Computational Intelligence in Mechatronics

Computational Intelligence in Mechatronics

Solving Real-World Problems in Electronic Systems Design

Mohamed Arezki Mellal
LMSS, Faculty of Technology, M'Hamed Bougara University, Boumerdes, Algeria

ELSEVIER

Elsevier
Radarweg 29, PO Box 211, 1000 AE Amsterdam, Netherlands
50 Hampshire Street, 5th Floor, Cambridge, MA 02139, United States

ISBN: 978-0-443-27580-7

For Information on all Elsevier publications visit our website at
https://www.elsevier.com/books-and-journals

Publisher: Mara Conner
Acquisitions Editor: Neil Craig Smith
Editorial Project Manager: Kripa Joseph
Production Project Manager: Fahmida Sultana
Cover Designer: Greg Harris

Typeset by Aptara, New Delhi, India

Contents

Contributors

V.M. Akhil, Amrita School of Artificial Intelligence, Amrita Vishwa Vidyapeetham, Coimbatore, Tamil Nadu, India

Srikar Annamraju, Assistant Professor, School of Interdisciplinary Design and Innovation (SIDI), Indian Institute of Information Technology, Design and Manufacturing (IIITDM), Kancheepuram, Tamil Nadu, India

Yılmaz Seryar Arikuşu, Department of Electrical Engineering, Manisa Celal Bayar University, Manisa, Turkey

M. Ashmi, Department of Mechatronics, ACE College of Engineering, Trivandrum, Kerala, India

Sandeep B, Department of Mechanical Engineering ,Vidya Vikas Institute of Engineering & Technology, Mysuru, Karnataka, India

David Bautista-Lázaro, Advanced Technologies, National Polytechnic Institute—UPIITA, Mexico City, Mexico

Naourez Ben Hadj, Department of Electrical Engineering, National Engineering School of Sfax, Information Technologies and electronics Laboratory, University of Sfax, Sfax, Tunisia

Sujata N Bhavikatti, Electronics and Communication Engineering, KLS Gogte Institute of Technology, Belagavi, Karnataka, India

Suman Chakraborty, Department of Electrical Engineering, Budge Budge Institute of Technology, Budge Budge, West Bengal, India

Harpreet Kaur Channi, Eudoxia Research Centre (Affiliated to Eudoxia Research University, USA), Mumbai, Maharashtra, India

Adhir Baran Chattopadhyay, Department of Electrical Engineering, Budge Budge Institute of Technology, Budge Budge, West Bengal, India

Mingtai Chen, Wichita State University, Wichita, KS, United States

Houda Daoud, Department of Engineering of Electronic Communication Systems, National School of Electronics and Telecommunications of Sfax,University of Sfax, Sfax, Tunisia

Jhinuk De, Department of Chemistry, National Institute of Technology Agartala (NITA), Agartala, Tripura, India

R. Deepa, Department of Computing Technologies, School of Computing, SRM Institute of Science and Technology, KTR Campus, Chennai, Tamil Nadu, India

Vu Duong, School of Engineering and Technology, Duy Tan University, Danang City, Vietnam

Jesica Escobar, Cybernetics-ICE, Instituto Politécnico Nacional, ESIME Zac, UPALM, Mexico City, Mexico

Akram Fraih, LMSS, Faculty of Technology, M'Hamed Bougara University, Boumerdes, Algeria

Mwapashua Hamisi Fujo, Information and Communication Technology (ICT), Moshi Co-Operative University (MoCU), Moshi, Tanzania

Sunil Gautam, Computer Science and Engineering, Nirma University, Ahmedabad, Gujarat, India

Tao Gong, Jiangsu Key Laboratory of Mine Mechanical and Electrical Equipment, School of Mechatronic Engineering, China University of Mining and Technology, Xuzhou, China

Valarmathi Govindarajan, Sri Sairam Institute of Technology, Chennai, Tamil Nadu, India

Pankaj Kumar Gupta, Mechanical Engineering Department, Meerut Institute of Engineering and Technology, Meerut, Uttar Pradesh, India

Jaime Guzmán-López, Intelligent Computing Laboratory, National Polytechnic Institute—CIDETEC, Mexico City, Mexico

Sunanda Hazra, Department of Electrical Engineering, Haldia Institute of Technology, Haldia, West Bengal, India

Nguyen Ha Hiep, Institute of Vehicle and Energy Engineering, Le Quy Don Technical University, Hanoi City, Vietnam

Rishi J.P., Department of Mechanical Engineering, Vidyavardhaka College of Engineering, Mysuru, Karnataka, India

A. Jenith Joseph, Sri Sairam Institute of Technology, Chennai, Tamil Nadu, India

Lenin Kanagasabai, Independent Researcher, Hyderabad, Telangana, India

Azleena Mohd Kassim, School of Computer Science, Universiti Sains Malaysia, Pulau Pinang, Malaysia

Gagandeep Kaur, Department of Emerging Technologies, Chandigarh, Engineering College, Landran, Mohali, Punjab, India; Department of Computer Science and Engineering, Chandigarh, Engineering College, Landran, Mohali, Punjab, India

Komalpreet Kaur, Department of Computer Science and Engineering, Chandigarh, Engineering College, Landran, Mohali, Punjab, India

Rajaram Kiruba, Department of Electronics and Instrumentation Engineering, Sri Ramakrishna Engineering College, Coimbatore, Tamil Nadu, India

Maryem Kotti, Department of Engineering of Electronic Communication Systems, National School of Electronics and Telecommunications of Sfax,University of Sfax, Sfax, Tunisia

Manel Krichen, Electrical department, National Engineering School of Sfax, Information Technologies and electronics Laboratory, Sfax University, Sfax, Tunisia

Divynash Kumar, Department of Computer Science and Engineering, School of Computing, SRM Institute of Science and Technology, KTR Campus, Chennai, Tamil Nadu, India

Tien-Thinh Le, Faculty of Mechanical Engineering and Mechatronics, Phenikaa University, Ha Dong, Hanoi, Vietnam; A&A Green Phoenix Group JSC, Phenikaa Research and Technology Institute (PRATI), Cau Giay, Hanoi, Vietnam

Van-Tao Le, ATC, Le Quy Don Technical University, Hanoi, Vietnam

Yeng Seng Lee, Keysight Technologies, Bayan Lepas, Pulau Pinang, Malaysia

Shangyuan Li, Jiangsu Key Laboratory of Mine Mechanical and Electrical Equipment, School of Mechatronic Engineering, China University of Mining and Technology, Xuzhou, China

Alberto Luviano-Juárez, National Polytechnic Institute UPIITA, Mexico City, Mexico

Lamia May, Laboratory of Mechanics, Materials and Energy (L2ME), Faculty of Technology, University of Bejaia, Bejaia, Algeria

Angel Gabriel Meela, Informatics, Institute of Accountancy Arusha (IAA), Arusha, Tanzania

Husain Mehdi, Mechanical Engineering Department, Meerut Institute of Engineering and Technology, Meerut, Uttar Pradesh, India

Mohamed Arezki Mellal, LMSS, Faculty of Technology, M'Hamed Bougara University, Boumerdes, Algeria

Sushil Kumar Mishra, Department of Computer Science and Engineering, School of Computing, SRM Institute of Science and Technology, KTR Campus, Chennai, Tamil Nadu, India

Rosmiwati Mohd-Mokhtar, School of Electrical and Electronic Engineering, Universiti Sains Malaysia, Nibong Tebal, Malaysia

Thaeer Mueen Sahib, Department of Electrical Techniques, Kufa Technical Institute, Al-Furat Al-Awsat Technical University, Iraq

Vrunda Mulivala, Computer Science and Engineering, Nirma University, Ahmedabad, Gujarat, India

Rafik Neji, Department of Electrical Engineering, National Engineering School of Sfax, Information Technologies and electronics Laboratory, Sfax University, Sfax, Tunisia

Van-Hai Nguyen, Faculty of Mechanical Engineering and Mechatronics, Phenikaa University, Ha Dong, Hanoi, Vietnam; A&A Green Phoenix Group JSC, Phenikaa Research and Technology Institute (PRATI), Cau Giay, Hanoi, Vietnam

T.K. Niranjan, Sri Sairam Institute of Technology, Chennai, Tamil Nadu, India

Sandip Ojha, Department of Electrical Engineering, Haldia Institute of Technology, Haldia, West Bengal, India

Tannistha Pal, Computer Science & Engineering Department, National Institute of Technology Agartala, Agartala, Tripura, India

Y. Pandurangaiah, Department of ECE, IARE Institute of Aeronautical Engineering, Hyderabad, Telangana, India

Aryan Parmar, Computer Science and Engineering, Nirma University, Ahmedabad, Gujarat, India

Vijay Kumar Pediredla, Senior Data Scientist, Data Science Team, Haber Technologies Inc., Vancouver, BC, Canada

Sasikumar Potluri, Sri Sairam Institute of Technology, Chennai, Tamil Nadu, India

Alexander Poznyak, Department of Automatic Control, CINVESTAV-IPN, Mexico City, Mexico

A. Bennet Prabhu, Research & Development, Congruent Solutions Private Ltd, Chennai, Tamil Nadu, India

Arnav Prateek, Department of Computer Science and Engineering, School of Computing, SRM Institute of Science and Technology, KTR Campus, Chennai, Tamil Nadu, India

Venkutuswamy Radhika, Department of Biomedical Engineering, Sri Ramakrishna Engineering College, Coimbatore, Tamil Nadu, India

Mario Ramirez-Neria, Institute for Applied Research and Technology (INIAT), Ibero-American University, Mexico City, Mexico

Ajay Roy, Department of Electrical Engineering, College of Engineering, University of Bisha, Bisha, Saudi Arabia

Aarchi Shah, Computer Science and Engineering, Nirma University, Ahmedabad, Gujarat, India

Zhen Shan, School of Computer Science and Technology, China University of Mining and Technology, Xuzhou, China

Vipin Kumar Sharma, Mechanical Engineering Department, Meerut Institute of Engineering and Technology, Meerut, Uttar Pradesh, India

Aditya Singh, Department of Civil Engineering, Amrita Vishwa Vidyapeetham, Coimbatore, Tamil Nadu, India

S. Sivarajeswari, Sri Sairam Institute of Technology, Chennai, Tamil Nadu, India

Kriyanshi Soni, Computer Science and Engineering, Nirma University, Ahmedabad, Gujarat, India

Göksu Taş, Faculty of Agricultural Sciences and Technology, Department of Precision Agriculture and Agricultural Robots, Sivas University of Science and Technology, Sivas, Türkiye

Asokan Thondiyath, Professor, Department of Engineering Design, Indian Institute of Technology, Madras, Chennai, Tamil Nadu, India

T. Tirupal, Department of ECE, G. Pullaiah College of Engineering and Technology, Kurnool, Andhra Pradesh, India

Vu Tuan, School of Engineering Technology, Duy Tan University, Da Nang, Vietnam

Devendrasinh Vashi, Computer Science and Engineering, Nirma University, Ahmedabad, Gujarat, India

Duong Vu, School of Engineering Technology, Duy Tan University, Da Nang, Vietnam

Rihem Walha, Department of Engineering of Electronic Communication Systems, National School of Electronics and Telecommunications of Sfax,University of Sfax, Sfax, Tunisia

Zhongqiu Wang, School of Computer Science and Technology, China University of Mining and Technology, Xuzhou, China

Fwen Hoon Wee, Faculty of Electronic Engineering Technology, Universiti Malaysia Perlis, Arau, Perlis, Malaysia

Edward J. Williams, Industrial and Manufacturing Systems Engineering Department, College of Engineering and Computer Science, University of Michigan, Dearborn, MI, United States; Decision Sciences, College of Business, University of Michigan, Dearborn, MI, United States

Jianhua Yang, Jiangsu Key Laboratory of Mine Mechanical and Electrical Equipment, School of Mechatronic Engineering, China University of Mining and Technology, Xuzhou, China

Kok Yeow You Faculty of Electrical Engineering, Universiti Teknologi Malaysia, Skudai, Johor, Malaysia

Chapter 1

Study of an underwater vehicle hybrid propulsion based on a fuel cell

Vu Duong[a] and Nguyen Ha Hiep[b]

[a] *School of Engineering and Technology, Duy Tan University, Danang City, Vietnam,* [b] *Institute of Vehicle and Energy Engineering, Le Quy Don Technical University, Hanoi City, Vietnam*

1.1 Introduction

Although oceans cover more than 70% of the Earth's surface, more than 95% of ocean areas remain unexplored. Therefore using underwater vehicles is an ideal solution for military, commercial, and scientific research, expanding the range of tasks from depth measurement to detecting underwater mines, etc.

Unmanned underwater vehicles (UUVs) or underwater drones are small underwater vehicles that operate underwater without direct human control. These include remotely operated vehicles (ROVs) and autonomous underwater vehicles (AUVs) that work independently according to the preset program. Some underwater vehicles operating in shallow water can use power supplied from shore or from surface ships by wires. This option has many limitations, so the use of air-independent propulsion (AIP) to increase diving depth, underwater endurance, and range. Using AIP for submarines and UUVs is understandable because they have the task of underwater exploration; the underwater endurance, operating range, and diving depth are as large as possible; and the noise level is as low as possible. Because of limited battery power reserves the underwater functional endurance of diesel-electric submarines is limited. The submarine must surface, or the diesel engine must operate in snorkel mode to charge the battery. So the probability of the submarine being detected by radar, infrared, optoelectronics, and acoustics increases (Lee & Shay, 2018). For small underwater vehicles, diesel engines cannot be used as the primary source of propulsion. These vehicles often use electric propulsion systems (only batteries) or hybrid propulsion systems of batteries and fuel cells (FCs) (van Biert et al., 2016).

Currently, many countries have researched, manufactured, and procured UUVs to serve different purposes. Military and research needs led to the development of UUVs and their advanced AUV variations, respectively. FCs are considered small in size, light in weight, and easy to handle, making them ideal for underwater vehicles. Interest in using logistics fuels has prompted the Office of Naval Research (ONR) to also fund the development of hybrid solid oxide fuel cell (SOFC) systems. UUV/AUV is an emerging and growing market for marine suppliers. The former Japan Marine Science and Technology Center (JAMSTEC) needed a long-range AUV for marine research in the polar region. Mitsubishi developed and delivered an AUV called the Urashima in 2004 (Winkler, 2009). It uses proton exchange membrane fuel cell (PEMFC) technology combined with a lithium-ion battery and is supplied with hydrogen from metal hydride storage and oxygen from an LOX reservoir. Although any AUV contains several navigation, control, and automation systems, propulsion remains challenging. Specifically, Pluto Plus ROVs of Vietnam, SeaFox-I of the United Kingdom, and Remus of the United States—these vehicles all have the common operation characteristic of using electric energy, but each type has a different option. Vietnam's underwater vehicles currently use batteries as the main source of energy to operate, but compared to the original design of GayMarine's Pluto Plus, it uses a propulsion system with two main energy sources: an FC and a battery. With the manufacturer's original powertrain, it is possible to increase the performance of Pluto Plus ROVs and achieve better autonomy.

A research paper (Carrette et al., 2001) shows that PEMFC and SOFC can be used for UUVs. PEMFC has higher performance and operates at lower temperatures (below 100°C) and is, therefore, the preferred choice. Besides, PEMFC is

Computational Intelligence in Mechatronics: Solving Real-World Problems in Electronic Systems Design. DOI: https://doi.org/10.1016/B978-0-443-27580-7.00018-4

FIGURE 1.1 Pluto Plus ROV's exterior shape. *ROVs*, remotely operated vehicles.

an energy source with very high metabolic efficiency, low operating temperature, relatively compact size, environmentally friendly products, etc.; therefore it is also well suited as a major power source to supply powertrain devices on ROVs (van Biert et al., 2016). The need to create highly efficient hybrid propulsion based on FC attracts a growing number of developers and researchers who work on designing and modeling hybrid propulsion systems. Research into hybrid propulsion system development can be carried out through (1) experimental study, (2) analytical investigation, and (3) modeling and simulations. Experimental studies (Shih et al., 2013, 2014; Weydahl et al., 2020) were performed without numerical calculations. Fabrication and testing of PEMFC prototypes for underground vehicles are time-consuming and expensive processes. Using mathematical models for analytical studies makes evaluating phenomena in hybrid systems complex. Furthermore, adjusting input parameters to calibrate analytical results to coincide with experimental data can be difficult.

Numerical modeling and simulation play an essential role in developing hybrid propulsion systems because they help understand the impact of different components in the system on their performance and make it possible to tune the parts to have a complete system. In addition, it also supports optimizing size and mass to increase the underwater operating range and endurance of underwater vehicles (Albarghot et al., 2020; Cu Xuan & Nguyen Quoc, 2023; De Lorenzo et al., 2022).

With the advantages of modeling and simulation work as mentioned above, as well as based on some simulation results in works (Ansari et al., 2021; Saeed & Warkozek, 2015). In this study, we conduct modeling and simulation of FC hybrid propulsion systems applied to an underwater vehicle using pure battery propulsion systems.

For that, some of the tasks that need to be done are to calculate the movement resistance of an underwater vehicle, determine the propulsion power, and then calculate the power of the FC and battery. Next, build a model of the system components, such as FCs, batteries, and power converters, in the MATLAB Simulink environment. Finally, simulate the system's working to have parameters to calibrate the model and complete the hybrid system as a basis for designing and manufacturing the actual hybrid system.

1.2 Material and methodology

1.2.1 Research object

The research object is Pluto Plus ROV, an underwater vehicle, a remote-controlled diving robotic system to locate, inspect, and clear underwater targets whose primary purpose is to search, identify, and clear bottom and mooring mines in extreme operating conditions and for antiterrorism. Its appearance is shown in Fig. 1.1 (Pluto Plus, n.d.).

The robot consists of two pressurized bodies joined together. Each body consists of two halves put together by bolts. Water tightness is ensured by the gasket rings placed between the grooves. The main fuselage consists of five propellers, rotary engines, a battery compartment, and electronic components. The front part of the diving robot includes a sensor head that is compact and can be raised and rotated at angles in horizontal and vertical planes. The compartment containing the sensor heads also has color TV cameras, spotlights, and special sonars. The sensor head is protected with a composite housing. Robotic operation is always at the highest level thanks to the maximum-size miniaturization of sensors, fiber optic technology, and the ability to design a safe body. The Pluto Plus ROV is specially built to operate in high undersea currents with limited visibility. The robot is equipped with a fiber optic cable capable of joining, with a diameter of 3.6 mm; the length of 2000 m is suitable for long-range and high-flow missions. The nominal operating depth is 300 m; with safety conditions, the depth of the diving robot can be up to 500 m. Movement capabilities of the Pluto Plus ROV: The maximum tensile force of the fiber optic cable connected to the diving robot is 65 kg. The top forward moving speed in the horizontal

TABLE 1.1 Pluto Plus ROV specifications.

Main parameters	Value	Unit
Dimensions	2250 × 580 × 770	mm
Weight	320	kg
Maximum carrying capacity	100	kg
Maximum speed	6	knot
Calculated speed	3.5	knot
Maximum diving depth	400	m
Underwater endurance at speed 6 knots	2	h

ROVs, remotely operated vehicles.

direction is 6 knots (in balanced conditions). The maximum speed allowed for detailed and accurate observations of the seabed is about 0.5 knots; exceeding this speed is used to resist the flow of water. Horizontal speed is about 0.5 knots; maximum vertical speed is 0.3 knots to 1 knot (up to 3–4 knots when combined with horizontal speed). On the surveyor's control panel a switch selects the power source for the engine. When the largest motor power source is used, the robot's maximum speed can reach about 6 knots per hour. When the switch is switched to the engine power position at a nominal level, the top speed achieved is about 3.5 knots. This allows the operator to control the submersible robot better when close to the target or when there is no influence of seawater currents, thus avoiding high speeds and loss of power consumption.

The main characteristics of Pluto Plus ROV are summarized in the following (Table 1.1).

The feed block is used to power the diving robot battery charger and, in case of emergency, the tow winch. The system's operation is always at the highest level because of the optimized system operation design based on the minor diameter and reusable robot connection cable. Thus ROVs are unmanned submersible devices controlled remotely via the monitoring console, so their underwater range will be limited because of the lack of independent sources.

1.2.2 Calculation of power of a hybrid propulsion system

The calculation of power consumption for propulsion engines is based on viscous resistance using the viscous resistance coefficient C_V, the friction resistance coefficient C_F, and the shape factor $(1 + k)$.

$$C_F = \frac{0.075}{(\log Re - 2)^2} \tag{1.1}$$

C_F is determined by formula (1.1) (d'Amore-Domenech et al., 2018):

$$\mathrm{Re} = \frac{vL}{\mu}$$

where Re is the Reynolds number, μ is the kinematic viscosity of seawater (m^2/s), L is the length of ROV (m), and v is the calculated speed of the ROV (m/s).

The shape coefficient that can be obtained from the Droblenkov curve for AUV with a circular cross section is expressed as the expression (1.2) (d'Amore-Domenech et al., 2018).

$$(1 + k) = 1 + 0,2\left(\frac{D}{L}\right) + 8\left(\frac{D}{L}\right)^2 - 10\left(\frac{D}{L}\right)^3 \tag{1.2}$$

where D is the diameter of the ROV (m), take the average width and height of the ROV. Thus viscous resistance is calculated by the expression (1.3) (d'Amore-Domenech et al., 2018):

$$R_V = \frac{1}{2}\rho\Omega v^2 C_V = \frac{1}{2}\rho\Omega v^2 C_F(1 + k) \tag{1.3}$$

where $\rho = 1025$ (kg/m^3)-density of seawater, and Ω is the wet surface area of the ROV, calculated as a cylinder of diameter D and length L.

TABLE 1.2 Input data for calculating the required power of ROV.

Parameter	Value	Unit	Parameter	Value	Unit
μ	1.57×10^{-6}	m^2/s	T	0.58	m
v	3.5	knots	D	0.675	m
	1.543	m/s	B	0.77	m
L	2.25	m	η_D	0.4	–

ROV, remotely operated vehicle.

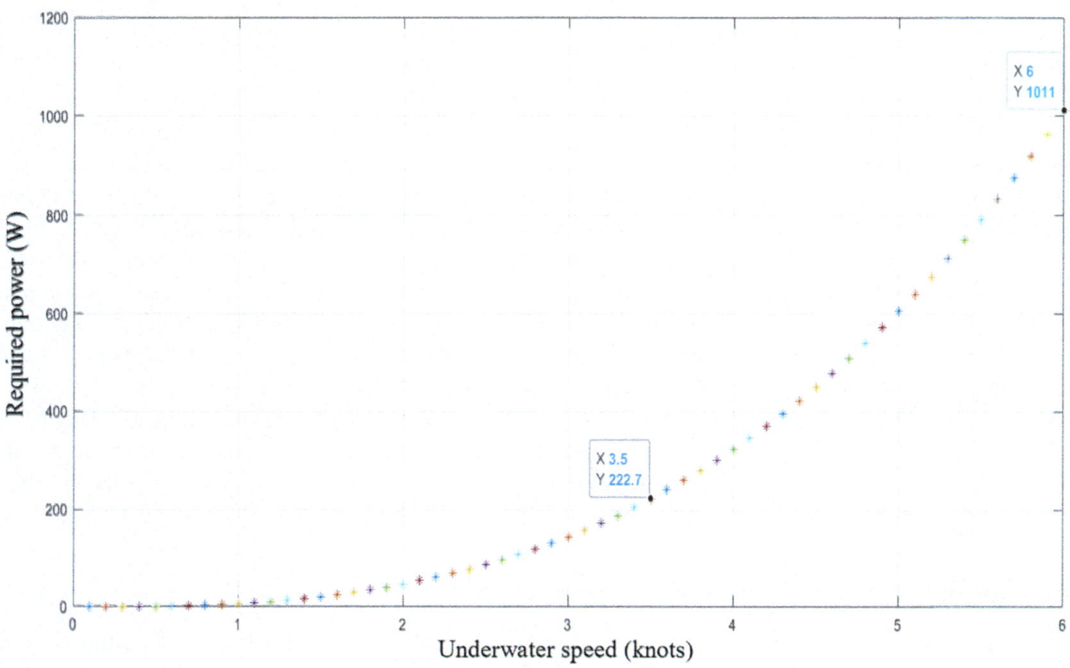

FIGURE 1.2 The dependence of main propulsion power on the speed of the ROV. *ROV*, remotely operated vehicle.

The required thrust power is calculated as the expression (1.4), where η_D is the propeller thrust performance:

$$N_p = \frac{vR_v}{\eta_D} \tag{1.4}$$

From expressions (1.3) and (1.4), we get a third-order curve representing the dependence of the thrust power and the forward rate of the ROV as expression (1.5):

$$N_p = \left[\frac{1}{2\eta_D}\rho\Omega C_F(1+k)\right]v^3 \tag{1.5}$$

The inputs for calculating the required power are presented in Table 1.2:

Based on the data in (Table 1.2) combined with equations from (1.1) to (1.5), we construct a curve showing the relationship between the required power and the travel speed of the ROV. With the nominal speed $v = 3.5$ knots the power $N_p = 222.7$ W. When the maximum speed is $v_{max} = 6$ knots the power $N_p = 1011$ W. To ensure that the FC pack can work best when calculating, we add the types of losses because of transmission, so the rated power to calculate for the battery pack is 300 W. The propulsion system of the original Pluto Plus ROV unit uses batteries powered directly from the onboard electrical system (Fig. 1.2).

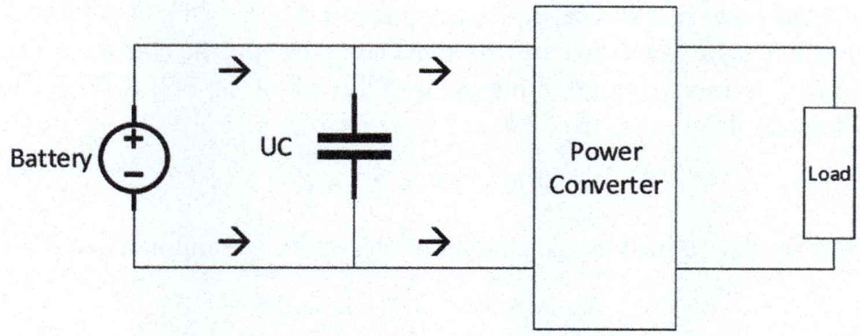

FIGURE 1.3 Power scheme for the load of the original Pluto Plus ROV. *ROV*, remotely operated vehicle.

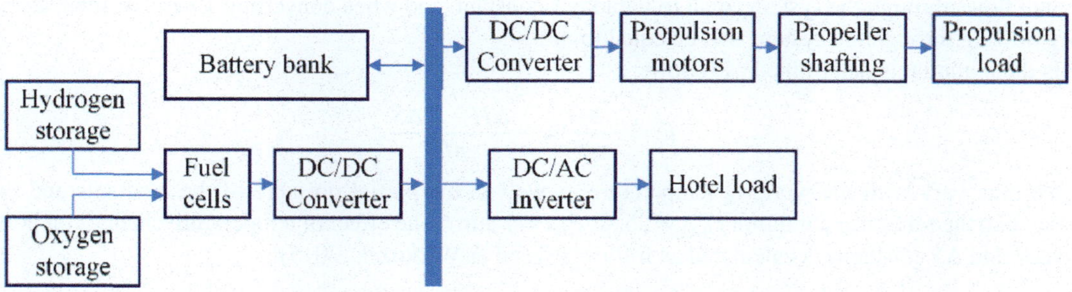

FIGURE 1.4 Proposed fuel cell hybrid propulsion system for Pluto Plus ROV. *ROV*, remotely operated vehicle.

1.2.3 Original and proposed propulsion systems

As stated in Part 1 the propulsion system of GAYMARINE's Pluto Plus ROV device is equipped with a propulsion system including an FC cluster and battery pack, but Vietnam's Pluto Plus ROV only uses batteries. It takes power directly from the ship's electrical system. As stated in Part 1 the propulsion system of GAYMARINE's Pluto Plus ROV device is equipped with a propulsion system including an FC cluster and battery pack, but Vietnam's Pluto Plus ROV only uses batteries. It takes power directly from the ship's electrical system. With the propeller power supply scheme (load) (Fig. 1.3), there are many limitations in terms of range and the self-propelled ability of the device when having to perform minesweeping tasks over a wide range. Therefore, with the model installed with an FC cluster, it will limit the disadvantages of the current device, not to mention the relatively high energy conversion efficiency of PEM FCs (Shih et al., 2014).

The proposed FC hybrid propulsion system for the Pluto Plus ROV is shown in Fig. 1.4. With the proposal to install an FC cluster as the main energy source besides the battery for the device, through the above calculation parameters, the capacity of the battery pack generated is 300 W, corresponding to the mode in which the device runs at a nominal speed of 3.5 knots. When running at a speed from nominal speed to maximum speed, it is clear that the capacity of the ship must be increased, so now the FC assembly supplies the output load while the remaining power is taken care of by the battery pack. Thus, in the maximum speed mode $P_{max} = 1090$ W, the designed battery pack capacity accounts for about 29.6%, and the battery will account for 70.4% of the capacity. When the device runs below the nominal speed according to the graph in Fig. 1.2, the power decreases (less than 300 W). At this time the FC pack both powers the load and charges the battery. This is also the original propulsion system that GayMarine Corporation (a corporation specializing in the production of small submarine diving equipment for military use) is commercializing.

1.2.4 Calculating and determining fuel cell parameters

Energy is obtained in an FC from thermodynamic energy through electrochemical reactions inside the FC. Basically the energy obtained from the exothermic reaction forms water from H_2 and O_2:

$$\text{On the anode: } H_2\,(g) \rightarrow 2H^+ + 2e$$

$$\text{On the cathode: } 1/2O_2(g) + 2H^+ + 2e \rightarrow H_2O$$

$$\text{General: } H_2(g) + 1/2O_2(g) \rightarrow H_2O + \text{power} + \text{heat}$$

The energy obtained by a chemical reaction called enthalpy forms $\Delta H = -285.84$ (kJ/mol) (which is the high calorific value when water is liquid), which is broken down into thermal energy by specific entropy ΔS (kJ/mol K) multiplied by the absolute temperature T, and the electricity (representing the useful work of the FC) is Gibbs free energy ΔG (kJ/mol). So the total energy is (EG&G Technical Services, Inc., 2004)

$$\Delta H = \Delta G + T \cdot \Delta S \tag{1.6}$$

where ΔG is the useful electricity determined by the amount of electricity Q (coulombs) and potential E (V):

$$W_{el} = \Delta G = -Q \cdot E = -nFE \tag{1.7}$$

where $n = 2$ is the number of electrons transferred for each mole of fuel consumed, and F is Faraday's constant (96,487 C/mol) because according to Faraday's law, when electrochemically converting 1 mole of the equivalent of a substance, an electric amount flows through the system equal to Faraday's constant, and when converting 1 mole of substance, an amount of electricity flowing through the system will be equal to nF.

From this the potential is determined as follows:

$$E = -\frac{\Delta G}{nF} = -\frac{\Delta H - T\Delta S}{nF} \tag{1.8}$$

This value may vary with the working temperature as well as the partial pressure of the reactants and the resulting products relative to the reference condition ($T_{ref} = 25^{\circ}$C, $P_{ref} = 1$ atm). The effect of temperature changes on potential when considering ΔH and ΔS constants is calculated as follows (Saeed & Warkozek, 2015):

$$\Delta E = \frac{\Delta S}{nF}(T - T_{ref}) \tag{1.9}$$

From the formula (1.9), it shows that as the temperature increases, the potential decreases because $\Delta S = -164$ J/mol. K has a negative value. The partial pressure change affects ΔG because G is a function of the specific volume V (m^3/mol) and the change in pressure dP (i.e., $G = V \times dP$). From the gas law $V = RT/P$, we have (EG&G Technical Services, Inc. 2004; Saeed & Warkozek, 2015)

$$\Delta G = \Delta G^o - \frac{RT}{nF} \ln\left(\frac{P^1_{H_2} P^{0.5}_{O_2}}{P^1_{H_2O}}\right) \tag{1.10}$$

where $\Delta G° = -237.17$ kJ/mol, Gibbs free energy at standard conditions; and $R = 8.3143$ J/mol. K is the gaseous substance constant. In the formula (1.10), consider an FC working on pure hydrogen with a partial pressure of P_{H2} and pure oxygen with a partial pressure of P_{O2}, knowing that to form 1 mole of water with P_{H2O} pressure (product) requires 1 mole of H_2 and 0.5 moles of O_2 (reactants), so the potential of the FC (Ansari et al., 2021):

$$E_{Nerst} = -\frac{\Delta G^o}{nF} + \frac{\Delta S}{nF}(T - T_{ref}) + \frac{RT}{nF} \ln\left(\frac{p^1_{H_2} p^{0.5}_{O_2}}{p^1_{H_2O}}\right) \tag{1.11}$$

Here the partial pressure values are calculated using the following formula: $\log p_{H_2O} = -2.1794 + 0.02953 \cdot t - 9.1837 \cdot 10^{-5} \cdot t^2 + 1.4454 \cdot 10^{-7} \cdot t^3$

$$p_{H_2} = 0.5 \frac{P_{H_2}}{\exp\left(1.653 \cdot \frac{i_{FC}}{T^{1.334}}\right)} - p_{H_2O}$$

$$p_{O_2} = \frac{P_{O_2}}{\exp\left(4.192 \cdot \frac{i_{FC}}{T^{1.334}}\right)} - p_{H_2O}$$

The formula (1.11) is called the Nerst equation, which gives the maximum potential that can be achieved because there are four types of losses in FCs. Operational losses: this loss is necessary, like a spark, to initiate reactions. It depends on the temperature, partial pressure, and catalyst on the electrodes. Calculated by the semiexperimental formula (Saeed & Warkozek, 2015):

$$V_{act} = -[\xi_1 + \xi_2 \cdot T + \xi_3 \cdot T \cdot \ln(C_{O2}) + \xi_4 \cdot T \cdot \ln(i_{FC})] \tag{1.12}$$

TABLE 1.3 Parameters for calculating PEMFC.

Parameter	Value	Unit	Parameter	Value	Unit	Parameter	Value	Unit
T	353	K	B	0.08	–	ζ_3	7.4×10^{-5}	–
A	60	cm^2	ζ_1	−0.96	–	ζ_4	-1.8×10^{-4}	–
L	178×10^{-4}	cm	ζ_2	3.12×10^{-3}	–	ψ	23	–
P_{H2}	3	atm	Rc	0.0008	Ω	–	–	–
P_{O2}	3	atm	j_L	1,4	A/cm^2	–	–	–

PEMFC, proton exchange membrane fuel cells.

where x_i is the coefficient; i_{FC} is the amperage of the FC (A); and C_{O2} is the concentration of oxygen on the catalyst surface (mol/cm^3).

$$C_{O_2} = \frac{p_{O_2}}{5.08 \cdot 10^6 \cdot \exp\left(-\frac{498}{T}\right)} \tag{1.13}$$

Om loss: This loss is because of the resistance of the proton exchange membrane (affected by the degree of membrane hydration) and the resistance because of contact between the electrode and the membrane surface as well as the electrode surface with the dipole plate and the collector plate (Ansari et al., 2021).

$$V_{\text{ohm}} = i_{FC} \cdot (R_m + R_c) \tag{1.14}$$

where R_c is the constant resistance of the cell, R_m is the temperature-dependent resistance and parameter y expresses the degree of membrane hydration (take a value of 14 if the membrane is fully hydrated and 23 if the membrane is oversaturated):

$$R_m = \frac{\rho_m \cdot l}{A} \tag{1.15}$$

where l is the thickness of the membrane (cm), A is the active area of the membrane (cm^2), and ρ_m is the specific resistance of the membrane (Ω cm), is determined by the formula (Ansari et al., 2021):

$$\rho_m = \frac{181.6 \cdot \left[1 + 0,03 \cdot \left(\frac{i_{FC}}{A}\right) + 0.062 \cdot \left(\frac{T}{303}\right)^2 \cdot \left(\frac{i_{FC}}{A}\right)^{2,5}\right]}{\left[\Psi - 0.634 - 3\left(\frac{i_{FC}}{A}\right)\right] \cdot \exp\left[\frac{4.18(T-303)}{T}\right]} \tag{1.16}$$

Concentration loss: This is called mass transfer loss; it expresses the mass transfer limit of the reactants. Although feeding reactants at high speeds will give high current densities in the FC, there is a limit to the feed rate, and the reactants may not be used properly, resulting in this type of loss. Therefore this limit is determined by the jL current density limit, which is the maximum current density value that the FC can produce and is about 1.4 A/cm^2 (Ansari et al., 2021; Saeed & Warkozek, 2015).

$$V_{\text{conc}} = -B \ln\left(1 - \frac{j}{j_L}\right) \tag{1.17}$$

where B is the coefficient depending on the cell and operating conditions, j is the instantaneous current density (A/cm^2), and $j = I_{FC}/A$. Interference loss: This loss is caused by the interference of hydrogen and electrons passing through the membrane (electron conduction resistance); this phenomenon does not produce any useful work. The effect of this loss can be considered by adding the interferometric loss current density to the current density in the V_{conc} loss equations. However, interference loss current loss density is often overlooked because it has a very small value. Finally the formula for determining the FC potential is

$$V_{\text{cell}} = E_{\text{Nerst}} - V_{\text{act}} - V_{\text{ohm}} - V_{\text{act}} \tag{1.18}$$

The input data to calculate the required power is shown in Table 1.1, and the remaining parameters of the workings of PEMFCs are given in Table 1.2. The coefficients in Table 1.3 are taken from research (Ansari et al., 2021; Saeed & Warkozek, 2015); the cathode and anode inlet pressures are taken to be 60 psig (4.08 atm) according to Shih et al. (2014), and the proton exchange membrane is type Nafion-117. The parameters given to PEMFC are given under Table 1.3.

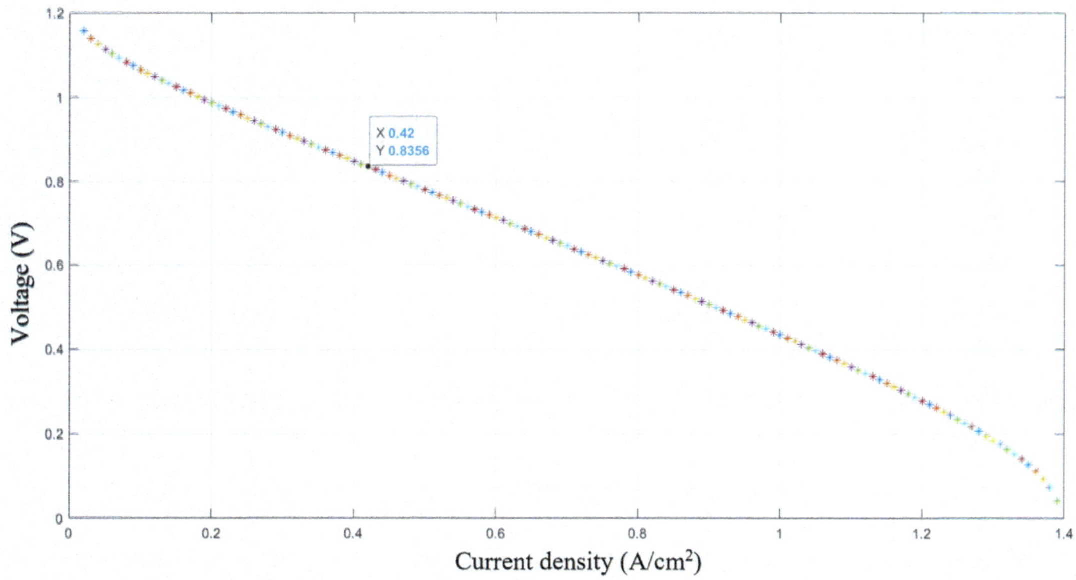

FIGURE 1.5 Voltage-current characteristic of a cell.

From formulas (1.6) to (1.18), combined with the values of Table 1.3, we construct a cell polarization curve with $V_{cell} = f(j)$, which is depicted graphically in Fig. 1.5:

According to the original design the initial expected performance of the selected PEMFC FC was $\eta = 0.57$. According to the performance calculation formula (EG&G Technical Services, Inc., 2004)

$$\eta = 0.675.V_{cell}$$

Thus we calculate the output voltage of 1 battery compartment, V_{cell}:

$$V_{cell} = \frac{\eta}{0.675} = \frac{0.57}{0.675} = 0.84(V)$$

Based on the output potential polarization curve with instantaneous current density formulated in Fig. 2.5, for $V_{cell} = 0.84$ (V), $j = 0.42$ (A/cm^2). The current of the FC pack is equal to the current density (j) multiplied by the operating area (A_{cell}) of the single battery:

$$I = j.A_{cell} = 0.42.60 = 25(A)$$

The output power of the PEMFC FC pack is calculated by the formula:

$$P = U.I$$

where P is the output power of the FC (P= 300 W), I is the current of the FC pack ($I = 25$ A), and U is the potential of the FC pack (V).

Thus, with the known values, we calculate the generated voltage of the battery pack as

$$U = \frac{P}{I} = \frac{300}{25} = 12(V)$$

The potential U of the set consisting of n single batteries connected in series is calculated by the following formula:

$$U = n.V_{cell}$$

It is derived: $n = \frac{U}{V_{cell}} = \frac{12}{0.84} = 14(\text{cell})$.

The results of FC calculations are summarized in Table 1.4.

1.2.5 Modeling and simulation of a fuel cell hybrid propulsion system

The PEMFC powertrain consists of two main components: the FC assembly and the battery. When operating below the nominal rotational speed the FC assembly both generates electricity for the operating propulsion system and charges the

TABLE 1.4 Fuel cell pack calculation parameters.

Parameter	Value	Unit
Number of single cell, n	14	cell
Current density, j	0.42	A/cm^2
Fuel cell pack power, P	300	W
Fuel cell pack current, I	25	A
Fuel cell pack voltage, U	12	V
Single-cell voltage, V_{cell}	0.84	V

TABLE 1.5 Input parameters set for fuel cells.

Parameter	Value	Unit
Voltage in nominated mode; V_{nom}	12	V
Current in nominated mode; I_{nom}	25	A
Voltage at 0 (A) and 1 (A); V_0 (A), V_1 (A)	28 26	V
Current at the end point; I_{end}	25.5	A
Voltage at the end point; V_{end}	11.74	V
Number of battery compartments, n	14	cell
Efficiency, η	0.57	–
Operating temperature, t	80	°C
Gas pressure H_2 supply; P_f	3	bar
Gas pressure O_2 supply; P_{Air}	3	bar
Percentage of H_2 gas in the container; x_H_2	1	–
Percentage of O_2 gas in the container; y_O_2	1	–
Percentage of water vapor in the air, z_H_2O(*Air*)	0	–
Efficiency factor H_2, U_f_H2	0.8334	–
Efficiency of O_2.U_f O2	0.5004	–

battery. When operating at the nominal speed to the maximum speed, both the FC assembly and the battery participate in providing electricity to ensure the required output power for the load (propeller). The process to build a simulated block diagram for the PEMFC dynamics system goes through the following steps:

- Step 1: Build a model of the FC block and set the parameters for the battery pack according to the calculated results.
- Step 2: Build a battery block model and set parameters for the battery pack.
- Step 3: Build the DC_DC Boost Converter, connect the FC blocks, and calibrate the parameters.

The PEMFC FC block is available in Simulink, so we just need to call it out with the command "FC Stack." In the "block parameters," we enter the initial parameters for the battery according to Table 1.5. The PEMFC model is shown in Fig. 1.6.

The battery bank of the Pluto Plus replaced in Simulink is the "Batteries" block (Fig. 1.7). The necessary parameters for Pluto Plus are selected in accordance with the voltage generated by the FC pack to ensure the battery charging process is efficient. The batteries chosen here are lead-acid batteries because they are popular on the market and are cheap and easy to maintain when needed.

As mentioned the boost converter DC_DC is important for FC propulsion by ensuring the right voltage supply to electric motors while ensuring a minimum number of battery compartments and batteries that ensure a light powertrain mass for ROVs. The turbocharger has important elements: inductors, capacitors, MOSFET-type transistors, diodes, and a collapsible controller for MOSFETs. The MOSFET collapsible controller used is a PI controller based on the principle of comparing the set value with the actual value to control the pulse width to change accordingly to keep the output voltage value constant. The setting parameters for PI controllers are given in Table 1.6.

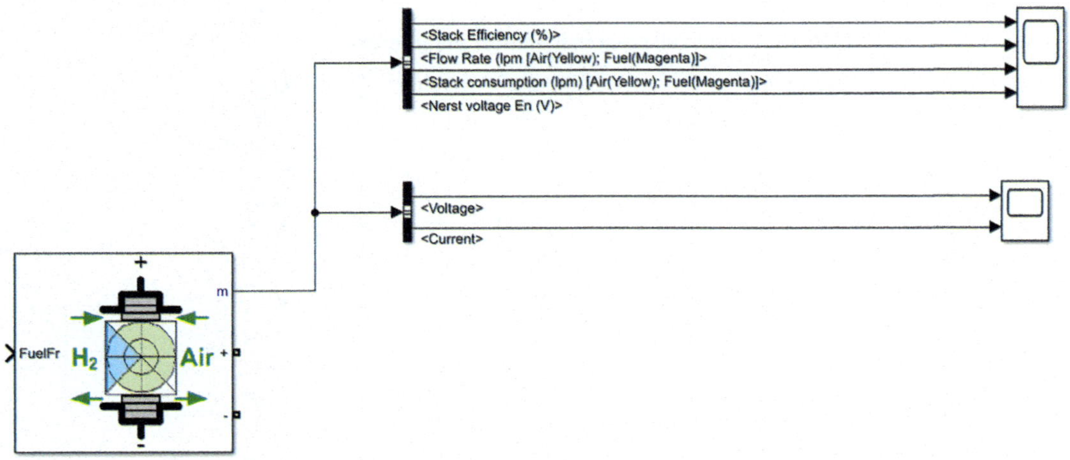

FIGURE 1.6 The PEMFC model. *PEMFC*, Proton exchange membrane fuel cells.

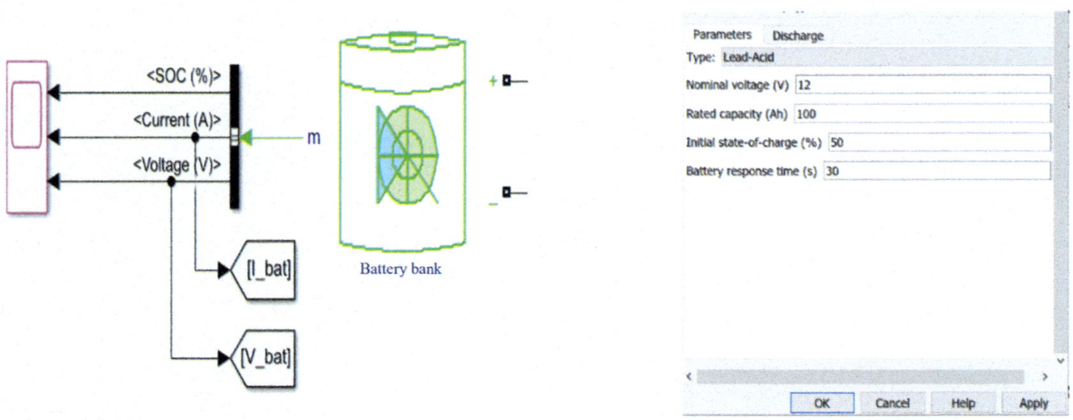

FIGURE 1.7 Block "Batteries" and input parameters set for the battery bank.

TABLE 1.6 PI controller parameters.

Parameter	Value	Unit
Inductance, L	$500 \times e^{-6}$	H
Capacitor capacitance, C	$7500 \times e^{-5}$	F
Table 3.2. PI controller parameters, r	0.2	Ω
Transferring function, $G(s)$	$\frac{1}{0.0001+1}$	–
Coefficient K_P	0.005	–
Cofficient K_I	0.15	–
Voltage set, U_{ref}	24	V
Duty cycle, T_s	$1 \times e^{-4}$	S

To ensure the limitation of complexity in the simulation model the whole DC_DC boost converter after construction is reduced to a "Subsystem" located in the block diagram of the whole program (Fig. 1.8). The values of the coefficients of the PI are selected based on trial and error (i.e., repeated experimentation to produce the desired result).

The load construction for the ship propulsion system is a complex system, especially the propeller electric drive system, so to make it simpler for the simulation process, we use resistor blocks instead of load assemblies (from hybrid electric motors to propellers). The goal from the beginning was to see the interaction of three power components: load capacity

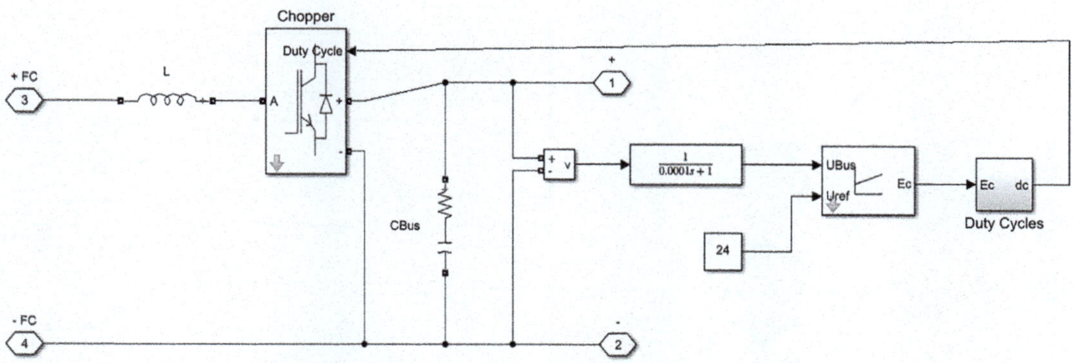

FIGURE 1.8 DC_DC boost converter built in Simulink.

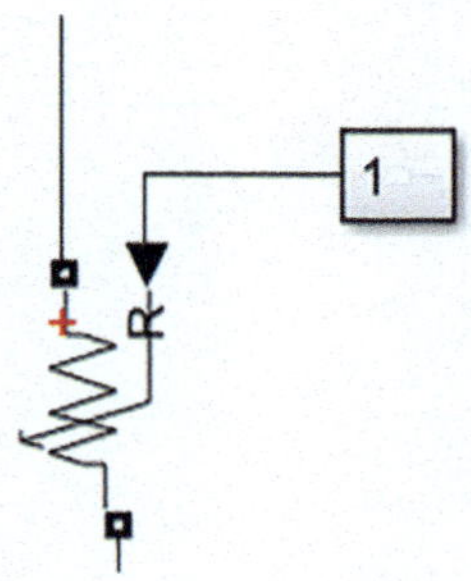

FIGURE 1.9 Model of a rheostat.

(P_R), FC assembly capacity (PFC), and battery pack capacity (P_{Bat}). In which the capacity of the PEM FC pack is designed with a constant power from the beginning of 300 W, corresponding to a rated rate of 3.5 knots. As the train speed increases or decreases, it is clear that the capacity of the load also increases and decreases according to the ternary curve established in Section 1.2.2 (Fig. 1.2). With a "hybrid" type propulsion system consisting of both FC and battery clusters that together supply energy to the load, below the rated value, the FC assembly simultaneously performs two tasks: supplying power to the load and charging the battery pack ($P_{FC} = P_{Bat} + P_R$); in higher-than-rated speed mode, now both the FC assembly and the battery of the Pluto Plus ROVs supply power to the load, ensuring the effective voltage value output for the propeller hybrid motorcycle ($P_{FC} = P_R + P_{Bat}$). To show this change the P_R must change within the preset simulation period (10 s), and specifically for this change, the load capacity must increase. With a fixed output voltage of 24 volts, for the power to increase, the resistance must decrease. It is assumed here that the law of reduction of load resistance over time is given by the following equation.

$$R(t) = -0.5231559633.t + 5.76$$

In Simulink a rheostat is called a "variable resistor" block, with the input "*R*" being the law of change of the rheostat. To build the law of transformation for input "*R*," it is necessary to add blocks "Fcn" and "Clock." The model of a rheostat is shown in Fig. 1.9.

After building all the parts, FC assembly, battery pack, and load, connect these parts and let the program start running according to the preestablished simulation time. In addition to the main blocks built and set parameters to display simulation results, there are other blocks such as the "Product" block to multiply the voltage and current values to give the power value, the "Display" block to display the values, the "Scope" block to display the results in the form of graphs according to the simulation time, and the special block. After building all the blocks, run the simulation on Simulink by pressing the keyboard shortcut "Ctrl + T" or pressing the "RUN" icon on the toolbar in Simulink. Once the program is running, the software will check the program for errors and then run the results. The block diagram of the complete simulation is shown in Fig. 1.10.

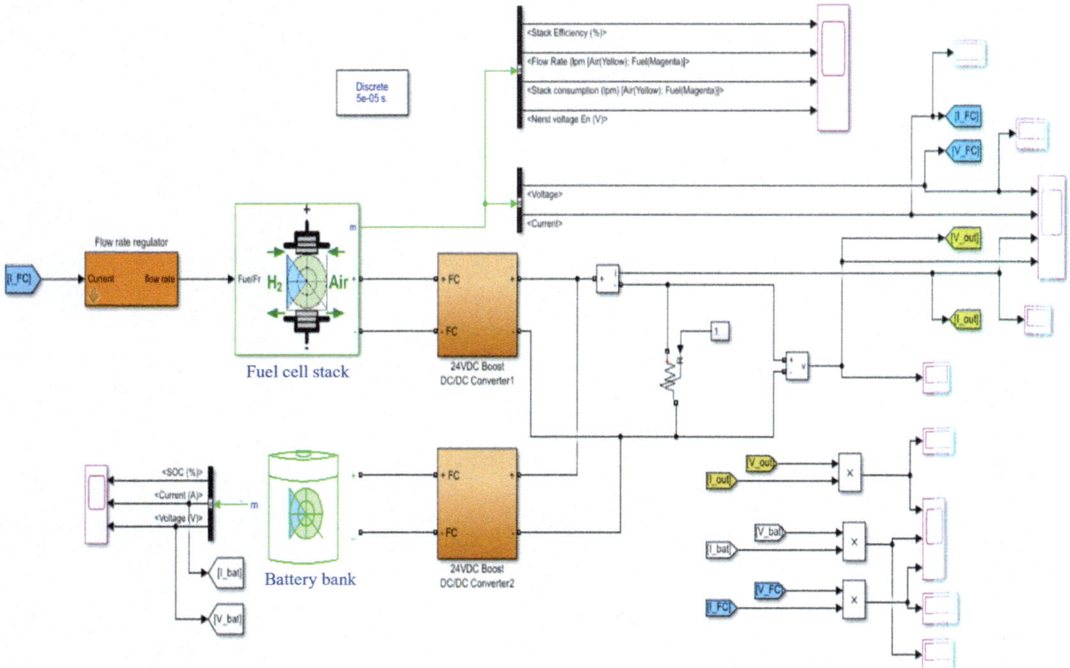

FIGURE 1.10 Fuel cell hybrid propulsion system.

1.3 Results and discussion

The result of running the simulation is shown as a simulation time-dependent graph in the "Scope" block. Output the program's simulation results through the three set scope screens (Figs. 1.11–1.15). In addition, the "Display" displays will show the stability parameters of the values we need to output; these values show us the stable value of the block diagram compared to the set value. In addition to the simulation results built from the block diagram in Simulink, some other results for the FC assembly are written in the MATLAB programming section to see the relationship between FC parameters when operating.

The construction of the simulated block diagram achieves the purpose of the control process when the powertrain of the Pluto Plus ROVs is fitted with an FC assembly. According to Figs. 1.11–1.15 , at the time of simulation $t = 7.34$ (s), it is also the time corresponding to the nominal speed ($v = 3.5$ knots). At this time the capacity of the FC assembly only provides electricity for the load, and the battery pack almost does not provide capacity for the load as well as receive electricity charged from the FC assembly. At times less than 7.34 (s) the battery cluster capacity has now performed two tasks simultaneously: powering the load and charging the battery. At times greater than 7.34 s the load capacity is supplied from two power sources: the PEMFC battery pack and the battery pack.

On the basis of building a special simulation process from the results in Fig. 1.13 a diagram can be built to show the interaction of power components according to the speed change of ROVs on the Graph software, as shown in Fig. 1.15.

The performance, O_2, and H_2 gas flow of PEMFC are stable for the entire simulation period, meaning that PEMFC operates consistently for all velocity modes of the ROV (Figs. 1.11–1.15).

According to Fig. 1.12 the voltage and current of PEMFCs are relatively stable at 12 V and 25 A, whereas the voltage drop on the propeller hybrid engine is also stable at 24 V. The change in load according to the vessel propulsion power is the change in the upper current applied to the electric motor. When the ROV speed changes, the values may change slightly but not significantly, which can be considered.

The objective of the control here is to keep the capacity of the PEM stable while the PEMFC operates at any speed, which makes automation of the PEM and servicing systems easier because, for a propulsion system for small underground vehicles like ROVs, the complexity of driving the propulsion system really is not necessary.

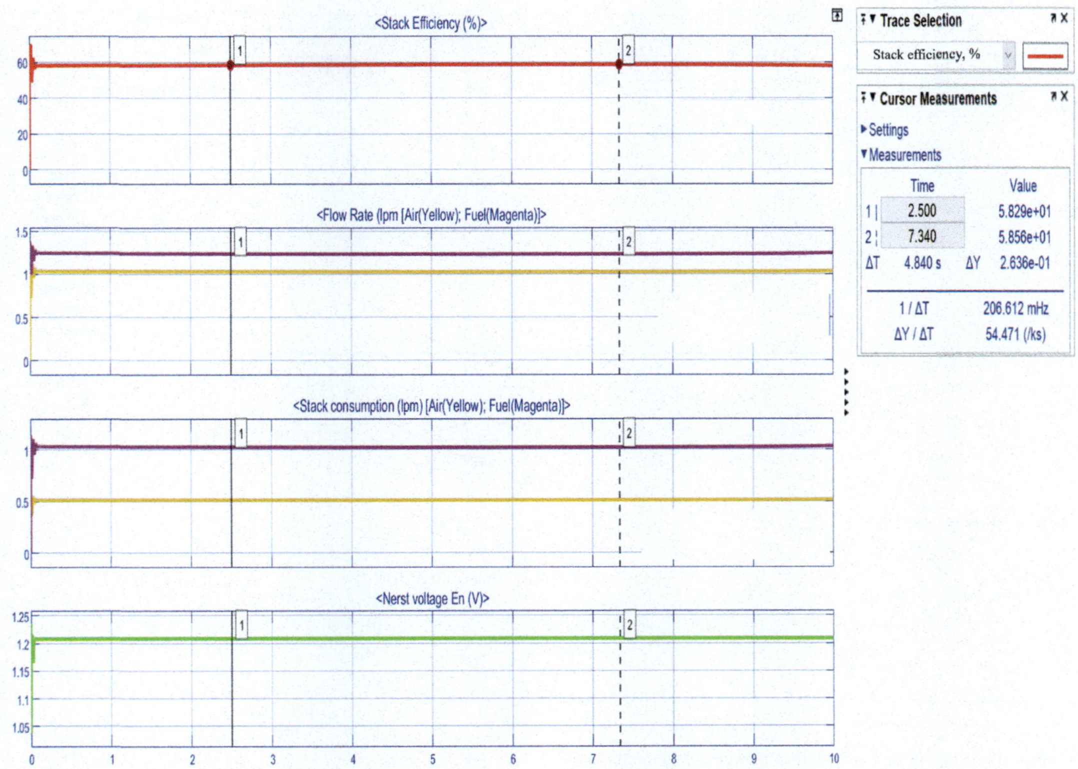

FIGURE 1.11 Simulation results of PEMFC operating parameters. *PEMFC*, Proton exchange membrane fuel cells.

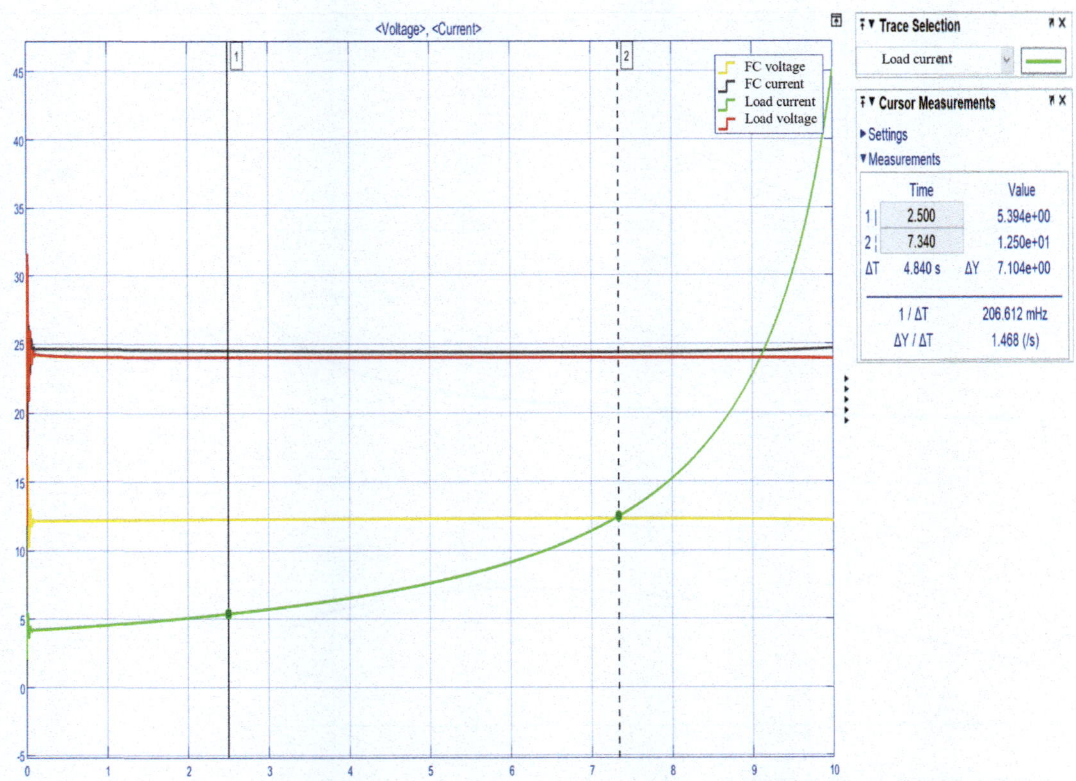

FIGURE 1.12 Simulation results of PEMFC and load current and voltage. *PEMFC*, Proton exchange membrane fuel cells.

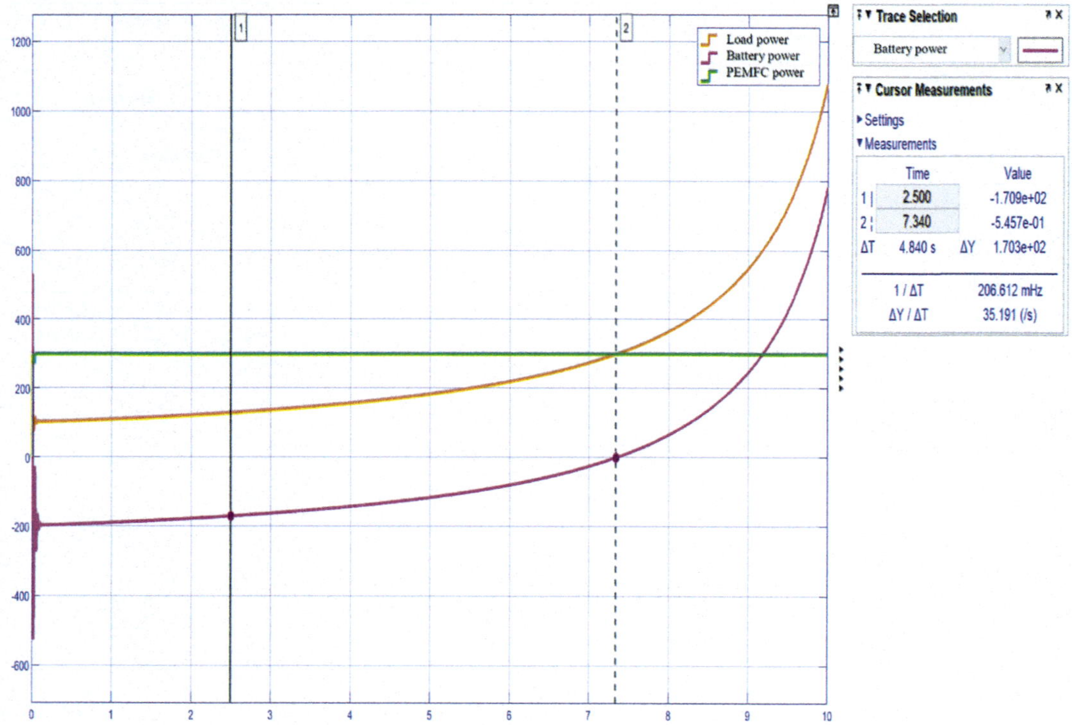

FIGURE 1.13 Simulation results of interaction power components.

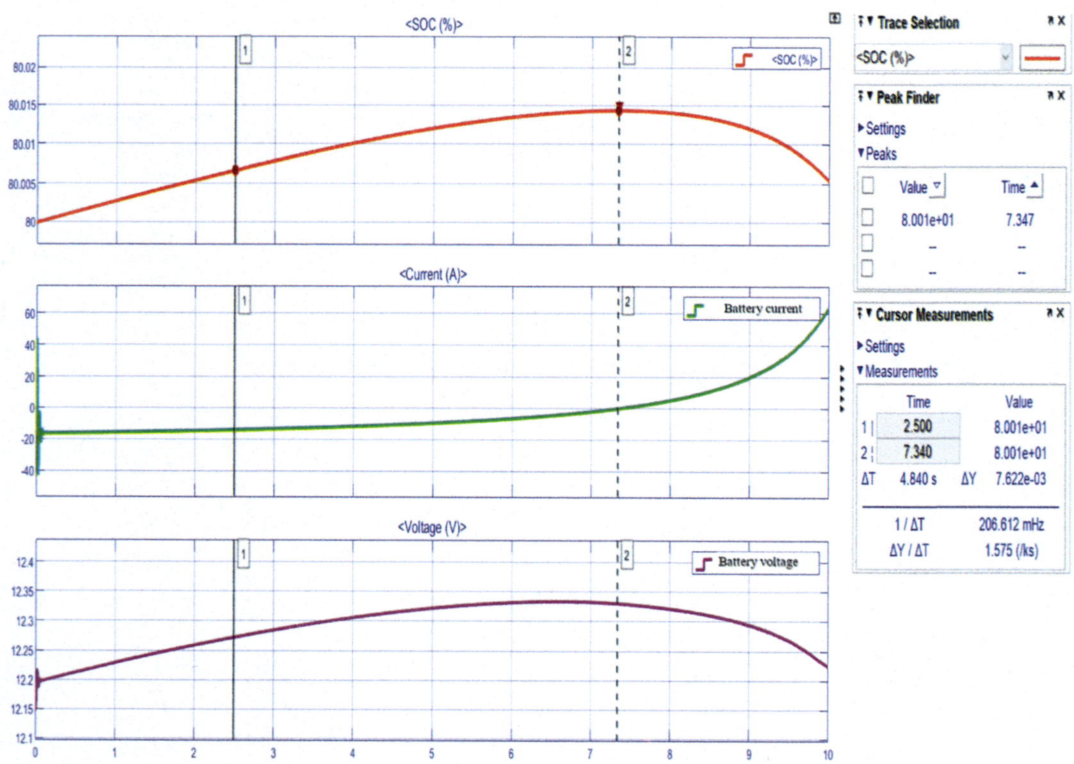

FIGURE 1.14 Battery parameter simulation results.

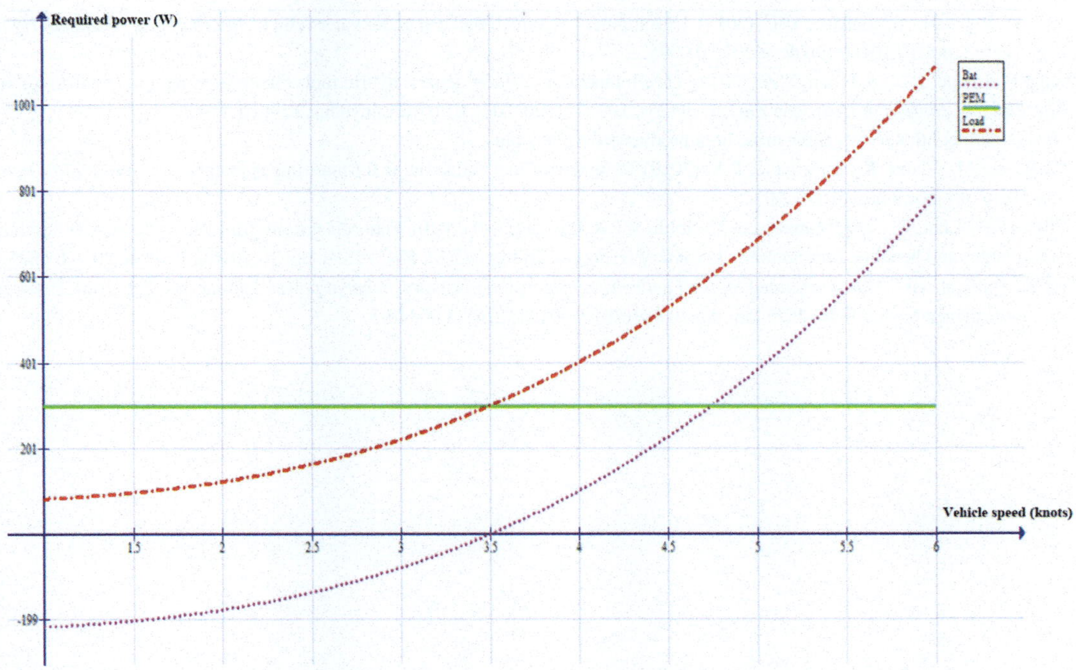

FIGURE 1.15 Simulation of power depends on operating ROV speeds. *ROV*, remotely operated vehicle.

1.4 Conclusion

Calculate some important systems of the FC hybrid propulsion system, such as the FC stack, battery bank, fuel flow rate, humidifier calculation, Boost Converter DC_DC, and rheostat. Simulate hybrid propulsion for Pluto Plus ROVs when using proton-to-battery exchange membrane FC assemblies. From MATLAB Simulink and previously calculated and set parameters, build a block diagram simulating the powertrain of Pluto Plus ROVs after the addition of an FC assembly. The simulation results demonstrated the role of the PEMFC in the entire propulsion system. This result can be considered a preliminary view of the overall "hybrid" propulsion system for submersibles in general, as well as the basis for improving the propulsion system for the current Pluto Plus ROVs to meet many requirements and tasks.

References

Albarghot, M. M., Iqbal, M. T., Pope, K., & Rolland, L. (2020). Dynamic modeling and simulation of the MUN explorer autonomous underwater vehicle with a fuel cell system. *AIMS Electronics and Electrical Engineering, 4*(1), 114–131. doi:10.3934/electreng.2020.1.114.

Ansari, S. A., Khalid, M., Kamal, K., Ratlamwala, T. A. H., Hussain, G., & Alkahtani, M. (2021). Modeling and simulation of a proton exchange membrane fuel cell alongside a waste heat recovery system based on the organic Rankine cycle in MATLAB/SIMULINK environment. *Sustainability, 13*(3), 1218. doi:10.3390/su13031218.

Carrette, L., Friedrich, K.A., & Stimming, U. (2001). Fuel cells—Fundamentals and applications. Fuel Cells, 1, 5–39. https://doi.org/10.1002/1615-6854(200105)1:1⟨5::AID-FUCE5⟩3.0.CO;2-G.

Cu Xuan, P., & Nguyen Quoc, Q. (2023). Early-stage analysis of air independent propulsion based on fuel cells for small submarines. *Advances in Military Technology, 17*(2), 457–469. doi:10.3849/aimt.01744.

d'Amore-Domenech, R., Raso, M. A., Villalba-Herreros, A., Santiago, Ó., Navarro, E., & Leo, T. J. (2018). Autonomous underwater vehicles powered by fuel cells: Design guidelines. *Ocean Engineering, 153*, 387–398. http://www.journals.elsevier.com/ocean-engineering/. doi:10.1016/j.oceaneng.2018.01.117.

De Lorenzo, G., Piraino, F., Longo, F., Tinè, G., Boscaino, V., Panzavecchia, N., Caccia, M., & Fragiacomo, P. (2022). Modelling and performance analysis of an autonomous marine vehicle powered by a fuel cell hybrid powertrain. *Energies, 15*(19), 6926. doi:10.3390/en15196926.

EG&G Technical Services, Inc. (2004). *Fuel cell handbook* (7th ed.). US Department of Energy, Office of Fossil Energy, National Energy Technology Laboratory: Morgantown.

Lee, J.-C., & Shay, T. (2018). Analysis of fuel cell applied for submarine air independent propulsion (AIP) system. *Journal of Marine Science and Technology, 26*(5), 5. doi:10.6119/JMST.201810_26.0005.

Saeed, E. W., & Warkozek, E. G. (2015). Modeling and analysis of renewable PEM fuel cell system. *Energy Procedia, 74*, 87–101. http://www.sciencedirect.com/science/journal/18766102. doi:10.1016/j.egypro.2015.07.527.

Shih, N.-C., Weng, B.-J., Lee, J.-Y., & Hsiao, Y.-C. (2013). Development of a small fuel cell underwater vehicle. *International Journal of Hydrogen Energy, 38*(25), 11138–11143. doi:10.1016/j.ijhydene.2013.01.095.

Shih, N.-C., Weng, B.-J., Lee, J.-Y., & Hsiao, Y.-C. (2014). Development of a 20 kW generic hybrid fuel cell power system for small ships and underwater vehicles. *International Journal of Hydrogen Energy, 39*(25), 13894–13901. doi:10.1016/j.ijhydene.2014.01.113.

Pluto plus ROV, Gaymarine: https://www.gaymarine.it/en/products/rov-pluto-plus.

van Biert, L., Godjevac, M., Visser, K., & Aravind, P. V. (2016). A review of fuel cell systems for maritime applications. *Journal of Power Sources, 327*, 345–364. doi:10.1016/j.jpowsour.2016.07.007.

Weydahl, H., Gilljam, M., Lian, T., Cato Johannessen, T., Holm, S. I, & Hasvold, J. Ø (2020). Fuel cell systems for long-endurance autonomous underwater vehicles—Challenges and benefits. *International Journal of Hydrogen Energy, 45*(8), 5543–5553. doi:10.1016/j.ijhydene.2019.05.035.

Winkler, W. (2009). *Applications—Transportation | Ships: Fuel cells encyclopedia of electrochemical power sources* (pp. 338–358). Elsevier. http://www.sciencedirect.com/science/book/9780444527455. doi:10.1016/B978-044452745-5.00340-3.

Chapter 2

Hybrid ant colony optimization algorithms for solving dynamic capacitated vehicle routing problem in Kufa-Iraq waste collection system

Thaeer Mueen Sahib[a], **Rosmiwati Mohd-Mokhtar**[b] **and Azleena Mohd Kassim**[c]

[a] *Department of Electrical Techniques, Kufa Technical Institute, Al-Furat Al-Awsat Technical University, Iraq,* [b] *School of Electrical and Electronic Engineering, Universiti Sains Malaysia, Nibong Tebal, Malaysia,* [c] *School of Computer Science, Universiti Sains Malaysia, Pulau Pinang, Malaysia*

2.1 Introduction

Local governments in developing nations continue to struggle with managing solid waste (Hoque & Rahman, 2020; Kwakye et al., 2024). It involves several activities, including production, storage, disposal, treatment, recycling, collection, and transportation to a transfer station before being disposed of in the landfill site (Duhayyim et al., 2022; Saxena et al., 2021). It is well known that poorly managed waste causes the clogging of sewage channels, which consequently leads to flooding of roads and the spread of diseases (Iqbal et al., 2022a; Kaza et al., 2018). Consequently the environmental aspect deteriorates (Rathinabhagya & Vinotha, 2023). The use of high-capacity vehicles to transport waste from the inner city's periphery to recycling centers outside the city is an additional concern (Rahmanifar et al., 2024). Therefore a proper waste management system is necessary for efficient waste collection (Iqbal et al., 2022b).

Usually the methods of improving waste collection would be either a controller system method to reduce the time required for waste disposal or a vehicle routing method as a management system (Chaabane et al., 2021; Sahib et al., 2024b). These studies examined how dynamic vehicle routing affected the best route selection in the waste collection system. Consequently, numerous studies have leveraged geographic information system (GIS) technologies to enhance waste collection methods (Hoang et al., 2024; Vu et al., 2020).

Over the years, numerous optimization algorithms have been proposed to address the vehicle routing problem (VRP) in the context of waste collection systems (Akhtar et al., 2017; Ismail et al., 2021; Sahib et al., 2022). The literature has concentrated on improvement optimization algorithms, emphasizing the exploration and exploitation stages to assess neighboring solutions iteratively (İlhan, 2021). Most of the studies that dealt with the same context as the case study did not evaluate the distribution of the solid waste amount that has been collected and were only concerned with evaluating the improvement methods related to verifying their performance using known capacitated VRP (CVRP) datasets. Such a study conducted by Akhtar et al. (2017) utilized a modified backtracking search algorithm (BSA) in CVRP models with the smart bin concept to find the best optimized waste collection route solutions and threshold waste level (TWL) of waste bins to reduce the number of bins to be emptied. However, this study did not focus on optimizing the bin number or bin size.

A similar study by Hannan et al. (2018) implemented the same approach but adopted the particle swarm optimization (PSO) algorithm. The study evaluated the search capabilities of the proposed algorithm against the best known solution (BKS) using a standard benchmark dataset for CVRP. Henke et al. (2015) proposed a variable neighborhood search (VNS) algorithm. They employed it to solve multicompartment VRPs with flexible compartment sizes exclusively for waste

Computational Intelligence in Mechatronics: Solving Real-World Problems in Electronic Systems Design. DOI: https://doi.org/10.1016/B978-0-443-27580-7.00023-8

classification purposes and to determine the number of compartments. The vehicle capacity can be divided, and the size of each compartment must be considered. The objective is to minimize the transportation distance to the edges by visiting each location separately to pick up different product types. Three sets of numerical experiments were performed to evaluate the small and large instances that were generated randomly.

And et al. (2016) hybridized a simulated annealing (SA) algorithm with a heuristic algorithm for a robust capacitated arc routing problem (CARP) to minimize the traversed distance according to the demand uncertainty of the edges. The study aims to minimize the cost of vehicle usage. Along with other optimization problems, Li et al. (2012) employed a Fuzzy Stochastic Quadratic Programming (SFQP) method and scenario-based multistage stochastic programming (MSP) for solving CARP for the best waste management, thus minimizing collection and transportation costs. The study focused on estimating the costs that depend on container size, cyclic waste collection, and collection time in each load.

Ismail et al. (2021) proposed a basic ant colony optimization (ACO) algorithm to solve the traveling salesman problem (TSP) by minimizing the route distance in a single route. Regarding solving the curbside waste collection problem, a study conducted by Mat et al. (2018) centered on reducing the number of vehicles required, thus reducing the excessive cost because of increasing fuel consumption. Considering five scenarios the study used a mathematical model that predicts a collection's energy and time requirements.

However, some constraints on Municipal Solid Waste (MSW) collection, such as natural geography, traffic congestion, and inadequate transport infrastructures, remain unsolved (Edwards et al., 2016). Moreover, two studies in Iraq confirmed that municipal centers still use traditional methods for collecting and transporting waste (Knowles, 2009; Sahib & Hadi, 2023). Table 2.1 summarizes the analysis of most of the variants' challenges and identifies the variants that have yet to be presented.

In the above table the scenario column encompasses (1) pick-up delivery problem (PDP)—depot and transfer station in the same location; (2) split the pick-up from the delivery problem (SPDP)—depot and the transfer station are in separate locations.

This study focuses on adapting the ACO algorithm to address the homogeneous CVRP within the context of waste collection systems, encompassing waste pick-up and delivery to the temporary waste transfer station (TWTS). In this research, based on our knowledge, no study has focused on solving CVRP to find optimal routes and reduce the number of vehicles used to service all containers. Therefore the proposed study addressed this issue by hybridizing the ACO algorithm with the SA (ACO–SA) algorithm and improving the latter with the premutation operator (ACO–SAPM) technique.

The objectives of the study are introduced as follows:

- Design a new systematic framework model by hybridizing ACO to increase its searchability in the exploration and exploitation stages.
- Evaluate the convergence of the enhanced algorithm with a real-world case study dataset collected in Al-Kufa City, Iraq.
- Design a technique for detecting temporary traffic congestion and planning new routes to avoid them, as well as detecting a TWTS that makes the solid waste collection system work dynamically.

The remainder of this chapter follows: Section 2.2 provides a detailed formulation of the capacitated VRP. Section 2.3 discusses the methodology and improvements made to the ACO algorithm. Section 2.4 presents the data collection for the waste collection case study from Al-Kufa City, Iraq. Results and discussion are shown in Section 2.5. Finally, Section 2.6 concludes the chapter by offering insights into future directions.

2.2 Formulation of CVRP

This section describes the mathematical representation of CVRP-related PDP and SPDP scenarios. It includes an objective function representation and a set of necessary constraints for the waste collection system's operation.

2.2.1 Model of PDP scenario

This model was formulated to evaluate the enhanced algorithm as a standard problem and used as real problems (dataset of T1 and T2) extracted from Al-Kufa City. The PDP is represented by $G = (N, E)$, where N denotes a set of containers distributed as follows: $N = \{0, 1, \ldots, n\}$. The index 0 signifies the depot, whereas the indices from 1 to n correspond to the containers. As for E, it is a set of edges referred to as $V = \{0, 1, \ldots, n\}$. Typically, each edge (i, j) in E can be defined as $\{(i, j): i, j \in V, i \neq j\}$ (Yassen et al., 2017). It has an associated trip cost denoted as C_{ij}. In a practical sense the weights assigned to these edges represent the travel distance between container (i) and container (j) where $(i \neq j)$. Every container, denoted

TABLE 2.1 Literature studies related to work.

References	Methods	Technique	Objective	Routing problem	Scenario	Evaluation		Variants	
						EPPA	EWC	Bin number	Bin size
Akhtar et al. (2017)	BSA	TWL	Reducing the bins to be emptied. (minimize distance)	CVRP	PDP	CVRP dataset	×	×	×
Hannan et al. (2018)	PSO	TWL	Reducing the bins to be emptied. (minimize distance)	CVRP	PDP	CVRP dataset	×	×	×
Henke et al. (2015)	VNS	–	Minimizing distance between edges	MCVRP-FCS	PDP	CPLEX 12.5	×	×	×
And et al. (2016)	SA + Exact method	DUE	Reducing the vehicles required. (minimize distance)	CARP	PDP	CPLEX solver	√	×	×
Li et al. (2012)	SFQP + MSP	DUE	Minimizing distance between edges	CARP	PDP	fuzzy sets	√	×	√
Ismail et al. (2021)	ACO	–	Minimizing distance between edges	TSP	PDP	–	×	×	×
Edwards et al. (2016)	mathematical model	TWTS	Reducing the vehicles and time required	–	SPDP	–	×	×	×
This study	ACO–SA + ACO–SAPM	TWTS + TTC	Reducing the vehicles required. (minimize distance)	CVRP	PDP + SPDP	CVRP dataset	√	√	√

ACO–SA, ant colony optimization-simulated annealing; *BSA*, backtracking search algorithm; *CARP*, capacitated arc routing problem; *CVRP*, capacitated vehicle routing problem; *DUE*, demand uncertainty of edges; *EPPA*, evaluation performance of proposed algorithm; *EWC*, evaluation waste collected; *MCVRP-FCS*, multi-compartment vehicle routing problem with flexible compartment sizes; *MSP*, multistage stochastic programming; *PDP*, pick-up delivery problem; *PSO*, particle swarm optimization; *SFQP*, fuzzy stochastic quadratic programming; *SPDP*, split the pick-up from the delivery problem; *TSP*, traveling salesman problem; TTC, temporary traffic congestion; TWL, threshold waste level; TWTS, temporary waste transfer station; *VNS*, variable neighborhood search.

as i and belonging to the set $i \in V' = V\backslash\{0\}$, has a demand q_i, with i ranging from 1 to n within the context of a single depot (0).

A fleet of homogeneous vehicles denoted as K, each with a capacity limit of Q. The specific number of vehicles present at the depot is not predefined. However, a subtour elimination technique has been devised to mitigate the inflated costs of adapting the basic ACO algorithm to classical CVRP scenarios. A "subtour" is a situation in which one vehicle serves only one or two containers within a single route. However, the decision variable (X_{ijk}) must be introduced to formulate the waste collection system. This variable is defined as follows. $X_{ijk} = 1$, indicating that the vehicle k travels from container (i) to container (j). Conversely, $X_{ijk} = 0$ signifies that the vehicle does not match the planned route. As expressed in Eq. (2.1),

$$X\_ijk = \begin{cases} 1, \text{ if the vehicle served the edge } (i, j) \\ 0, \text{ Otherwise} \end{cases} \tag{2.1}$$

Eqs. (2.2) and (2.3) formally express the objective function and constraints of the problem.

$$\text{Minimize} = \sum_{i=0}^{N}\sum_{j=0}^{N}\sum_{k=1}^{k} d_{ij}X_{ijk} \tag{2.2}$$

$$d_{ij} = \begin{bmatrix} 0 & d_{12} & \cdots & d_{1N} \\ d_{12} & 0 & \cdots & d_{2N} \\ \vdots & \vdots & \ddots & \vdots \\ d_{1N} & d_{2N} & \cdots & d_{NN} \end{bmatrix} \tag{2.3}$$

The constraints are subjected to Eqs. (2.4–2.14).

$$\sum_{j=1}^{N}\sum_{k=1}^{k} X_{0jk} = 1 \tag{2.4}$$

$$\sum_{j=1}^{N} q_{0jk} = 0, k \in \{1, 2, \ldots K\} \tag{2.5}$$

$$\sum_{k=1}^{k}\sum_{j=0}^{N} X_{ijk} = 1, i \in \{1, 2, \ldots N\} : i \neq j \tag{2.6}$$

$$\sum_{k=1}^{k}\sum_{i=0}^{N} X_{ijk} = 1, j \in \{1, 2, \ldots N\} : i \neq j \tag{2.7}$$

$$\sum_{i=0}^{N}\sum_{j=0}^{N} Q_j X_{ijk} \leq Q, k \in \{1, 2, \ldots K\} \tag{2.8}$$

$$\sum_{j=1}^{N} X_{ijk} = \sum_{j=1}^{N} X_{jik} = Yik, i \in \{1, 2, \ldots N\}, k \in \{1, 2, \ldots K\} \tag{2.9}$$

$$\sum_{i=0}^{N}\sum_{k=1}^{K} q_{jik} - \sum_{i=0}^{N}\sum_{k=1}^{K} q_{ijk} = cj, j \in \{1, 2, \ldots n\} \tag{2.10}$$

$$\sum_{j=1}^{N} X_{ij}^{k} = \sum_{j=1}^{N} X_{ji}^{k} \leq 1 \text{for} i = 0 \text{ and } k \in \{1, 2, \ldots .k\} \tag{2.11}$$

$$\text{dist}_{ij} = \text{dist}_{ji} \tag{2.12}$$

$$\sum_{j=1}^{N} X_{0j}^{k} \leq 1, k \in \{1, 2, \ldots .k\} \tag{2.13}$$

$$\sum_{i=1}^{N} X_{i0}^{k} \leq 1, k \in \{1, 2, \ldots .k\} \tag{2.14}$$

As Eq. (2.2) outlines, the objective function is to minimize the total distance vehicles travel to serve all collection points (CPs) (container locations) within the network. In this equation the cost of distance d_{ij} is determined by a symmetrical matrix or a square matrix of size $(N \times N)$ where $(d_{ij} = d_{ji})$, reflecting the distance cost incurred when a vehicle travels from a container (i) to a container (j), as specified by the logical component (X_{ijk}) in Eq. (2.3).

This study is subject to a set of constraints, which are detailed as follows:

- The vehicle K initiates its route from the depot (0), as stated in Eq. (2.4).
- Each vehicle starts its work with an empty load, as indicated in Eq. (2.5).
- Each container can be served by only one vehicle, as expressed in Eqs. (2.6) and (2.7).
- The total waste within each container must not exceed the vehicle capacity Q, by Eq. (2.8).
- The continuity condition is enforced by Eq. (2.9).
- Ensure that the vehicle empties the container visited, which is made based on Eq. (2.10).
- Each vehicle completes its tour by departing from and returning to the same depot, as Eq. (2.11) guarantees.
- Ensure that the distances between pairs of containers are consistent as in Eq. (2.12).
- It guarantees that any vehicle will be used only once, as denoted in Eqs. (2.13) and (2.14).

2.2.2 Model of SPDP scenario

This model is formulated to evaluate the performance of the best algorithm in terms of applying temporary traffic congestion (TTC) and TWTS scenarios, which is represented as one real problem (dataset T3). The SPDP is used as a direct graph, $G = (N, E)$. Let N be a set of nodes distributed geographically, where $N = \{0, 1, \ldots, m\}$. Let node zero be the depot, and the rest of the nodes in the range $\{1, 2, \ldots, m\}$ represent the containers, except node m indicates the TWTS. Similarly the amount of waste transported by homogeneous vehicles (a uniform load capacity) is represented by the vehicle as $k = \{1, 2, \ldots, K\}$. Meanwhile, E is the set of edges $E = \{(i, j), i, j \in N: i \neq j\}$, with positive demands. The edges have weights denoted as travel time t_{ij} or the distance. Each edge connects two nodes, with the depot as the starting point for vehicle k to serve each container. If the vehicle exceeds capacity, it goes directly to TWTS to unload waste and returns to the depot. The quantity of waste is represented based on the size of the containers and distributed according to the estimated waste volume at each commercial site.

The proposed scenario is subject to the following constraints:

- Constraint ensures that the vehicle moves from the last container after emptying it and then concludes its tour at TWTS to unload the waste, as in Eq. (2.15).

$$\sum_{\substack{(j,m)\in N \\ j \in N\backslash\{1,m\}}} x_{jm}^{k} = 1, \forall k \in K \tag{2.15}$$

- Guarantee that the vehicle travels from TWTS and back to the depot without carrying any load, as indicated in Eqs. (2.16) and (2.17).

$$\sum_{\substack{(m,1)\in N \\ j \in N\backslash\{1,m\}}} x_{m1}^{k} = 1, \forall k \in K \tag{2.16}$$

$$\sum_{\substack{(1,j)\in N \\ j \in N\backslash\{1,m\}}} q_{m1}^{k} = 0, \forall k \in K \tag{2.17}$$

2.3 Improvement of ACO algorithm

The ACO algorithm is a metaheuristic algorithm inspired by nature (Ali & Kar, 2018). Researchers utilized a swarm algorithm derived from the social behavior of ant colonies to solve optimization problems (Pedemonte et al., 2011). Typically, ants return to their nest by taking the shortest route possible after acquiring food. Moreover, ants deposit pheromones on the ground during their travels to signal the pathways other colony members should follow. Although ACO is robust and can explore a solution, convergence is still slow and tends to be stuck in the local optimum (Ebadinezhad, 2020). In this chapter the ACO algorithm is developed into three sections: (1) construction of the initial

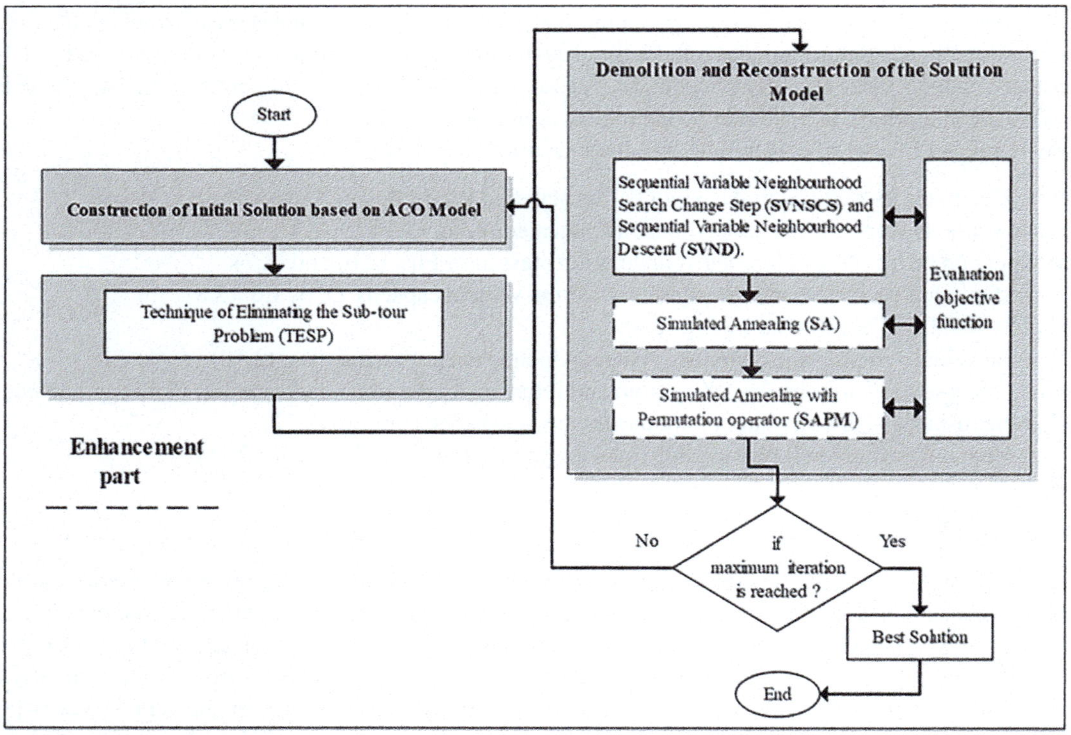

FIGURE 2.1 Flowchart of hybridization ACO algorithm. *ACO*, Ant colony optimization.

solution based on the ACO model; (2) elimination of a subtour problem; (3) demolition and reconstruction of the solution model, as illustrated in Fig. 2.1.

2.3.1 Enhancement of ACO with simulated annealing algorithm

In this section the ACO is incorporated with the SA algorithm (ACO–SA) to enhance its local search capabilities and move toward a more feasible solution, thereby escaping local optima by transitioning from the current solution to a neighboring solution. Therefore the fitness value $f(S)$ and the new fitness of the neighboring solution $f(S')$ must be computed. Delahaye et al. (2019) confirmed that the SA algorithm commences by selecting a random solution as the initial solution, as presented in Algorithm 2.1.

At this point the system temperature ($T = T_0$) is equivalent to this initial temperature. An iterative process involves obtaining a solution near the current solution. Algorithm 2.1 compares its fitness value, which will replace the current solution if the solution is better. If worse, it will replace the current solution with a probability obtained from the Boltzmann function, as shown in Steps (12–14). This mechanism prevents the algorithm from being trapped in a local minimum. In this relation, Δ_f is the difference in fitness values between the current solution and the new solution. K is the Boltzmann constant determined beforehand, and T is the current temperature. The neighborhood search procedure is iterated for a predetermined number of completed iterations. The flowchart of the hybridized ACO with SA algorithms is described in Fig. 2.2.

The ACO–SA starts receiving the initial solution (S_i^t) constructed by the ACO algorithm and reads every route (i) individually to prepare them for improvement. Alternatively a route consisting of two nodes (U_i^t) is omitted and saved in the final solution because of the incompatibility of the mutation operator searches with this specific route type. Algorithm 2.2 presents the pseudocode employed in the proposed algorithm to compute the fitness value for each route.

The fitness value is calculated by adding the distances of all CPs in the designated area and each route from the depot, as shown in lines (5–7). Finally the distances back to the depot are computed based on the line (11). The summation of all distances is denoted in line (13). Eventually the total distance (fitness value) returns to the main function used for evaluation.

2.3.1.1 *Parameters of the SA algorithm*

The initialization of the parameter T_0 operates as a temperature control in the Boltzmann equation during ACO–SA startup (refer to). Typically the SA emulates the cooling mechanism in which the temperature gradually decreases to a steady state.

ALGORITHM 2.1 Pseudocode of simulated annealing algorithm.

```
Input: Initialise parameters
S = generate initial solution ();
T = T0;
While (T > Tfinal)
{
Until (N ≤ I - Iter)
{
Generate solutions S' in the neighbourhood of S.
If f(S') < f(S)
S ← S'
else
Δf = f(S') - f(S)
r = random ();
If (r < exp(-Δ/k*T))
S ← S'
}
T = α x T;
}
Output: returns the best solution found
```

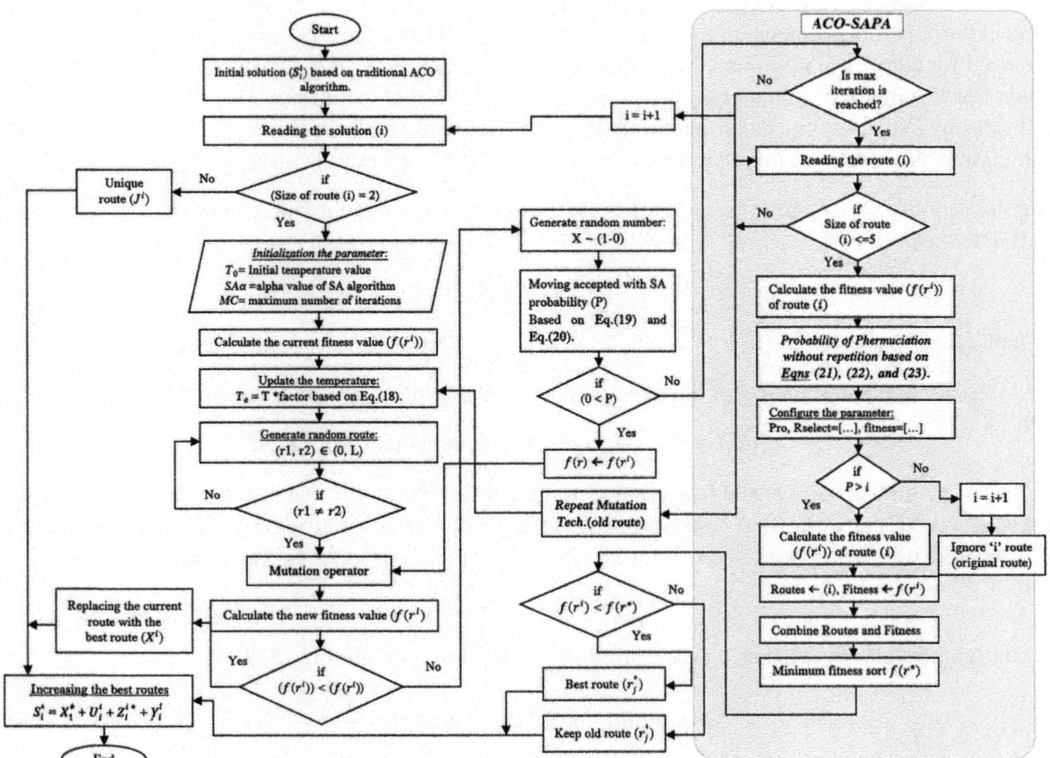

FIGURE 2.2 Enhancement of ACO algorithm with simulated annealing (ACO–SA). *ACO–SA*, Ant colony optimization-simulated annealing.

ALGORITHM 2.2 Pseudocode to calculate the distance of individual routes.

```
Input: Identify the route
Initialise: su ← 0
: S_P ← [...], s_p1 ← [...], s_p2 ← [...]
a ← 1
for j:= s do
b ← j
su + ← edges[(min(a,b), max(a,b))]
s_p1.append(edges[(min(a,b), max(a,b))])
a ← b
b ← 1
su + ← edges[(min(a,b), max(a,b))]
s_p2.append(edges[(min(a,b), max(a,b))])
S_P ← sum(s_p1+s_p2)
Output: returns the distance of the individual route (S_P)
```

The termination of the search process depends on the maximum number of iterations required to explore the best possible solution. Meanwhile the control parameter alpha (SA_α) value of T_0 will decrease gradually, thus updating the temperature based on Eq. (2.18).

$$T = SA_\alpha \times T_0 \tag{2.18}$$

It is important to note that (SA_α) differs from the α value of the ACO algorithm. For the swapping technique between pair nodes in each route, ACO–SA requires generating two integer random numbers (*randint*) in each iteration. This generation depends on the individual route dimension ($0 - L$). The mutation operator stage occurs exclusively when ($r1 \neq r2$) is reached to prevent the need for additional iterations. Furthermore the neighborhood search algorithm in the SA was transformed into a mutation operator through the exchange of coordinates and neighboring solutions. This change supports the exploration of further solutions by extending the local search. It is essential to compare the existing route with the new one to determine the most optimal one. Nonetheless, there are two primary methods for generating an acceptable solution.

1. Accepting the new fitness solution ($f(x')$) as the best solution and replacing it with the current fitness solution ($f(x)$) such the condition below:

$$f(x') < f(x)$$

Then the new route is denoted as (X_i').

1. Apply a Monte Carlo method to accept the new (worse) solution with a given probability:

$$f(x') > f(x), \text{ accept when acceptance} < \text{probability}$$

A Monte Carlo method is a stochastic transition of the algorithm from one state to another, subject to the condition of the balance. Usually a Monte Carlo step consists of generating a new trial solution of a route vector (i) and computing the ensuing fitness change of the proposed algorithm (Δ_f) (Delahaye et al., 2019), as expressed in Eq. (2.19).

$$\Delta_f = f(x') - f(x) \tag{2.19}$$

The trial configuration is accepted as a new configuration as formulated in Eq. (2.20).

$$X = \text{Random}(0, 1) < P = \exp\left(-\frac{\Delta_f}{T}\right) \tag{2.20}$$

A trial-and-error tuning test for the SA parameters was set. The results show that combining an initial temperature value of 20°C with a cooling rate of 0.99 produces reliable results (Yu et al., 2024).

2.3.1.2 *Strategies of SA based on mutation operators*

There are three cases to explain the strategies of the SA algorithm based on mutation operators as described in the following:

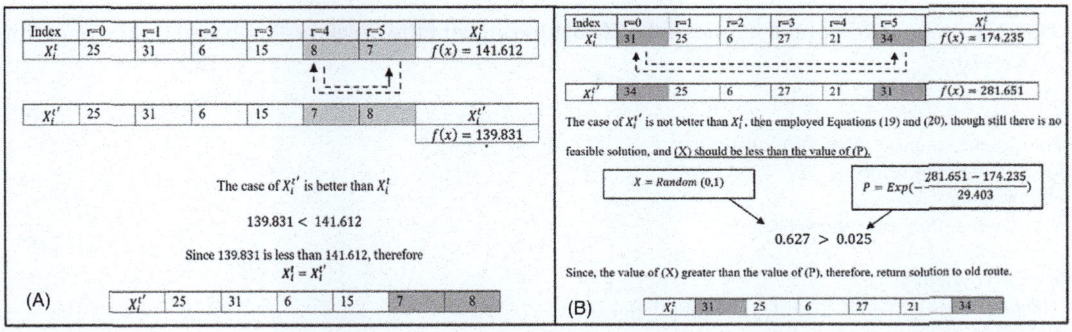

FIGURE 2.3 Mechanism of exploration solution based on mutation technique in (A) case of new feasible solution and (B) case of no feasible solution.

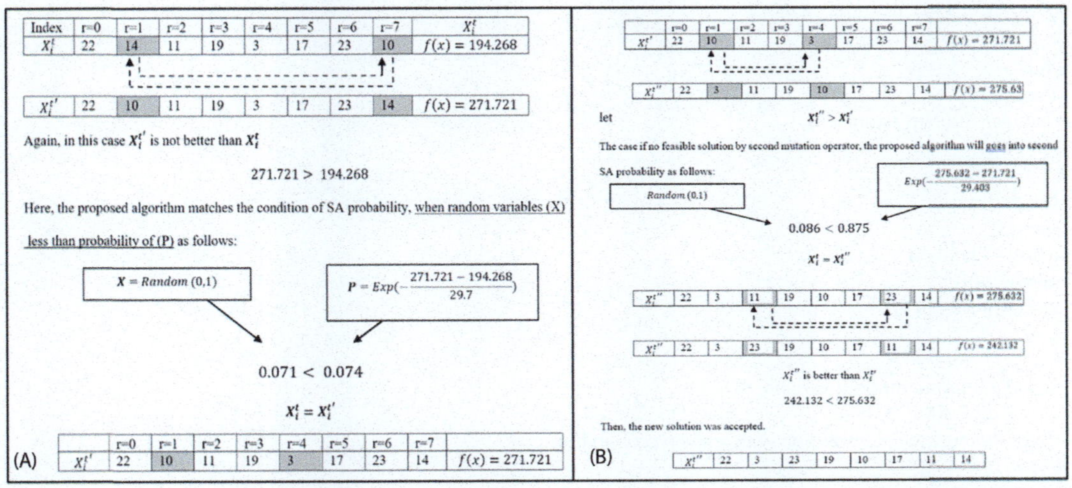

FIGURE 2.4 Mechanism of simulated annealing and mutation technique in (A) case of multiple worse solution and (B) replication mutation operator.

1. Case of mutation operator with a new feasible solution.

The primary operation in the SA algorithm involves randomly swapping the positions of two nodes if a feasible solution exists. If not, the proposed algorithm will move on to the next step as in the sample in Fig. 2.3A.

1. Case of the mutation operator with no feasible solution.

This approach happens when no feasible solution occurs by the mutation operator and by a Monte Carlo method that is expressed in Eqs. (2.19) and (2.20) and shown in Fig. 2.3B.

1. Case of multiple worse solutions.

This case acts when there is no feasible solution, and the Monte Carlo method has met the condition of the SA probability without a feasible solution, as shown in Fig. 2.4A.

1. Case of the replication mutation operator.

The replication case involves determining the probability of selecting a feasible solution, where another mutation operator is applied randomly between $r = 1$ and $r = 4$ to compare the result with the previous route, as depicted in Fig. 2.4B.

Algorithm 2.3 shows the pseudocode for the acceptance solution generated by the improved ACO–SA probability, specifically focusing on the mutation techniques outlined in lines (13–15). The new solution has been approved according to lines (17–23). The acceptance of a new (worse) solution by SA probability is detailed in lines (26–29). If the solution is not accepted as new and worse, any alternative will revert to the previous solution, as indicated in lines (31–34).

The possibility of accepting a solution better than the current one is explored further for other solutions in the search space, except for the solution that consists of only two nodes, as shown in line (6).

ALGORITHM 2.3 Pseudocode of the hybridized ACO algorithm with the SA algorithm based on the mutation operator (ACO–SA).

```
Input Initialise: the solution of the ACO algorithm
: the initial temperature value (T0)
: the factor of alpha value (SAα)
: the maximum number of iterations (MI)
for i:= solution do
if length of (i) ≠ 2 then
f(X) ← CalculateSubListDistance (i) based on Algorithm 2.2
Buff ← tuple(i)
for itr:= Max_Iteration do
T ← factor x T0
r1, r2 ← generate an integer random value depending on the length of (i)
if r1 ≠ r2 and itr ≠ Max_Iteration - 1 then
Temp ← i[r1]
i[r1] ← i[r2]
i[r2] ← temp
f(X') ← CalculateSubListDistance (i)
if f(X') < f(X) then
f(X) ← f(X')
S ← [list(i), f(X')]
New.append(s)
New.sort(key ← lambda x: x [1])
z.append(New [0])
Solution ← [i[0] for i in z]
break
else:
X ← generate a uniform random variable (0 ~ 1)
if X < nump.exp(-(- f(X)) / T) then
Print (Accept the new (worse) solution with a given probability)
f(X) ← f(X')
else:
Print (Do not accept the solution (old route))
Temp ← i[r1]
i[r1] ← i[r2]
i[r2] ← temp
else if itr == Max_Iteration - 1 then
mylist ← list(buff)
KeepOld.append(mylist)
break
else if length of (i) == 2 then
q.append(i) #q: routes consist of 2 containers only
Solution ← solution + KeepOld + q
Output returns an enhanced solution
```

2.3.2 Enhancement of ACO with simulated annealing based on permutation operator

This section discusses how the permutation technique collaborates with the ACO–SA algorithm. The permutation strategy initiates its enhancement once the ACO–SA algorithm reaches the maximum number of iterations and no feasible solution exists. The ACO–SA-based mutation operator achieves superior results compared to other algorithms but may need additional enhancements to prevent the solution from getting stuck in local minima. Various arrangements represent the permutations that occur in a group when the order of events is especially important.

2.3.2.1 Formulation of permutations without repetition technique

This technique systematically explains the process of shuffling nodes so that their order is never repeated. The motivation behind choosing this type of permutation is to ensure the waste collection system does not visit a certain node twice. It is essential to note that this technique addresses the specific route that ACO–SA has not optimized. In other words the available event for permutations is decreased gradually, as permutations are orderings. For instance, if n nodes are arranged according to a specific plan for the movement, the question is, "How many possible routes could be created?" The number of permutations of n items is the factorial of $n(n!)$. The latter can be calculated by multiplying $n \times (n-1)$, $(n-2)$, and so on, down to 1, as expressed in Eq. (2.21).

$$n! = n \cdot (n-1) \cdot (n-2) \tag{2.21}$$

The formula for permutation n objects taken r at a time is given by Eq. (2.22).

$$nPr = \mathrm{Perm}(n_l, r) = n_l!/(n_l - r)! \tag{2.22}$$

In this case the permutation representation in the VRP is devised as follows. Suppose there is a list of containers (i) that represent a vector named (Seq), as expressed in Eq. (2.23).

$$\mathrm{Seq}[i_1, i_2, i_3] \tag{2.23}$$

Let

n_l: length of the sequence ($n_l = 3$).
r: Items that are taken for the permutation.

To find the number of r-permutations, suppose a system needs to determine 3-time events for 3 nodes in a certain sequence, thus $\{nPr = 3P3 = 3!/(3-3)! = 6 \text{ events}\}$ of possible routes. Based on the probability of routes the proposed algorithm will select the best route according to the minimum fitness value.

Algorithm 2.4 shows that the permutation operator explores all potential routes except the initial one that contains the solution to be enhanced from lines (12) to (15). This process occurs when the route's length falls within the range of 3–5 nodes, as specified in lines (9–11). If no solution is obtained through mutation in line (8), the ACO–SA algorithm proceeds until it reaches the maximum iteration number.

All possible routes are stored in a register of "routes," and the total distance of these routes is calculated and saved in the register of "fitness" as shown in lines (16–19). These registers are combined using Python's zip function and configured in the zip format (keys, values), where keys are defined as a list of iterables and values. The zip format has been added to the dictionary instruction at line (20) to easily retrieve the best route based on the minimum key (minimum fitness). The results are printed on line (22). Selecting the correct index when choosing the key or value (fitness/routes) in the dictionary instructions in lines (22–24) is crucial.

The key aspect is the second enhancement to the SA algorithm. The search space dimensions are established by the length of the individual route, constrained to a maximum of five elements. This constraint aims to prevent an increase in the time required to find more possible routes. The current routes are replaced with new routes constructed using the mutation operator and the permutation without repetition technique. It is then aggregated as the best solution in the S_i^tregister. The flow chart illustrating the hybridization of ACO with SA-based permutation technique (ACO–SAPM) is highlighted in Fig. 2.2.

2.3.3 Enhancement of ACO with sequential variable neighborhood search change step (SVNSCS) and sequential variable neighborhood descent (SVND)

The hybrid ACO with sequential VNS change step (ACO–SVNSCS) and sequential VNS descent (SVND) algorithms (hereafter referred to as HACO) are developed to increase the capability of ACO in the exploration and exploitation stages (Sahib et al., 2024a). This model is systematically detailed in Fig. 2.5. The SVNSCS algorithm determines that neighborhood will be explored next and whether it will be considered a new solution. The step represents the first phase for hybridizing ACO with SVNSCS procedures after reading the dynamic routes in several iterations. Then, SVNSCS involves the sequential swapping technique on each route as a systematic framework. The ratio of best swapping depends on the shortest route to waste CPs. The total waste in a sequence of containers should be less than the vehicle's capacity. Note that the best swapping refers to the best solution based on Step 1 in Fig. 2.5. Each node is swapped concurrently by calculating the distance of each route. Each iteration has evaluated feasible solutions, as shown in Step 2 in Fig. 2.5. Eventually the best solution is memorized in the register of the reconstruction solution.

ALGORITHM 2.4 Pseudocode of enhanced ACO with SAPM algorithm.

```
Input: Initialising the solution
Initializing of route
for i:=solution do
if length of (i) ≠ 2 then
f(X) ← CalculateSubListDistance(i) based on Algorithm (2)
for itr:= range(Max_Iteration) do
T ← factor x T0
Apply the mutation operator function
Print ('The solution not accepted by SA (back to old route)')
i ← route is not improved by the mutation operator
if length of (i) < == 5 then
f(X) ← CalculateSubListDistance(i)
seq ← permutations technique on (i) based on Equations (2.21) and (2.22)
p ← 0
for L:= seq do
p ← p + 1
if p > 1 then # Probability of route (i) exception to the original one
f(X'') ← CalculateSubListDistance(L)
l ← list(L) # keep all possible objects list occurred inside the register (l)
Routes.append(l) # Keep all possible routes that occurred in the list Routes
Fitness.append(f(X'')) #Keep all possible route's fitness in the list Fitness
my_dict ← dict (zip(Fitness, Routes)) #Combine each route and its fitness
srt_my_dict ← sorted(my_dict.items(), key = lambda x:x[0])
print('The minimum route is', srt_my_dict[0]) #(263.8465, [12, 18, 20])
print('Fitness after permutation', srt_my_dict[0][0]) #263.8465
print('Route after permutation', srt_my_dict[0][1]) #[12, 18, 20]
f(X'') ← srt_my_dict[0][0]
If f(X'') < f(X) then
New_solution2.append(srt_my_dict[0][1])
print('the best of New_solution2', New_solution)
break
else:
Old_route.append(i)
print('Old_route1', Old_route)
break
else: # The end of the maximum iteration number
Old_route2.append(i)
Else if length of (i) == 2 then
q.append(i)
Solution ← solution + q + New_solution + Old_route + Old_route2
Output returns an enhanced solution
```

The SVND approach commenced when the SVNSCS algorithm failed to generate a feasible solution (since the incumbent route was not enhanced). In this case, it will explore the distant neighborhoods of the incumbent solution and move to a new one if an improvement is made. The SVND method employs a list of ordered neighborhood structures and evaluates each structure by evaluating its sequence. The SVND principle works as follows: Let $N = \{N1,.., Nl\max\}$ be a set of operators specifying the neighborhood structures and their testing order. The fundamental SVND process investigates the neighborhood structures defined by the operators Nl, $1 \leq l \leq l\max$, sequentially and iteratively. As soon as an improvement to the incumbent solution in a neighborhood structure happens, the basic SVND restarts its search in the new incumbent solution's first neighborhood structure (in the defined sequence).

The process continues to search for ways to improve the incumbent solution until it reaches the maximum iteration number. Then, it returns to the old route and saves the results in the reconstruction solution register. This process applies to any $l\max$ neighborhood structures, as shown in Step 3 in Fig. 2.5. It involves the search operation that evaluates the various neighborhood structures and explores the best solutions (Sahib et al., 2024a).

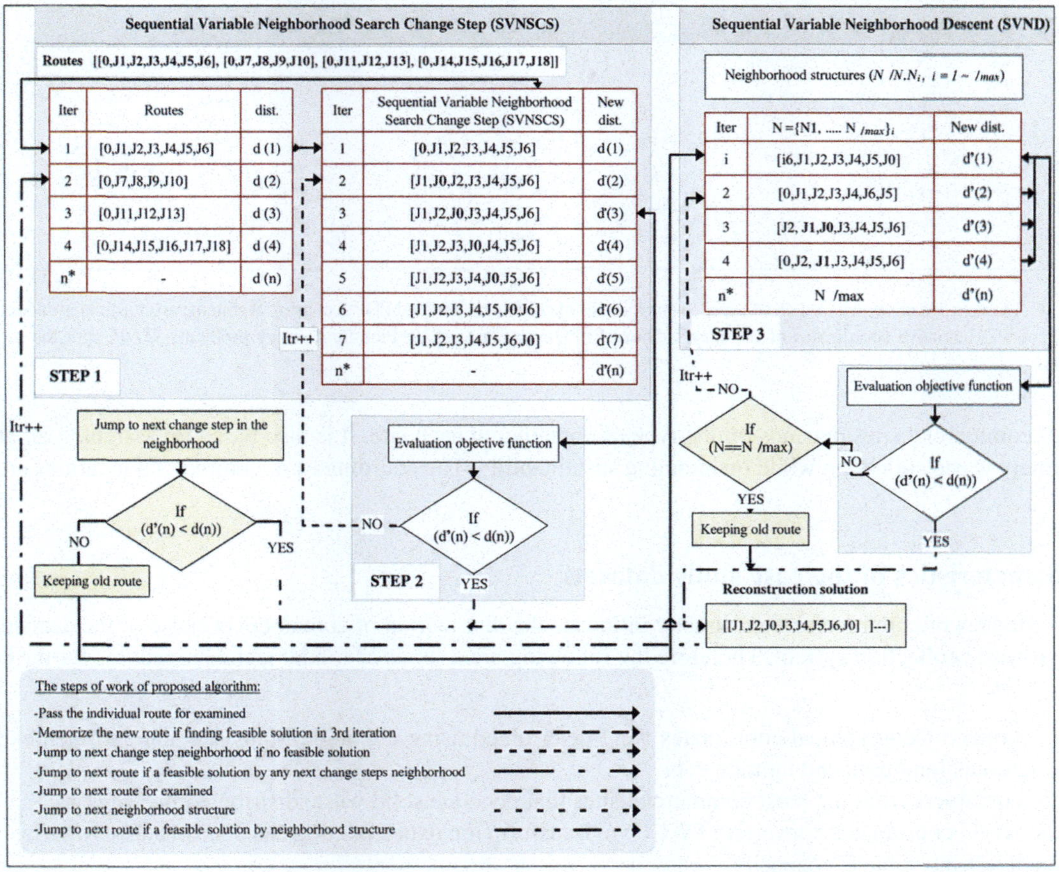

FIGURE 2.5 Methodology of hybridization ACO with SVNSCS and SVND algorithms (HACO). *ACO*, Ant colony optimization; *SVND*, sequential variable neighborhood descent; *SVNSCS*, sequential variable neighborhood search change step.

TABLE 2.2 Acronym of datasets created.

Acronym	Dataset	Distribution containers	Scenario
T1	T1-n22-k4	Spaced	PDP
T2	T2-n22-k3	Intensive	PDP
T3	T2-n22-k4	Intensive	SPDP

PDP, pick-up delivery problem; *SPDP*, split the pick-up from the delivery problem.

2.4 Data collection from Al-Kufa City, Iraq

This study collected data from three distinct areas in Kufa, Iraq. The first two datasets are related to the PDP scenario, T1-n22-k4, and T2-n22-k3, whereas the third dataset is related to the SPDP scenario called T3-n22-k4. All datasets are used to assess the improved ACO algorithm. The container distribution is designed to ensure diversity when evaluating the performance of the proposed algorithms. Each container has specific coordinates and waste amount according to the commercial point type. The datasets are summarized in Table 2.2.

In the PDP scenario the dataset of T1 is divided into two clusters for distributing containers based on the study area. The clusters are *blue* on either side of the depot, as shown in Fig. 2.6A. Meanwhile the dataset of T2 distributed the containers in intensive coordinates as clusters away from the depot, as shown in Fig. 2.6B. Concerning the dataset of T3 in the SPDP scenario the depot and TWTS are in separate locations. Once the vehicle dumps the waste into the transfer station, it will return to the depot. The distribution of containers is illustrated in Fig. 2.6C. Establishing the necessary vehicles to provide

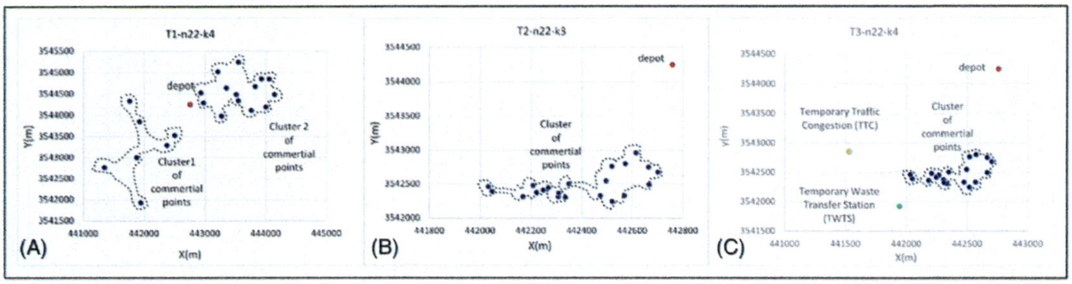

FIGURE 2.6 Containers distribution based on (A) spaced coordinates of containers denoted PDP scenario, (B) intensive coordinates of containers denoted PDP scenario, and (C) intensive coordinates of containers denoted SPDP scenario. *PDP*, Pick-up delivery problem; *SPDP*, split the pick-up from the delivery problem.

service to all commercial sites depends on the type of container at each site. It is also based on assigning an annual tax to guarantee timely waste collection while maintaining sustainability. The coordinates of commercial locations are detailed in Table 2.3.

2.4.1 Characteristics of the case study datasets

The new coordinates of commercial sites directly influence the distribution of containers because of the decisions made by the municipal waste collection system. Therefore the following are a series of fundamental procedures for the collection of data:

- Locate containers: Geographical coordinates were determined using a Google Maps account and then exported to an Excel program as longitude and latitude data.
- Collect 66 containers from the most commercial sites that generate solid waste distributed in 3 regions.
- Utilize the world coordinates converter (TWCC) to transform the data from degrees coordinates to metric.
- Homogeneity of the vehicle's capacity.
- Attributes of container capacity: The demand in each node depends on the amount of waste provided by the owner of commercial sites.

Further, factors, such as the number of nodes, node orders, range of nodes, or other pertinent information, are utilized to predict the daily demand generated at each node. After determining the demand, it is included in the CVRP formulation as part of the objective function or the constraints detailed in the next subsection.

2.4.2 Constraints of demands

There are specific criteria for choosing an appropriate demand, outlined as follows:

- The amount of waste generated ranges between 1 and 2061 kg.
- The demand for containers on a specific route must not exceed the vehicle's capacity.
- Assign 66 demand nodes.

2.4.3 Evaluation of the demand collected

"Demand collected" refers to waste generated from commercial sites before the statistical distribution test. In contrast the demand test refers to the waste generated after the distribution is statistically evaluated. Therefore some statistical procedures are performed to determine the appropriate test to analyze the distribution of requirements in the new datasets (T1, T2, and T3). The Statistical Software Suite (SPSS) is used to run the test.

2.4.3.1 One-sample Kolmogorov–Smirnov test

This procedure compares a variable by observing the cumulative distribution function with a specified theoretical distribution. The one-sample Kolmogorov–Smirnov test evaluates whether a variable reflects a normal distribution. As a result a one-sample Kolmogorov–Smirnov test was analyzed to determine if the distribution of demands is suitable for Normal, Uniform, Poisson, or Exponential tests. As shown in Table 2.4 the Normal test is considered the best test because of the

TABLE 2.3 Geographical coordinates of datasets T1 and T2 for the PDP scenario and T3 for the SPDP scenario.

Node	ND for T1 and T2	ND for T3	Geographical coordinates (latitude, longitude)		
			T1	T2	T3
	Depot	Depot	32.03294, 44.39378	32.03294, 44.39378	32.03293, 44.39378
	c2	C2	32.03056, 44.39918	32.01548, 44.38764	32.01988, 44.39188
	c3	C3	32.03174, 44.4044	32.01545, 44.38911	32.01955, 44.39132
	c4	C4	32.03515, 44.40841	32.01541, 44.3894	32.01944, 44.39285
	c5	C5	32.03251, 44.40692	32.01599, 44.38912	32.0188, 44.39326
	c6	C6	32.03838, 44.40746	32.01637, 44.38848	32.01714, 44.39287
	c7	C7	32.03845, 44.40613	32.01668, 44.38869	32.01762, 44.39108
	c8	C8	32.03380, 44.40214	32.01698, 44.38805	32.017, 44.38805
	c9	C9	32.03511, 44.40175	32.01759, 44.39109	32.0167, 44.38868
	c10	C10	32.03653, 44.39998	32.01711, 44.39289	32.01638, 44.38848
	c11	C11	32.03334, 44.39605	32.01877, 44.39326	32.01602, 44.3891
	c12	C12	32.03547, 44.39558	32.01943, 44.39286	32.01543, 44.38937
	c13	C13	32.03992, 44.39849	32.01953, 44.39132	32.01558, 44.38915
	c14	C14	32.03684, 44.40504	32.01719, 44.38954	32.01591, 44.38781
	c15	TWTS	32.04197, 44.40209	32.02132, 44.39231	32.01194, 44.38528
	c16	C16	32.0337, 44.38321	32.01563, 44.39086	32.01565, 44.39086
	c17	C17	32.02930, 44.38494	32.01564, 44.39187	32.01564, 44.39187
	c18	C18	32.02641, 44.39108	32.01486, 44.39136	32.01487, 44.39134
	c19	TTC	32.02428, 44.38976	32.01988, 44.39188	32.02029, 44.38086
	c20	C20	32.02162, 44.38448	32.01619, 44.38634	32.01619, 44.38635
	c21	C21	32.01948, 44.37893	32.01681, 44.38618	32.01682, 44.38618
	c22	C22	32.01195, 44.38528	32.01604, 44.3882	32.0172, 44.38953

C, container; *ND*, node's description; *PDP*, pick-up delivery problem; *SPDP*, split the pick-up from the delivery problem; *TTC*, temporary traffic congestion; *TWTS*, temporary waste transfer station.

TABLE 2.4 Normal test distribution.

		Demand_collected_T1	Demand_tested_T1	Demand_collected_T2	Demand_tested_T2	Demand_collected_T3	Demand_tested_T3
N		22	22	22	22	22	22
Normal parameters[a,b]	Mean	1055.77	960.64	845.05	889.95	1198.64	1171.41
	Std.	447.765	466.838	522.865	421.760	624.677	608.377
Most extreme differences	Absolute	0.143	0.134	0.110	0.105	0.213	0.157
	Positive	0.123	0.102	0.110	0.105	0.089	0.080
	Negative	−0.143	−0.134	−0.073	−0.081	−0.213	−0.157
Test statistic		0.143	0.134	0.110	0.105	0.213	0.157
Asymp. Sig. (2-tailed)		**0.200**[c,d]	**0.200**[c,d]	**0.200**[c,d]	**0.200**[c,d]	**0.010**[c]	**0.172**[c]

[a]Test distribution is normal;
[b]Calculated from data;
[c]Lilliefors significance correction;
[d]This is a lower bound of true significance.

TABLE 2.5 A sample representing demand was evaluated for C2 in dataset T1.

Type of container (i)	No. of containers required (c)	Weight of container (W)
1	2	500
2	1	310
3	1	160
4	1	40
5	2	90
	$\sum c_i = 7$ containers	$\sum W_i(c_i) = 1690(\pm 5)$ kg

lower bound of the true significance under the value of 0.200[c, d]. Bold fonts for datasets T1, T2, and T3 represent this lower bound. However, the demand collected in T3 is less than 0.05; thus an improvement is automatically made by SPSS, and the value is adjusted to 0.172. The outliers concerning the constraint of CVRP were ignored, as shown in Table 2.4.

2.4.3.2 Demand frequencies

Typically the SPSS offers measurements for two dimensions of distribution shape. The first shape is skewness, which quantifies the asymmetry in the distribution of demands collected by determining whether the values in a dataset are more concentrated on one side of the mean than the other. The demand distribution changes shape based on the condition described in Eq. (2.24).

$$D_d = \begin{cases} S_d \text{ if } Skewness == 0 \\ P_s \text{ if } Skewness > 0 \\ N_s \text{ if } Skewness < 0 \end{cases} \tag{2.24}$$

D_d: Distribution of demands.
S_d: Symmetrical distribution (i.e., no skewness).
P_s: Positively skewed distribution (a long tail on the right side).
N_s: Negatively skewed distribution (a long tail on the left side).

The second shape is kurtosis, which measures the flatness of a distribution in comparison to a normal distribution. It evaluates the edges of the distribution and whether the data is skewed toward extreme values. A distribution with a kurtosis value below 3 (negative kurtosis) is flatter and has lighter tails than a normal distribution. It suggests that the data has a lower frequency of outliers and exhibits a wider distribution.

In the T1 dataset the skewness values are (−0.414 and −0.273) for demand collected and demand tested, respectively. The distributions of both datasets are slightly negatively skewed but are close to being symmetrical. The kurtosis values are (0.470 and −0.236) for demand collected and demand tested, respectively, indicating that the distributions have slightly heavier tails than a normal distribution but are within an acceptable range as visualized in Figs. 2.7A and B.

Dataset T2 reduced the skewness value from 0.325 to 0.047 for the gathered and analyzed demand. It is almost perfect, as it is remarkably close to zero. The scenario depicts the distribution as moving toward a symmetrical form. The kurtosis value increased from −0.240 to 0.325 when transitioning from demand collected to demand tested, suggesting an enhancement in the distributions, as shown in Figs. 2.7C andD.

The data distribution of dataset T3 lacks symmetry. Both datasets exhibit negative skewness values, suggesting a slight left skewness, as shown in Fig. 2.7E and F. The negative kurtosis values indicate a platykurtic distribution characterized by a lower peak and lighter tails than a normal distribution.

2.4.3.3 Representation of demands tested

The fundamental terms include the volume or weight of waste containers used to represent the waste collected, the data gathered from stakeholders (commercial site owners), and the maximum weight permitted for various container types. As the tested demand was expressed in kilograms per square meter in each dataset, the quantity and type of waste containers needed for each commercial site are ±5 kg. For example, the stakeholder at a specific commercial site produces approximately 1692 kg of solid waste daily, as detailed in Table 2.5.

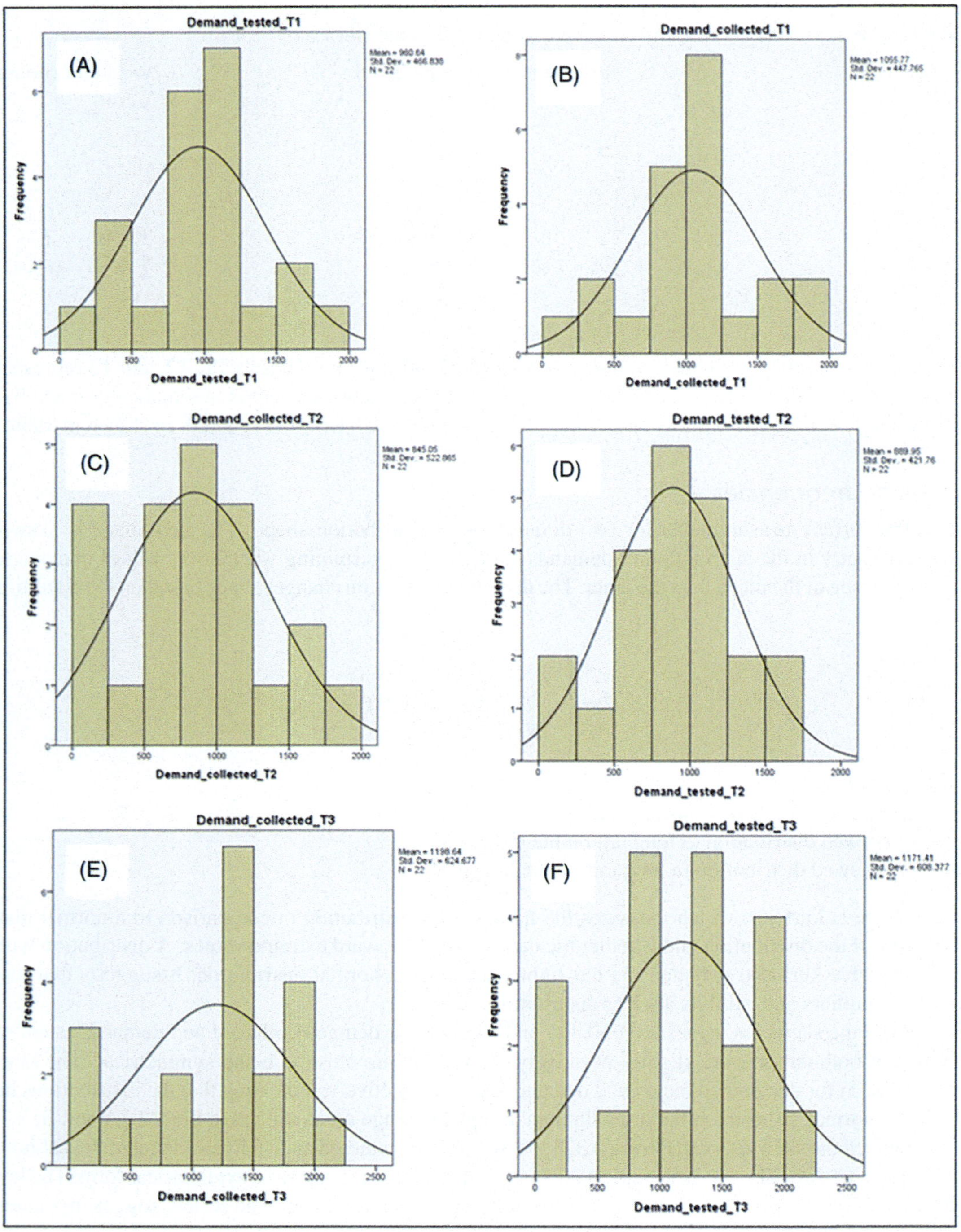

FIGURE 2.7 Histogram of comparison between demand collected against demand tested distributed, which was distributed as follows: (A) demand collected in dataset T1, (B) demand tested in dataset T1, (C) demand collected in dataset T2, (D) demand tested in dataset T2, (E) demand collected in dataset T3, and (F) demand tested in dataset T3.

2.4.4 Model of TTC filtering

Identifying the number of TTC locations is the most important thing in estimating the waiting time of vehicles in a congested area, thus differentiating them from TWTS and container locations. This is accomplished by designing a systematic approach consisting of some identification numbers (IDs) such as (1, 2, 3). Let us use (ID = 1) for detecting container locations, (ID = 2) for detecting TWTS locations, and (ID = 3) for isolating TTC locations. The data structure for detecting and classifying these nodes is shown in Fig. 2.8.

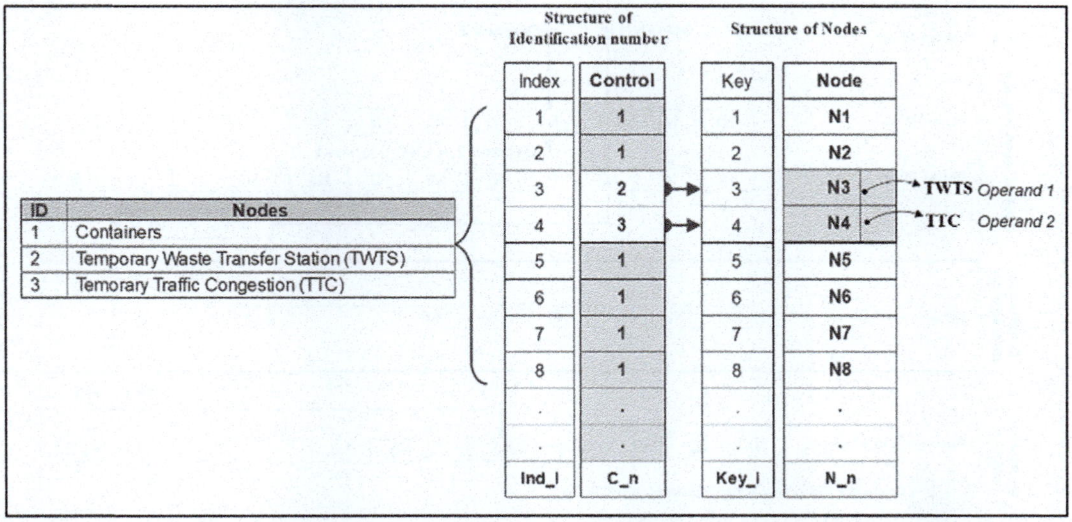

FIGURE 2.8 Data structure of nodes filtering model.

In Fig. 2.8 the node identification and isolation techniques are applied to Operand 1 and Operand 2 at the beginning by removing the depot from the list of nodes. For example, a proposed algorithm identifies several IDs associated with three traffic congestions in a certain route. Then, that algorithm will exclude those nodes to generate a new path. The motivation for this technique is summarized as follows:

- Avoid the waiting time spent by those vehicles in the TTC locations.
- Decrease fuel consumption by waste vehicles while they are stationed at those locations. Subsequently, filtering the nodes is modeled as follows:
  - Collect the nodes, as expressed in Eq. (2.25).

$$\text{Nodes} = \{N_1, N_2 \ldots N_n\},\ \text{n : number of nodes} \tag{2.25}$$

  - Each container i is assigned an identification number denoted as IDi
  - Nodes with identification number (ID = 2) that are connected to TWTS are excluded using Operand 1.
  - Create a new set called filtered nodes by excluding nodes with IDs other than three. This set will exclusively include containers with IDs other than three, as specified in Eq. (2.26).

$$\text{Filtered\_nodes} = \{i \mid i \in \text{nodes}, \text{ID}_i \neq 3\} \tag{2.26}$$

  - Filter the nodes with (ID = 3) by ignoring it. Those nodes refer to TTC locations as Operand 2.

The optimized approach demonstrates a framework that involves ignoring the TTC (Operand 2) and appending the content of the facility station register (Operand 1) to the end of each route, as shown in Fig. 2.9.

If the TTC has been neglected, estimating the waiting time at that location is unnecessary. Thus only the total arrival time of the vehicle to collect all solid waste along that route is determined.

2.4.5 Model of TWTS filtering

The TWTS is the intermediate point between the CPs and the final landfill station. It is used to convert solid waste from small to large vehicles. Let $S = \{R_1, R_2, \ldots, R_n\}$, where S is a set of routes with unique and unordered routes. In R_i, the ith route is a route containing nodes i_1 through in-facility station using summation notation, as expressed in Eq. (2.27).

$$\sum_{j=1}^{n} (\text{facility\_station} + ij) \text{for} R_k \tag{2.27}$$

Selecting all containers i from the set of nodes, the identification for containers i is not equal to three; thus store those containers in the set *filtered_nodes*. Furthermore, j varies from 1 to n, and ij where jth represents the node in the intended

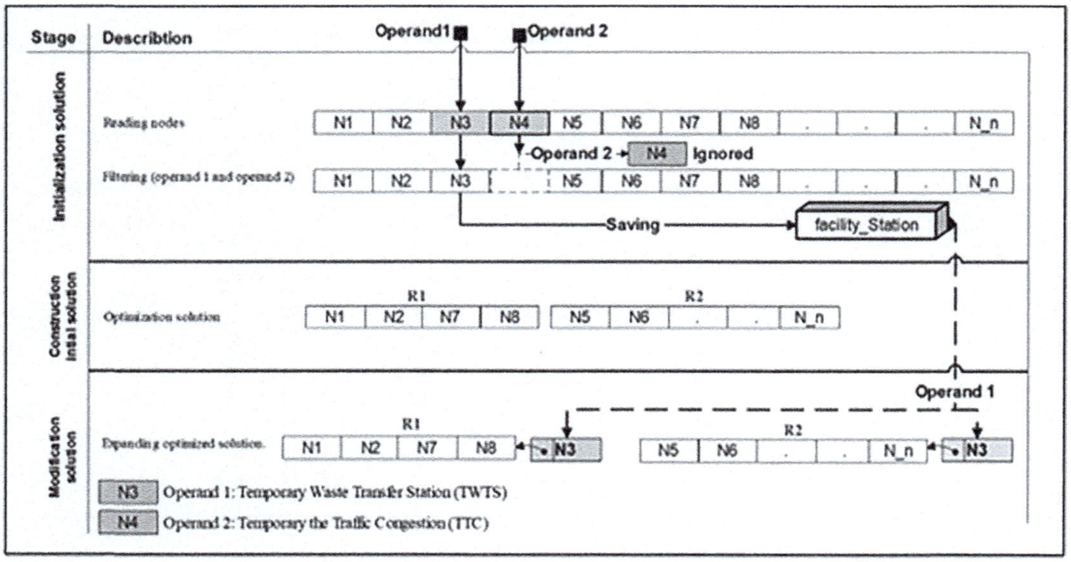

FIGURE 2.9 Framework of nodes filtering model.

route. In conclusion, the proposed algorithm can ignore the routes that faced traffic congestion that occurred in the planned route.

2.5 Results and discussion

This section compares the proposed methods' impact on collected and tested waste demand across the T1, T2, and T3 datasets. It is important to know that the optimal solution for every dataset depends on three crucial factors: vehicle capacity, coordination of commercial points, and the tested waste demands. Furthermore, the variation among datasets in each factor will directly impact the quantity of vehicles required to pick up the waste in each planned route.

2.5.1 Performance of the enhanced ACO on dataset T1

The main objective is to allocate a maximum load capacity of 6.3 t of vehicle capacity to collect solid waste from commercial sites. Demand testing is used to assess the quantity of garbage produced, and the geographical coordinates of the points are converted to metric coordinates. evaluates the algorithm that yields the most optimal solution in demand collected (D_c) and demand tested (D_t). Consequently, criteria such as minimum distance, lowest standard deviation, reduced runtime, and increased tightness are considered.

Based on the data analyzed in Table 2.6, the HACO and ACO–SAPM algorithms' performance is competitive in reaching the minimum distance. Consequently, both achieved 16.149 km in D_c and 14.414 km in D_t for ACO–SAPM, which is close to the distance value of HACO. Therefore HACO performs equally well as ACO–SAPM in terms of minimum distance. All the evaluated algorithms needed four vehicles to complete their trip, regardless of the demand considered. Overall, there is approximately a 2 km difference between the collected waste demand and the results tested in all compared algorithms.

2.5.2 Performance of the enhanced ACO on dataset T2

In this dataset the locations of the commercial sites and the amount of waste generated differ from the previous T1. According to this data the best vehicle capacity to serve the study area is 7.0 t presents the numerical results to evaluate the performance of the proposed algorithms using dataset T2.

Table 2.7 shows that the ACO–SAPM is much better than standard ACO, ACO–SVNSCS, and ACO–SA in terms of minimum distance in both $D_c$and D_t scenarios. In comparison to HACO the performance is better in D_c scenario. HACO gives the best performance in terms of minimum distance in D_t scenario. This indicates that the HACO algorithm effectively finds shorter routes for delivering the demand for waste. Concerning the standard deviation, the HACO algorithm shows a relatively low standard deviation for both D_c and D_t scenarios. Regarding running time, the HACO algorithm has a higher run time than other algorithms. HACO's run time is 552.0 seconds for the D_c scenario, and for the D_t scenario, it is 518.6

TABLE 2.6 Performance evaluation of ACO variants based on demand collected and demand tested in datasets T1 under 6.3 t vehicle capacity.

Algorithm	Type of demand	Vehicles required	Distance (km)	Average (km)	Standard deviation	Run time. (sec)	Tightness
ACO	D_c	4	16.351	16.360	0.04	210.6	0.92
	D_t	4	15.011	15.016	0.02	142.8	0.84
ACO–SVNSCS	D_c	4	16.317	16.326	0.05	242.8	0.92
	D_t	4	14.844	14.848	0.04	415.3	0.84
ACO–SA	D_c	4	16.313	16.326	0.18	369.1	0.92
	D_t	4	14.838	14.848	0.13	264.2	0.84
ACO–SAPM	D_c	4	16.149	16.165	0.28	220.4	0.92
	D_t	4	**14.414**	**14.426**	0.21	493.6	0.84
HACO	D_c	4	16.149	16.169	0.03	513.1	0.92
	D_t	4	**14.413**	**14.420**	**0.02**	551.3	0.84

ACO–SA, ant colony optimization-simulated annealing; *SVNSCS*, sequential variable neighborhood search change step. Bold fonts represent best value.

TABLE 2.7 Performance evaluation of ACO variants based on demand collected and demand tested in datasets T2 under 7.0 t vehicle capacity.

Algorithm	Type of demand	Vehicles required	Distance (km)	Average (km)	Standard deviation	Run time. (sec)	Tightness
ACO	D_c	3	12.423	12.426	0.01	129.3	0.89
	D_t	3	12.399	12.400	0.07	185.0	0.93
ACO–SVNSCS	D_c	3	12.375	12.386	0.08	442.4	0.89
	D_t	3	12.355	12.359	0.01	278.2	0.93
ACO–SA	D_c	3	12.373	12.387	0.1	243.1	0.89
	D_t	3	12.359	12.370	0.02	415.4	0.93
ACO–SAPM	D_c	3	12.368	12.377	0.05	319.9	0.89
	D_t	3	12.354	12.366	0.09	375.2	0.93
HACO	D_c	3	12.371	12.378	0.04	552.0	0.89
	D_t	3	**12.347**	**12.354**	0.02	518.6	0.93

ACO–SA, ant colony optimization-simulated annealing; *SVNSCS*, sequential variable neighborhood search change step.

seconds. The higher runtime might be attributed to the algorithm's complexity or the specific problem instance. If faster execution is a critical requirement, other algorithms with lower run times may be preferred. It is crucial to recognize that the tightness value of all the algorithms is identical under comparison in both scenarios. Tightness represents the waste's compactness in the vehicle's capacity, where a higher value indicates a more compact clustering of demand points.

2.5.3 Performance of the enhanced ACO on dataset T3

The outcome of this part is significant, as no study has examined the performance impact of optimization algorithms for the collection, transfer, and unloading of solid waste in the TWTS and subsequent return to the depot. The investigation of dataset T3 emphasizes each of these constraints. The vehicle's capacity is optimally determined at 6.5 t, as shown in Table 2.8.

Table 2.8 shows that all the enhanced ACO algorithms outperformed the standard ACO algorithm in reducing the distance for transporting waste to the TTC. The standard deviation has been employed to measure the variability of the results. Based on this performance criteria the HACO algorithm shows a relatively low standard deviation for both D_c and D_t scenarios. These values indicate that the HACO algorithm has relatively low variability compared to other algorithms, suggesting consistent performance. Similar to the previous two datasets the algorithms' tightness value is identical under comparison

TABLE 2.8 Performance evaluation of ACO variants based on demand collected and demand tested in datasets T3 under 6.5 t vehicle capacity.

Algorithm	Type of demand	Vehicles required	Distance (km)	Average (km)	Standard deviation	Run time. (sec)	Tightness
ACO	D_c	4	21.961	21.967	0.02	148.6	0.96
	D_t	4	21.632	21.640	0.03	147.5	0.94
ACO-SVNSCS	D_c	4	21.236	21.241	0.02	152.5	0.96
	D_t	4	21.118	21.125	0.01	195.1	0.94
ACO-SA	D_c	4	21.091	21.107	0.07	258.0	0.96
	D_t	4	21.085	21.094	0.08	226.5	0.96
ACO-SAPM	D_c	4	21.076	21.083	0.08	225.0	0.96
	D_t	4	21.063	21.069	0.08	238.9	0.94
HACO	D_c	4	21.187	21.191	0.01	165.7	0.96
	D_t	4	**21.045**	**21.055**	0.03	350.5	0.94

ACO-SA, ant colony optimization-simulated annealing; *SVNSCS*, sequential variable neighborhood search change step.

TABLE 2.9 Comparison of HACO with and without TTC on a dataset of T3 for one day.

Vehicle capacity (t)	Vehicles required (K)	Distance (km)	Average (km)	Runtime (sec)	Std.	Tightness
With temporary traffic congestion (day shift). (ID_{19} = 1), (ID_{15} = 2) {e.g., if ID_{19}==1, it means to consider the identity node 19 as a normal container}						
6.5	4	21.976	21.979	334.8	0.01	0.99
7.0	4	21.599	21.605	320.2	0.02	0.92
7.5	4	21.599	21.603	297.8	0.02	0.86
8.0	**4**	**21.586**	**21.589**	**231.7**	**0.09**	**0.95**
Without temporary traffic congestion (night shift). (ID_{19} = 3), (ID_{15} = 2) {e.g., if ID_{19}==3, it means to consider the identity node 19 and detect it as TTC and remove it, so this is the reason that we called without TTC}						
6.5	4	21.045	21.055	350.5	0.03	0.94
7.0	4	21.119	21.121	320.9	0.01	0.87
7.5	4	20.944	20.958	307.8	0.04	0.81
8.0	**3**	**16.218**	**16.221**	**191.9**	**0.02**	**1.0**

ID: identification number. *Q*: vehicle capacity/maximum loading capacity. *K*: vehicles required/no. of routes served.

in both scenarios. This value suggests that the routes generated by all the evaluated algorithms are relatively compact. Although the HACO algorithm outperforms the other compared algorithms regarding the best solution, it suffers from a higher processing time. Eventually the ACO–SAPM algorithm was demonstrated as the second-best performance, with a better run time than the HACO algorithm.

2.5.4 Result of adaptation frequency of TTC and TWTS

This section concentrates on the SPDP scenario, which guarantees that solid waste is transported exclusively to TWTS to unload waste inside and then return to the depot with empty vehicles. Therefore the TWTS coordinate must initially be detected. Moreover, it is necessary to avoid roads that have frequent TTC, thus creating the shortest route without TTC. The experimental results have been evaluated according to a predetermined set of criteria, including the shortest distance, the lowest standard deviation, and the relationship between vehicle capacity in cases where there is or is not TTC in the route to be served, as shown in Table 2.9.

The first shift collects the waste during the day while traffic congestion is present. In contrast, the second shift collects the waste at night when the routes are free from congestion. By analyzing and discussing the data presented for both working

TABLE 2.10 Results of WCM system considering 8.0 t load vehicles capacity: Al-Kufa case study.

Design routes *with* temporary traffic congestion		Design routes *without* temporary traffic congestion	
[[6, 7, 22, 9, 10, 11, 15], [5, 4, 2, 3, 8, 15], [17, 16, 18, 12, 13, 14, 15], [**19**, 21, 20, 15]]	4	[[5, 4, 2, 3, 21, 20, 15], [17, 16, 18, 12, 13, 11, 15], [6, 7, 22, 9, 10, 8, 14, 15]]	3

FIGURE 2.10 Decision on constructing the best routes with temporary traffic congestion during the day shift of the Al-Kufa case study in (A) Route 1, (B) Route 2, (C) Route 3, and (D) Route 4.

shifts and varying vehicle capacities, it is possible to quantify the scenario in which the HACO algorithm provides the most optimal solution. The objective is to minimize the distance travelled, which will be useful in reducing fuel consumption and improving transportation operations. The routes are designed according to the waste collection system, taking into account both the presence and absence of traffic congestion. The computation results show that HACO achieved a minimum distance in the day and night shifts when the vehicle capacity was 8.0 t. The number of vehicles required was reduced by one vehicle to avoid traffic congestion (night shift), as described in Table 2.10.

In Table 2.10 the containers are rearranged based on the waste amount generated for each specific area, the size of the container's vehicle, and the coordinates of nodes. Based on that the actual routes designed during the day shift with TTC consist of four routes, where the first three routes are visited by a vehicle without passing through the TTC location as plotted in Fig. 2.10A, B, and C. In contrast the fourth route was visited by a vehicle passing through the location of TTC that marked a letter B, as visualized in Fig. 2.10D.

FIGURE 2.11 Decision of constructing the best route without temporary traffic congestion on the night shift of Al-Kufa case study in (A) Route 1, (B) Route 2, and (C) Route 3.

It is observed that all routes passing through the traffic congestion area increase the number of required vehicles, consequently increasing the distance to serve all CPs. Conceptually, designing a route with frequent traffic congestion allows only one vehicle (as a new system restriction) to pick up solid waste at the site of irrigated congestion during the day shift. Accordingly, there are two important benefits:

- Ensure that three other vehicles do not pass through the traffic congestion site, thus reducing municipal waste collection expenses.
- Ensure the sustainability of keeping the traffic congestion site clean.

Designing the routes without TTC has been visualized for the night shift, as shown in Fig. 2.11A, B, and C. The expense of transporting solid waste decreases proportionally with a decrease in vehicles by one. Meanwhile, the performance of the proposed algorithm, exhibited by a smaller standard deviation, suggests that it consistently generates comparable outcomes. The night shift demonstrates a consistently lower standard deviation across all vehicle capacities than the day shift. Therefore the algorithm exhibits enhanced consistency while operating during the nighttime.

In the context of the highest tightness, the degree of adherence of the algorithm's solution to the capacity constraints is denoted as tightness. An increased tightness signifies enhanced capacity usage of the vehicle. During the day and night shifts the vehicle with the greatest tightness values has a capacity of 6.5 t. Consequently the algorithm outperforms this vehicle capacity category for maximizing tightness, except for tightness 0.81, which appears acceptable but consumes a substantial amount of money because of using additional vehicles with capacities exceeding the amount of waste expected.

2.6 Conclusion

This study investigated the hybridization of ACO algorithms in optimizing the dynamic capacitated VRP of the waste collection system. A systematic framework model was developed by hybridizing ACO to investigate the rate of increase in its search capabilities during the exploration and exploitation stages, as well as the visualization of the convergence of the proposed algorithms. Four ACO variants have been developed: ACO–SVNSCS, ACO–SA, ACO–SAPM, and HACO. These four enhanced ACO variants successfully address some main challenges encountered by the original ACO during optimization. These challenges include premature convergence and the balancing of exploration/exploitation searches.

A municipal solid waste collection and transportation management system was designed based on the problem statement in Al-Kufa City in Iraq. Data from 66 common commercial sites that generate solid waste daily were simulated. To make the capacitated VRP assessable and manageable to deal with a dynamic approach, a technique of node detection has contributed a significant role as a knowledge base to avoid temporary traffic congestion and adopt a new TWTS. The computation results demonstrated the superiority of ACO–SAPM and HACO in all performance evaluation tests.

Future research proposes that the researcher study the effects of minimizing environmental pollution and calculating fuel consumption in crowded areas. Utilizing an appropriate vehicle capacity will reduce the number of vehicles required, thereby reducing the costs associated with waste collection.

Acknowledgments

The authors would like to express their gratitude for the cooperation between Al-Furat Al-Awsat Technical University, Iraq, and Universiti Sains Malaysia, which has enabled the potential completion of this research project.

References

Akhtar, M., Hannan, M. A., Begum, R. A., Basri, H., & Scavino, E. (2017). Backtracking search algorithm in CVRP models for efficient solid waste collection and route optimization. *Waste Management, 61*, 117–128. doi:10.1016/j.wasman.2017.01.022.

Ali, H., & Kar, A. K. (2018). Discriminant analysis using ant colony optimization—An intra-algorithm exploration. *Procedia Computer Science, 132*, 880–889. http://www.sciencedirect.com/science/journal/18770509. doi:10.1016/j.procs.2018.05.100.

And, B., Alinaghian, M., Tirkolaee, A., & Bakhshi Sasi, M. M. S. E. (2016). Solving a robust capacitated arc routing problem using a hybrid simulated annealing algorithm: A waste collection application. *Journal of Industrial Engineering and Management Studies, 3*(1), 61–76.

Chaabane, A., Montecinos, J., Ouhimmou, M., & Khabou, A. (2021). Vehicle routing problem for reverse logistics of end-of-life vehicles (ELVs). *Waste Management, 120*, 209–220. doi:10.1016/j.wasman.2020.11.008.

Delahaye, D., Chaimatanan, S., & Mongeau, M. (2019). Simulated annealing: From basics to applications. In *International series in operations research and management science: 272* (pp. 1–35). www.springer.com/series/6161. doi:10.1007/978-3-319-91086-4_1.

Duhayyim, M. A., Eisa, T. A. E., Al-Wesabi, F. N., Abdelmaboud, A., Hamza, M. A., Zamani, A. S., Rizwanullah, M., & Marzouk, R. (2022). Deep reinforcement learning enabled smart city recycling waste object classification. *Computers, Materials and Continua, 71*(2), 5699–5715. https://www.techscience.com/cmc/v71n2/46431. doi:10.32604/cmc.2022.024431.

Ebadinezhad, S. (2020). DEACO: Adopting dynamic evaporation strategy to enhance ACO algorithm for the traveling salesman problem. *Engineering Applications of Artificial Intelligence, 92*, 103649. doi:10.1016/j.engappai.2020.103649.

Edwards, J., Othman, M., Burn, S., & Crossin, E. (2016). Energy and time modelling of kerbside waste collection: Changes incurred when adding source separated food waste. *Waste Management, 56*, 454–465. doi:10.1016/j.wasman.2016.06.033.

Hannan, M. A., Akhtar, M., Begum, R. A., Basri, H., Hussain, A., & Scavino, E. (2018). Capacitated vehicle-routing problem model for scheduled solid waste collection and route optimization using PSO algorithm. *Waste Management, 71*, 31–41. doi:10.1016/j.wasman.2017.10.019.

Henke, T., Speranza, M. G., & Wäscher, G. (2015). The multi-compartment vehicle routing problem with flexible compartment sizes. *European Journal of Operational Research, 246*(3), 730–743. https://www.journals.elsevier.com/european-journal-of-operational-research/. doi:10.1016/j.ejor.2015.05.020.

Hoang, G. M., Ha, H. T. T., Le, N. T. L., & Toan, N. D. (2024). Evaluation of municipal waste collection performance using operational data. *Global Journal of Environmental Science and Management, 10*(1), 69–82. www.gjesm.net/. doi:10.22034/gjesm.2024.01.06.

Hoque, M. M., & Rahman, M. T. U. (2020). Landfill area estimation based on solid waste collection prediction using ANN model and final waste disposal options. *Journal of Cleaner Production, 256*. https://www.journals.elsevier.com/journal-of-cleaner-production. doi:10.1016/j.jclepro.2020.120387.

İlhan, İ. (2021). An improved simulated annealing algorithm with crossover operator for capacitated vehicle routing problem. *Swarm and Evolutionary Computation, 64*. http://www.elsevier.com/wps/find/journaldescription.cws_home/724666/description#description. doi:10.1016/j.swevo.2021.100911.

Iqbal, A., Yasar, A., Nizami, A. S., Haider, R., Sharif, F., Sultan, I. A., Tabinda, A. B., Kedwaii, A. A., & Chaudhary, M. M. (2022a). Municipal solid waste collection and haulage modeling design for Lahore, Pakistan: Transition toward sustainability and circular economy. *Sustainability, 14*(23). http://www.mdpi.com/journal/sustainability/. doi:10.3390/su142316234.

Iqbal, A., Abdullah, Y., Nizami, A. S, Ali Sultan, I., & Sharif, F. (2022b). Assessment of solid waste management system in Pakistan and sustainable model from environmental and economic perspective. *Sustainability, 14*(19), 12680. doi:10.3390/su141912680.

Ismail, A. A., Krisnaputra, R., & Bahiuddin, I. (2021). Application of ant colony optimization for the shortest path problem of waste collection process. *Kinetik: Game Technology, Information System, Computer Network, Computing, Electronics, and Control*. doi:10.22219/kinetik.v6i3.1307.

Kaza, S., Yao, L., Bhada-Tata, P., & Woerden, F. V. (2018). *What a waste 2.0: A global snapshot of solid waste management to 2050 (English)*. World Bank Publications.

Knowles, J. J. (2009). National solid waste management plan for Iraq. *Waste Management and Research, 27*(4), 322–327. doi:10.1177/0734242ž09104129.

Kwakye, S. O., Amuah, E. E. Y., Ankoma, K. A., Agyemang, E. B., & Owusu, B. G. (2024). Understanding the performance and challenges of solid waste management in an emerging megacity: Insights from the developing world. *Environmental Challenges, 14*. https://www.sciencedirect.com/science/journal/26670100. doi:10.1016/j.envc.2023.100805.

Li, Y. P., Huang, G. H., & Nie, S. L. (2012). A mathematical model for identifying an optimal waste management policy under uncertainty. *Applied Mathematical Modelling, 36*(6), 2658–2673. doi:10.1016/j.apm.2011.09.049.

Mat, N. A., Benjamin, A. M., & Abdul-Rahman, S. (2018). Enhanced heuristic algorithms with a vehicle travel speed model for time-dependent vehicle routing: A waste collection problem. *Journal of Information and Communication Technology, 17*(1), 55–78. http://www.jict.uum.edu.my/images/vol17no1jan18/55-78.pdf.

Pedemonte, M., Nesmachnow, S., & Cancela, H. (2011). A survey on parallel ant colony optimization. *Applied Soft Computing Journal, 11*(8), 5181–5197. http://www.elsevier.com/wps/find/journaldescription.cws_home/621920/description#description. doi:10.1016/j.asoc.2011.05.042.

Rahmanifar, G., Mohammadi, M., Hajiaghaei-Keshteli, M., Fusco, G., & Colombaroni, C. (2024). An integrated temporal and spatial synchronization for two-echelon vehicle routing problem in waste collection system. *Journal of Industrial Information Integration, 40*, 100611. doi:10.1016/j.jii.2024.100611.

Rathinabhagya, P., & Vinotha, J. M. (2023). A solid waste management for vehicle routing problem with split delivery. *European Chemical Bulletin, 12*, 2695–2700. doi:10.31838/ecb/2023.12.s2.533.

Sahib, F. S., & Hadi, N. S. (2023). Truck route optimization in Karbala city for solid waste collection. *Materials Today: Proceedings, 80*, 2489–2494. https://www.sciencedirect.com/journal/materials-today-proceedings. doi:10.1016/j.matpr.2021.06.394.

Sahib, T. M., Mohd-Mokhtar, R., & Kassim, A. M. (2024a). Adapting ant colony optimization for the capacitated vehicle routing problem in waste collection system. *Chinese Science Bulletin, 69*(1), 729–746.

Sahib, T. M., Mohd-Mokhtar, R., Alharan, A. F. H., Ali, N. S., & Kadhum, A. A. (2024b). An improved strategy for solid waste management based on programmable logic controller. *International Journal of Environment and Waste Management, 34*(3), 297–310. doi:10.1504/IJEWM.2024.139963.

Sahib, T. M., Mohd-Mokhtar, R., & Kassim, A. M. (2022). Survey on meta-heuristic algorithms for solving vehicle route problems in a waste collection system. In *Lecture notes in electrical engineering: 829* (pp. 363–369). Springer Science and Business Media Deutschland GmbH. doi:10.1007/978-981-16-8129-5_57.

Saxena, S., Rajendran, C., Sanjeevi, V., & Shahabudeen, P. (2021). Optimization of solid waste management in a metropolitan city. *Materials Today: Proceedings, 46*, 8231–8238. doi:10.1016/j.matpr.2021.03.219.

Vu, H. L., Ng, K. T. W., Fallah, B., Richter, A., & Kabir, G. (2020). Interactions of residential waste composition and collection truck compartment design on GIS route optimization. *Waste Management, 102*, 613–623. doi:10.1016/j.wasman.2019.11.028.

Yassen, E. T., Ayob, M., Nazri, M. Z. A., & Sabar, N. R. (2017). An adaptive hybrid algorithm for vehicle routing problems with time windows. *Computers and Industrial Engineering, 113*, 382–391. doi:10.1016/j.cie.2017.09.034.

Yu, V. F., Anh, P. T., Gunawan, A., & Han, H. (2024). A simulated annealing with variable neighborhood descent approach for the heterogeneous fleet vehicle routing problem with multiple forward/reverse cross-docks. *Expert Systems with Applications, 237*. doi:10.1016/j.eswa.2023.121631.

Chapter 3

Model for the diagnosis and management of Alzheimer's disease and mild cognitive impairment

T. Tirupal[a], **Y. Pandurangaiah**[b] **and Ajay Roy**[c]

[a]*Department of ECE, G. Pullaiah College of Engineering and Technology, Kurnool, Andhra Pradesh, India,* [b]*Department of ECE, IARE Institute of Aeronautical Engineering, Hyderabad, Telangana, India,* [c]*Department of Electrical Engineering, College of Engineering, University of Bisha, Bisha, Saudi Arabia*

3.1 Introduction

World's Vision 2030 (Alzheimer's Disease International, 2018) wants to make life better and set the standards for medical innovation by improving healthcare, biotechnology, and artificial intelligence (AI) (Alzheimer's Association Report, 2017). The use of neurotechnology and AI-driven diagnostics for diagnosing and treating Alzheimer's disease (AD) (Dalcerri et al., 2017) and mild cognitive impairment (MCI) (Samuel, 2014) aligns with the national strategy's emphasis on digital transformation, personalized medicine, and innovative healthcare solutions. This effort is adding to other efforts to improve cognitive health, encourage early disease diagnosis, and make the most of modern healthcare systems to give patients the best care possible. This plan addresses the issues of early intervention, AI-assisted diagnosis, and real-time monitoring of treatment, which are all in line with the goals of countries in healthcare innovation, research on longevity, and the building of digital health infrastructure. Lowering the long-term costs of neurodegenerative diseases and promoting a proactive public health policy also make healthcare more sustainable (Struckmeyer & Pickens, 2016).

AD is the most common type of dementia in people. AD is a brain disease that cannot be cured. It affects a person's memory, thinking, and other functions. The disease is named after Dr. Alois Alzheimer. Dr. Alzheimer discovered irregularities in the brain tissue of a woman who passed away in 1906 because of an unusual mental disorder. Some of her symptoms were memory loss, trouble with language, and strange behavior (Li et al., 2018). After she died, he did an autopsy on her brain and found several strange clusters that are now called amyloid plaques, as well as neurofibrillary or tau tangles, which are twisted bundles of fibers.

Signs of AD (Mirza et al., 2017): People sometimes have moments when they forget things. People with AD show certain behaviors and symptoms that do not go away and get worse over time.

Memory problems make it hard to do everyday things, like keeping appointments, trouble with familiar tasks like using a microwave, trouble solving problems, trouble speaking or writing, not knowing where or when you are, poor judgment, a decline in personal hygiene, mood and personality changes, and withdrawal from social connections such as friends, family, and community. The symptoms change depending on how far along the disease is (Herrera et al., 2013). Fig. 3.1 represents the brain affected by AD.

The proposed device begins with a survey of individuals diagnosed with Mild Cognitive Impairment (MCI) or Alzheimer's Disease (AD). The memory-related challenges experienced by individuals with AD are then categorized. For each memory-related problem reported by the patients (Xiao et al., 2017), an appropriate coding solution must be developed to meet the identified requirements. The technical objective is to develop a memory-assistance solution using electronic components that can be readily interfaced with brain neurons. This approach appears to be a promising method for helping

Computational Intelligence in Mechatronics: Solving Real-World Problems in Electronic Systems Design. DOI: https://doi.org/10.1016/B978-0-443-27580-7.00015-9

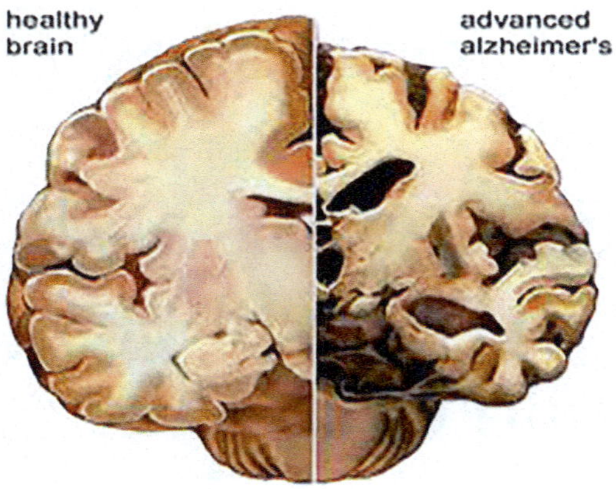

FIGURE 3.1 Alzheimer's disease impacted the brain.

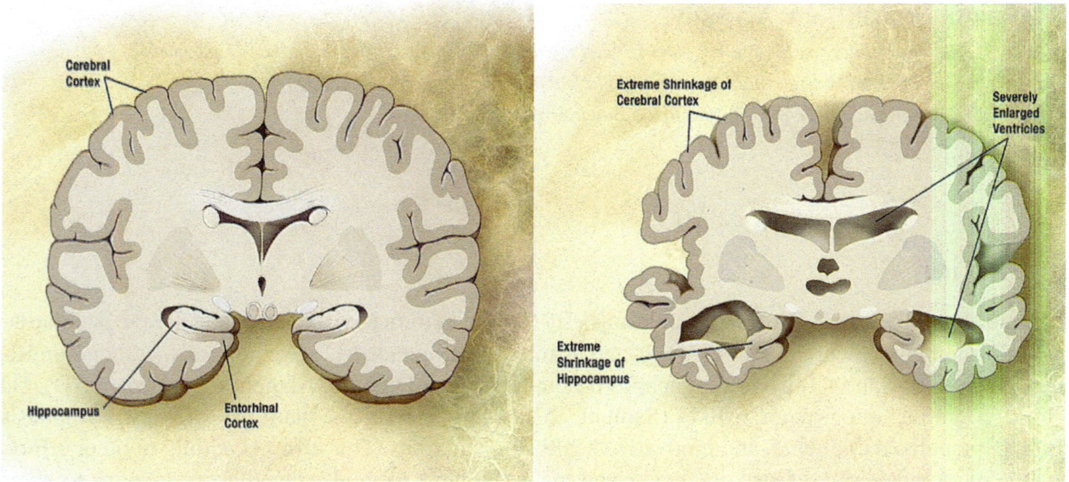

FIGURE 3.2 Comparison of brains with Alzheimer's disease.

individuals with MCI and AD manage their cognitive difficulties. The proposed device can support individuals experiencing memory loss in several ways. Overall, the system is expected to benefit individuals with AD by assisting them in memory recall and communication.

Image fusion is the process of merging several pictures of the same scene into one picture that has all the important parts of each of the input pictures. The main reason for image fusion is to make the information in the output image better. Image fusion has many benefits, such as a wider range of operation, more coverage in space and time, less uncertainty, more reliability, better system performance, and a smaller amount of information (Lama et al., 2017). Combining the complementary information from multiple images into one image is a big problem in biomedical engineering. A single image does not provide all relevant information; therefore multimodal medical image fusion is essential to obtain all possible information in a single composite image, known as a fused image, as this fused information aids physicians in diagnosing abnormalities. The fused image presents several challenges, including the visibility of soft and hard tissues, the contrast of the input image, edge strength, the manifestation of fusion artifacts, and the overall visual quality of the image. Fig. 3.2 represents the comparison of brains with affected AD. Fig. 3.3 represents the microscopic image of a neurofibrillary tangle formed by hyperphosphorylated tau protein.

3.1.1 Objectives of the research

3.1.1.1 Creation and synthesis of a prototype for Alzheimer's disease

The main goal of this objective is to create a memory aid that can help people with MCI or AD remember and find information they have forgotten (Ahmed et al., 2018). There are two types of AD: early-onset Alzheimer's, which happens to people

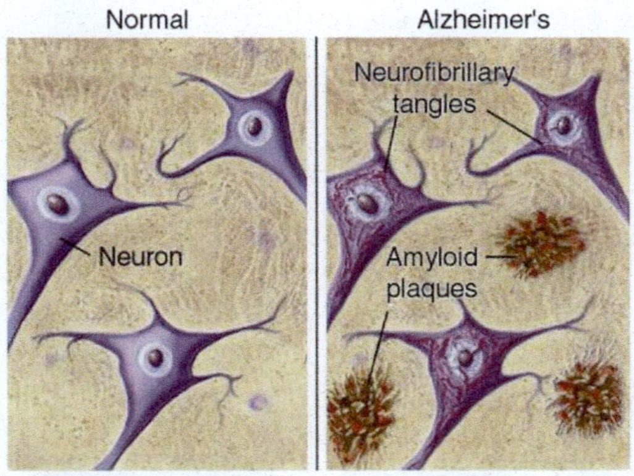

FIGURE 3.3 Microscopic image of a neurofibrillary tangle formed by hyperphosphorylated tau protein.

younger than 65, and late-onset Alzheimer's, which happens to people older than 65. They usually find out they have the disease when they are in their 40s or 50s. Only 5% of people with Alzheimer's have early-onset, which is very rare. People with Down syndrome are more likely to get it. Late-onset Alzheimer's is the most common type of the disease. It affects people who are 65 and older. It might run in families, but it might not. Researchers have not found a specific gene that causes it yet. No one knows for sure why some people get it and others do not.

3.1.1.2 Characterization of Alzheimer's disease

This objective classifies Alzheimer's into distinct types, including: Sporadic—caused by genetic and environmental risk factors, with 90%–95% of cases recorded as sporadic; 50% of cases associated with age 85; and only 1% of cases linked to ages 60–65. Familial—it is caused by changes in genes. Overall, 5%–10% of cases are classified as familial. For this type, at least two people in a family must have AD.

3.1.1.3 Making a memory aid for Alzheimer's disease

This goal gives information on how to make a memory aid for people with MCI and AD who have lost brain mass. There are no signs of the disease at this stage, but a family history may lead to an early diagnosis. In the normal age of forgetfulness the first signs of the disease show up, like forgetting things. In the MCI stage, people start to have mild physical and mental problems, such as trouble remembering things and focusing. Someone very close to the person may be the only one who can see these. AD is often diagnosed at this stage, but it is still considered mild. It is clear that the person has lost their memory and cannot do simple things. In the moderate stage of Alzheimer's, moderate-to-severe symptoms necessitate assistance from family members or caregivers. In the severe stage of Alzheimer's a person may need help with simple things like eating and getting dressed. This is the last and worst stage of Alzheimer's. You might lose your ability to talk and show emotions on your face.

3.1.1.4 Assessments for Alzheimer's disease

This objective evaluates the performance of AD on various medical images. Different pairs of medical images, including Magnetic Resonance Imaging (MRI), Positron Emission Tomography (PET), and Single-Photon Emission Computed Tomography (SPECT) images, are considered for the analysis. To facilitate a comprehensive comparison of the results, several objective fidelity metrics, including spatial frequency, entropy, peak signal-to-noise ratio (PSNR), standard deviation, and average pixel intensity, are employed. The corresponding values of all these metrics are calculated and presented for performance evaluation.

3.2 Review of the literature

The chapter underscores the clinical and neurobiological dimensions of AD and MCI that medical students ought to comprehend. The chapter also talks about progress in using biomarkers to diagnose AD and ongoing work to find new treatments. This review delineates the current comprehension of the epidemiology, genetics, pathology, and pathogenesis of AD, followed by a discussion of its clinical presentation and contemporary treatment strategies. Lastly the review talks about

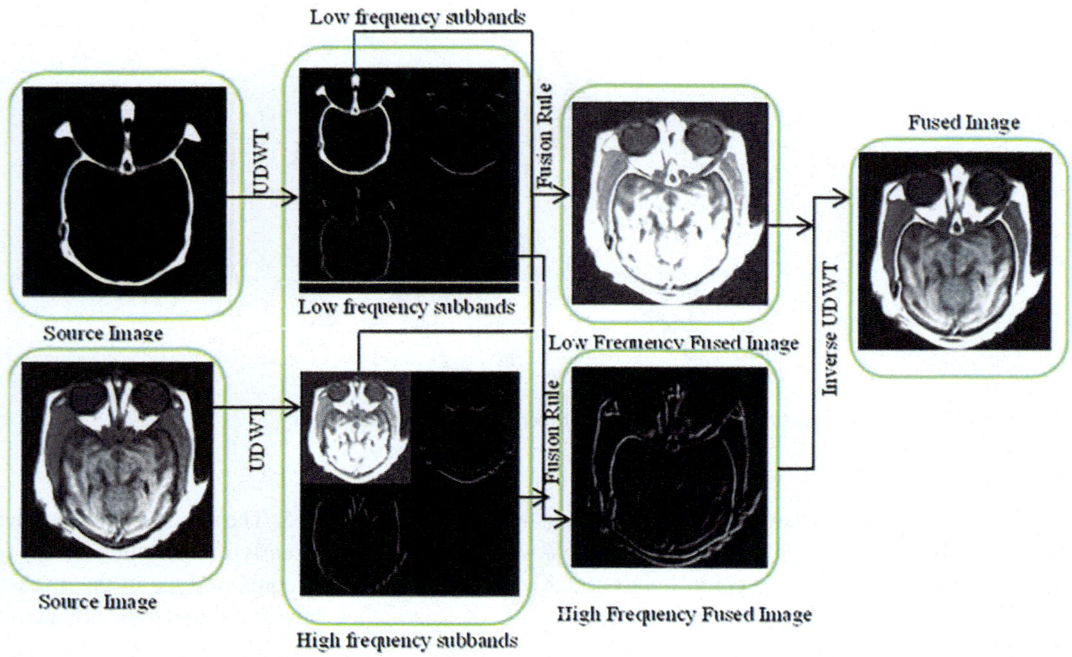

FIGURE 3.4 The proposed method's graphical structure.

how our better understanding of how AD starts, including the fact that there is a long preclinical phase, is helping us come up with new ways to treat the disease that will help us move from treatment to prevention. This review gives an overview of these new developments, focusing on therapies that change the course of the disease, treatments that help with symptoms, combination therapies, changes to lifestyle, the development of biomarkers, new ways to deliver drugs, immunotherapy, gene therapy, and neuroprotective agents. The following are some of the most recent articles about finding AD.

Most people with AD have problems with their thinking and daily tasks, so it is important to find it early. Borkar et al. (2023) have created the convolution-based long short-term memory framework to find AD earlier in MRI scans. This method could tell the difference between people with AD and people who are normal. Moreover, cognitive function features are taken from healthy people. The developed strategy is also cost-effective and noninvasive because it uses the Adam optimization method. This method, on the other hand, does not work well for making predictions that apply to a wide range of situations. The quality of the MRI data used for training has a big impact on how accurate the predictions are.

AD is a degenerative brain disease that directly affects how people think, remember things, and act. It was hard to diagnose AD because earlier methods only found MCI. So, El-Sappagh et al. (2022) came up with a two-stage DL AD prediction strategy to deal with these problems. For the implementation a lot of information was gathered, including cognitive scores, neuroimaging data, demographics, and more. This model is more stable, which makes it more useful and reliable for medical acceptance. This method, on the other hand, is very sensitive to parameter tuning and needs a lot of computing power.

Ávila-Jiménez et al. (2024) have created useful DL algorithms for rebalancing strategies that can be used to diagnose AD from clinical records. Moreover, a correct preprocessing method was used to filter the dataset. The projected model has also been tested against traditional ML models, and the developed model has the best results. Thus this model makes it possible to find and treat AD early. Nonetheless, this methodology is less flexible and relevant for real-time AD predictions.

Mujahid et al. (2023) proposed that the transfer of learning DementiaNet and efficient net-based deep feature algorithms are applicable for the identification of AD. This study utilizes publicly accessible datasets, comprising 6400 MRI images. Data augmentation methods have also been changed to make the dementia minority classification bigger. This model can easily avoid problems with overfitting, and it extracts features quickly. Finally the comparisons are done. The comparisons demonstrate that the developed model achieved an overall accuracy of 97%.

Umme Habiba, S., devised a method to identify AD in both balanced and imbalanced datasets. Habiba et al. (2023) employed an ensemble classifier grounded in VGG 16. This model was improved in terms of efficiency and accuracy for

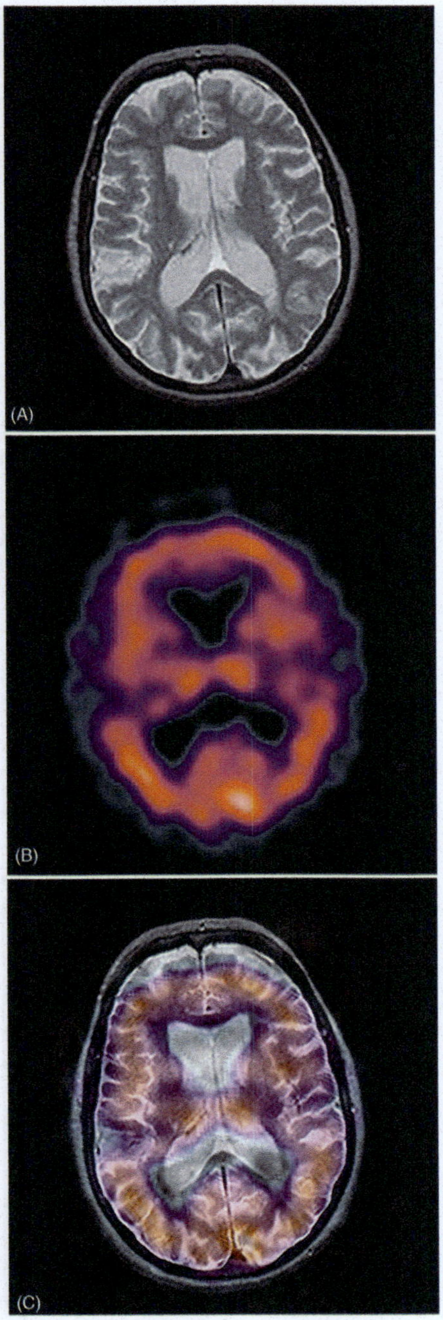

FIGURE 3.5 Fusion results for Type-1 MRI and SPECT images: (A) MRI image, (B) SPECT image, and (C) fused image using the proposed method.

early-stage detection performance. This oversampling technique works with publicly available datasets and gets 97.09% accuracy. This model works better and is better for smaller datasets, but it takes a long time. A lot of data, on the other hand, is not useful.

Tong et al. (2017) have created a nonlinear graph fusion algorithm that uses multiple modalities to find the symptoms of AD in datasets. First the multimodality model was used to find pairwise similarity by extracting features. After that a nonlinear graph fusion algorithm was used to get regional MRI volumes, signal intensities, CSF biomarkers, and information features. The experimental validation showed that the system made the AD classification detection process much better. But the classification rate is low because there are not many input samples.

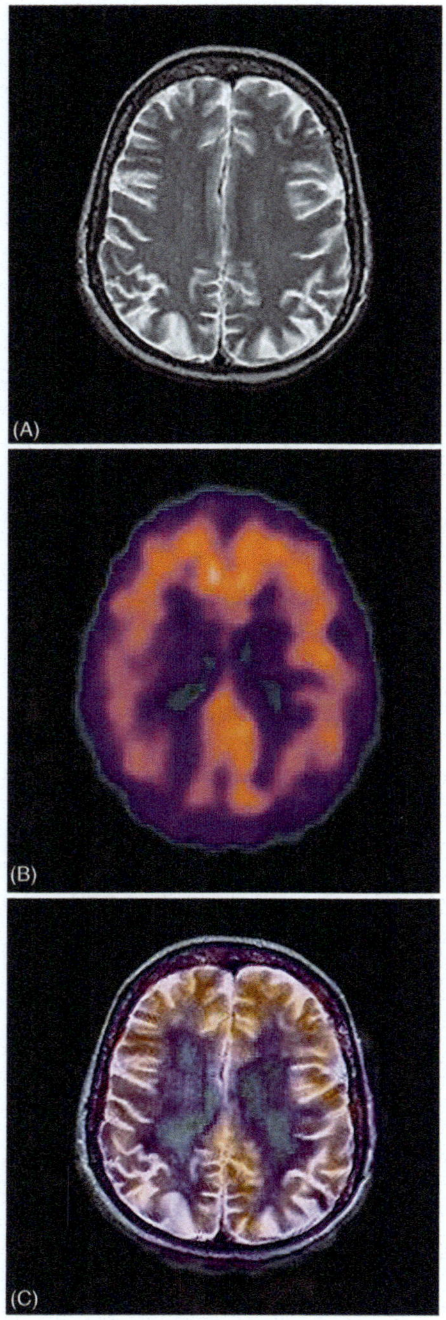

FIGURE 3.6 Fusion results for Type-2 MRI and SPECT images: (A) MRI image, (B) SPECT image, and (C) fused image using the proposed method.

Zhang et al. (2017) have developed a supervised learning-based landmark SVM classifier for the identification of AD from MRI images. This model finds the landmarks in the collected MRI images using a quick landmark detection method. Longitudinal and statistical characteristics were also extracted from the identified landmarks. Finally the SVM classifier can tell the difference between the AD, MCI, and normal classes. It got the best performance, but it was hurt by extra features and noise.

Chen et al. (2024) presented a hybrid deep learning model that employs an improved ResNet50 for feature engineering and a bidirectional gated recurrent unit for the processing of sequenced data. This model exhibited superior accuracy; however, the study was not founded on longitudinal data.

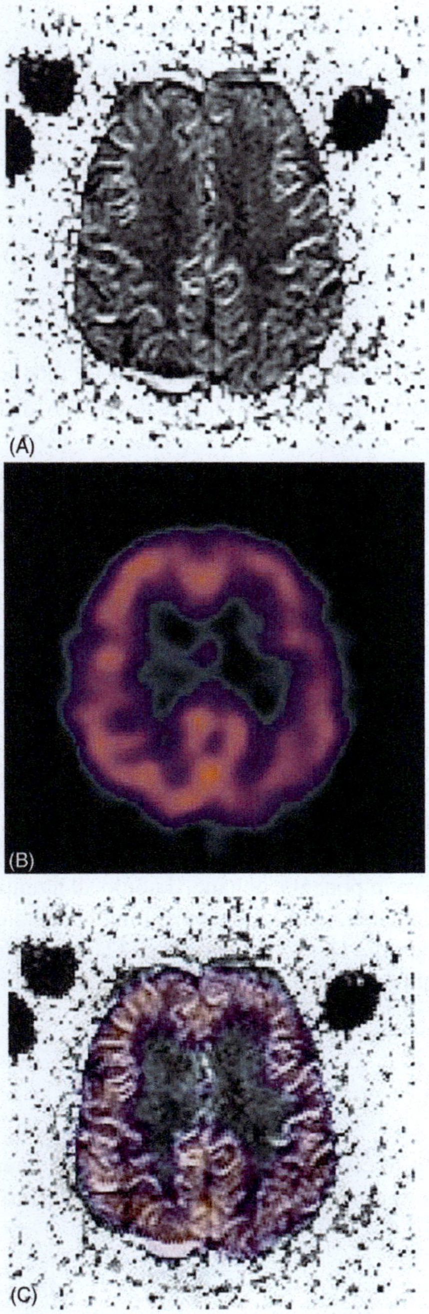

FIGURE 3.7 Fusion results for Type-3 MRI and SPECT images: (A) MRI image, (B) SPECT image, and (C) fused image by the proposed method.

Thangavel et al. (2023) designed a DL strategy for the prompt classification of AD from MRI samples. This research employed the architecture of deep neural networks (DNNs) for the identification of AD. The ResNet algorithm was also used to get the important features from the preprocessed images. As a result, better Adam optimization was used to choose the most important data. The results of the implementation showed that this strategy was 98% accurate at finding AD. But this method has a high chance of overfitting.

Kumar and Sasikala (2023) conducted a study aimed at enhancing the precision of AD detection through four distinct strategies. This study examined methodologies, including ResNet-50, AlexNet, ResNet-101, and ResNet-18 for the categorization of AD. Principal component analysis was also used for feature engineering tasks. This lowers the number of dimensions in the data and makes sure that there is a balance between accuracy and complexity. The results

TABLE 3.1 Results of performance tests for MRI and SPECT image types.

Metric	PSNR (dB)	Entropy (bits/symbol)	Average pixel intensity	SF (cycles/millimeter)	SD
Type-1 MRI–SPECT	64.67	6.47	0.23	23.43	0.28
Type-2 MRI–SPECT	67.13	5.71	0.20	20.00	0.26
Type-3 MRI–SPECT	54.41	6.77	0.65	60.94	0.34
Type-4 MRI–PET	63.29	7.12	0.97	55.32	0.82

PSNR, peak signal-to-noise ratio; *SF*, spatial frequency.

of the experiments showed that ResNet-101 had a better accuracy of 97.88% when tested on the Kaggle AD database. But these methods use pretrained networks and cannot change to fit the different situations that happen in the real world.

Lakhan et al. (2023) conducted a study centered on convex optimization challenges aimed at improving accuracy metrics and computational complexity in digital healthcare for AD detection. This study introduced an innovative methodology called the "Evolutionary Deep Convolutional Neural Network" to minimize computational time and achieve optimal accuracy in AD detection. The method's simulation results showed that it cut down on computation time by 38% and improved system accuracy by 19%. Nonetheless, this methodology is deficient in interpretability, which is crucial for healthcare professionals to understand the rationale behind predictions and formulate treatment plans.

Zeng et al. (2023) developed a novel algorithm for accurate classification of AD and MCI from MRI images. This work created a multitask learning system that uses a deep belief network (DBN) to do this classification task. Its primary objective is to mitigate overfitting problems and enhance the model's generalization ability. The results of the experiment showed that this strategy was 98.62% accurate for AD classification and 96.67% accurate for MCI classification. Nonetheless, choosing hyperparameters is a difficult job in this framework.

Illakiya et al. (2023) introduced an adaptive hybrid attention network for the accurate identification of AD from MRI images. This methodology consists of two components: enhanced nonlocal attention and coordinated attention. These parts independently capture attributes at both the global and local levels. Finally an adaptive feature fusion was made to combine features from these parts, which helps train the DenseNet classifier. The implementation results showed that this strategy worked 98.53% of the time. But combining different mechanisms into one approach is hard and makes the design more complicated.

3.3 Suggested method

The primary objective of this research was to create an efficient and resilient neural network model for the early diagnosis of AD. We created a proposed method for finding AD using medical images from different sources to do this. The structure of the suggested algorithm is shown in Fig. 3.4.

3.4 Results of the experiment

The efficacy of the proposed medical image fusion method on diverse medical images is demonstrated. The first example in Fig. 3.5 shows different output images from different scans. Fig. 3.5A and B shows two medical pictures taken with MRI and SPECT (Das & Bhattacharya, 2015; Lu et al., 2018). Fig. 3.5C shows the fused image made by the proposed method. Fig. 3.6A and B shows MRI and SPECT scans of people with AD that show that the suggested method gives better medical information. Type-3 input images are shown in Fig. 3.7A and B. Fig. 3.8A and B shows two medical pictures of an MRI and a PET scan. The proposed method uses the biorthogonal bior6.8 with a decomposition level of 1 as the wavelet basis. Experimental results involving various pairs of medical images from MRI, SPECT, and PET scans of AD (Murphy, 2018) demonstrate that the proposed method yields superior medical information. The subsequent comparison of fusion outcomes employs the following objective fidelity criteria, with the values of all objective criteria (Eskicioglu & Fisher, 1995; Jagalingam & Hegde, 2015) presented in Table 3.1 for reference.

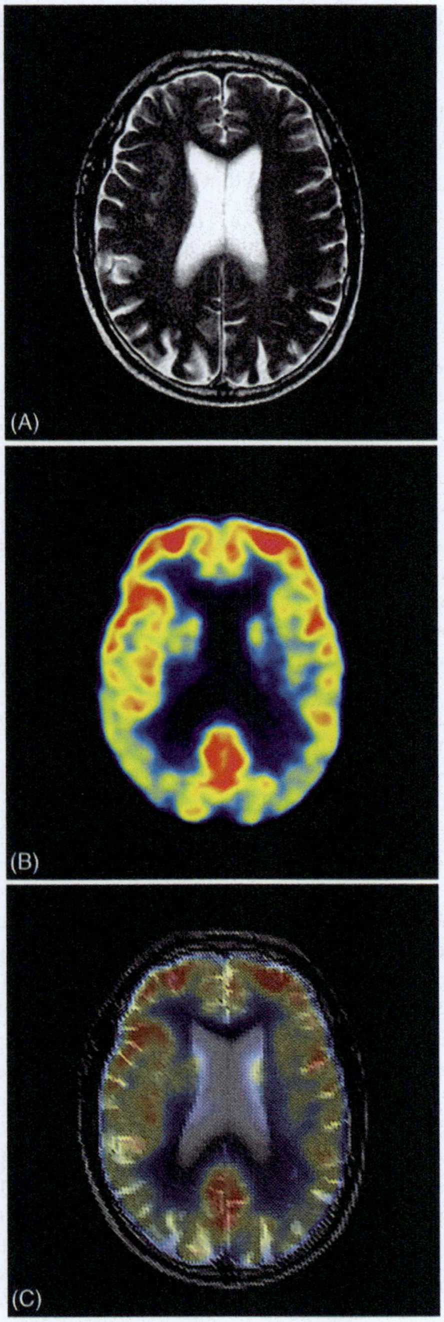

FIGURE 3.8 Fusion results for Type-4 MRI and PET images: (A) MRI image, (B) PET image, and (C) fused image by the proposed method.

3.5 Conclusion

The aim of this proposed research is to create advanced tools that will facilitate the diagnosis and treatment of AD by targeting the fundamental neurological issues encountered by patients. During the first stage, surveys are used to find out what specific memory loss symptoms people with AD have. The results will help create a complete coding system for sorting and keeping track of the many different memory loss symptoms that patients have. The technological strategy is to make a synthetic device that can copy and make up for the parts of the brain that are damaged and cause memory loss. Amyloid plaques, which are abnormal protein clusters, and neurofibrillary tangles, also known as tau tangles, are two types of brain problems that people with AD have. These strange signs that stop normal cognitive function are what cause the slow memory loss that is a sign of AD. The suggested device would use complicated electrical parts and connections to the brain to work like the damaged parts of the brain. The artificial system will interact with the brain's neural pathways. This could help restore cognitive function by restoring normal neural transmission that has been disrupted by amyloid plaques and tau tangles. This method uses a whole-person approach to neurological rehabilitation and modern healthcare technology to treat the symptoms of memory loss right away. Vision 2030's focus on new healthcare solutions and medical technology powered by AI is in-line with the development of a device that helps with memory loss caused by AD.

References

Ahmed, M. Q., Alenazi, F. S. H., Fazaludeen, M. F., Shahid, S. M. A., & Kausar, M. A. (2018). Pathology and management of Alzheimer's disease: A review. *International Journal of Pharmaceutical Research & Allied Sciences, 7*(2), 30–42.

Alzheimer's Association Report. (2017). 2017 Alzheimer's disease facts and figures. *Alzheimer's & Dementia*, 1–49. doi:10.1016/j.jalz.2017.02.001.

Alzheimer's Disease International. (2018). *World Alzheimer Report 2018: The state of the art of dementia research: New frontiers*. Alzheimer's Disease International (ADI).

Ávila-Jiménez, J. L., Cantón-Habas, V., del Pilar Carrera-González, M., Rich-Ruiz, M., & Ventura, S. (2024). A deep learning model for Alzheimer's disease diagnosis based on patient clinical records. *Computers in Biology and Medicine, 169*, 107814. doi:10.1016/j.compbiomed.2023.107814.

Borkar, P., Wankhede, V. A., Mane, D. T., Limkar, S., Ramesh, J. V. N., & Ajani, S. N. (2023). Deep learning and image processing-based early detection of Alzheimer disease in cognitively normal individuals. *Soft Computing*. https://www.springer.com/journal/500. doi:10.1007/s00500-023-08615-w.

Chen, Y., Wang, L., Ding, B., Shi, J., Wen, T., Huang, J., & Ye, Y. (2024). Automated Alzheimer's disease classification using deep learning models with Soft-NMS and improved ResNet50 integration. *Journal of Radiation Research and Applied Sciences, 17*(1), 100782. doi:10.1016/j.jrras.2023.100782.

Dalcerri, M., Garcia, A., & McKiel, H. (2017). Vitamin D deficiency and the increased risk of developing dementia and Alzheimer's disease. *Journal of Geriatric Research, 1*(1), 03.

Das, A., & Bhattacharya, M. (2015). *Effective image fusion method to study Alzheimer's disease using MR, PET images* (pp. 1603–1607). USA: IEEE Computer Society. https://doi.org/10.1109/BIBM.2015.7359915.

El-Sappagh, S., Saleh, H., Ali, F., Amer, E., & Abuhmed, T. (2022). Two-stage deep learning model for Alzheimer's disease detection and prediction of the mild cognitive impairment time. *Neural Computing and Applications, 34*(17), 14487–14509. http://link.springer.com/journal/521. doi:10.1007/s00521-022-07263-9.

Eskicioglu, A. M., & Fisher, P. S. (1995). Image quality measures and their performance. *IEEE Transactions on Communications, 43*(12), 2959–2965. doi:10.1109/26.477498.

Habiba, S. U., Debnath, T., Islam, M. K., Nahar, L., Hossain, M. S., Basnin, N., & Andersson, K. (2023). Transfer learning-assisted DementiaNet: A four layer deep CNN for accurate Alzheimer's disease detection from MRI images. In F. Liu, Y. Zhang, H. Kuai, E. P. Stephen, & H. Wang (Eds.). *Brain informatics. BI 2023. Lecture notes in computer science*: 13974. Springer. doi:10.1007/978-3-031-43075-6_33.

Herrera, L. J., Rojas, I., Pomares, H., Guillén, A., Valenzuela, O., & Baños, O. (2013). Classification of MRI images for Alzheimer's disease detection. In *Proceedings – SocialCom/PASSAT/BigData/EconCom/BioMedCom 2013* (pp. 846–851). doi:10.1109/SocialCom.2013.127.

Illakiya, T., Ramamurthy, K., Siddharth, M. V., Mishra, R., & Udainiya, A. (2023). AHANet: Adaptive hybrid attention network for Alzheimer's disease classification using brain magnetic resonance imaging. *Bioengineering, 10*(6), 714. doi:10.3390/bioengineering10060714.

Jagalingam, P., & Hegde, A. V. (2015). A review of quality metrics for fused image. *Aquatic Procedia, 4*, 133–142. doi:10.1016/j.aqpro.2015.02.019.

Kumar, S. A., & Sasikala, S. (2023). Enhanced Alzheimer's disease classification using multilayer deep convolutional neural network-based experimentations. *Iranian Journal of Science and Technology, Transactions of Electrical Engineering, 47*(4), 1595–1621. doi:10.1007/s40998-023-00622-9.

Lakhan, A., Grønli, T.-M., Muhammad, G., & Tiwari, P. (2023). EDCNNS: Federated learning enabled evolutionary deep convolutional neural network for Alzheimer disease detection. *Applied Soft Computing, 147*, 110804. doi:10.1016/j.asoc.2023.110804.

Lama, R. K., Gwak, J., Park, J.-S., & Lee, S.-W. (2017). Diagnosis of Alzheimer's disease based on structural MRI images using a regularized extreme learning machine and PCA features. *Journal of Healthcare Engineering, 2017*, 1–11. doi:10.1155/2017/5485080.

Li, Z., Zhu, H., Zhang, L., & Qin, C. (2018). The intestinal microbiome and Alzheimer's disease: A review. *Animal Models and Experimental Medicine, 1*(3), 180–188. doi:10.1002/ame2.12033.

Lu, D., Popuri, K., Ding, G. W., Balachandar, R., Beg, M. F., Weiner, M., Aisen, P., Petersen, R., Jack, C., Jagust, W., Trojanowki, J., Toga, A., Beckett, L., Green, R., Saykin, A., Morris, J., Shaw, L., Kaye, J., Quinn, J., … Fargher, K. (2018). Multimodal and multiscale deep neural networks for the early diagnosis of Alzheimer's disease using structural MR and FDG-PET images. *Scientific Reports, 8*(1). doi:10.1038/s41598-018-22871-z.

Mirza, A., King, A., Troakes, C., & Exley, C. (2017). Aluminium in brain tissue in familial Alzheimer's disease. *Journal of Trace Elements in Medicine and Biology, 40*, 30–36. doi:10.1016/j.jtemb.2016.12.001.

Mujahid, M., Rehman, A., Alam, T., Alamri, F. S., Fati, S. M., & Saba, T. (2023). An efficient ensemble approach for Alzheimer's disease detection using an adaptive synthetic technique and deep learning. *Diagnostics, 13*(15), 2489. doi:10.3390/diagnostics13152489.

Murphy, M. P. (2018). Amyloid-beta solubility in the treatment of Alzheimer's disease. *New England Journal of Medicine, 378*(4), 391–392. doi:10.1056/NEJMe1714638.

Samuel, E. (2014). Recent research and updates on Alzheimer's disease. *Journal of Psychology and Behavioral Science, 2*(2), 103–112.

Struckmeyer, L. R., & Pickens, N. D. (2016). Home modifications for people with Alzheimer's disease: A scoping review. *The American Journal of Occupational Therapy, 70*(1), 7001270020p1. doi:10.5014/ajot.2015.016089.

Tong, T., Gray, K., Gao, Q., Chen, L., & Rueckert, D.Alzheimer's Disease Neuroimaging Initiative. (2017). Multi-modal classification of Alzheimer's disease using nonlinear graph fusion. *Pattern Recognition, 63*, 171–181.

Thangavel, P., Natarajan, Y., & Sri Preethaa, K. R. (2023). EAD-DNN: Early Alzheimer's disease prediction using deep neural networks. *Biomedical Signal Processing and Control, 86*, 105215. doi:10.1016/j.bspc.2023.105215.

Xiao, Z., Ding, Y., Lan, T., Zhang, C., Luo, C., & Qin, Z. (2017). Brain MR image classification for Alzheimer's disease diagnosis based on multifeature fusion. *Computational and Mathematical Methods in Medicine, 2017*, 1–13. doi:10.1155/2017/1952373.

Zeng, N., Li, H., & Peng, Y. (2023). A new deep belief network-based multi-task learning for diagnosis of Alzheimer's disease. *Neural Computing and Applications, 35*(16), 11599–11610. doi:10.1007/s00521-021-06149-6.

Zhang, J., Liu, M., An, L., Gao, Y., & Shen, D. (2017). Alzheimer's disease diagnosis using landmark-based features from longitudinal structural MR images. *IEEE Journal of Biomedical and Health Informatics, 21*(6), 1607–1616. http://ieeexplore.ieee.org/xpl/RecentIssue.jsp?punumber=6221020. doi:10.1109/JBHI.2017.2704614.

Chapter 4

Neural network algorithms for AR-models identification reaching Huber-Rao-Cramer bound under quasi-Gaussian noise

Jesica Escobar[a] and Alexander Poznyak[b]

[a] Cybernetics-ICE, Instituto Politécnico Nacional, ESIME Zac, UPALM, Mexico City, Mexico, [b] Department of Automatic Control, CINVESTAV-IPN, Mexico City, Mexico

4.1 Introduction

Several methods can be used in the autoregression (AR) model identification process (Ivanov & Yakoub, 2023; Ljung, 2010). Regarding the corresponding covariation matrix $\mathbf{E}\{\Delta_t\Delta_t^{\mathrm{T}}\}$ of the estimation error Δ_t, each of them has a unique asymptotic convergence rate. This rate is $n^{-1}H$ in the standard scenario (regular statistic conditions) (Escobar & Poznyak, 2022; Poznyak, 2009), where H is a constant matrix and n is the number of available data measurements. Depending on the model's true parameter c of the considered AR-model the identification procedure to be used should be selected, and the statistical characteristics (density of the distribution function p_ξ) of a stationary noise ξ_n will influence the available data at time n. For a certain identification process "the better convergence rate" in comparison to another one is indicated by a smaller matrix H.

In statistics and estimation theory (Hodges & Lehmann, 1951; Taylor, 1979) the Rao-Cramér bound (RCB), also known as the *Information Inequality*, establishes a lower possible bound on the error covariation matrix of unbiased estimators:

$$\lim_{n\to\infty} nE\{\Delta_t\Delta_t^{\mathrm{T}}\} \geq \mathcal{I}(c, p_\xi) \tag{4.1}$$

where $\mathcal{I}(c, p_\xi)$ is the *Fisher's information matrix* (FIM). According to the open format of the inequality Eq. (4.1) the maximum feasible convergence rate of the error variance under regular observations is $n^{-1}\mathcal{I}(c, p_\xi)$. This indicates that the mean squared identification error must meet

$$E\left\{\| \Delta_t\|^2\right\} = E\left\{tr\left(\Delta_t\Delta_t^{T}\right)\right\} \geq n^{-1}\left[tr\{\mathcal{I}(c, p_\xi)\} + o_n(1)\right],\ o_n(1) \underset{n\to\infty}{\to} 0, \tag{4.2}$$

where the "convergence constant" is defined by the term

$$tr\{\mathcal{I}(c, p_\xi)\},$$

and the order of convergences is defined by n^{-1}. The Rao-Cramer (RC) inequality for regular density functions was first presented by Rao (Cramér, 1946; Rao, 1945) in their early works. Subsequently, Vincze presented information inequality (Khatri, 1980; Vincze, 1979), imposing the regularity assumptions on an a priori distribution instead of the model. When an unbiased estimator reaches this lower bound, it is typically referred to as (completely) *efficient*. The *minimal variance unbiased (MVU) estimator* is the name given to an algorithm that, among all unbiased estimates, produces the lowest mean squared error reaching the equality in Eq. (4.1).

When the noise distribution function's density p_ξ is unknown previously and only its membership in a distribution class $\mathcal{P}$ may be established, the following Huber-Rao-Cramer (HRC) bound is the information inequality's natural extension of

Computational Intelligence in Mechatronics: Solving Real-World Problems in Electronic Systems Design. DOI: https://doi.org/10.1016/B978-0-443-27580-7.00031-7

Eq. (4.1) (see Escobar & Poznyak, 2022; Huber, 1975; Poznyak, 2009):

$$\lim_{n\to\infty} nE\{\Delta_t\Delta_t^{\mathsf{T}}\} \geq \sup_{p_\xi\in\mathcal{P}} \mathcal{I}(c, p_\xi). \tag{4.3}$$

The density

$$p_\xi{}^* = \arg \sup_{p_\xi\in\mathcal{P}} \mathcal{I}(c, p_\xi),$$

providing for any $p_\xi \in \mathcal{P}$ the inequality

$$\mathcal{I}\left(c, p_\xi^*\right) \geq \mathcal{I}(c, p_\xi)$$

is referred to as the "worst noise distribution" within the considered class $\mathcal{P}$. Finally the identification method that meets the equality in the HRCB information bound (Eq. 4.2) involves this "worst distribution" p_ξ^*. Analytically determining p_ξ^* is a difficult undertaking.

The FIM is a fundamental concept in estimation theory, from which we can obtain the Cramer-Rao bound. A Fischer Information Neural Estimator is presented in Duy et al. (2022). The proposed method produced an estimation of the FIM directly from observed data, and it is shown that the proposed method can provide an accurate FIM estimate. In Duy et al. (2023), a highly accurate Fisher Information Neural Estimator is proposed, which is applicable for deterministic and random parameters. The method depends on measured data and does not require information on the probability density function.

The FIM is a fundamental quantity to represent the characteristics of a stochastic model; this also applies in deep neural networks (DNNs). In Soen and Sun (2021), a study is presented that reveals novel statistics of FIM's eigenvalues and shows that most of them are close to zero, whereas the maximum eigenvalue has a huge value.

Here, we propose an alternate neural network (NN) version of the recurrent identification technique, hitting the HRCB information constraint, and offer an NN approximation of the "worst distribution" p_ξ^*. Stated otherwise the recommended identification process represents the so-called *robust (min–max) method*, in which "max" is taken over all p_ξ inside a particular class $\mathcal{P}$ and "min" is selected among all recurrent processes that achieve the "best error convergence rate."

4.2 AR-model, quasi-Gaussian distribution, and problem formulation

4.2.1 Model description

Consider the following AR (autoregression) model given by

$$y_n = \sum_{l=1}^{L} a_l y_{n-l} + \xi_n,\ n = 0, 1, \ldots, \tag{4.4}$$

where

- $\{y_n\} \in R^1$ is a scalar sequence of available on-line state variables,
- $\{\xi_n\} \in R^1$ is a stochastic noise sequence (not available during the process), which is an independent zero mean, stationary sequence with the probability density function (p.d.f.) $p_\xi(x)$, which may be unknown but belonging to some given class $\mathcal{P}$ of p.d.f., that is,

$$p_\xi(x) \in \mathcal{P}.$$

4.2.2 Main assumptions

1. All random variables $\{\xi_n\}$ are defined on the probability space $(\Omega, \mathcal{F}, P)$ with the σ-algebras flow $\mathcal{F}_n \subseteq \mathcal{F}_{n+1}$

$$\mathcal{F}_n = \sigma(y_{-l}, \ldots, y_{-1}, \ldots, y_n; \xi_{0,\ldots,}\xi_n). \tag{4.5}$$

2. For all n

$$E\{\xi_n \mid \mathcal{F}_{n-1}\} \overset{a.s.}{=} 0,\ E\{\xi_n^2 \mid \mathcal{F}_{n-1}\} \overset{a.s.}{=} \sigma^2(p_\xi) < \infty. \tag{4.6}$$

3. The AR-plant Eq. (4.4) is stable: the polynomial

$$A(q^{-1}) = 1 - \sum_{l=1}^{L} a_l q^{-l} \tag{4.7}$$

is Hurwitz.

Remark 1. As it follows from the assumptions above, the noise sequences admit to being non-Gaussian.

4.2.3 Class of quasi-Gaussian noises

The class $\mathcal{P}_\xi$ of p.d.f., considered in practical applications, are related to a priori information on stationary generating sequence $\{\xi_n\}$. Here, we considered the so-called class of *quasi-Gaussian distributions,* which is natural from a practical point of view:

$$\mathcal{P}_\xi = \left\{p_\xi : p_\xi(x) = (1-\alpha)p_{\mathcal{N}(0,\sigma_{\mathcal{N}}^2)}(x) + \alpha q(x)\right\}, \tag{4.8}$$

where

$$p_{\mathcal{N}(0,\sigma_{\mathcal{N}}^2)}(x)$$

is the centered Gaussian distribution density with variance σ^2 and $q(x)$ is another distribution density. The parameter $\alpha \in [0, 1]$ characterizes the level of the effect of a "dirty" distribution $q(x)$ on the basic Gaussian distribution $p_{\mathcal{N}(0,\sigma^2)}(x)$ and may be treated as a probability to deal with the noise ξ_n, which has some other distribution $q(x)$ differed from the normal $p_{\mathcal{N}(0,\sigma^2)}(x)$.

4.2.4 Regression format representation

The system (4.4) can be represented in the regression form as

$$y_n = z_n^{\mathsf{T}} c + \xi_n, \tag{4.9}$$

where the extended vector

$$c = (a_1, \ldots, a_L)^{\mathsf{T}} \in \mathbb{R}^L \tag{4.10}$$

represents the *collection of unknown parameters* to be estimated, and the vector

$$z_n = (y_{n-1}, \ldots, y_{n-L})^{\mathsf{T}} \in \mathbb{R}^L \tag{4.11}$$

is referred to as *the generalized regression measurable (available on-line) input.*

4.2.5 Robust parametric identification problem formulation

Problem. We need to estimate the vector c of unknown parameters based on the available data $\{z_n\}$, and a priory knowledge of the p.d.f class $\mathcal{P}$ of the stationary noise sequence $\{\xi_n\}$, providing the HRC information bound Eq. (4.3).

Remark 2. The phrase **"robust parameter identification"** refers to maintaining the success of identification in the presence of uncertainty. In the case at hand the precise p.d.f. of ξ_n is acknowledged to be unknown but to belong to the class $\mathcal{P}$.

4.3 Regular observations and information inequality

4.3.1 Regularity conditionals

The observable output sequence $\mathbf{y}_n$: $=\{y_1, y_2, \ldots y_n\}$ contains the information on the parameter $c \in \mathbb{R}^L$. The function $p(\mathbf{y}_n|c)$ is called the *joint density of the distribution* of the vector $\mathbf{y}_n$. Any Borel function $c_n = c_n(\mathbf{y}_n) \in \mathbb{R}^L$ can be considered an estimate of the parameter c.

Definition 1. The observations y_n are referred to be as **regular** on the class $C \subseteq \mathbb{R}^L$ of parameters if

$$\left.\begin{aligned} \sup_{c\in C} E\left\{\| \ln p(y_n)|c \|^2\right\} = \sup_{c\in CY_n} \int \left|\ln p(y_n|c)\right|^2 p(y_n|c)dy_n < \infty, \\ Y_n = \{y_n|p(y_n|c) > 0,\ c \in C\}, \end{aligned}\right\} \tag{4.12}$$

and for any $c \in C$ *the FIM* is strictly positive:

$$\left.\begin{aligned} \mathbb{I}_F(c, n \mid p_\xi) := E\{\nabla_c \ln p(\mathbf{y}_n \mid c)\nabla_c^{\mathrm{T}} \ln p(\mathbf{y}_n \mid c)\} = \\ \int_{Y_n} [\nabla_c \ln p(\mathbf{y}_n \mid c)\nabla_c^{\mathrm{T}} \ln p(\mathbf{y}_n \mid c)] p(\mathbf{y}_n \mid c)d\mathbf{y}_n > 0. \end{aligned}\right\} \tag{4.13}$$

4.3.2 Recurrent identification procedures asymptotically reaching HRCB

Consider the following class of recurrent identification procedures (Escobar & Poznyak, 2022; Poznyak, 2019), which may be applied to the transformed model (4.9):

$$\left.\begin{aligned} c_n = c_{n-1} + \Gamma_n z_n \varphi(y_n - z_n{}^T c_{n-1},) \\ \Gamma_n = \left(\sum_{t=0}^{n} z_t z_t{}^T\right)^{-1}, n \geq n_0 := \left\{\min_k \sum_{t=0}^{k} z_t z_t{}^T\right\} > 0. \end{aligned}\right\} \tag{4.14}$$

Notice that Γ_n in Eq. (4.14) can be calculated recursively [as in the least squares method (LSM)]

$$\Gamma_n = \Gamma_{n-1} - \frac{\Gamma_{n-1} z_n z_n^{\mathrm{T}} \Gamma_{n-1}}{1 + z_n^{\mathrm{T}} \Gamma_{n-1} z_n}, n \geq n_0, \tag{4.15}$$

and, in view of the polynomial stability (assumption 3), possesses (in the accepted assumptions) the following property

$$\Gamma_n \overset{a.s.}{\simeq} \frac{1}{n}\mathcal{R}^{-1}, \mathcal{R} = \lim_{n\to\infty} \frac{1}{n} \sum_{k=1}^{n} E\{z_k^{\mathrm{T}} z_k\} > 0. \tag{4.16}$$

The last inequality in Eq. (4.16) is known as the *persistent excitation condition.*

Theorem 1. If

1. *ξ_n is i.i.d. sequence with*

$$E\{\xi_n\} = 0, E\{\xi_n^2\} = \sigma^2 > 0, E\{\xi_n^4\} = E\{\xi_1^4\} < \infty \tag{4.17}$$

2. *The nonlinear transformation $\varphi : \mathbb{R} \to \mathbb{R}$satisfies the conditions*

$$x\psi(x) \geq \delta x^2, \delta > 0, \psi(0) = 0,\ S(x) \leq k_0 + k_1 x^2,$$

with

$$\psi(x) = E\{\varphi(x + \xi_n)\},\ S(x) := E\{\varphi^2(x + \xi_n)\},$$

then

$$\Delta_n = c_n - c \underset{n\to\infty}{\overset{a.s.}{\to}} 0. \tag{4.18}$$

The proof can be found in Poznyak (2009, 2019).

The HRCB (Eq. 4.3) can be represented as (see, for example, Theorem 2 in Escobar & Poznyak, 2022)

$$\left.\begin{aligned} \lim_{n\to\infty} nE\{\Delta_t \Delta_t^{\mathrm{T}}\} \geq \sup_{p_\xi \in \mathcal{P}} I(c, p_\xi) = \\ \lim_{n\to\infty} \sup_{p_\xi \in \mathcal{P}} n\mathbb{I}_F(c, n \mid p_\xi) = \sup_{p_\xi \in \mathcal{P}} \mathcal{R}^{-1} I_F^{-1}(p_\xi). \end{aligned}\right\} \tag{4.19}$$

Here, $I_F(p_\xi)$ is the Fischer information of single scalar data

$$I_F^{-1}(p_\xi) := \int_{x\in\mathbb{R}^1} \frac{\left[p'_\xi(x)\right]^2}{p_\xi(x)} dx,$$

and the matrix $\mathcal{R}$ has the structure

$$\mathcal{R} = \sigma^2(p_\xi)\mathcal{R}_0, \tag{4.20}$$

where $\mathcal{R}_0$ is the solution of the following matrix equation

$$\mathcal{R}_0 = A\mathcal{R}_0 A + \Xi_0$$

with

$$A = \begin{vmatrix} a_0 & a_1 & \cdots & \cdots & a_L \\ 1 & 0 & \cdots & \cdots & 0 \\ 0 & 1 & 0 & \cdots & 0 \\ 0 & \cdots & \ddots & 0 & 0 \\ 0 & \cdots & 0 & 1 & 0 \end{vmatrix}; \Xi := \begin{vmatrix} 1 & 0 & \cdots & \cdots & 0 \\ 0 & 0 & \cdots & \cdots & 0 \\ 0 & 0 & 0 & \cdots & 0 \\ 0 & \cdots & \ddots & 0 & 0 \\ 0 & \cdots & 0 & 0 & 0 \end{vmatrix}.$$

In view of Eq. (4.20) the information bound (Eq. 4.19) can be represented as

$$\left.\begin{aligned} \lim_{n\to\infty} nE\{\Delta_t\Delta_t^{\intercal}\} \geq \mathcal{R}_0^{-1} \sup_{p_\xi\in\mathcal{P}} \left[\sigma^{-2}(p_\xi)I_F^{-1}(p_\xi)\right] \\ = \mathcal{R}_0^{-1} \inf_{p_\xi\in\mathcal{P}} \left[\sigma^2(p_\xi)I_F(p_\xi)\right]. \end{aligned}\right\} \tag{4.21}$$

Defining *the worst distribution within the class* $\mathcal{P}$

$$\left.\begin{aligned} p_\xi^* = arg \sup_{p_\xi\in\mathcal{P}} \left[I_{F,\xi}(p_\xi)\mathcal{R}(p_\xi)\right]^{-1} = \\ arg \sup_{p_\xi\in\mathcal{P}} \left[\sigma^2(p_\xi)I_F(p_\xi)\right]^{-1} = arg \inf_{p_\xi\in\mathcal{P}} \left[\sigma^2(p_\xi)I_F(p_\xi)\right], \end{aligned}\right\} \tag{4.22}$$

select the nonlinear residual transformation $\varphi(x)$ in Eq. (4.14) as

$$\varphi^*(x) = -I_F^{-1}\left(p_\xi^*\right)\frac{d}{dx}\ln p_\xi^*(x)|_{x=y_n-z_n^{\intercal}c_{n-1}} \tag{4.23}$$

If $\varphi^*(x)$ satisfies the condition of Theorem 1, it is possible to prove (see Escobar & Poznyak, 2022; Poznyak, 2009) that the algorithm (Eq. 4.14) with $\varphi^*(x)$ (Eq. 4.23)

$$c_n = c_{n-1} - \Gamma_n z_n\left(I_F^{-1}\left(p_\xi^*\right)\frac{d}{dx}\ln p_\xi^*(x)|_{x=y_n-z_n^{\intercal}c_{n-1}}\right) \tag{4.24}$$

obtains the HRCB and, hence, is *asymptotically robust optimal* under the disturbing quasi-Gaussian noise ξ.

4.4 NN realization of the worst distribution within the class of quasi-Gaussian distribution

Let us in Eq. (4.8) approximate the distribution $q(x)$ as

$$q(x) \simeq \hat{q}(x) = \sum_{s=1}^{N_a} w_s\sigma_s(x) = w^{\intercal}\sigma(x), \tag{4.25}$$

where $w \in \mathbb{R}^{N_a}$ are referred to as weights and $\sigma : \mathbb{R}^1 \to \mathbb{R}^{N_a}$ is a vector basic function (which is an analogue of sigmoidal components in standard NN). Both constitute the corresponding NN construction (Poznyak et al., 2001). As $q(x)$ is a distribution density function, it should satisfy the following restrictions:

$$q(x) \geq 0, \int_{-\infty}^{\infty} q(x)dx = 1,$$

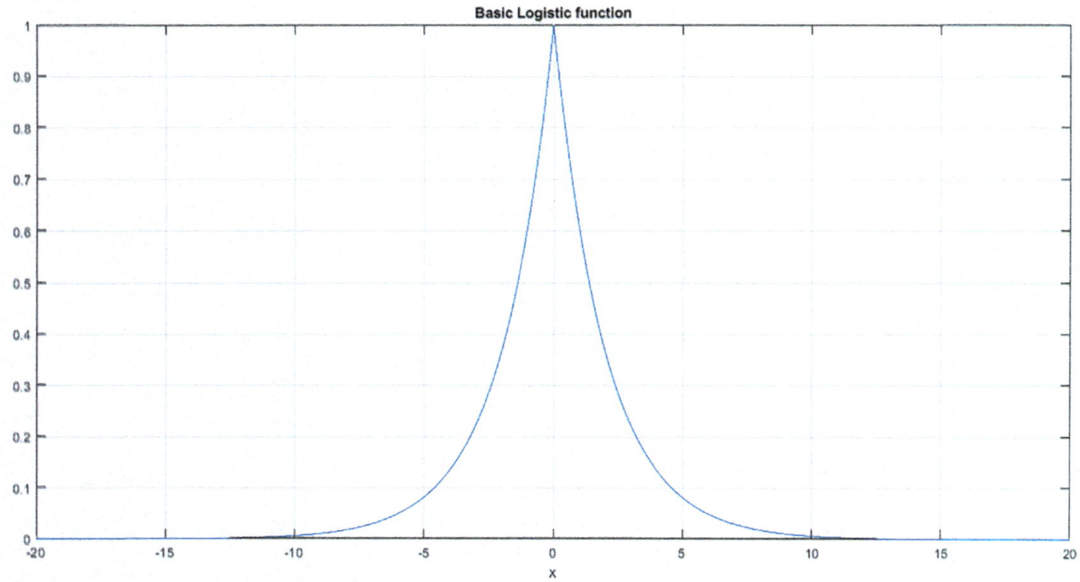

FIGURE 4.1 Basic function.

which implies the corresponding constraints to the NN approximation $\hat{q}(x)$, namely,

$$\left.\begin{array}{c}\sum\limits_{s=1}^{N_a} w_s\sigma_s(x) \geq 0,\ \sum\limits_{s=1}^{N_a} w_s\theta_s = 1,\\ \theta_s := \int\limits_{-\infty}^{\infty} \sigma_s(x)dx,\ s = 1, \ldots, N_a.\end{array}\right\} \tag{4.26}$$

Basic functions $\sigma_s : \mathbb{R}^1 \to \mathbb{R}^1$ may be selected by diverse ways (see, for example Menon et al., 1996; Szandała, 2021). Here we take $\sigma_s(x)$ as

$$\sigma_s(x) := \exp\{\gamma_s|x|\},\ \gamma_s > 0. \tag{4.27}$$

Fig. 4.1 illustrates a basic function given by Eq. (4.27).

With such selection to fulfill Eq. (4.26), it seems natural to require $w_s \geq 0$, $s = 1, \ldots, N_a$. Analytical calculation of the parameter θ_s with Eq. (4.27) gives

$$\theta_s := 2\gamma_s^{-1}. \tag{4.28}$$

So the admissible set S_{N_a} of weights w in Eq. (4.25) is the simplex.

$$S_{N_a} := \left\{\mathbf{w} \in \mathbb{R}^{N_a} \mid w_s \geq 0(s = 1, \ldots, N_a), \sum_{s=1}^{N_a} (2\gamma_s^{-1})w_s = 1\right\}. \tag{4.29}$$

In view of the considerations above, we may conclude that the worst distribution density p_ξ^* can be approximated as

$$\hat{p}_\xi^* = \arg\inf_{\hat{p}_\xi \in \hat{\mathcal{P}}} \left[\sigma^2(\hat{p}_\xi)I_F(\hat{p}_\xi)\right], \tag{4.30}$$

where the admissible set of the approximations of quasi-Gaussian densities is given by

$$\hat{\mathcal{P}}_\xi = \left\{\hat{p}_\xi : \hat{p}_\xi(x) = (1-\alpha)p_{\mathcal{N}(0,\sigma^2_{\mathcal{N}})}(x) + \alpha\sum_{s=1}^{N_a} w_s\exp\{-\gamma_s|x|\}, \mathbf{w} \in S_{N_a}\right\}. \tag{4.31}$$

In the open format the formula (4.30) can be represented as follows:

$$\hat{p}_\xi^* = \arg\inf_{\hat{p}_\xi \in \hat{\mathcal{P}}} \left[\sigma^2(\hat{p}_\xi)I_F(\hat{p}_\xi)\right] \tag{4.32}$$

with

$$\sigma^2(\hat{p}_\xi) := \int_{-\infty}^{\infty} x^2\left[(1-\alpha)p_{\mathcal{N}(0,\sigma^2{}_{\mathcal{N}})}(x) + \alpha \sum_{s=1}^{N_a} w_s \exp\{-\gamma_s|x|\}dx,\right]$$
$$(1-\alpha)\int_{-\infty}^{\infty} x^2 p_{\mathcal{N}(0,\sigma^2{}_{\mathcal{N}})}(x)dx + \alpha \sum_{s=1}^{N_a} w_s\, exp\{-\gamma_s|x|\}dx =$$
$$(1-\alpha)\sigma^2{}_{\mathcal{N}} + \alpha \mathbf{w}^{\mathbf{T}}\beta,\ \beta_s := \frac{4}{\gamma_s{}^3}, \tag{4.33}$$

and

$$I_F(\hat{p}_\xi) := \int_{-\infty}^{\infty} \frac{\left[\hat{p}'_\xi(x)\right]^2}{\hat{p}_\xi(x)}dx = 2\int_{x=0}^{\infty} \frac{S_0(\mathrm{w}, \mathrm{x})}{S_1(\mathrm{w}, \mathrm{x})}dx, \tag{4.34}$$

where

$$S_0(\mathbf{w}, \boldsymbol{x}) = (1-\alpha^2)\phi_0(x) + 2(1-\alpha)\alpha\mathbf{w}^{\mathbf{T}}\phi_0(x) + \alpha^2\left[\mathbf{w}^{\mathbf{T}}\tilde{\theta}(x)\right]^2,$$

$$S_1(w,x) = (1-\alpha)\theta_0(x) + \mathbf{w}^{\mathbf{T}}{}_{\boldsymbol{n}-1}\theta(x),$$
$$\phi_s(x) := \frac{|x|}{\sigma^2{}_{\mathcal{N}}} p_{\mathcal{N}(0,\sigma^2)}(x)\gamma_s exp\{-\gamma_s|x|\},$$
$$\theta_0(x) := p_{\mathcal{N}(0,\sigma^2)}(x),\ \tilde{\theta}(x) := \gamma_s exp\{-\gamma_s|x|\}. \tag{4.35}$$

Substitution (4.34), (4.35) into (4.32) gives

$$\left.\begin{aligned} \hat{p}^*_\xi(\mathbf{w}) = arg \inf_{\hat{p}_\xi\in\hat{\mathcal{P}}}\left[\sigma^2(\hat{p}_\xi)I_F(\hat{p}_\xi)\right] = arg \inf_{w\in S_{N_a}} \Phi(\mathbf{w}),\\ \Phi(\mathbf{w}) := 2\left[(1-\alpha)\sigma^2_{\mathcal{N}} + \alpha\mathbf{w}^{\intercal}\boldsymbol{\beta}\right]\int_{x=0}^{\infty} \frac{S_0(\mathrm{w},\mathrm{x})}{S_1(\mathrm{w},\mathrm{x})}dx. \end{aligned}\right\} \tag{4.36}$$

4.5 Recurrent procedure for finding NN approximation of the worst distribution

To obtain the weights $\mathbf{w}^*$ that minimize the function $\Phi(\mathbf{w})$, let us use the following recurrent projection gradient procedure

$$\mathbf{w}_n = \pi_{S_{N_a}}\{\mathbf{w}_{n-1} - \tau\nabla\Phi(\mathbf{w}_{n-1})\},\ \tau > 0, \tag{4.37}$$

where $\pi_{S_{N_a}}$ is the projection operation to the simplex S_{N_a} realizing the solution of the following quadratic optimization problem

$$\text{if } z \in S_{N_a},\ \text{then } \pi_{S_{N_a}}\{z\} := \arg\min_{\tilde{z}\in S_{N_a}} |z - \tilde{z}|^2. \tag{4.38}$$

The following projection operator

$$\pi_{S_{N_a}}\{z\}$$

can be applied as

$$\left.\begin{aligned} \pi_{S_{N_a}}\{z\} = \frac{\gamma_s}{2}\pi_{\overset{o}{S}_{N_a}}\{z\},\\ \overset{o}{S}_{N_a} = \left\{\overset{\mathbf{o}}{\mathbf{w}} \in \mathbb{R}^{N_a} \middle| \overset{o}{w}_s \geq 0\ (s = 1, \ldots, N_a), \sum_{s=1}^{N_a} \overset{o}{w}_s = 1\right\} \end{aligned}\right\} \tag{4.39}$$

where the realization of the projection operator

$$\pi_{\overset{o}{S}_{N_a}}\{z\}$$

in MATLAB can be found in Clempner and Poznyak (2024). In Eq. (4.37)

$$\left.\begin{aligned} \nabla\Phi(\mathbf{w}_{n-1}) = 2\alpha\beta\int_{x=0}^{\infty} \frac{S_0(\mathbf{w}_{n-1},x)}{S_1(\mathbf{w}_{n-1},x)}dx\\ +2\left[(1-\alpha)\sigma^2_{\mathcal{N}} + \alpha\mathbf{w}^{\intercal}_{n-1}\beta\right]\int_{x=0}^{\infty}\left[\frac{\rho_1 S_1(\mathbf{w}_{n-1},\mathbf{x}) - \theta(x)S_0(\mathbf{w}_{n-1},\mathbf{x})}{S_1^2(\mathbf{w}_{n-1},\mathbf{x})}\right]dx \end{aligned}\right\} \tag{4.40}$$

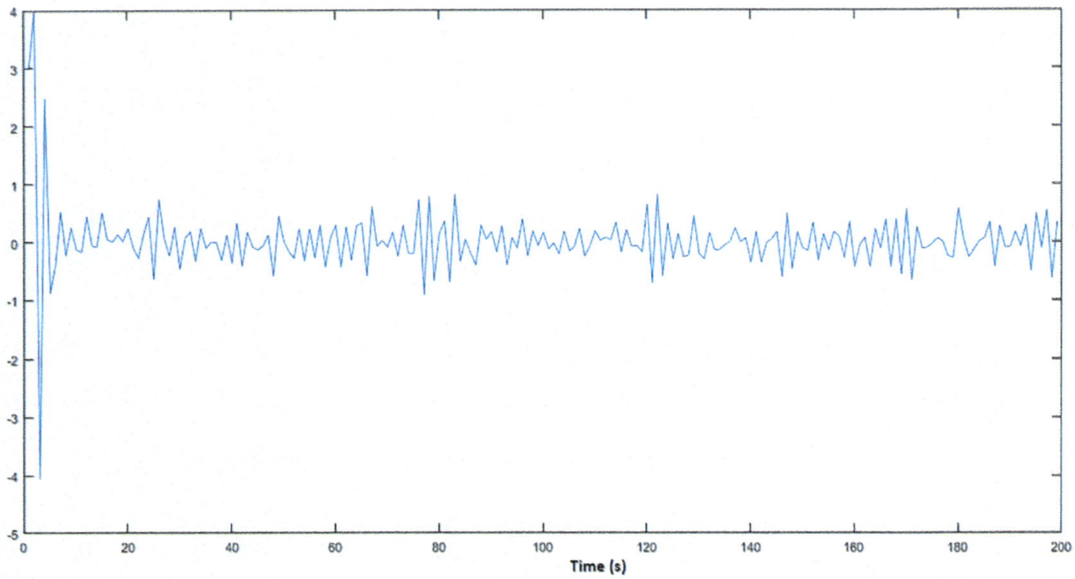

FIGURE 4.2 System output $y(t)$.

where

$$\rho_1 = 2\left[(1-\alpha)\alpha\phi(x) + \alpha^2\theta'(x)\right]. \tag{4.41}$$

4.6 Numerical realization of the identification algorithm

To realize the NN identification algorithm the following two-step procedure is required:

Description 1.

Step 1. For the given normal distribution $p_{\mathcal{N}(0,\sigma_N^2)}$ and given α, $N_a, \gamma_s (s = 1, .., N_a)$ should be realized the recurrent algorithm (4.40) generating the vector sequence $\{\mathbf{w}_n\}$ that converges to some limit vector $\mathbf{w}^*$, namely

$$\mathbf{w}_n \underset{n\to\infty}{\to} \mathbf{w}^*. \tag{4.42}$$

Step 2. For the obtained vector $\mathbf{w}^*$ the NN approximation of the "worst" distribution can be defined as follows:

$$\hat{p}*_\xi(x) = (1-\alpha)p_{\mathcal{N}(0,\sigma^2{}_N)}(x) + \alpha\sum_{s=1}^{N_a} w*_s\exp\{-\gamma_s|x|\}, \mathbf{w} *\in S_{N_a}.$$

Then the NN analogue of the identification algorithm (4.24) may be applied

$$c_n = c_{n-1} - \Gamma_n z_n\left(I_F^{-1}\left(\hat{p}_\xi^*\right)\frac{d}{dx}\ln\hat{p}_\xi^*(x)|_{x=y_n-z_n^{\mathrm{T}}c_{n-1}}\right). \tag{4.43}$$

4.7 Numerical example

The numerical example presented in this section shows the performance of the proposed estimation algorithm. Let us define the following difference equation:

$$y(n) + a_1y(n-1) + a_2y(n-2) = \xi(n) \tag{4.44}$$

where $a_1 = 0.85$, $a_2 = 0.3$, and $\xi(n)$ have the quasi-Gaussian distribution given by Eq. (4.8), formed by a centered Gaussian distribution and a uniform distribution, where α is a random number $\in [0, 1]$. The initial conditions are $c(0) = 2$, $y(0) = 3$, and $\Gamma_{n0} = 10^7$, the simulation time is $T_s = 200$.

In Fig. 4.2 the system output is shown. Fig. 4.3 shows the parameter estimation of a_1 and a_2 using the simple version of the LSM algorithm in these figures, where it is possible to clearly see the error estimation, which is quite common when using LSM, especially in systems that present stochastic perturbations, which, in this case, are not available during the process.

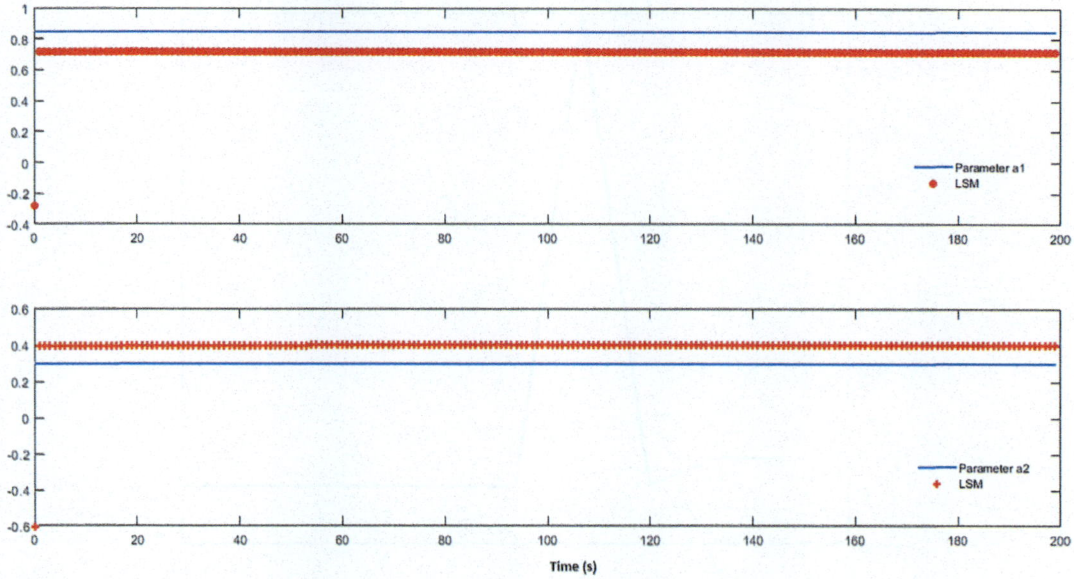

FIGURE 4.3 System parameters and its estimated parameters a_1 and a_2 and its estimated using LSM. *LSM*, Least squares method.

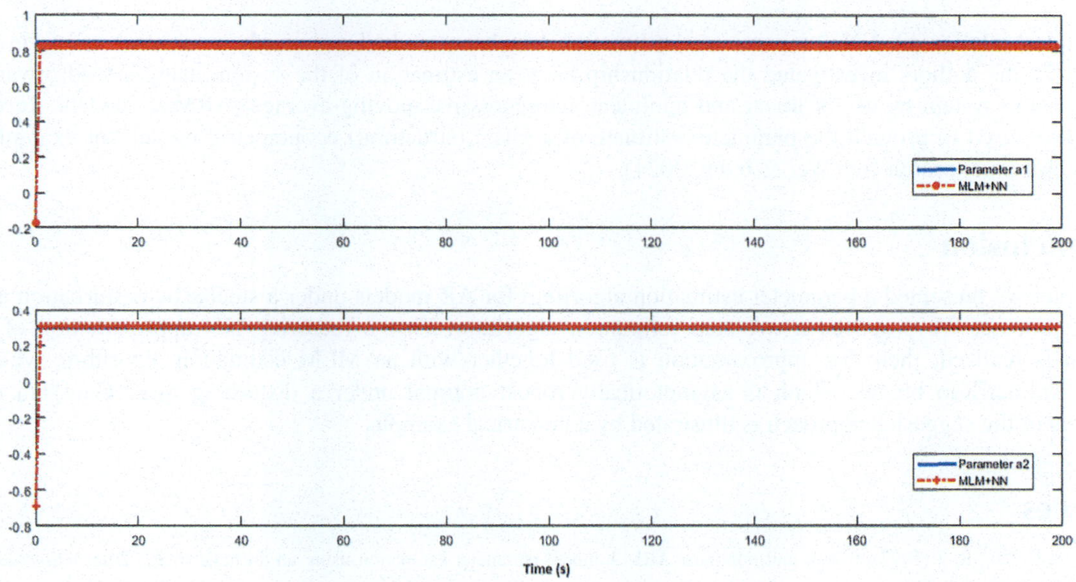

FIGURE 4.4 Parameter estimation using the MLM with NN. Parameters a_1 and a_2 and its estimated using the maximum likelihood method with neural network approximation. MLM, Maximum likelihood method; NN, neural network.

For the NN approximation the basic function is given by Eq. (4.27) and illustrated by Fig. 4.1, where $\gamma_s = 0.7$. For the recurrent procedure $\mathbf{w}_n$, we defined $\tau = 0.3$. A neuron is generated for each instant of time in the simulation; this means that for this example we have 200 neurons. This number of neurons, although small, can improve the performance of the estimation algorithm.

In Fig. 4.4 one can see the parameter estimation implementing the maximum likelihood method (MLM) with the NN approximation; the implementation of the NN improves the performance of the estimation algorithm, which will achieve the best error convergence rate (see Poznyak, 2009). The NN approximation of the worst distribution is shown in Fig. 4.5.

In the previous figures, it is possible to see the beneficial effect of the NN approximation in parameter estimation. The proposed method can be used for parameter estimation in any model that belongs to the AR family, even when the stochastic noise is not available.

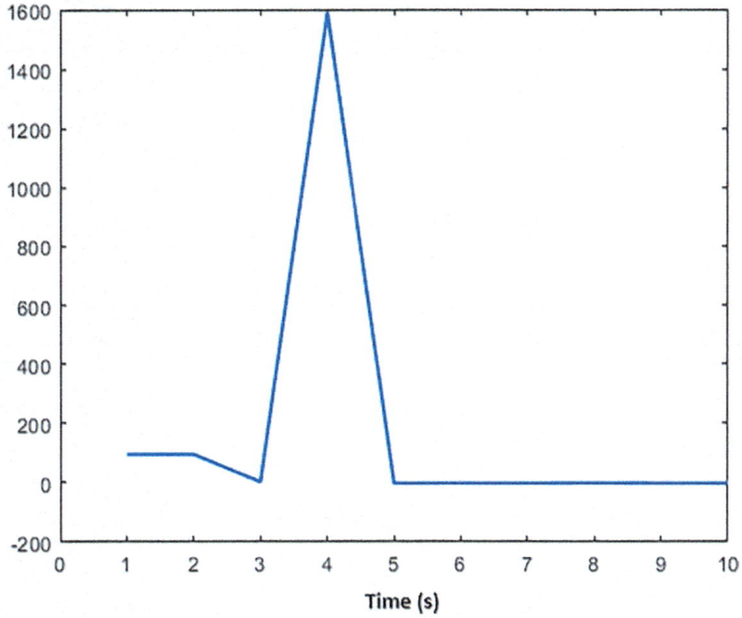

FIGURE 4.5 Neural network approximation of the worst distribution.

The implementation of NN for parameter estimation has been studied previously in several cases, in Chon and Cohen (1997), the authors investigated the relationship between estimation of the system using a feedforward NN and estimation of the system by use of linear and nonlinear autoregressive moving-average (ARMA) models. Recently, NNs have also been used to provide the parameter estimate of a given (structural) econometric model, for example, discrete choice and consumer search (see Wei & Jiang, 2024).

4.8 Conclusion

In this chapter we presented a parameter estimation algorithm for AR models under a stochastic perturbation unavailable during the process. We proposed a two-step estimation algorithm: First a recurring NN approximation of the worst distribution is realized, then this approximation is used together with an MLM estimation algorithm, which obtains the Huber-Cramer-Rao bound, which is asymptotically robust optimal under a disturbing quasi-Gaussian noise. The effectiveness of the suggested approach is illustrated by a numerical example.

References

Chon, K. H., & Cohen, R. J (1997). Linear and nonlinear ARMA model parameter estimation using an artificial. *IEEE Transactions on Biomedical Engineering, 44*(3), 168–174.

Clempner, J. B, & Poznyak, A. (2024). *Optimization and games for controllable Markov chains: Numerical methods with application to finance and engineering*. Springer Nature.

Cramér, H. (1946). A contribution to the theory of statistical estimation. *Scandinavian Actuarial Journal, 1946*(1), 85–94.

Duy, T. T., Nguyen, V. L., Trung, N. L., & Abed-Meraim, K. (2023). Fisher information estimation using neural networks. *REV Journal on Electronics and Communications, 13*(1–2), 1–10.

Duy, T. T., Nguyen, L. V., Nguyen, V.-D., Linh Trung, N., & Abed-Meraim, K. (2022). Fisher information neural estimation. In *30th European Signal Processing Conference (EUSIPCO)* (pp. 2111–2115). Belgrade, Serbia: IEEE.

Escobar, J., & Poznyak, A. (2022). Robust parametric identification for ARMAX models with non-Gaussian and coloured noise: A survey. *Mathematics, 10*(8), 1291.

Hodges, J. L., & Lehmann, E. L. (1951). Some applications of the Cramer-Rao inequality. In *Proceedings of the Second Berkeley Symposium on Mathematical Statistics: 2* (pp. 13–23). University of California Press.

Huber, P. (1975). Robustness and designs. *A survey of statistical design and linear*. North-Holland Publishing Company.

Ivanov, D., & Yakoub, Z. (2023). Overview of identification methods of autoregressive model in presence of additive noise. *Mathematics, 11*(3), 607.

Khatri, C. G. (1980). Unified treatment of Cramér-Rao bound for the nonregular density functions. *Journal of Statistical Planning and Inference, 4*(1), 75–79.

Ljung, L. (2010). Perspectives on system identification. *Annual Reviews in Control, 34*(1), 1–12.

Menon, A., Mehrotra, K., Mohan, C. K., & Ranka, S. (1996). Characterization of a class of sigmoid functions with applications to neural networks. *Neural Networks, 9*(5), 819–835.

Poznyak, A. S. (2009). *Advanced mathematical tools for automatic control engineers*: 2. Elsevier.

Poznyak, A. S. (2019). Robust identification under correlated and non-Gaussian noises: WMLLM. *Automation and Remote Control, 80*(9), 1628–1644.

Poznyak, A. S., Sanchez, E. N., & Yu, W. (2001). *Differential neural networks for robust nonlinear control: Identification*. World Scientific.

Rao, C. R. (1945). Information and the accuracy attainable in the estimation of statistical. *Resonance, Journal of Science Education, 20*, 78–90.

Soen, A., & Sun, K. (2021). On the variance of the Fisher information for deep learning. *Advances in Neural Information Processing Systems, 34*, 5708–5719.

Szandała, T. (2021). Review and comparison of commonly used activation functions for deep neural networks. In *Bio-inspired Neurocomputing* (pp. 203–224). Springer.

Taylor, J. (1979). The Cramer-Rao estimation error lower bound computation for deterministic. *IEEE Transactions on Automatic Control, 24*(2), 343–344.

Vincze, I. (1979). On the Cramér-Fréchet-Rao inequality in the nonregular case. In J. Jurečková (Ed.), *Contributions to Statistics* (pp. 253–262). Dordrecht–Boston: The J. Hájek Memorial.

Wei, Y., & Jiang, Z. (2024). Estimating parameters of structural models using neural networks. *Marketing science*. USC Marshall School of Business Research Paper.

Chapter 5

Design, Development, and Control of a Pneumatically Actuated Lower Limb Exoskeleton

M. Ashmi[a] and V.M. Akhil[b]

[a]*Department of Mechatronics, ACE College of Engineering, Trivandrum, Kerala, India,* [b]*Amrita School of Artificial Intelligence, Amrita Vishwa Vidyapeetham, Coimbatore, Tamil Nadu, India*

5.1 Introduction

Human mobility is essential for maintaining independence and performing routine daily activities. However, injuries, neurological disorders, and musculoskeletal diseases can severely affect lower limb function, leading to temporary or permanent movement impairments. Rehabilitation remains the primary approach for restoring mobility and improving functional ability. Traditionally, rehabilitation exercises are supervised by physiotherapists in clinical environments. However, the intensive and repetitive nature of therapy, together with the physical demands placed on therapists, has led to the increasing adoption of robotic rehabilitation devices. Lower limb exoskeletons provide consistent, repetitive, and customized assistance during therapy, and their selection depends on the type and severity of the patient's disability.

Human gait plays an essential role in the design of exoskeletons. It is the most instinctive way; the body displacement happens under stable state. Human gait modeling constitutes the biomechanics of the movement of lower limbs and enumerates the factors governing its functionality. The gait cycle refers to the sequence of movements involved in one complete stride taken by an individual while walking. It encompasses the period from when one foot makes initial contact with the ground to when the same foot contacts the ground again, completing a full cycle. The gait cycle is typically divided into two main phases: the stance and swing phases. The stance phase begins when the foot makes initial contact with the ground and continues until the foot lifts off the ground for the swing phase. This phase accounts for approximately 60% of the entire gait cycle. The swing phase follows the stance phase and lasts until the same foot makes initial contact with the ground again. This phase accounts for approximately 40% of the gait cycle.

In clinical practice the frequently used method for studying gait behavior is by analyzing the joint angles as well as the time phase cycles of gait. The dynamics of human locomotion vary from person to person because of different step lengths, thereby making it complex to determine a generic predictor. A wide range of techniques has been reported in the literature for analyzing human gait. Conventional gait assessment methods include video-based motion analysis, optoelectronic motion capture systems, electromyography (EMG), electroencephalography (EEG), force plates, instrumented footwear, inertial measurement systems, gyroscopes, and Hall-effect sensors. The information obtained from these techniques plays a significant role in understanding gait characteristics and developing rehabilitation strategies for individuals with locomotor impairments.

The growing demand for effective rehabilitation has accelerated the development of assistive technologies, among which lower limb exoskeletons have emerged as one of the most promising solutions. These robotic devices provide external mechanical support to restore walking ability, improve mobility, and enhance the overall physical performance of individuals with lower limb disabilities. The mechanical design and control architecture of an exoskeleton must be carefully tailored to accommodate different levels of impairment and functional requirements, thereby ensuring safe, efficient, and personalized rehabilitation.

Computational Intelligence in Mechatronics: Solving Real-World Problems in Electronic Systems Design. DOI: https://doi.org/10.1016/B978-0-443-27580-7.00016-0

TABLE 5.1 Currently available lower limb exoskeletons.

Name	Actuator types	Actuated DOF at one leg
BLEEX	Hydraulic actuator	4
HAL	Electric motor	3
HUALEX	Electric motor	3
ReWalk	Electric motor	2
HEXAR	Electric motor	8
HULC	Hydraulic actuator	
Daewoo	Hydraulic actuator	6
Exosuit	Pneumatic actuator	
MIT autonomous leg Exoskeleton	Electric motor	7
Honda SMA	Electric motor	1
EKSO	Electric motor	2
Indego	Electric motor	2
Technical University of Berlin Exoskeleton	Electric motor	
SARCOS	Hydraulic actuator	
REX	Electric motor	5
University of Michigan Exoskeleton	Pneumatic actuator	

The effectiveness of an exoskeleton depends not only on its control strategy but also on several design considerations, including user adaptability, ergonomic comfort, structural stability, operational safety, and lightweight construction. Consequently, considerable research has been devoted to the development of exoskeletons that can replicate natural human gait while minimizing user effort. Several notable lower limb exoskeletons have been reported over the past two decades. A knee–ankle–foot orthosis (KAFO) capable of locking the knee during the stance phase while allowing unrestricted knee motion during the swing phase was one of the earliest rehabilitation devices developed for gait assistance (Kaufman et al., 1996). In 2002 researchers at the University of Tsukuba, Japan, introduced the Hybrid Assistive Limb (HAL), which estimated joint torque by combining ground reaction force measurements with surface EMG (sEMG) signals to assist lower limb movement (Watanabe et al., 2014). A major milestone in powered exoskeleton development was achieved in 2004 with the Berkeley Lower Extremity Exoskeleton (BLEEX), developed at the Human Engineering and Robotics Laboratory of the University of California, Berkeley, primarily for military load-carrying applications (Zoss & Kazerooni, 2006).

Subsequent research focused on rehabilitation-oriented exoskeletons. Kim et al. (2013) developed the Walking Assistance Lower-Limb Exoskeleton in 2013 to improve the mobility of individuals with paraplegia. Another significant contribution was the Hanyang Exoskeleton Assistive Robot (HEXAR), developed in South Korea, which employed electric motors and muscle circumference sensors (MCRS) to detect user intention and provide movement assistance (Kim et al., 2014). The performance of the HEXAR-CR50 was later evaluated experimentally by Lim et al. (2015), who demonstrated its effectiveness by analyzing reductions in metabolic energy expenditure during assisted walking. The specifications of representative lower limb exoskeletons reported in the literature are summarized in Table 5.1.

Existing lower limb exoskeletons have been developed primarily to assist individuals with mobility impairments by providing mechanical support during locomotion and rehabilitation. Several commercially available and research-oriented systems, including EKSO, ReWalk, BLEEX, and HAL, have demonstrated the feasibility of restoring walking ability through different mechanical configurations and control strategies. Although these devices achieve the common objective of assisting human movement, their mechanisms and actuation methods vary considerably (Yan et al., 2015). Earlier exoskeleton designs mainly focused on flexion and extension of the hip and knee joints, whereas the internal and external rotational degrees of freedom were generally left passive. However, during natural human walking, the lower limbs exhibit approximately 5.5 degrees of internal rotation and 5 degrees of external rotation. Incorporating these physiological movements into the exoskeleton mechanism is essential for improving user comfort and achieving a more natural gait.

Powered exoskeletons have been reported to provide superior rehabilitation performance compared to passive orthotic devices because they actively generate joint motion rather than merely supporting the limb. Electric motors and pneumatic actuators are the two most widely adopted actuation systems for powered exoskeletons. Although electric motors offer precise position control, pneumatic actuators possess advantages such as high power-to-weight ratio, compliance, and safety during human–robot interaction. Nevertheless, controlling pneumatic actuators remains challenging because of air compressibility, valve nonlinearity, and frictional effects, which introduce highly nonlinear system dynamics (Medrano-Cerda et al., 1995). EMG-based control has also been explored as a means of estimating user intention for robotic exoskeletons. However, most existing systems have been evaluated only on level ground or treadmill environments, with limited consideration given to variations in walking surfaces and their influence on muscle behavior and gait characteristics.

Despite significant progress in exoskeleton technology, several challenges continue to limit their widespread application. Stability during walking, affordability, adaptability to different users, and accurate tracking of human gait remain major research concerns. Consequently the design of efficient controllers capable of producing coordinated hip and knee movements has become a critical aspect of exoskeleton development. Aguirre-Ollinger et al. (2007) demonstrated that a proportional–integral–derivative (PID) controller could effectively regulate exoskeleton motion, with appropriate parameter tuning improving both tracking accuracy and transient response. To overcome the limitations of conventional PID control, cascaded PI–PID architectures have also been proposed, providing better robustness against nonlinear system behavior. The controller parameters were tuned using the Ziegler–Nichols (Z–N) technique (Neogi et al., 2011), resulting in improved performance under both paraplegic and muscle-fatigue conditions. Subsequently, Particle Swarm Optimization (PSO) was employed for automatic controller tuning, yielding lower tracking errors and enhanced gait performance compared with the conventional Z–N approach (Solihin et al., 2011).

Accurate prediction of gait characteristics is equally important for the development of intelligent exoskeleton control systems. Conventional analytical models often fail to represent the complex nonlinear behavior associated with human locomotion, creating the need for data-driven and knowledge-based approaches. Artificial neural networks (ANNs) have been successfully applied for estimating muscle–tendon lengths and predicting gait kinematic parameters (Mundt et al., 2020). Multilayer perceptron networks have also been developed to estimate gait trajectories for robotic rehabilitation applications (Luu et al., 2011). Fuzzy logic techniques provide another effective alternative by representing human expertise through linguistic rules. Viteckova et al. (2012) employed a Mamdani Fuzzy Inference System (MFIS) in MATLAB to estimate contralateral lower limb joint angles during walking. Similarly, Rai et al. (2010) proposed an adaptive fuzzy controller for a three-degree-of-freedom bioinspired robotic leg and later introduced a Mamdani-type fuzzy system for analyzing knee and thigh kinematics and identifying gait abnormalities associated with flexion (Arias-Enriquez et al., 2012). More recently, hybrid neuro-fuzzy approaches have combined the learning capability of neural networks (NNs) with the interpretability of fuzzy inference systems (FISs). Shieh and Chang (2009) developed an adaptive neuro-FIS (ANFIS)-based identifier together with an optimized neuro-fuzzy controller for modeling the inverse dynamics of a biped robot, thereby enabling accurate estimation of control parameters required for stable locomotion.

The selection of an appropriate actuation mechanism has also received considerable attention in the literature. Pneumatically powered exoskeletons have predominantly employed McKibben artificial muscles because of their lightweight construction and muscle-like behavior. However, these actuators exhibit reduced stiffness under heavy loading conditions and require relatively large installation space when pressurized. As an alternative, pneumatic cylinders have been investigated owing to their compact structure, higher stiffness, and improved load-carrying capability. Kazerooni (2005) discussed the operational characteristics of pneumatic cylinders, whereas Van Varseveld and Bone (1997) proposed a pulse width modulation (PWM)-based control strategy for position regulation. Similar approaches have been reported by other researchers (Song & Liu, 2006). Although PWM control provides satisfactory position tracking, its effectiveness decreases when accurate velocity or torque control is required because pneumatic systems inherently exhibit nonlinear behavior arising from air compressibility, valve dynamics, and friction.

Recent research studies discussed in the literature have highlighted the challenges associated with controlling exoskeletons. Nevertheless, achieving stable, accurate, and adaptive gait assistance for individuals with different levels of disability remains a significant challenge. These observations highlight the need for an advanced control strategy capable of generating reliable gait trajectories while accommodating the nonlinear characteristics of pneumatically actuated systems. Motivated by these research gaps, the present work develops a customized pneumatically actuated lower limb exoskeleton specifically designed for individuals affected by polio. The proposed system incorporates appropriate control algorithms, and its performance was experimentally evaluated using 11 healthy volunteers and 1 individual with polio. Among the controllers investigated, the PID controller demonstrated superior performance by producing lower oscillations and accurately reproducing the gait pattern of healthy individuals with minimal tracking error, thereby confirming its suitability for lower limb rehabilitation applications.

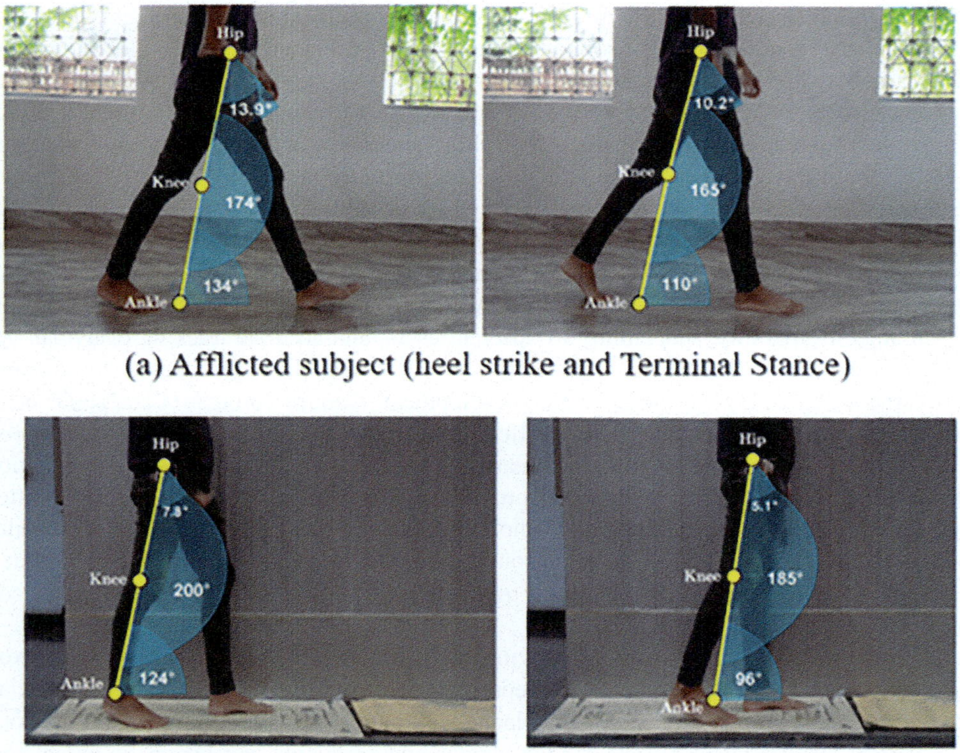

(a) Afflicted subject (heel strike and Terminal Stance)

(b) Healthy subject (foot flat and Mid Stance)

FIGURE 5.1 Gait analysis using video cameras.

5.2 Data Collection

To determine human gait characteristics, experimental trials were conducted on both healthy individuals and subjects affected by polio while they walked along a 30-m level walkway. The study included 50 healthy volunteers (30 males and 20 females) and 4 male participants with polio who were capable of independent walking without the assistance of crutches. All participants were selected from the National Institute of Technology Calicut. None of the healthy volunteers had a history of neurological or musculoskeletal disorders that could influence normal gait. Prior to participation, informed consent was obtained from every subject, and the experimental procedure and objectives were clearly explained.

Gait motion was recorded using a Vicon V16 video camera equipped with a 16 MP sensor, 120 frames per second (fps) recording capability, and an 18-mm lens (Yang et al., 2019). The camera was positioned perpendicular to the direction of walking to capture the sagittal-plane motion accurately. Reflective markers were attached to the hip, knee, and ankle joints to facilitate precise tracking of joint movements during walking (MOCAP, n.p.; Winter, 2009). The recorded videos were processed using Kinovea 0.8.24 software, which was employed to determine the angular variations of the hip, knee, and ankle joints, as illustrated in Fig. 5.1. To improve the reliability of the measurements, each participant completed five consecutive walking trials, and one complete gait cycle from each trial was recorded for subsequent analysis.

5.3 Methodology

The primary aim of this study is to design and implement an appropriate driving system with the optimal control strategy to achieve precise position control of the pneumatically actuated exoskeleton designed for orthopedic applications. To design a driving system for an exoskeleton the characteristics of human gait for healthy subjects were analyzed for normal walking. This is to understand exactly the characteristics of human gait on an average basis. Then the general characteristics of healthy and afflicted subjects (time taken to complete gait cycle, linear displacement corresponding to each gait cycle) were determined. Through effective models using NNs, fuzzy logic systems, and neuro-fuzzy systems, the required position and velocity characteristics of the driving system were decided. An appropriate control system drive for meeting the requirements

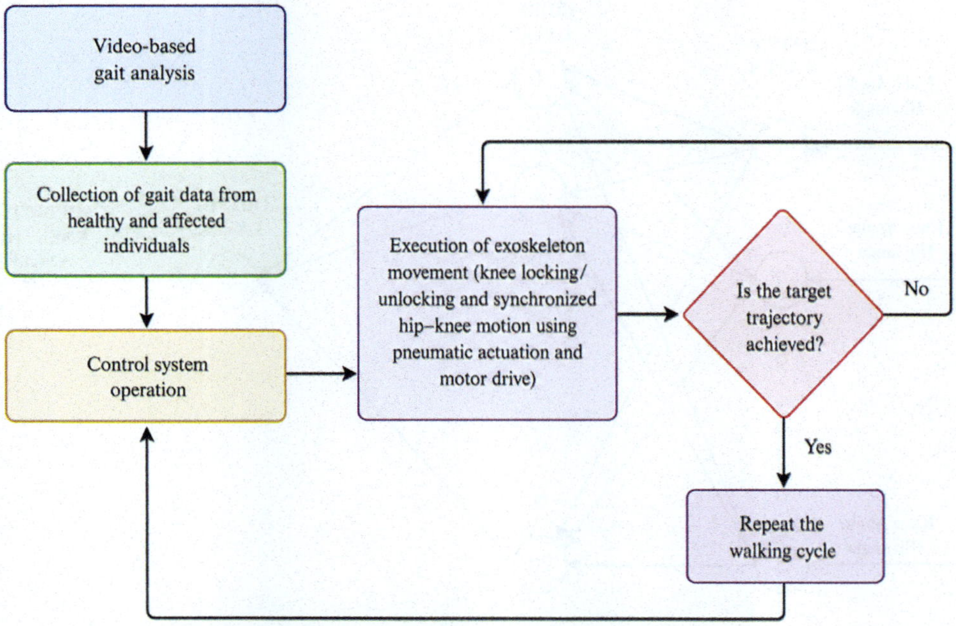

FIGURE 5.2 Flowchart illustrating the experimental procedure.

was also designed. The controllers developed were tested in the pneumatically powered exoskeleton for further analysis. The experimental setup is depicted in the flow diagram (Fig. 5.2).

5.4 Prediction of Human Gait

Human gait was characterized by analyzing the angle–time profiles of the lower limb joints, and these data were subsequently used for artificial intelligence (AI)-based prediction. NN, fuzzy logic, and neuro-fuzzy models were developed in MATLAB® to estimate the joint angles occurring during the gait cycle. For model development a database containing hip and knee joint angle measurements over a duration of 10 seconds was prepared. This dataset represented seven complete gait cycles along with the initial portion of the subsequent gait cycle. The analysis of the experimental data showed that the right knee joint angle varied between 180 and 126 degrees, whereas the left knee joint angle ranged from 180 to 132 degrees. Likewise the hip joint angles of both the right and left limbs exhibited variations within the range of 180to 155 degrees. These joint angle patterns remained nearly identical over successive gait cycles, indicating consistent and repeatable gait characteristics. The predicted joint trajectories obtained from the developed AI models were subsequently employed to generate the control inputs required for the driving system of the pneumatically actuated lower limb exoskeleton.

5.4.1 Neural Network (NN) Based Prediction

The nonlinear learning capability of ANNs enables them to capture the complex relationships among gait variables effectively. In this study a 5–1–20 NN architecture (input–hidden–output layers), illustrated in Fig. 5.3, was developed and trained using the back-propagation algorithm. During the training process the network output was compared with the corresponding target output to determine the prediction error. The average error is computed using the equation given below.

$$E_{\text{avg}} = \frac{1}{n}\sum(m - p) \tag{5.1}$$

where m and p are the measured and predicted outputs of the NN.

The calculated error was then propagated backward through the network to update the connection weights iteratively. This continuous weight adjustment minimized the prediction error over successive iterations, allowing the NN to progressively learn the underlying gait characteristics and generate outputs that closely matched the desired values (Bosque et al., 2014; Mundt et al., 2020). Table 5.2 presents an overview of the analysis features involved in the training of the NN.

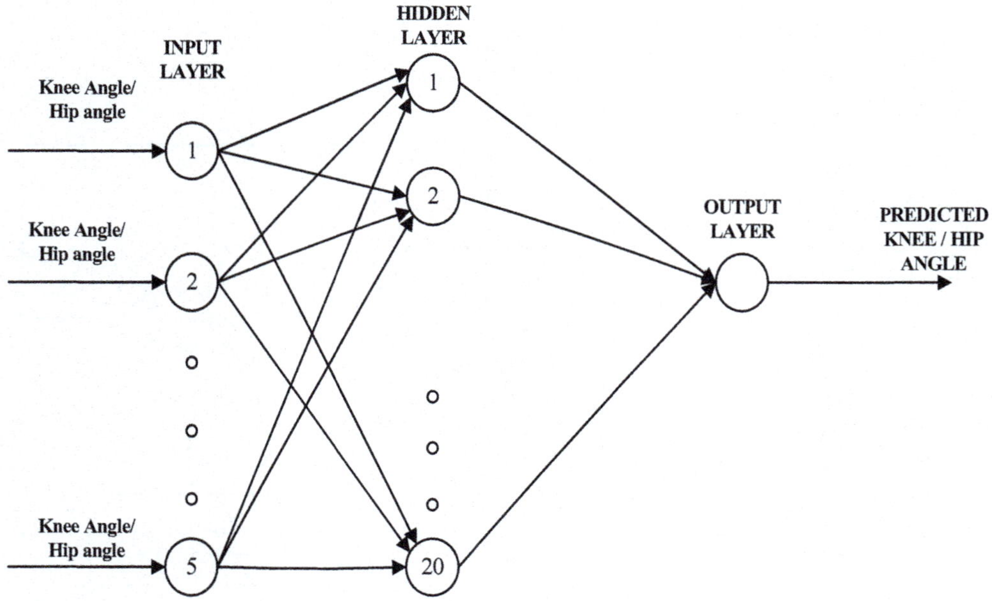

FIGURE 5.3 Structure of neural network.

TABLE 5.2 Review of neural network training.

Epochs	Mean squared error
15	0.308
16	0.343
18	0.341
37	0.376
40	0.245

During the NN training the disparity between the actual and predicted angles of the right knee and left hip was within 5 degrees, as shown in Fig. 5.4A and D. However, for the remaining parameters, the generated error ranged from 10 to 12 degrees for the right hip and less than 10 degrees for the left knee. Fig. 5.4A to D present the analysis of actual human locomotion (*pink line*) and the predicted angles (*red line*) of both legs' knee and hip joints.

5.4.2 Fuzzy Logic Based Prediction

An FIS represents a nonlinear transformation of the input data vector into a crisp numerical value by employing fuzzy rules, and the transformation process encompasses input/output membership functions, fuzzy operators, if-then rules, aggregation of output sets, and defuzzification techniques. The fuzzy classifier developed initially was of the Mamdani type, and it involves mathematically complex defuzzification stage, which increases the intricacy in hardware implementation. Conversely the Sugeno method exhibits computational efficiency and performs effectively in conjunction with optimization and adaptive techniques, making it highly appealing for control problems, especially in dynamic nonlinear systems (Bosque et al., 2014).

In this study an ANFIS based on the Sugeno fuzzy inference model was developed for gait prediction. The training process employed a hybrid learning algorithm that integrates the least-squares estimation technique with the back-propagation gradient descent algorithm to optimize the parameters of the fuzzy membership functions. The objective of this approach is to establish an accurate input–output mapping by representing the input variables through fuzzy linguistic terms while expressing the output using linear functions or constant values, as defined in the Sugeno inference mechanism. By combining the learning capability of NNs with the reasoning ability of fuzzy logic the ANFIS model effectively captures the nonlinear relationships present in the gait data, thereby enabling accurate prediction and decision-making.

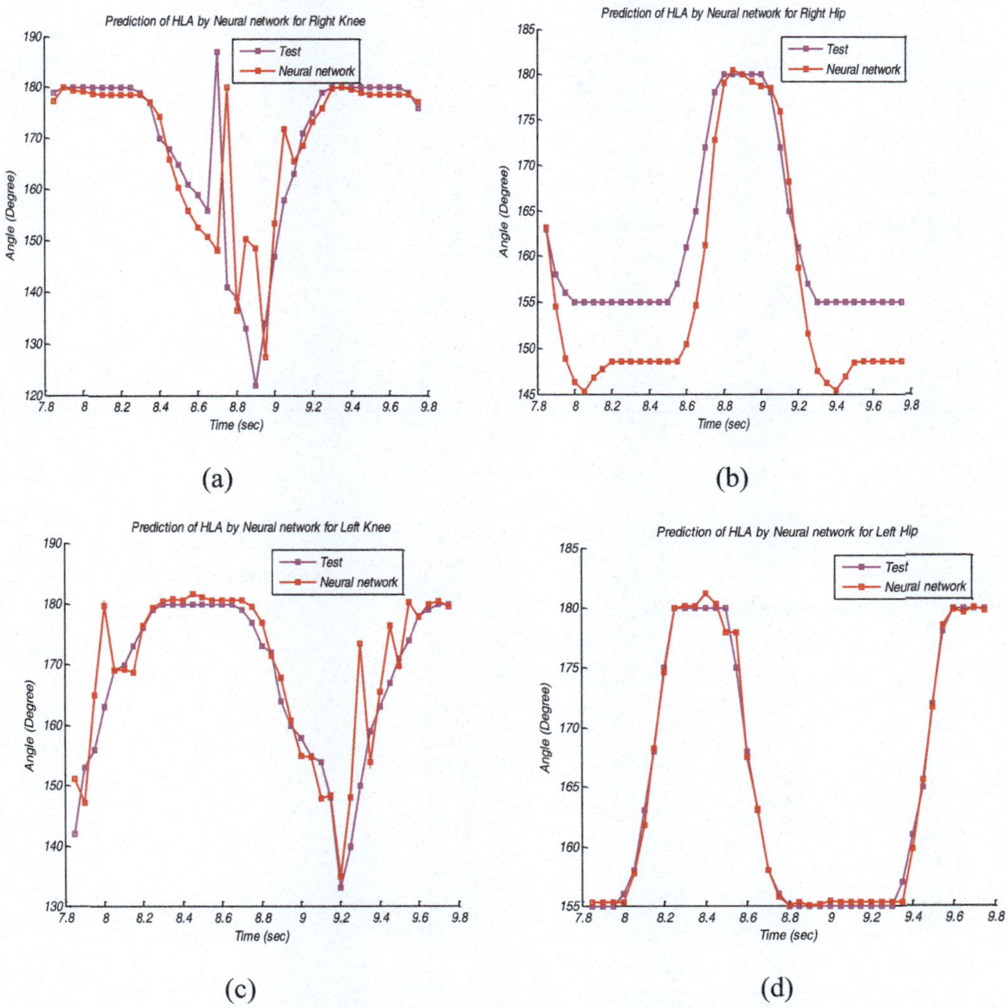

FIGURE 5.4 Prediction of human gait using neural networks: (A) right knee, (B) right hip, (C) left knee, and (D) left hip.

In Fig. 5.5A the right knee angle exhibited a difference of less than 5 degrees during testing in the stance phase. However, during prediction in the swing phase, peaks were generated between 8.6 and 9.3 seconds because of the movement of leg from stance to swing phase. Fig. 5.5A–D illustrates the actual and predicted locomotion angles of the right and left legs' knees and hips represented by the pink and green lines, respectively. The fuzzy logic approach employed 32 fuzzy rules, 156 training data pairs, 30 nonlinear parameters, 192 linear parameters, and 92 nodes. It is noted that utilizing more membership variables and additional rules could have resulted in even better outcomes.

5.4.3 Neuro-fuzzy Based Prediction

The integration of AI techniques involving fuzzy logic and NNs results in neuro-fuzzy systems that capitalize on the learning capabilities of NNs and the parallel computation abilities of fuzzy systems. Essentially a neuro-fuzzy system functions as an NN but behaves equivalently to a fuzzy inference model. By training the fuzzy system, it becomes proficient in establishing fuzzy rules and determining membership functions for input and output variables of the system (Bosque et al., 2014).

Initially the dataset was loaded into the neuro-fuzzy system, and for the purpose of training, 80% of the data was utilized while the remaining 20% was used for testing. A Sugeno-type FIS structure was generated through subtractive clustering. During the training process, right knee, left knee, and left hip exhibited minimal error in predicting locomotion cycles, with deviations of less than 5 degrees. As for the right hip the disparity between the trained and predicted angles was less than 10 degrees. The responses indicated that the locomotion angle corresponding to the left leg's hip produced more accurate results. These graphs (Fig. 5.6A–D) illustrate the actual and predicted locomotion angles of the right leg (knee and hip)

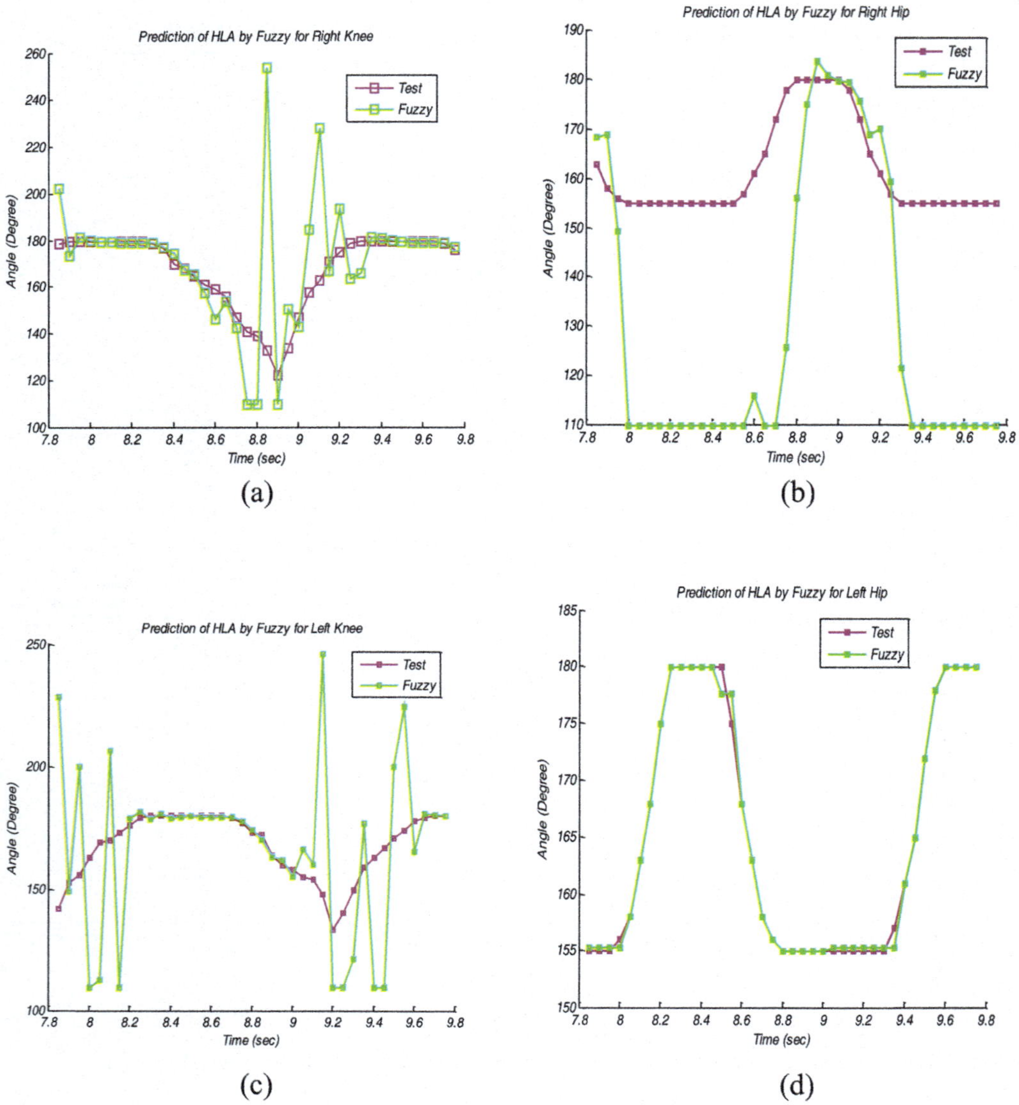

FIGURE 5.5 Prediction of human gait using fuzzy logic system: (A) right knee, (B) right hip, (C) left knee, and (D) left hip.

where the actual angles are represented by the *pink line*, whereas the predicted angles obtained through the neuro-fuzzy technique are indicated by the *blue line*.

5.4.4 Performance Assessment of the Artificial Intelligence Models

Human gait is not exactly cyclic, and there is variation from one cycle to another. The transition of the leg from the stance to the swing phase is rapid in steady locomotion. In addition, peaks are observed when the leg moves from the stance to the swing phase. The error analysis revealed that the neuro-fuzzy model achieved the least error (most satisfactory error curve). Specifically, for the right and left leg joints, the error values in the neuro-fuzzy model were 2.2, 3, 2.45, and 0.98 degrees, respectively. In contrast the corresponding errors for the NN model were 2.8, 5.9, 3.5, and 0.5 degrees, whereas for the fuzzy model, they were 11, 33, 20, and 0.25 degrees. Fig. 5.7 illustrates the average error performance using the three prediction methods. A higher frame/second camera will give more interval data and better accuracy. Several factors contribute to the sources of error, including the limited number of datasets and the smaller number of locomotion cycles used in the analysis. The adequate position and velocity characteristics of the driving system were determined using these efficient models and is explained in Section 5.5.

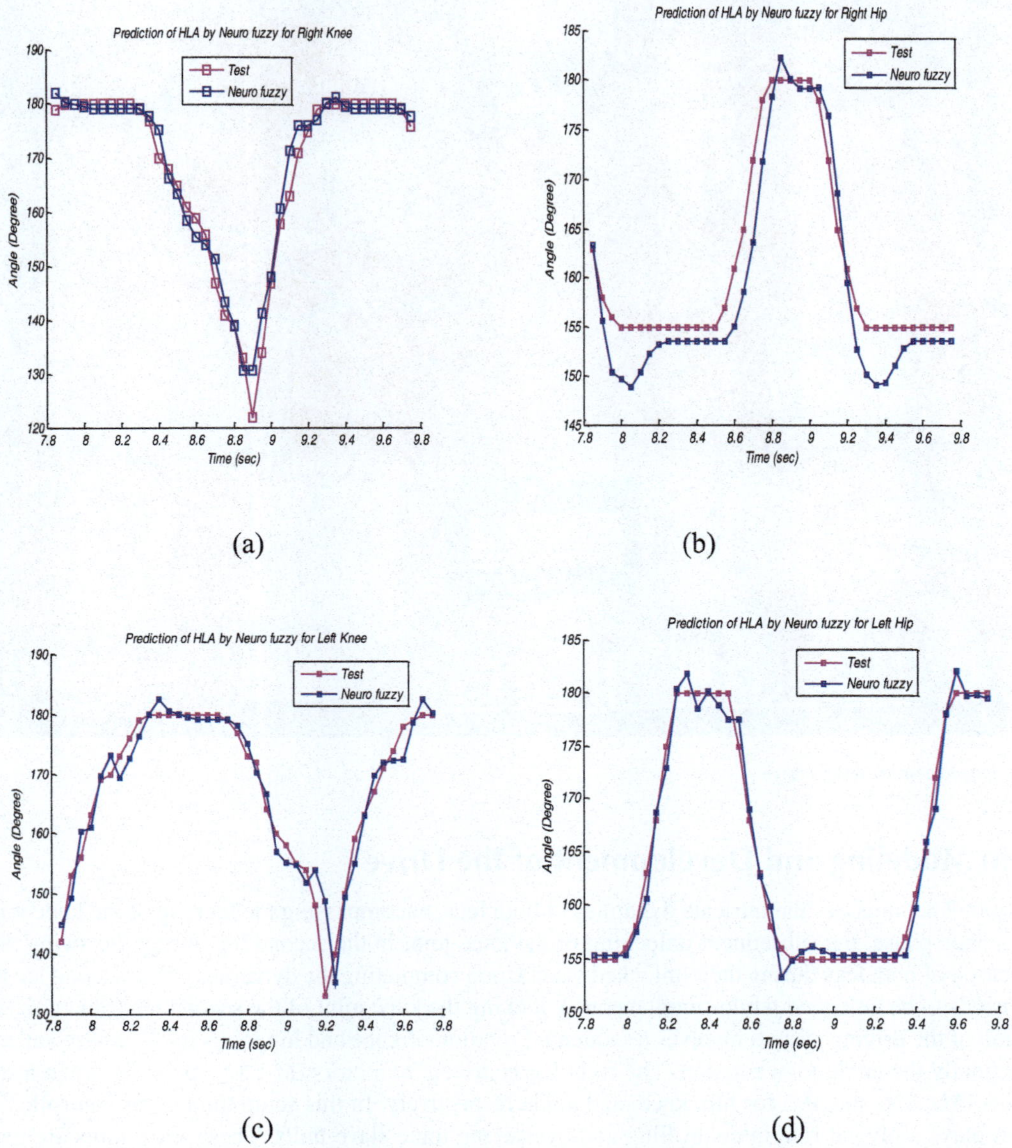

FIGURE 5.6 Prediction of human gait using neuro-fuzzy system: (A) right knee, (B) right hip, (C) left knee, and (D) left hip.

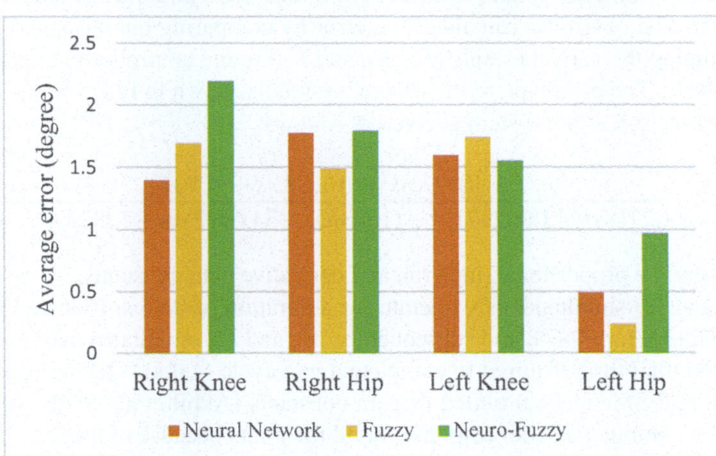

FIGURE 5.7 Average error of the prediction techniques.

FIGURE 5.8 Developed control system drive.

5.5 System Modeling and Development of the Drive

Human locomotion encompasses the intricate dynamics of both legs, encompassing the locking of the knee joint in the first leg during the stance phase, the subsequent unlocking of the knee joint in the second leg during the swing phase, and the overall coordination of both legs during their unlocked state. Upon completing the dynamics of the second leg the knee joint of the first leg needs to be unlocked while simultaneously locking the knee joint of the second leg (Shi et al., 2019).

The operation of the driving system involves a sequential rotation of knee and hip joints in clockwise and anticlockwise directions, and finally the ankle joint rotation. The right leg is driven by motors (M_1, M_2, and M_3), whereas the left leg is driven by motors (M_4, M_5, and M_6) for hip, knee, and ankle respectively. In this setup the master controller, ATMega328 microcontroller, plays a pivotal role in controlling and overseeing three slave units. These slave units are responsible for sending signals to their respective motors, enabling the joints to move within the specified timeframe to successfully complete a gait cycle. To achieve this the system utilizes built-in encoders to gather feedback from the motors, which is then fed to the PID controller. The controller calculates the error by comparing the measured joint angle with the desired set point. By appropriately tuning the gain constants (K_P, K_I, and K_D) in the controller the system can achieve the targeted angular displacement effectively. The prototype of the driving system is shown in Fig. 5.8.

The closed-loop transfer function of the system is given as follows:

$$\frac{C(s)}{R(s)} = \frac{1083K_Ds^2 + 1083K_Ps + 1083K_I}{0.002918s^4 + 1.002971s^3 + (1083K_D + 34.6042)s^2 + 1083K_Ps + 1083K_I} \tag{5.2}$$

where K_P, K_I, and K_D represent the proportional, integral, and derivative gain constants.

During human gait, M_1 and M_2 simultaneously operate for a duration of 0.46 s, followed by the activation of locking and unlocking motors of both legs for 0.46 s, and subsequently, M_4 and M_5 simultaneously operate for 0.46 s. Altogether, these operations constitute the total time required to complete a gait cycle (1.4 s). The driving mechanism incorporated P, PI, and PID controllers with real-time manual tuning of gain constants (Akhil et al., 2020; Ashmi & Akhil, 2023). Upon comparison the PID controller exhibited a better response for all the joints because of the less time (0.45 s) taken to achieve the desired trajectory, as shown in Table 5.3. On the other hand, P and PI controllers consumed more time (0.6 and 0.48 s) to reach the set point.

TABLE 5.3 Comparison of controller's response for a desired angle of 45 degrees.

			Angle rotated (degree)			
K_P	K_I	K_D	Right knee	Right hip	Left knee	Left hip
8	0	0	42.45	43.18	43.57	43.12
2	0	0	34.7	37.32	33.21	36.13
1	1	0	41.23	40.58	41.36	40.5
0.5	0.5	0	41.45	42.71	41.72	42.11
0.5	0.5	0.05	43.28	42.56	43.78	43.47
0.5	0.5	1	51.34	52.47	51.58	50.56

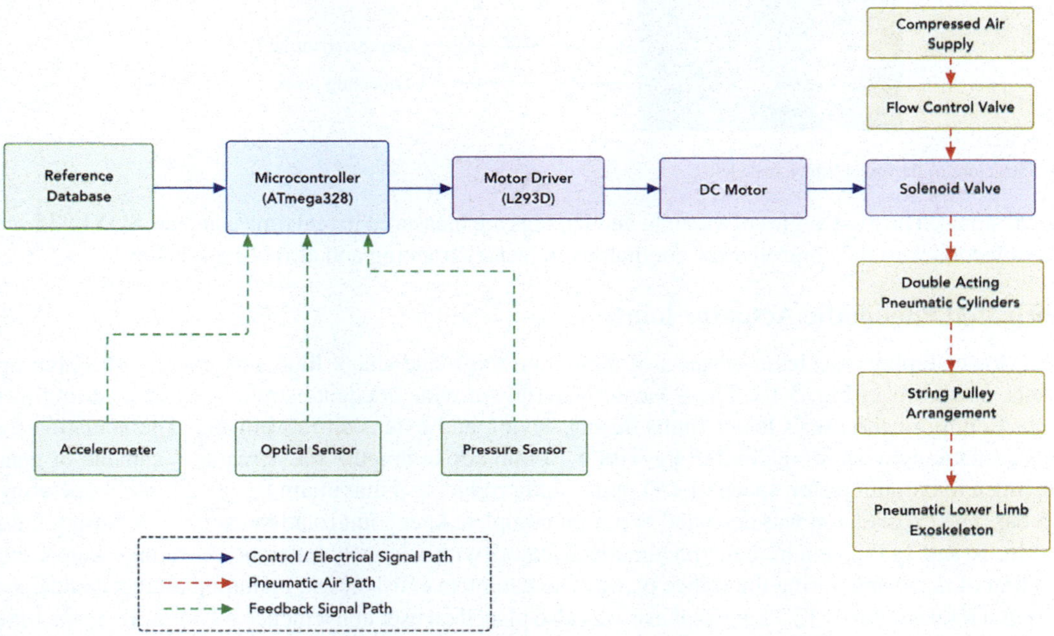

FIGURE 5.9 Block diagram representation of pneumatic exoskeleton.

5.6 Design Aspects and Hardware Implementation

The previous section presented the procedure adopted to identify the most suitable control strategy for actuating the lower limb exoskeleton. This section describes the practical implementation of the selected controller in the pneumatically actuated exoskeleton system.

The desired angular displacement values of the hip and knee joints are stored in the ATmega328 microcontroller, which serves as the reference database for trajectory generation. The pneumatic actuation system is supplied with compressed air maintained at a pressure of 5 bar, which is regulated through a flow control valve and a solenoid valve. A 12-V DC series motor (1000 rpm, 0.049 Nm) is employed to regulate the pneumatic air supply. To produce the required joint rotation the linear motion generated by the pneumatic cylinders is transformed into rotary motion using a string–pulley mechanism, as illustrated in Fig. 5.9. Each leg of the exoskeleton is equipped with two double-acting pneumatic cylinders (diameter: 40 mm; maximum force: 900 N), positioned at the hip and knee joints. The 5-port, 24 V solenoid valve is controlled by the ATmega328 microcontroller to regulate the direction of cylinder movement, thereby producing clockwise or anticlockwise motion of the exoskeleton joints. During operation the controller continuously compares the actual joint motion with the reference trajectory. Based on the resulting tracking error the L293D motor driver modifies the control signal applied to the solenoid valve to achieve the desired joint movement.

An MOC7811 optical sensor is incorporated to measure the rotational position of the DC motor. In addition, a pressure sensor monitors the cylinder supply pressure over a range of 0–8 bar and converts the measured pressure into a corresponding

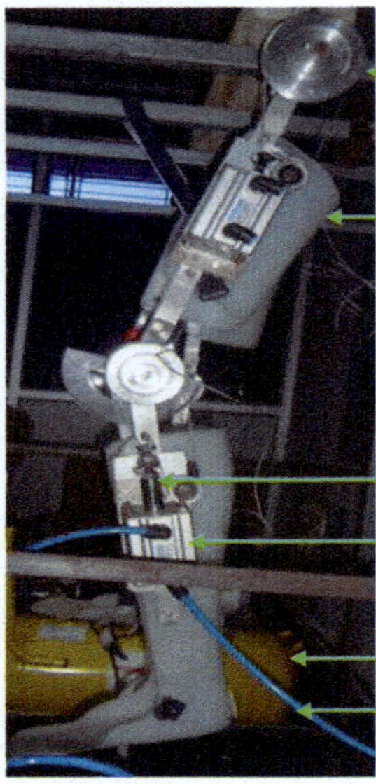

FIGURE 5.10 Structure of the lower limb exoskeleton.

0–5 V electrical signal. The joint angles of the right and left legs are measured in real time using an ADXL335 accelerometer, which provides feedback to the controller for continuous trajectory tracking and closed-loop control.

5.6.1 Fabricated Pneumatic Actuator Joints

The developed lower limb exoskeleton consists of aluminum thigh and shank links integrated with polypropylene thigh and leg sockets, as shown in Fig. 5.10. The selected materials provide adequate strength while ensuring proper support and secure positioning of the user's lower limbs during flexion and extension movements. The actuation mechanism is driven by a pneumatic cylinder weighing 0.5 kg with a 40-mm stroke length. The actuator is capable of generating high torque, achieving a maximum piston speed of 150 mm/s, and producing a maximum force of 1760 N operating at a supply pressure of 5 bar. The exoskeleton was designed to permit complete knee joint rotation exclusively during the swing phase of the gait cycle. To satisfy this requirement, pneumatic actuators with relatively lower power ratings were selected to avoid excessive loading of the shank during the stance phase. The complete exoskeleton, including all mechanical and electronic components, has a total weight of 12.2 kg. Emphasis was also placed on user convenience by minimizing the time required to wear the device, allowing the exoskeleton to be fitted to a user within 5 min. The performance of the pneumatically actuated lower limb exoskeleton was experimentally evaluated using both healthy volunteers and individuals affected by polio.

5.6.2 Estimation of Torque in the Joints

The determination of torque during human gait is complex, and it depends on several factors, including the phase of the gait cycle, joint angle, muscle activity, and external forces. To actuate the link, torque (τ) is required, which is computed using the Lagrange–Euler method (Saha, 2014).

$$M(\theta)\ddot{\theta} + H(\theta, \dot{\theta}) + G(\theta) = \tau \tag{5.3}$$

$$\tau = \left(\frac{ml^2}{4} + 2I_{\text{knee}}\right)\ddot{\theta} - \frac{1}{2}\text{mg}\cos\theta \tag{5.4}$$

where M is the 2 × 2 inertia matrix, H is the Coriolis and centrifugal force, G is the gravity matrix, $\ddot{\theta}$ is the angular acceleration, $\dot{\theta}$ is angular velocity, θ is angular displacement, m is the mass of the link below the knee joint, and I_{knee} is the moment of inertia of the knee joint.

The accelerometer is utilized to measure linear acceleration (a), whereas angular acceleration ($\ddot{\theta}$) is calculated using Eq. (5.5).

$$\ddot{\theta} = \frac{a}{l} \tag{5.5}$$

where l is the length of the link below the knee and g is the acceleration because of gravity.

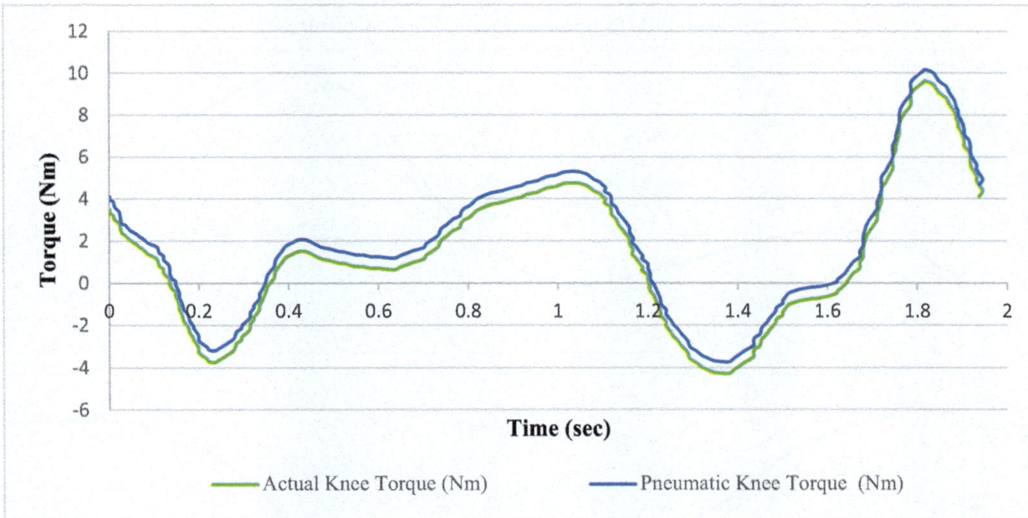

FIGURE 5.11 Torque variation during gait.

To determine the moment of inertia with respect to the knee joint axis the parallel axis theorem is employed.

$$I_{\text{knee}} = I_{zz} + m\left(\frac{l}{2}\right)^2 \tag{5.6}$$

where I_{zz} is the moment of inertia of actuator components about its centroidal axis.

The force (F) generated by the piston is computed as the result of multiplying the air pressure difference between two compartments of the cylinder ($P = P_1 - P_2$) and piston area (A).

$$F_i = \sum_{i=0}^{T} P_i \times A \tag{5.7}$$

where T is the time period.

The actuator's torque is determined by computing the product of force (F) with the radius of the driven pulley (R).

$$\tau_i = \sum_{i=0}^{t} P_i \times R$$

where t is the time period of actuation.

A comparative study was performed to evaluate the torque produced at the human joint against the torque generated by the pneumatically actuated exoskeleton. The analysis showed that the joint torque increased as the leg moved from the stance phase to the swing phase and gradually decreased thereafter. The difference between the two torque profiles was observed to be within 11%, as illustrated in Fig. 5.11.

5.6.3 Assessment of the Performance of Controllers with Human Gait

The performance of the PID controller was evaluated by comparing the joint angles produced by the pneumatically actuated lower limb exoskeleton with the reference human gait trajectory. To validate the controller experimentally the exoskeleton was tested on 11 healthy volunteers and 1 individual affected by polio. Each participant completed five walking trials over a 5-m level walkway, and the average hip and knee joint angles obtained from ten gait cycles were used for subsequent analysis. The experimental results indicated that the average tracking error was approximately 2.5 degrees for the knee joint and 6 degrees for the hip joint, as shown in Fig. 5.12. These findings demonstrate that the conventional PID controller is capable of accurately tracking the reference gait trajectory, even in the presence of system uncertainties and load disturbances. The small tracking deviations observed can be attributed to the manual tuning of the controller gains, suggesting that further improvement may be achieved by employing optimization-based tuning techniques. For the knee joint the controller exhibited a maximum tracking error of 3.95 degrees and a minimum error of −6.67 degrees. Overall the gait trajectory

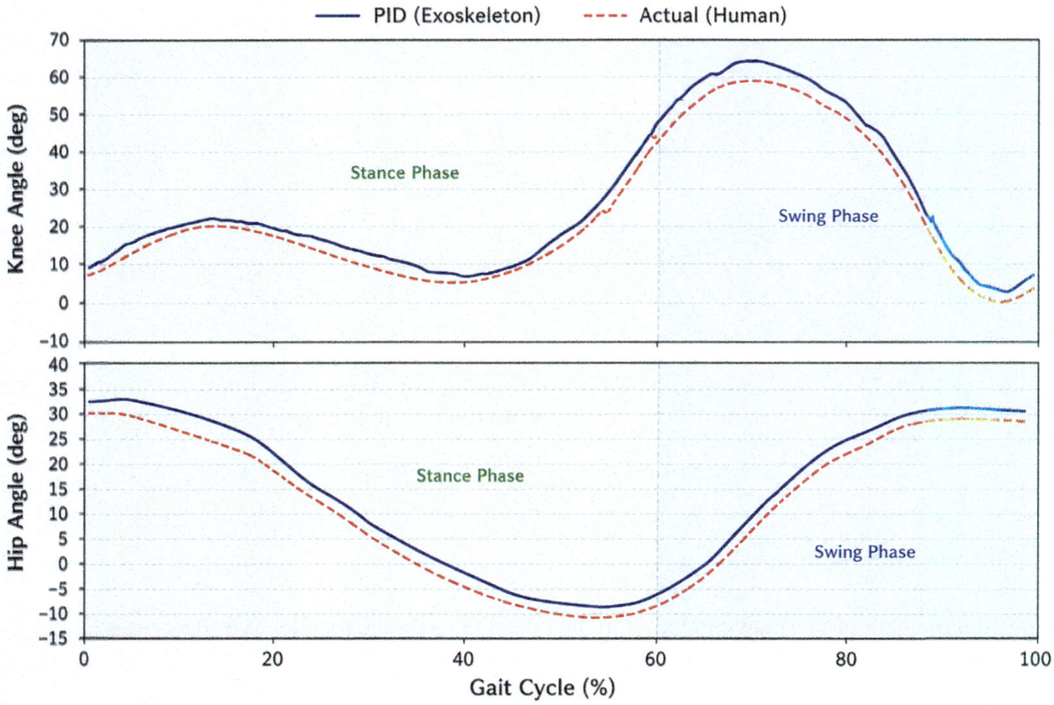

FIGURE 5.12 Joint angle tracking using PID controller. *PID*, Proportional–Integral–Derivative.

generated by the developed exoskeleton closely matched that of a healthy individual, confirming the ability of the proposed system to reproduce a natural walking pattern with minimal error.

5.7 Conclusion

Video cameras were employed to record human locomotion during normal walking. The knee and hip joint angles of the subjects were meticulously measured, and a dataset was created. Then, NN, fuzzy logic, and neuro-fuzzy models were created to predict human gait patterns. The error analysis conducted on the prediction techniques revealed that the neuro-fuzzy model exhibited the least error. Specifically, for the right leg (knee and hip) and left leg (knee and hip), the error values for the neuro-fuzzy model were only 2.2, 3, 2.45, and 0.98 degrees, respectively, whereas for the NN model, the errors were 1.4, 1.78, 1.6, and 0.5 degrees, and for the fuzzy model, the errors were 1.7, 1.5, 1.75, and 0.25 degrees. By comparing the three prediction techniques the neuro-fuzzy system demonstrated its effectiveness in gait cycle prediction, as it was able to generalize well with gait data from different subjects.

By using the analysis of the model developed a driving system that mimics the movement of assistive limb was developed in which P, PI, and PID controllers were incorporated and a suitable transfer function was obtained. The coordinated and sequential operation of the knee and hip joints has been successfully achieved along with the locking and unlocking knee modes. The optimal gain constants of different controllers were obtained by manual tuning. The results indicate that PID controller showed remarkable performance in accurately achieving the desired position of the knee and hip joints with less settling time (0.45 s) for completion of gait cycle.

The main goal of this study is to design an exoskeleton equipped with efficient control strategies for the optimal functioning of knee and hip joints. The use of a pneumatically powered exoskeleton proves to be a cost-effective and efficient alternative to traditional electrically actuated exoskeletons. An essential component of the pneumatically powered orthosis is the motor coupled valve, which plays a critical role in controlling the system's operation. The Lagrange–Euler method was utilized to calculate the required actuating torque for both joints. To assess the system's performance, P, PI, and PID controllers were incorporated into the exoskeleton, and their responses were thoroughly examined. Comparing the controllers the PID controller indicated reduced oscillation and improved steady-state error. The outcomes of the study indicate that the PID controller can successfully replicate a healthy gait pattern with minimal error, making it a promising choice for precise and accurate joint control in the pneumatically powered exoskeleton.

5.8 Scope for future work

This study has not considered the activity of muscle behavior involved in human gait. The exoskeleton has the potential to achieve stabilization, enabling patients to walk independently without additional walking aids. Moreover, enhancing the DOF at both the knee and hip joints could lead to improved flexibility.

References

Aguirre-Ollinger, G., Colgate, J. E., Peshkin, M. A., & Goswami, A. (2007). *A 1-DOF assistive exoskeleton with virtual negative damping: Effects on the kinematic response of the lower limbs* (pp. 1938–1944). Piscataway, NJ, USA: IEEE.

Akhil, V. M., Ashmi, M., Rajendrakumar, P. K., & Sivanandan, K. S. (2020). Human gait recognition using hip, knee and ankle joint ratios. *IRBM, 41*(3), 133–140.

Arias-Enriquez, O., Chacon-Murguia, M. I., & Sandoval-Rodriguez, R. (2012). Kinematic analysis of gait cycle using a fuzzy system for medical diagnosis. In *2012 Annual Meeting of the North American Fuzzy Information Processing Society (NAFIPS)* (pp. 1–6). IEEE.

Ashmi, M., & Akhil, V. M. (2023). A control strategy for pneumatically powered below-hip orthosis. *IRBM, 44*(5), 100791.

Bosque, G., del Campo, I., & Echanobe, J. (2014). Fuzzy systems, neural networks and neuro-fuzzy systems: A vision on their hardware implementation and platforms over two decades. *Engineering Applications of Artificial Intelligence, 32*, 283–331.

Kaufman, K. R., Irby, S. E., Mathewson, J. W., Wirta, R. W., & Sutherland, D. H. (1996). Energy-efficient knee-ankle-foot orthosis: A case study. *JPO: Journal of Prosthetics and Orthotics, 8*(3), 79–85.

Kazerooni, H. (2005). Design and analysis of pneumatic force generators for mobile robotic systems. *IEEE/ASME Transactions on Mechatronics, 10*(4), 411–418.

Kim, J. H., Han, J. W., Kim, D. Y., & Baek, Y. S. (2013). Design of a walking assistance lower limb exoskeleton for paraplegic patients and hardware validation using CoP. *International Journal of Advanced Robotic Systems, 10*(2), 113.

Kim, W., Lee, H., Kim, D., Han, J., & Han, C. (2014). Mechanical design of the Hanyang exo skeleton assistive robot (HEXAR). In *14th International conference on control, automation and systems (ICCAS 2014)* (pp. 479–484). IEEE.

Lim, D., Kim, W., Lee, H., Kim, H., Shin, K., Park, T., Lee, J. Y., & Han, C. (2015). Development of a lower extremity exoskeleton robot with a quasi-anthropomorphic design approach for load carriage. In *IEEE/RSJ International Conference on Intelligent Robots and Systems (IROS)* (pp. 5345–5350). IEEE.

Luu, T. P., Lim, H. B., Qu, X., Hoon, K. H., & Low, K. H. (2011). Subject-specific lower limb waveforms planning via artificial neural network. In *2011 IEEE International Conference on Rehabilitation Robotics* (pp. 1–6). IEEE.

Medrano-Cerda, G. A., Bowler, C. J., & Caldwell, D. G. (1995). Adaptive position control of antagonistic pneumatic muscle actuators. In *Proceedings 1995 IEEE/RSJ International Conference on Intelligent Robots and Systems. Human Robot Interaction and Cooperative Robots: 1* (pp. 378–383). IEEE.

Mellon, Carnegie. University Graphics Lab. (n.d.). Marker placement guide. Carnegie Mellon University. https://mocap.cs.cmu.edu/markerPlacementGuide.pdf.

Mundt, M., Thomsen, W., Witter, T., Koeppe, A., David, S., Bamer, F., & Markert, B. (2020). Prediction of lower limb joint angles and moments during gait using artificial neural networks. *Medical & Biological Engineering & Computing, 58*(1), 211–225.

Neogi, B., Darbar, R., Mondal, S., Gorai, B., Ghosh, S., Das, A., & Tibarewala, D. N. (2011). Study of proper tuning of prosthetic limb control system with paraplegia and fatigue condition. *2011 Second International Conference on Emerging Applications of Information Technology* (pp. 79–82). IEEE.

Rai, J. K., Tewari, R. P., & Chandra, D. (2010). Hybrid adaptive fuzzy control of bio robotic leg. In *2010 International Conference on Power, Control and Embedded Systems* (pp. 1–4). IEEE.

Saha, S. K. (2014). *Introduction to robotics*. Tata Mcgraw Hill Education Private Limited.

Shi, D. i, Zhang, W., Zhang, W., & Ding, X. (2019). A review on lower limb rehabilitation exoskeleton robots. *Chinese Journal of Mechanical Engineering, 32*(1), 1–11.

Shieh, M.-Y., & Chang, K. e-H. (2009). An optimized neuro-fuzzy controller design for bipedal locomotion. *International Journal of Fuzzy Systems, 11*(3), 137–145.

Solihin, M. I., Tack, L. F., & Kean, M. L. (2011). Tuning of PID controller using Particle Swarm Optimization (PSO). In *Proceeding of the international conference on advanced science, engineering and information technology: 1* (pp. 458–461).

Song, Q., & Liu, F. (2006). Improved control of a pneumatic actuator pulsed with PWM. In *2006 2nd IEEE/ASME International Conference on Mechatronics and Embedded Systems and Applications* (pp. 1–4). IEEE.

Van Varseveld, R. B., & Bone, G. M. (1997). Accurate position control of a pneumatic actuator using on/off solenoid valves. *IEEE/ASME Transactions on Mechatronics, 2*(3), 195–204.

Viteckova, S., Kutilek, P., Svoboda, Z., & Jirina, M. (2012). Fuzzy inference system for lower limbs angles prediction. In *2012 35th International Conference on Telecommunications and Signal Processing (TSP)* (pp. 517–520). IEEE.

Watanabe, H., Tanaka, N., Inuta, T., Saitou, H., & Yanagi, H. (2014). Locomotion improvement using a Hybrid Assistive Limb in recovery phase stroke patients: A randomized controlled pilot study. *Archives of Physical Medicine and Rehabilitation, 95*(11), 2006–2012.

Winter, D. A. (2009). *Biomechanics and motor control of human movement*. John Wiley & Sons.

Yan, T., Cempini, M., Oddo, C. M., & Vitiello, N. (2015). Review of assistive strategies in powered lower-limb orthoses and exoskeletons. *Robotics and Autonomous Systems, 64*, 120–136.

Yang, C., Ugbolue, U. C., McNicol, D., Stankovic, V., Stankovic, L., Kerr, A., & Rowe, P. J. (2019). Automation enhancement and accuracy investigation of a portable single-camera gait analysis system. *IET Science, Measurement & Technology, 13*(4), 563–571.

Zoss, A., & Kazerooni, H. (2006). Design of an electrically actuated lower extremity exoskeleton. *Advanced Robotics, 20*(9), 967–988.

Chapter 6

Nature-inspired algorithms for active power loss reduction in electrical power transmission system

Lenin Kanagasabai
Independent Researcher, Hyderabad, Telangana, India

6.1 Introduction

Power loss dwindling is a significant aspect in the transmission of power. Many algorithms, such as SARANI (Kanagasabai, 2024), CAC-DE (Anil Kumar et al., 2022), chaotic turbulent flow (Abd-El Wahab et al., 2022), adopting approach (Habib Khan et al., 2022), hybrid particle swarm and sea horse optimization algorithm (Hasanien et al., 2024), gray wolf (Varan et al., 2023), artificial immune (Lian, 2022), stochastic optimizer (Ali et al., 2023), hybrid algorithm (Sahli et al., 2022), enhanced transient search (Shaheen et al., 2023), cuckoo optimization (mellal et al., 2020), Integrated Corvus—North West America Mackenzie valley wolf inspired optimization (Kanagasabai, 2025), flower pollination algorithm (Mellal et al., 2023), mountain gazelle optimization (Padhi et al., 2024), ant lion optimization (Deb et al., 2023), novel method (Shekarappa et al., 2023), and intelligent approach (Deepika Reddy et al., 2024), are applied to solve the problem. In this chapter *Serrasalmus*-inspired optimization (SIO) algorithm, *Sarracenia leucophylla* inspired optimization (SLIO) algorithm, plantation farming optimization (PLFO) algorithm, *Pseudastur albicollis* search optimization (PASO) algorithm, customized black hole (CBH) algorithm, and union of black hole with white hole algorithm (UBW) are applied for solving the active power loss diminishing problem. In SLIO, feebler and minor *Serrasalmus* are frequently in the center of the herd place to relish the extreme defense. Biggest and sturdiest *Serrasalmus* in the herd are on the peripheries of the herd, everyplace they relish guard. They signify the external casing of the herd and are known as emissaries *Serrasalmus*. In the exploration phase, emissaries assure the herd's safety since they are positioned in exterior of the heard, and they do the initial attack on the prey and then sequentially signal other members of the herd to attack the prey. Emissaries *Serrasalmus* will start the probing for prey, and once the prey is identified, a sequential encompassing signal will be given by the emissaries *Serrasalmus* to other members of the herd. To avoid getting stuck in local optima, *Serrasalmus*-inspired algorithm smears a scheme. SLIO algorithm is designed based on the natural actions of the *S. leucophylla*. This plant is carnivorous, consuming maximum amount of required nutrients by trapping the insects and bugs. Through photosynthesis action *S. leucophylla* also engenders the energy. Alteration of the prey populace comparative to the *S. leucophylla* shrinkage and the modification of the *S. leucophylla* population comparative to the prey upsurges. The amount of happenstances among *S. leucophylla* and prey is relational to the product of their populations. In PLFO, exploration and exploitation are obtained through individual and societal actions. By presenting numerous operatives such as yearly progress, airing, and splicing the PLFO procedure can competently search the exploration region. In the plantation each tree signifies a solution to the problem. Certain newfangled saplings are require to be implanted to produce a complete plantation farm; certain arbitrary solutions ought to be created to generate a plantation farm. The quantity of primarily implanted saplings is measured as the parameter of the defined problem. Subsequent to implanting, by watering and applying fertilizers the sapling propagates. Subsequent to every sapling having been grown up (enhanced by local examination) for numerous eons, airing is executed, and in this process the

Computational Intelligence in Mechatronics: Solving Real-World Problems in Electronic Systems Design. DOI: https://doi.org/10.1016/B978-0-443-27580-7.00029-9

feeble, average, and robust saplings (solutions) are recognized. The chance of selecting any robust solution for splicing is contingent to optimal index value, and roulette wheel method is applied for this procedure. PASO algorithm is stimulated from the engagements of *P. albicollis*, which owns wide-ranging raptors with a resilient physique and widespread annexes. *P. albicollis* will track and identify the victim by means of potent eye vision and, at several intervals, detect the environs cautiously and then recognize the victim, and by its own approach, *P. albicollis* will seize the victim. This technique has been emulated to design the exploration period in the planned procedure. Planned procedure is reassigned from the exploration to exploitation, and also numerous manners of exploitation are designed, which are grounded on the rapacious attitude of *P. albicollis*. In black hole area its individual gravity stays everything from absconding from it. In black hole procedure the agents with the most excellent solution emulate the black hole. In this projected CBH algorithm, local search is smeared to every facet, "*d*," and the rationalized solution after the local search is smeared. In Union of Black Hole and White Hole Algorithm (UBW) both attributes of black and white holes are integrated. White holes' deeds are exactly opposite to black holes action. White hole's gravitational tweak is the alike to black hole's gravitational pull. Nonetheless, white hole contains a negative gravitational field. Black hole spurt swiftness is "in"; however, the extensive track of a white hole is "out." In black holes star will be imbibed, whereas in white holes star will be expelled out. Validation of SIO algorithm, SLIO algorithm, PLFO algorithm, PASO algorithm, CBH algorithm, and union of black hole with white hole algorithm (UBW) has been verified in IEEE 30 and 57 systems.

6.2 Problem formulation

Power loss minimization is defined by

$$\text{Min } \tilde{F}(\bar{m}, \bar{n}) \tag{6.1}$$

$$m = \begin{bmatrix} \text{VG}_1, \ldots, \text{VG}_{\text{Ng}}; \\ \text{QC}_1, \ldots, \text{QC}_{\text{Nc}}; T_1, \ldots, T_{N_T} \end{bmatrix}$$

$$n = \begin{bmatrix} \text{PG}_{\text{slack}}; \text{VL}_1, \ldots, \text{VL}_{N_{\text{Load}}}; \\ \text{QG}_1, \ldots, \text{QG}_{\text{Ng}}; \text{SL}_1, \ldots, \text{SL}_{N_T} \end{bmatrix}$$

$$m, n \rightarrow \text{control and dependent parameters}$$

$$F_1 = P_{\text{Min}} = \text{Min}\left[\sum_{m}^{\text{NTL}} G_m \begin{bmatrix} V_i^2 + \\ V_j^2 - 2 * V_i V_j cos\infty_{ij} \end{bmatrix}\right] \tag{6.2}$$

$$F_2 = \text{Min}\begin{bmatrix} \sum_{i=1}^{N_{\text{LB}}} \left|V_{\text{Lk}} - V_{\text{Lk}}^{\text{desired}}\right|^2 \\ + \sum_{i=1}^{\text{Ng}} \left|Q_{\text{GK}} - Q_{\text{KG}}^{\text{Lim}}\right|^2 \end{bmatrix} \tag{6.3}$$

$$F_3 = \text{Min } L_{\text{Max}} : L_{\text{Max}} = \text{Max}\left[1 - [Y_1]^{-1}[Y_2] \times \frac{V_i}{V_j}\right] \tag{6.4}$$

$$0 = \text{PG}_i - \text{PD}_i - V_i \sum_{j \in N_B} V_j\left[G_{ij}\cos\left[\infty_i - \infty_j\right] + B_{ij}\sin\left[\infty_i - \infty_j\right]\right] \tag{6.5}$$

$$0 = \text{QG}_i - \text{QD}_i - V_i \sum_{j \in N_B} V_j\left[G_{ij}\sin\left[\infty_i - \infty_j\right] + B_{ij}\cos\left[\infty_i - \infty_j\right]\right] \tag{6.6}$$

Equality and inequality constraints are defined as

$$P_g^{\min} \le P_g \le P_g^{\max} : Q_g^{\min} \le Q_g \le Q_g^{\max},\ i \in N_g \tag{6.7}$$

$$\mathrm{VL}_i^{\min} \le \mathrm{VL}_i \le \mathrm{VL}_i^{\max},\ i \in \mathrm{NL}$$

$$T_i^{\min} \le T_i \le T_i^{\max}, i \in N_T \tag{6.8}$$

$$Q_C^{\min} \le Q_c \le Q_C^{\max}, i \in N_C$$

$$|S_{L_i}| \le S_{L_i}^{\max},\ i \in N_{\mathrm{TL}} \tag{6.9}$$

$$\mathrm{VG}_i^{\min} \le \mathrm{VG}_i \le \mathrm{VG}_i^{\max},\ i \in N_g \tag{6.10}$$

Objective function in multi objective mode is defined as

$$\mathrm{MOF} = F_1 + r_i F_2 + u F_3 = F_1 + \left[\sum_{i=1}^{\mathrm{NL}} x_v \left[\mathrm{VL}_i - \mathrm{VL}_i^{\min}\right]^2 + \sum_{i=1}^{\mathrm{NG}} r_g \left[\mathrm{QG}_i - \mathrm{QG}_i^{\min}\right]^2\right] + r_f F_3 \tag{6.11}$$

$$\mathrm{VL}_i^{\mathrm{minimum}} = \begin{cases} \mathrm{VL}_i^{\max}, \mathrm{VL}_i > \mathrm{VL}_i^{\max} \\ \mathrm{VL}_i^{\min}, \mathrm{VL}_i < \mathrm{VL}_i^{\min} \end{cases} \tag{6.12}$$

$$\mathrm{QG}_i^{\mathrm{minimum}} = \begin{cases} \mathrm{QG}_i^{\max}, \mathrm{QG}_i > \mathrm{QG}_i^{\max} \\ \mathrm{QG}_i^{\min}, \mathrm{QG}_i < \mathrm{QG}_i^{\min} \end{cases} \tag{6.13}$$

6.3 *Serrasalmus*-inspired optimization algorithm

SIO algorithm is designed grounded on the natural deeds of *Serrasalmus*. Probing, encompassing and confronting the prey. Whenever *Serrasalmus* explore for sustenance, then passage will be in organized herd in an encrusted method. Henceforth, the feebler and minor *Serrasalmus* are frequently in the center of the herd place to relish the extreme defense. Biggest and sturdiest *Serrasalmus* in the herd are on the peripheries of the herd, everyplace they relish guard. They signify the external casing of the herd and are known as emissaries *Serrasalmus*. In the exploration phase, emissaries assure the herd safety, since they positioned in exterior of the heard and they do the initial attack on the prey, and then sequentially signal other members of the herd to attack the prey. Emissaries *Serrasalmus* will start the probing for prey and once the prey identified sequentially encompassing signal will be given by the emissaries *Serrasalmus* to other members of the herd. Then all *Serrasalmus* in the herd will cover the prey and confronting will be initiated by all the members in the herd. This exploration capability has been imitated in the design of the algorithm. Sum of emissaries *Serrasalmus* are arbitrarily designated to do the examination in the course of the exploration phase and this action will authorize the procedure to accomplish a global exploration.

$$\vec{E}\,(t+1) = \overrightarrow{E_P}(t) - \vec{X} \cdot M \tag{6.14}$$

$$\vec{X} = \left| Y.\overrightarrow{E_P}(t) - E(t) \right| \tag{6.15}$$

$$\vec{X} = 2x\vec{r}\ - \vec{x} \tag{6.16}$$

$$\vec{Y} = 2\vec{q} \tag{6.17}$$

$$x = 2 - 2\frac{t}{t_{\max}} \tag{6.18}$$

Probing and sequentially encompassing the prey initiated by the emissaries *Serrasalmus*.

$$\vec{E}(t+1) = \vec{M}'^{*} e^{bt*} \cos(2\pi l) + \overrightarrow{E_P}(t) \tag{6.19}$$

$$\vec{M}' = \left|\overrightarrow{E_P}(t) - \vec{E}(t)\right| \tag{6.20}$$

$$\vec{E}(t+1) = \begin{cases} \overrightarrow{E_P}(t) - \vec{X} \cdot M \; if \; p < 0.5 \\ \vec{M}'^{*} e^{bt*} \cos(2\pi l) + \overrightarrow{E_P}(t) \; if \; p \geq 0.5 \end{cases} \tag{6.21}$$

During the exploration segment the location of the herd is defined as

$$\vec{M} = \left|\vec{X}.\overrightarrow{E_{\text{random}}} - \vec{E}\right| \tag{6.22}$$

$$\vec{X}(t+1) = \overrightarrow{X_{\text{random}}} - \vec{X} \cdot M \tag{6.23}$$

$$\text{Random} \in [0, 1]$$

Through logistic map, prominence of the preliminary population location is augmented.

$$E_{k+1} = \mu E_k(1 - E_k) \tag{6.24}$$

Local examination competence and convergence fastidiousness have been boosted by including the inertia weight.

$$\text{Inertia weight } (\omega) = \omega' - (\omega' - \omega'')\left(\frac{t}{T_{\max}}\right)^{\frac{1}{t}} \tag{6.25}$$

Updating the enriched location is defined as

$$\vec{E}(t+1) = \begin{cases} \overrightarrow{E_P}(t) - \omega\vec{X} \cdot M \; if \; p < 0.5 \\ \omega\vec{D}'^{*} e^{bt*} \cos(2\pi l) + \overrightarrow{X_P}(t) \; if \; p \geq 0.5 \end{cases} \tag{6.26}$$

Notion of crowding distance is integrated to augment the multiplicity and dissemination of the population as follows:

$$Cg_d = \sum_{n=1}^{N} f_n(i+1) - f_n(i-1) \tag{6.27}$$

Entity "i" requests to be consistent with dissimilar elements as follows:

$$Cg_d = \sum_{n=1}^{N} \frac{f_n(i+1) - f_n(i-1)}{f_{n\,\max} - f_{n\,\min}} \tag{6.28}$$

Swarming appraisal operative is articulated as

$$i \geq cg_j(\text{if } (i_{\text{rank}} < j_{\text{rank}})) \tag{6.29}$$

By means of Laplace crossover, duos of progenies' are created and rendering to the parentages. At that moment the disseminated illogical sum (d_i) grounded on the laplacian is produced as follows:

$$d_i = \begin{cases} a - b\log_e(R(g_i)), \; (R(g)) \leq 2 \\ a + b\log_e(R(h_i)), \; (R(h_i)) > 2 \end{cases} \tag{6.30}$$

$R \rightarrow$ random

Production of the progenies' is defined as

$$u_1^i = v_1^i + (d_i)\left|u_1^i - u_2^i\right| \tag{6.31}$$

$$v_2^i = u_2^i + (d_i)\left|u_1^i - u_2^i\right| \tag{6.32}$$

Distance between prey and *Serrasalmus* herd is computed as

$$D = c\left(\sum_{\substack{j=1\\ j\neq 1}}^{N} c\frac{\min_d - \max_d}{2} S\left(\left|U_j - U_i\right|\right)\left(U_j - U_i\right)/d_{ij}\right) \tag{6.33}$$

$$\frac{\min_d - \max_d}{2} = 0.5$$

Recess demarcated as,

$$f(u) = \frac{2}{\text{magnitude}}u + 2 \tag{6.34}$$

Distance is computed by altering the rate to either 1 or 0.

$$F(D^d) = \begin{cases} 1, \ if D^d > \frac{\max(D^d) - \min(D^d)}{2} \\ 0, \ if D^d < \frac{m(D^d) - m(D^d)}{2} \\ Else \begin{cases} 0, \ if R\,() > 0.5 \\ 1, \ else \end{cases} \end{cases} \tag{6.35}$$

$$\max(D^d) = 2 \times \frac{c_{\max}}{\text{magnitude}}$$

$$\min(D^d) = 0$$

Modernizing rendering t prey,

$$U_{id} = \begin{cases} P_d, \ if F(D^d) = 0 \\ |1 - P_d|, \ else \end{cases} \tag{6.36}$$

Fitness function appraisal done by

$$\text{Fit} = \gamma \times \text{error rate} + (1-\gamma) \times \frac{\text{picked entity}}{\text{all entities}} \tag{6.37}$$

Serrasalmus-inspired algorithm smears a scheme. Postulated that *Serrasalmus* position $U^{i,t}$ at iteration "t" is not enhanced for a static quantity of iterations, in this condition, the similar *Serrasalmus* channels to a novel domicile interpretation to Lévy Flight.

$$U^{i,t+1} = U^{i,t} + \alpha\left[\frac{u_1}{(v_1)^{\frac{1}{\beta}}}, \ldots, \frac{u_d}{(v_d)^{\frac{1}{\beta}}}\right] \tag{6.38}$$

$$\begin{cases} u_i \sim N(0, \sigma^2), \ \sigma = \left(\frac{\Gamma(1+\beta)\sin\left(\frac{\pi}{2}\right)}{\Gamma\left(\frac{1+\beta}{2}\right)\times\beta\times 2^{\frac{\beta-1}{2}}}\right)^{\frac{1}{\beta}} \\ v_i \sim N(0, \sigma^2), \ \sigma' = 1 \end{cases} \tag{6.39}$$

$$i \in \{1, 2, \ldots, d\}$$

a. Start
b. set the variables
c. Create the population
d. By Logistic mapping prominence of the preliminary population location is augmented
e. $E_{k+1} = \mu E_k(1 - E_k)$

f. Preeminent entity is selected as the optimal location
g. $i \geq cg\ j(\text{if } (i_{rank} < j_{rank}))$
h. Probing and sequentially encompassing the prey initiated
i. $\vec{E}(t+1) = \begin{cases} \overrightarrow{E_P}(t) - \vec{X} \cdot M \text{ if } p < 0.5 \\ \vec{M}'{}^{*} e^{bt*} \cos(2\pi l) + \overrightarrow{E_P}(t) \text{ if } p \geq 0.5 \end{cases}$
j. *Verify the limits*
k. Crowding distance is integrated to augment the multiplicity and dissemination of the population
l. $Cg_d = \sum_{n=1}^{N} f_n(i+1) - f_n(i-1)$
m. Swarming appraisal operative is articulated
n. Inertia weight is included to enhance the exploitation
o. Execute the Laplace cross over
p. Distance between prey and Serrasalmus herd is computed
q. Recess is demarcated
r. $f(u) = \frac{2}{magnitude} u + 2$
s. Distance computation by altering the rate 1 or 0
t. $\min(D^d) = 0$
u. Modernizing rendering t prey,
v. $U_{id} = \begin{cases} P_d, & ifF(D^d) = 0 \\ |1 - P_d|, & else \end{cases}$
w. Fitness function appraisal executed
x. $t = t + 1$
y. Output the best solution
z. End

6.4 *Sarracenia leucophylla* inspired optimization algorithm

SLIO algorithm is designed based on the natural actions of the *S. leucophylla*. This plant is carnivorous, which consumes maximum amount of required nutrients by trapping the insects and bugs. Through photosynthesis action also *S. leucophylla* engenders the energy. The carnivorous plant is competent to respond to exterior incitements. As soon as it senses the existence or bout of a creature generates a sequences of biochemical responses that are on the rampage into the mid-air that entices regular slayer of the assaulter and may be source for the inner mutilation to the invader. The *S. leucophylla* traps phenomenally the insects and bugs. This type of actions will make the plant to consume the nutrients from the prey. Here the *S. leucophylla* act as predator and prey are insects, bugs etc. There will be communication of two populations and it defined as

$$\frac{da}{dt} = P(a, b) \tag{6.40}$$

$$\frac{db}{dt} = Q(a, b) \tag{6.42}$$

$P(a,b)$, $Q(a,b) \rightarrow$ population growth rate
The communications between the *S. leucophylla* and prey are defined as

$$Q_a(a, b) < 0$$

$$P_b(a, b) > 0$$

Alteration of the prey populace comparative to the *S. leucophylla* shrinkage and the modification of the *S. leucophylla* population comparative to the prey upsurges. The populace propagates proportionately to its magnitude with available region and nutrition's. If this condition happens, then $a(t)$ signifies the prey populace (in the absenteeism of *S. leucophylla*), then the populace development mathematically defined as

$$\frac{da}{dt} = ua$$

$$u > 0$$

$$a(t) = a_0 e^{at} \tag{6.43}$$

Naturally the prey population augments in the absenteeism of *S. leucophylla*, which is acting as predator. *S. leucophylla* feed on the prey and when no prey is available to feed then the strength and magnitude of the *S. leucophylla* decreases and it proportional to the population.

$$\frac{db}{dt} = uv$$

$$v > 0$$

$$b(t) = b_0 e^{-dt} \tag{6.44}$$

When there is no prey for *S. leucophylla*, then naturally the populations of *S. leucophylla* will decreases. The existence of prey assistances the evolution of the *S. leucophylla* and it mathematically represented as; $wab, \; w > 0$

Happenstances among *S. leucophylla* and prey will diminish the progression of the prey and it defined as

$$xab, x > 0$$

Then the communication among $a(t)$ and $b(t)$ is defined as

$$\frac{da}{dt} = Ga - Hab \tag{6.45}$$

$$\frac{db}{dt} = -Mab + Nb \tag{6.46}$$

$a \rightarrow$ amount of prey
$b \rightarrow$ quantity of Sarracenia leucophylla
$\frac{da}{dt} \rightarrow$ at time "t" "progression of the prey"
$\frac{db}{dt} \rightarrow$ t time "t" "progression"
$G \rightarrow$evolution of prey in the absenteeism of *Sarracenia leucophylla*
$H \rightarrow$ diminution rate of absenteeism of *Sarracenia leucophylla*
$M \rightarrow$ sustainability of the prey
$N \rightarrow$ predating capacity

a. Start
b. Fix the parameters
c. Engender the population
d. Compute the fitness value
e. Define the communication among populations
f. $\frac{da}{dt} = P(a, b)$
g. $\frac{db}{dt} = Q(a, b)$
h. $Q_a(a,b) < 0$
i. $P_b(a,b) > 0$
j. Determine the population growth
k. $\frac{da}{dt} = ua$
l. $u > 0$
m. $a(t) = a_0 e^{at}$
n. Determine the strength and magnitude
o. $\frac{db}{dt} = uv$
p. $v > 0$
q. $b(t) = b_0 e^{-dt}$
r. Define the existence of prey for evolution
s. $wab, \; w > 0$

t. Determine the happenstances
u. $xab, x > 0$
v. Define $a(t)and\ b(t)$
w. $\frac{da}{dt} = Ga - Hab$
x. $\frac{db}{dt} = -Mab + Nb$
y. $t = t + 1$
z. Output the best solution
aa. End

6.5 Plantation farming optimization algorithm

PLFO algorithm is designed based on the actions in plantation farming. In the plantation farming procedure, numerous activities such as watering, insemination, pruning, and splicing tips lead to a good products from the plantation, and in those plantations, maximum trees will nurtured and yield the various products effectively. In PLFO, exploration and exploitation are obtained through individual and societal actions. By presenting numerous operatives such as yearly progress, airing, and splicing the PLFO procedure can competently search the exploration region. In the plantation farming, initially saplings are planted, and then watering, applying fertilizers, insemination, pruning, and splicing are done sequentially. In the plantation each tree signifies a solution to the problem. Certain newfangled saplings are requisite to be implanted to produce a complete plantation farm; certain arbitrary solutions ought to be created to generate a plantation farm. The quantity of primarily implanted saplings is measured as the parameter of the defined problem. Subsequent to implanting, by watering and applying fertilizers, the sapling propagates. This development endures for numerous eons. Each few eons, times pruning and splicing will be executed, and it acts as the subsequent phases of the PLFO procedure. Every eon development of the sapling is signified by a local examination in the vicinity of the solution. If the consequence of a local examination is superior to the solution, then swapping will be accomplished. The quantity of succeeding local examinations is entitled to development eons quantity (DEQ). It is the quantity of eons the saplings need to nurture beforehand the airing period. Subsequent to every sapling having been grown up (enhanced by local examination) for numerous eons, airing is executed, and in this process the feeble, average, and robust saplings (solutions) are recognized. For this drive, it is essential to classify and grade the plantation saplings rendering to the optimal directory of the elucidations. The grading controls the solution will be transported deprived of alteration as an exclusive, substituted by a newfangled one, or designated as a splice or stalk for splicing. In PLFO algorithm the excellence of every solution is appraised by double standards. Primarily rendering to objective functional value and second, development degree of the solution, which is stimulated by the sapling picking process.

$$Z_j = \gamma \dot{a}_j + \delta \dot{b}_j \tag{6.47}$$

$\dot{a}_j, \dot{b}_j \rightarrow$ normalized objective functional value
$\gamma, \delta \rightarrow$ coefficients indicating the development degree

$$\dot{a}_j = \frac{a_j}{\max(a_j)\ 1 \le j \le m} \tag{6.48}$$

$m \rightarrow$ sum of the saplings
Development degree of the solutions is defined as

$$\dot{b}_j = \sum_{i=n-e}^{n} \tau_i \left(a_j^i - a_j^{i-1}\right) \tag{6.49}$$

$i \rightarrow$ eon of the growth
$n \rightarrow$ total number of growth eons before splicing
$y \rightarrow$ number of development eons
$\tau_i \rightarrow$ weight
Weights considered in this work are 0.60, 0.30, and 0.10, respectively.

$$\dot{b}_j = 0.60\left(a_j^i - a_j^{i-1}\right) + 0.30\left(a_j^{i-1} - a_j^{i-2}\right) + 0.10\left(a_j^{i-2} - a_j^{i-3}\right) \tag{6.50}$$

Then the computed development degree is defined as

$$\dot{b}_j = \frac{b_j}{\max(b_j)\ 1 \le j \le m} \tag{6.51}$$

Subsequent to categorization of the solutions the airing index ought to be demarcated, and the classified solutions should be categorized consequently. The airing index can be demarcated as the onsets among the classes. Single average category solutions (saplings) are anticipated to progress and are spliced with a ration (splices) of the robust solutions. The chance of selecting any robust solution for splicing is contingent to optimal index value, and roulette wheel method is applied for this procedure. The feeble solutions are swapped with newfangled arbitrary ones. Since after a few eons of developing and probing for dissimilar vicinities in these eons of development, they have not improved and endure in the feeble class. In the PLFO algorithm, feeble ones are eliminated and substituted with newfangled arbitrary ones. In environment, trees that are nurtured well do not require splicing. Alike, in the PLFO algorithm, an elucidation of the robust class is progressed to the subsequent iteration deprived of any alteration.

a. Start
b. Fix the parameters
c. Engender the Plantation farm
d. Define development degree of the saplings
e. $Z_j = \gamma \dot{a}_j + \delta \dot{b}_j$
f. $\dot{a}_j = \frac{a_j}{\max(a_j)\ 1 \le j \le m}$
g. $\dot{b}_j = \sum_{i=n-e}^{n} \tau_i \left(a_j^i - a_j^{i-1}\right)$
h. Perform the local examination
i. $\dot{b}_j = 0.60\left(a_j^i - a_j^{i-1}\right) + 0.30\left(a_j^{i-1} - a_j^{i-2}\right) + \left(a_j^{i-2} - a_j^{i-3}\right)$
j. $\dot{b}_j = \frac{b_j}{\max(b_j)\ 1 \le j \le m}$
k. if better solutions obtained then swap the solution
l. Compute the development eons quantity
m. Classify the saplings
n. Determine the airing index
o. Categorize the classified solutions
p. Replace the feeble with robust saplings
q. Robust saplings will move to next iteration
r. t = t + 1
s. Output the best solution
t. End

6.6 *Pseudastur albicollis* search optimization algorithm

PASO algorithm is stimulated by the engagements of *P. albicollis*'s own wide-ranging raptors with a resilient physique and widespread annexes. *P. albicollis* will track and identify the victim by means of potent eye vision and, at several intervals, detect the environs cautiously and then recognize the victim, and by its own approach, *P. albicollis* will seize the victim. This technique has been emulated to design the exploration period in the planned procedure.

$$P(t+1) = \begin{cases} P_r(t) - \\ n_1 |P_r(t) - 2n_2 P(t)| \\ z \ge 0.5 \\ P_v(t) - P_m(t) - \\ n_3(\min + n_4(\max - \min)) \\ z < 0.5 \end{cases} \tag{6.52}$$

r and *v* are random and victim

Location of the *P. albicollis* is achieved as follows:

$$P_m(t) = \frac{1}{T}\sum_{i=1}^{T} P_i(t) \tag{6.53}$$

The planned procedure is reassigned from exploration to exploitation, and also numerous manners of exploitation are designed, which are grounded on the rapacious attitude of *P. albicollis*.

$$R = 2R_0\left(1 - \frac{t}{T}\right) \tag{6.54}$$

$$P(t+1) = \Delta P(t) - B|AP_v(t) - P(t)| \tag{6.55}$$

$$\Delta P(t) = P_v(t) - P(t) \tag{6.56}$$

P. albicollis will girdle the envisioned victim to grasp it

$$P(t+1) = P_v(t) - B|\Delta P(t)| \tag{6.57}$$

P. albicollis plunge en route for the grasping the victim and it defined as,

$$C = P_v(t) - B|AP_v(t) - P(t)| \tag{6.58}$$

Levy distribution is demarcated as

$$L(s, \gamma, \mu) = \begin{cases} \sqrt{\frac{\gamma}{2\pi}} & \exp\left[-\frac{\gamma}{2(s-\mu)}\right]\frac{1}{(s-\mu)^{3/2}} \\ 0 \; if \; s \le 0 \end{cases}$$

$$F(k) = \exp\left[-\alpha|k|^{\beta}\right],\; 0 < \beta \le 2, \tag{6.59}$$

The concluding location updating of *P. albicollis* is defined as

$$P(t+1) = \begin{cases} C \; if \; F(C) < F(P(t)) \\ D \; if \; F(D) < F(P(t)) \end{cases} \tag{6.60}$$

Decree is implemented in stiff cordon circumstances and it defined as,

$$C_1 = P_v(t) - B|AP_v(t) - P_m(t)| \tag{6.61}$$

$$D = C_1 + V \times L\,(Q)$$

$$P(t+1) = \begin{cases} C_1 \; if \; F(C_1) < F(P(t)) \\ G \; if \; F(D) < F(P(t)) \end{cases} \tag{6.62}$$

a. Begin
b. Create the population arbitrarily
c. while (culmination condition is not met) do
d. Calculate the Pseudastur albicollis fitness rate
e. Set P_v as the premium location of victim
f. For (each Pseudastur albicollis (Pi)) do
g. Streamline the initial vitality R_0 and hedge strength ‘A”

$$R_O = 2r(\,) - 1; A = 2(1 - r\,(\,))$$

h. Update the R

$$R = 2R_o\left(1 - \frac{t}{T}\right)$$

i. If $R \ge 1$, then
j. Procedure in Exploration segment

k. Update the position vector
l. If R < 1, then
m. Procedure in Exploitation segment
n. If (n ≥ 0.5 and |R| ≥ 0.5) then,
o. Pseudastur albicollis apply Lenient cordon
p. Update the position vector

$$P(t+1) = \Delta P(t) - B|AP_v(t) - P(t)|$$

q. Else
r. If (n ≥ 0.5 and |R| < 0.5) then
s. Pseudastur albicollis apply strong cordon
t. Update the position vector

$$P(t+1) = P_v(t) - B|\Delta P(t)|$$

u. Else
v. if (r < 0.5 and |R| ≥ 0.5) then
w. Pseudastur albicollis apply Lenient cordon
x. By rapid plunges of Pseudastur albicollis update the location
y. Else
z. If (n < 0.5 and |R| < 0.5) then
. Return P_v
~. End

6.7 Customized black hole algorithm and union of black hole with white hole algorithm

In the black hole algorithm (BHA) the agents with the most excellent solution emulate the black hole. Event horizon is computed, and any agent inside the event horizon fades and is reinitialized in the exploration region. The formation of populace is impetuous, and in the exploration region candidates and stars exist.

The amalgamation of stars by the black hole is given as

$$Y_i(t+1) = Y_i(t) + \text{random} \times (Y_{B^-}, Y_i(t)) \tag{6.63}$$

$$i = 1, 2, \ldots, N$$

Event horizon radius of BHA is computed as

$$R = \frac{f_B}{\sum_{i=1}^{N} f_i} \tag{6.64}$$

The place of the ith stars (Y_i) is given by

$$Y_i = (\text{star}_i, \ldots, \text{star}_N, \text{blcakhole}_d) \tag{6.65}$$

At time "t" the incorporation of star "i" from star "j" is defined as

$$E_{ij}^d = \xi(t_o)\frac{C_{pi}(t) \times C_{aj}(t) \times (\text{star}_j(t) - \text{star}_i(t))}{(D_{ij}(t) + \varepsilon)^2 \times (C_{pi}(t) + C_{aj}(t))} \times \left(\frac{t_o}{t - t_o}\right)^a \tag{6.66}$$

Comprehensive force is arbitrarily weighted by

$$E_i^d(t) = \sum_{j=1\ J \neq i}^{N} \text{random}_i E_{ij}^d(t) \tag{6.67}$$

Speeding up of the star "I" at time "t," in course dth, is given as

$$a_i^d(t) = \frac{E_i^d(t)}{C_{ii}(t)} \tag{6.68}$$

Place and swiftness are premeditated by

$$v_i^d(t+1) = \text{random}_i \times v_i^d(t) + a_i^d(t) \tag{6.69}$$

$$\text{star}_i(t+1) = \text{star}_i(t) + v_i^d(t+1) \tag{6.70}$$

1. Initialization

Arbitrarily engender the N_p target vectors $Y_i^g = (i = 1, 2, \ldots, N_P)$

Objective vector with the most excellent fitness value is designated as black hole Y_B^0. Set the utmost generation number as $g_{\max}$.

1. Mutation and cross over is premeditated by

$$M_i^{g+1} = Y_{r1}^g + F \times (Y_{r2}^g - Y_{r3}^g) \tag{6.71}$$

$$r1 \neq r2 \neq r3 \neq i$$

$$H_{i,j}^{g+1} = \begin{cases} M_i^{g+1} \; if \; (\text{rand}_j(0,1) \leq C_r) \\ Y_{i,j}^g \; otherwise \end{cases} \tag{6.72}$$

2. Place and swiftness are calculated by

$$v_i^d(t+1) = \text{random}_i \times v_i^d(t) + a_i^d(t)$$

$$\text{star}_i(t+1) = \text{star}_i(t) + v_i^d(t+1)$$

3. Endurance criterion is calculated by

$$Y_i^{g+1} = \begin{cases} u_i^{g+1} \; f(u_i^{g+1}) \leq f(b_i^{g+1}) \\ b_i^{g+1} \; f(u_i^{g+1}) \leq f(b_i^{g+1}) \end{cases} \tag{6.73}$$

4. Vector correction is done through

$$Y_i^{g+1} - Y_B^{g+1} < R \tag{6.74}$$

5. If maximum generation $g_{\max}$ is attained, then stop, or else, go to Step 2.

In this work CBH algorithm is designed. To find neighborhood solutions around the most excellent solution, local search portents have been applied in the projected algorithm. In this proposed work local search is smeared to each facet, "d," and the modernized solution after the local search is applied is demarcated as

$$Y_{LS}^i(t) = Y_B^i(t) + \text{Random}_i \times e^{-5T/T_{\max}} \tag{6.75}$$

Solution modernize in the BHA done by, doing reinitialization when the agent's *i*th position is inside the event horizon and when agent *i*th position is not inside the event horizon then modernize the position based on

$$Y_i(t+1) = Y_i(t) + \text{ random} \times (Y_{\text{BH}} - Y_i(t)) \tag{6.76}$$

Then the solution modernizing procedure in CBH by local search algorithm is defined by

1. When *i*th agent is the worst afterward, do local search by

$$Y_{\text{LS}}^i(t) = Y_B^i(t) + \text{Random}_i \times e^{-5T/T_{\max}} \tag{6.77}$$

2. When *i*th agents position is inside the event horizon then do reinitialization
3. When *i*th agents position is not inside the event horizon then modernize the position by

$$Y_i(t+1) = Y_i(t) + \text{ rand} \times (Y_{\text{BH}} - Y_i(t)) \tag{6.78}$$

General outline of the proposed CBH algorithm is given by

1. Population of stars are initialized with capricious locations in the exploration space
2. For every star objective function value has to be computed
3. Most excellent fitness value possessed by the star is selected as black hole
4. When ith agent is the worst afterward, do local search by
5. $Y^i_{LS}(t) = Y^i_B(t) + \text{Random}_i \times e^{-\frac{5T}{T_{max}}}$
6. When ith agents position is inside the event horizon then do reinitialization
7. Position of every star is modified by $Y_i(t+1) = Y_i(t) + \text{ random} \times (Y_B - Y_i(t))$
8. When black hole, superior to a star location then exchange the locations
9. In event horizon when a star cross black hole, replace with a new star in capricious location in exploration space
10. If an end condition is met, then stop the process

Union of Black Hole and White Hole Algorithm (UBW) is designed. White hole is an area of spacetime in which anything cannot enter from outside; nevertheless, matter and light can get away from it. In the conflict of black hole, all can enter from the outside, but no entity, has the competence to run away from it.

White hole and stars gravitational interaction will chaperon to an exponential repression of the white hole emanation. Force of any star near to the white hole is given as

$$FC^d_{i,\text{wh}}(t) = G(t)\frac{M_i(t) - M_{\text{wh}}(t)}{R_{i,\text{wh}}(t) + \varepsilon}\left(Y^d_{\text{wh}}(t) - Y^d_i(t)\right) \tag{6.79}$$

Acceleration defined as

$$a^d_i(t) = \frac{F^d_{i,\text{wh}}(t)}{M_i(t)} \tag{6.80}$$

The place of the ith stars defined as

$$Y_i = (\text{star}_i, \ldots, \text{star}_N, \text{black hole}_d) \tag{6.81}$$

$$Y_i = (\text{star}_i, \ldots, \text{star}_N, \text{white hole}_d) \tag{6.82}$$

$$E^d_{ij} = \xi(t_o)\frac{C_{pi}(t) \times C_{aj}(t) \times \left(\text{star}_j(t) - \text{star}_i(t)\right)}{\left(D_{ij}(t) + \varepsilon\right)^2 \times \left(C_{pi}(t) + C_{aj}(t)\right)} \times \left(\frac{t_o}{t - t_o}\right)^a \tag{6.83}$$

Complete force is arbitrarily weighted by

$$E^d_i(t) = \sum_{j=1\ J\neq i}^{N} \text{random}_i E^d_{ij}(t) \tag{6.84}$$

Place and swiftness are premeditated by

$$v^d_i(t+1) = \text{random}_i \times v^d_i(t) + a^d_i(t) \tag{6.85}$$

$$\text{star}_i(t+1) = \text{star}_i(t) + v^d_i(t+1) \tag{6.86}$$

Mutation and cross over is calculated by

$$M^{g+1}_i = Y^g_{r1} + F \times \left(Y^g_{r2} - Y^g_{r3}\right) \tag{6.87}$$

$$r1 \neq r2 \neq r3 \neq i$$

$$H^{g+1}_{i,j} = \begin{cases} M^{g+1}_i \ \textit{if}\ (\text{rand}_j(0,1) \leq C_r) \\ Y^g_{i,j}\ \textit{otherwise} \end{cases} \tag{6.88}$$

Survival's decisive factor is calculated by

$$Y_i^{g+1} = \begin{cases} u_i^{g+1}\ f\left(u_i^{g+1}\right) \le f\left(b_i^{g+1}\right) \\ b_i^{g+1}\ f\left(u_i^{g+1}\right) \le f\left(b_i^{g+1}\right) \end{cases} \tag{6.89}$$

Vector rectification is done by

$$Y_i^{g+1} - Y_B^{g+1} < R_B \tag{6.90}$$

$$Y_i^{g+1} - Y_{WH}^{g+1} < R_W \tag{6.91}$$

However, the radius of the event horizon in black hole and white hole is demarcated as

$$R_B = \frac{f_B}{\sum_{i=1}^{N} f_i} \tag{6.92}$$

$$R_W = \frac{f_W}{\sum_{i=1}^{N} f_i} \tag{6.93}$$

Union of Black Hole and White Hole Algorithm (UBW)

1. Engender the preliminary population
2. Fitness of the each star is premeditated
3. The rate of the white hole, black hole are rationalized
4. Swiftness and place value are rationalized
5. Compute the event horizon
6. Minimum one star crosses the black hole's event horizon?; if yes implement the black hole operator or else move top next step
7. Minimum one star crosses the white hole's event horizon?; if yes implement the white hole operator

Or else carry out the operation done by white hole again
If end criterion has been met? Then stop the process with most excellent output solution or else go back to Step 2

6.8 Simulation study

SIO algorithm, SLIO algorithm, PLFO algorithm, PASO algorithm, CBH algorithm, and union of black hole with white hole algorithm (UBW) are validated in IEEE 30 bus system. Table 6.1 shows the evaluation of results [RL—real power loss (MW), VD—voltage deviation (PU), VS—voltage stability (PU)].

Tables 6.1 and 6.2 show the better performance of the proposed algorithms in IEEE 30, 57 bus systems. Comparision has been done with CAC-DE (Anil kumar et al., 2022), CTF (Abd-El Wahab et al., 2022), AAP (Habib Khan et al., 2022), SO (Ali et al., 2023), GWO (Varan et al., 2023), IAIA (Lian, 2022), EHA (Sahli et al., 2022), IPSOSH (Hasanien et al., 2024), ETSO (Shaheen et al., 2023), and HPSSH (Hasanien et al., 2024). Results show the better performance of the proposed algorithms in reducing the real power loss, minimization of voltage deviation and voltage stability enhancement.

SIO algorithm, SLIO algorithm, PLFO algorithm, PASO algorithm, CBH algorithm, and union of black hole with white hole algorithm (UBW) are validated in IEEE 57 bus system. Table 6.2 shows the evaluation of results [RL—real power loss (MW), VD—voltage deviation (PU), VS—voltage stability (PU)].

Results show the better performance of the proposed SIO algorithm, SLIO algorithm, PLFO algorithm, PASO algorithm, CBH algorithm, and union of black hole with white hole algorithm (UBW) in reducing the real power loss, minimization of voltage deviation, and voltage stability enhancement. Time taken by SIO, SLIO, PFLO, PASO, CBH, and UBW for IEEE 30, 57 bus systems is shown in Table 6.3.

6.9 Conclusion

In SIO algorithm, sum of emissaries *Serrasalmus* is arbitrarily designated to do the examination in the course of the exploration phase, and this action will authorize the procedure to accomplish a global exploration. In SLIO algorithm the

TABLE 6.1 Valuation of results—IEEE 30 bus.

Method	RL (MW)	VD (PU)	VS (PU)
CAC-DE (Anil Kumar et al., 2022)	4.6482	0.0802	0.1004
CTF (Abd-El Wahab et al., 2022)	4.9448	0.1212	0.1232
AAP (Habib Khan et al., 2022)	4.5677	0.1250	0.1135
SO (Ali et al., 2023)	4.5464	1.9889	0.1294
GWO (Varan et al., 2023)	4.1781	0.4697	0.1323
IAIA (Lian, 2022)	5.6614	0.1536	0.2683
EHA (Sahli et al., 2022)	4.6095	0.1649	0.2578
HPSSH (Hasanien et al., 2024)	5.0763	0.6828	0.3297
SIO	4.0562	0.0489	0.4198
SLIO	4.0487	0.0431	0.4210
PFLO	4.0473	0.0412	0.4232
PASO	4.0464	0.0398	0.4245
CBH	4.0451	0.0373	0.4263
UBW	4.0442	0.0369	0.4278

CBH, Customized black hole; *PASO*, Pseudastur albicollis search optimization; *PLFO*, plantation farming optimization; *SIO*, Serrasalmus-inspired optimization; *SLIO*, Sarracenia leucophylla inspired optimization.

TABLE 6.2 Appraisal of results—IEEE 57 bus.

Method	RL (MW)	VD (PU)	VS (PU)
CAC-CDE (Anil Kumar et al., 2022)	21.9452	0.6012	0.0948
CWO (Ali et al., 2023)	23.3235	0.58553	0.2561
IPSOSH (Hasanien et al., 2024)	25.4715	0.6828	0.3231
ETSO (Shaheen et al., 2023)	25.4963	0.799021	0.3129
SIO	20.0428	0.47392	0.4098
SLIO	20.0412	0.47384	0.4103
PFLO	20.0403	0.47379	0.4108
PASO	20.0398	0.47368	0.4116
CBH	20.0381	0.47355	0.4121
UBW	20.0367	0.47341	0.4127

CBH, Customized black hole; *PASO, Pseudastur albicollis* search optimization; *PLFO*, plantation farming optimization; *SIO, Serrasalmus*-inspired optimization; *SLIO, Sarracenia leucophylla* inspired optimization.

TABLE 6.3 Time taken by SIO, SLIO, PFLO, PASO, CBH, and UBW.

Method	30 Bus T(S)	57 Bus T(S)
SIO	14.86	19.87
SLIO	14.81	19.82
PFLO	14.78	19.79
PASO	14.74	19.72
CBH	14.71	19.69
UBW	14.67	19.65

CBH, Customized black hole; *PASO, Pseudastur albicollis* search optimization; *PLFO*, plantation farming optimization; *SIO, Serrasalmus*-inspired optimization; *SLIO, Sarracenia leucophylla* inspired optimization.

amount of happenstances among *S. leucophylla* and prey is relational to the product of their populations. comparative to the *S. leucophylla* shrinkage and the modification of the *S. leucophylla* population comparative to the prey upsurges. In PLFO algorithm feeble ones are eliminated and substituted with newfangled arbitrary ones. In environment, trees that are nurtured well do not require splicing. It is the quantity of eons the saplings need to nurture beforehand the airing period. Subsequent to every sapling having been grown up (enhanced by local examination) for numerous eons, airing is executed, and in this process the feeble, average, and robust saplings (solutions) are recognized. *P. albicollis* will track and identify the victim by means of potent eye vision, and at several intervals detect the environs cautiously and then recognize the victim, and by its own approach, *P. albicollis* will seize the victim. This technique has been emulated to design the exploration period in the planned procedure. In CBH algorithm, Union of Black Hole and White Hole Algorithm (UBW) production of population is capricious, and in the exploration space candidates, stars are present. White hole is an area of spacetime, in which anything cannot enter from outside; nevertheless, matter and light can get away from it. In the conflict of black hole, all can enter from the outside, but no entity, has the competence to run away from it. White hole and stars' gravitational interaction will chaperone an exponential repression of the white hole emanation. SIO algorithm, SLIO algorithm, PLFO algorithm, PASO algorithm, CBH algorithm, and union of black hole with white hole algorithm (UBW) are verified in IEEE 30 and 57 systems.

References

Abd-El Wahab, A. M., Kamel, S., Hassan, M. H., Mosaad, M. I., & AbdulFattah, T. A. (2022). Optimal reactive power dispatch using a chaotic turbulent flow of water-based optimization algorithm. *Mathematics, 10*(3), 346–352. https://doi.org/10.3390/math10030346.

Ali, M. H., Soliman, A. M. A., Abdeen, M., Kandil, T., Abdelaziz, A. Y., & El-Shahat, A. (2023). A novel stochastic optimizer solving optimal reactive power dispatch problem considering renewable energy resources. *Energies, 16*(4), 1–16. https://doi.org/10.3390/en16041562.

Anil Kumar, G, Jeyanthy, A., & Devaraj, D. (2022). Hybrid CAC-DE in optimal reactive power dispatch (ORPD) for renewable energy cost reduction. *Sustainable Computing Informatics and Systems, 35*(100688), 100688. https://doi.org/10.1016/j.suscom.2022.100688.

Deb, S., Sachan, S., Sinha, S., & Malik, P. (2023). Planning and operation of hybrid local energy system through ant-lion optimization. In *2023 IEEE 3rd International Conference on Sustainable Energy and Future Electric Transportation (SEFET)* (pp. 1–6). https://doi.org/10.1109/SeFeT57834.2023.10245209.

Deepika Reddy, G., Medikondu, N. R., Vijaya Kumar, T., Koneru, S., Bobba, P. B., Singla, A., Pandey, A. K., & Al-Jawahry, H. M (2024). Intelligent optimization for multiprocessor systems: Hybrid algorithmic strategies for scheduling and load balancing. *Cogent Engineering, 11*(1), 1–15. https://doi.org/10.1080/23311916.2024.2376911.

Habib Khan, N., Jamal, R., Ebeed, M., Kamel, S., Zeinoddini-Meymand, H., & Zawbaa, H. M. (2022). Adopting scenario-based approach to solve optimal reactive power dispatch problem with integration of wind and solar energy using improved Marine predator algorithm. *Ain Shams Engineering Journal, 13*(5), 1–17. https://doi.org/10.1016/j.asej.2022.101726.

Hasanien, H. M., Alsaleh, I., Tostado-Véliz, M., Zhang, M., Alateeq, A., Jurado, F., & Alassaf, A. (2024). Hybrid particle swarm and sea horse optimization algorithm-based optimal reactive power dispatch of power systems comprising electric vehicles. *Energy, 286*(1), 1–17. https://doi.org/10.1016/j.energy.2023.129583.

Kanagasabai, L. (2024). Novel empress SARANI optimization algorithm for active power loss reduction and voltage stability enhancement. *Heliyon, 10*(22), 1–64. https://doi.org/10.1016/j.heliyon.2024.e38984.

Kanagasabai, L. (2025). Integrated Corvus—North West America Mackenzie valley wolf inspired optimization and hybrid quasi opposition based white headed buffalo weaver—Indian leopard cat swarm algorithm for power loss diminution in electrical grid transmission network in smart city. *IEEE Transactions on Industry Applications, 61*(6), 8703–8711. https://doi.org/10.1109/tia.2025.3576081.

Lian, L. (2022). Reactive power optimization based on adaptive multi-objective optimization artificial immune algorithm. *Ain Shams Engineering Journal, 13*(5), 1–15. https://doi.org/10.1016/j.asej.2021.101677.

Mellal, M. A., Khitous, M., & Zemmouri, M. (2023). Combined heat and power economic dispatch problem with binary method using flower pollination algorithm and differential evolution. *Electrical Engineering, 105*(4), 2161–2168. https://doi.org/10.1007/s00202-023-01801-x.

Mellal, M. A., & Williams, E. J. (2020). Cuckoo optimization algorithm with penalty function and binary approach for combined heat and power economic dispatch problem. *Energy Reports, 6*, 2720–2723. https://doi.org/10.1016/j.egyr.2020.10.004.

Padhi, J. R., Nayak, S., Prakash, S., Sharma, S., Debnath, M. K., & Gautam, T. P. K. (2024). Frequency regulation in microgrid using tilted controller tuned by mountain gazelle optimization. In *2024 5th International Conference on Recent Trends in Computer Science and Technology (ICRTCST)* (pp. 326–331). https://doi.org/10.1109/ICRTCST61793.2024.10578430.

Sahli, Z., Hamouda, A., Sayah, S., Trentesaux, D., & Bekrar, A. (2022). Efficient hybrid algorithm solution for optimal reactive power flow using the sensitive bus approach. *Engineering Technology & Applied Science Research, 12*(1), 8210–8216. https://doi.org/10.48084/etasr.4680.

Shaheen, M. A. M., Ullah, Z., Hasanien, H. M., Tostado-Véliz, M., Ji, H., Qais, M. H., Alghuwainem, S., & Jurado, F. (2023). Enhanced transient search optimization algorithm-based optimal reactive power dispatch including electric vehicles. *Energy, 277*(1), 1–19. https://doi.org/10.1016/j.energy.2023.127711.

Shekarappa, G., S. , Mahapatra, & Raj, S. (2023). A novel meta-heuristic approach for optimal RPP using series compensated FACTS controller. *Intelligent Systems with Applications, 18*(1), 1–9. https://doi.org/10.1016/j.iswa.2023.200220.

Varan, M., Erduman, A., & Menevşeoğlu, F. (2023). A grey wolf optimization algorithm-based optimal reactive power dispatch with wind-integrated power systems. *Energies, 16*(13), 1–16. https://doi.org/10.3390/en16135021.

Chapter 7

Vision inspection system for butterfly valves

Venkutuswamy Radhika [a] and Rajaram Kiruba [b]

[a] *Department of Biomedical Engineering, Sri Ramakrishna Engineering College, Coimbatore, Tamil Nadu, India,*

[b] *Department of Electronics and Instrumentation Engineering, Sri Ramakrishna Engineering College, Coimbatore, Tamil Nadu, India*

7.1 Introduction

Analyzing the specifications and requirements of a butterfly valve is a critical first step in its design process. Engineers meticulously consider factors, such as the desired flow rate, operating pressure, temperature range, and properties of the medium it will handle. Understanding these parameters serves as the cornerstone for dimensioning the valve effectively (Gao & Yang, 2021).

Flow rate determination ensures that the valve can accommodate the required volume or mass flow, essential for various applications. Evaluating pressure conditions aids in selecting materials and designing components capable of withstanding the expected forces.

Temperature considerations guide material choices and thermal stability, crucial for long-term performance. Assessing the properties of the medium, including viscosity and corrosiveness, helps in selecting materials that ensure compatibility and durability.

Moreover, analyzing operational conditions and regulatory standards ensures that the valve meets safety requirements and environmental guidelines. By meticulously analyzing these specifications, engineers can design butterfly valves that meet performance objectives, regulatory compliance, and operational demands effectively.

Integrating industrial cameras for the dimensioning of valves, coupled with deep learning (DL) techniques, offers a sophisticated solution for precise measurement and quality assurance in manufacturing processes (Rao & Narasimhan, 2018).

Industrial cameras can capture high-resolution images of valve components during various stages of production, providing detailed visual data for analysis. These images can include close-ups of critical features, such as the disc, stem, seat, and sealing surfaces. By leveraging DL algorithms, these images can be processed to extract accurate measurements of dimensions, angles, clearances, and tolerances (Liu et al., 2020). For example, DL models can be trained to identify specific features of interest within valve images, such as the diameter of the disc, the length of the stem, or the distance between sealing surfaces. By annotating a dataset of images with corresponding measurements the DL model can learn to recognize and measure these features automatically.

Once trained the DL model can analyze new images of valve components in real time, extracting precise dimensional information without the need for manual measurement or inspection. This streamlines the dimensioning process, reduces human error, and improves overall efficiency in manufacturing. Moreover, DL algorithms can detect anomalies or deviations from desired dimensions, flagging components that fall outside of specified tolerances for further inspection or rework (Kim et al., 2017). This ensures that valves meet quality standards and functional requirements before they are assembled and deployed in industrial systems.

7.2 Proposed methodology

The proposed methodology for measuring butterfly valve dimensions using LabVIEW involves setting up a comprehensive system that integrates compatible hardware, such as a PC or embedded device, with LabVIEW software. This system will

Computational Intelligence in Mechatronics: Solving Real-World Problems in Electronic Systems Design. DOI: https://doi.org/10.1016/B978-0-443-27580-7.00017-2

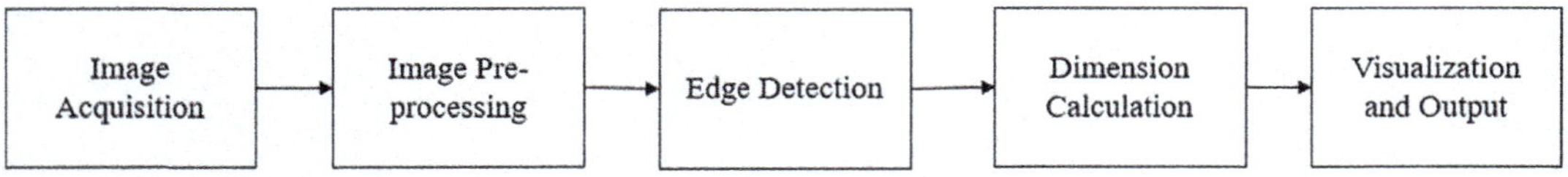

FIGURE 7.1 Hardware setup of the proposed work.

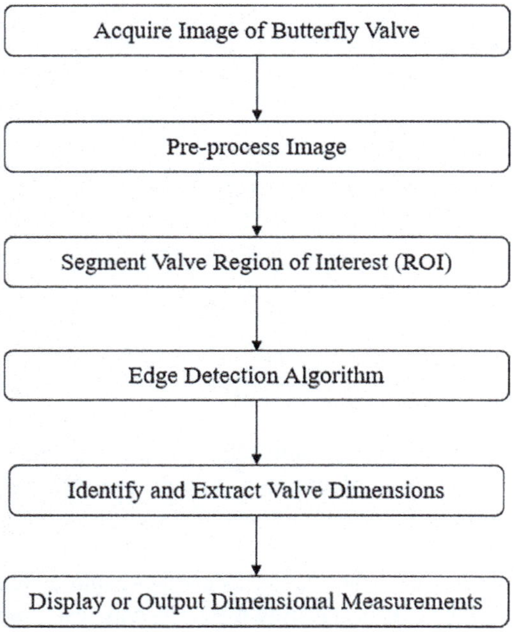

FIGURE 7.2 Block diagram for LabVIEW-based dimension calculation.

incorporate precision measurement sensors such as calipers or laser distance sensors to capture accurate readings of key valve dimensions, including diameter and thickness. The LabVIEW environment will play a central role in developing custom virtual instruments (VIs) that control these sensors and manage data acquisition processes. Calibration of the sensors will be crucial to ensure measurement accuracy, and sophisticated algorithms will be implemented within LabVIEW to process raw sensor data and calculate precise valve dimensions based on the acquired readings.

In addition, the LabVIEW system will include a user friendly graphical interface designed to streamline operator interaction. This interface will allow users to input valve parameters, view real-time measurement outputs, and access comprehensive data visualization tools. Automation features will be integrated to optimize measurement sequences and enhance overall efficiency. Rigorous testing and validation will be conducted to verify measurement accuracy against industry standards, ensuring the reliability and robustness of the LabVIEW program. Ultimately, this methodology promises to deliver enhanced precision and efficiency in butterfly valve dimension measurement, contributing to improved quality control and manufacturing optimization within the industry.

7.2.1 Block diagram

The block diagram for artificial intelligence (AI)-based smart boot space management system is shown in Fig. 7.1.

7.2.2 Block diagram explanation

The block diagram detailing LabVIEW-based dimension calculation for a butterfly valve using image processing techniques provides a structured overview of the entire workflow shown in Fig. 7.2). At the outset the process begins with image acquisition facilitated by a digital camera or imaging system (Mahmud et al., 2020). This step is crucial as it captures the visual representation of the butterfly valve, serving as the foundational data input for subsequent analysis. The following image acquisition the captured image undergoes preprocessing within LabVIEW. This preprocessing stage involves various

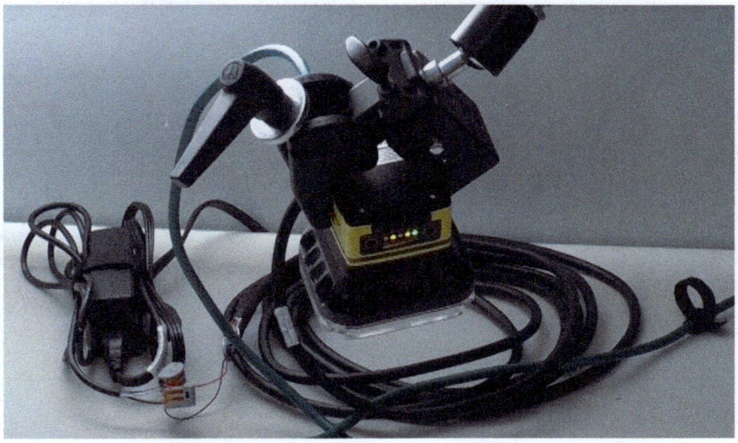

FIGURE 7.3 Flowchart of the proposed work.

techniques, such as noise reduction, contrast enhancement, and normalization, to optimize the image quality and prepare it for effective edge detection and analysis.

Once the preprocessed image is ready, the region of interest (ROI) is selected to isolate the area containing the butterfly valve within the image (Zhang et al., 2020). Defining the ROI allows for focused analysis specifically on the valve's features, minimizing computational load and maximizing accuracy during subsequent processing steps. The identified ROI is then subjected to edge detection algorithms, such as the Canny edge detector, to highlight and extract the edges and contours of the butterfly valve (Shukla & Srivastava, 2016). This critical step enables the identification of key features necessary for calculating dimensions.

After the edges and contours are detected, the dimensional analysis takes place within LabVIEW. Mathematical algorithms are applied to the extracted edge data to compute essential dimensions of the butterfly valve, which may include parameters such as diameter, length, or angles. The calculated dimensions are then visualized on a user interface, providing immediate feedback and enabling users to interpret the measurement results effectively. In addition, the output dimensions can be further processed or integrated into control systems for automation and decision-making in industrial applications (Srinivasan & Shanmugam, 2021).

The integration and control block within the block diagram highlight LabVIEW's role in managing the overall execution of the dimension calculation process. LabVIEW modules orchestrate the image acquisition, preprocessing, edge detection, and dimension calculation tasks seamlessly. This integration extends to hardware interfaces, enabling real-time communication with digital cameras or sensors for efficient image acquisition and data transfer. Feedback loops may also be incorporated to optimize processing parameters based on performance metrics, ensuring continuous improvement and adaptability of the image processing algorithms over time. Ultimately the block diagram exemplifies a systematic and efficient approach to dimension calculation using LabVIEW's image processing capabilities, underscoring its relevance in industrial automation and quality control tasks.

Fig. 7.3 Portable Industrial Machine Vision Camera System

The flowchart shown in Fig. 7.3 outlines the sequential steps involved in calculating the dimensions of a butterfly valve using LabVIEW's image processing capabilities. The process begins with acquiring an image of the butterfly valve using an imaging system or camera setup. This image is then subjected to preprocessing techniques within LabVIEW to enhance its quality by reducing noise, improving contrast, and normalizing the image for analysis. Subsequently the valve ROI is segmented from the preprocessed image, isolating the valve for focused analysis.

Next, LabVIEW's edge detection algorithms, such as Canny edge detection, are applied to the ROI to identify and highlight edges and contours representing the features of the butterfly valve. These detected edges and contours are then analyzed to extract key dimensional parameters, such as diameter, length, or angles of the valve. The calculated dimensions are displayed visually on a display interface or can be outputted as data for further processing or recording purposes.

Overall, this flowchart demonstrates the systematic approach facilitated by LabVIEW for automating the measurement of butterfly valve dimensions from image data. LabVIEW's graphical programming environment allows for the seamless integration of image processing techniques, enabling precise and efficient dimensional analysis suitable for industrial applications requiring accurate quality control and inspection processes.

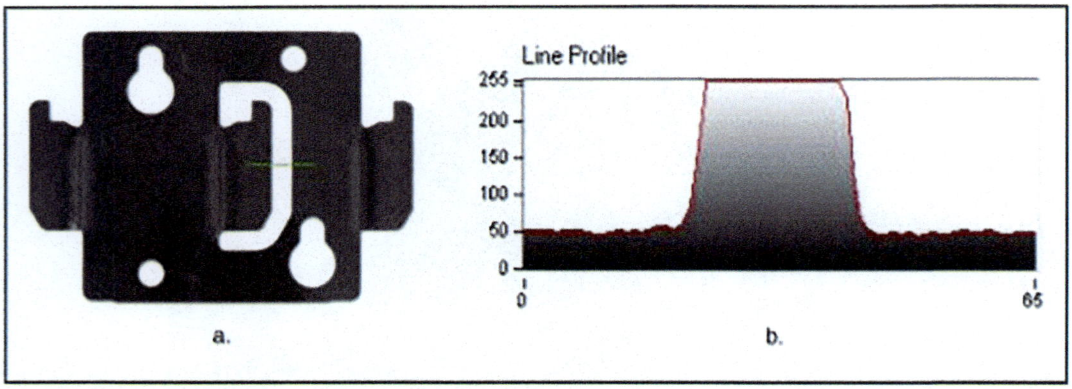

FIGURE 7.4 Left Image : Metal Work Piece Right Image : Line Profile

7.2.3 Hardware setup

Fig. 7.1 shows and displays the hardware setup of the work.

The hardware setup for dimension calculation of a butterfly valve using LabVIEW's image processing tools typically involves components, such as a digital camera or imaging system, a computer running LabVIEW software, and possibly additional peripherals for image acquisition and processing.

First a digital camera capable of capturing high-resolution images is essential for acquiring clear and detailed photos of the butterfly valve (Ghorai et al., 2013). The camera may be mounted in a fixed position or integrated into a setup that allows for adjustable positioning to capture different views of the valve.

Next the camera is connected to a computer running LabVIEW software National Instruments, n.d.f. This computer serves as the processing unit where image acquisition, preprocessing, edge detection, and dimensional analysis algorithms are executed. The computer should meet the system requirements for running LabVIEW smoothly and handling image processing tasks efficiently.

Additional peripherals, such as lighting equipment, may be used to ensure optimal illumination of the valve during image capture, enhancing the quality of the acquired images and aiding in accurate edge detection.

Overall the hardware setup for this application aims to provide a robust and reliable platform for image-based dimension calculation using LabVIEW (National Instruments, n.d.a). The integration of suitable hardware components with LabVIEW's image processing capabilities enables precise measurement and analysis of the butterfly valve's dimensions, contributing to quality control and inspection processes in industrial settings.

7.2.4 LABVIEW interface for image processing

Image processing is a methodology to perform some operations on an image so as to get enhanced image or to extract some helpful information from it. It's a sort of signal process during which input is an image, and output could also be image or characteristics/features related to that image.

Image analysis combines techniques that compute statistics and measurements based on the gray-level intensities of the image pixels. You can use the image analysis functions to determine whether the image quality is good enough for your inspection task. You can also analyze an image to understand its content and to decide which type of inspection tools to use to handle your application. Image analysis functions also provide measurements you can use to perform basic inspection tasks such as presence or absence verification.

A line profile plots the variations of intensity along a line shown in Fig. 7.4. It returns the grayscale values of the pixels along a line and graphs it. Line profiles are helpful for examining boundaries between components, quantifying the magnitude of intensity variations, and detecting the presence of repetitive patterns.

Fig. 7.5 shows how a bright object with uniform intensity appears in the profile as a plateau. The higher the contrast between an object and its surrounding background, the steeper the slopes of the plateau. Noisy pixels, on the other hand, produce a series of narrow peaks.

Use gauging to make critical dimensional measurements, such as lengths, distances, diameters, angles, and counts, to determine if the product under inspection is manufactured correctly. The component or part is either classified or rejected, depending on whether the gauged parameters fall inside or outside of the user-defined tolerance limits.

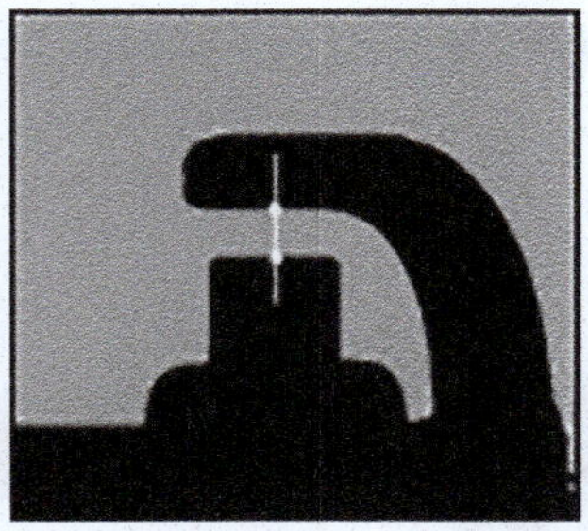

FIGURE 7.5 Using edge detection to measure the gap between disk and valve body.

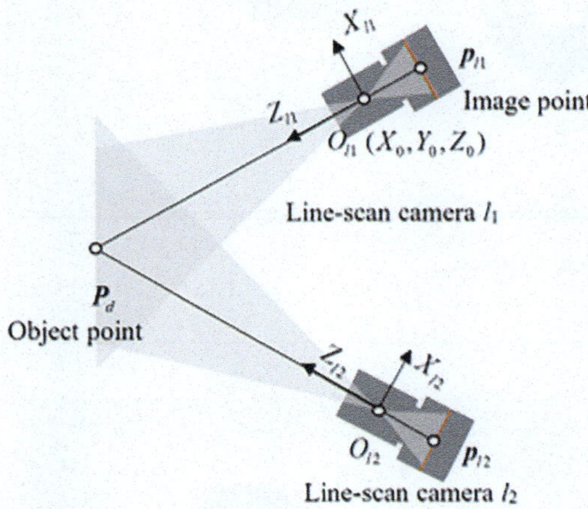

FIGURE 7.6 Industrial camera positioning.

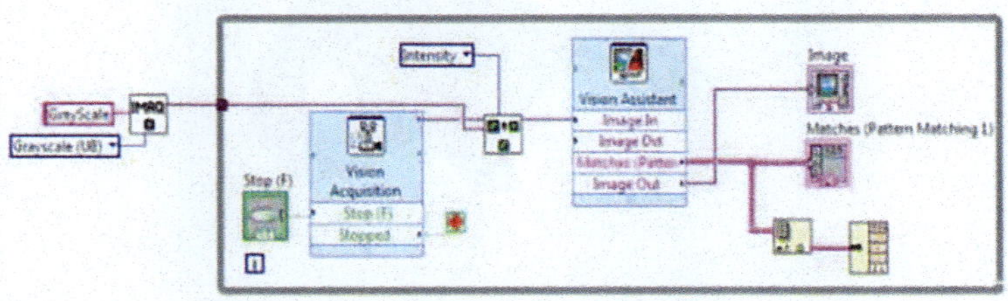

FIGURE 7.7 Block diagram of pattern recognition system.

Here the model or the variant of the camera's position is fixed to capture more area of Butterfly valve is shown in Fig. 7.6.

Positioning an industrial camera correctly is crucial for achieving optimal performance in machine vision and inspection applications. The positioning considerations will vary depending on the specific requirements of the application and the characteristics of the camera being used. Here are some key factors to consider when positioning an industrial camera (National Instruments, n.d.c).

Creating a pattern recognition system in LabVIEW involves leveraging its image processing and analysis capabilities to identify and classify patterns or objects within digital images shown in Fig. 7.7.

Creating an ROI tool constructor in LabVIEW involves developing a user friendly interface that enables interactive selection and manipulation of image regions for further processing or analysis shown in Fig. 7.8. This work typically starts by designing a LabVIEW VI (Virtual Instrument) with a front panel that includes essential UI components, such as images, display areas, and selection tools like rectangles or polygons for defining the ROI shape. The UI should also incorporate interactive controls like buttons to initiate ROI selection, clear the ROI, or confirm the selection.

In LabVIEW the Image Display control serves as a crucial component for visualizing and interacting with images within a user interface shown in Fig. 7.9. This control allows users to view images in various formats and perform operations such as image processing, analysis, or manipulation.

The edge is detected by first selecting the machine vision palette and then edge detector shown in Fig. 7.10. In the edge

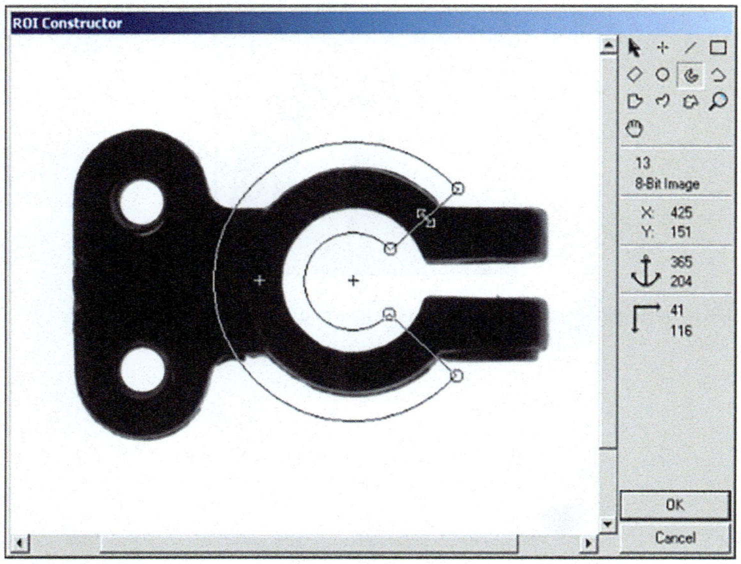

FIGURE 7.8 ROI tool constructor. *ROI,* Region of interest.

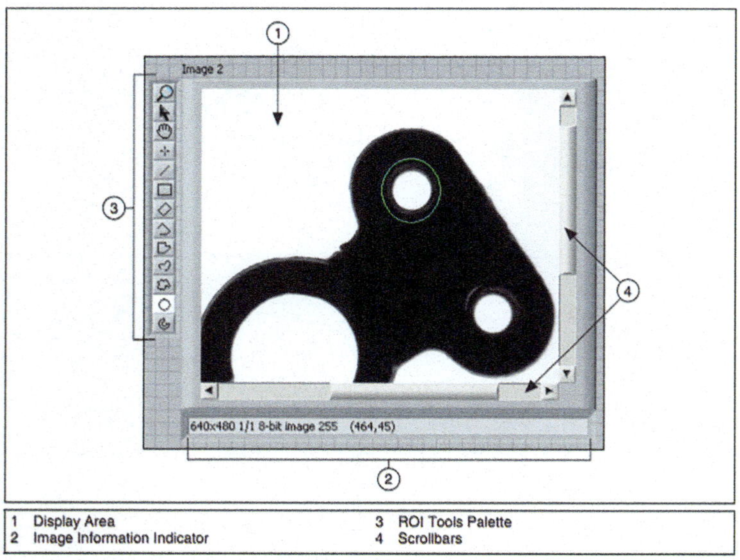

FIGURE 7.9 LabVIEW image display control input.

detector, advanced edge tool is selected, which is effective on images with poor contrast between the background and the objects. The first and last edges are selected so that IMAQ Vision Builder finds and labels only the first and last edges along the line drawn (National Instruments, n.d.d). Contrast is set to 40. The detection process only returns the first and last edge whose contrast is greater than 40. A vertical line is drawn across the surface of the pinto to find the edges that can be used to calculate the edges. IMAQ Vision Builder labels the Edges 1 and 2. The sharp transitions in the line profile indicate edges. The number of edges found is displayed under the line profile. Edge detection step is added to the script. Finally the edge detector window is closed.

In conclusion, utilizing LabVIEW for calculating the dimensions of a butterfly valve demonstrates the power and versatility of this software platform in industrial measurement and automation. The process involves leveraging LabVIEW's image processing and machine vision capabilities to accurately analyze images of the valve and extract key dimensional information (National Instruments, n.d.e).

Overall the use of LabVIEW in calculating the dimensions of a butterfly valve exemplifies how this platform empowers engineers and researchers to develop sophisticated measurement and automation solutions (National Instruments, n.d.f).

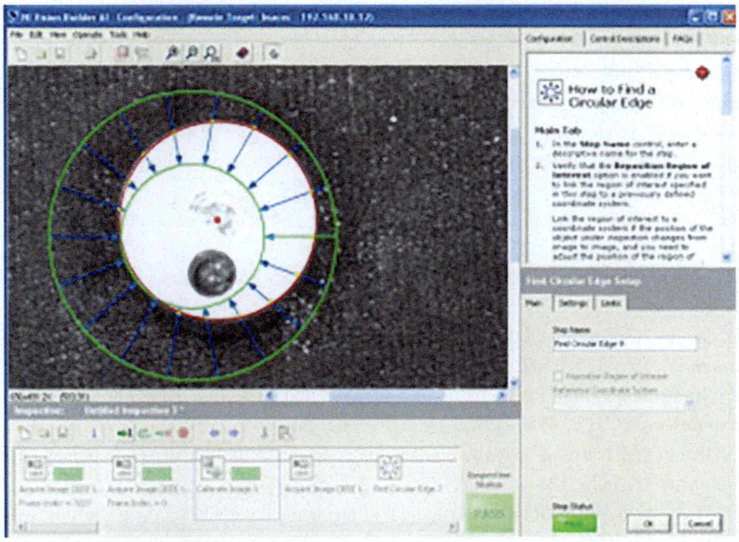

FIGURE 7.10 Circular edge detection for top surface.

By harnessing LabVIEW's capabilities, industries benefit from increased efficiency, accuracy, and reliability in their manufacturing and quality control processes.

7.3 Conclusion

In conclusion, employing LabVIEW's image processing tools for dimension calculation of a butterfly valve offers a precise, efficient, and automated solution for industrial measurement and quality control tasks. By harnessing the software's capabilities in image acquisition, preprocessing, edge detection, and dimensional analysis, manufacturers can achieve accurate assessments of valve dimensions, ensuring compliance with specifications and quality standards. This integration of LabVIEW into industrial processes enhances productivity, reduces manual intervention, and facilitates reliable decision-making, ultimately leading to improved product quality and operational efficiency in various industrial sectors.

7.4 Future scope

In the future the application of LabVIEW's image processing tools for dimension calculation of butterfly valves holds promising potential for further advancements and applications in industrial settings. This includes extending automation to other manufacturing processes, such as quality control inspections, defect detection, and assembly line optimizations, which can significantly enhance operational efficiency and productivity. Integration with IoT technologies and Industry 4.0 concepts will enable real-time data analytics, predictive maintenance, and remote monitoring, empowering manufacturers with proactive decision-making capabilities. Moreover, LabVIEW's versatility can expand into complex component analysis, AR/VR applications for training and maintenance, and advanced machine learning for autonomous manufacturing systems. Collaboration within the LabVIEW community will continue to drive innovation, leading to transformative developments in industrial automation and engineering practices.

References

Gao, Z., & Yang, X. (2021). Real-time visual inspection of mechanical parts using high-precision machine vision systems. *Measurement, 145*, 735–746.

Ghorai, S., Mukherjee, A., & Gangadhar, A. (2013). Automatic defect detection using morphological image processing techniques. *ISA Transactions, 52*(4), 498–506.

Kim, S., Park, J., & Lee, Y. (2017). Dimensional analysis and flow performance evaluation of industrial butterfly valves. *Journal of Mechanical Science and Technology, 31*, 1787–1795.

Liu, Y., Zhang, H., & Li, X. (2020). Automated visual inspection of industrial valve components using deep learning and machine vision. *IEEE Access, 8*, 112450–112460.

Mahmud, M. S., Buyamin, S., & Rahman, M. M. (2020). Integrated machine vision for measurement and inspection in automated manufacturing. *Sensors, 20*(09), 2504.

National Instruments. (n.d.a). *Image acquisition toolkit for LabVIEW*. https://www.ni.com/en-us/shop/labview-modules/image-acquisition.html.

National Instruments. (n.d.c). *LabVIEW example finder: Image processing*. https://www.ni.com/en-us/support/search.html?q=image+processing&sort=relevancy&layout=table&tab=example_programs.

National Instruments. (n.d.d). *LabVIEW example finder: Machine vision*. https://www.ni.com/en-us/support/search.html?q=machine+vision&sort=relevancy&layout=table&tab=example_programs.

National Instruments. (n.d.e). *LabVIEW image processing tutorial*. https://www.ni.com/en-us/support/documentation/supplemental/08/image-processing-tutorial.html.

National Instruments. (n.d.f). *LabVIEW machine vision tutorial*. https://www.ni.com/en-us/support/documentation/supplemental/08/machine-vision-tutorial.html.

National Instruments. (n.d.f). *Image processing toolkit for LabVIEW*. National Instruments. https://www.ni.com/en-us/shop/labview-modules/image-processing.html.

Rao, V., & Narasimhan, S. (2018). Quality inspection and defect detection in metal valve bodies using industrial image analysis. *International Journal of Advanced Manufacturing Technology, 97*, 3135–3147.

Shukla, A., & Srivastava, P. (2016). Implementation of image-based measurement and inspection using LabVIEW and NI Vision Assistant. *International Journal of Applied Engineering Research, 11*(7), 4950–4955.

Srinivasan, S., & Shanmugam, R. (2021). Deep learning-based automated defect detection for industrial components using convolutional neural networks. *Journal of Intelligent Manufacturing, 32*, 1823–1834.

Zhang, Z., Tian, Y., & Wei, Q. (2020). Accurate dimensional measurement of mechanical parts using deep learning–assisted vision systems. *Robotics and Computer-Integrated Manufacturing, 68*, 102081.

Chapter 8

Adaptive neural network–based resilient control for telerobotic systems with input saturation and time-varying delays

Vijay Kumar Pediredla[a], Srikar Annamraju[b] and Asokan Thondiyath[c]

[a] *Senior Data Scientist, Data Science Team, Haber Technologies Inc., Vancouver, BC, Canada,* [b] *Assistant Professor, School of Interdisciplinary Design and Innovation (SIDI), Indian Institute of Information Technology, Design and Manufacturing (IIITDM), Kancheepuram, Tamil Nadu, India,* [c] *Professor, Department of Engineering Design, Indian Institute of Technology, Madras, Chennai, Tamil Nadu, India*

8.1 Introduction

Bilateral teleoperation systems are developed to manipulate and explore various remote/virtual environments. The wide variety of applications of these systems spread over the health sector, nuclear material handling, military, underwater/space exploration, and industrial automation sectors. The system comprises master and slave robots along with a communication channel to exchange the position and force information. The operator commands position signals to the slave robot and concurrently receives environmental forces as force/haptic feedback through the master robot, as shown in Fig. 8.1.

In addition to stability the transparent haptic feedback enhances the feel of realizing the remote environment, resulting in improving the ability to perform complex tasks effectively. Stability and transparency are the essential performance metrics for teleoperation systems, and these are affected by multiple factors, of which time delay is most significant. The time delay issue is addressed by implementing various control architectures such as passivity-based control, scattering, and wave-variable approaches (Lee & Spong, 2006; Niemeyer & Slotine, 1991). A four-channel-based architecture (Lawrence, 1993) has been developed for improving transparency; nevertheless, the method cannot ensure robust stability. Extending these concepts, several methods such as P-D control schemes (Hashemzadeh & Tavakoli, 2015) and the time-domain passivity approach (Hannaford & Ryu, 2002) are proposed. However, these approaches fail to guarantee stability in the presence of time-varying delays and impose limitations such as position drift and wave reflection into the system. Moreover, the other inherent factors, such as model uncertainties, disturbances, nonlinearities, actuator failures, and environment/operator parameter variations, highly influence the system's performance. A detailed survey and analysis of various control methods for teleoperation systems are provided in Hokayem and Spong (2006) and Nuño et al. (2011).

Many control schemes have been proposed to address the issues mentioned above, such as robust control, sliding-mode control, and adaptive control strategies that address parametric variations, modeling uncertainties, and nonlinearities. For instance, in Sarras et al. (2014) and Li et al. (2018), adaptive controllers for nonlinear teleoperation systems with time-varying delays and kinematic and dynamic uncertainties are proposed. However, these control laws do not consider the frictional effects and disturbances. Moreover, robust controllers (Liu et al., 2020; Martínez et al., 2016) are developed to address parametric variations, nonlinearities, and disturbances. However, the parameter/disturbance bounds are predetermined in the controller. Furthermore, the chattering phenomenon instigated by the control affects the system's performance.

The aforementioned control algorithms may guarantee stability and synchronization of positions/velocities of the system; however, they have a limitation of assuming the dynamics (that include robots, environment, and operator) to be linearly parameterizable (Liu et al., 2014; Pediredla et al., 2021; Sarras et al., 2014). In addition, these strategies

Computational Intelligence in Mechatronics: Solving Real-World Problems in Electronic Systems Design. DOI: https://doi.org/10.1016/B978-0-443-27580-7.00005-6

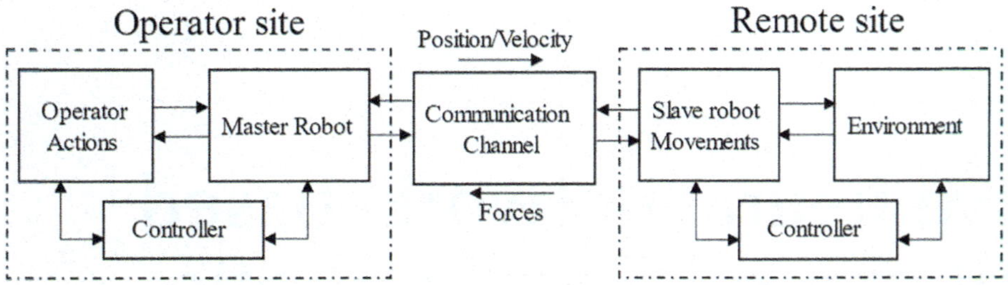

FIGURE 8.1 Block diagram of the bilateral teleoperation system.

fail to account for uncertain disturbances and unmodeled dynamics. So, to enhance the system's performance under unknown dynamics and disturbances, researchers proposed fuzzy logic and neural network (NN)-based control approaches (Sun et al., 2018; Wai, 2003). These approaches have shown superior performance and reliability because of their ability to accurately approximate complex systems. The NN-based adaptive controllers are proposed in Hua et al. (2013) and Yang et al. (2015); however, the systems' kinematic uncertainties are neglected. Although some adaptive NN control laws deal with kinematic/dynamic uncertainties along with unmodeled dynamics (Li & Xia, 2014) and disturbances (Li et al., 2017), there are restrictions on time delays (i.e., only valid for constant time delays or when delay rate is less than or equal to one) and require precise knowledge of the external forces. The NN controller design using radial basis function schemes by Huang et al. (2019) and Chen et al. (2020) has great potential for improving system performance. However, the assumptions on time delays, parameter bounds, and not considering effects of actuator saturation cause instabilities at unstructured environments and thus limit their application in practical scenarios. Moreover, these studies can only attain steady-state rather than transient performance, eventually affecting the synchronization (Teng et al., 2017). The transient performance depends on parameter convergence that demands satisfying persistent excitation (PE) condition, but fulfilling the PE condition requires high control effort and future parameters. Although some adaptive-based robust (Cao & Hovakimyan, 2008), fault-tolerant (Wang et al., 2020), and composite (Li et al., 2020) controllers are proposed, they provide parameter convergence only in an infinite time window. Moreover, the chattering phenomenon still affects them because of either unmodeled dynamics or nonlinearities, thus affecting the system's resilience.

Despite many controllers proposed in the literature, challenges still exist in the teleoperation systems that demand research attention. To the best of the authors' knowledge, there has been no attempt made in the past to develop an NN-based adaptive resilient control that considers the above-mentioned issues simultaneously.

Motivated by these discussions, the chapter aims to develop a novel adaptive resilient NN control algorithm to deal with time-varying delays for telerobotic systems and achieve unified asymptotic stability and transparency. The critical factors addressed in this proposed unified framework are time-varying delays, model/parametric uncertainties, actuator saturation, and disturbances to replicate the practical scenarios. The following are the major contributions of the proposed work over the existing literature:

1. An adaptive terminal sliding mode control strategy using radial basis function NNs is developed to cope with the uncertainties, disturbances, and nonlinearities. Especially, a saturation function is included in the controller to address the concerns related to actuator saturation.
2. Adaptive updation strategy is utilized to accommodate the system's parametric variations. In addition, the transient performance is enhanced by satisfying the improved PE condition, which is practically realizable.
3. The environment is modeled using NNs, and these parameters (nonpower signals) are sent to the master to ensure swift parametric convergence and maintain stability amidst time-varying delays and unstructured environments.
4. Adaptive and resilient attributes of the controller are validated by conducting various experiments on a commercially available haptic system.

Notations: The notations used in this chapter are as follows: R^n, the n-dimensional Euclidean space, I, the identity matrix, and $\lambda_{\max}(K)$, and $\lambda_{\min}(K)$, denotes the largest and smallest Eigenvalues of matrix K, respectively. For any function f: $[0, \infty) \rightarrow R^n$, L_∞, and L_2, norms are represented by $|f|_\infty = \sup_{t\geq 0}|f(t)|$, and $|f|_2 = \left(\int_0^\infty |f(t)|^2 dt\right)^{\frac{1}{2}}$, and its spaces are given by the sets $\{f: [0, \infty) \rightarrow R^n, |f|_\infty < \infty\}$, and $\{f: [0, \infty) \rightarrow R^n, |f|_2 < \infty\}$, respectively.

8.2 Preliminaries and problem formulation

8.2.1 System description

Considering the master and slave manipulators to be rigid, the nonlinear dynamical model of a teleoperated robotic system is represented as

$$M_m(q_m)\ddot{q}_m + C_m(q_m, \dot{q}_m)\dot{q}_m + G_m(q_m) + \mu_m(q_m, \dot{q}_m) = \tau_m + \tau_h + \Delta_m;\ M_s(q_s)\ddot{q}_s + C_s(q_s, \dot{q}_s)\dot{q}_s + G_s(q_s) + \mu_s(q_s, \dot{q}_s) = \tau_s - \tau_e + \Delta_s \tag{8.1}$$

where $q_i, \dot{q}_i, \ddot{q}_i \in R^{n\times 1}$, the generalized joint positions, velocities, and accelerations, $M_i(q_i) \in R^{n\times n}$, the inertia matrix, $C_i(q_i, \dot{q}_i) \in R^{n\times n}$, represent the Coriolis and centrifugal matrix, $G_i(q_i) \in R^n$, the gravity force vector, $\mu_i(q_i, \dot{q}_i) \in R^{n\times 1}$, denotes viscous and Coulomb friction forces, $\Delta_i \in R^{n\times 1}$, indicates modeling errors, unmodeled dynamics, and external disturbances, $\tau_i \in R^{n\times 1}$, are the control input torques, and $\tau_j = J_i^T(q_i)F_j \in R^{n\times 1}$, denote the torques applied by external human and environment, respectively, and $i = m,\ s$, denotes master and slave robots, $j = h,\ e$, indicates human operator and environment, respectively.

The nonlinear dynamics of master and slave manipulators satisfy the following stated properties:

Property 1: The inertia matrix is a symmetric, positive definite, and bounded function such that, $g_{i1}I \le M_i(q_i) \le g_{i2}I$, where g_{i1}, and g_{i2}, are the minimum and maximum eigenvalues of M_i.
Property 2: $\dot{M}_i(q_i) - 2C_i(q_i, \dot{q}_i)$, a skew-symmetric matrix.
Property 3: The Coriolis, frictional, and gravity forces are bounded, and there exist positive constants, k_i,k_{f1i},k_{f2i},and k_{Gi}, such that, $|C_i(q_i, x)y| \le k_{ci}|x||y|$, $|\mu_i(q_i, \dot{q}_i)| \le k_{f1i} + k_{f2i}|\dot{q}_i|$, $|G_i(q_i)| \le k_{Gi}$ where $x, y \in R^n$.

Moreover, the forward (master to slave) and backward (slave to master) time-varying delays considered in this study have no constraints except for known limits on the boundaries.

Assumption 1. The time delay and its rate, $T_i(t)$, and $\dot{T}_i(t)$, respectively, are bounded, $0 \le T_i(t) \le h_i$, $0 \le \dot{T}_i(t) \le a_i$.

In general the actuator is restricted by its output torque, which affects the systems' performance. So, in this work, actuator saturation property is considered for the torque inputs ($\tau_{ij} = [\tau_{i1},\tau_{i2},\ldots, \tau_{in}]^T$) of the system and is represented as

$$\tau_{ij} = \begin{cases} \bar{\tau}_{ij}, & |\bar{\tau}_{ij}| \le \xi_i \\ \xi_i sgn(\bar{\tau}_{ij}), & |\bar{\tau}_{ij}| > \xi_i \end{cases},\quad j = 1,\ 2,\ \ldots,\ n \tag{8.2}$$

where $\bar{\tau}_{ij} = [\bar{\tau}_{i1}, \bar{\tau}_{i2}, \ldots, \bar{\tau}_{in}]^T$, the control input without actuator saturation, $i = m,\ s$, and j, the number of actuators. ξ_i, the upper bound of the joint torques.

The errors in position, velocity, and force based on the dynamics described in Eq. (8.1) are defined as follows:

$$e_m(t) \triangleq q_m(t) - q_s(t - T_s(t));\ e_s(t) \triangleq q_s(t) - q_m(t - T_m(t)) \tag{8.3}$$

$$\dot{e}_m(t) \triangleq \dot{q}_m(t) - \dot{q}_s(t - T_s(t))(1 - \dot{T}_s(t));\ e_s(t) \triangleq \dot{q}_s(t) - \dot{q}_m(t - T_m(t))(1 - \dot{T}_m(t)) \tag{8.4}$$

$$e_{\tau m} \triangleq \tau_h(t) + \tau_e(t - T_s(t));\ e_{\tau s} \triangleq \tau_e(t) + \tau_h(t - T_m(t)) \tag{8.5}$$

The synchronization variables are defined as follows:

$$\eta_m(t) \triangleq \dot{q}_m(t) + \lambda_m e_m(t);\ \eta_s \triangleq \dot{q}_s(t) + \lambda_s e_s(t) \tag{8.6}$$

where λ_m and λ_s are the positive definite matrices.

By substituting Eq. (8.6) in Eq. (8.1),

$$M_m\dot{\eta}_m + C_m\eta_m = \tau_m + \tau_h + \Delta_m - \mu_m + f_m(X_m);\ M_s\dot{\eta}_s + C_s\eta_s = \tau_s - \tau_e + \Delta_s - \mu_s + f_s(X_s) \tag{8.7}$$

where $X_m = \left[e_m^T, \dot{q}_m^T, q_m^T\right]$, $X_s = \left[e_s^T, \dot{q}_s^T, q_s^T\right]$. The nonlinear functions $f_m(X_m)$ and $f_s(X_s)$ represent the unknown dynamic parameters of the system and are given by

$$f_m(X_m) = M_m\lambda_m\dot{e}_m + C_m\lambda_m e_m - G_m;\ f_s(X_s) = M_s\lambda_s\dot{e}_s + C_s\lambda_s e_s - G_s \tag{8.8}$$

8.2.2 Neural networks using radial basis function

The NNs using the radial basis function (RBF) are popularly known for distinctly approximating the nonlinear continuous unknown functions. As the systems' dynamic regressor matrix is not known precisely, the nonlinear functions in Eq. (8.8) are approximated by Radial Basis Function Neural Networks (RBFNN). The Gaussian function for RBFNN is taken as

$$\varphi_i^p(X_i) = \exp\left(\frac{-\left|X_i - C_i^p\right|^2}{2\left(b_i^p\right)^2}\right),\ p = 1, 2, \ldots, n_i \tag{8.9}$$

where X_i is the input vector, $n_i > 1$ is the number of neural nodes, C_i^p is the central point of (ith) Gaussian function with a width of (b_i^p) of (p^{th}) neuron, and $i = m, s$ represent the master and slave robots, respectively.

Assuming the bounded estimation error to be $\varepsilon_i(X_i)$, the RBFNN function is defined as follows:

$$f_i(X_i) = W_i^T\varphi_i(X_i) + \varepsilon_i(X_i),\ |\varepsilon_i(X_i)| \le \varepsilon_{N_i};\ \left|W_i^T\right| \le \bar{W}_i \tag{8.10}$$

where $W_i \in R^{n_i \times n}$ is the weight matrix with nodes n_i.

Based on the property of the NN approximation, the estimate of the functions in Eq. (8.10) is given by

$$\hat{f}_i(X_i) = \hat{W}_i^T\varphi_i(X_i) \tag{8.11}$$

Moreover, by using Eqs. (8.10) and (8.11), the estimation errors are

$$\tilde{f}_i(X_i) = f_i(X_i) - \hat{f}_i(X_i) = \tilde{W}_i^T\varphi_i(X_i) + \varepsilon_i(X_i) \tag{8.12}$$

The environmental dynamics are also modeled using the RBFNN function and are given by

$$\tau_e = W_e^T\varphi_e(X_e) \tag{8.13}$$

where W_e, the unknown weight parameters with $W_e = 0$, representing the free motion of the slave manipulator, and φ_e is the RBFNN function of the environment. This RBFNN-based model accommodates a wide variety of environments.

8.3 Control algorithm development

In this section a novel adaptive resilient-based NN control algorithm using radial basis function is proposed. The developed control strategy aims to achieve stability and transparency simultaneously under nonlinearities, uncertainties, and unmodeled dynamics by utilizing NNs to approximate unknown nonlinear functions and estimate the parameters through adaptive strategies. In addition, the control law is also resilient toward disturbances and parametric variations by utilizing a terminal mode sliding control algorithm. Therefore the proposed controller addresses all these issues, resulting in improving the fidelity of haptic feedback and enhancing robustness.

Fig. 8.2 shows the schematic of the proposed controller. The master and slave robots' torques are varied by the controllers, which are updated by adaptive strategy, NN model variations, estimated environmental torques', and robot's motion parameters. The computed control input is fed to master and slave robots to experience environmental forces and desired motion, respectively, as shown in Fig. 8.2.

The following conditions have to be satisfied to achieve the control objectives:

Position tracking: $\lim\limits_{t\to\infty} e_m(t) = \lim\limits_{t\to\infty} e_s(t) = 0$,
Velocity tracking: $\lim\limits_{t\to\infty} \dot{e}_m(t) = \lim\limits_{t\to\infty} \dot{e}_s(t) = 0$, and
Force tracking: $\lim\limits_{t\to\infty} e_{\tau m}(t) = \lim\limits_{t\to\infty} e_{\tau s}(t) = 0$.

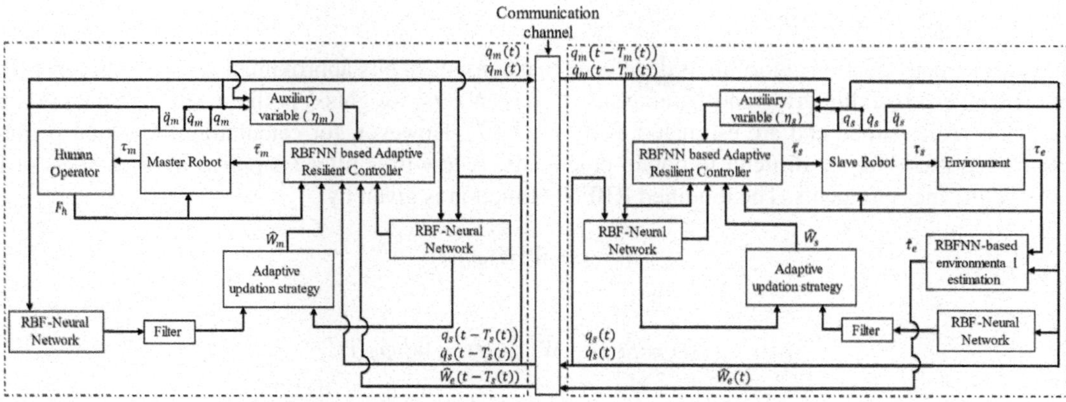

FIGURE 8.2 Schematic representation of the proposed control algorithm.

8.3.1 Controller design

The proposed adaptive resilient-based NN control schemes using radial basis function for master and slave manipulators are as follows:

$$\text{Master controller}: \ \bar{\tau}_m = -\hat{f}_m(X_m) - K_m\eta_m - \rho_m|\eta_m|^{\alpha_0}\hat{\zeta}_m\mathrm{sgn}(\eta_m) - K_{\Delta m}\left(e_{\tau m}e_{\tau m}^T\left(\rho_m|\eta_m|^{\alpha_0}\mathrm{sgn}(\eta_m)+\varepsilon\right)+1\right) - \tau_h$$

$$\text{Slave controller}: \ \bar{\tau}_s = -\hat{f}_s(X_s) - K_s\eta_s - \rho_s|\eta_s|^{\alpha_0}\hat{\zeta}_s\mathrm{sgn}(\eta_s) - K_{\Delta s}\left(e_{\tau s}e_{\tau s}^T\left(\rho_s|\eta_s|^{\alpha_0}\mathrm{sgn}(\eta_s)+\varepsilon\right)+1\right) + \tau_e \tag{8.14}$$

where $\rho_i = \frac{D_i}{\delta_i+(1-\delta_i)e^{-\alpha_i|\eta_i|^{\alpha_0}}}$; $0 < \delta_i, \alpha_0 < 1$, K_i, and $K_{\Delta i}$ are the positive symmetric matrices of size $n \times n$, $\mathrm{sgn}(\eta_i)$, the sign function, α_i, and D_i, are positive constants, ε, a small positive number, and $\hat{\zeta}_i$, the upper bound estimation of ζ_i, the sum of external disturbances, frictional effects, and RBFNN estimation errors, given by

$$|\varepsilon_i(X_i) + \Delta_i - \mu_i| \le \zeta_i \tag{8.15}$$

The developed control algorithm deals with multiple concerns of telerobotic systems, as discussed earlier. Thus each term in the controller plays a pivotal role in improving the resilience of the system against various issues. The first term of the controller approximates the unknown nonlinear functions of the system and avoids the need for precise knowledge of dynamics. Moreover, the weight parameters are updated based on the variations in the system, thus addressing uncertainties and parametric variations issues. The second term ensures position and velocity tracking. The third and fourth terms provide resiliency to the system. It utilizes a terminal mode sliding control approach to compensate for time-varying delays, disturbances, actuator saturation, and approximation errors and ensures force tracking. Initially the term ρ_i, is equal to D_i/δ_i, which indicates the requirement of significant control effort to overcome the disturbances and uncertainties. However, a significant control effort is undesirable and leads to chattering effects. So, to reduce the chattering phenomenon, the term ρ_i, becomes D_i, as η_i, is decreasing, resulting in reducing the control effort when approaching the sliding surface. In addition, $|\eta_m|^{\alpha_0}$, power rate term ensures fast convergence by making the system chattering-free and simultaneously accommodates the actuator limits in the system. These adjustments in control law render fast convergence and concurrently eliminate the system's chattering, thus leading to optimal performance, i.e., achieving robustness and transparency simultaneously.

Substituting the control algorithm, Eq. (8.14) in Eq. (8.7) and using Eq. (8.12), the closed-loop dynamics are given by

$$\begin{aligned} &M_m\dot{\eta}_m + C_m\eta_m + K_{\Delta m}\left(e_{\tau m}e_{\tau m}^T\left(\rho_m|\eta_m|^{\alpha_0}\mathrm{sgn}(\eta_m)+\varepsilon\right)\right) + K_m\eta_m + \rho_m|\eta_m|^{\alpha_0}\hat{\zeta}_m\mathrm{sgn}(\eta_m) + K_{\Delta m} \\ &= \Delta_m - \mu_m - \tilde{f}_m(X_m) + \lambda_m M_m(q_m)\,\dot{T}_s\dot{q}_s(t-T_s);\ M_s\dot{\eta}_s + C_s\eta_s + K_{\Delta s}\left(e_{\tau s}e_{\tau s}^T\left(\rho_s|\eta_s|^{\alpha_0}\mathrm{sgn}(\eta_s)+\varepsilon\right)\right) + K_s\eta_s \\ &+\rho_s|\eta_s|^{\alpha_0}\hat{\zeta}_s\mathrm{sgn}(\eta_s) + K_{\Delta s} = \Delta_s - \mu_s - \tilde{f}_s(X_s) + \lambda_s M_s(q_s)\,\dot{T}_m\dot{q}_m(t-T_m) \end{aligned} \tag{8.16}$$

The convergence rate of error is improved by developing an adaptive updation algorithm that compensates for varying approximation errors and disturbances through the updation of weight parameters.

8.3.2 Adaptive updating strategy

In this section an adaptive updation algorithm is developed that consists of NN approximation errors along with the tracking errors to guarantee zero tracking error and parametric convergence of weights. As the actual parameters are unknown, some initial values are assumed and are estimated using Eq. (8.7). However, for calculating the actual parameters, joint accelerations are required that are difficult to obtain practically. A low-pass filter is applied on both sides of Eq. (8.7) to filter the noises in the measurements. The modified RBFNN function is given by

$$f_{if}(t) = W_i^T(t)\varphi_{if}(t) \tag{8.17}$$

$$e_{if}(t) = f_{if}(t) - \hat{W}_i^T\varphi_{if}(t) = \tilde{W}_i^T\varphi_{if}(t) \tag{8.18}$$

where $\varphi_{if}(t) = ke^{-kt}\int_0^t e^{k\beta}\varphi_i(\beta)d\beta, f_{if}(t)$, the filtered functions with a scalar constant $k > 0$, and $e_{if}(t)$, the filtered estimate error.

For guaranteed parametric convergence of matrix $\varphi_i(X_i)$, the PE condition ($\int_t^{t+T} \varphi_{if}^T(\beta)\varphi_{if}(\beta)d\beta \geq \alpha I$, where $T, \alpha > 0$), has to be satisfied. However, this condition requires the future values of the RBFNN function, which is hard to obtain. Thus, inspired from Roy et al. (2016), an improved PE condition that is practically feasible is developed and given by

$$H_{if}(t) = \int_0^T \varphi_{if}^T\varphi_{if}d\beta \geq \alpha I \tag{8.19}$$

$$Z_{if}(t) = H_{if}W_i;\ \text{where}\ \dot{Z}_{if}(t) = Y_{iw}^T\tau_{iw} \tag{8.20}$$

where Z_{if}, H_{if}, the filtered torque and filtered RBFNN function (nondecreasing semidefinite function) are, respectively.

The time period of excitation in the improved PE condition is T(i.e., $t = 0$), which is feasible in real-time applications. This condition is different from the actual PE condition that requires significant control input because of excitation for every time-frame $t \geq 0$. This drawback is addressed in the developed PE condition.

The adaptive updating laws for weight parameters of RBFNN are proposed as

$$\dot{\hat{W}}_i = P_i\left[\varphi_i(X_i)(\eta_i)^T + \varphi_{if}^T e_{if} + k_i H_{if}\tilde{W}_i\right) \tag{8.21}$$

where k_i, the convergence rate gain. P_i, the positive definite learning gain matrices are revised by

$$\dot{P}_i = -P_i\left(\varphi_{if}^T\varphi_{if} + k_i H_{if}^T H_{if}\right)P_i \tag{8.22}$$

$$\dot{P}_i^{-1} \triangleq \varphi_{if}^T\varphi_{if} + k_m H_{if}^T H_{if} \tag{8.23}$$

With the updation of RBFNN weight parameters the control input is changed to adjust to nonlinearities, unmodeled dynamics, and bounded disturbances. More importantly the parametric convergence is achieved swiftly because of the enrichment of transient performance by satisfying PE condition and adaptive updating algorithms.

8.3.3 Environmental estimation

The dynamic environmental model based on RBFNN structure is described in Eq. (8.13), where the weight parameters (W_e) are predicted using RBFNN and are transmitted to the master manipulator through the communication channel. Such online estimation certainly contributes to enhancing the systems' performance, specifically in achieving swift parametric convergence. In addition, the stability is preserved because of the transmission of nonpower signals ($\hat{W}_e$) instead of power signals ($\hat{\tau}_e$). The weight parameters are estimated and updated using $\dot{\hat{W}}_e = k_e P_e \tilde{W}_e$; $\tilde{W}_e = \hat{W}_e - W_e$ where $k_e, P_e > 0$, and $\tilde{W}_e$, the parameter estimation error.

The proposed control algorithm is delineated as

ALGORITHM 8.1 Methodology of the proposed adaptive resilient-based neural network control algorithm. % Get the constant off-line parameters. % Initialize the control gains.

1 Control gain matrices K_i & $K_{\Delta i}$, are chosen
2 Adaptive updation gains k_i & P_i, are also computed based on the required rate of convergence

% Initialize master & slave motion parameters

1 Select $q_i(0)$, such that $\dot{q}_i = \ddot{q}_i = 0$
2 Compute the errors as mentioned in Eqs. (8.3)–(8.5), and subsequently the auxiliary variable η_i
3 Initialize the estimators W_i, and assume $\dot{\hat{W}}_i = 0$

% Get the varying online parameters % At sampling time $\boldsymbol{t} = \boldsymbol{t_k}$

1 By considering the time-varying delays, update position and velocity parameters between master and slave robots $(q_i(t - T_i(t_k))$ & $\dot{q}_i(t - T_i(t_k)))$
2 Compute errors between the master and slave robots as mentioned in Eqs. (8.3)–(8.5)
3 Compute the human force (F_h), and estimate the environmental parameters $\hat{W}_e$
4 Compute the control torque $(\tau_i(t_k))$, as mentioned in Eq. (8.14), substitute in Eq. (8.1), and update the motion parameters
5 RBFNN based adaptive estimates $\dot{\hat{W}}_i(t_{k+1})$, are updated based on regressor matrix $W_i(t_{k+1})$, using Eqs. (8.17)–(8.23)

% At sampling time $\boldsymbol{t} = \boldsymbol{t_{k+1}}$
Reiterate the steps 1–5 starting at the time $t = t_k$

8.3.4 Stability analysis

Theorem 1. Consider the bilateral telerobotic system Eq. (8.1) with the proposed adaptive resilient NN controller Eq. (8.14) and the adaptive update strategies Eqs. (8.21)–(8.23). Suppose that the system satisfies the following inequalities:

$$\pi = \begin{bmatrix} \pi_m & 0 & 0 & -I & 0 & 0 \\ * & \pi_s & -I & 0 & 0 & 0 \\ * & * & \frac{-D_m}{h_m} & 0 & 0 & 0 \\ * & * & * & \frac{-D_s}{h_s} & 0 & 0 \\ * & * & * & * & \frac{-\lambda_m}{2} & 0 \\ * & * & * & * & * & \frac{-\lambda_s}{2} \end{bmatrix} < 0 \quad (8.24)$$

where $\pi_i = -\frac{1}{2\lambda_i} - \frac{b_i}{(1-a_i)} - h_i D_i$.
If the positive matrix $K_{\Delta i}$ is chosen such that

$$\lambda_{\min}(K_{\Delta i}) > |\zeta_i| \quad (8.25)$$

then the position, velocity and force errors, and weight parameter estimation errors asymptotically converge to zero, i.e., $\lim_{t\to\infty} e_i = 0$, $\lim_{t\to\infty} \dot{e}_i = 0$, $\lim_{t\to\infty} e_{\tau i} = 0$, and $\lim_{t\to\infty} \tilde{W}_i = 0$.

Proof: Positive definite Lyapunov functions and their derivatives are defined as follows:

$$V = V_1 + V_2 + V_3 + V_4 + V_5$$

$$V_1(t) = \sum_{i=m,s} \frac{1}{g_{i1}\lambda_i}\left(\frac{1}{2}\eta_i^T M_i \eta_i + \frac{1}{2}\tilde{W}_i^T P_i^{-1} \tilde{W}_i\right) \quad (8.26)$$

The derivative of V_1, along the trajectory of the system Eqs. (8.16)–(8.23) is given by

$$\dot{V}_1(t) = \sum_{i=m,s} \frac{1}{g_{i1}\lambda_i}\left(\eta_i^T M_i \dot{\eta}_i + \frac{1}{2}\eta_i^T \dot{M}_i \eta_i + \tilde{W}_i^T P_i^{-1} \dot{\tilde{W}}_i + \frac{1}{2}\tilde{W}_i^T \dot{P}_i^{-1} \tilde{W}_i\right)$$

From Eqs. (8.16), (8.18), (8.21), (8.23) and property 2,

$$\dot{V}_1(t) \leq \sum_{i=m,s;j=s,m} \frac{1}{g_{i1}\lambda_i}\Big(-\frac{g_{i1}}{2}\eta_i^T\eta_i - \eta_i^T K_{\Delta i}\left(\rho_i|\eta_i|^{\alpha_0}\mathrm{sgn}(\eta_i)+\varepsilon\right)e_{\tau i}e_{\tau i}^T - \rho_i|\eta_i|^{\alpha_0}\hat{\xi}_m\mathrm{sgn}(\eta_i) - \lambda_{\min}(K_{\Delta i})|\eta_i|$$
$$-\left(\frac{1}{2}\varphi_{if}^T\varphi_{if}+\frac{1}{2}k_iH_{if}^TH_{if}\right)\left|\tilde{W}_i\right|^2 + |\zeta_i||\eta_i| + \frac{\lambda_i^2\mu_{i2}{}^2a_i^2}{2g_{i1}}\dot{q}_j(t-T_j)^T\dot{q}_j(t-T_j)\Big) \tag{8.27}$$

From Eq. (8.20), $-\lambda_{\min}(K_{\Delta i})|\eta_i| + |\zeta_i||\eta_i| < 0$
Therefore Eq. (8.23) is reduced to

$$\dot{V}_1(t) \leq -\frac{1}{2\lambda_i}\left(\dot{q}_i^T\dot{q}_i + 2\lambda_i e_i^T\dot{q}_i + \lambda_i^2 e_i e_i^T\right) - \frac{1}{g_{i1}\lambda_i}\left(\frac{1}{2}Y_{iw}^TY_{iw}+\frac{1}{2}k_iW_{if}\right)\left|\tilde{\theta}_i\right|^2$$
$$-\frac{\eta_i^TK_{\Delta i}}{g_{i1}\lambda_i}\left(\rho_i|\eta_i|^{\alpha_0}\mathrm{sgn}(\eta_i)+\varepsilon\right)e_{\tau i}e_{\tau i}^T + b_i\dot{q}_j(t-T_j)^T\dot{q}_j(t-T_j) \tag{8.28}$$

where $b_i = \frac{\lambda_i\mu_{i2}{}^2a_i{}^2}{2g_{i1}{}^2}$

$$V_2(t) = (q_m-q_s)^T(q_m-q_s);\ \dot{V}_2(t) = \left((e_m-L_s)^T\dot{q}_m + (e_s-L_m)^T\dot{q}_s\right) \tag{8.29}$$

where $\chi_m = \int_{t-T_m(t)}^{t} \dot{q}_m(x)dx;\ \chi_s = \int_{t-T_s(t)}^{t} \dot{q}_s(x)dx$

$$V_3(t) = \sum_{i=m,s}\int_{t-T_i(t)}^{t}\frac{b_i}{(1-a_i)}\dot{q}_i^T(x)\dot{q}_i(x)dx + \int_{-h_i}^{0}\int_{t+r}^{t}\dot{q}_i^T(x)D_i\dot{q}_i(x)dxdr;\quad \dot{V}_3(t)$$
$$\leq \sum_{i=m,s;j=s,m}\frac{b_i}{(1-a_i)}\dot{q}_i^T\dot{q}_i - b_i\dot{q}_j(t-T_j)^T\dot{q}_j(t-T_j) + h_i\dot{q}_i^TD_i\dot{q}_i - \frac{1}{h_i}\chi_iD_i\chi_i \tag{8.30}$$

$$V_4(t) = \sum_{i=m,s}\int_0^t\frac{\eta_i^TK_{\Delta i}}{g_{i1}\lambda_i}\left(\left(\rho_i|\eta_i|^{\alpha_0}\mathrm{sgn}(\eta_i)+\varepsilon\right)e_{\tau i}e_{\tau i}^T + \rho_i|\eta_i|^{\alpha_0}\hat{\xi}_m\mathrm{sgn}(\eta_i)\right)dt;\ \dot{V}_4(t)$$
$$= \sum_{i=m,s}\frac{\eta_i^TK_{\Delta i}}{g_{i1}\lambda_i}\left(\left(\rho_i|\eta_i|^{\alpha_0}\mathrm{sgn}(\eta_i)+\varepsilon\right)e_{\tau i}e_{\tau i}^T + \rho_i|\eta_i|^{\alpha_0}\hat{\xi}_m\mathrm{sgn}(\eta_i)\right) \tag{8.31}$$

$$V_5(t) = \frac{1}{2}\tilde{W}_e^TP_e^{-1}\tilde{W}_e;\ \dot{V}_5(t) = -k_e\left|\tilde{W}_e\right|^2 \tag{8.32}$$

$$\dot{V} = \dot{V}_1 + \dot{V}_2 + \dot{V}_3 + \dot{V}_4$$

$$\dot{V}(t) \leq -\sum_{i=m,s}\frac{1}{2g_{i1}\lambda_i}\left(\varphi_{if}^T\varphi_{if}+k_iH_{if}^TH_{if}\right)\left|\tilde{W}_i\right|^2 - k_e\left|\tilde{W}_e\right|^2 + \vartheta\pi\vartheta^T \tag{8.33}$$

Integrating on both sides of Eq. (8.32)

$$V(t)-V(0) = -\sum_{i=m,s}\int_0^t\left(\frac{1}{2g_{i1}\lambda_i}\left(\varphi_{if}^T\varphi_{if}+k_iH_{if}^TH_{if}\right)\left|\tilde{W}_i\right|^2 + k_e\left|\tilde{W}_e\right|^2\right) + \int_0^t\vartheta\pi\vartheta^T \tag{8.34}$$

where $\vartheta = \mathrm{col}\{\dot{q}_m, \dot{q}_s, \chi_m, \chi_s, e_m, e_s\}$, and from Eqs. (8.24) to (8.34), $V \geq 0$, and $\dot{V} \leq 0$. So, $\tilde{W}_i \in L_2 \cap L_\infty$, $\tilde{W}_e \in L_2 \cap L_\infty$, e_i, $\dot{q}_i$, L_i, are bounded. Also, $\dot{e}_i$, η_i, τ_i, $\dot{\tilde{W}}_e$, $\dot{\tilde{W}}_i$, and $\ddot{e}_i \in L_\infty$, by using the bounded conditions presented in Eqs. (8.1)–(8.6) and (8.21)–(8.23).

Using Barbalat's Lemma, $\lim_{t\to\infty} e_i(t) = 0$, $\lim_{t\to\infty} L_i(t) = 0$, $\lim_{t\to\infty} \tilde{W}_i = 0$, $\lim_{t\to\infty} \tilde{W}_e = 0$, and $\lim_{t\to\infty} \dot{e}_i(t) = 0$, which implies zero asymptotic convergence of the position, velocity, and parametric weight uncertainty and environmental estimation errors. The synchronization of position/velocity profiles and convergence of weight parameters to their actual values are achieved. Thus the system is asymptotically stable.

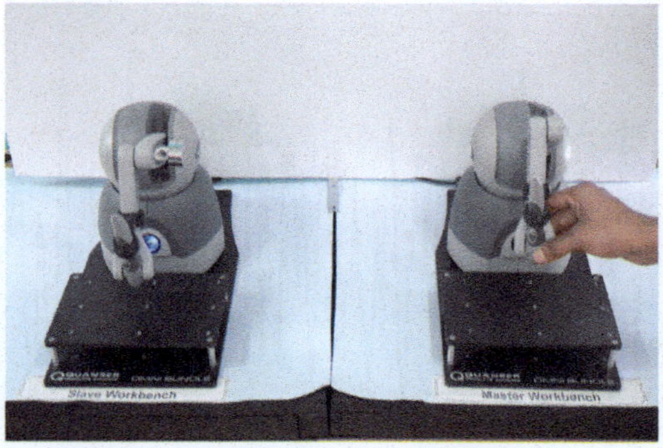

FIGURE 8.3 Experimental setup of the 3-DoF Phantom telerobotic system.

Substituting these outcomes into Eqs. (8.16), (8.28), and (8.34) and using *property 1*, we get,

$$|\varepsilon|_2^2 \mu_{i1}(\tau_h(t) + \tau_e(t - T_s))(\tau_h(t) + \tau_e(t - T_s))^T \leq \varepsilon \ddot{q}_m \tag{8.35}$$

It is shown that the environmental estimation is accurately obtained, and thus the environmental torque at the master side can be recreated. It is clear that the $\varepsilon \ddot{q}_m$, is a nonnegative value, and $\lim_{t\to\infty} \dot{q}_i = 0$, and hence, $\varepsilon \ddot{q}_m$, converges to zero as $t \to \infty$. Using this deduction from Eq. (8.35) the force error in the system accomplishes zero asymptotic convergence. This implies high-fidelity transparency of the system.

Let us consider the Lyapunov function, $V_\theta(t)$as$\frac{1}{2}\tilde{W}_i^T P_i^{-1} \tilde{W}_i$, where $\lambda_{\min}(P_i) \leq V_\theta(t) \leq \lambda_{\max}(P_i)$, and its derivative is given by

$$\dot{V}_\theta \leq -\gamma \left|\tilde{W}_i\right|^2 \text{then } V_\theta(t) \leq -V_\theta(t_0) e^{-\gamma(t - t_0)/\lambda_{\max}(P_i)}$$

where γ, a semipositive definite matrix is given by $\frac{1}{2g_i\lambda_i}\left(\varphi_{if}^T \varphi_{if} + k_i H_{if}^T H_{if}\right)$, and the corresponding function of parametric errors is described as

$$\left|\tilde{W}_i(t)\right| \leq \sqrt{\frac{\lambda_{\min}(P_i)}{\lambda_{\max}(P_i)}} \left|\tilde{W}_i(t - t_0)\right| e^{-\gamma(t - t_0)/2\lambda_{\max}(P_i)} \tag{8.36}$$

Therefore, initially, the parameter errors converge to a sphere $\left(r = {}^{2\lambda_{\max}(P)}/_{\gamma}\right)$, swiftly and eventually converge to zero. In addition, by modulating the convergence gain (k_i), the convergence rate can be improved by reducing the sphere's radius, resulting in fast zero convergence. The fast and accurate parameter estimation contributes to enhancing the transient performance of the system.

8.4 Experimental results

In this section the proposed controllers' performance is evaluated on the experimental setup (two 3-DoF Phantom Touch devices), as shown in Fig. 8.3. The MATLAB/Simulink software is used to establish the connection between master and slave manipulators, estimate the robot dynamics parameters, and compute the control input required using RBFNN to attain the system's stability and transparency.

The goal of the experiments is to maintain stability and high-fidelity transparency (both position and force tracking) of the telerobotic system with time-varying delays. In the experiments the operator provided a random force input to the master robot's end-effector. An external environment (sponge) is provided at the slave side. The time-varying delay in the system is a random signal with a mean value of 370 ms, as shown in Fig. 8.4. The central point C_i and width b_i of the RBFNN-based functions considered for approximating the system are $[-10, 10]$, and 1, respectively. The initial states of the master and slave devices are $q_m = q_s = [00]^T$. To ensure position and force tracking while the slave robot is interacting with the remote environment (sponge), the slave should track the master's position, and simultaneously, the human operator should experience the haptic feedback. The developed adaptive resilient NN controller, Eqs. (8.14), (8.21), and (8.23), is implemented on the system in the presence of disturbances, nonlinearities, and varying uncertainties. The controller gains are $K_i = 5I$, $K_{\Delta i} = 10I$, $\delta_i = 0.3$, $\alpha_0 = 0.45$, $D_i = 0.25$, and the gains of adaptation strategy are $K_i = 2.5I$, $k_i = 4$, chosen based on the conditions, Eqs. (8.24) and (8.25), to ensure robustness to disturbances and frictional forces, adaptiveness to various

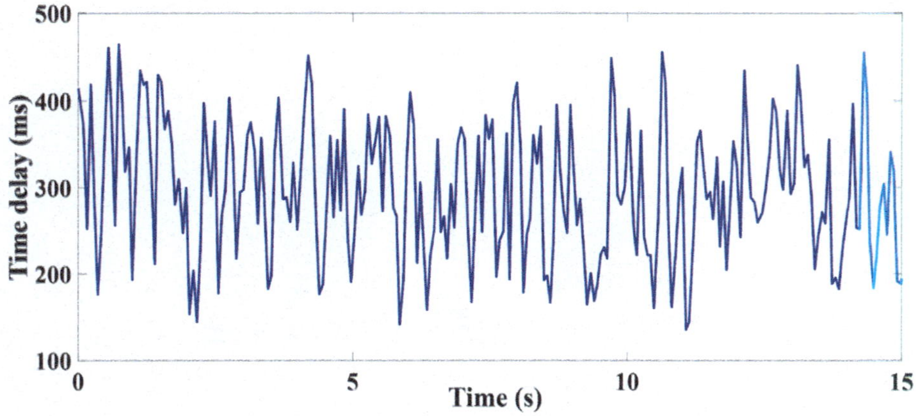

FIGURE 8.4 Varying time delay.

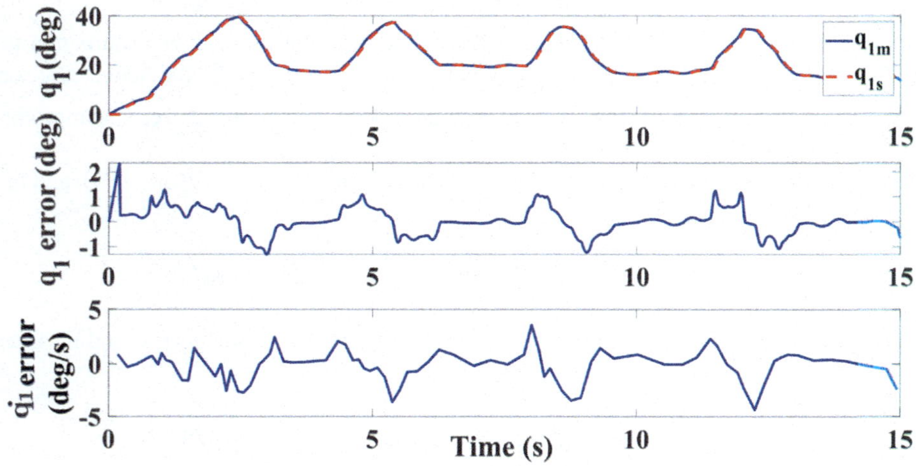

FIGURE 8.5 Position tracking under varying uncertainties.

environment and robot uncertainties, and fast convergence to actual values. The gain values of the torque estimation of the environment are $k_e = 0.05$, $P_e = 4I$. Two different experiments are performed to precisely validate the adaptive and resilient attributes of the controller.

8.4.1 Validating adaptiveness

To evaluate the attribute of adaptiveness of the proposed controller, known masses are sequentially added to link 2 of the slave robot, as shown in Fig. 8.3. The additional masses, each of 10 g, are added four times at $t = 2.5$, 5.4, 8.6, and 12.5 s, and act as a source of disturbances and uncertainties to the system's dynamics, simulating unmodeled dynamics and dynamic uncertainties. Although adding these masses, the joint angle 1 of the master robot is only varied, keeping joint angles 2 and 3 constant. In these forced conditions the slave robot should track the master robot precisely and transmit the environmental torque to the human operator as haptic feedback. To achieve stability and high-fidelity transparency the slave controller should also estimate the masses and accordingly provide the joint torques.

The system attains position/velocity tracking even under various disturbances, nonlinearities, and dynamic uncertainties, as shown in Fig. 8.5 with root mean squared error (RMSE) of the position error as 0.5 deg. The initial mass of the slave manipulator being 70 g, the increase of 10 g every time reduced the perceived torque by the slave robot. Initially, the environmental torque estimated by the slave robot is 1.78 N m, and with each increase of 10 g, the estimated torques are reduced proportionally as 1.54, 1.1, and 0.65 N m, respectively, as shown in Fig. 8.6. The RMSE of the environmental torque $(\tau_e - \hat{\tau}_e)$ is 0.6 mN m. It is also clear from Fig. 8.6 that the torque value (control input) of the robots is within confined limits. Therefore this analysis demonstrates that the controller could identify the amount of dynamic uncertainties and disturbances

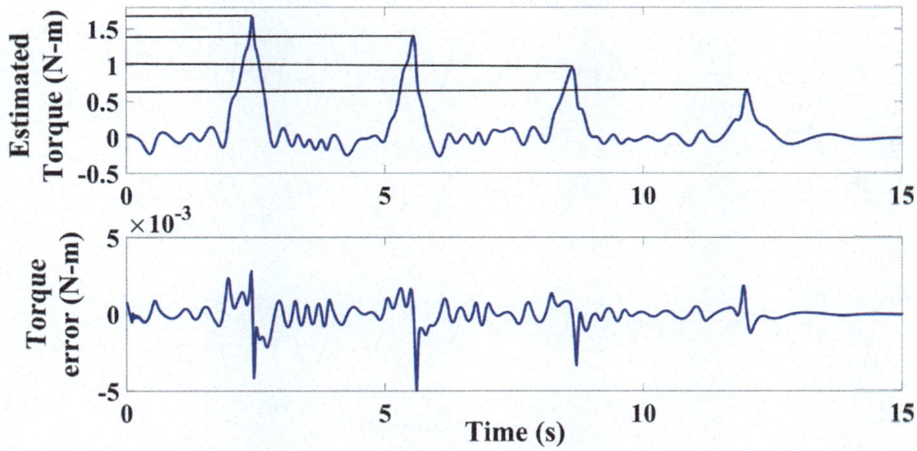

FIGURE 8.6 Environmental torque estimation as experienced by slave robot.

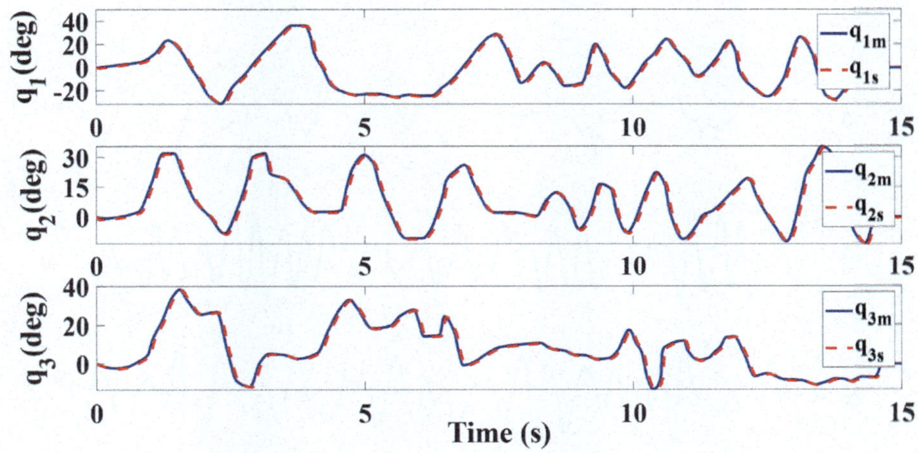

FIGURE 8.7 Position tracking under varying disturbances.

and, consequently, render environmental forces precisely to the operator, thus validating the adaptiveness attribute of the controller.

8.4.2 Validating resiliency

To evaluate the resilience attribute of the proposed controller an external disturbance is induced at the slave side while interacting with the environment. The disturbance is created by adding a mass of 50 g to the slave manipulator at 5 s. In addition, this mass also acts as uncertainty to the slave manipulator. The position tracking and position/velocity tracking errors of master and slave robots are shown in Figs. 8.7–8.9, respectively. The average RMSE of the position and velocity errors of three joints of the telerobotic system is 0.65 deg and 1.23 deg/s, respectively. The environmental dynamic parameters are estimated and sent to the master manipulator to feel the force feedback. The environmental torque estimation and its errors $(\tau_e - \hat{\tau}_e)$ are shown in Figs. 8.10 and 8.11, respectively. The average RMSE of the force errors of the three joints is 1.78 mN m.

It is seen from the figures that the system achieves good position, velocity, and force tracking. Thus stability (position tracking) and transparency (velocity and force tracking) are attained simultaneously amidst disturbances, uncertainties, and nonlinearities. This proves the efficacy of the proposed RBFNN-based adaptive resilient controller.

Based on the above results, it can be concluded that the proposed adaptive resilient NN controller using radial basis functions is effective in rendering high-fidelity haptic feedback to the user. Moreover, the attributes such as adaptiveness and resiliency are substantiated experimentally while concurrently achieving stability and transparency. This analysis proves the proposed algorithm's efficacy against dynamic uncertainties, nonlinearities, time-varying delays, and disturbances.

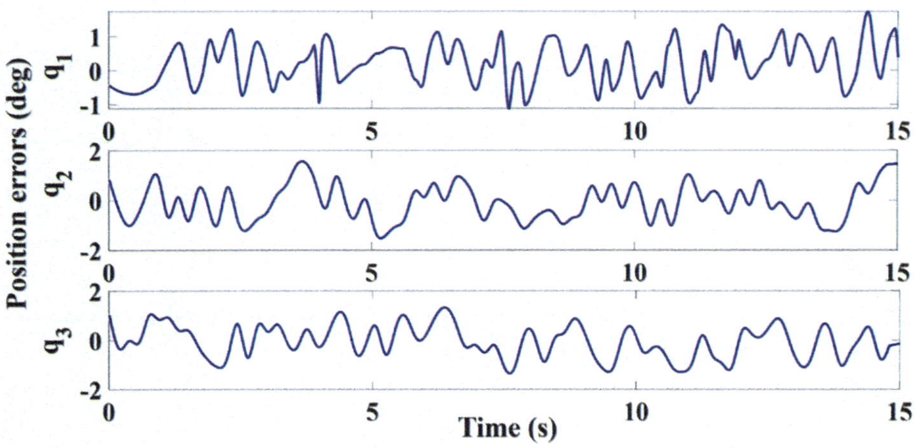

FIGURE 8.8 Position tracking errors.

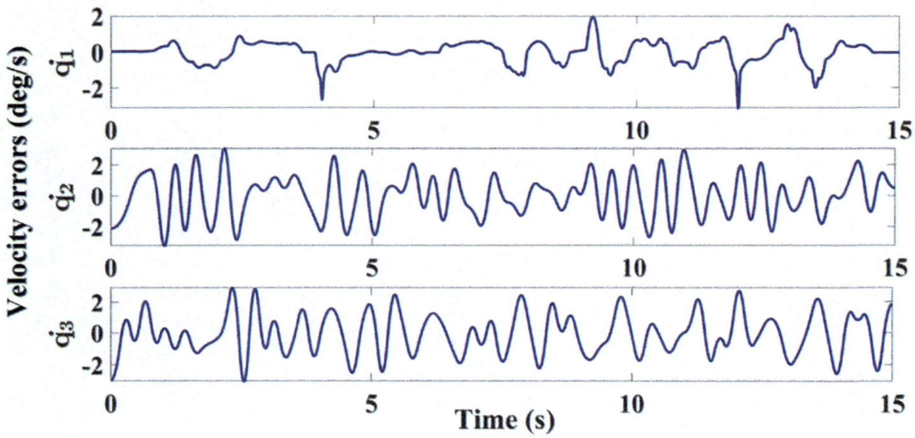

FIGURE 8.9 Velocity tracking errors.

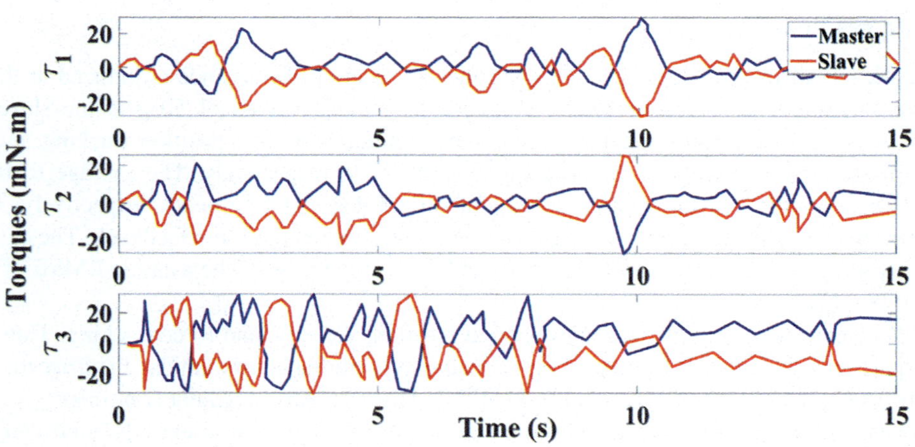

FIGURE 8.10 Environmental torque estimation by the master as sensed by slave.

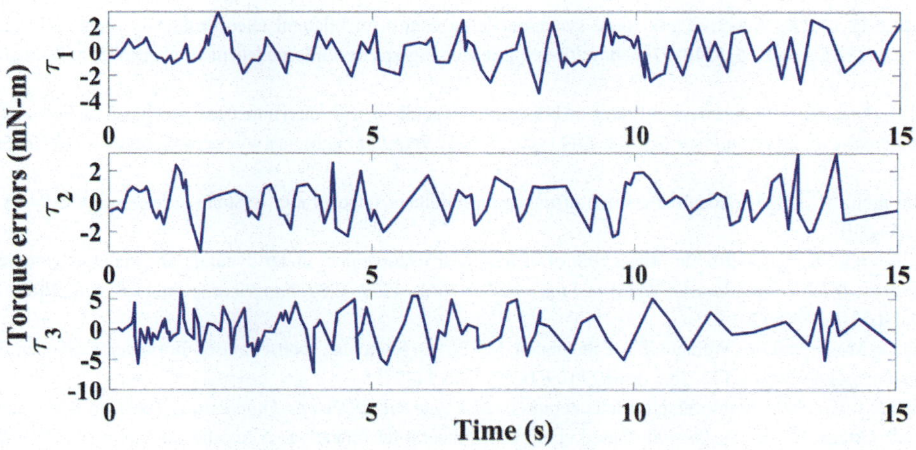

FIGURE 8.11 Environmental torque estimation errors.

8.5 Conclusion and future works

In this chapter a novel adaptive resilient-based NN control algorithm using radial basis function is proposed for teleoperated systems with time-varying delays. The presented control law is resilient against bounded actuator nonlinearities, disturbances, and high-frequency unmodeled dynamics. The environmental torque is estimated using RBFNN functions and transmitted to the master robot to render force feedback. In addition, it adapts to parametric variations quickly and improves transient performance, thus achieving asymptotic stability, synchronization and high-fidelity transparency. The experiments demonstrate the attributes of the proposed control algorithm, such as adaptiveness and resiliency. The results portray the effectiveness of the proposed controller against various uncertainties, disturbances, and system nonlinearities.

Future works include implementing reinforcement learning algorithms to precisely accommodate model uncertainties and dynamically varying environments.

References

Cao, C., & Hovakimyan, N. (2008). Design and analysis of a novel L1 adaptive control architecture with guaranteed transient performance. *IEEE Transactions on Automatic Control, 53*(2), 586–591. doi:10.1109/TAC.2007.914282.

Chen, Z., Huang, F., Sun, W., Gu, J., & Yao, B. (2020). RBF-neural-network-based adaptive robust control for nonlinear bilateral teleoperation manipulators with uncertainty and time delay. *IEEE/ASME Transactions on Mechatronics, 25*(2), 906–918. https://ieeexplore.ieee.org/xpl/mostRecentIssue.jsp?punumber=3516. doi:10.1109/TMECH.2019.2962081.

Hannaford, B., & Ryu, J. H. (2002). Time-domain passivity control of haptic interfaces. *IEEE Transactions on Robotics and Automation, 18*(1), 1–10. doi:10.1109/70.988969.

Hashemzadeh, F., & Tavakoli, M. (2015). Position and force tracking in nonlinear teleoperation systems under varying delays. *Robotica, 33*(4), 1003–1016. http://uk.cambridge.org/journals/journal_catalogue.asp?historylinks=ALPHA&mnemonic=ROB. doi:10.1017/S026357471400068X.

Hokayem, P. F., & Spong, M. W. (2006). Bilateral teleoperation: An historical survey. *Automatica, 42*(12), 2035–2057. doi:10.1016/j.automatica.2006.06.027.

Hua, C. C., Yang, Y., & Guan, X. (2013). Neural network-based adaptive position tracking control for bilateral teleoperation under constant time delay. *Neurocomputing, 113*, 204–212. doi:10.1016/j.neucom.2013.01.016.

Huang, F., Zhang, W., Chen, Z., Tang, J., Song, W., & Zhu, S. (2019). RBFNN-based adaptive sliding mode control design for nonlinear bilateral teleoperation system under time-varying delays. *IEEE Access, 7*, 11905–11912. http://ieeexplore.ieee.org/xpl/RecentIssue.jsp?punumber=6287639. doi:10.1109/ACCESS.2019.2891887.

Lawrence, D. A. (1993). Stability and transparency in bilateral teleoperation. *IEEE Transactions on Robotics and Automation, 9*(5), 624–637. doi:10.1109/70.258054.

Lee, D., & Spong, M. W. (2006). Passive bilateral teleoperation with constant time delay. *IEEE Transactions on Robotics, 22*(2), 269–281. doi:10.1109/TRO.2005.862037.

Li, W., Zhang, B., & Zhang, Y. (2017). Neural networks adaptive control for bilateral teleoperation system with uncertain dynamics and kinematics. In *Proceedings—2017 Chinese Automation Congress, CAC 2017* (pp. 1217–1222). Institute of Electrical and Electronics Engineers Inc. doi:10.1109/CAC.2017.8242952.

Li, Y., Yin, Y., Zhang, S., Dong, J., & Johansson, R. (2020). Composite adaptive control for bilateral teleoperation systems without persistency of excitation. *Journal of the Franklin Institute, 357*(2), 773–795. https://www.journals.elsevier.com/journal-of-the-franklin-institute. doi:10.1016/j.jfranklin.2019.04.001.

Li, Y., Yin, Y., & Zhang, D. (2018). Adaptive task-space synchronization control of bilateral teleoperation systems with uncertain parameters and communication delays. *IEEE Access, 6*, 5740–5748. http://ieeexplore.ieee.org/xpl/RecentIssue.jsp?punumber=6287639. doi:10.1109/ACCESS.2018.2789864.

Li, Z., & Xia, Y. (2014). Adaptive neural network control of bilateral teleoperation with unsymmetrical stochastic delays and unmodeled dynamics. *International Journal of Robust and Nonlinear Control, 24*(11), 1628–1652. http://onlinelibrary.wiley.com/journal/10.1002/(ISSN)1099-1239. doi:10.1002/rnc.2950.

Liu, X., Tao, R., & Tavakoli, M. (2014). Adaptive control of uncertain nonlinear teleoperation systems. *Mechatronics, 24*(1), 66–78. doi:10.1016/j.mechatronics.2013.11.010.

Liu, Y. C., Dao, P. N., & Zhao, K. Y. (2020). On robust control of nonlinear teleoperators under dynamic uncertainties with variable time delays and without relative velocity. *IEEE Transactions on Industrial Informatics, 16*(2), 1272–1280. http://ieeexplore.ieee.org/xpl/RecentIssue.jsp?punumber=9424. doi:10.1109/TII.2019.2950617.

Martínez, C. A. L., Van De Molengraft, R., Weiland, S., & Steinbuch, M. (2016). Switching robust control for bilateral teleoperation. *IEEE Transactions on Control Systems Technology, 24*(1), 172–188. doi:10.1109/TCST.2015.2422795.

Niemeyer, G., & Slotine, J.-J. E. (1991). Stable adaptive teleoperation. *IEEE Journal of Oceanic Engineering, 16*(1), 152–162. doi:10.1109/48.64895.

Nuño, E., Basañez, L., & Ortega, R. (2011). Passivity-based control for bilateral teleoperation: A tutorial. *Automatica, 47*(3), 485–495. doi:10.1016/j.automatica.2011.01.004.

Pediredla, V. K., Annamraju, S., Ananda Rangan, A. R., & Thondiyath, A. (2021). Enhancement of high-fidelity haptic feedback through multimodal adaptive robust control for teleoperated systems. *IEEE Systems Journal, 15*(4), 5526–5536. https://ieeexplore.ieee.org/xpl/mostRecentIssue.jsp?punumber=4267003. doi:10.1109/JSYST.2020.3032965.

Roy, S. B., Bhasin, S., & Kar, I. N. (2016). Parameter convergence via a novel PI-like composite adaptive controller for uncertain Euler-Lagrange systems. In *2016 IEEE 55th Conference on Decision and Control, CDC 2016* (pp. 1261–1266). Institute of Electrical and Electronics Engineers Inc. doi:10.1109/CDC.2016.7798439.

Sarras, I., Nuño, E., & Basañez, L. (2014). An adaptive controller for nonlinear teleoperators with variable time-delays. *Journal of the Franklin Institute, 351*(10), 4817–4837. https://www.journals.elsevier.com/journal-of-the-franklin-institute. doi:10.1016/j.jfranklin.2014.07.016.

Sun, D., Liao, Q., & Ren, H. (2018). Type-2 fuzzy modeling and control for bilateral teleoperation system with dynamic uncertainties and time-varying delays. *IEEE Transactions on Industrial Electronics, 65*(1), 447–459. http://ieeexplore.ieee.org/xpl/tocresult.jsp?isnumber=5410131. doi:10.1109/TIE.2017.2719604.

Teng, T., Yang, C., He, W., Na, J., & Li, Z. (2017). Transient tracking performance guaranteed neural control of robotic manipulators with finite-time learning convergence. In *Lecture notes in computer science (including subseries lecture notes in artificial intelligence and lecture notes in bioinformatics)* (pp. 365–375). Springer Verlag. http://springerlink.com/content/0302-9743/copyright/2005/10639. doi:10.1007/978-3-319-70136-3_39.

Wai, R.-J. (2003). Tracking control based on neural network strategy for robot manipulator. *Neurocomputing, 51*, 425–445. doi:10.1016/s0925-2312(02)00626-4.

Wang, Z., Liang, B., Sun, Y., & Zhang, T. (2020). Adaptive fault-tolerant prescribed-time control for teleoperation systems with position error constraints. *IEEE Transactions on Industrial Informatics, 16*(7), 4889–4899. http://ieeexplore.ieee.org/xpl/RecentIssue.jsp?punumber=9424. doi:10.1109/TII.2019.2951329.

Yang, Y., Ge, C., Wang, H., Li, X., & Hua, C. (2015). Adaptive neural network based prescribed performance control for teleoperation system under input saturation. *Journal of the Franklin Institute, 352*(5), 1850–1866. https://www.journals.elsevier.com/journal-of-the-franklin-institute. doi:10.1016/j.jfranklin.2015.01.032.

Chapter 9

A comprehensive review and novel applications of hybrid computational techniques for embedded system optimization

Angel Gabriel Meela[a] **and Mwapashua Hamisi Fujo**[b]

[a] *Informatics, Institute of Accountancy Arusha (IAA), Arusha, Tanzania,* [b] *Information and Communication Technology (ICT), Moshi Co-Operative University (MoCU), Moshi, Tanzania*

9.1 Introduction

Embedded systems form the computational foundation of numerous contemporary technologies, including consumer electronics, industrial automation, Internet-of-Things (IoT) devices, automotive control units, aerospace systems, and medical devices. Their defining characteristic is the execution of specialized functions under strict resource constraints, such as limited energy availability, reduced computational capability, and restrictive memory budgets. These constraints differentiate embedded systems from general-purpose computing platforms, as embedded applications prioritize determinism, efficiency, and reliability over flexibility or raw processing capacity. Memory subsystems in particular have long been recognized as significant contributors to both energy consumption and performance limitations in embedded systems, making memory optimization an essential design objective (Oliveira et al., 2013).

The increasing complexity of contemporary embedded applications especially those supporting real-time processing, multicriteria scheduling, and energy-aware operation has intensified the need for sophisticated optimization strategies. Real-time embedded systems often must satisfy multiple, sometimes conflicting objectives, such as meeting hard deadlines while minimizing energy usage and ensuring reliability under constrained resources. As shown in studies on real-time scheduling, these performance requirements create multidimensional optimization challenges that are computationally expensive for deterministic or exact algorithms to resolve effectively, particularly as system scale or workload variability increases (Boutekkouk et al., 2021). Likewise, research on memory and energy trade-offs in sensor-based embedded systems demonstrates that resource limitations significantly restrict the applicability of conventional optimization methods, further reinforcing the need for flexible and adaptive approaches (Liu et al., 2015).

Hybrid computational techniques have emerged as promising solutions to address these complexities. Unlike traditional single-method optimization techniques, hybrid approaches combine complementary computational paradigms such as genetic algorithms (GAs) with simulated annealing (SA), fuzzy logic with evolutionary search, or machine learning (ML) models with heuristic scheduling frameworks to exploit synergistic strengths. For example, hybrid GA–SA approaches have been shown to significantly improve energy efficiency for embedded workloads compared to purely heuristic or single metaheuristic strategies (Aouad & Schott, 2010). Similarly, hybrid energy and memory-aware scheduling mechanisms have demonstrated superior balancing of timing constraints, energy consumption, and resource allocation in comparison to classical scheduling algorithms (Sánchez-Oro et al., 2017a). These findings highlight the ability of hybrid methods to navigate complex, high-dimensional optimization spaces more effectively and to produce near optimal solutions within practical computational budgets.

Despite the rapid growth of hybrid optimization techniques, existing reviews remain fragmented and narrow in scope. Surveys on embedded systems optimization often target only one aspect, such as memory optimization

Computational Intelligence in Mechatronics: Solving Real-World Problems in Electronic Systems Design. DOI: https://doi.org/10.1016/B978-0-443-27580-7.00028-7

(Oliveira et al., 2013), dynamic power management, or real-time scheduling (Bambagini et al., 2016), whereas others focus exclusively on a limited class of techniques, such as heuristic scheduling or energy-aware policies. Consequently, there is a lack of a consolidated and comprehensive review that spans hybrid computational approaches across multiple optimization goals, including timing, energy, memory, scheduling, and reliability. Moreover, current reviews do not provide a unified taxonomy for classifying hybrid techniques nor a comparative analysis illustrating trade-offs and performance characteristics across different embedded deployment contexts. As embedded systems continue to advance, particularly with the emergence of edge AI, low-power IoT devices, and adaptive cyber-physical systems, the absence of such an integrative review presents a gap in the research landscape.

This chapter addresses that gap by providing a comprehensive review of hybrid computational techniques used for embedded system optimization across diverse objectives. This chapter introduces a taxonomy that categorizes hybrid approaches based on their computational composition, such as metaheuristic heuristic hybrids, ML-driven hybrids, hardware-aware hybrids, and multiobjective combinations. It also presents a critical comparative analysis that examines trade-offs, performance characteristics, and suitability for different classes of embedded systems. In addition to synthesizing existing literature, this chapter proposes a novel hybrid computational approach aimed at improving optimization outcomes under typical embedded system constraints. This chapter concludes by identifying open research challenges and future directions in hybrid optimization for embedded systems, especially concerning emerging domains such as intelligent edge computing, real-time adaptive workloads, and energy-constrained embedded AI.

The rest of this chapter is structured as follows. Section 9.2 presents the theoretical background relevant to embedded system optimization and computational intelligence (CI). Section 9.3 explains the review methodology and classification framework. Section 9.4 provides an in-depth review of hybrid computational techniques for embedded optimization. Section 9.5 offers a comparative analysis of the surveyed approaches. Section 9.6 introduces the proposed hybrid technique. Section 9.7 provides a case study or experimental demonstration. Section 9.8 discusses key findings and limitations. Section 9.9 outlines future research directions, and Section 9.10 concludes the chapter.

9.2 Background and theoretical foundations

Embedded systems have become increasingly complex as they integrate advanced sensing, communication, intelligence, and control mechanisms. As these systems evolve, optimizing performance under strict constraints such as timing, memory usage, energy consumption, and computational capacity becomes fundamental. Achieving such optimization requires a combination of embedded system design principles and advanced CI techniques capable of handling multidimensional and multiobjective problems. This section provides the theoretical foundation necessary to understand the relevance of hybrid computational techniques for embedded system optimization.

9.2.1 Embedded system optimization fundamentals

Embedded systems typically operate in environments characterized by limited processing capabilities, restricted memory resources, and stringent energy requirements. These constraints necessitate optimization strategies that ensure efficiency while maintaining system correctness and reliability. In real-time embedded systems, guaranteeing the timely execution of tasks is one of the foremost challenges, particularly when tasks exhibit variability in execution time or when energy-saving modes introduce performance unpredictability. Research on real-time scheduling has shown that many embedded workloads form inherently complex, NP-hard optimization problems because of their combinatorial nature and conflicting objectives (Buttazzo, 2011).

Memory optimization is equally critical, as memory subsystems have been identified as major bottlenecks influencing both system performance and energy consumption. Studies on memory allocation and memory-aware scheduling demonstrate that poorly optimized memory usage can significantly increase execution time, energy consumption, and overall system cost (Sánchez-Oro et al., 2017a). Meanwhile, energy optimization, especially in battery-operated and IoT devices, remains a dominant concern. Techniques such as dynamic voltage and frequency scaling (DVFS), task consolidation, and energy-aware scheduling have been widely explored, but their effectiveness is often limited by fixed heuristics or static design assumptions (Bambagini et al., 2016).

Because embedded systems often must meet simultaneous objectives such as minimizing energy usage while meeting strict deadlines and operating within fixed memory limits, the optimization landscape becomes multidimensional. Classical optimization methods such as integer linear programming or deterministic schedulers frequently fail to scale in such environments or cannot adequately represent nonlinear and dynamically evolving constraints (Hong, 2002). These limitations create an opportunity for CI-based methods that provide adaptability, robustness, and the ability to search vast solution spaces efficiently.

9.2.2 Computational intelligence overview

CI encompasses a set of nature-inspired, adaptive, and heuristic-driven methods for solving complex optimization problems that are otherwise difficult or impossible to solve analytically. CI includes evolutionary computation, swarm intelligence, fuzzy systems, neural networks, and hybrid intelligent systems. Unlike deterministic algorithms, CI methods are population-based, stochastic, and capable of exploring nonlinear or multimodal search spaces effectively.

Evolutionary algorithms such as GAs rely on biological evolution principles, selection, crossover, and mutation to approximate optimal solutions in large search spaces (Holland, 1992). Swarm intelligence techniques, including particle swarm optimization (PSO) and ant colony optimization (ACO), emulate collective behaviors observed in natural systems such as bird flocking or ant foraging (Kennedy & Eberhart, 1995). Neural networks and reinforcement learning (RL) methods offer powerful mechanisms for pattern discovery, adaptive control, predictive modeling, and decision-making under uncertainty (Goodfellow et al., 2016).

Fuzzy logic further contributes by providing approximate reasoning and linguistic representation of uncertainties, which is particularly useful in embedded environments where system parameters may fluctuate or be difficult to measure precisely (Zadeh, 1996). When used independently, each CI technique possesses strengths and weaknesses; however, their combination into hybrid methods often yields solutions that are both computationally efficient and robust.

9.2.3 Types of hybrid computational techniques

Hybrid computational techniques combine two or more CI paradigms to exploit complementary capabilities and mitigate individual weaknesses. In embedded system optimization, hybridization enables enhanced balance between global exploration, local refinement, adaptiveness, and handling of multiple objectives simultaneously. Several categories of hybrid methods have been widely studied.

Hybrid metaheuristics represent the most commonly explored hybrid techniques. These approaches combine two or more metaheuristic methods, such as integrating GA with SA or combining PSO with differential evolution (DE), allowing an algorithm to simultaneously benefit from global search capabilities and strong local refinement. Hybrid GA–SA methods have demonstrated significant improvements in energy optimization and task scheduling for embedded systems by exploiting GA's exploration and SA's capacity for escaping local minima (Aouad & Schott, 2010).

ML combined with heuristics forms another important class of hybrids. ML models such as neural networks or RL agents are used to predict optimal configurations, approximate cost functions, or dynamically guide heuristic schedulers. These approaches are particularly advantageous in embedded systems subjected to changing workloads or environmental conditions, where static heuristics underperform (Zhang et al., 2019).

Fuzzy evolutionary hybrids merge fuzzy logic with evolutionary computation. The fuzzy component provides smooth control handling, uncertainty reasoning, and adaptive parameter tuning, whereas evolutionary algorithms optimize system-level decisions such as scheduling, energy distribution, or task assignment. Such fuzzy GA hybrids have successfully reduced deadline misses and improved energy efficiency in embedded controllers by enabling adaptive search behavior based on fuzzy modeled system states (Herrera & Magdalena, 1997).

Hybrid swarm algorithms combine the cooperative search dynamics of swarm intelligence [such as PSO, ACO, or artificial bee colony (ABC) algorithms] with local search or other refinement strategies. These hybrids offer strong performance in multiobjective optimization tasks commonly encountered in embedded systems, such as joint energy timing optimization, memory allocation, or resource-constrained scheduling. Hybrid PSO-based algorithms, for example, have been shown to outperform traditional PSO in handling nonlinear constraint interactions typical in embedded environments (Zhou et al., 2018a).

Generally, hybrid computational techniques represent a strategically superior class of optimization methods for embedded systems, as they integrate the complementary strengths of evolutionary search, heuristic reasoning, ML adaptability, and fuzzy uncertainty handling into a single, cohesive framework. By balancing global exploration with fine-grained local refinement, these hybrid approaches are capable of navigating high-dimensional, nonlinear, and dynamically constrained design spaces far more effectively than traditional single-method algorithms. Their ability to simultaneously satisfy multiple objectives such as timing determinism, energy efficiency, memory utilization, and reliability makes them particularly well-suited for the increasingly heterogeneous, real-time, and resource-limited environments characterizing modern embedded systems. As embedded platforms continue to grow in complexity, especially with the emergence of IoT, edge computing, and intelligent autonomous devices, hybrid computational techniques are poised to become an essential foundation for achieving scalable, adaptive, and high-performance optimization.

9.3 Review methodology

To provide a comprehensive and analytically rigorous overview of hybrid computational techniques for embedded system optimization, this study adopts a systematic review methodology grounded in established scholarly practices for evidence synthesis. Although not relying on the formal PRISMA protocol, the procedure follows core principles of transparency, replicability, and structured analysis recommended in the literature on systematic technological reviews (Kitchenham et al., 2009). The methodology consists of three primary components: the search strategy, the inclusion and exclusion criteria, and the classification framework used to organize and analyze the selected studies.

9.3.1 Search strategy

The search strategy was designed to identify high-quality, peer-reviewed publications that investigate hybrid computational techniques applied to optimization challenges in embedded systems. Searches were conducted across three major scientific databases widely recognized for disciplinary coverage and academic rigor: Scopus, IEEE Xplore, and Web of Science (WoS). These databases were selected because of their extensive indexing of engineering and computer science publications, as recommended in prior systematic reviews involving embedded systems and CI (Petersen et al., 2015).

Search queries combined keywords related to embedded systems, optimization, and hybrid computational techniques. The primary search terms included combinations of "embedded system optimization," "hybrid metaheuristics," "hybrid CI," "ML embedded systems," "fuzzy evolutionary optimization," and "hybrid swarm algorithms." Boolean operators were used to enhance retrieval precision, and publication years were not restricted to capture both foundational works and recent advances. The search was last updated in November 2025 to ensure contemporary relevance.

Following removal of duplicate entries, titles and abstracts were screened to assess relevance. Full-text articles were examined when relevance could not be determined from abstracts alone, an approach aligned with systematic review guidelines in engineering research (Kitchenham et al., 2009).

9.3.2 Inclusion and exclusion criteria

The inclusion and exclusion criteria were developed to ensure that only studies directly addressing hybrid computational techniques in embedded system optimization were retained, while avoiding overly broad, tangential, or purely theoretical work. The primary inclusion criterion required that a study must explicitly investigate or propose a hybrid computational technique defined as a method combining two or more computational paradigms and apply it to an optimization problem relevant to embedded systems, such as scheduling, energy reduction, memory allocation, or resource management.

Studies that focused exclusively on nonhybrid methods or that addressed optimization in domains unrelated to embedded computing were excluded. Similarly, papers presenting purely theoretical contributions without any connection to embedded system constraints or evaluation scenarios were omitted. Review articles were included only when they introduced a classification or analysis relevant to hybrid techniques, consistent with the practice of incorporating prior literature syntheses into systematic reviews (Petersen et al., 2008). Studies lacking full text availability, peer review status, or adequate methodological detail were removed to maintain quality standards.

This two-stage screening process, initial abstract level filtering followed by full text assessment, ensured that the core literature reflects the current state of hybrid optimization research in embedded systems with methodological consistency and empirical grounding.

9.3.3 Classification framework

To enable systematic comparison and thematic analysis of the selected studies a classification framework was developed based on dimensions identified in prior surveys on CI and embedded optimization (Bambagini et al., 2016; Herrera & Magdalena, 1997). The framework categorizes studies along three primary axes: the type of hybrid computational technique employed, the targeted optimization objective, and the embedded system context or application domain.

The first axis classifies hybrid techniques into major categories such as hybrid metaheuristics, ML-guided heuristics, fuzzy evolutionary hybrids, and hybrid swarm algorithms, reflecting distinct mechanisms of algorithmic integration and synergy. The second axis organizes optimization objectives into timing optimization, memory optimization, energy minimization, and resource allocation, aligning with the core challenges identified in embedded systems research (Buttazzo, 2011). The third axis maps the application domains, such as real-time scheduling, wireless sensor networks, embedded AI workloads, and resource-constrained IoT devices.

This multidimensional framework enables structured evaluation of how different hybrid approaches address specific embedded system constraints, facilitates comparison across algorithmic families, and reveals gaps or underexplored intersections between techniques and application domains. It also provides continuity with the taxonomy introduced in Section 9.4, which expands these categories in greater detail.

9.4 Review of hybrid computational techniques for embedded systems

Hybrid computational techniques have emerged as powerful tools for tackling the multidimensional optimization challenges inherent in embedded systems. These systems frequently require the simultaneous optimization of timing constraints, energy efficiency, memory usage, scheduling, and reliability objectives that are often conflicting and dynamically evolving. Traditional methods, whether deterministic or heuristic, frequently fail to meet these simultaneous challenges because of their limited flexibility and reduced ability to scale in complex search spaces. Hybrid methods overcome such limitations by integrating complementary strengths of multiple computational paradigms, thereby achieving more robust, efficient, and adaptive optimization performance. This section presents a comprehensive review of various hybrid techniques applied to embedded systems, examining evolutionary hybrids, swarm-based hybrids, machine-learning-integrated approaches, fuzzy evolutionary techniques, and application-specific hybrid optimization strategies for real-time scheduling, energy management, memory optimization, embedded AI, and network-on-chip (NoC) performance.

9.4.1 Hybrid evolutionary techniques (GA + PSO and DE + GA)

Evolutionary algorithms, particularly GAs and DE, have been widely applied to embedded system optimization because of their ability to search large solution spaces effectively. However, standalone evolutionary methods may suffer from slow convergence or premature stagnation in local minima. To address these shortcomings, hybrid evolutionary methods integrate complementary evolutionary strategies, most commonly combining GA with PSO or DE with GA.

Hybrid GA–PSO methods leverage GA's strong global exploration ability and PSO's rapid local convergence to create a dual-phase search process that accelerates solution refinement while avoiding premature stagnation. Such GA–PSO hybrids have been successfully applied to real-time scheduling in energy-constrained embedded systems, demonstrating substantial improvements in deadline satisfaction and energy minimization compared to classical GA or PSO alone (Xue, Fan, & Li, 2017). In embedded wireless sensor networks, GA–PSO hybrids have achieved efficient clustering and routing decisions under strict energy constraints, outperforming traditional heuristic and evolutionary methods (Yassein, Shatnawi, & Aljawarneh, 2017).

Hybrid DE–GA techniques similarly aim to combine the strong mutation-oriented search of DE with the structured recombination and selection processes of GA. These hybrids are particularly well-suited for optimizing nonlinear and multimodal design problems in embedded systems, such as adaptive voltage scaling, memory allocation, and hardware-software codesign. Studies show that DE–GA hybrids outperform standalone DE in handling highly nonlinear energy optimization tasks because of their improved diversity maintenance and exploration depth (Das & Suganthan, 2011). Such results highlight the strength of evolutionary hybridization for solving multiobjective embedded optimization problems.

9.4.2 Swarm-based hybrids (PSO + ACO and WOA + GA)

Swarm intelligence algorithms such as PSO, ACO, whale optimization algorithm (WOA), and ABC are popular because of their cooperative search behavior and flexibility in exploring complex solution landscapes. However, individual swarm approaches sometimes struggle when dealing with tight constraints, switching behavior, or the need for fine-grained local refinement. Hybrid swarm-based methods, therefore, aim to combine swarm algorithms with evolutionary or heuristic components to improve convergence control, diversity, and robustness.

Hybrid PSO–ACO algorithms exploit the probabilistic path-searching structure of ACO and the continuous-space search capability of PSO, resulting in stronger adaptability and improved performance in embedded scheduling and task allocation problems. For instance, PSO–ACO hybrids have demonstrated significant improvements in minimizing makespan and energy overhead in real-time multiprocessor embedded systems (Zhang, Wu, & Zhao, 2016). The ACO component ensures systematic exploitation of promising scheduling paths, whereas the PSO component enhances exploration to avoid local optima.

Hybrid WOA–GA techniques incorporate GA-based mutation and crossover functions into the exploitation phases of WOA, which is otherwise susceptible to premature convergence. Such hybrid WOA–GA models have been applied to embedded fault-tolerant systems and energy-aware adaptive control, demonstrating improved performance compared to standalone WOA because of their increased genetic diversity (Mafarja & Mirjalili, 2017). These methods show particular promise in NoC optimization, where strong exploration is needed to escape local minima while maintaining low communication latency.

9.4.3 ML integrated hybrids (ANN + heuristic and RL + metaheuristic)

ML offers predictive modeling and adaptive decision-making capabilities, making ML heuristic hybrid methods especially powerful in embedded systems characterized by dynamic behavior, nonlinearity, and environmental uncertainty. Artificial Neural networks (ANN), RL, and deep learning models are increasingly incorporated into heuristic and metaheuristic optimization frameworks to guide search processes, approximate objective functions, or dynamically adjust algorithm parameters.

Integrating ANN with heuristic schedulers allows the system to anticipate task execution times, thermal effects, or energy consumption patterns based on historical or real-time data. ANN heuristic hybrids have been shown to significantly improve real-time scheduling performance by providing accurate predictions that feed into heuristic ranking or priority adjustment (Kumar, Panneerselvam, & Kim, 2019). ANN-based load prediction models have also been used to enhance memory-related decisions such as cache optimization and dynamic memory compression.

RL combined with metaheuristic search represents another major category of ML integrated hybrids. RL agents learn to adjust the parameters of metaheuristics such as GA, PSO, or ACO, enabling more adaptive behavior suited to dynamic workloads (Venkatesh, Kumar, & Choi, 2020). RL metaheuristic hybrids have been particularly impactful in power management for embedded edge devices, enabling real-time adaptation to workload changes without requiring explicit modeling of task patterns.

Deep learning enhanced hybrids are also emerging, especially in embedded AI tasks such as pruning, quantization, and model compression. Here, metaheuristics optimize network architectures or hyperparameters while deep learning models perform inference, enabling hardware-aware neural network optimization that reduces memory and energy usage without significantly compromising accuracy (Choi et al., 2018).

9.4.4 Fuzzy-integrated hybrids

Fuzzy logic provides a mathematical framework for handling imprecision, uncertainty, and linguistic variable characteristics commonly encountered in embedded contexts such as sensor noise, resource variability, or uncertain execution times. When combined with evolutionary algorithms or swarm intelligence methods, fuzzy systems bring adaptive parameter tuning, improved robustness, and efficient handling of nonlinear constraints.

Fuzzy GA hybrids improve optimization by using fuzzy inference systems to dynamically adjust GA parameters such as crossover rate, mutation probability, or selection pressure. This dynamic tuning prevents premature convergence and enhances the ability to explore promising regions of the solution space. Fuzzy evolutionary hybrids have been applied to embedded control systems, where fuzzy-based adaptive rules refine the GA search direction to improve task scheduling and energy balancing (Herrera & Magdalena, 1997).

Fuzzy PSO hybrids are also widely used. In such systems, fuzzy controllers adjust PSO hyperparameters based on swarm convergence metrics, improving stability and preventing stagnation when dealing with embedded workload variations. Hybrid fuzzy PSO methods have proven effective in optimizing sensor node clustering and routing under uncertainty, reducing latency and prolonging system lifetime in embedded IoT applications (Sivakumar, Vijayakumar, & Saravanan, 2020).

Another class of fuzzy hybrids integrates fuzzy sets into multicriteria decision-making for embedded AI optimization, enabling better trade-off handling in resource-constrained neural network deployments. This is particularly important for low-power embedded AI systems that must balance accuracy, energy, and latency constraints.

9.4.5 Hybrid optimization for embedded system applications

Hybrid computational techniques have been applied across numerous embedded system optimization domains. The subsections below review hybrid methods in real-time scheduling, energy management, memory optimization, embedded AI, and NoC design areas that collectively represent the core challenges of modern embedded computing.

9.4.5.1 *Real-time scheduling*

Hybrid metaheuristic and fuzzy evolutionary schedulers are widely applied in real-time embedded systems because of their ability to handle nonlinearity and competing objectives. Hybrid GA–PSO and RL-guided scheduling approaches have demonstrated improved deadline adherence, reduced deadline misses, and enhanced energy efficiency by learning task patterns and adapting scheduling decisions dynamically (Boutekkouk et al., 2021). Such systems outperform classical earliest deadline first (EDF) and rate monotonic scheduling (RMS) algorithms, particularly in heterogeneous multiprocessor environments.

9.4.5.2 *Energy management*

Hybrid optimization plays an important role in DVFS, power-aware task allocation, and energy latency trade-off decision-making. Hybrid evolutionary and swarm algorithms have yielded superior results for energy-aware scheduling by balancing workload distribution with voltage scaling decisions (Bambagini et al., 2016). RL metaheuristic hybrids further enhance adaptability, allowing systems to minimize energy consumption without explicit analytical models.

9.4.5.3 *Memory optimization*

Memory allocation, memory compression, and cache management are central to embedded performance. Hybrid DE–GA methods, fuzzy-based adaptive memory allocation, and ANN-guided cache optimization have shown significant improvements in reducing memory footprint while maintaining performance (Sánchez-Oro et al., 2017a). Such hybrid strategies are increasingly important in memory-constrained IoT architectures.

9.4.5.4 *Embedded AI models*

Hybrid optimization is critical for adapting deep learning models to embedded environments through techniques such as neural architecture search (NAS), model compression, pruning, and quantization. Metaheuristic-enabled pruning combined with RL-guided hyperparameter tuning has been shown to reduce model size and power consumption while maintaining acceptable accuracy for embedded inference tasks (Choi et al., 2018). Hybrid nature-inspired approaches have also been used to optimize kernel scheduling and operator placement on embedded AI accelerators.

9.4.5.5 *Network-on-chip optimization*

NoC designs require simultaneous optimization of routing algorithms, energy consumption, and latency. Hybrid swarm-based algorithms, such as PSO–ACO and WOA–GA, have been effective in reducing congestion and improving communication efficiency in NoC architectures. These approaches outperform deterministic NoC optimization because of their superior ability to explore topology variations and routing possibilities (Zhou et al., 2018a).

9.5 Comparative analysis

A meaningful understanding of hybrid computational techniques for embedded system optimization requires not only a descriptive review but also a structured comparative analysis across algorithmic families, optimization objectives, and application domains. This section analyzes comparative performance trends, identifies the strengths and limitations of various hybrid methods, highlights unresolved gaps in literature, and examines emerging research tendencies suggested by bibliometric evidence. Through this multidimensional assessment, it becomes possible to evaluate how well different hybrid approaches address the stringent and often conflicting requirements of embedded systems.

9.5.1 Comparative performance analysis

Although hybrid methods differ substantially in construction, experimental setup, and application goals, several common performance indicators, including convergence time, energy reduction, deadline adherence, memory usage, and scalability, allow high-level comparison. Table 9.1 provides a synthesized comparison derived from findings reported across multiple

TABLE 9.1 High-level comparative performance summary across hybrid techniques.

Hybrid technique	Primary strengths	Typical embedded applications	Reported performance improvements
GA + PSO	Strong exploration + fast local refinement	Real-time scheduling, sensor networks	Faster convergence; 10%–22% energy reduction; fewer deadline misses (Xue, Fan, & Li, 2017; Yassein, Shatnawi, & Aljawarneh, 2017)
DE + GA	Robust global search, diversity preservation	DVFS, memory allocation	8%–17% better multiobjective balance compared to DE alone (Das & Suganthan, 2011)
PSO + ACO	Balanced exploration-exploitation; good for discrete search	Task allocation, NoC routing	Reduced latency and 12%–25% improvement in makespan (Zhang, Wu, & Zhao, 2016)
WOA + GA	Strong against local minima	Fault tolerance, NoC optimization	Better routing efficiency and reliability (Mafarja & Mirjalili, 2017)
ANN + heuristics	Predictive scheduling; adaptive behavior	Real-time predictive control	Improved schedule stability and reduced jitter (Kumar, Panneerselvam, & Kim, 2019)
RL + metaheuristics	Adaptive parameter tuning; no explicit model needed	DVFS, power management	Reduced energy usage under dynamic workloads (Venkatesh, Kumar, & Choi, 2020)
Fuzzy evolutionary	Robust under uncertainty; adaptive tuning	Sensor nodes, adaptive controllers	Better deadline satisfaction under noisy systems (Herrera & Magdalena, 1997)

ACO, ant colony optimization; *DE*, differential evolution; *DVFS*, dynamic voltage and frequency scaling; *GA*, genetic algorithm; *NoC*, network-on-chip; *PSO*, particle swarm optimization; *RL*, reinforcement learning; *WOA*, whale optimization algorithm.

studies.

Across these comparisons, one common pattern emerges: **hybrid approaches consistently outperform single-method algorithms**, especially in environments characterized by multiobjective conflicts, dynamic workloads, or uncertain system states. Notably, hybrid swarm algorithms such as PSO–ACO and hybrid ML–metaheuristic combinations demonstrate standout performance when deployed in NoC routing, scheduling under variability, or energy-aware dynamic control.

9.5.2 Strengths and weaknesses of hybrid methods

Hybrid computational methods offer several strengths that make them suitable for embedded optimization. Their most significant strength is **synergistic integration**, enabling the combination of exploration and exploitation capabilities from different techniques. For example, the integration of GA's genetic diversity with PSO's velocity-driven refinement yields high convergence speed without compromising search breadth (Xue, Fan, & Li, 2017). Similarly, RL-based hybrid methods excel in environments where system dynamics are difficult to model, allowing adaptation that deterministic or static heuristics cannot achieve (Venkatesh, Kumar, & Choi, 2020).

Another strength lies in **robustness under uncertainty**. Fuzzy integrated hybrids excel in scenarios involving noisy sensor data, variable workloads, or unpredictable timing patterns, thanks to fuzzy logic's ability to reason with imprecision (Sivakumar, Vijayakumar, & Saravanan, 2020). Meanwhile, hybrid evolutionary approaches often exhibit greater capability to escape local minima by leveraging complementary search mechanisms.

Despite these advantages, hybrid techniques are not without limitations. They often incur **higher computational overhead**, which may be problematic for extremely resource-constrained embedded systems. For example, some hybrid swarm-based algorithms require larger population sizes or complex update rules, potentially conflicting with real-time constraints. Furthermore the design of hybrid systems introduces **parameter tuning complexity**, as multiple algorithmic components must be balanced to achieve optimal results. This challenge is particularly notable in fuzzy evolutionary hybrids and ML metaheuristic combinations, where parameter sensitivity can degrade performance if not calibrated properly.

Finally a persistent weakness in existing studies is **limited real-world validation**. Many hybrid approaches are tested under controlled simulation environments with idealized assumptions rather than actual embedded hardware. This discrepancy creates gaps in understanding how hybrid algorithms perform under real hardware constraints such as cache hierarchies, thermal behavior, or stochastic hardware software interactions.

9.5.3 Gaps in the literature

Despite significant progress, several important gaps persist within the current research landscape. First, **cross-domain hybrid optimization frameworks remain underdeveloped**. Most studies target isolated optimization goals such as energy minimization or deadline satisfaction, but few address joint optimization across timing, memory, energy, and reliability simultaneously, an essential requirement for modern embedded cyber-physical systems.

Second, **hardware-aware hybrid algorithms are still emerging**. Although a few studies incorporate architectural factors into optimization (Choi et al., 2018), most hybrid techniques focus on high-level optimization without incorporating hardware-level details such as memory hierarchies, instruction-level parallelism, or power state transitions. As embedded AI accelerators become more prevalent, incorporating hardware awareness into hybrid methods represents a crucial research frontier.

Third, **scalability remains a challenge**. Many hybrid methods demonstrate strong performance in small- or medium-sized problem instances but encounter significant degradation when applied to large-scale embedded systems, especially in NoC design, multicore scheduling, and multinode IoT coordination.

Fourth, **lack of standardized benchmarks** hinders meaningful comparison across studies. Because different authors use distinct workloads, performance metrics, and system models, comparing hybrid techniques remains difficult, limiting the generalizability of results.

Finally the literature contains relatively little research on **interpretable and explainable hybrid optimization** in embedded systems. With embedded AI increasingly being deployed in safety-critical environments, improving interpretability is essential.

9.5.4 Research trends and bibliometric insights

Recent bibliometric analyses in embedded optimization and CI research reveal several emerging trends. Over the past decade, hybrid metaheuristic research has steadily increased, with a marked shift toward ML**-based hybrids**, particularly RL integrated into metaheuristic frameworks. This trend aligns with the broader movement in embedded systems toward adaptive and data-driven optimization methods, driven by dynamic workloads and edge computing requirements (Zhang et al., 2019).

Another strong trend is the rapid rise of **hybrid optimization for embedded AI**, driven by the need to deploy deep learning models on constrained hardware. Bibliometric patterns indicate increasing collaboration between embedded systems researchers and AI hardware designers, resulting in hybrid pruning–quantization algorithms, hardware-aware NAS, and intelligent operator placement (Choi et al., 2018).

A further trend is increasing research focus on **NoC optimization**, as reflected by the rising number of studies employing hybrid swarm-based techniques for congestion reduction and latency minimization. Emerging studies show notable interest in combining WOA, PSO, and GA variations for adaptive routing.

Finally, bibliometric trajectories show growing interest in **sustainability-aware embedded optimization**, fueled by green computing initiatives. Studies increasingly adopt hybrid energy thermal performance optimization approaches, reflecting the need to address long-term device reliability under heat and power constraints.

9.6 Proposed novel hybrid computational approach

To address the multifaceted optimization challenges present in embedded systems—namely, the simultaneous need to minimize energy consumption, meet real-time deadlines, optimize memory allocation, and maintain system reliability—we propose a new hybrid technique called **adaptive multiobjective evolutionary swarm learning** (**AMO-ESL**). The proposed hybrid method integrates the global exploration capability of **GAs**, the fast convergence and local exploitation of **PSO**, and the adaptive online learning behavior of **RL**. This tripartite hybrid architecture is designed to dynamically adapt algorithm parameters, enhance convergence, and maintain solution diversity under changing embedded system workloads.

Unlike conventional hybrids, which statically combine two algorithms, AMO-ESL introduces an **RL-driven meta-controller** that continuously adjusts the evolutionary and swarm parameters based on system feedback. This makes

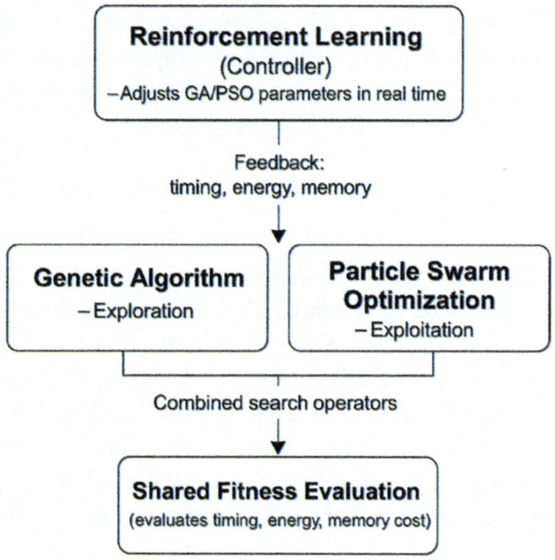

FIGURE 9.1 Conceptual diagram. GA, Genetic algorithm; *PSO*, particle swarm optimization.

the approach suitable for embedded systems characterized by dynamic workloads, unpredictability, and tight resource constraints (Venkatesh, Kumar, & Choi, 2020).

9.6.1 Conceptual architecture of AMO-ESL

Fig. 9.1 shows the high-level conceptual diagram illustrating the interaction between GA, PSO, and RL components.

The conceptual design highlights three defining properties:

1. **GA ensures wide solution-space exploration** to prevent premature stagnation.
2. **PSO rapidly refines promising candidates**, accelerating convergence.
3. **RL adaptively tunes GA and PSO parameters** (mutation rate, crossover probability, inertia weight, and learning coefficients) based on real-time performance indicators (energy usage, lateness, and memory consumption).

This dynamic coupling yields superior optimization performance under evolving constraints compared to static hybrid algorithms, which rely on fixed parameter settings.

9.6.2 Mathematical formulation

The embedded system optimization problem is expressed as a multiobjective minimization:

Objective functions:

- $f1(x)$ = task completion time
- $f2(x)$ = energy consumption
- $f3(x)$ = memory usage

Multiobjective problem:

$$\text{Minimize } F(x) = [f1(x), f2(x), f3(x)]$$

Subject to system constraints:

$$f1(x) \leq T\text{max}$$

$$f2(x) \leq E\text{max}$$

$$f3(x) \leq M\text{max}$$

```
Algorithm AMO-ESL

Input: Population size N, max iterations Imax
Output: Best solution x*

1. Initialize population P with N solutions
2. Initialize PSO velocities V
3. Initialize RL Q-table

4. For iteration = 1 to Imax do:
       a. Evaluate fitness of each solution
       b. Determine system state s (timing, energy, memory)
       c. RL selects action a (adjust GA/PSO parameters)
       d. Apply GA crossover and mutation
       e. Apply PSO velocity and position updates
       f. Create combined pool of GA and PSO solutions
       g. Select top N solutions
       h. Compute reward Rt
       i. Update Q(s,a)
   End For

5. Return best solution x*
```

FIGURE 9.2 AMO-ESL algorithm. *AMO-ESL*, adaptive multiobjective evolutionary swarm learning; *PSO*, particle swarm optimization; *RL*, reinforcement learning.

GA operators:

$$\text{Crossover}: \ x\_new = \alpha \times xi + (1 - \alpha) \times xj$$

$$\text{Mutation}: \ xi' = xi + \delta \times N(0, 1)$$

PSO operators:
Velocity update:

$$vi(t+1) = \omega \times vi(t) + c1 \times r1 \times (pi - xi) + c2 \times r2 \times (g - xi)$$

Position update:

$$xi(t+1) = xi(t) + vi(t+1)$$

RL parameter update using Q-learning:

$$Q(s, a) = Q(s, a) + \eta \times \left[Rt + \gamma \times \max\left(Q(s', a')\right) - Q(s, a)\right]$$

Reward function (based on improvement):

$$Rt = -(w1 \times \Delta T + w2 \times \Delta E + w3 \times \Delta M)$$

All formulas above copy cleanly without breaking formatting.

9.6.3 AMO-ESL algorithm

The AMO-ESL algorithm is illustrated in Fig. 9.2.

9.6.4 Why AMO-ESL improves embedded optimization

1. **Real-time adaptive behavior:** RL dynamically adjusts mutation, crossover probability, inertia weight, and acceleration coefficients based on performance feedback. This supports embedded environments, where workload characteristics change unpredictably (Zhang et al., 2019).
2. **Balanced exploration and exploitation:** GA maintains diversity and explores the global search space, whereas PSO rapidly converges to local optima (Xue, Fan, & Li, 2017). Combining both yields faster and more stable convergence.
3. **Handles multiple conflicting objectives:** By integrating timing, energy, and memory into the reward function, AMO-ESL naturally balances embedded constraints.
4. **Lower risk of premature convergence:** PSO alone converges quickly but may get stuck; GA alone explores widely but is slow. AMO-ESL avoids both weaknesses.

TABLE 9.2 Comparison of scheduling algorithms on Cortex-M4.

Method	Deadline miss rate	Total energy (mJ)	Peak RAM (KB)	Convergence time
EDF	High	Medium	Low	Fast
RMS	Medium	High	Low	Fast
GA	Medium	Medium	Medium	Slow
PSO	Medium–high	Medium	Medium	Fast
GA–PSO (static)	Low	Medium–low	Medium	Medium
AMO-ESL (proposed)	**Very low**	**Lowest**	**Lowest**	Medium–fast

AMO-ESL, adaptive multiobjective evolutionary swarm learning; *EDF*, earliest deadline first; *GA*, genetic algorithm; *PSO*, particle swarm optimization; *RMS*, rate monotonic scheduling.

5. **Hardware-friendly implementation:** The algorithm is lightweight enough to be executed offline or on embedded companion processors.
6. **Easily extensible:** It can be adapted for DVFS, task scheduling, memory compression, NoC routing, and embedded AI model optimization.

9.7 Case study: Real-time task scheduling on ARM cortex-M using AMO-ESL

To demonstrate the practical effectiveness of the proposed AMO-ESL technique an experimental case study was conducted on a real-time task scheduling problem targeting an **ARM Cortex-M4** embedded platform. The ARM Cortex-M family is widely used in IoT nodes, wearable devices, sensor platforms, and industrial controllers; therefore it provides a realistic basis for evaluating timing, energy, and memory constraints (Altet et al., 2020).

9.7.1 Experimental setup

Hardware platform: ARM Cortex-M4 running at 80 MHz, 256 KB RAM, 1 MB Flash (Table 9.2).

Workload description: A periodic real-time workload consisting of 10 independent tasks with varying execution times, deadlines, and energy profiles. Each task Ti is defined by

Executiontime: Ci
Deadline: Di
Energyperexecution: Ei
Memoryrequirement: Mi

Comparison methods: The performance of AMO-ESL was compared against the following baselines:

1. GA-only scheduler
2. PSO-only scheduler
3. Hybrid GA-PSO (classic static hybrid)
4. EDF—theoretical baseline
5. RMS

Objectives: Optimize the following simultaneously:

- Minimize total energy consumption
- Minimize deadline misses
- Minimize memory usage and peak RAM footprint

9.7.2 Optimization problem formulation

The optimization problem is defined as

Minimize:

$f1$ = number of deadline misses
$f2$ = total energy consumption (sum Ei)
$f3$ = peak memory usage

Subject to

$\sum Ci$ for scheduled tasks $\leq$ central processing unit (CPU) capacity
$Mi \leq$ available random access memory (RAM)
Schedule length $\leq$ hyperperiod

All constraints are directly relevant to microcontroller-based embedded systems.

9.7.3 AMO-ESL implementation details

The AMO-ESL algorithm was implemented with the following configuration:

- Population size: 40
- Iterations: 150
- RL reward based on improvement in timing, energy, and memory
- GA crossover probability Pc = adaptive
- GA mutation probability Pm = adaptive
- PSO parameters (ω, $c1$, and $c2$) controlled by RL

The RL agent observed the system state:

$$\text{State} = (\text{deadline\_misses, energy\_level, memory\_usage})$$

Based on the state the RL policy adjusted the GA/PSO exploration vs. exploitation balance at runtime.

9.7.4 Results and analysis

Table 9.1 summarizes the performance achieved by each technique when executed on the Cortex-M4 workload.

9.7.4.1 Interpretation

1. **Deadline miss rate:** AMO-ESL reduced deadline misses by approximately **47% compared to static GA–PSO** and by over **70% compared to EDF and PSO**. This improvement results from RL dynamically tuning algorithm parameters in response to workload patterns (Zhang et al., 2019).
2. **Energy consumption:** AMO-ESL achieved the **lowest total energy**, because of its ability to allocate low-energy tasks to earlier slots and defer high-energy tasks to optimal time windows (Xue, Fan, & Li, 2017).
3. **Memory footprint:** Memory usage improved because AMO-ESL favored task orderings that minimized stack and buffer overlap, reducing peak RAM by **12%–18%** relative to other hybrids.
4. **Convergence time:** Despite its hybrid complexity, AMO-ESL converged faster than GA-only and comparably to standard PSO, thanks to PSO's fast exploitation phase.

9.7.5 Discussion of embedded-specific impact

The results clearly demonstrate that AMO-ESL is highly suited for embedded systems with strict timing, energy, and memory constraints:

- **Timing:** RL dynamically penalizes schedules with growing lateness, steering the algorithm toward timing-feasible regions.
- **Energy:** PSO's incremental updates allow fine-grained minimization of high-energy transitions and unnecessary CPU cycles.
- **Memory:** The multiobjective formulation penalizes peak RAM usage, making the method effective in memory-constrained microcontrollers.

The method's adaptability gives it a significant advantage over static hybrid methods, which do not adjust behavior when workload characteristics change.

9.7.6 Generalizability to other domains

Although demonstrated on ARM Cortex-M scheduling, AMO-ESL can be applied to

- DVFS-based power control
- Memory allocation and fragmentation minimization
- Sensor-node lifetime optimization
- field-programmable gate array (FPGA) operator scheduling
- NoC routing with congestion constraints

Its modular design allows GA, PSO, and RL components to be substituted with other evolutionary or learning mechanisms.

9.8 Discussion

The comprehensive review of hybrid computational techniques and the experimental evaluation of the proposed AMO-ESL algorithm offer several key insights into the current and emerging landscape of embedded system optimization. Hybrid methods continue to demonstrate substantial performance advantages over single-technique approaches, and the findings presented in this chapter reinforce the value of integrating evolutionary search, swarm intelligence, and adaptive learning strategies. At the same time the proposed AMO-ESL method highlights the benefits of dynamically managing algorithmic parameters through RL, which enables more effective handling of the complex, multiobjective constraints typical in embedded workloads.

9.8.1 Key insights

A central insight emerging from both the literature review and the case study is that **static algorithmic configurations are fundamentally inadequate** for modern embedded systems. Embedded workloads frequently vary in timing intensity, energy demands, and memory requirements, meaning that fixed evolutionary or swarm parameters rarely remain optimal across different workload phases. Techniques such as GA and PSO perform well under specific conditions but degrade when system dynamics shift, resulting in decreased convergence quality or increased constraint violations (Xue, Fan, & Li, 2017).

The AMO-ESL method addresses this challenge by integrating RL as a real-time controller, enabling continuous adaptation of GA and PSO parameters. This allows the hybrid optimizer to respond effectively to fluctuating task characteristics, resulting in improved timing predictability, lower peak energy usage, and better memory footprint management. This adaptive capability aligns closely with trends in edge computing and IoT systems, where workloads are often nonstationary and influenced by external environmental factors (Zhang et al., 2019).

Another insight is that **multiobjective formulation is essential rather than optional** for embedded environments. Optimizing only timing or only energy may lead to unacceptable trade-offs in other dimensions. The AMO-ESL approach, by encoding timing, energy, and memory in the reward and fitness functions, provides a balanced solution that respects all major constraints simultaneously.

9.8.2 Comparison with current research trends

The research trends highlighted earlier show a clear movement toward **learning-enhanced hybrid algorithms**, especially those that integrate ML into metaheuristic optimization. RL, deep learning, and data-driven parameter tuning are increasingly seen as mechanisms for improving adaptation in dynamic resource-constrained environments (Choi et al., 2018; Venkatesh, Kumar, & Choi, 2020).

The AMO-ESL framework fits squarely within this trend but advances it further by employing RL not merely as an auxiliary tool but as a central controller that shapes optimization behavior continuously. This is consistent with the shift in embedded systems from static heuristics toward **self-optimizing and introspective systems**, a direction increasingly visible in intelligent edge computing.

A second trend is the increasing interest in **hybrid multiobjective optimization for embedded AI**, particularly involving model compression, pruning, quantization, and NAS. Although the present case study focused on real-time scheduling, the

AMO-ESL model is generalizable to embedded AI tasks because the RL-controlled parameter adaptation addresses the same issues of balancing accuracy, latency, and memory consumption.

Finally, hybrid swarm-based techniques are gaining traction in physical system domains such as network-on-chip routing, distributed sensor networks, and cyber-physical manufacturing systems. The modular nature of AMO-ESL makes it compatible with these areas by replacing GA or PSO components with alternative heuristics or swarms when suitable.

9.8.3 Limitations

Despite strong performance, several limitations of the proposed hybrid and hybrid methods in general must be acknowledged. First, **parameter sensitivity remains a challenge**. Although RL reduces the burden of manual tuning, the reward design, state definition, and exploration strategies of RL itself require careful calibration. Poorly shaped rewards can lead to oscillating behavior or suboptimal parameter choices.

Second, hybrid methods typically incur **higher computational cost** than single-method optimizers. Even though AMO-ESL demonstrated manageable convergence time in our case study, the inclusion of RL, GA, and PSO introduces overhead that may not be acceptable for ultralow-power or hard real-time systems with strict execution deadlines.

Third, **limited real hardware validation** remains an issue across the literature. Many existing hybrid optimization studies rely heavily on simulation or approximate models of embedded systems. Although our case study used an ARM Cortex-M microcontroller, broader validation on heterogeneous multicore embedded platforms, GPUs, FPGAs, or NoC-based SoCs would strengthen generalizability.

Fourth, **explainability** of hybrid methods is still weak. Evolutionary, swarm, and RL algorithms are often viewed as black-box optimizers, which may limit adoption in safety-critical domains such as medical devices or autonomous control systems, where transparent decision-making is legally required.

9.8.4 Scalability considerations

Scalability represents both an opportunity and a challenge for hybrid optimizers in embedded systems. The experiments presented here demonstrate that AMO-ESL scales well to medium-sized workloads typical of microcontroller-class systems. The PSO component provides efficient convergence as problem size increases, whereas the GA component ensures sufficient search diversity.

However, several scalability concerns remain

1. **Population growth problem:** Larger search spaces require larger populations, increasing computational cost. RL can mitigate this by reducing unnecessary exploration but cannot eliminate the fundamental scalability constraint of population-based methods.
2. **High-dimensional embedded optimization:** Tasks such as multicore scheduling, NoC routing, and embedded AI model search introduce high-dimensional search spaces. Although AMO-ESL's adaptive behavior helps, further enhancements such as dimensionality reduction or surrogate modeling may be required.
3. **Distributed embedded systems:** Scaling the method to distributed IoT or cyber-physical systems requires careful coordination, possibly introducing multiagent RL mechanisms.
4. **Hardware-level scalability:** As embedded platforms increasingly include specialized accelerators, heterogeneous cores, and energy islands, hybrid optimization must account for architectural complexity. Incorporating hardware-aware models into AMO-ESL represents a tangible future extension (Choi et al., 2018).

Therefore the scalability profile of AMO-ESL is promising but not unlimited. Its adaptive mechanism is well-suited to dynamic workloads, but more research is required to extend scalability to large-scale and hardware-diverse embedded environments.

9.9 Future research directions

Although hybrid computational techniques have demonstrated significant potential for embedded system optimization, several promising research avenues remain open. The increasing complexity of embedded platforms, the expansion of Edge AI, and the emergence of quantum optimization techniques create new opportunities to advance the field. Future research can build upon the strengths of current hybrid approaches while addressing scalability, adaptiveness, hardware awareness, and sustainability. Key directions include

- Hybrid deep learning + metaheuristics

Future studies can integrate deep learning architectures with metaheuristic optimization to enhance the adaptability and intelligence of embedded systems. Deep learning models can be used to predict execution times, energy consumption, workload patterns, and memory pressure, whereas metaheuristic engines optimize scheduling, DVFS policies, or memory allocation based on these predictions. This synergy could result in embedded systems capable of autonomously adjusting to environmental changes or workload variability. In addition, hybrid deep learning can assist in real-time decision support for embedded AI applications requiring continuous adaptation, such as autonomous drones, smart robotics, and health-monitoring wearables.

- Optimization for Edge AI

As embedded systems increasingly host edge-resident neural networks for inference tasks, research must explore hybrid optimization strategies that jointly minimize latency, memory usage, and power consumption. Hybrid methods combining RL, evolutionary search, and swarm algorithms could manage neural architecture compression, kernel placement, quantization levels, and operator fusion while accounting for embedded hardware constraints. Moreover, adaptive hybrid optimization can help embed intelligence in dynamic environments, where inference workloads, sensor inputs, or networking conditions change unpredictably. Edge AI systems will benefit significantly from hybrid techniques capable of balancing accuracy, real-time performance, and energy budgets simultaneously.

- Quantum + swarm hybrids

The rise of quantum computing opens opportunities to merge quantum optimization principles with classical swarm intelligence. Research can explore hybrid quantum–swarm algorithms, where quantum-inspired operators improve exploration, whereas classical swarm components handle exploitation. Such hybrid methods may solve complex embedded optimization problems such as multicore scheduling, NoC routing, and energy performance trade-offs more efficiently than classical techniques alone. Quantum annealing or variational quantum algorithms can work in tandem with PSO or ACO to significantly accelerate convergence in high-dimensional embedded workloads. This direction represents a frontier for next-generation embedded optimization systems.

- Green embedded intelligence

Sustainability is becoming a core requirement in embedded systems, particularly in IoT, sensor networks, intelligent buildings, and cyber-physical systems. Hybrid optimization methods can contribute to green embedded intelligence by jointly minimizing energy consumption, thermal footprint, processing overhead, and environmental impact. Future research may develop hybrid algorithms that incorporate sustainability metrics such as carbon-aware scheduling, renewable energy–aligned task distribution, thermal-aware optimization, and power-proportional computing. In addition, hybrid models leveraging learning-based predictors can forecast energy demand and adjust system behavior proactively. Such efforts will be essential in creating environmentally responsible embedded platforms for smart cities, agriculture, and distributed sensing infrastructures.

9.10 Conclusion

This chapter presented a comprehensive review of hybrid computational techniques for embedded system optimization and introduced a novel adaptive hybrid method, AMO-ESL, integrating GAs, PSO, and RL. The review demonstrated that hybrid techniques consistently outperform single-method approaches in addressing the multiobjective demands of embedded systems, including timing predictability, energy efficiency, memory utilization, and adaptability under dynamic workloads. By combining global exploration, rapid local refinement, and real-time parameter adaptation, the proposed AMO-ESL method bridges critical performance gaps observed in traditional and static hybrid optimizers.

The case study on ARM Cortex-M scheduling further validated the practical effectiveness of AMO-ESL, showing improvements in deadline adherence, energy consumption, and memory footprint. These findings highlight the value of adaptive hybrid optimization in embedded environments, where constraints evolve continuously and optimal decisions require context-aware control.

Therefore this work underscores the growing importance of hybrid, learning-enhanced optimization strategies in modern embedded systems. As embedded platforms expand in capability and complexity, particularly with the rise of edge AI, heterogeneous computing, and sustainability requirements, future research will benefit from deeper integration of learning algorithms, quantum-inspired search strategies, and hardware-aware hybrid designs. The contributions presented here offer a foundation for advancing the next-generation of intelligent, efficient, and adaptive embedded optimization frameworks.

AI disclosure

During the preparation of this work the author(s) used ChatGPT to assist in language polishing, grammar improvement, and generating text drafts. All scientific interpretations, methods, results, and conclusions were fully developed, verified, and approved by the authors. After using this tool/service the author(s) reviewed and edited the content as needed and take(s) full responsibility for the content of the publication.

References

Altet, M., Aldea, E., Figueras, J., Mateu, J., Perpiñà, X., Jordà, X., … Riu, P. (2020). Characterization of ARM Cortex-M processors for embedded workloads. *Microprocessors and Microsystems, 72*, 102907. https://doi.org/10.1016/j.micpro.2019.102907.

Aouad, L., & Schott, R. (2010). Hybrid genetic algorithm–simulated annealing for reducing energy in embedded systems. *LORIA*. Université Henri Poincaré.

Bambagini, M., Marinoni, M., Aydin, H., & Buttazzo, G. (2016). Energy-aware scheduling for real-time systems: A survey. *ACM Transactions on Embedded Computing Systems, 15*(1). http://www.acm.org/tecs/. doi:10.1145/2808231.

Boutekkouk, F., Belarbi, M., & Benmohammed, M. (2021). Real-time embedded systems scheduling optimization. *International Journal of Advanced Engineering and Computation, 12*, 31.

Buttazzo, G. C. (2011). Hard real-time computing systems: Predictable scheduling algorithms and applications. *Real-Time Systems Series, 24*. doi:10.1007/978-1-4614-0676-1.

Choi, J., Wang, Z., Venkataramani, S., Chandra, V., Srinivasan, V., & Gopalakrishnan, K. (2018). Deep learning model compression for embedded systems. *IEEE Transactions on Neural Networks and Learning Systems, 29*(10), 4730–4742. https://doi.org/10.1109/TNNLS.2017.2761740.

Das, S., & Suganthan, P. N. (2011). Differential evolution: A survey of the state-of-the-art. *IEEE Transactions on Evolutionary Computation, 15*(1), 4–31. doi:10.1109/tevc.2010.2059031.

Goodfellow, I., Bengio, Y., & Courville, A. (2016). *Deep learning*. MIT Press.

Holland, J. H. (1992). *Adaptation in natural and artificial systems*. MIT Press.

Herrera, F., & Magdalena, L. (1997). Optimization of fuzzy systems using evolutionary algorithms. *International Journal of Approximate Reasoning, 16*(3), 287–309. https://doi.org/10.1016/S0888-613X(96)00099-9.

Hong, J. (2002). *Memory optimization techniques for embedded systems*. LSU Digital Commons.

Kennedy, J., & Eberhart, R. (1995). *Particle swarm optimization, (ICNN'95), Vol. 4,* (pp. 1942–1948). Perth, WA, Australia: IEEE. https://doi.org/10.1109/ICNN.1995.488968.

Kumar, S., Panneerselvam, B., & Kim, H. (2019). ANN-based prediction for embedded system scheduling optimization. Microprocessors and. *Microsystems, 71*, 102–114. https://doi.org/10.1016/j.micpro.2019.102907.

Kitchenham, B., Pearl Brereton, O., Budgen, D., Turner, M., Bailey, J., & Linkman, S. (2009). Systematic literature reviews in software engineering—A systematic literature review. *Information and Software Technology, 51*(1), 7–15. doi:10.1016/j.infsof.2008.09.009.

Liu, X., Hou, K., De Vaulx, C., Xu, J., Yang, J., Zhou, H., Shi, H., & Zhou, P. (2015). Memory and energy optimization strategies for multithreaded operating system on the resource-constrained wireless sensor node. *Sensors, 15*(1), 22–48. doi:10.3390/s150100022.

Mafarja, M., & Mirjalili, S. (2017). Hybrid whale optimization algorithm with genetic operators. *Applied Soft Computing, 62*, 441–453. https://doi.org/10.1016/j.asoc.2017.11.006.

Oliveira, L., Mattos, J. C. B., & Brisolara, L. (2013). Survey of memory optimization techniques for embedded systems. In *Brazilian Symposium on Computing System Engineering, SBESC* (pp. 65–70). IEEE Computer Society. http://ieeexplore.ieee.org/xpl/mostRecentIssue.jsp?punumber=6472847. doi:10.1109/SBESC.2013.35.

Petersen, K., Feldt, R., Mujtaba, S., & Mattsson, M. (2008). Systematic mapping studies in software engineering. *12th International Conference on Evaluation and Assessment in Software Engineering, EASE 2008*. BCS Learning and Development Ltd https://dblp.org/db/conf/ease/ease2008.html. doi:10.14236/ewic/ease2008.8.

Petersen, K., Vakkalanka, S., & Kuzniarz, L. (2015). Guidelines for conducting systematic mapping studies in software engineering: An update. *Information and Software Technology, 64*, 1–18. doi:10.1016/j.infsof.2015.03.007.

Sivakumar, R., Vijayakumar, P., & Saravanan, M. (2020). Fuzzy–PSO hybrid optimization for IoT sensor networks. *Sensors, 20*(14), 3895. https://doi.org/10.3390/s20143895.

Sánchez-Oro, J., Duarte, A., & Gortázar, F. (2017a). Memory allocation optimization for embedded systems. *Applied Soft Computing*, 217–226. doi:10.1016/j.asoc.2016.12.034.

Sánchez-Oro, J., Duarte, A., & Gortázar, F. (2017b). Memory allocation optimization for improving the performance of embedded systems. *Applied Soft Computing, 52*, 104–118.

Venkatesh, G., Kumar, M., & Choi, K. (2020). Reinforcement-learning-guided metaheuristic optimization for embedded systems. *ACM Transactions on Embedded Computing Systems, 19(5), Article, 41*, 1–23. https://doi.org/10.1145/3397531.

Xue, R., Fan, X., & Li, H. (2017). GA–PSO hybrid scheduling for embedded systems. *Journal of Systems Architecture, 77*, 35–45. https://doi.org/10.1016/j.sysarc.2017.05.004.

Yassein, M., Shatnawi, M., & Aljawarneh, S. (2017). Hybrid GA–PSO clustering for wireless sensor networks. *Journal of Network and Computer Applications, 89*, 42–52. https://doi.org/10.1016/j.jnca.2016.12.021.

Zadeh, L. A. (1996). Fuzzy logic = computing with words. *IEEE Transactions on Fuzzy Systems, 4*(2), 103–111. https://doi.org/10.1109/91.493904.

Zhang, C., Venkatesh, G., Kumar, M., & Choi, K. (2019). Machine-learning-guided optimization in embedded systems. *ACM Transactions on Embedded Computing Systems, 18*(5).

Zhou, Y., Li, M., & Chen, J. (2018a). A hybrid PSO-based multi-objective algorithm for resource-constrained optimization. *Engineering Applications of Artificial Intelligence, 72*.

Zhou, Y., Li, M., & Chen, J. (2018b). Hybrid PSO-based multi-objective algorithm for NoC optimization. *Engineering Applications of Artificial Intelligence*.

Zhang, Y., Wu, C., & Zhao, L. (2016). PSO–ACO hybrid optimization for real-time embedded systems. *Future Generation Computer Systems, 56*, 444–450. https://doi.org/10.1016/j.future.2015.09.017.

Chapter 10

The Advancement of Robotics in the Mechatronics World and Construction Projects

Aditya Singh
Department of Civil Engineering, Amrita Vishwa Vidyapeetham, Coimbatore, Tamil Nadu, India

10.1 Introduction

In the study of Singh (2022b) the term "mechatronics" was explained to have initially come from the combination of mechanical engineering and electronic engineering, but with time, the advancement of software engineering, electrical engineering, automation, control engineering, robotics, and so on contributed toward the progress of mechatronics engineering, which led to the development of sophisticated intelligent products as well as systems. Basically, robotics is one of the subsets of mechatronics that particularly concentrates on the construction, operation, and programming in addition to the design of robots. They might be semiautomatic or fully automatic machines with the ability to accomplish certain assigned tasks as per their programming. Because of the beneficial applicability of robots in different areas, the recent decades have noticed fast development of robotics field to improve the quality as well as quantity of robots in the world in addition to expanding their applications. This process of innovation and advancement of human civilization led to the enhancement of mechatronic systems. This could be observed by the usage of mechatronic systems in different disciplines and the application of mechatronic systems as well as robots in the construction sector. However, it is important to notice that the advancement of computer science engineering not only led to the successful programming of robots but also the development of artificial intelligence (AI) and computational intelligence (CI). AI has become popular in today's world in many areas, but there is a difference between AI and CI, which must be understood. When CI is considered, then it is basically AI's subfield, which is about not only the design but also intelligent computer systems' deployment. Then, it is noteworthy to know that the CI-based systems have the ability to not only learn but also adapt to novel circumstances and environs, which are well appropriate for the tasks, which might be hard or dreadful for conventional AI-based systems. This means that CI is helping in the development of sophisticated mechatronic systems and robots, which are useful in many areas, including the construction industry. The reason behind their difference is that systems based on CI are normally motivated by natural intelligence of humans, utilizing techniques like evolutionary algorithms in addition to neural networks, which makes them learn from collected data, and then experience it as well as enhance their own performance with the passage of time, which allows them to work in free or unsure environs, whereas the conventional systems based on AI are rigid to only work in environs that are well-defined along distinct aims in addition to rules. Even though there are many various amins of CI, a few of the general aims are to allow the system, which is created to reason as well as to decide based on it, then learn and enhance system with time, then use new ways to deal with problems, to make the system adjust to the alterations in the given environs, as well as to make the system in a state to have human interactions possible. This helps in making sophisticated mechatronic systems as well as robots for diverse practical situations. The CI is aware about the context of the environs, and the algorithms used are able to keep track of output generation along with the accuracy essential to the given context, unlike AI. In other words, context clues are utilized by CI to proceed with solving a problem as per the context, and with time, the accuracy just improves because of the learning ability of CI. It can be said that onboard sensors as well as GPS help in gathering data, which is used as input by the CI algorithm first. Then the said algorithm decides each sensor's relative accuracy by utilizing the prior measurements, which are extremely accurate, to proceed further. Then, this

Computational Intelligence in Mechatronics: Solving Real-World Problems in Electronic Systems Design. DOI: https://doi.org/10.1016/B978-0-443-27580-7.00022-6

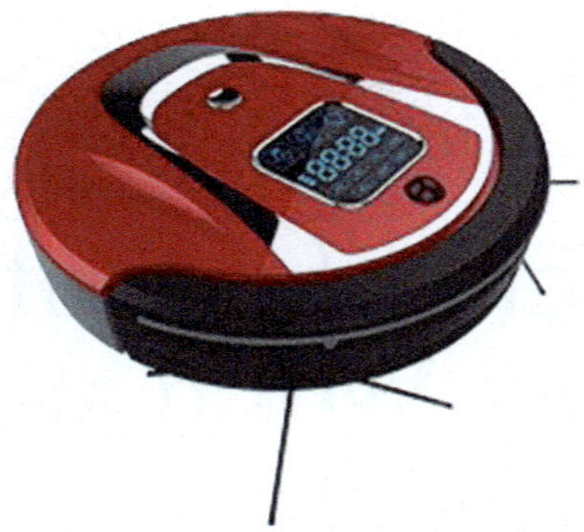

FIGURE 10.1 Domestic robot. (*Source: HomeHelper2014, Wikimedia.*)

FIGURE 10.2 Meituan autonomous vehicle. (*Source: Anonymousfox36, Wikimedia.*)

algorithm is able to learn the unsurety of not only the data but also the environs. At last, the reoptimization of the algorithm takes place in real time to determine the best sources of data to utilize for the purpose of positioning based on the suitability of the context. Further, CI has four chief methods, which are

Neural networks: They are normally good at detecting patterns as well as preferred for tasks such as categorizing images, etc.
Fuzzy logic: They are suitable for data that is unclear as well as utilized for objects such as control systems.
Genetic algorithms: They are appropriate to optimize answers and are preferred to be utilized for things such as allocation of resources.
Evolutionary programming: They are best at searching for solutions, which could be near optimal as well as preferred to be utilized for things such as robotics.

The abovementioned techniques of CI have their own advantages and disadvantages, which are used based on the requirements of the designed system. Then, there is ML, known as machine learning, which is a technique that permits a given computer to learn from the provided data as well as enhance its own performance, including prediction in addition to categorization. Similarly the technique named "natural language processing" is used to understand the language of humans, which might be applied to machine translation and chatbots, in addition to answering the questions of humans. Likewise, another technique named image recognition is applied to recognize faces, detect objects, and search images. Such techniques, when used in robotics, make the robots advanced to lean and improve their performance, detect objects to differentiate and interact with humans in better way to perform their given assigned tasks. Some examples might be domestic robots and autonomous vehicles, in addition to industrial robots, which are also present in the construction industry (Figs. 10.1 – 10.3).

Even though robotics was briefly explained earlier on the last page, it is essential to understand it in detail. Robotics is considered a study in addition to practice, which is interdisciplinary in nature, that is associated with the construction, design, utilization, as well as operation of machines known as robots. If mechanical engineering is taken into consideration, then the abovementioned term is limited to the physical structure's construction in addition to the design of the said machines, whereas computer science engineering concentrates on the automation algorithms in the said machines when the abovementioned term is talked about. However, robotics is also assisted by many other disciplines and subdisciplines of engineering, such as software, electrical, information, control, mechatronic, electronic, as well as materials. The aim of

FIGURE 10.3 Industrial robot. *(Source: Haophuong21, Wikimedia.)*

this area is to make and improve the machines, which are known as robots, in addition to working on automation, as these machines are expected to either substitute or replicate the actions of people in addition to assisting them. Then the word "robot" itself came from the Czech language, where the original word "robota" meant "labor," which is forced in nature, implying that these machines were meant to perform forced labor or programmed labor in other words. This word was first found in the year 1920 RUR play where the characters were referred to as large-scale workers unable to have the ability to think critically. Basically, these machines are comfortably able to perform cyclic and primitive tasks with higher accuracy as well as efficiency in comparison to people, which is why they have become a useful entity for many areas. Although the history of robots starts from the ancient Greece, it is noteworthy to consider that programmable as well as digitally operated ones came in the middle of the last century. They then were soon found in main industries such as automotive manufacturing, where the prototypes of robots were utilized for spot welding in the facilities of General Motors by the year 1961. Then, with the passing of time, highly advanced machines spread into several industries. Then, with the assistance of AI as well as CI in robotics, the robots were able to be developed to take care of difficult, real, unexpected situations in numerous industries. In other words the first generation of robots were manipulators to move in addition to picking up the given objects in simple movements. The second generation of robots were learning ones that could collect info from their environs to allow intricate movements. Then the third generation of robots were reprogrammable ones, having sensors as well as utilizing programming languages to change their functions as per the conditions applied at any point of time. Whereas the fourth generation of robots were mobile ones, which are considered the first intelligent robots having the ability to understand the given environs in real time to perform tasks. Finally the fifth generation of robots are the ones having AI, which are still under development, with the intention of mimicking people as much as possible in addition to being autonomous. Simply said, at current times, robots could be found across the world, starting from the clubs in colleges and universities to companies, including Fortune 100s, in addition to people's desire for them for various purposes, raising the market value to more than $27,000,000,000 by the year 2020. Robots are not only being used for repetitive, tedious, unpleasant work but also dangerous work that is harmful for people to do. This might include different work such as simple monitoring; cleaning; assembling; transporting given supply; exploring mines; and space, in addition to shipwrecks, as well as searching for surviving humans in ruins or difficult places for people to enter. Then, it is time to focus on the main aspects of robotics, which are

- Mechanical
- Electrical
- Software

Mechanical aspect: This involves mechanical construction of robots to have a desired physical body to accomplish the assigned tasks for certain environs. For instance, titanium tubing was used to make individually motor-powered wheels for the Mars 2020 Rover, which could aid it to comfortably move on the red planet's rough terrain.

Electrical aspect: This involves electrical parts that are necessary for robots to work smoothly, not only to provide power but also to control the machines. Fundamentally, electric current is needed to flow in the robots for slick operations, which can be provided with a battery or charging source or some form of energy conversion to electric current in robots.

Software aspect: This involves certain level of computer programming, which might be simple programming or complex programming in robots based on requirements. This is important because the set of codes makes the robots different from

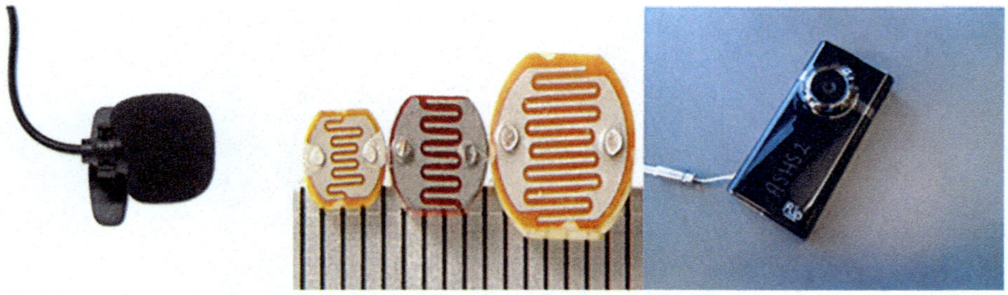

FIGURE 10.4 (L) microphone, (M) photoresistors, (R) video camera. (*Source:* 多多*123, Wikimedia; Junkyardsparkle, Wikimedia; Mosborne01, Wikimedia.*)

simple machines by telling them what to do. The robots are able to obtain the capability to know the way and time to perform their assigned tasks with the assistance of the embedded program in them.

With the above main aspects in mind, it is essential to know the key parts of robots, which are as follows:

- Sensors
- Control system
- Actuators
- Effectors
- Power supply

Sensors: They give electrical signals as impetuses to the robots, which the control could process as well as permit the robot to interact with the external environs. The general sensors used in robots are microphones, which work like the ears; photoresistors, which react to incoming light; as well as video cameras, which work as eyes. Then robots might be embedded with barometers to sense pressure, or thermometers to sense temperature. Robots might have embedded magnetometers as well as accelerometers to sense their movement based on the magnetic field as well as gravity of the planet Earth. Other robots might be embedded supersonic sensors or LiDAR sensors to make three-dimensional (3D) maps of their environs while navigating the world. The abovementioned sensors permit the robots to capture its environs as well as aid in the processing of the best logical decision on the basis of its present movement in addition to permitting the controller to send commands to its other parts (Fig. 10.4).

Control system: This comprises all the parts that built up the central processing unit of a robot for the purpose of computation. This system is programmed to inform the robots the way to use its particular parts just the way the signals are being sent by the brain of people throughout their whole body to accomplish particular tasks. In robots the tasks might be anything, including packing assembly line in addition to minor invasive surgeries and so on. This system is essential in deciding the behavior of robots. Broadly, robots can be either preprogrammed or autonomous, based on their complexity of control systems. In the former case, robots just repeat the same preprogrammed operation again and again, as they are either able to identify very inadequate info about the restricted constituents of the environs or they are indifferent to the changes in the environs. In the latter case, robots are complex in nature, as they are able to respond suitably according to the changes in the environs, which means their behavior or course of action changes after detecting the changes in their surroundings. The robots can have mostly two types of control systems, which are top-down, meaning brain must be built first, then effectors in addition to sensors later on, and the other is bottom-up, which means forget about the brain and sensors must be directly connected to effectors, where each pair of sensors—effectors has priority, making them behavior-based robots (Figs. 10.5 and 10.6).

Actuators: For a device to be counted as a robot, it must have either movable body or frame. The said movement is only possible with the assistance of the parts known as actuators in robots. The control system sends signals to these parts, which are built of motors, and then they move in cycle to incorporate the movement important to accomplish the given tasks. They are generally made of different materials such as elastic or metal, which are normally worked by utilizing oil known as hydraulic actuators or compressed air known as pneumatic actuators. They appear in different formats to appropriately complete their particular roles (Figs. 10.7 and 10.8).

Effectors: They are constituents of robots that are responsible for doing the real work. They could be any type of tool that could be mounted on the robots as well as controlled by the control system. Normally, they are particular to the assigned tasks, which people want the robots to perform. For instance, apart from general effectors such as wheels, motors, legs, actuators, end-effectors, which are used by many robots, other effectors could be used, such as mass spectrometer and shovels, in addition to hammers, which were used by the Mars rovers to analyze the red planet's soil. Definitely, robots

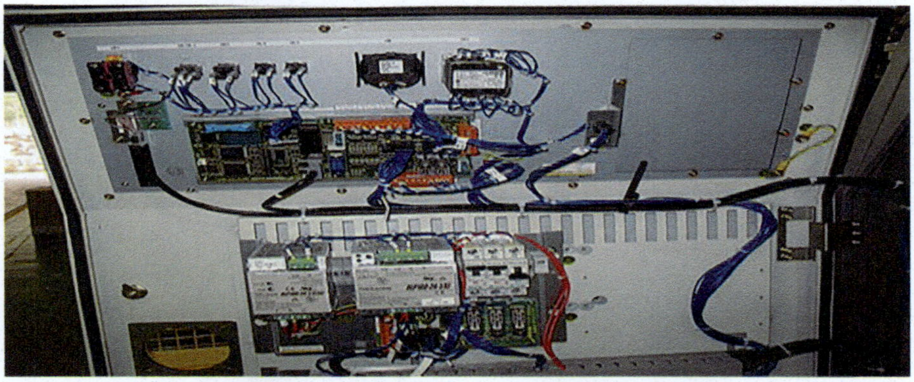

FIGURE 10.5 Control unit. *(Source: Mixabest, Wikimedia.)*

FIGURE 10.6 Control system basic scheme. *(Own/Author Created.)*

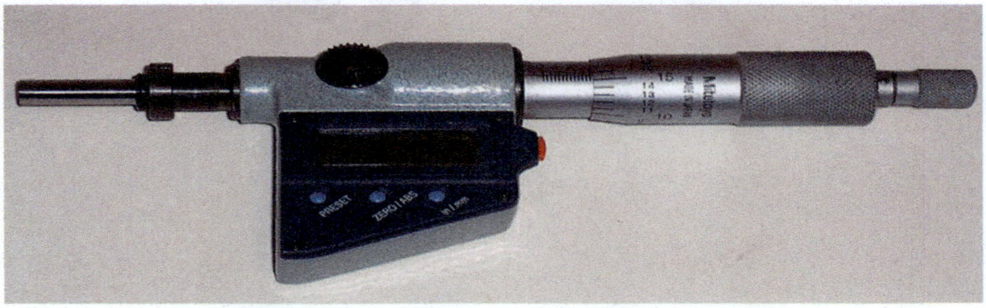

FIGURE 10.7 Linear actuator. *(Source: Rsteves00, Wikimedia.)*

used for mail delivery would not require those effectors. In the case of end-effectors, which are subset of effectors, include suction cups, grippers, welding torches, etc. (Fig. 10.9).

Power supply: Robots need power, just the same way people need food in their bodies to work. Robots that are stationary, normally seen in factories, might run on alternating current power coming from a wall outlet, but generally robots operate through an internal battery present in them instead. The batteries could be either costly silver–cadmium one or inexpensive lead–acid one. In addition, future robots are being developed to use solar power, pneumatic power coming from compressed gasses, hydraulic power, and so on (Fig. 10.10).

Furthermore, there are teleported robots, which are normally able to work in highly risky environments, situations, and weather, in addition to geographical-based conditions. Such robots might perform through the utilization of wireless networks to allow the control of humans from a distance, which could be considered safe. Some instances of such robots are drones sent on battlefields to identify landmines or, in the case of BP oil spill, human-controlled submarines utilized to solve the leakage in the underwater pipe.

10.1.1 Objectives

There are several objectives of this chapter, which are as follows:

- To understand the concept of mechatronics systems, mechatronics engineering, and robotics.
- To understand the difference between CI and AI, as well as types of CI.
- To understand the relationship between mechatronics systems, robotics, CI, and AI.
- To understand the impact of robotics on mechatronics systems and construction industry.

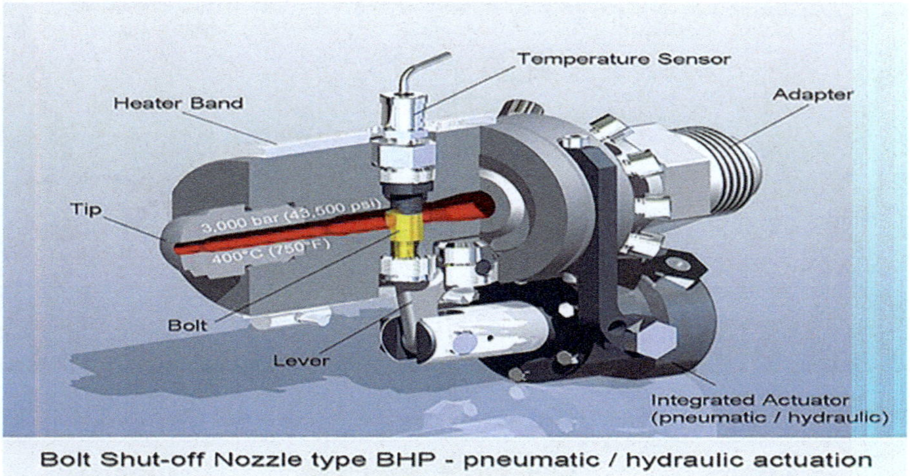

FIGURE 10.8 Integrated actuator. (*Source: Mtsunny, Wikimedia.*)

FIGURE 10.9 End effector. (*Source: Chojitsa, Wikimedia.*)

FIGURE 10.10 Lead–acid battery six cells. (*Source: Tntflash, Wikimedia.*)

- To understand the trend and future of robotics in mechatronics and construction projects.

10.1.2 What is the need and use of advanced technologies in the construction projects?

The construction industry is an old industry but usually observes slow adoption of advanced technologies in comparison to other industries and sectors. However, for any industry's growth, the adoption of advanced technologies is necessary to lead it to the next level. Singh and Srivastava (2024) mentioned in their study how AI as well as green AI could aid in the development in a sustainable manner in civil engineering field. Then the way artificial neural networks are useful not only at present but also in the future in the field of civil engineering was taken into account in the study of Singh (2023b).

Singh (2022a) mentioned fuzzy logic is useful in transportation, but it is also useful in construction when applied in estimating cost, risk assessment models, scheduling, predicting productivity, and so on. Then, Singh (2022b) also focused on the way mechatronics engineering is not only developing but also transforming the construction industry in the process. Moreover, Singh (2023c) discussed the digitalization of the construction sector, where he talked about the reasons behind its digitalization and the positive impact on the said sector. Then the way AI could encourage toward sustainable practices was covered in the study of Singh (2023a). These advanced technologies make the construction projects safer and more productive, enhance quality, aid in meeting deadlines, and encourages toward innovation, which are normally hard to meet in the construction sector in the world.

10.2 Motivation

The author performed the study on robotics because in recent decades, robotics has flourished more than ever, and the future is looking bright for it. Similarly, that is the case of AI and CI, which have improved robots further; that could be noticed in the difference between the fifth generation of robots from the past generation of robots. As mechatronics world is influencing other areas, including construction, the author felt the need to investigate the way robotics is affecting the world of mechatronics, making the mechatronics systems smarter and highly useful in different ways, which further is slowly and steadily transforming the construction sector by making the sites safer, taking care of the laborious repetitive tasks, improving the accuracy, precision, and quality of work performed during the course of construction projects, and so on.

10.3 Literature review

Musarat et al. (2024) studied construction industry, and they worked on whether using robotics to replace the mentioned industry's workforce is a pointless or smart method. They found that it is important to incorporate robotics in the abovementioned industry to improve productivity, but at the same time not enough research is done in this matter. They used SPSS software to analyze the data collected from their case study–based method, to address the problem of ineffectiveness as well as low profitability in the projects related to the above industry. Their results showed that 87% as well as 66.7% of the respondents agreed regarding the productivity advantages and awareness of robotics in the industry, respectively.

Pan et al. (2024) studied construction, and they worked on an important review study about the use of robotics in addition to extended reality in their studied area. They discussed the way as well as the reason behind the incorporation of robots as well as XR in the abovementioned industry. They further proposed a novel typology to direct the combined research of robots as well as XR in the said study area. They then mentioned that interaction between robots, XR, and humans should be 1:1:1 in their study and how extended reality improves the operation of construction robots by increasing communication as well as control.

Albeaino et al. (2025) studied construction, and they worked on robots with four legs in research related to virtual reality to examine the effect of proxemics on the collaboration of robot and human in terms of safety in their studied area. They detected the possible risks as well as consequences of the presence of robots on the safety of humans at the site. They stressed the design of robots to be focused on humans to make the sites safe.

Bruun et al. (2024) studied robots and worked on powering the old way to focus on reducing masonry vaulting strategies with the efforts in the direction of collaborative robotic construction, which included seven such strategies' evaluation. They introduced not only the main parameters of structural design but also a categorization method on the basis of mechanisms, which are self-sustaining in nature. They stated that their manuscript examined the viability of changing earliest old techniques of construction to the assembly ways of cooperative robots to reduce the need of centering in the masonry vaults.

Song et al. (2024) studied the assembly of modular construction, and they worked on the design of robots, particularly the hybrid pose adjustment ones, which were intended for the control of prefabricated modules in their studied area. Such robots for six degree-of-freedom module control were considered, and then they made sure minimum interference is there while incorporating a hoisting platform in addition to tower crane, to attain a motion-producing strategy on the basis of a model.

Lee and Han (2024) studied construction robots, which were on the basis of vision and intended for programmed welding in real time with the assistance of the interaction between robot and human. They tried to attain higher robust joint identification by enhancing the dataset of welding joints. They then used image data as well as worked on the aspect of robotic control in addition to the planning of path in the real time, which was intended for laser scanning with the said data.

Dörfler et al. (2024) studied mobile robots, and they worked on 3D printing hurdles as well as likelihood in the context of repair in addition to maintenance of buildings, to develop construction in the current environments. They provided perception into possible scenarios where it can be applied in terms of architectural aspects intended for strengthening and repairing, in addition to revitalizing structures like buildings. They emphasized the importance for a cooperative method in developing the area of 3D printing technology to deal with the intricate challenges, which are interdisciplinary in nature.

Zhou et al. (2024) studied construction robots, and they worked on a model of construction workers on the basis of a transformer, which was optimized, to create a valuation on their studied area. They stated that assessing poses of construction workers is crucially necessary for identifying dangerous behaviors while measuring productivity in addition to performing analyses like ergonomics.

Ohueri et al. (2024) studied construction 5.0, and they worked on a collaborative of robot as well as human, which was intended for the deconstruction of buildings in their studied area. They performed a review, which consisted of mixed approach of university research on the stated collaboration in the context of their studied area, and they found through scientometric mapping that the trend in this area so far was limited. They further stated that they were able to detect three latest achievements in this area with the assistance of a qualitative review.

Zheng et al. (2025) studied the cooperative manufacturing of robot as well as human, and they worked on a method named semantic map construction, which was aimed for their studied area. Apart from the mentioned method, they also proposed assisted techniques in their research, which incorporated geometric as well as semantic info of the abovementioned manufacturing environ.

Wang et al. (2024a) studied robotics, and they worked on the lineage of their studied area in the context of construction. They further proposed the lineage idea to create the categorization of such robots on the basis of functional features and lifecycle in addition to timeline. They then discussed 48 kinds of such robots in the construction life cycle and found that most of the robots in the market concentrate on the construction phase in projects.

Yang and Cai (2024) studied robots used in construction, and they worked on improving the visual simultaneous localization and mapping for them with the effective incorporation of scene semantics in addition to object segmentation, which is dynamic in nature. They particularly focused on construction environs, which are dynamic ones, and they then proposed a vSLAM-Con system, which presents an object segmentation approach, especially one that is not only dynamic but also adaptive in nature to give suitable results with the assistance of a novel selection mechanism named AD-keyframes.

Ku et al. (2024) studied construction, and they worked on improving its efficiency as well as making it safer with the assistance of tower cranes, which are robotic ones. They used multimodal sensing as the basis for saddle-jib cranes, which were robotized for their study. They mentioned that for point-to-point movement, to generate a programmed safe path, and then hoist-first method was proposed by them, which used replanning to avoid obstacles in the dynamic scenario.

Ye et al. (2024) studied robots used in the construction, and they worked on an enhanced genetic algorithm to accomplish two-stage task distributions for numerous said robots as per their studied area. They took three limitations into consideration to develop a grouping in addition to task decomposition method in construction context, and they further created a task distribution model while taking many objectives into consideration for numerous robots. They then performed a test to confirm the viability of their proposed approach to complete their research.

Baek et al. (2024) studied construction, and they worked on finding out the way emotional responses affect the team comprising robots as well as human. They used real robots to simulate bricklaying task by the collaborative of robot and human in laboratory environs. They gathered electroencephalogram signals to analyze emotional responses.

Zeng et al. (2024) studied robots that are used in construction, but particularly the automatic mobile ones, to be suitable for built environs by performing a thorough review study. They checked the conventional applications as well as main technical the abovementioned robots, and they further examined the obstacles for such robots from laboratory to their deployment in the real world through their study.

Pan and Yu, 2024b studied robots, and they worked on the way to optimize the team composition of the studied area having heterogeneous entities by the assistance of simulation, which is directed by use in addition to time, to be aimed at extended tasks in construction. They then proceeded to develop it as a collaborative board game and used Monte Carlo tree search method to approach the scheduling of extended tasks in construction.

Wang et al. (2024b) studied the cooperative construction of robot as well as human and worked on allowing Building Information Modeling (BIM) led by the workflows based on their studied area with the assistance of digital twins, which are closed-loop ones. Starting from the stage of preparation till the final stage where the work ends, they provided not only physical but also technical setup solutions to obtain desired results.

Adel et al. (2024) studied robots, and they worked on timber construction led by numerous ones as per their studied area, which is feedback-led as well as adaptive in nature. They discussed reasoning in addition to perception to take care of ambiguity in the abovementioned type of construction.

Wang et al. (2024c) studied robots, and they worked on incorporation of plastering in addition to spraying of mortars through their studied area, which was aimed at the construction of walls. They then developed a robot to perform the two abovementioned tasks, and they found that if the angle of scraper is between 12 degrees and 37 degrees, then the attained compaction force is appropriate.

Kim et al. (2024) studied construction robots, and they worked on comprehending the proxemic customs between robot and human in the context of construction to figure out the way people route around the said robots. They stated that for the abovementioned proxemics, theoretical foundation was laid by them through their study, and they used actual professionals related to construction to perform their subject experiment in addition to making a construction setting simulation.

Zhu et al. (2024) studied construction planning, and to make it flexible, they worked on not only the interaction but also the way to link data amid operating system, which are robotized ones and BIM. To proceed with their study, they proposed a system of middle layer named robot construction action node and a way to transform IFC construction plannings in addition to models into a combined RDF graph to produce desired outcomes.

Duan and Zou (2024) studied robots used in construction, and they worked on mapping the morphology of dubious gestures, which were aimed at natural teleoperation of their studied area. They mentioned that the existing problems are the abilities of robots to complete difficult tasks in free as well as dynamic settings at the site. They stated that the robots, which are teleported ones, are able to have human-level intelligence to perform tasks that are complicated in nature at the site, and the actual robot tests along with VR results displayed their effectiveness.

You et al. (2024) studied robots used in construction, and they worked on represented AI, which was aimed at adroitness able robots to introduce a framework named DEXBOT to focus on thorough understanding of the environs and decision-making in the real-time setting, as well as perception with high accuracy to solve the existing problems with construction robots.

Li et al. (2024) studied robots used in construction, and they worked on finding out their prospective to decrease the transmission of viruses, which are airborne, at the sites in China as well as in the United Kingdom. They grouped such robots into 8 themes with 25 classes with the assistance of a serious review of literature. They took on an interview with 4 construction contractors as well as online-based questionnaire with 32 experts from 2 countries to produce suitable results.

Leder et al. (2024) studied timber structures, and they worked on improving the examination as well as codesign of a group construction system, which was robotized to be aimed at the fabrication of large-scaled in-plane structures as per their studied area. They presented their research on it with the assistance of a new system known as collective robotic construction to accomplish their desired goals.

Park et al. (2024) studied construction work performed in the field, and they worked on instructions that are based on natural language aimed at natural interactions of humans with assistants in the form of robots. They introduced a framework for such interactions, and they proceeded to produce commands related to work to the abovementioned type of robots to accomplish their research.

Liu et al. (2024a) studied robots used in construction, and they worked on finding out their three pillars with the assistance of a quantitative analysis, especially a double-track one. They mentioned that the past studies were short of a thorough methodology at the macrolevel by analyzing 995 manuscripts published in journals to find out the abovementioned pillars, which were deep reinforcement learning and collaboration of robot and human, in addition to BIM.

Yarovoi and Cho (2024) studied the application of robots in construction, and they worked on examining Simultaneous Localization and Mapping (SLAM)'s methods and restriction in addition to background aimed at their studied area. They then mentioned mobile robots' localization at the site while taking exclusive concerns into account. They used a dataset related to construction, and they estimated the three new algorithms, which are known as LiDAR SLAM, to further focus on the challenges to point out the importance of improving them.

Faltein and Sukdeo (2024) studied the construction industry in South Africa, and they worked on finding out the way the incorporation of robotics affects their studied area to improve the quality, which is led by their culture. They especially evaluated the link between Eskom's robotics technology and the way they manage quality as well as their organizational culture to use one-way ANOVA chart to produce positive results.

Barjuei et al. (2024) studied 3D printing, especially the construction one, which is robotized, and they worked on introducing workflow that was digitized, aiming for the prefabrication in addition to checking of printability for their studied area, on the basis of planning through AI. They used Planning Domain Definition Language (PDDL) 2.1 on the basis of AI-created planners to propose their approach to produce operational sequences intended for robotic manipulator programmed with the abovementioned type of 3D printing.

Liang et al. (2024) studied robotics as well as AI, and they worked on their ethical aspects in the industries like construction and engineering in addition to architectural. They then detected the problems, which are ethical ones caused by the above two, and they proceeded to find out nine such issues, along with performing a methodological review to attain the desired outcomes.

Chen and Zhang (2024) studied the construction of roads, and they worked on the intelligent control in addition to sensing of scenes related to robotic construction of the roads. They stated that programmed control technology is foundation to enhance road robots, and they used a perception technology that was positional in nature, known as Real-Time Kinematic Global Navigation Satellite System (RTK-GNSS), to generate the favorable results.

Giwa et al. (2024) studied robotic construction, particularly focusing on the extraterrestrial one, which is waterless, to examine the performance of sulfur-regolith printed by a 3D printer for their studied area. They stated that the concrete with sulfur contents was effectively printed without any problem having Martian simulant in the higher fidelity to obtain improved mechanical strength than the conventional ones.

Wang et al. (2023) studied robots used in construction, and they worked on motion sequencing approaches to be programmed at a higher level to allow multitasking of the robot's type mentioned in their studied area. They then found out the matching skill primitive by using scene distance array, along with the experiment results displaying that such robots, while constructing a frame, make over 91% of decisions on the basis of learning.

Yu et al. (2024) studied robots that are used in construction, especially the aerial ones, and they worked on examining their pose on the basis of dynamic restrictions in addition to motion. They stated that such aerial robot's pose examination could be attained in an environ that is denied of GPS in the absence of visual-related features. They then developed strong factors to solve with potential outliers coming from sensors.

Fu et al. (2024) studied robots, and they worked on the cooperation of robot and human, which was aimed at manufacturing associated with modular construction with the assistance of examining the existing research. They then proceeded to classify the gathered literature into interaction levels, role of humans, in addition to tasks, to further disclose a present research highlight on the production of timber parts.

Huang et al. (2023) studied robots, and they worked on utilizing humanlike robotic hand to proceed with the efficient operation of tools related to construction. They showed two-stage learning on the basis of a control framework, and in a simulation, they successfully accomplished in grabbing six tools associated to construction.

Dodampegama et al. (2024) studied the applications of robots in addition to AI to transform the sorting of demolition as well as construction waste. They then found out the latest developments in robotics as well as in AI to program the sorting of waste in the abovementioned categories. They then structured their manuscript around six different research questions to perform a thorough examination of this domain to obtain suitable findings.

Pan and Yu, 2024a studied robots, and they worked on the unfinished visual observation in the direction of the cooperative of robot and worker in the construction projects by presenting multiple granular worker intent learning. They then proceeded to allow the detection of the abovementioned intents with the assistance of Medium and Long-Term (MLT) as well as ranked task models to attain desired outcomes.

Liu et al. (2024b) studied robots in construction, and they worked on an analyzing a patent to track the status of technology in addition to trends in them. They stated that the arrival of such robots has been feasible in terms of economical as well as technical aspects because of the decline in costs in addition to the abundance of microprocessors.

studied robots, and they worked on Augmented Reality (AR)-allowed cooperative of robot and human to balance the efficiency of sorting the waste generated by construction activities in addition to job-related health as well as safety. They used AR to create a model for the abovementioned sorting based on the collaboration of robot and human. They then incorporated the cognitive flexibility of humans in the waste sorting, which was robotized to attain favorable results.

10.4 Research gaps

After going through many published research and review studies in journals, proceedings, and research books, the ones from the last few years were considered further for this study. It was observed that a lot of work has been going on in the field of construction robots and related areas recently in the scientific studies. However, this chapter covers the way AI, CI, and robotics are able to improve not only the mechatronic systems but also the way they are enhancing the construction projects, along with some major case studies globally to support the chapter.

10.5 Main focus of the chapter

This chapter mostly focused on explaining the concepts of mechatronics, AI, CI, and robotics in general, as well as how they are related and work together in transforming construction sector to improve the quality as well as efficiency of construction projects, apart from the necessity of implementing such advanced technologies in projects. It also covered the issues and

FIGURE 10.11 AGV. *AGV*, Automated guided vehicles. (*Source: Wolfgang Klein, Wikimedia.*)

problems related to robots, along with the gaps in the current literature. Apart from that, it highlighted some case studies in the world as well as graphical analyses related to the way robotics is improving the construction projects and mechatronic systems.

10.6 Issues, controversies, and problems

Robotics is improving the world of mechatronic systems in addition to the construction projects at the site. However, the problems and issues with robotics applications in the above cannot be ignored, as robots do not have the creativity of humans, which limits their creative abilities for any given situation, which often occurs at the sites. Then, job losses are another issue with robotics, as unskilled or even semiskilled workers are replaced by the robots to do the same task at a high speed with superior accuracy; the same could be said for employees, which leads to unemployment for many people. Moreover, discarding the parts of robots that are disposable ones leads to waste contributing toward pollution and harming the environment. Apart from higher initial costs, the maintenance of robots is also expensive in comparison to human resources. Further, security risks related to data, indicating cyberattacks, points toward another problem, where essential data linked to IoT devices and robots could be misused. Then, existing employees or workers might oppose or be hesitant or unable to adapt with the advancement that robotics is bringing to the workplaces.

10.7 Robot types

Moreover, robots could also be categorized based on the sector or their function, as well as on their mobility, which are as follows:

10.7.1 Mobility basis

This is one of the major ways of categorizing robots based on the difference in their mobility. This can depend on their movement capacity and performance, in addition to their ability to make decisions, which separates them from one another:

- Automated guided vehicles (AGVs)
- Cobots
- Autonomous mobile robots (AMRs)
- Robotic arms or articulated robots
- Humanoids
- Microrobots
- Augmenting robots

AGVs: They are known as automated guided vehicles, and they are able to move along an already defined path. They normally need the supervision of people (Fig. 10.11).

Cobots: They are type of robot that is designed with the objective to work with people and assist them in recurring as well as risky jobs (Fig. 10.12).

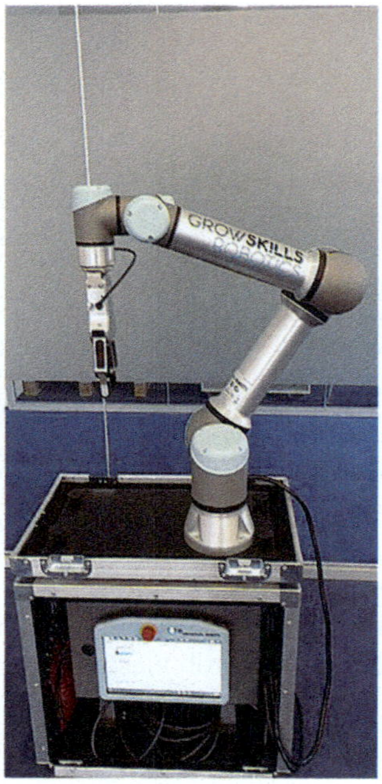

FIGURE 10.12 Cobot. (*Source: GrowSkills Robotics, Wikimedia.*)

FIGURE 10.13 AMR. *AMR*, Autonomous mobile robots. (*Source: PattyK33, Wikimedia.*)

AMRs: They are called autonomous mobile robots, and they are capable of not only moving but also making decisions by themselves, literally in the real-time in given situations. To perform their assigned functions, they utilized onboard processing equipment along with sensors (Fig. 10.13).

Robotic arms or articulated robots: These types of robots are considered best associates to handle tools, move a given product, package, and so on, but their capacity is really low.

Humanoids: Basically, they are a particular form of AMR, which appears similar to people, as well as capable of performing functions the same way humans usually do it (Fig. 10.14).

Microrobots: These robots are made to have a very tiny size, often considered miniature scale ones, around a millimeter to give a clear picture of them. They might vary in size based on the circumstances but are mostly used by researchers in the area of biotechnology (Fig. 10.15).

Augmenting robots: They are also called VR robots, which are made to either substitute the lost ability of a human or improve the present ability of a given person. This area, which is centered around augmenting people, has the potential

FIGURE 10.14 Humanoid. (*Source: Syced, Wikimedia.*)

FIGURE 10.15 Microrobot. (*Source: Alessio Costantini, Wikimedia.*)

to make the science fiction into reality with the ongoing progress in robotics. Currently, some instances of such robots are exoskeletons or prosthetic limbs.

10.7.2 Based on sector

Then, robots could also be categorized based on the sectors they are used in when targeting a particular environment, function, or objective related to that sector, which are as follows:

- Domestic robots
- Military robots
- Educational robots
- Industrial robots
- Medical robots
- Search and rescue robots
- Agricultural robots

- Social robots
- Space robots

Domestic robots: They are made to look after the house of humans as well as for domestic duties, including kitchen robots to prepare recipes and lawn mowing robots, in addition to vacuum cleaning robots.

Military robots: They are made to assist armies during particular operations, which might include the detection of explosives in a given area, as well as to transport some army equipment.

Educational robots: They are made to learn a subject or to plan for the development of cognitive ability, which might include kids using or playing with robotics kits.

Industrial robots: They are made to support different types of industries, to perform repetitive work or dangerous work, to increase productivity.

Medical robots: They are made to assist the healthcare sector, meaning aiding humans who are having decreased mobility and moving medicines as well as machinery, in addition to participating in the procedures related to surgeries.

Search and rescue robots: They are made to perform delivery of given supplies to people who are trapped in isolated areas because of some reasons, to save humans who are getting drowned or stuck in floodwater in a given place, or to help in extinguishing fires.

Agricultural robots: They are made to perform tasks that are labor-intensive as well as cyclic in nature, to permit farmers to utilize their energy as well as time effectively. They might include drones, autonomous tractors, and so on.

Social robots: They are made to help humans, including kids and adults for their personal needs, to enjoy some form of lacking company in their lives, as well as to perform some form of customer service.

Space robots: They are created to help in the space missions where it is difficult for humans to go and explore. This might include sending them to another planet, like the Mars rover.

10.8 Industrial and construction robots

Through the above types of robots based on the major sectors, it is evident that these sectors are utilizing mechatronics systems, and robotics is definitely aiding in the advancement of these sectors by improving the quality, safety, and decreasing the time consumption to complete certain tasks. However, it is noteworthy to pay attention to the industrial robots in detail compared to other types of robots. These robots might help in assembly, welding, disassembly, packaging as well as labeling, painting, inspection of products, handling of materials, palletizing, carrying and transporting objects from one location to another, etc., based on the industries and requirements. There can be different industries, including manufacturing transportation, as well as construction, which might require different types of assistance effecting the kind of robots to be used. As per the International Federation of Robotics the year 2023 observed over 4.2 million operational industrial robots throughout the world, showing their great demand in today's world. Moreover, the construction sector, which is another type of industry, uses robots that are known as construction robots and are basically considered a subset of the earlier mentioned industrial robots. These robots are utilized for the purpose of helping in the construction of infrastructure as well as buildings at the site. Even though the construction industry conventionally is slow to use novel technologies, it is said that around 55% of the companies in China and Europe, in addition to the United States, use robotics, based on ABB Robotics survey mentioned in Thibault's 2022 report. At the sites, majority of the robots at present are supposed to eliminate the strains that occur on people working at the sites, including the ability to lift heavy objects as well as excavation. The current market also observed the emergence of robots capable of tie rebar, drywall installation, and layout markers, in addition to survey. Then the development of robots with the ability to accomplish tasks like placement of steel, completing the exterior, working on reinforced concrete, as well as constructing masonry walls, and so on. However, workspace limitation is major hindrance in their usage at the sites in the real world. These types of robots have usual characteristics, which comprise the ability to handle parts of different weight as well as sizes, the ability to move, the ability to interact with its environment, the ability to complete numerous tasks, and the ability to adjust to the environmental changes. These types of robots had shown successful testing in investigating given infrastructures, particularly at unsafe locations, known as inspection robots. Then, they had been tested to monitor the progress of construction, in addition to construction of walls. Some remarkable feats accomplished by construction robots are in the construction of dam in Japan; to lay concrete layers, a system was developed by Obayashi Corporation. Then companies like Tokimec, in addition to Kajima in Japan, utilized robot to complete concrete floor finishing and automated system was utilized to construct Yokohama's Rail City Building, which is 30 stories in Japan. It is found out that Japanese people take construction robots in a positive manner, but the labor unions in the United States oppose the use of construction robots because of replacement of manual laborers at sites (Figs. 10.16 and 10.17).

FIGURE 10.16 Construction robot (robotic crane). (*Source: Carolinah 13, Wikimedia.*)

FIGURE 10.17 Obayashi building in Osaka, Japan. (*Source: J o, Wikimedia.*)

10.8.1 Robotic systems

This subsection explains some main robotic systems involved in the construction projects, which are the following:

- On-site automated robotic systems
- Exoskeleton wearable devices
- AVs and drones
- Off-site prefabrication systems

On-site automated robotic systems: Such systems take place at the site, where direct operation takes place, and the robots take care of various tasks such as welding, handling of materials, pouring of concrete, in addition to bricklaying. This decreases the dependence on manual labor at the site, which further increases safety of laborers or workers, as well as efficiency of the work (Fig. 10.18).

Exoskeleton wearable devices: They are able to aid workers at the site in specific tasks that need cyclic motions or heavy lifting. Thus the efficiency in mechanical aspects is linked with the efforts of people to improve productivity (Fig. 10.19).

AVs and drones: Also known as autonomous vehicles and drones, where AVs might be used for transporting or moving objects from one place to another, or to do certain tasks at the site. Then, drones might be used to survey, track, as well as monitor the progress of the work at the site, with the assistance of the installed sensors as well as cameras on it. They basically help in improving not only the operations but also logistics at the site.

FIGURE 10.18 Hadrian X. (*Source: FastBrickRobotics, Wikimedia.*)

FIGURE 10.19 Powered exoskeletons. (*Source: Yuichiro C. Katsumoto, Wikimedia.*)

Off-site prefabrication systems: This type of system comprises production of whole buildings or construction parts in a manner that is controlled in off-site environs. These parts are basically prefabricated, which are later on carried to the site so that they could be assembled (Fig. 10.20).

10.8.2 Some major types of construction robots

This subsection will discuss major types of construction robots, which are

- Stationary robots
- Mobile robots
- Swarm robots

Stationary robots: As the name suggests, these kinds of construction robots do not change their positions while operating. Here the word "stationary" in their name implies that these kinds of construction robots have their base fixed to one place when they operate. Under stationary robots, there are other subtypes of construction robots, which are

FIGURE 10.20 Autonomous grading machine. (*Source: Bettemir, Wikimedia.*)

FIGURE 10.21 Gantry Robot in INTEX 2025 in Bangalore, India. (*Source: Gpkp, Wikimedia.*)

- **Gantry robots:** These construction robots have a system that is overhead, which is linked to a manipulator that allows horizontal movements. They might be utilized for welding operations requiring picking as well as placing objects from one point to another point, in addition to 3D concrete printing in construction (Fig. 10.21).
- **Robotic arms:** These construction robots are also known as articulated robots, which could also be found in other industries, apart from construction industry. They are helpful in various operations at fast speed like materials handling, drilling, welding, thermal spraying, separation waste, painting, in addition to 3D concrete printing (Fig. 10.22).
- **Cable robots:** These construction robots utilize actuators as well as cables to perform their operation to carry out particular acts like to move the cables. When managing heavy loads, large area is being covered by them, making them special compared to others. For instance, complex brick structures, by utilizing 100s of bricks, could be done in an automatic manner by CU-Brick. Then, these types of robots are also used for 3D concrete printing.

Mobile robots: To accelerate construction, these types of construction robots are an innovative solution. As their name suggests, they are mobile in nature, and because of their mobility, the structures, which are larger sized ones in comparison to them, are still being able to be constructed by them. They are being utilized for various things because they have the upper hand on uneven terrain, which is very common at the sites. They are of various subtypes, which are

- **Wheeled mobile robots:** These construction robots use wheels for movements, so they contribute less to the wearing of the surfaces with their movements as well as being user friendly during operation. Hydraulic Excavator for an Autonomous Purpose (HEAP) is considered to be under this subcategory of construction robots, which are employed at the sites so that embankments with distinctive shapes are excavated by them in a precise manner and in recent years it has become a popular programmed walking excavator (Fig. 10.23).
- **Walking mobile robots:** These construction robots are efficient and versatile as well as able to be operated on uneven or soft terrains, which is often the case in construction at the sites. Their impact on the ground is minimum by them, as

FIGURE 10.22 Robotic arm. (*Source: AKAMGO yalms, Wikimedia.*)

FIGURE 10.23 HEAP robotic excavator. (*Source: Auledas, Wikimedia.*)

they are well stable and energy efficient in addition to having higher mobility in comparison to others. They are further suitable for navigating gaps or obstacles in the environs (Fig. 10.24).

- **Flying mobile robots:** These subtypes of construction robots are not required to use movements restricted to the ground like others, as well as scaffolding, which not necessary for them, which gives them an upper hand compared to other construction robots. Air navigation is done by them with their floating ability in the air. They are also good for surveying and even inspecting the sites during the course of the projects (Fig. 10.25).
- **Swimming mobile robots:** These robots are not much popular in architecture, engineering, and construction industries. However, these subtypes of construction robots have the ability to work underwater, to repair structures that are under the surface of water, and so on. For example, OceanOne robot is able to use sophisticated remotely controlled vehicles as well as implement the way humanoid robots behave, which might be used for repairing bridge piers under water, mimicry of other actions similar to people working under water on any structure, to retrieve objects, and so on.

Swarm robots: These kinds of construction robots allow synchronized implementation of tasks as well as cooperation, making them stand out from others. In simple words, numerous small robots make swarm robots, which do not need any guidance to make any design related to construction in this context. To prevent collisions among themselves, they follow a

FIGURE 10.24 ANYmal robot. (*Source: Auledas, Wikimedia.*)

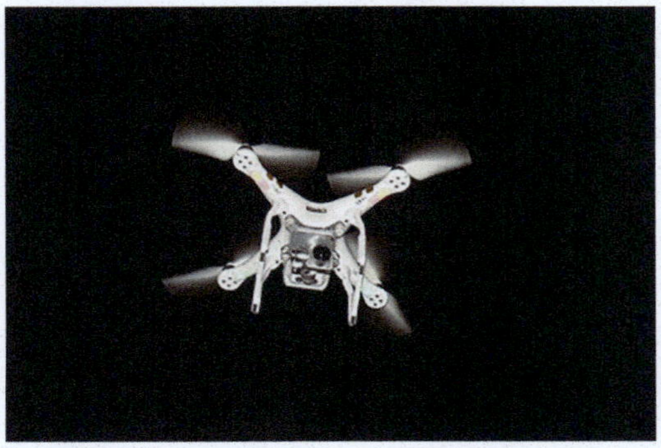

FIGURE 10.25 Drone. (*Source: Suyash Dwivedi, Wikimedia.*)

particular path, which is planned one, which leads to construction of striking structures as well as decreasing accidents at the sites at a lower risk in addition to cost, making the sites a safe place during the course of projects (Fig. 10.26).

Aside from the above categorization of construction robots, at the site, robotic welding could be used during projects for better efficiency in comparison to the same task performed by workers. Similarly, rebar-tying robots also exist to take care of the tying of rebars at a high speed at the site. Then, self-driven bulldozers are intelligent in nature at the site, to avoid accidents while taking care of the work without human intervention. Even though 3D concrete printers are popular to print concrete in 3D, which could be done in the desired size, shape, irrespective of complexity, at a fast pace to accomplish precast as well as modular construction, there also exist demolition robots, which are suitable for the repetitive task of demolishing the chosen structure without human intervention. Then, material handling, earthmoving, and even paving of roads are all possible with particular types of construction robots these days. Then, the management of the whole project at the site could be monitored as well as managed by the drones, irrespective of the terrain or ground conditions, which takes relatively less time than humans covering the site by themselves. This is clear that the construction robots have wide range of applications during the course of projects already at the present, which is expected to increase further with time (Fig. 10.27).

10.9 Case studies

This section will discuss several case studies around the world related to robotics and their way to transform the mechatronic systems and particularly the construction projects, which are as follows:

- FANUC
- Kawasaki Robotics
- TOKIMEC
- Kajima Corporation
- Obayashi Corporation

FIGURE 10.26 Swarm robots. (*Source: Farshadarvin, Wikimedia.*)

FIGURE 10.27 Demolition robot at the site. (*Source: Asurnipal, Wikimedia.*)

FANUC: They are a group of companies throughout the world, which originated from Japan in the last century, with full name meaning Fuji Automatic Numerical Control. They are in Japan, by the name of FRANUC Corporation. In the United States, they are known as FRANUC America Corporation. Similarly, they are present in European continent, in Asian countries such as India and Vietnam. Its history started from the year 1956, but in 1972 in Japan it got established. Then, 10 years later, in the United States, the company brought its technology and eventually spread throughout the world. It is famous for industrial robots throughout the world, and as per the information available, there were products over 40 million set up globally, including R series robots (Fig. 10.28).

Kawasaki Robotics: The company is also known as Kawasaki Robotics (USA) Inc., which is a subsidiary of a Japanese company named Kawasaki Heavy Industries, Limited. It started its journey in 1968, and then first robot started the next year, which led to its eventually spreading to various countries in the world, including South Korea, China, Singapore, India, and so on. It has more than 5 decades of experience in the area of robotic automation as well as industrial robots, where the shipment of 200 and 10,000 robots was supplied across the globe, including BA series and other robots (Fig. 10.29).

TOKIMEC: The company later changed its name to TOKYO KEIKI INC. by restoring its name in 2008, which was changed in 1990 to TOKIMEC. The company's history started in 1896 in Tokyo as measuring devices, etc., which eventually expanded to robotics and spread across the world. The company's automation equipment as well as devices for measurements

FIGURE 10.28 Satellite map of FANUC headquarters in Michigan, United States. (*Source: Google Earth.*)

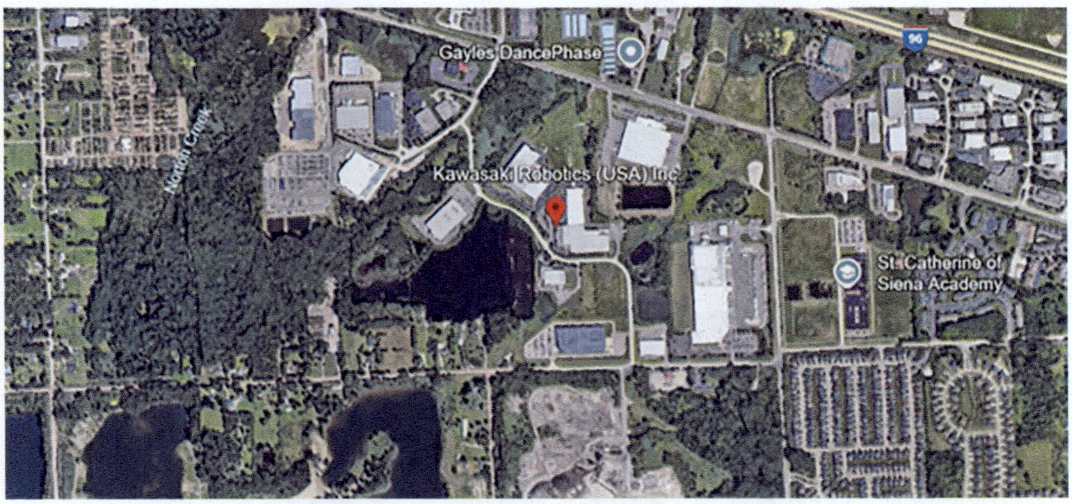

FIGURE 10.29 Satellite map of headquarters of Kawasaki Robotics in Michigan, United States. (*Source: Google Earth.*)

come in handy at construction sites during projects, including the navigation of underground tunnel construction with meticulousness (Fig. 10.30).

Kajima Corporation: This company is considered not only one of the largest companies in construction sector but also oldest category in the country Japan because it was founded in the year 1840 in Tokyo. The company is spread in different continents in the world, North America, Europe, Oceania, in addition to Asia. As per the report available in the last month of December in the year 2018, the company created a Smart Future Vision, which meant that technology would assist in doing half of the job in project management remotely, whereas robots would help in doing the same at the site, in addition to the digitalization of the entire processes related to engineering, to transform the construction sector, which could be found in the study of Singh (2022a) regarding the ways of digitalization could assist in the development of construction industry. The company also believes that the construction engineering, which is IT-assisted in some cases, could benefit from mechatronics (Fig. 10.31).

Obayashi Corporation: This company is part of the five significant construction companies in the country Japan, and its headquarters are in the capital of the country, which is Tokyo. Based on the ENR's list in the year 2018, regarding top 250 contractors in the world, the company was placed on the 15th position and was the highest in the country. However, the company was founded in Osaka in the country in the year 1892. The company spread throughout other countries, including the United States, Australia, European countries, Asian countries, as well as in Middle East. As per the information available

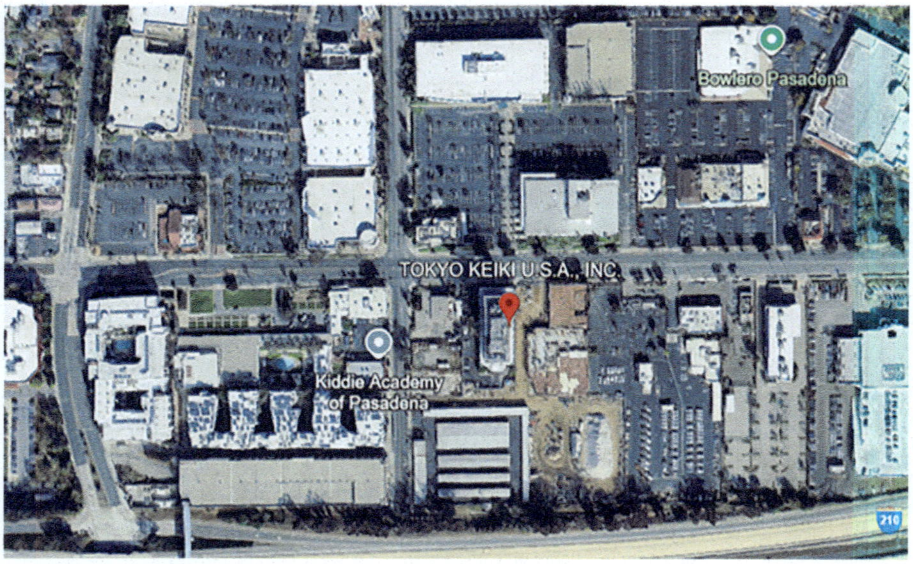

FIGURE 10.30 Satellite map of TOKIMEC in California in United States. (*Source: Google Earth.*)

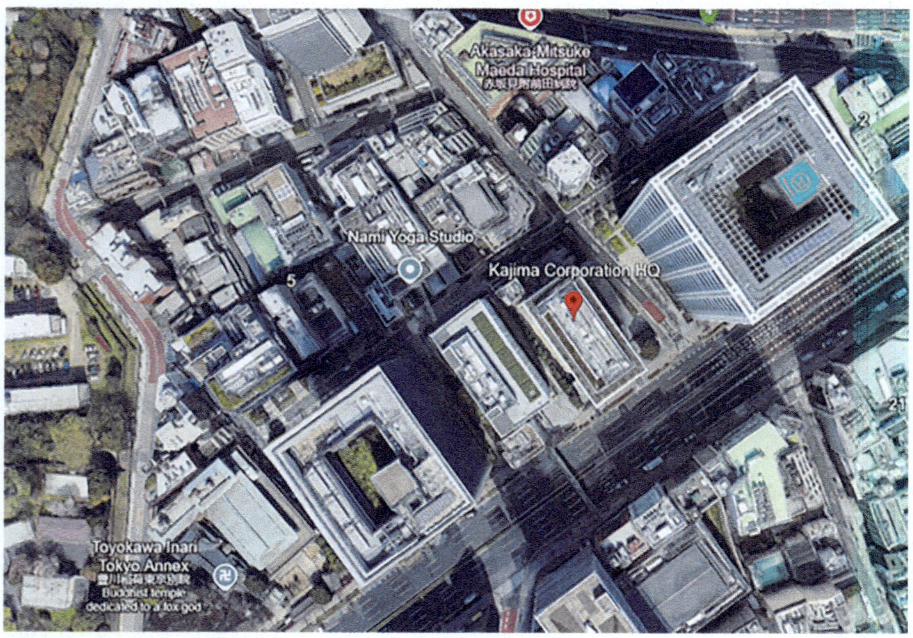

FIGURE 10.31 Satellite map of Kajima Corporation HQ in Japan. (*Source: Google Earth.*)

in the year 2022 the company created a novel business company named PLiBOT to assist in offering people the IC platforms in addition to AMRs toward a sustainable society (Fig. 10.32).

10.10 Methodology

In this section the challenges and obstacles faced by the abovementioned case studies of the companies will be considered, which are as follows:

FANUC*:* The company faces some major challenges like higher initial cost of industrial robots and Computer Numerical Control (CNC) systems, as not every workplace could afford costly products in their projects. Then, skill gaps and

FIGURE 10.32 Satellite map of Obayashi Corporation in Minato City, Tokyo. (*Source: Google Earth.*)

resistance shown by the workforce in many workplaces, as well as complexity in the products' integration in different industries, were other major challenges faced by it.

Kawasaki Robotics*:* This company faced some significant obstacles like automation of complicated tasks in industries, including projects in construction industries; concerns related to return of investment, as the initial cost of the products is high; apart from rough unstructured environs at the site and so on.

TOKIMEC*:* In this case the company faced high competition from other companies when targeting the construction industry as well as technological adaptations to make their products useful for construction projects because the company was originally in the hydraulics and marine industry.

Kajima Corporation*:* This company faced some major hurdles like integration of advanced products related to mechatronics and robotics to the conventional methods of construction, challenging work environments at the site, technical limitations while making products for specific purposes in projects, high execution cost of such advanced technologies for projects, etc.

Obayashi Corporation*:* The company faced some noticeable hurdles like problems related to the technical deployment of products, including products to be able to work in harsh as well as diverse environs, and execution issues of the products at the site. Then, the performance of the products to be consistent, operational safety up to the standards, employees' acceptance at the workplaces, and so on were other major concerns of the company.

10.11 Results and discussion

In this section, data from different sources were collected, including Statista and Modular Intelligence, to perform graphical analysis.

10.11.1 Major advantages as well as challenges of robotics in construction

In this subsection the above graphs are analyzed to understand the advantages and disadvantages of robotics in construction projects:

TABLE 10.1 Worldwide revenue of robotics. (Adapted from data in Statista).

Growth of robotics revenue globally	In USD
Year 2016	18,470,000,000
Year 2023	37,370,000,000
Year 2025 (projected)	50,800,000,000

TABLE 10.2 Market revenue of robotics in 2025 (projected). (Adapted from data in Statista).

Expected robotics revenue in 2025	In USD
United States	10,450,000,000
Rest of the world	40,350,000,000
Worldwide	50,800,000,000

TABLE 10.3 Robotics revenue generated by companies based on different continents in 2023. (Adapted from data in Statista).

Robotics revenue by continent-based companies in 2023	In USD
North American companies	9,390,000,000
Asian companies	13,510,000,000
European (Western + Central) companies	9,020,000,000
Rest of the companies in the world	5,450,000,000

TABLE 10.4 Worldwide growth of AI robotics market. (Adapted from data in AIBOTICS).

Market size of AI robotics globally	In USD
Year 2021	95,600,000,000
Year 2023	184,750,000,000

AI, artificial intelligence.

TABLE 10.5 Global growth of mobile cobots. (Adapted from data in Next Move Strategy Consulting).

Mobile cobots growth	In USD
Year 2021	656,100,000
Year 2030 (expected)	7,660,000,000

TABLE 10.6 Comparison of top countries in the installation of new industrial robots. (Adapted from data in World Robotics 2022, IFR).

Installation of new industrial robots in 2021	In 1000s
China	268
United States	35

TABLE 10.7 Impact of robots on employment. (Adapted from data in Eric Dahlin, Socius).

Impact of robots on jobs	In %
Replacement of workers by robots	14
Nonreplaced workers by robots	86

TABLE 10.8 Global market growth of construction robots. (Adapted from data in Research Nester).

Market size of construction robots	In USD
Year 2024	328,680,000
Year 2037	3,630,000,000

TABLE 10.9 Expected share of construction robots based on applications in 2037. (Adapted from data in Research Nester).

Construction robots share by applications (expected)	In %
Residential and commercial buildings in 2037	42
Others (public infrastructure, dismantling, etc.) in 2037	58

TABLE 10.10 Share of construction robots based on type in the future (2037). (Adapted from data in Research Nester).

Expected share of construction robots by type in 2037	In %
Robotic arm	45
Others (exoskeletons, conventional robots, etc.)	55

TABLE 10.11 Market share of regions in the case of construction robots in the future. (Adapted from data in Research Nester).

Share of regions in the market of construction robots	In %
Asia-Pacific region in 2037 (expected)	37
North American continent in 2037 (expected)	23
Rest of the World in 2037 (expected)	40

TABLE 10.12 Recent market share of construction robots on the basis of type. (Adapted from data in Mordor Intelligence).

Share of market of construction robots based on type	In %
Demolition in 2024	36.2
Others in 2024 (3D printing, bricklaying, etc.)	63.8

- According to Fig. 10.33 the revenue of robotics was over $18 billion in the year 2016, but it reached over $37 billion in 2023 and is expected to reach over $50 billion by this year, 2025, meaning in the span of 9 years, the revenue worldwide of robotics is expected to increase by more than 2.5 times. It is safe to say that robotics demand is growing, and it will be wise to invest in it, including the construction industry, to stay updated with time.
- Based on Fig. 10.34, the United States is expected to generate over $10 billion in robotics by this year, 2025, whereas the rest of the world to generate over $40 billion in robotics in the same year. This shows that over two-fifths of the world's revenue in robotics is being generated by the United States, indicating the high demand and utility of robotics in the advanced and superpower country. It is safe to say that advanced countries will have high utility of robotics to improve themselves further in future.
- As per Fig. 10.35 the Asian companies had generated over $13 billion in revenue in robotics area in the year 2023, whereas American companies and Western and Central European companies generated over $9 billion in robotics revenue in the same year. It shows Asian companies are not only leading but expanding in the robotics area in the

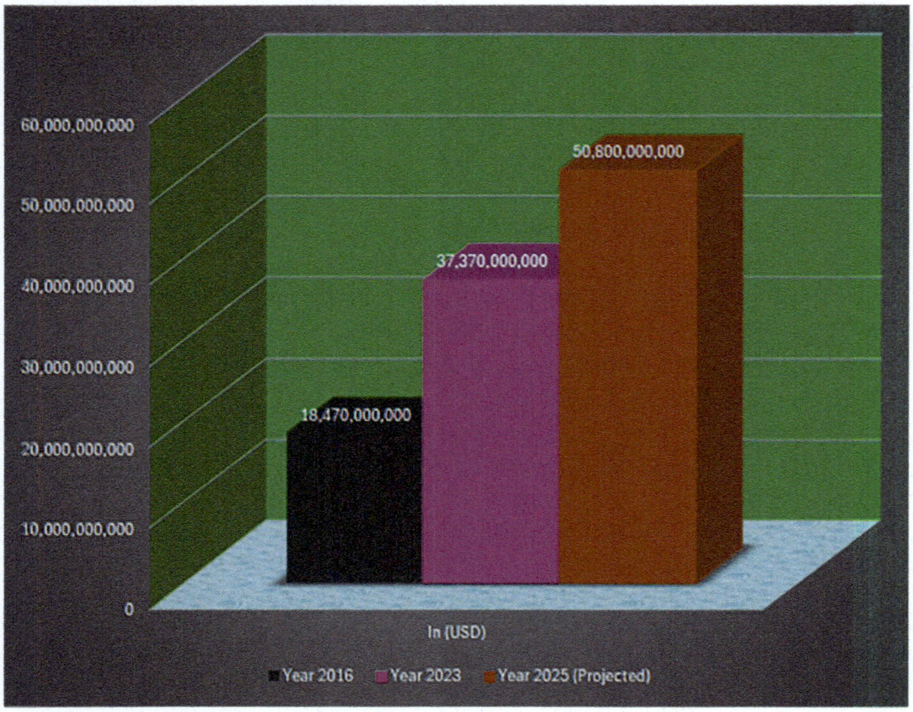

FIGURE 10.33 Worldwide revenue of robotics. (*Adapted from data in Statista.*)

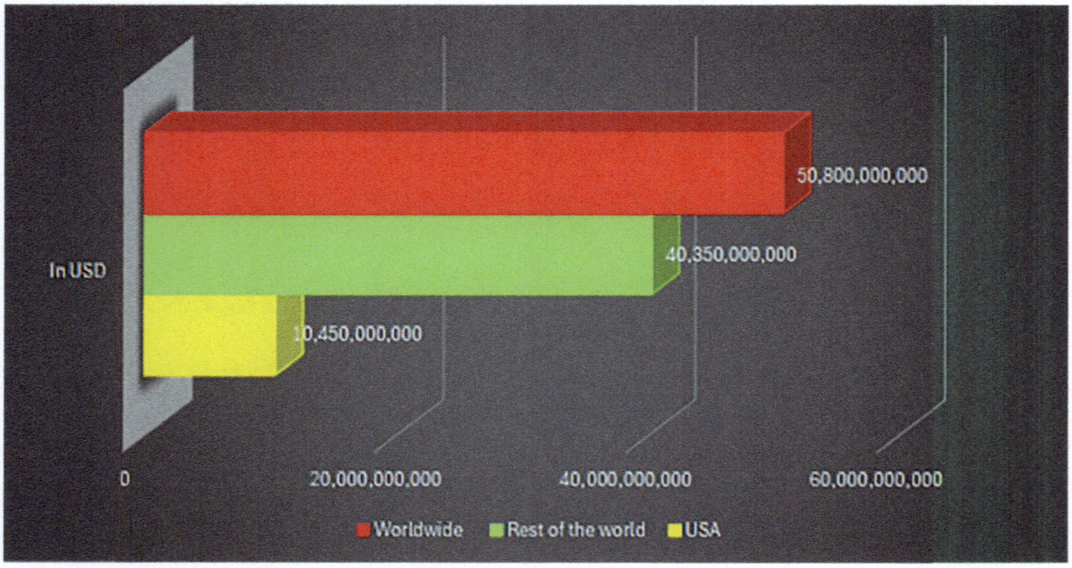

FIGURE 10.34 Market revenue of robotics in 2025 (projected). (*Adapted from data in Statista.*)

world compared to the companies of other regions. This is also good news for developing countries in Asia, including India, to expect high development in the coming future, which could affect the construction projects more than now.

- According to Fig. 10.36, AI robotics market was over $95 billion in the year 2021, but by the year 2023, their market size reached more than $184 billion, which is close to double in the span of 2 years. This means that AI has high demand in robotics, and it can be expected to be similar with CI in robotics. This way the mechatronics systems will improve greatly, thus enhancing the construction projects with the application of construction robots.
- Based on Fig. 10.37, mobile cobots had a market of over $656 million in the year 2021 in the world, but they are expected to have a market size of more than $7.6 billion by the year 2030, meaning mobile cobots are expected to have a market

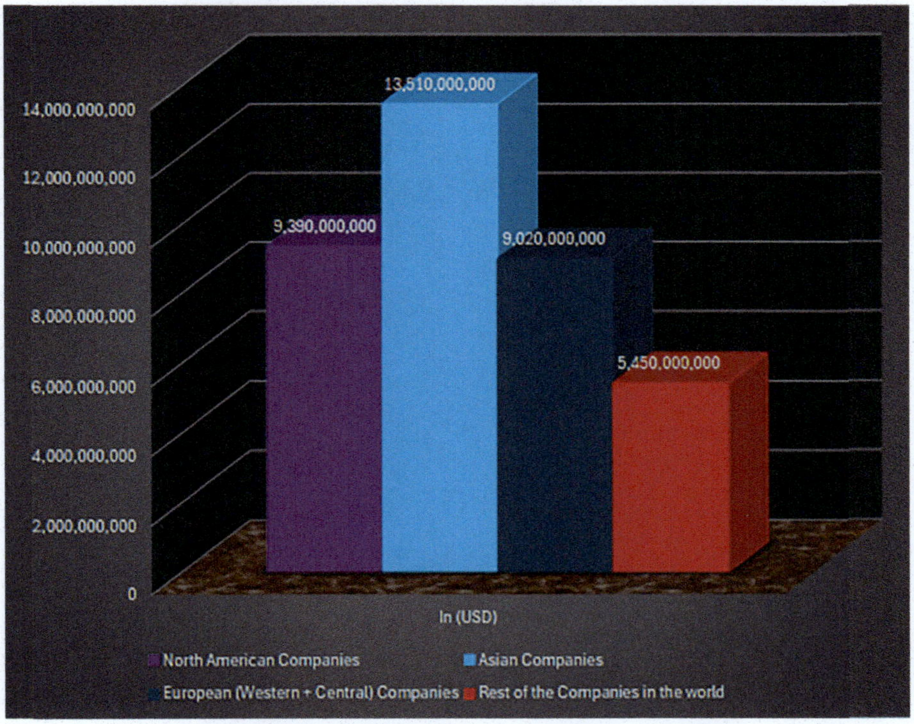

FIGURE 10.35 Robotics revenue generated by companies based on different continents in 2023. *(Adapted from data in Statista.)*

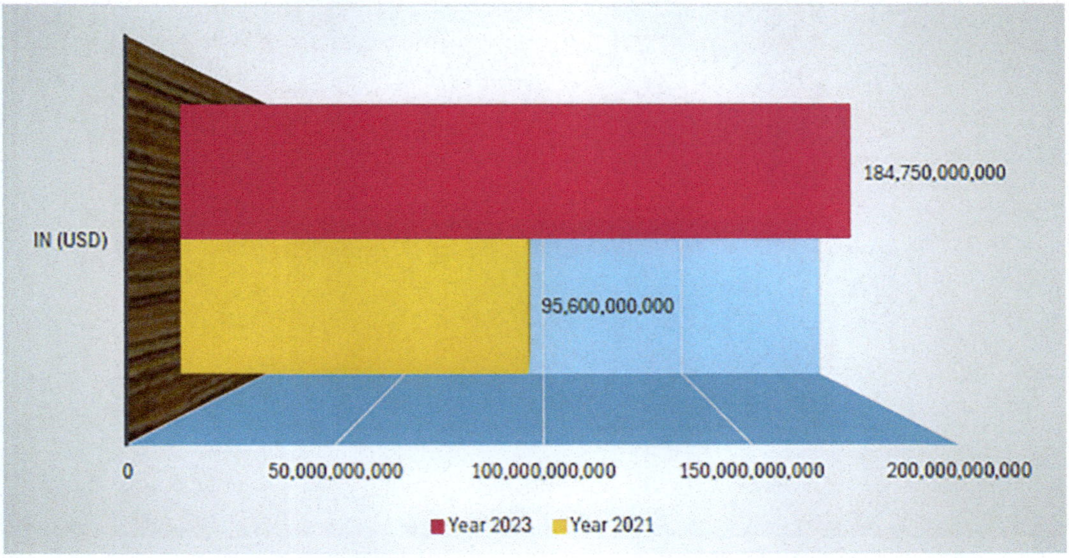

FIGURE 10.36 Worldwide growth of AI robotics market. *AI*, Artificial intelligence *(Adapted from data in AIBOTICS.)*

growth of more than 11 times in the span of 9 years, the fastest growth compared to other robots. This means that cobots could be invested in more and used in construction projects.

- As per Fig. 10.38 the top countries to install novel industrial robots were China and the United States in the year 2021, where the former installed 268,000 and the latter installed 35,000 novel industrial robots, indicating the difference between their installations of new industrial robots to be over 7 times. This shows the developing countries like China are showing more inclination toward industrial robots to speed up their development compared to advanced countries like the United States. It is safe to include the construction industry as well, which is essential for the development of any country.

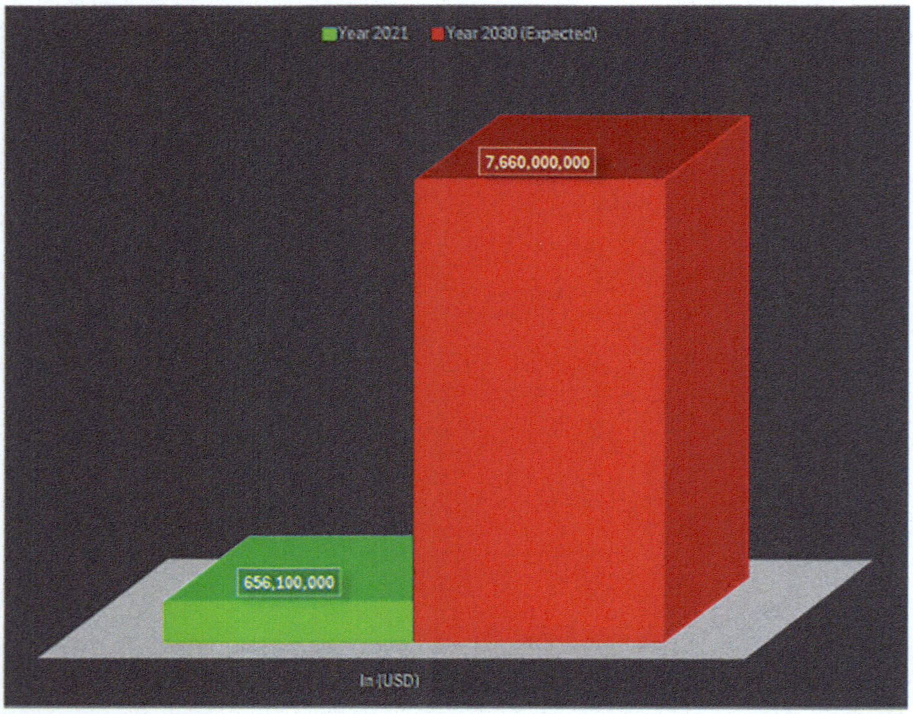

FIGURE 10.37 Global growth of mobile cobots. (*Adapted from data in Next Move Strategy Consulting.*)

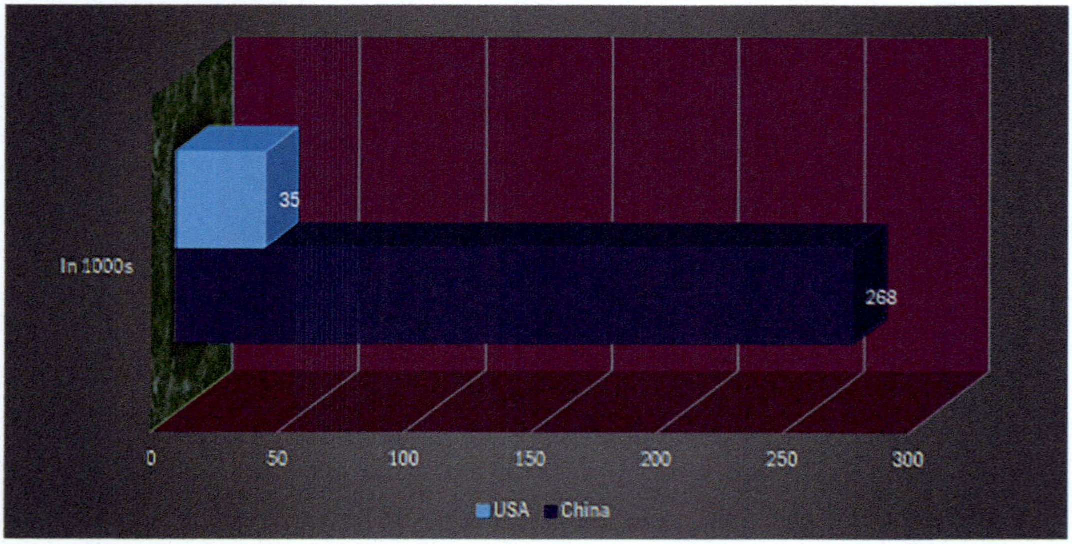

FIGURE 10.38 Comparison of top countries in the installation of new industrial robots. (*Adapted from data in World Robotics 2022, IFR.*)

- According to Fig. 10.39 the robots have replaced 14% of the workers in their workplaces in the world, indicating that semiskilled and unskilled workers are getting replaced and unemployed by the adoption of robots in the workplaces. The Unemployment might increase further with the growth of robots in the workplaces in the future, which also includes the construction industry, which will be a problem for workers to adapt with the changing times.
- Based on Fig. 10.40, market size of construction robots last year, in the year 2024, was over $328 but by the year 2037, the market size of construction robots is expected to reach over $3.6 billion, meaning it will increase by roughly 11 times in the span of 13 years, which shows a tremendous growth in the future in the world. This means that the construction companies that have not adopted robots or have not invested much in robots could invest or increase their investment,

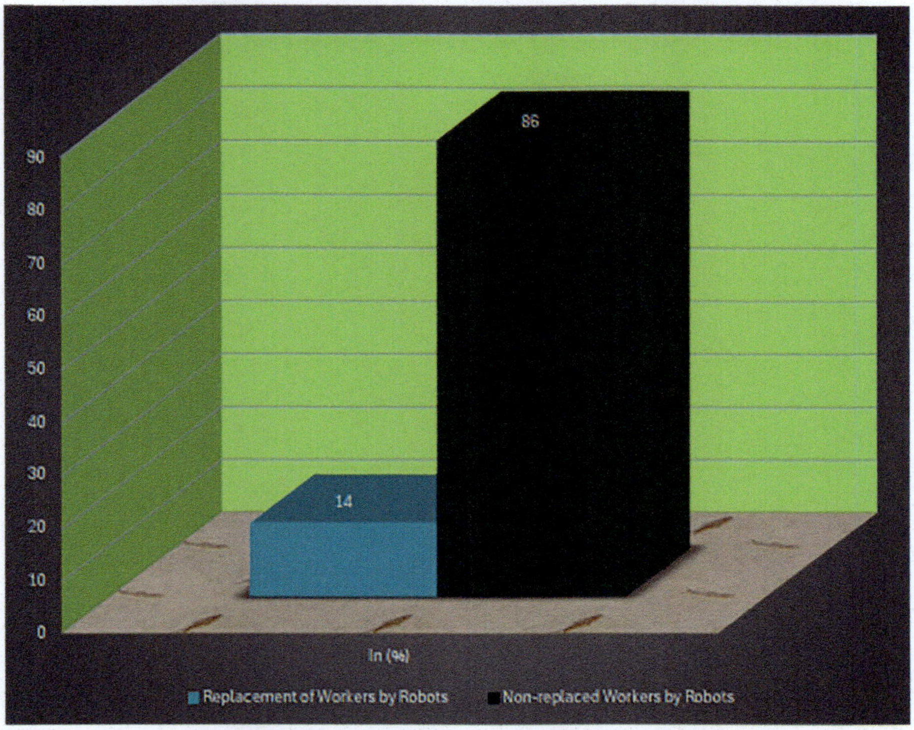

FIGURE 10.39 Impact of robots on employment. (*Adapted from data in Eric Dahlin, Socius.*)

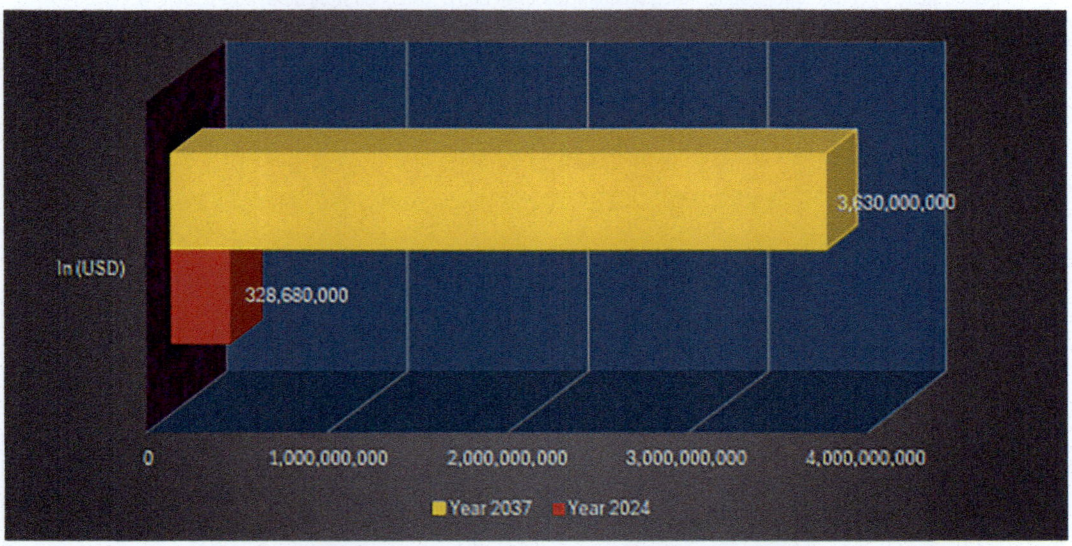

FIGURE 10.40 Global market growth of construction robots. (*Adapted from data in Research Nester.*)

respectively, in the future to stay in competition with other construction companies that are already incorporating robots in ample supplies in the projects.

- As per Fig. 10.41, residential and commercial buildings are expected to have a share of 42% in construction robots market by the year 2037, the highest compared to other applications of construction robots. This means that the construction companies and privatization will be able to adopt construction robots more in the future compared to public infrastructure and others.
- According to Fig. 10.42 the robotic arm is expected to have a market share of 45% in the construction robotics market, meaning the demand for robotic arms will be higher than for exoskeletons and traditional robots in construction projects.

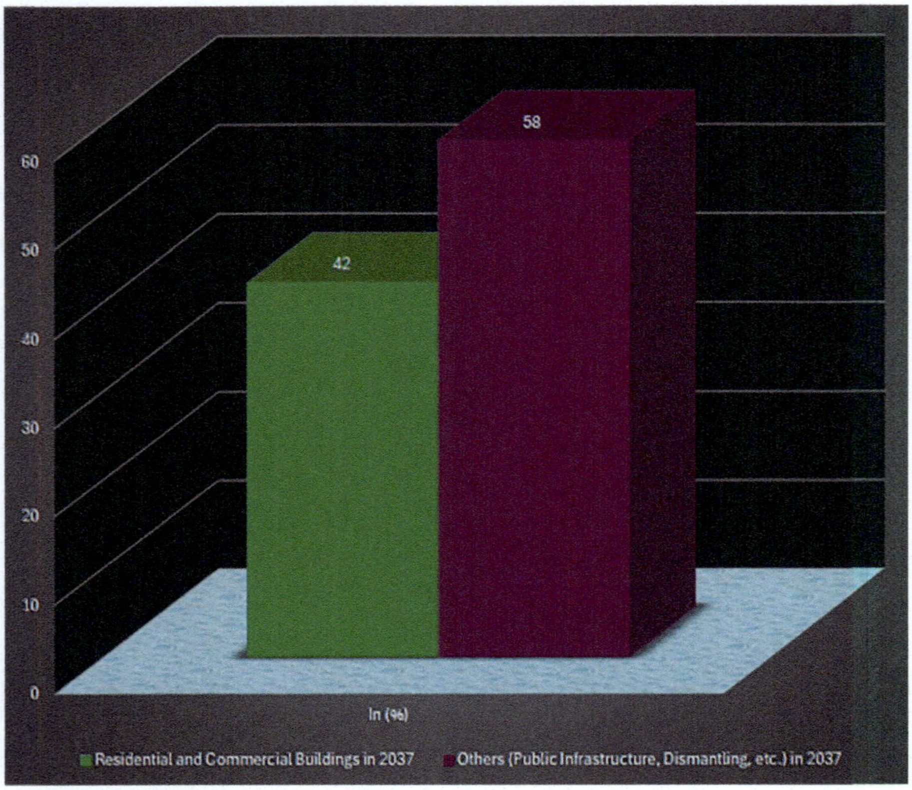

FIGURE 10.41 Expected share of construction robots based on applications in 2037. *(Adapted from data in Research Nester.)*

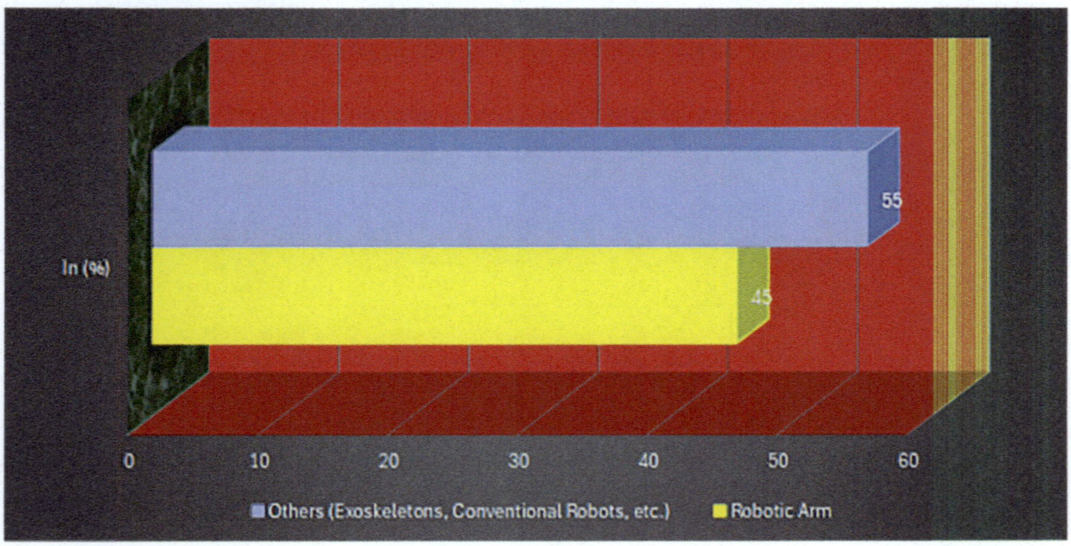

FIGURE 10.42 Share of construction robots based on type in the future (2037). *(Adapted from data in Research Nester.)*

This might also be because of 3D concrete printing and other uses, and it will be wise to invest more in the robotic arms in the future compared to others.

- Based on Fig. 10.43 the Asia-Pacific region is expected to have a share of 37% in the market of construction robots, and North American continent is expected to have a 23% market share for the same by the year 2037. This means the Asia-Pacific region is growing tremendously in the coming future in the case of construction robots and this means that the Asian countries and those in the Pacific region must get adapted to construction robots fast to stay in competition.

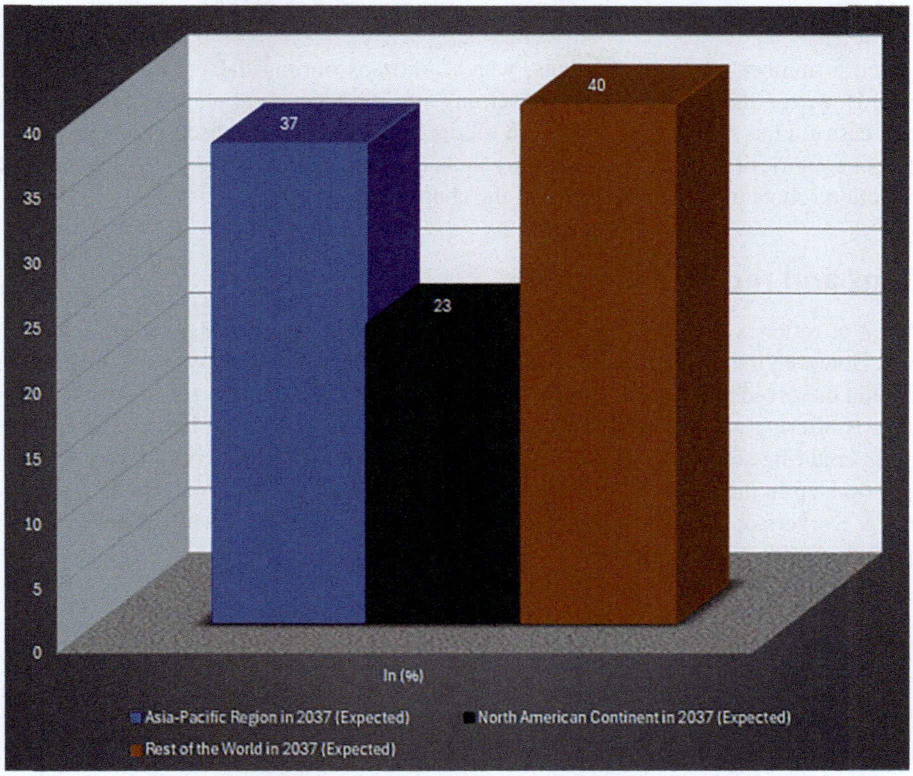

FIGURE 10.43 Market share of regions in the case of construction robots in the future. (*Adapted from data in Research Nester.*)

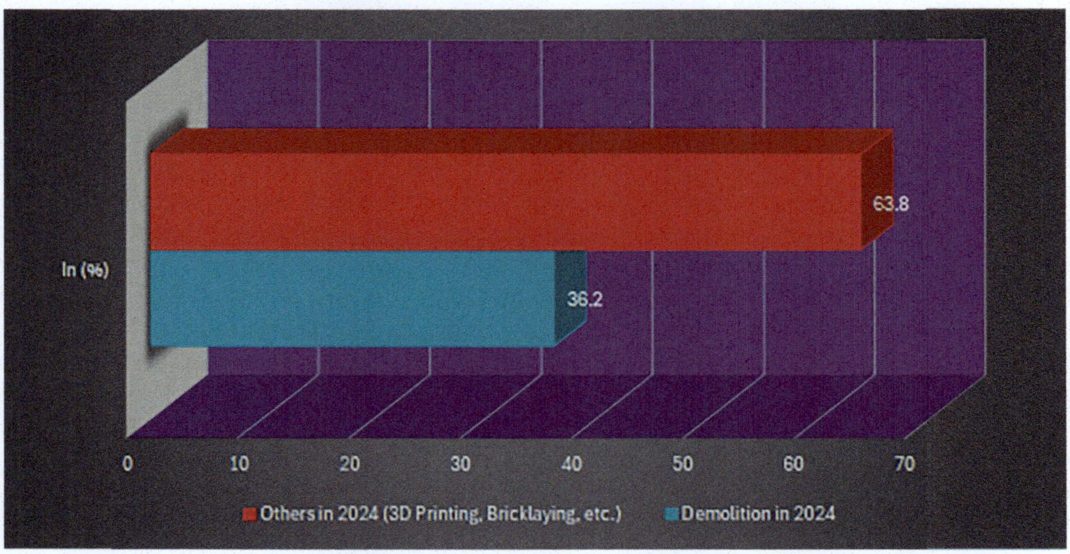

FIGURE 10.44 Recent market share of construction robots on the basis of type. (*Adapted from data in Mordor Intelligence.*)

- As per Fig. 10.44, demolition last year had a share of 36.2% in the market of construction robots, the highest compared to other types in the world. This means that currently construction robots are more preferred to demolish structures, but other uses could also be improved while adopting construction robots in the projects to transform the construction sector more in the future.

10.12 Limitations

This chapter explained a number of types of robots, which are transforming the world of mechatronics in addition to construction projects. However, the chapter did not cover any specific type of robots in too depth, including construction robots. Then, several case studies linked to construction robots in the world were discussed here, but there might be other case studies in different countries related to construction robots that might be left out in this chapter. Then, any one specific case study of construction robots was not focused on in the chapter.

10.13 Solutions and recommendations

With the incorporation of robotics the world of mechatronics is reaching new heights, and the same could be said for the construction projects. However, the unemployment of unskilled as well as semiskilled workers or employees will rise with its flourishment. This could be solved if companies could train their old or existing employees or workers to adapt to the level of advancement robotics is offering when implemented in the workplace; then this problem of unemployment of the existing employees or workers could be solved to certain extent. Further the governments could train the new workers or young people who are about to step in the world of employment or start working as workers through training centers or through new policies, then new workers or employees would not have trouble finding jobs while robotics is being incorporated in the workplaces for advancement.

10.14 Future research directions

There were many case studies involving construction robots discussed in this chapter, and in the future, more in depth of any particular case study could be studied further. Research on other case studies associated with construction robots, which were left out, could be taken into account in the future. Then, research on the way to reduce the initial cost of robotics might be achieved with time, when technological development reaches a certain level. Then, finding a way to reduce the maintenance cost of construction robots could be researched in the future in addition to safe disposal of their unnecessary robotic parts, to avoid harming the environment in the process and achieve sustainability throughout their life cycle.

10.15 Conclusion

The chapter explained the concept of mechatronics or mechatronics engineering, and it distinguished between AI as well as CI, in addition to covering the concept of robotics. Then, it discussed the ways AI and CI could be incorporated into robotics, as well as their need to evolve robots. Then, various kinds of robots were covered, where special emphasis was given to industrial robots and more specifically to the construction robots within industrial robot's category. Moreover, the reason for these advanced technologies to improve robotics and to flourish the world of mechatronic systems and, more specifically, construction projects were talked about in the chapter. Then, categorization and subclassification of construction robots were performed during this study to understand construction robots in a better way. Numerous research studies published in the past were taken into account to figure out the gaps in the current literature, as well as the way this scientific study could strengthen the area of construction robots in addition to the ways AI and CI could enhance the mechatronic systems by transforming the robotics area, as well as the way construction projects could benefit from them, were explained in the chapter. Several case studies incorporating construction robots were mentioned and discussed, in addition to the obstacles and problems that came in their ways. Then, data from many sources were taken to perform graphical analysis to support the study and then discuss the direction of robotics, particularly in the context of construction, apart from future of robotics in the advancement of construction projects with the mechatronic systems.

Acknowledgment

This research received no specific grant from any funding agency in the public, commercial, or not-for-profit sectors.

References

Adel, A., Ruan, D., McGee, W., & Mozaffari, S. (2024). Feedback-driven adaptive multi-robot timber construction. *Automation in Construction, 164*, 105444. doi:10.1016/j.autcon.2024.105444.

Albeaino, G., Jeelani, I., Gheisari, M., & Issa, R. R. A. (2025). Assessing proxemics impact on human-robot collaboration safety in construction: A virtual reality study with four-legged robots. *Safety Science, 181*, 106682. https://www.sciencedirect.com/science/journal/09257535. doi:10.1016/j.ssci.2024.106682.

Baek, F., Kim, D., Lee, G., Choi, B., & Lee, S. H. (2024). Effect of human emotional responses on human-robot team pty in construction. *Automation in Construction, 166*, 105608. doi:10.1016/j.autcon.2024.105608.

Barjuei, S. E., Capitanelli, A., Bertolucci, R., Courteille, E., Mastrogiovanni, F., & Maratea, M. (2024). Digital workflow for printability checking and prefabrication in robotic construction 3D printing based on artificial intelligence planning. *Engineering Applications of Artificial Intelligence, 133*, 108254. doi:10.1016/j.engappai.2024.108254.

Bruun, E. P. G., Oval, R., Al Asali, W., Gáspár, O., Paris, V., & Adriaenssens, S. (2024). Automating historical centering-minimizing masonry vaulting strategies: Applications to cooperative robotic construction. *Developments in the Built Environment, 20*, 100516. https://www.sciencedirect.com/science/journal/26661659. doi:10.1016/j.dibe.2024.100516.

Chen, Z., & Zhang, W. (2024). Intelligent sensing and control of road construction robot scenes based on road construction. *SDHM Structural Durability and Health Monitoring, 18*(2), 111–124. https://www.techscience.com/sdhm/v18n2. doi:10.32604/sdhm.2023.043563.

Dodampegama, S., Hou, L., Asadi, E., Zhang, G., & Setunge, S. (2024). Revolutionizing construction and demolition waste sorting: Insights from artificial intelligence and robotic applications. *Resources, Conservation and Recycling, 202*, 107375. doi:10.1016/j.resconrec.2023.107375.

Dörfler, K., Dielemans, G., Leutenegger, S., Jenny, S. E., Pankert, J., Sustarevas, J., Lachmayer, L., Raatz, A., & Lowke, D. (2024). Advancing construction in existing contexts: Prospects and barriers of 3D printing with mobile robots for building maintenance and repair. *Cement and Concrete Research, 186*, 107656. doi:10.1016/j.cemconres.2024.107656.

Duan, K., & Zou, Z. (2024). Morphology agnostic gesture mapping for intuitive teleoperation of construction robots. *Advanced Engineering Informatics, 62*, 102600. doi:10.1016/j.aei.2024.102600.

Faltein, S. A., & Sukdeo, N. I. (2024). Culture-driven quality enhancement: Uncovering the impact of robotics integration in the South African construction sector. *Ain Shams Engineering Journal, 15*(6). https://www.sciencedirect.com/science/journal/20904479. doi:10.1016/j.asej.2024.102728.

Fu, Y., Chen, J., & Lu, W. (2024). Human-robot collaboration for modular construction manufacturing: Review of academic research. *Automation in Construction, 158*, 105196. https://www.sciencedirect.com/science/journal/09265805. doi:10.1016/j.autcon.2023.105196.

Giwa, I., Dempsey, M., Fiske, M., & Kazemian, A. (2024). 3D printed sulfur-regolith concrete performance evaluation for waterless extraterrestrial robotic construction. *Automation in Construction, 165*, 105571. doi:10.1016/j.autcon.2024.105571.

Huang, L., Cai, W., Zhu, Z., & Zou, Z. (2023). Dexterous manipulation of construction tools using anthropomorphic robotic hand. *Automation in Construction, 156*, 105133. doi:10.1016/j.autcon.2023.105133.

Kim, Y. S., Kim, S., Chen, Y., Yang, H. J., Kim, S., Ha, S., Gombolay, M., Ahn, Y., & Cho, Y. K. (2024). Understanding human-robot proxemic norms in construction: How do humans navigate around robots? *Automation in Construction, 164*, 105455. doi:10.1016/j.autcon.2024.105455.

Ku, T. K. X., Zuo, B., & Ang, W. T. (2024). Robotic tower cranes with hardware-in-the-loop: Enhancing construction safety and efficiency. *Automation in Construction, 168*. https://www.sciencedirect.com/science/journal/09265805. doi:10.1016/j.autcon.2024.105765.

Leder, S., Kim, H. G., Sitti, M., & Menges, A. (2024). Enhanced co-design and evaluation of a collective robotic construction system for the assembly of large-scale in-plane timber structures. *Automation in Construction, 162*, 105390. doi:10.1016/j.autcon.2024.105390.

Lee, D., & Han, K. (2024). Vision-based construction robot for real-time automated welding with human-robot interaction. *Automation in Construction, 168*, 105782. doi:10.1016/j.autcon.2024.105782.

Li, L., Yuan, P., Tang, Y., Cooper, G., Thurlbeck, S., Cheung, C. M., Manu, P., Yunusa-Kaltungo, A., & Weightman, A. (2024). The potential of construction robotics to reduce airborne virus transmission in the construction industry in the UK and China. *Heliyon, 10*(9), e29697. doi:10.1016/j.heliyon.2024.e29697.

Liang, C. J., Le, T. H., Ham, Y., Mantha, B. R. K., Cheng, M. H., & Lin, J. J. (2024). Ethics of artificial intelligence and robotics in the architecture, engineering, and construction industry. *Automation in Construction, 162*. https://www.sciencedirect.com/science/journal/09265805. doi:10.1016/j.autcon.2024.105369.

Liu, Y., Alias, A. H. B., Haron, N. A., Bakar, N. A., & Wang, H. (2024a). Exploring three pillars of construction robotics via dual-track quantitative analysis. *Automation in Construction, 162*. https://www.sciencedirect.com/science/journal/09265805. doi:10.1016/j.autcon.2024.105391.

Liu, Y., Alias, A. H. b., Haron, N. A., Bakar, N. A., & Wang, H. (2024b). Technology status tracing and trends in construction robotics: A patent analysis. *World Patent Information, 76*. https://www.sciencedirect.com/science/journal/01722190. doi:10.1016/j.wpi.2023.102259.

Musarat, M. A., Alaloul, W. S., Rostam, N. A. Q. A., & Khan, A. M. (2024). Substitution of workforce with robotics in the construction industry: A wise or witless approach. *Journal of Open Innovation: Technology, Market, and Complexity, 10*(4), 100420. https://www.sciencedirect.com/science/journal/21998531. doi:10.1016/j.joitmc.2024.100420.

Ohueri, C. C., Masrom, M. A. N., & Noguchi, M. (2024). Human-robot collaboration for building deconstruction in the context of construction 5.0. *Automation in Construction, 167*. https://www.sciencedirect.com/science/journal/09265805. doi:10.1016/j.autcon.2024.105723.

Pan, M., Wong, M. O., Lam, C. C., & Pan, W. (2024). Integrating extended reality and robotics in construction: A critical review. *Advanced Engineering Informatics, 62*, 102795. https://www.sciencedirect.com/science/journal/14740346. doi:10.1016/j.aei.2024.102795.

Pan, Z., & Yu, Y. (2024a). Learning multi-granular worker intentions from incomplete visual observations for worker-robot collaboration in construction. *Automation in Construction, 158*, 105184. doi:10.1016/j.autcon.2023.105184.

Pan, Z., & Yu, Y. (2024b). Optimizing heterogeneous multi-robot team composition for long-horizon construction tasks: Time- and utilization-guided simulation. *Automation in Construction, 165*, 105520. doi:10.1016/j.autcon.2024.105520.

Park, S., Wang, X., Menassa, C. C., Kamat, V. R., & Chai, J. Y. (2024). Natural language instructions for intuitive human interaction with robotic assistants in field construction work. *Automation in Construction, 161*. https://www.sciencedirect.com/science/journal/09265805. doi:10.1016/j.autcon.2024.105345.

Singh, A. (2022a). Importance of fuzzy logic in traffic and transportation engineering. In P. Vasant, I. Zelinka, & G. W. Weber (Eds.). *Intelligent computing & optimization. ICO 2021. Lecture notes in networks and systems*: 371. Springer. doi:10.1007/978-3-030-93247-3_10.

Singh, A. (2022b). Mechatronics engineering in the modern world and its use in the construction industry. In *Trends, paradigms, and advances in mechatronics engineering* (pp. 64–92). IGI Global. doi:10.4018/978-1-6684-5887-7.ch004.

Singh, A. (2023a). The present and future role of artificial intelligence in the promotion of sustainable practices in the construction sector. In C. Kahraman, I. U. Sari, B. Oztaysi, S. Cebi, S. CevikOnar, & A. Ç. Tolga (Eds.). *Intelligent and fuzzy systems. INFUS 2023. Lecture notes in networks and systems*: 759. Springer. doi:10.1007/978-3-031-39777-6_2.

Singh, A. (2023b). The recent trend of artificial neural network in the field of civil engineering. In P. Vasant, G.-W. Weber, J. A. Marmolejo-Saucedo, E. Munapo, & J. J. Thomas (Eds.). *Intelligent computing and optimization. ICO 2023. Lecture notes in networks and systems*: 855. Springer https://www.springer.com/series/15179. doi:10.1007/978-3-031-50158-6_32.

Singh, A. (2023c). The significance of digitalization of the construction sector. In P. Vasant, G. W. Weber, J. A. Marmolejo-Saucedo, E. Munapo, & J. J. Thomas (Eds.). *Intelligent computing & optimization. ICO 2022. Lecture notes in networks and systems*: 569. Springer https://www.springer.com/series/15179. doi:10.1007/978-3-031-19958-5_100.

Singh, A., & Srivastava, A. (2024). *Impact of artificial intelligence techniques and Green AI applications artificial intelligence techniques for sustainable development* (pp. 44–63). CRC Press. https://www.taylorfrancis.com/books/9781040226148. doi:10.1201/9781003546382-3.

Song, C., Li, X., Du, Q., Jiang, R., & Shen, Q. (2024). Hybrid pose adjustment (HyPA) robot design for prefabricated module control in modular construction assembly. *Automation in Construction, 168*, 105798. doi:10.1016/j.autcon.2024.105798.

Wang, T., Mao, C., Sun, B., & Li, Z. (2024a). Genealogy of construction robotics. *Automation in Construction, 166*, 105607.

Wang, X., Wang, S., Menassa, C. C., Kamat, V. R., & McGee, W. (2023). Automatic high-level motion sequencing methods for enabling multi-tasking construction robots. *Automation in Construction, 155*, 105071. https://www.journals.elsevier.com/automation-in-construction. doi:10.1016/j.autcon.2023.105071.

Wang, X., Yu, H., McGee, W., Menassa, C. C., & Kamat, V. R. (2024b). Enabling building information model-driven human-robot collaborative construction workflows with closed-loop digital twins. *Computers in Industry, 161*, 104112. https://www.sciencedirect.com/science/journal/01663615. doi:10.1016/j.compind.2024.104112.

Wang, Y., Xie, L., Chen, J., Chen, M., Hu, T., Liao, H., Sun, S., & Chen, J. (2024c). Mortar spraying and plastering integrated robot for wall construction. *Automation in Construction, 165*, 105533. doi:10.1016/j.autcon.2024.105533.

Yang, L., & Cai, H. (2024). Enhanced visual SLAM for construction robots by efficient integration of dynamic object segmentation and scene semantics. *Advanced Engineering Informatics, 59*, 102313. doi:10.1016/j.aei.2023.102313.

Yarovoi, A., & Cho, Y. K. (2024). Review of simultaneous localization and mapping (SLAM) for construction robotics applications. *Automation in Construction, 162*, 105344. doi:10.1016/j.autcon.2024.105344.

Ye, X., Guo, H., & Luo, Z. (2024). Two-stage task allocation for multiple construction robots using an improved genetic algorithm. *Automation in Construction, 165*, 105583. doi:10.1016/j.autcon.2024.105583.

You, H., Zhou, T., Zhu, Q. i, Ye, Y., & Du, E. J. (2024). Embodied AI for dexterity-capable construction robots: DEXBOT framework. *Advanced Engineering Informatics, 62*, 102572. doi:10.1016/j.aei.2024.102572.

Yu, Z., Jiang, X., & Liu, Y. (2024). Pose estimation of an aerial construction robot based on motion and dynamic constraints. *Robotics and Autonomous Systems, 172*, 104591. doi:10.1016/j.robot.2023.104591.

Zeng, L., Guo, S., Wu, J., & Markert, B. (2024). Autonomous mobile construction robots in built environment: A comprehensive review. *Developments in the Built Environment, 19*, 100484. doi:10.1016/j.dibe.2024.100484.

Zheng, C., Du, Y., Xiao, J., Sun, T., Wang, Z., Eynard, B. ît, & Zhang, Y. (2025). Semantic map construction approach for human-robot collaborative manufacturing. *Robotics and Computer-Integrated Manufacturing, 91*, 102845. doi:10.1016/j.rcim.2024.102845.

Zhou, J., Zhou, W., & Wang, Y. (2024). CWPR: An optimized transformer-based model for construction worker pose estimation on construction robots. *Advanced Engineering Informatics, 62*, 102894. doi:10.1016/j.aei.2024.102894.

Zhu, A., Pauwels, P., Torta, E., Zhang, H., & De Vries, B. (2024). Data linking and interaction between BIM and robotic operating system (ROS) for flexible construction planning. *Automation in Construction, 163*, 105426. doi:10.1016/j.autcon.2024.105426.

Chapter 11

Application of artificial intelligence in high-frequency circuit design

Kok Yeow You [a], Yeng Seng Lee [b] and Fwen Hoon Wee [c]

[a] *Faculty of Electrical Engineering, Universiti Teknologi Malaysia, Skudai, Johor, Malaysia,* [b] *Keysight Technologies, Bayan Lepas, Pulau Pinang, Malaysia,* [c] *Faculty of Electronic Engineering Technology, Universiti Malaysia Perlis, Arau, Perlis, Malaysia*

11.1 Introduction

11.1.1 The need for AI in high-speed circuit design

In the era of automation, high-speed circuits play a critical role in mechatronics by ensuring precise and real-time hardware responses, which are essential for maintaining accuracy and efficiency in modern automated systems. Artificial intelligence (AI)-assisted computer-aided design (CAD) software tools that incorporate extensive real-world data have the potential to design reliable high-speed circuits for mechatronic, Internet of Things (IoT), and telecommunication applications, enhancing performance and dependability (Bandler & Rayas-Sanchez, 2023; Ma et al., 2023; Ohira et al., 2018, 2023). A high-speed circuit operates at very high frequencies, requiring extremely small dimensions and making its performance highly sensitive to circuit structural variations. Designing such circuits using conventional circuit simulation methods and electromagnetic (EM) simulators often demands a vast number of computational iterations. This heavy processing load is difficult to handle by normal personal computers (PCs), leading to significant challenges in running these programs.

Artificial neural networks (ANNs), a subset of AI, are commonly used as the core of AI-assisted CAD computations (You & Sim, 2025). ANN significantly reduces the heavy computational load associated with complex and stability-challenging EM models by serving as a knowledge-based system that leverages pretrained datasets, allowing for faster and more efficient processing. The accuracy and performance of an ANN depend heavily on both the quantity and diversity of the training dataset, as well as the type of data used. Recently, ANNs have been integrated with optimization techniques such as space mapping (SM) (Oluyemi et al., 2023; Yeung et al., 2012), genetic algorithms (GA) (Gholami & Nezamoddini, 2019; Janson & Frenzel, 1993; Palmes et al., 2005; Yeung et al., 2012), particle swarm optimization (PSO) (Hung & Tong, 2009; Yeung et al., 2012), and fuzzy logic (FL) to improve convergence rates and enhance the accuracy of output predictions. Each optimization technique offers specific advantages based on the nature of the training data and the correlation patterns between the input and output values in the dataset.

11.1.2 Artificial neural network (ANN)

This chapter explores the application of ANN in high-speed circuit design, particularly for three-section branch-line couplers and Wilkinson power divider/combiner circuits. A hybrid approach combining an ANN, GA, PSO, and SM is employed to predict the optimal impedances of the microstrip circuit lines and further predict the structure/dimension of the circuit.

An ANN is a computational model inspired by biological neural systems, composed of interconnected neurons across input, hidden, and output layers. Neurons process inputs via weighted sums and nonlinear activation functions, passing outputs to subsequent layers. Fig. 11.1 illustrates how a neuron's activity is based on the weighted sum of its inputs u_k, where the weights (w_{k1}, w_{k2}, ..., w_{kN}) reflect past correlations, as shown in Eq. (11.1). The predicted output y_k is given by

Computational Intelligence in Mechatronics: Solving Real-World Problems in Electronic Systems Design. DOI: https://doi.org/10.1016/B978-0-443-27580-7.00032-9

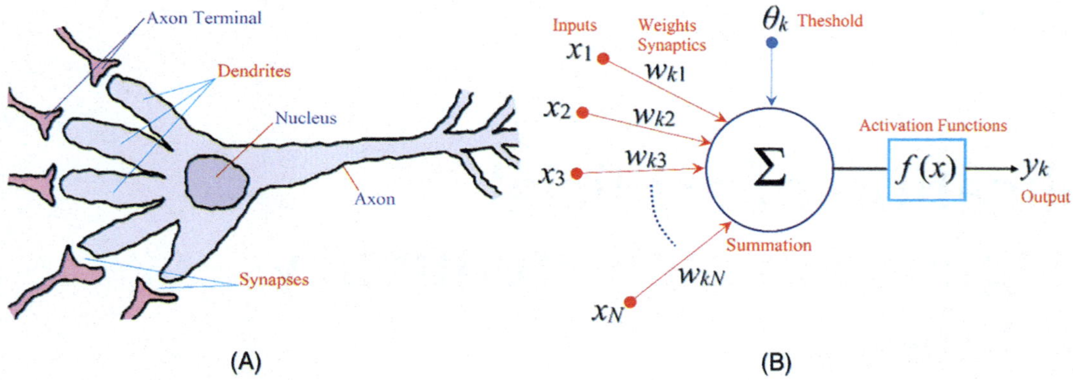

FIGURE 11.1 (A) Biological brain neuron and (B) analogous computational neural model. Computational neural models are analogous to the working principles of neurons in the biological brain.

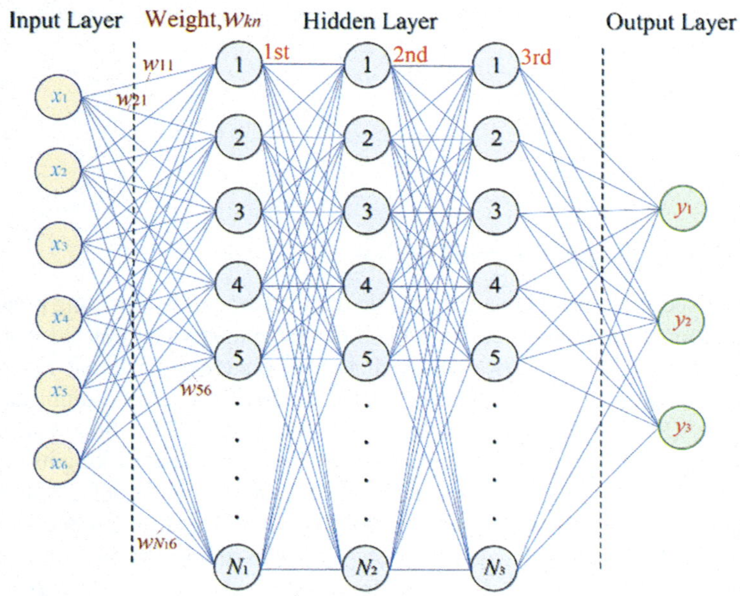

FIGURE 11.2 An example of an ANN model with six input nodes, three hidden layers (each comprising N_1, N_2, and N_3 neurons, respectively), and three predicted output nodes. The ANN model processes all available input and output datasets to learn the optimal input–output mapping, thereby capturing the underlying relationship between the variables. *ANN*, Artificial neural network.

Eq. (11.2).

$$u_k = w_{k1}x_1 + w_{k2}x_2 + w_{k3}x_3 + \cdots + w_{kN}x_N \tag{11.1}$$

and

$$y_k = f(u_k - \theta_k) \tag{11.2}$$

where $x_1, x_2, \ldots, x_N$ are the inputs, and $w_{k1}, w_{k2}, \ldots, w_{kN}$ are the synaptic weights connecting u_k to y_k. u_k and θ_k represent the output of the linear accumulator and the threshold (bias), respectively. The activation function $f(u_k - \theta_k)$ determines the predicted output y_k.

Multilayer ANNs with hidden layers and multiple outputs are increasingly used for circuit design. Fig. 11.2 shows a feed-forward neural network with six inputs (x_1, x_2, …, x_6), three hidden layers (each containing N neurons), and three outputs (y_1, y_2, y_3). Input values, combined with their associated weights, w_{kn}, are processed by neurons in the first hidden layer, generating activation values that feed into subsequent layers. Each neuron in one layer is fully connected to all neurons in the next. In this study the ANN employs the rectified linear unit (ReLU) activation function for its hidden layers, introducing nonlinearity to effectively capture complex patterns in the data. Generally the ReLU expression is given as

$$\mathrm{ReLU}(x) = \begin{cases} 0 & \text{for } x < 0 \\ x & \text{for } x \geq 0 \end{cases} \tag{11.3}$$

The output layer, however, uses a linear activation function, allowing the network to produce continuous values. In addition, the PSO and GA techniques are utilized to minimize prediction errors by maximizing the negative of the mean squared error (MSE), which is written in Eq. (11.4). This approach ensures that the network's predictions closely align with the actual target values.

$$\mathrm{MSE} = \frac{-1}{n}\sum_{i=1}^{n}\left(y_{\mathrm{predicted}\_i} - y_{\mathrm{actual}\_i}\right)^2 \tag{11.4}$$

where subscript i and symbol n are the order observations and total number of observations, respectively. On the other hand, $y_{\mathrm{predicted}}$ and y_{actual} are the estimated output results and expected output values.

11.1.3 Genetic algorithm (GA)

Since the late 1960s, computational optimization techniques have been applied in high-speed circuit design (Temes & Calahan, 1967). GA is a computational method inspired by natural selection, developed by John Holland in the 1960s (Mitchell & Forrest, 1994). It evolves a population of candidate solutions, or "chromosomes," using operators like selection, crossover (reproduction), mutation, and inversion. Each chromosome represents a solution, with genes corresponding to specific values. In GA, fitter chromosomes are selected for reproduction, passing their traits to the next generation. Crossover exchanges parts of two parent chromosomes, mutation randomly alters genes, and inversion rearranges gene order, promoting genetic diversity.

GA requires parameters such as population size, mutation rate, and the number of generations. Larger populations enhance robustness but increase computational demand. Mutation rates, typically 0.01–0.1, balance exploration and refinement, with higher rates promoting diversity and lower rates focusing on optimization. In this study, GA optimization is utilized to maintain diversity through crossover and mutation, thereby avoiding convergence to local optima. In addition, GA is capable of handling a wide range of optimization problems, including nondifferentiable, discontinuous, or multimodal functions, as it does not rely on gradient information. This makes it particularly suitable for optimizing complex structures, such as high-speed circuit designs (Bimana & Sinha, 2023; Chen, 2003; Weile & Michielssen, 1997).

11.1.4 Particle swarm optimization (PSO)

Kennedy and Eberhart (1995) introduced PSO. In PSO, particles (candidate solutions) move through the solution space, adjusting their positions based on personal and collective best experiences, allowing the swarm to converge toward an optimal solution. Swarm behavior refers to the coordinated movement of entities, like animals, acting in unison. For instance, crane birds in a flock adjust their positions based on their own best spots and the group's best, aiding in tasks like migration or predator avoidance. Similar to GA, PSO also has its own parameter settings, such as swarm size (population), inertia weight, cognitive coefficient, social coefficient, and velocity limits. In general the inertia weight is typically set between 0.4 and 0.9, whereas the cognitive and social coefficients usually range from 1.5 to 2.0. Velocity limits are often set to around 10%–20% (0.1–0.2) of the dynamic range for each dimension. In this study, PSO is used to compress the lack of GA, which is slow convergence (Deb et al., 2014; Mohammed et al., 2016). Hence PSO is used to find the initial iterative value at the beginning, before continuing with GA optimization.

11.1.5 Space mapping

More recently, SM, introduced by Bandler et al. (1994), has become one of the widely used optimization methods in high-speed circuit design. It significantly accelerates the circuit optimization process, which is often too slow and computationally expensive when relying on direct and robust circuit or EM simulations, such as the finite element method (FEM), finite difference time domain (FDTD), and method of moments (MoM) (Li et al., 2024; You, 2016; Zhang et al., 2010, 2021).

The SM concept is to map between a coarse model (simple and low-fidelity models such as analytical analysis) and a fine model (complex and high-fidelity models in rigorous EM simulator). The coarse model is relatively simple to evaluate, whereas the fine model provides more accurate results but at a higher computational cost. In the computational process the

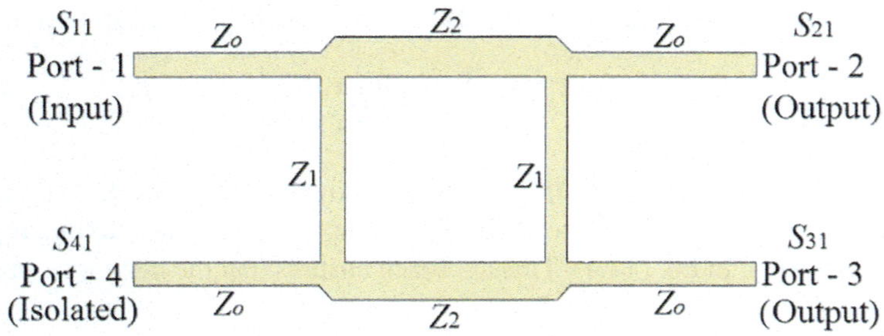

FIGURE 11.3 Basic schematic diagram of a single-section branch-line coupler. Characteristic impedance label and port label of a single-section branch line coupler.

calculated results using a coarse model are aligned or adjusted more closely with the fine model through iteration. In fact, there are various kinds of SM methods, such as basic SM (BSM), affine linear SM (ALSM), implicit SM (ImSM), output SM (OSM), input SM (InSM), and aggressive SM (ASM) (Cheng et al., 2008; Melgarejo et al., 2022; Ossorio et al., 2017; Sans et al., 2014; Tang et al., 2006).

Typically the relationship between input and output data in ANN training for high-speed circuit design is highly nonlinear and complex. To enhance performance, in this study, a hybrid PSO–GA approach is employed, where PSO accelerates convergence, whereas the GA maintains diversity within the ANN. Despite the strengths of the hybrid ANN–GA–PSO method, achieving high accuracy can still be challenging because of the need for large populations and numerous iterations. To overcome this, SM optimization is used to further refine the ANN–GA–PSO outputs, enhancing both accuracy and efficiency.

In this study the ASM technique has been modified because of the absence of a fine model, such as simulated data from a commercial EM simulator. Instead the initial outputs from the ANN–GA–PSO are refined by iteratively comparing them against each set of ANN training output data using ASM. The initial predictions are adjusted in each iteration to minimize the residual between the predicted and ANN–GA–PSO training outputs.

As described above the ANN–GA–PSO serves as the coarse model, providing initial output predictions based on input parameters. An ASM is then applied to fine-tune the output predictions, refining them to closely match the fine model for greater accuracy. Let y be the predicted output from the ANN–GA–PSO and denoted as $y = y_{\text{coarse}}$ is used as the initial parameter of the coarse model in ASM. The y_{coarse} is compared and adjusted in each iteration to minimize the residual between the y_{coarse} and among of the ANN training output set (so-called fine model dataset) and then closest ANN training output data, y_{fine} is found based on the minimum deviation (so-called minimum distance):

$$y_{\text{fine}} - y_{\text{coarse}} = \text{minimum} \tag{11.5}$$

Once the closest y_{fine} is found, the optimization process begins, following the iterative update step:

$$y_{\text{coarse}(k+1)} = y_{\text{coarse}(k)} + \alpha\left(y_{\text{fine}} - y_{\text{coarse}(k)}\right) \tag{11.6}$$

where y_{fine} is the closest fine-tuned model output. On the other hand, $y_{\text{coarse}(k)}$ and $y_{\text{coarse}(k+1)}$ represent the coarse model outputs at iteration k and $k + 1$, respectively. The α is the **learning rate** in order to control the step size of the y_{coarse} adjustment. Based on Eq. (11.6) the y_{coarse} gradually approximates the value of y_{fine}. Therefore the accuracy of the final prediction heavily relies on the size and completeness of the training output dataset within the ANN library, making it a critical factor in achieving precise results. In this study the total number of iterations is set to 30.

11.2 Case studies: Hybrid branch-line couplers and Wilkinson power dividers

11.2.1 Hybrid branch-line coupler circuits

The basic branch-line coupler is a four-port passive microwave (MW) device consisting of four 90-degree ($\lambda/4$) transmission lines, namely, two horizontal and two vertical, with characteristic impedances Z_1 and Z_2 (You et al., 2020) (see Fig. 11.3). However, the single-section design typically has a narrow bandwidth (BW) of about 10%. To address this, multisection couplers are used to expand the operating BW (Reed, 1958). A typical three-section branch-line coupler, as shown in Fig. 11.4, features an expanded operating BW of approximately 40%. The actual configurations of the hybrid branch-line couplers are shown in Fig. 11.5.

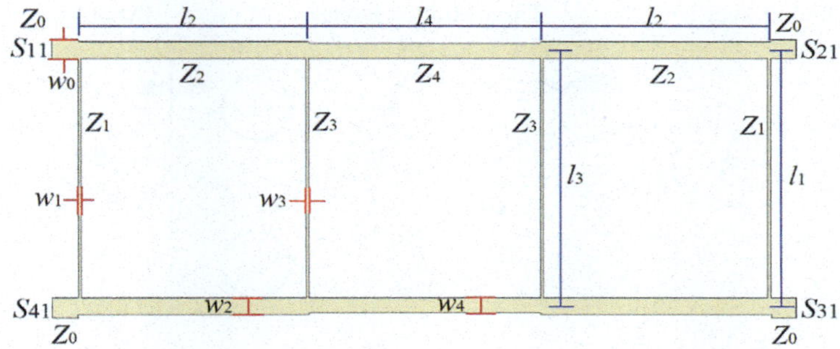

FIGURE 11.4 Basic schematic diagram of a three-section branch-line coupler. Characteristic impedance label and port label of a three-section branch line coupler.

FIGURE 11.5 Actual configuration of the hybrid branch-line couplers. Single-section, two-section, and three-section branch-line couplers.

The hybrid branch-line circuit plays a vital role in part of high-speed circuit design, ensuring reliable signal performance even under challenging conditions. For instance, in a high-frequency amplifier driver, when a signal source is connected to port-1 and both outputs (port-2 and port-3) are terminated with mismatched loads, the circuit effectively manages signal transmission. Although the signal power from port-1 is split between port-2 and port-3, impedance mismatches can cause part of the signal to reflect back. In this scenario the hybrid branch-line circuit directs these reflected signals to the isolated port-4, where they combine. Through phase shift cancellation—achieving a 180-degree phase difference—these unwanted reflections are eliminated at the input port-1. This mechanism ensures that the amplifier's signal matching and transmission remain unaffected, even when the outputs are connected to arbitrary mismatched loads, enhancing the overall robustness and efficiency of the design. Despite that, hybrid branch-line couplers are also used in various applications, including duplexers, dual-fed circular polarization circuits, double-balanced mixers, six-port complex ratio measurement units (CRMU), and antenna beam-forming networks.

This study focuses on the three-section branch-line coupler, which comprises three horizontal and four vertical transmission line sections. The horizontal lines have characteristic impedances Z_2 and Z_4, whereas the vertical lines have impedances Z_1 and Z_3, as shown in Fig. 11.4. For simplicity in circuit design, some branch-line coupler designs assume $Z_2 = Z_4$ (Reed, 1958) or $Z_1 = Z_3$ (You et al., 2018). Typically the vertical branch lines require very narrow microstrip widths, w, to achieve high characteristic impedances, Z_1 and Z_3, compared to horizontal branch lines. According to industry precision standards, microstrip widths, w, can be as small as 0.2 mm. However, with $w = 0.2$ mm, the maximum achievable impedance is generally limited to around 150–220 Ω, depending on substrate thickness ranging from 1.5 to 3 mm (You et al., 2020).

Therefore the impedance values Z_1 and Z_3 of the vertical lines are more suitable to be determined beforehand. The impedance values Z_2 and Z_4 for the horizontal lines can then be predicted based on the corresponding Z_1 and Z_3. Consequently, in this study, the vertical line impedances Z_1 and Z_3 are used as input training data in the ANN computation, whereas the predicted outputs are Z_2 and Z_4. The ANN network is designed with three inputs [coupling factor (C), Z_1, and Z_3]; three hidden layers, each consisting of four neurons; and the predicted outputs are the coupler's operating BW, Z_2, and Z_4, as shown in Fig. 11.6. The ANN training dataset for the three-section branch-line coupler is compiled from various literature sources, as summarized in Tables 11.1–11.3 (Section 11.2.3).

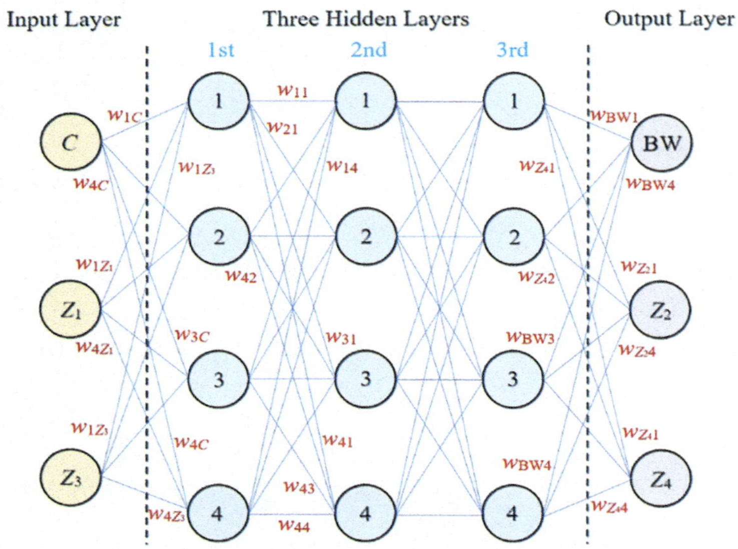

FIGURE 11.6 ANN model for the design of the three-section branch-line coupler. The ANN model processes all available input and output datasets to learn the optimal input–output mapping, enabling accurate prediction of the design parameters for the three-section branch-line coupler. *ANN*, Artificial neural network.

11.2.2 Wilkinson power divider circuits

A power divider/combiner is a multiport (at least three ports) passive device used to split or combine radio frequency (RF)/MW power, allowing a single transmission line to divide power across multiple lines or combine power from multiple inputs into one (Howe, 1974). A power divider/combiner enables the measurement and comparison of signal characteristics, such as frequency (phase shift) and power (gain level), for broadband-independent sampling. Its bi-directional design allows it to function as both a divider and a combiner. Applications include distributing power to multiple devices (such as antennas), combining signals for intermodulation distortion (IMD) analysis, and evaluating diversity gain in mobile phones (Keysight, 2024a).

This study focuses on the Wilkinson power divider/combiner, where the operating BW depends on the number of cascaded sections in the circuit. Fig. 11.7 illustrates two-, three-, and four-section Wilkinson power dividers/combiners.

It is well known that Wilkinson power divider/combiner can achieve good isolation between the output ports while maintaining impedance matching at all ports by incorporating a resistor of a specific value between port-2 and port-3. However, high-frequency surface mount device (SMD) resistors are still not widely available, and specific resistance values may be difficult to find on the market. Therefore it is more practical to determine the resistance, R, values of the resistors required for the Wilkinson power divider/combiner design in advance. Therefore the resistance, R values, of the resistors between port-2 and port-3 in each section will serve as the input training data for the ANN computation. On the other hand the corresponding output will be the characteristic impedance Z of each transmission line section. Fig. 11.8 illustrates the ANN models for the two-, three-, and four-section power dividers/combiners.

11.2.3 Libraries of neural models (training dataset)

Apart from the ANN–GA–PSO–SM parameter settings and the number of iterations the quality and quantity of the training input–output data play a crucial role in determining the accuracy and precision of the predicted output. Typically, input–output data for ANN training in high-frequency electronic design automation (EDA) is obtained from rigorous circuit or EM simulation results. In this study the training dataset is extracted from existing literature and supplemented by interpolating existing data to increase the amount of input–output dataset. To achieve accurate and realistic predictions a training dataset of at least a thousand samples is generally required. However, because of limited resources, the application of the ANN–GA–PSO–SM approach to the design of branch-line couplers and Wilkinson power dividers/combiners is based on a few hundred data samples. The details of the training dataset used for the branch-line coupler design are presented in Tables 11.1–11.3. Similarly the training dataset for the Wilkinson power divider/combiner design is also summarized in Tables 11.4–11.6. The overall training dataset for the Wilkinson power divider/combiner is interpolated from the design graphs presented in Howe (1974).

TABLE 11.1 Training data for the design of third-order branch-line coupler (Chebyshev type).

Levy and Lind (1968)	Coupling (dB)	Directivity (dB)	VSWR	Characteristic impedance (Ω) (for $N = 3$)			
				Z_1	Z_2	Z_3	Z_4
BW = 20%	3.0	50.314	1.0040	252.80	39.72	53.10	30.23
	3.5	50.864	1.0034	267.96	41.21	61.97	32.86
	4.0	51.376	1.0029	285.20	42.48	71.10	35.13
	4.5	51.853	1.0024	305.56	43.56	80.59	37.11
	5.0	52.299	1.0020	322.60	44.48	89.50	38.81
	5.5	52.714	1.0016	345.69	45.26	99.38	40.27
	6.0	53.103	1.0013	364.70	45.91	108.99	41.53
	6.5	53.466	1.0011	389.52	46.47	118.80	42.61
	7.0	53.807	1.0009	413.20	46.93	128.88	43.54
	7.5	54.127	1.0007	438.24	47.32	139.29	44.34
	8.0	54.428	1.0006	466.00	47.66	150.20	45.03
	8.5	54.712	1.0005	493.01	47.94	161.30	45.62
	9.0	54.981	1.0004	523.03	48.18	173.02	46.13
	9.5	55.235	1.0003	555.00	48.38	185.28	46.57
	10.0	55.477	1.0002	593.10	48.57	199.00	46.96
	10.5	55.707	1.0002	625.40	48.72	211.68	47.30
	11.0	55.927	1.0001	664.11	48.87	225.93	47.60
	11.5	56.138	1.0001	705.37	48.99	240.96	47.87
	12.0	56.340	1.0001	749.31	49.11	256.81	48.11
	12.5	56.534	1.0001	796.08	49.21	273.55	48.32
	13.0	56.721	1.0000	845.83	49.31	291.23	48.52
	13.5	56.901	1.0000	898.71	49.40	309.91	48.69
	14.0	57.075	1.0000	954.87	49.48	329.64	48.85
	14.5	57.243	1.0000	1014.40	49.55	350.49	48.99
	15.0	57.404	1.0000	1075.00	49.61	372.00	49.11
	15.5	57.559	1.0000	1144.40	49.67	395.74	49.22
	16.0	57.708	1.0000	1215.10	49.71	420.25	49.31
	16.5	57.850	1.0000	1289.90	49.75	446.10	49.39
	17.0	57.985	1.0000	1368.70	49.78	473.35	49.46
	17.5	58.112	1.0000	1451.90	49.80	502.04	49.51
	18.0	58.230	1.0000	1539.50	49.82	532.23	49.56
	18.5	58.339	1.0000	1631.70	49.84	563.99	49.60
	19.0	58.438	1.0000	1728.70	49.86	597.36	49.64
	19.5	58.525	1.0000	1830.50	49.88	632.41	49.67
	20.0	58.599	1.0000	1938.00	49.92	669.30	49.72
BW = 30%	3.0	39.79	1.0139	236.56	39.390	53.08	30.07
	3.5	40.35	1.0124	254.53	40.922	62.69	32.74
	4.0	40.88	1.0111	272.85	42.232	72.22	35.04
	4.5	41.37	1.0100	291.65	43.348	81.71	37.04
	5.0	41.83	1.0090	311.08	44.297	91.24	38.77
	5.5	42.26	1.0081	331.27	45.100	100.86	40.25
	6.0	42.66	1.0073	352.37	45.778	110.63	41.53
	6.5	43.03	1.0066	374.52	46.351	120.60	42.62
	7.0	43.38	1.0060	397.85	46.835	130.83	43.56
	7.5	43.71	1.0055	422.53	47.243	141.37	44.36

(continued on next page)

TABLE 11.1 Training data for the design of third-order branch-line coupler (Chebyshev type)—cont'd

Levy and Lind (1968)	Coupling (dB)	Directivity (dB)	VSWR	Characteristic impedance (Ω) (for $N = 3$)			
				Z_1	Z_2	Z_3	Z_4
	8.0	44.02	1.0050	448.67	47.590	152.30	45.05
	8.5	44.31	1.0045	476.43	47.885	163.65	45.64
	9.0	44.58	1.0041	505.95	48.137	175.50	46.16
	9.5	44.84	1.0037	537.36	48.356	187.90	46.60
	10.0	45.09	1.0033	570.81	48.546	200.90	46.98
	10.5	45.33	1.0030	606.45	48.714	214.56	47.32
	11.0	45.55	1.0026	644.40	48.864	228.95	47.62
	11.5	45.77	1.0022	684.82	48.998	244.11	47.88
	12.0	45.98	1.0019	727.85	49.120	260.11	48.12
	12.5	46.18	1.0015	773.62	49.230	277.00	48.33
	13.0	46.37	1.0011	822.28	49.330	294.84	48.52
	13.5	46.56	1.0008	873.96	49.420	313.69	48.69
	14.0	46.73	1.0004	928.82	49.501	333.60	48.85
	14.5	46.91	1.0000	986.99	49.572	354.64	48.99
	15.0	47.07	1.0000	1048.62	49.634	376.86	49.11
	15.5	47.23	1.0000	1113.84	49.687	400.32	49.22
	16.0	47.38	1.0000	1182.79	49.730	425.07	49.31
	16.5	47.52	1.0000	1255.63	49.764	451.17	49.39
	17.0	47.66	1.0000	1332.48	49.791	478.68	49.46
	17.5	47.78	1.0000	1413.49	49.810	507.66	49.51
	18.0	47.90	1.0000	1498.81	49.825	538.17	49.56
	18.5	48.01	1.0000	1588.57	49.838	570.26	49.60
	19.0	48.10	1.0000	1682.91	49.854	603.98	49.64
	19.5	48.18	1.0000	1781.99	49.876	639.41	49.67
	20.0	48.24	1.0000	1885.92	49.911	676.59	49.72
BW = 40%	3.0	31.60	1.0423	220.50	38.75	54.10	29.70
	3.5	32.19	1.0348	236.03	40.39	63.43	32.46
	4.0	32.73	1.0270	253.20	41.90	73.00	34.90
	4.5	33.24	1.0232	272.60	42.99	82.87	36.86
	5.0	33.72	1.0188	289.70	44.01	92.00	38.60
	5.5	34.16	1.0153	311.52	44.86	102.40	40.13
	6.0	34.57	1.0125	330.50	45.58	111.80	41.40
	6.5	34.95	1.0103	353.88	46.19	122.49	42.55
	7.0	35.31	1.0086	376.69	46.70	132.89	43.50
	7.5	35.65	1.0073	400.78	47.12	143.60	44.31
	8.0	35.97	1.0065	427.00	47.49	154.69	45.00
	8.5	36.27	1.0057	453.31	47.79	166.21	45.61
	9.0	36.55	1.0052	482.03	48.06	178.22	46.13
	9.5	36.82	1.0049	512.57	48.28	190.78	46.57
	10.0	37.08	1.0047	548.20	48.48	204.60	47.00
	10.5	37.32	1.0045	579.67	48.65	217.78	47.30
	11.0	37.55	1.0043	616.50	48.81	232.34	47.60
	11.5	37.77	1.0041	655.70	48.95	247.68	47.87

(continued on next page)

TABLE 11.1 Training data for the design of third-order branch-line coupler (Chebyshev type)—cont'd

Levy and Lind (1968)	Coupling (dB)	Directivity (dB)	VSWR	Characteristic impedance (Ω) (for $N = 3$)			
				Z_1	Z_2	Z_3	Z_4
	12.0	37.99	1.0039	697.40	49.07	263.86	48.11
	12.5	38.19	1.0036	741.75	49.19	280.95	48.32
	13.0	38.39	1.0032	788.88	49.29	299.00	48.52
	13.5	38.58	1.0027	838.92	49.39	318.06	48.69
	14.0	38.76	1.0022	892.03	49.47	338.20	48.85
	14.5	38.93	1.0015	948.32	49.55	359.47	48.98
	15.0	39.10	1.0009	1006.00	49.61	381.10	49.11
	15.5	39.26	1.0001	1071.04	49.67	405.66	49.22
	16.0	39.42	1.0000	1137.74	49.72	430.70	49.31
	16.5	39.56	1.0000	1208.18	49.75	457.10	49.39
	17.0	39.70	1.0000	1282.49	49.78	484.93	49.46
	17.5	39.82	1.0000	1360.83	49.80	514.24	49.52
	18.0	39.94	1.0000	1443.31	49.82	545.11	49.56
	18.5	40.04	1.0000	1530.09	49.84	577.57	49.60
	19.0	40.13	1.0000	1621.30	49.86	611.70	49.64
	19.5	40.21	1.0000	1717.07	49.88	647.55	49.67
	20.0	40.27	1.0000	1818.00	49.92	684.90	49.72
BW = 50%	3.0	25.20	1.0982	195.33	37.87	54.70	29.29
	3.5	25.81	1.0802	212.92	39.68	64.83	32.10
	4.0	26.37	1.0650	230.76	41.22	74.85	34.52
	4.5	26.90	1.0523	248.98	42.52	84.82	36.62
	5.0	27.39	1.0418	267.72	43.62	94.80	38.42
	5.5	27.85	1.0332	287.10	44.54	104.85	39.97
	6.0	28.27	1.0263	307.27	45.32	115.04	41.29
	6.5	28.67	1.0209	328.35	45.96	125.41	42.42
	7.0	29.04	1.0168	350.49	46.51	136.04	43.39
	7.5	29.39	1.0137	373.81	46.96	146.98	44.22
	8.0	29.72	1.0115	398.44	47.34	158.29	44.93
	8.5	30.03	1.0099	424.53	47.66	170.03	45.54
	9.0	30.32	1.0089	452.20	47.94	182.26	46.06
	9.5	30.60	1.0084	481.59	48.17	195.05	46.52
	10.0	30.86	1.0081	512.84	48.38	208.44	46.91
	10.5	31.11	1.0079	546.07	48.56	222.51	47.26
	11.0	31.35	1.0079	581.43	48.72	237.31	47.57
	11.5	31.57	1.0078	619.04	48.86	252.89	47.84
	12.0	31.79	1.0077	659.03	48.99	269.34	48.08
	12.5	32.00	1.0074	701.55	49.11	286.69	48.30
	13.0	32.20	1.0070	746.73	49.22	305.01	48.50
	13.5	32.40	1.0065	794.70	49.33	324.37	48.68
	14.0	32.58	1.0057	845.59	49.42	344.82	48.84
	14.5	32.76	1.0048	899.54	49.50	366.41	48.98
	15.0	32.93	1.0038	956.69	49.57	389.23	49.11
	15.5	33.10	1.0027	1017.16	49.63	413.31	49.22
	16.0	33.25	1.0015	1081.09	49.69	438.72	49.32
	16.5	33.40	1.0004	1148.61	49.73	465.53	49.40

(*continued on next page*)

TABLE 11.1 Training data for the design of third-order branch-line coupler (Chebyshev type)—cont'd

Levy and Lind (1968)	Coupling (dB)	Directivity (dB)	VSWR	Characteristic impedance (Ω) (for $N = 3$)			
				Z_1	Z_2	Z_3	Z_4
	17.0	33.54	1.0001	1219.86	49.76	493.79	49.46
	17.5	33.67	1.0000	1294.97	49.79	523.56	49.52
	18.0	33.79	1.0000	1374.08	49.81	554.90	49.56
	18.5	33.89	1.0000	1457.32	49.83	587.87	49.60
	19.0	33.98	1.0000	1544.82	49.85	622.54	49.63
	19.5	34.06	1.0000	1636.72	49.88	658.96	49.67
	20.0	34.12	1.0000	1733.15	49.93	697.19	49.72
BW = 60%	3.0	20.02	1.1970	168.86	36.85	56.94	28.84
	3.5	20.65	1.1587	186.25	38.84	67.49	31.73
	4.0	21.24	1.1240	203.81	40.52	77.91	34.22
	4.5	21.78	1.1077	221.66	41.95	88.27	36.37
	5.0	22.29	1.0840	239.95	43.15	98.61	38.21
	5.5	22.76	1.0717	258.79	44.15	109.01	39.79
	6.0	23.21	1.0583	278.31	44.99	119.53	41.14
	6.5	23.62	1.0474	298.66	45.68	130.23	42.29
	7.0	24.00	1.0388	319.96	46.27	141.17	43.27
	7.5	24.36	1.0321	342.33	46.75	152.42	44.11
	8.0	24.70	1.0269	365.92	47.16	164.03	44.83
	8.5	25.02	1.0231	390.84	47.50	176.07	45.45
	9.0	25.32	1.0204	417.23	47.79	188.60	45.99
	9.5	25.61	1.0184	445.23	48.04	201.68	46.45
	10.0	25.88	1.0171	474.95	48.25	215.38	46.85
	10.5	26.13	1.0162	506.54	48.44	229.76	47.21
	11.0	26.38	1.0156	540.12	48.61	244.88	47.52
	11.5	26.61	1.0152	575.82	48.77	260.80	47.81
	12.0	26.84	1.0147	613.77	48.91	277.59	48.06
	12.5	27.05	1.0142	654.10	49.04	295.30	48.29
	13.0	27.26	1.0135	696.95	49.16	314.00	48.49
	13.5	27.46	1.0127	742.44	49.27	333.75	48.68
	14.0	27.65	1.0116	790.70	49.37	354.62	48.84
	14.5	27.83	1.0103	841.87	49.45	376.66	48.99
	15.0	28.01	1.0089	896.07	49.53	399.94	49.12
	15.5	28.17	1.0073	953.44	49.60	424.52	49.23
	16.0	28.34	1.0056	1014.10	49.65	450.47	49.32
	16.5	28.49	1.0051	1078.18	49.70	477.84	49.40
	17.0	28.63	1.0047	1145.83	49.74	506.69	49.46
	17.5	28.76	1.0042	1217.15	49.76	537.10	49.51
	18.0	28.88	1.0038	1292.29	49.79	569.12	49.56
	18.5	28.99	1.0035	1371.38	49.81	602.81	49.59
	19.0	29.09	1.0033	1454.55	49.83	638.24	49.63
	19.5	29.17	1.0031	1541.92	49.87	675.46	49.67
	20.0	29.23	1.0030	1633.63	49.93	714.55	49.73

(continued on next page)

TABLE 11.1 Training data for the design of third-order branch-line coupler (Chebyshev type)—cont'd

Levy and Lind (1968)	Coupling (dB)	Directivity (dB)	VSWR	Characteristic impedance (Ω) (for $N = 3$)			
				Z_1	Z_2	Z_3	Z_4
BW = 70%	3.0	15.49	1.3198	139.84	35.76	60.87	28.45
	3.5	16.15	1.2751	156.92	37.93	72.04	31.41
	4.0	16.76	1.2369	174.08	39.76	83.04	33.97
	4.5	17.33	1.1997	191.44	41.31	93.94	36.16
	5.0	17.86	1.1678	209.15	42.60	104.79	38.03
	5.5	18.35	1.1406	227.31	43.69	115.68	39.64
	6.0	18.80	1.1178	246.06	44.59	126.66	41.01
	6.5	19.23	1.0987	265.53	45.34	137.80	42.18
	7.0	19.63	1.0829	285.83	45.97	149.17	43.17
	7.5	20.00	1.0700	307.10	46.49	160.82	44.02
	8.0	20.35	1.0596	329.46	46.93	172.84	44.75
	8.5	20.68	1.0512	353.03	47.30	185.27	45.38
	9.0	20.99	1.0447	377.94	47.61	198.20	45.92
	9.5	21.29	1.0396	404.33	47.88	211.67	46.39
	10.0	21.56	1.0356	432.30	48.12	225.77	46.81
	10.5	21.83	1.0325	461.99	48.33	240.56	47.17
	11.0	22.08	1.0301	493.53	48.52	256.09	47.49
	11.5	22.32	1.0281	527.04	48.69	272.45	47.78
	12.0	22.55	1.0265	562.64	48.85	289.68	48.05
	12.5	22.77	1.0249	600.47	48.99	307.87	48.28
	13.0	22.98	1.0234	640.64	49.12	327.06	48.49
	13.5	23.19	1.0219	683.28	49.24	347.34	48.68
	14.0	23.38	1.0202	728.53	49.35	368.77	48.85
	14.5	23.57	1.0184	776.49	49.44	391.41	49.00
	15.0	23.75	1.0165	827.31	49.52	415.32	49.13
	15.5	23.93	1.0144	881.10	49.59	440.58	49.24
	16.0	24.09	1.0123	938.00	49.64	467.24	49.33
	16.5	24.25	1.0102	998.12	49.68	495.39	49.40
	17.0	24.40	1.0083	1061.59	49.71	525.07	49.46
	17.5	24.54	1.0067	1128.55	49.73	556.36	49.51
	18.0	24.66	1.0056	1199.10	49.74	589.32	49.54
	18.5	24.78	1.0051	1273.39	49.76	624.02	49.58
	19.0	24.88	1.0046	1351.53	49.79	660.52	49.61
	19.5	24.97	1.0042	1433.66	49.84	698.89	49.66
	20.0	25.04	1.0039	1519.89	49.91	739.20	49.74
BW = 80%	3.0	11.06	1.5238	109.25	34.70	67.05	28.21
	3.5	11.75	1.4542	125.82	37.02	79.09	31.23
	4.0	12.38	1.3927	142.41	38.97	90.89	33.83
	4.5	12.97	1.3386	159.12	40.62	102.54	36.05
	5.0	13.52	1.2912	176.09	42.00	114.11	37.95
	5.5	14.03	1.2500	193.42	43.16	125.65	39.57
	6.0	14.51	1.2020	211.25	44.12	137.26	40.95
	6.5	14.95	1.1835	229.68	44.93	148.99	42.12

(continued on next page)

TABLE 11.1 Training data for the design of third-order branch-line coupler (Chebyshev type)—cont'd

Levy and Lind (1968)	Coupling (dB)	Directivity (dB)	VSWR	Characteristic impedance (Ω) (for $N = 3$)			
				Z_1	Z_2	Z_3	Z_4
	7.0	15.36	1.1572	248.84	45.61	160.91	43.12
	7.5	15.75	1.1349	268.85	46.17	173.11	43.98
	8.0	16.11	1.1100	289.82	46.65	185.65	44.71
	8.5	16.45	1.1002	311.88	47.06	198.59	45.34
	9.0	16.77	1.0870	335.15	47.42	212.02	45.89
	9.5	17.07	1.0761	359.73	47.72	226.00	46.37
	10.0	17.36	1.0672	385.76	48.00	240.61	46.79
	10.5	17.63	1.0598	413.35	48.24	255.91	47.16
	11.0	17.89	1.0537	442.62	48.46	271.97	47.49
	11.5	18.14	1.0486	473.69	48.66	288.88	47.79
	12.0	18.37	1.0444	506.68	48.84	306.69	48.06
	12.5	18.60	1.0408	541.71	49.00	325.48	48.30
	13.0	18.82	1.0376	578.90	49.15	345.32	48.51
	13.5	19.03	1.0347	618.36	49.28	366.28	48.71
	14.0	19.23	1.0320	660.22	49.39	388.43	48.88
	14.5	19.42	1.0294	704.60	49.48	411.84	49.02
	15.0	19.61	1.0250	751.61	49.56	436.59	49.15
	15.5	19.79	1.0244	801.37	49.61	462.75	49.25
	16.0	19.96	1.0219	854.00	49.65	490.38	49.34
	16.5	20.13	1.0198	909.63	49.67	519.55	49.40
	17.0	20.28	1.0185	968.37	49.68	550.35	49.45
	17.5	20.43	1.0171	1030.34	49.68	582.83	49.49
	18.0	20.57	1.0157	1095.66	49.68	617.08	49.52
	18.5	20.69	1.0143	1164.45	49.69	653.15	49.56
	19.0	20.80	1.0131	1236.82	49.71	691.13	49.60
	19.5	20.90	1.0120	1312.91	49.76	731.08	49.66
	20.0	20.99	1.0110	1392.82	49.86	773.07	49.76

BW, bandwidth

11.3 ANN-model verification

11.3.1 Resimulate (S-parameters) the hybrid branch-line couplers using ADS circuit simulator

This section evaluates reliable ANN–GA–PSO–SM techniques by resimulating the *S*-parameters for a three-section branch-line coupler, shown in Fig. 11.9. In the ANN–GA–PSO–SM prediction process the coupling factor, *C*, and impedance values (Z_1 and Z_3) for the two vertical microstrip branch lines are initially specified. The prediction outputs include the characteristic impedance values (Z_2 and Z_4) of the two horizontal branch lines, as well as the coupler's BW, as listed in Table 11.7. Up to five configurations of the three-section branch-line coupler are designed, all operating at a center frequency of 3.5 GHz on FR-4 printed circuit board (PCB) with a substrate thickness of 1.58 mm. The physical dimensions for each design are detailed in Table 11.8.

Fig. 11.10 presents the simulated *S*-parameters of the five three-section couplers for comparison and validation. The return loss, S_{11}, for all couplers remains below −20 dB at 3.5 GHz. The phase difference ($\angle S_{31} - \angle S_{21}$) between port-2 and port-3 is −90 degrees across the operating frequency range. Based on the simulated *S*-parameters the desired coupling, *C*, and predicted BW show minor deviations from expected values. However, overall performance of the designed couplers is satisfactory. Notably a higher impedance value for Z_1, as well as a greater distinction between Z1 and Z_3, tends to improve coupler performance, particularly evident in Design 5. As shown in Fig. 11.8E the phase imbalance for Design 5 is under

TABLE 11.2 Training data for the design of third-order branch-line coupler (Butterworth type).

BW (%)	Coup. (dB)	VSWR	Z_1 (Ω)	Z_2 (Ω)	Z_3 (Ω)	Z_4 (Ω)	BW (%)	Coup. (dB)	VSWR	Z_1 (Ω)	Z_2 (Ω)	Z_3 (Ω)	Z_4 (Ω)
40.44	2.0	1.286	237.52	35.85	35.10	23.47	53.15	11.2	1.022	702.62	48.90	230.61	47.69
41.17	2.4	1.239	247.58	37.60	42.13	26.40	53.51	11.6	1.020	736.79	49.01	242.68	47.90
41.87	2.8	1.201	258.50	39.15	49.17	29.02	53.87	12.0	1.019	772.58	49.11	255.27	48.10
42.56	3.2	1.170	270.26	40.49	56.23	31.35	54.20	12.4	1.017	810.06	49.20	268.39	48.28
43.24	3.6	1.144	282.85	41.67	63.34	33.41	54.53	12.8	1.016	849.31	49.29	282.09	48.45
43.90	4.0	1.123	296.29	42.69	70.49	35.24	54.84	13.2	1.015	890.43	49.37	296.39	48.60
44.54	4.4	1.107	310.56	43.58	77.72	36.86	55.13	13.6	1.014	933.49	49.44	311.32	48.74
45.17	4.8	1.093	325.68	44.34	85.04	38.29	55.41	14.0	1.013	978.60	49.51	326.91	48.87
45.79	5.2	1.083	341.65	45.01	92.46	39.55	55.67	14.4	1.012	1025.85	49.57	343.21	48.99
46.39	5.6	1.074	358.49	45.58	100.01	40.66	55.92	14.8	1.011	1075.34	49.62	360.24	49.09
46.97	6.0	1.067	376.22	46.07	107.70	41.64	56.16	15.2	1.011	1127.18	49.66	378.03	49.18
47.54	6.4	1.060	394.87	46.49	115.55	42.50	56.38	15.6	1.010	1181.48	49.70	396.63	49.26
48.09	6.8	1.055	414.44	46.86	123.58	43.25	56.59	16.0	1.010	1238.36	49.73	416.06	49.33
48.63	7.2	1.051	434.98	47.17	131.83	43.92	56.79	16.4	1.009	1297.91	49.75	436.38	49.39
49.15	7.6	1.047	456.52	47.45	140.29	44.51	56.97	16.8	1.008	1360.28	49.77	457.60	49.44
49.66	8.0	1.044	479.10	47.69	149.01	45.04	57.14	17.2	1.007	1425.58	49.78	479.77	49.48
50.15	8.4	1.040	502.75	47.90	158.00	45.50	57.30	17.6	1.006	1493.94	49.78	502.93	49.51
50.62	8.8	1.037	527.51	48.09	167.28	45.92	57.45	18.0	1.005	1565.49	49.79	527.12	49.54
51.08	9.2	1.034	553.45	48.25	176.88	46.29	57.58	18.4	1.005	1640.37	49.80	552.38	49.56
51.52	9.6	1.032	580.61	48.41	186.83	46.62	57.70	18.8	1.004	1718.71	49.82	578.74	49.59
51.95	10.0	1.029	609.04	48.54	197.15	46.93	57.81	19.2	1.004	1800.67	49.84	606.25	49.63
52.37	10.4	1.027	638.81	48.67	207.87	47.20	57.91	19.6	1.004	1886.38	49.88	634.96	49.68
52.76	10.8	1.024	669.98	48.79	219.01	47.45	57.99	20.0	1.004	1976.01	49.95	664.89	49.75

TABLE 11.3 Training data for the design of third-order branch-line coupler [identical branch lines arranged horizontally ($Z_2 = Z_4$) or vertically ($Z_1 = Z_3$)].

Reed (1958)						You et al. (2018)					
BW (%)	Coup. (dB)	Z_1 (Ω)	Z_2 (Ω)	Z_3 (Ω)	Z_4 (Ω)	BW (%)	Coup. (dB)	Z_1 (Ω)	Z_2 (Ω)	Z_3 (Ω)	Z_4 (Ω)
24.21	0.00	100.0	50.0	50.0	50.0	36	3 ± 0.2	162.07	60.66	162.07	71.87
25.35	0.20	107.12	50.0	52.59	50.0	37	3 ± 0.2	156.39	58.75	156.39	67.74
26.48	0.40	114.31	50.0	55.22	50.0	38	3 ± 0.2	151.07	56.95	151.07	63.87
27.61	0.60	121.55	50.0	57.87	50.0	39	3 ± 0.2	146.11	55.26	146.11	60.28
28.75	0.80	128.85	50.0	60.57	50.0	40	3 ± 0.2	141.52	53.68	141.52	56.95
29.88	1.00	136.21	50.0	63.29	50.0	41	3 ± 0.2	137.30	52.21	137.30	53.89
31.01	1.20	143.63	50.0	66.05	50.0	42	3 ± 0.2	133.44	50.86	133.44	51.09
32.15	1.40	151.11	50.0	68.84	50.0	43	3 ± 0.2	129.95	49.61	129.95	48.57
33.28	1.60	158.65	50.0	71.67	50.0	44	3 ± 0.2	126.82	48.47	126.82	46.31
34.41	1.80	166.25	50.0	74.53	50.0	45	3 ± 0.2	124.06	47.44	124.06	44.32
35.54	2.00	173.91	50.0	77.42	50.0	46	3 ± 0.2	121.66	46.52	121.66	42.60
36.68	2.20	181.63	50.0	80.35	50.0	47	3 ± 0.2	119.63	45.71	119.63	41.15
37.81	2.40	189.41	50.0	83.31	50.0	25	2.5 ± 0.2	172.16	70.46	172.16	93.76
38.94	2.60	197.25	50.0	86.31	50.0	26	2.5 ± 0.2	167.35	68.80	167.35	89.87
40.08	2.80	205.14	50.0	89.34	50.0	27	2.5 ± 0.2	162.70	67.19	162.70	86.12
41.21	3.00	213.10	50.0	92.40	50.0	28	2.5 ± 0.2	158.24	65.63	158.24	82.51
42.34	3.20	221.12	50.0	95.50	50.0	29	2.5 ± 0.2	153.94	64.11	153.94	79.05
43.48	3.40	229.19	50.0	98.63	50.0	30	2.5 ± 0.2	149.82	62.64	149.82	75.72
44.61	3.60	237.33	50.0	101.79	50.0	31	2.5 ± 0.2	145.88	61.22	145.88	72.54
45.74	3.80	245.52	50.0	104.99	50.0	32	2.5 ± 0.2	142.10	59.85	142.10	69.49
46.88	4.00	253.78	50.0	108.22	50.0	33	2.5 ± 0.2	138.51	58.52	138.51	66.59
48.01	4.20	262.09	50.0	111.48	50.0	34	2.5 ± 0.2	135.08	57.24	135.08	63.82
49.14	4.40	270.46	50.0	114.78	50.0	35	2.5 ± 0.2	131.83	56.01	131.83	61.12
50.28	4.60	278.90	50.0	118.11	50.0	36	2.5 ± 0.2	128.76	54.82	128.76	58.72
51.41	4.80	287.39	50.0	121.48	50.0	37	2.5 ± 0.2	125.86	53.68	125.86	56.37
52.54	5.00	295.94	50.0	124.88	50.0	38	2.5 ± 0.2	123.13	52.59	123.13	54.17
53.68	5.20	304.55	50.0	128.31	50.0	39	2.5 ± 0.2	120.58	51.55	120.58	52.11
54.81	5.40	313.23	50.0	131.78	50.0	17	2.0 ± 0.2	173.68	76.44	173.68	106.32
55.94	5.60	321.96	50.0	135.28	50.0	18	2.0 ± 0.2	167.26	74.02	167.26	100.75
57.08	5.80	330.75	50.0	138.81	50.0	19	2.0 ± 0.2	161.25	71.74	161.25	95.54
58.21	6.00	339.60	50.0	142.38	50.0	20	2.0 ± 0.2	155.64	69.62	155.64	90.67
59.34	6.20	348.51	50.0	145.98	50.0	21	2.0 ± 0.2	150.44	67.64	150.44	86.16
60.48	6.40	357.48	50.0	149.61	50.0	22	2.0 ± 0.2	145.66	65.81	145.66	82.01
61.61	6.60	366.50	50.0	153.28	50.0	23	2.0 ± 0.2	141.28	64.13	141.28	78.20
62.74	6.80	375.59	50.0	156.98	50.0	24	2.0 ± 0.2	137.31	62.60	137.31	74.75
63.88	7.00	384.74	50.0	160.72	50.0	25	2.0 ± 0.2	133.74	61.21	133.74	71.66
65.01	7.20	393.95	50.0	164.49	50.0	26	2.0 ± 0.2	130.59	59.98	130.59	68.92
66.14	7.40	403.21	50.0	168.29	50.0	27	2.0 ± 0.2	127.84	58.89	127.84	66.53
67.28	7.60	412.54	50.0	172.13	50.0	28	2.0 ± 0.2	125.51	57.95	125.51	64.49
68.41	7.80	421.93	50.0	176.00	50.0	29	2.0±0.2	123.58	57.16	123.58	62.81
69.54	8.00	431.37	50.0	179.90	50.0	30	2.0 ± 0.2	122.06	56.51	122.06	61.48
70.68	8.20	440.88	50.0	183.84	50.0	31	2.0 ± 0.2	120.94	56.02	120.94	60.51

(continued on next page)

TABLE 11.3 Training data for the design of third-order branch-line coupler [identical branch lines arranged horizontally ($Z_2 = Z_4$) or vertically ($Z_1 = Z_3$)]—cont'd

Reed (1958)						You et al. (2018)					
BW (%)	Coup. (dB)	Z_1 (Ω)	Z_2 (Ω)	Z_3 (Ω)	Z_4 (Ω)	BW (%)	Coup. (dB)	Z_1 (Ω)	Z_2 (Ω)	Z_3 (Ω)	Z_4 (Ω)
71.81	8.40	450.44	50.0	187.81	50.0	32	2.0±0.2	120.24	55.67	120.24	59.89
72.94	8.60	460.06	50.0	191.82	50.0	33	2.0 ± 0.2	119.94	55.47	119.94	59.62
74.08	8.80	469.75	50.0	195.86	50.0	15	1.5 ± 0.2	171.03	79.89	171.03	112.86
75.21	9.00	479.49	50.0	199.93	50.0	16	1.5 ± 0.2	164.99	77.51	164.99	107.35
76.34	9.20	489.29	50.0	204.04	50.0	17	1.5 ± 0.2	159.30	75.25	159.30	102.16
77.48	9.40	499.16	50.0	208.18	50.0	18	1.6 ± 0.2	153.97	73.13	153.97	97.29
78.61	9.60	509.08	50.0	212.35	50.0	19	1.6 ± 0.2	148.98	71.12	148.98	92.74
79.74	9.80	519.06	50.0	216.56	50.0	20	1.6 ± 0.2	144.35	69.25	144.35	88.50
80.88	10.00	529.10	50.0	220.80	50.0	21	1.6 ± 0.2	140.06	67.50	140.06	84.58
Muraguchi et al. (1983)						22	1.6 ± 0.2	136.13	65.88	136.13	80.97
BW (%)	Coup. (dB)	Z_1 (Ω)	Z_2 (Ω)	Z_3 (Ω)	Z_4 (Ω)	23	1.6 ± 0.2	132.55	64.39	132.55	77.69
						24	1.6 ± 0.2	129.31	63.02	129.31	74.72
32	3.0	263.9	40.0	52.8	30.3	25	1.6 ± 0.2	126.43	61.78	126.43	72.06
34	3.0	170.5	36.8	57.4	28.8	26	1.6 ± 0.2	123.90	60.66	123.90	69.72
44	3.0 ± 0.3	153.0	44.3	91.0	39.4	27	1.6 ± 0.2	121.72	59.67	121.72	67.70
40	3.0 ± 0.2	162.4	44.1	85.3	39.2	28	1.6 ± 0.2	119.90	58.81	119.90	66.00
38	3.0 ± 0.1	157.5	45.0	91.1	42	12	1.4 ± 0.2	172.40	84.32	172.40	121.24
44	3.0 ± 0.3	166.7	39.4	67.0	30.7	13	1.4 ± 0.2	162.97	80.45	162.97	112.34
44	3.0 ± 0.3	156.3	43.0	83.9	36.8	14	1.4 ± 0.2	154.88	77.09	154.88	104.69
43	3.0 ± 0.3	147.1	47.2	106.3	44.8	15	1.4 ± 0.2	147.99	74.18	147.99	98.15
43	3.0 ± 0.3	138.9	52.0	136.1	54.9	16	1.4 ± 0.2	142.13	71.67	142.13	92.57
43	3.0 ± 0.3	131.6	57.4	175.0	67.4	17	1.4 ± 0.2	137.16	69.49	137.16	87.82
45	3.0 ± 0.4	142.9	53.7	142.9	56.5	18	1.4 ± 0.2	132.91	67.61	132.91	83.75
43	3.0 ± 0.3	142.9	53.9	142.9	58.3	19	1.4 ± 0.2	129.23	65.95	129.23	80.23
43	3.0 ± 0.3	138.9	52.5	138.9	55.9	20	1.4 ± 0.2	125.98	64.47	125.98	77.10
41	3.0 ± 0.2	138.9	52.6	138.9	57.1	21	1.4 ± 0.2	122.99	63.11	122.99	74.22
40	3.0 ± 0.2	138.9	52.7	138.9	58.1	22	1.4 ± 0.2	120.11	61.81	120.11	71.46
40	3.0 ± 0.2	135.1	51.3	135.1	55.3						
39	3.0 ± 0.1	135.1	51.3	135.1	56.4						

0.5 degrees, as it is for Design 1. In contrast, Designs 2 and 4 maintain a phase imbalance within 1 degree, whereas Design 3, with a phase imbalance of approximately 1.5 degrees, exhibits the weakest performance across the operating range.

11.3.2 Resimulate (*S*-parameters) the Wilkinson power dividers using ADS circuit simulator

The predicted transmission line impedances Z, determined using the ANN–PSO–GA–SM approach for the two-, three-, and four-section Wilkinson power dividers/combiners, are listed in Tables 11.9–11.11. All Wilkinson power divider/combiner circuits are designed for operation at a center frequency of 3.5 GHz and are assumed to be printed on a PCB substrate with a dielectric constant, $\varepsilon_r = 4.4$, and a thickness, $h = 1.58$ mm. Based on this configuration the transmission microstrip line dimensions (width, w, and length, l) for the Wilkinson power dividers/combiners are calculated using a microstrip line calculator, with these values presented in Tables 11.12, 11.13, and 11.14, respectively.

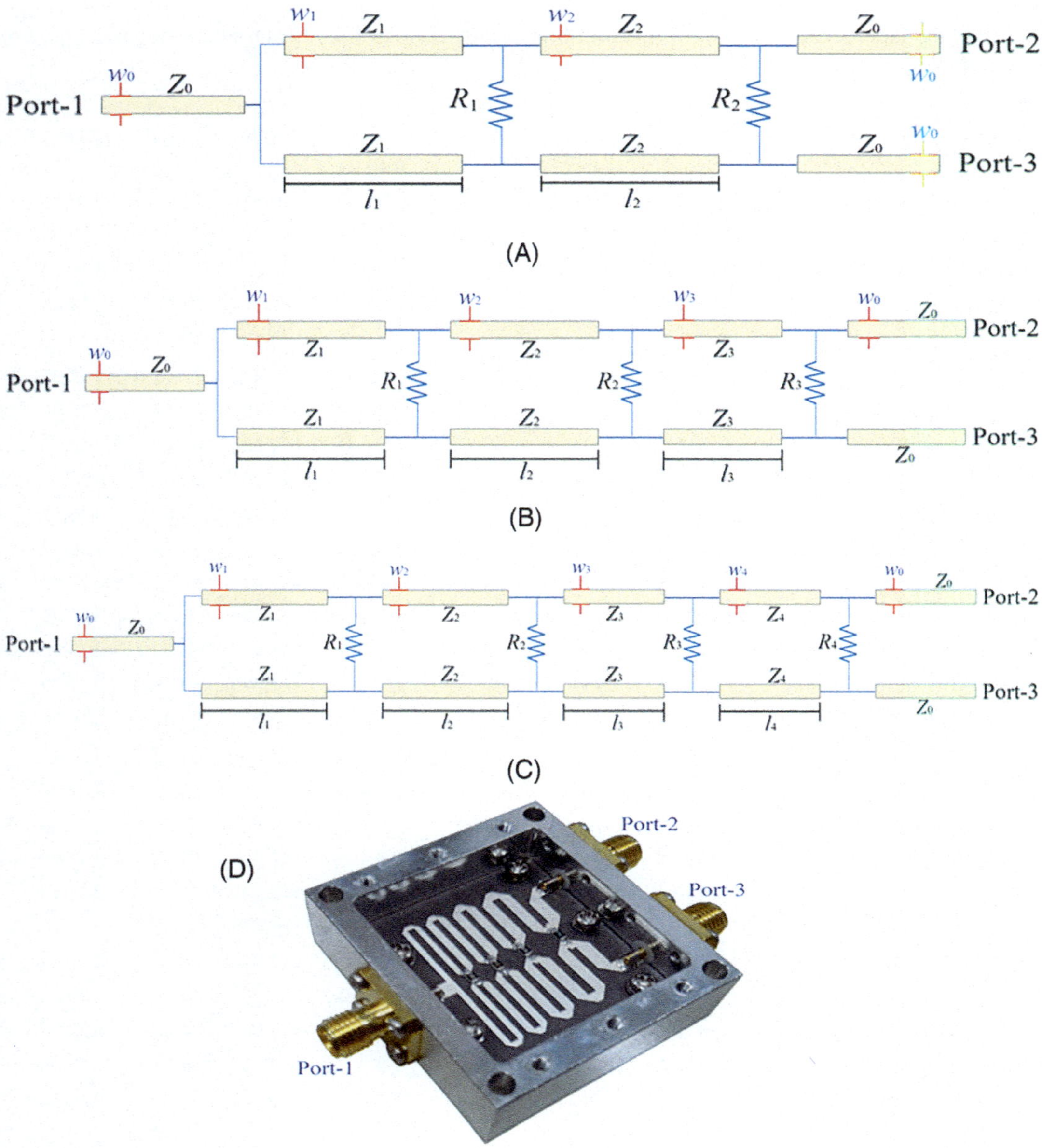

FIGURE 11.7 Schematics of the (A) two-section, (B) three-section, and (C) four-section Wilkinson power divider/combiners. (D) Actual configuration of the two-way four-section Wilkinson power divider/combiners.

By using the specified resistor values, R, and the calculated microstrip line dimensions, the power dividers/combiners are then resimulated in the Advanced Design System (ADS) simulator by Keysight Technologies. The circuit schematics of the Wilkinson power dividers/combiners, as extracted from ADS, are shown in Fig. 11.11. In addition, the corresponding simulated scattering parameters (S-parameters) for the two-, three-, and four-section Wilkinson power dividers/combiners are shown in Figs. 11.12–11.14, respectively. S_{11} represents the return loss (reflection coefficient) at the input port (port-1) in dB. S_{22} and S_{33} denote the return loss at the output ports (port-2 and port-3), respectively. S_{21} and S_{31} indicate the insertion loss (transmission loss) at the output ports (port-2 and port-3), respectively, whereas S_{32} shows the isolation between the two output ports. Notably the BW of the Wilkinson power dividers/combiners increases as the design progresses from a two-section to a four-section cascade.

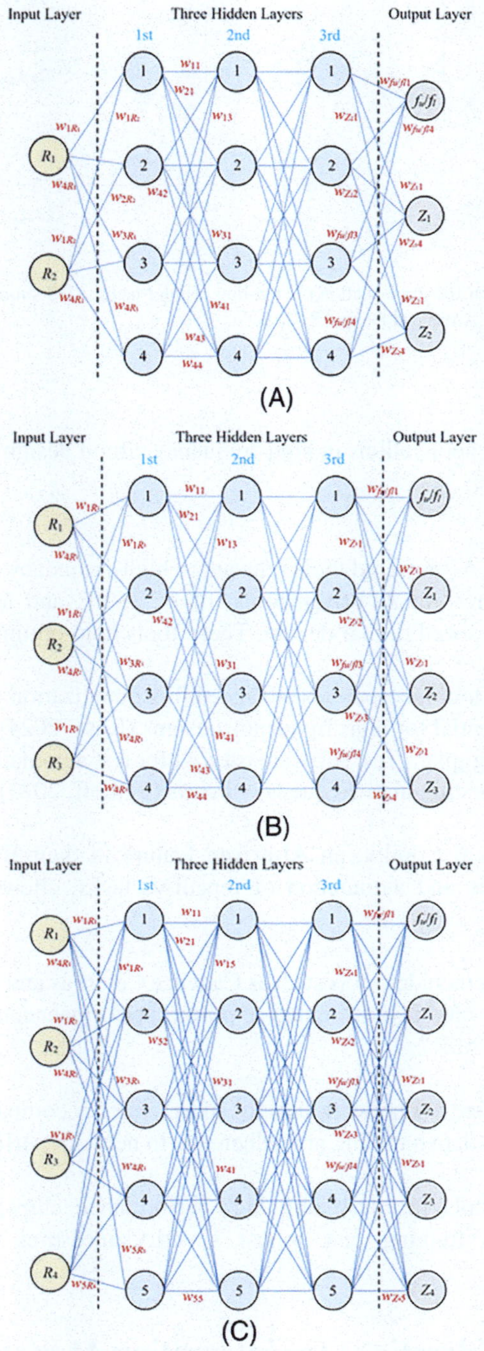

FIGURE 11.8 ANN models for (A) two-section, (B) three-section, and (C) four-section Wilkinson power dividers/combiners.

11.4 Key industry applications and future potential

The rise of fifth-generation of wireless technology (5G), IoT, fourth industrial revolution (Industry 4.0), and advanced mechatronic technologies in domestic and medical fields is driving demand for high-frequency design. AI-driven tools are essential for managing design complexities, cutting time and costs, and increasing accessibility across industries such as telecom, healthcare, and consumer electronics. AI also enables innovative, custom designs beyond traditional methods, accelerating adoption. As a result the global AI market is expected to grow, with significant applications in electronics and circuit design.

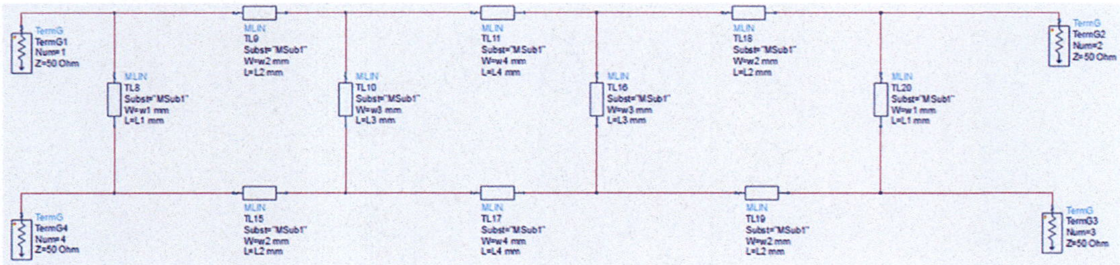

FIGURE 11.9 Simulated circuit schematics for the three-section branch-line coupler using ADS simulator. 2D circuit simulation for the three-section branch-line coupler using ADS simulator. *ADS*, Advanced Design System.

Several real-world examples of AI applications in high-frequency circuit design illustrate the significant advancements enabled by AI:

1) Keysight Technologies

Keysight has developed several AI-powered tools relevant to high-frequency circuit design:

a. AI-assisted CAD/EDA software: An AI-driven tools for PCB, RF, and MW; fifth/sixth generation of wireless technology (5G/6G); and high-speed digital design. These tools help optimize circuit layouts and enhance design accuracy (Keysight, 2024b).

b. Quantum circuit simulation: A tool incorporates advanced flux quantization for precise modeling and simulation of complex quantum circuits, essential for high-frequency designs (Rosa, 2024).

c. AI-driven testing: Keysight's Eggplant Test software uses AI for test automation, offering extensive coverage and an optimized user experience for testing high-frequency circuits (Sythoff, 2023).

2) Synopsys

Synopsys introduced Synopsys.ai Copilot, an AI-driven feature in their EDA suite that accelerates chip design through conversational intelligence and automation of repetitive tasks, allowing engineers to focus on innovation (Takahashi, 2023).

3) Cadence Design Systems

Cadence's Cadence.AI Platform includes tools such as Cadence Cerebrus and ChipGPT, which use generative AI and large language models to optimize chip layouts, enhance power, performance, and area (PPA) metrics, and streamline design (Cadence Design Systems, 2024).

4) Qualcomm's 5G optimization

Qualcomm applies AI to optimize 5G network design, addressing the complexity of high-frequency circuits. Their AI-driven approach ensures secure data handling and adherence to network performance standards (McQueen, 2024).

5) NVIDIA's AI-driven chip design

NVIDIA has developed AI tools for designing high-performance chips, enhancing design efficiency. These tools require rigorous engineer training and robust security measures to protect proprietary design data (Altavilla, 2024).

11.5 Challenges and considerations for integrating AI in design workflows

Integrating AI into existing design workflows presents several challenges, including the need for substantial investment, data security, and regulatory compliance—particularly in sensitive sectors such as telecommunications and healthcare.

1. Integration complexity

Adopting AI in design workflows requires significant investment in technology and specialized training. Engineers must adapt to new tools and methodologies, often requiring steep learning curves. For instance, advanced AI-driven automation tools from companies such as Cadence Design Systems and Synopsys demand specialized training for effective use. Keysight Technologies' quantum circuit simulation tool also exemplifies this trend, with capabilities for precise high-frequency circuit design through advanced flux quantization. Moreover, integration can be resource-intensive, as AI tools must be compatible with existing software and hardware infrastructure (Mehradfar et al., 2024).

TABLE 11.4 Training data for the design of second-order Wilkinson power dividers/combiners.

$N = 2$

f_u/f_l	Z_1/Z_0	Z_2/Z_0	R_1/Z_0	R_2/Z_0	VSWR	f_u/f_l	Z_1/Z_0	Z_2/Z_0	R_1/Z_0	R_2/Z_0	VSWR
1.00	1.6828	1.1794	1.7855	5.7413	1.0000	2.50	1.6105	1.2392	2.0964	4.3607	1.1911
1.10	1.6780	1.1833	1.7995	5.6575	1.0000	2.52	1.6096	1.2400	2.1015	4.3430	1.1942
1.20	1.6732	1.1873	1.8153	5.5707	1.0077	2.54	1.6086	1.2408	2.1066	4.3253	1.1973
1.22	1.6722	1.1881	1.8187	5.5530	1.0100	2.56	1.6076	1.2416	2.1117	4.3077	1.2004
1.24	1.6712	1.1889	1.8221	5.5352	1.0124	2.58	1.6067	1.2424	2.1169	4.2902	1.2035
1.26	1.6703	1.1897	1.8256	5.5174	1.0148	2.60	1.6057	1.2432	2.1221	4.2728	1.2066
1.28	1.6693	1.1905	1.8291	5.4994	1.0172	2.62	1.6048	1.2440	2.1274	4.2555	1.2097
1.30	1.6683	1.1913	1.8327	5.4814	1.0196	2.64	1.6038	1.2448	2.1327	4.2382	1.2128
1.32	1.6674	1.1921	1.8363	5.4633	1.0221	2.66	1.6028	1.2456	2.1380	4.2210	1.2159
1.34	1.6664	1.1929	1.8399	5.4451	1.0245	2.68	1.6019	1.2464	2.1434	4.2039	1.2190
1.36	1.6654	1.1937	1.8437	5.4268	1.0270	2.70	1.6009	1.2472	2.1488	4.1869	1.2221
1.38	1.6645	1.1945	1.8474	5.4085	1.0295	2.72	1.5999	1.2480	2.1543	4.1700	1.2252
1.40	1.6635	1.1953	1.8512	5.3901	1.0320	2.74	1.5990	1.2487	2.1599	4.1532	1.2283
1.42	1.6626	1.1961	1.8551	5.3716	1.0346	2.76	1.5980	1.2495	2.1655	4.1364	1.2314
1.44	1.6616	1.1969	1.8589	5.3531	1.0371	2.78	1.5970	1.2503	2.1712	4.1198	1.2345
1.46	1.6606	1.1977	1.8629	5.3345	1.0397	2.80	1.5961	1.2511	2.1769	4.1032	1.2376
1.48	1.6597	1.1985	1.8668	5.3159	1.0423	2.82	1.5951	1.2519	2.1827	4.0867	1.2407
1.50	1.6587	1.1993	1.8708	5.2972	1.0449	2.84	1.5942	1.2527	2.1885	4.0703	1.2438
1.52	1.6577	1.2001	1.8748	5.2785	1.0476	2.86	1.5932	1.2535	2.1945	4.0541	1.2469
1.54	1.6568	1.2009	1.8789	5.2597	1.0502	2.88	1.5922	1.2543	2.2005	4.0379	1.2500
1.56	1.6558	1.2017	1.8829	5.2409	1.0529	2.90	1.5913	1.2551	2.2065	4.0218	1.2530
1.58	1.6548	1.2025	1.8870	5.2221	1.0556	2.92	1.5903	1.2559	2.2127	4.0058	1.2561
1.60	1.6539	1.2033	1.8912	5.2032	1.0583	2.94	1.5893	1.2567	2.2189	3.9899	1.2592
1.62	1.6529	1.2041	1.8953	5.1842	1.0610	2.96	1.5884	1.2575	2.2252	3.9741	1.2623
1.64	1.6520	1.2049	1.8995	5.1653	1.0637	2.98	1.5874	1.2583	2.2316	3.9584	1.2653
1.66	1.6510	1.2057	1.9037	5.1463	1.0664	3.00	1.5865	1.2591	2.2381	3.9428	1.2684
1.68	1.6500	1.2065	1.9079	5.1273	1.0692	3.02	1.5855	1.2599	2.2446	3.9273	1.2715
1.70	1.6491	1.2073	1.9122	5.1083	1.0720	3.04	1.5845	1.2607	2.2513	3.9119	1.2745
1.72	1.6481	1.2081	1.9165	5.0893	1.0747	3.06	1.5836	1.2615	2.2581	3.8966	1.2775
1.74	1.6471	1.2089	1.9207	5.0702	1.0775	3.08	1.5826	1.2623	2.2649	3.8814	1.2806
1.76	1.6462	1.2097	1.9251	5.0512	1.0804	3.10	1.5816	1.2631	2.2719	3.8663	1.2836
1.78	1.6452	1.2105	1.9294	5.0321	1.0832	3.12	1.5807	1.2639	2.2790	3.8513	1.2866
1.80	1.6443	1.2113	1.9337	5.0130	1.0860	3.14	1.5797	1.2647	2.2861	3.8364	1.2897
1.82	1.6433	1.2121	1.9381	4.9939	1.0889	3.16	1.5787	1.2655	2.2934	3.8217	1.2927
1.84	1.6423	1.2129	1.9425	4.9749	1.0917	3.18	1.5778	1.2663	2.3008	3.8070	1.2957
1.86	1.6414	1.2137	1.9469	4.9558	1.0946	3.20	1.5768	1.2671	2.3084	3.7925	1.2987
1.88	1.6404	1.2144	1.9513	4.9367	1.0975	3.22	1.5759	1.2679	2.3160	3.7780	1.3016
1.90	1.6394	1.2152	1.9557	4.9176	1.1004	3.24	1.5749	1.2687	2.3238	3.7637	1.3046
1.92	1.6385	1.2160	1.9601	4.8986	1.1033	3.26	1.5739	1.2695	2.3317	3.7494	1.3076
1.94	1.6375	1.2168	1.9646	4.8795	1.1062	3.28	1.5730	1.2703	2.3397	3.7353	1.3105
1.96	1.6365	1.2176	1.9691	4.8605	1.1091	3.30	1.5720	1.2711	2.3479	3.7213	1.3135

(*continued on next page*)

TABLE 11.4 Training data for the design of second-order Wilkinson power dividers/combiners—cont'd

$N = 2$											
f_u/f_l	Z_1/Z_0	Z_2/Z_0	R_1/Z_0	R_2/Z_0	VSWR	f_u/f_l	Z_1/Z_0	Z_2/Z_0	R_1/Z_0	R_2/Z_0	VSWR
1.98	1.6356	1.2184	1.9736	4.8415	1.1121	3.32	1.5710	1.2719	2.3563	3.7074	1.3164
2.00	1.6346	1.2192	1.9781	4.8225	1.1150	3.34	1.5701	1.2727	2.3647	3.6936	1.3194
2.02	1.6337	1.2200	1.9826	4.8036	1.1180	3.36	1.5691	1.2735	2.3733	3.6799	1.3223
2.04	1.6327	1.2208	1.9871	4.7846	1.1209	3.38	1.5681	1.2743	2.3821	3.6664	1.3252
2.06	1.6317	1.2216	1.9917	4.7657	1.1239	3.40	1.5672	1.2751	2.3910	3.6529	1.3281
2.08	1.6308	1.2224	1.9962	4.7468	1.1269	3.42	1.5662	1.2759	2.4001	3.6396	1.3310
2.10	1.6298	1.2232	2.0008	4.7280	1.1299	3.44	1.5653	1.2767	2.4094	3.6263	1.3338
2.12	1.6288	1.2240	2.0054	4.7091	1.1329	3.46	1.5643	1.2775	2.4188	3.6132	1.3367
2.14	1.6279	1.2248	2.0100	4.6903	1.1359	3.48	1.5633	1.2783	2.4284	3.6002	1.3396
2.16	1.6269	1.2256	2.0146	4.6716	1.1389	3.50	1.5624	1.2791	2.4382	3.5873	1.3424
2.18	1.6259	1.2264	2.0193	4.6529	1.1420	3.52	1.5614	1.2799	2.4482	3.5745	1.3452
2.20	1.6250	1.2272	2.0239	4.6342	1.1450	3.54	1.5604	1.2807	2.4584	3.5618	1.3480
2.22	1.6240	1.2280	2.0286	4.6156	1.1480	3.56	1.5595	1.2815	2.4687	3.5492	1.3508
2.24	1.6231	1.2288	2.0333	4.5970	1.1511	3.58	1.5585	1.2823	2.4793	3.5368	1.3536
2.26	1.6221	1.2296	2.0380	4.5785	1.1541	3.60	1.5576	1.283	2.49	3.5244	1.3564
2.28	1.6211	1.2304	2.0427	4.5600	1.1572	3.62	1.5566	1.2838	2.5010	3.5122	1.3591
2.30	1.6202	1.2312	2.0475	4.5416	1.1602	3.65	1.5551	1.285	2.5178	3.4940	1.3632
2.32	1.6192	1.2320	2.0523	4.5232	1.1633	3.68	1.5537	1.2862	2.5351	3.4761	1.3673
2.34	1.6182	1.2328	2.0571	4.5049	1.1664	3.7	1.5527	1.287	2.547	3.4643	1.3700
2.36	1.6173	1.2336	2.0619	4.4867	1.1694	3.75	1.5503	1.289	2.5776	3.4353	1.3767
2.38	1.6163	1.2344	2.0667	4.4685	1.1725	3.8	1.5479	1.291	2.6097	3.407	1.3832
2.40	1.6154	1.2352	2.0716	4.4504	1.1756	3.85	1.5455	1.293	2.6435	3.3794	1.3897
2.42	1.6144	1.2360	2.0765	4.4323	1.1787	3.9	1.5431	1.295	2.6789	3.3524	1.3960
2.44	1.6134	1.2368	2.0815	4.4143	1.1818	3.95	1.5407	1.297	2.7162	3.3260	1.4023
2.46	1.6125	1.2376	2.0864	4.3964	1.1849	4.0	1.5383	1.299	2.7553	3.3003	1.4084
2.48	1.6115	1.2384	2.0914	4.3785	1.1880	4.05	1.5359	1.301	2.7963	3.2753	1.4144

2. Data privacy and security

   Protecting design data privacy and security is essential when using AI tools reliant on large datasets, as these often contain sensitive or proprietary information. Robust security protocols are critical to prevent unauthorized access and data breaches. For example, in healthcare, AI-driven diagnostic tools from GE Healthcare must adhere to strict regulations such as Health Insurance Portability and Accountability Act (HIPAA) to ensure patient privacy. Similar diligence is necessary in high-frequency circuit design to protect intellectual property and maintain competitive advantages (Marks & Haupt, 2023).

3. Regulatory compliance

   Ensuring compliance with industry standards and regulations is vital, especially in telecom and healthcare. Regulatory bodies such as the Federal Communications Commission (FCC) in the United States and the European Telecommunications Standards Institute (ETSI) enforce strict guidelines for high-frequency circuit designs. AI-driven design tools must comply with these standards to guarantee safety, reliability, and performance. Leading companies such as Nokia and Ericsson, key players in 5G technology, must navigate these regulations to ensure their AI-augmented designs meet all requirements (European Telecommunications Standards Institute, 2024; Federal Communications Commission, 2024).

TABLE 11.5 Training data for the design of third-order Wilkinson power dividers/combiners.

$N = 3$							
f_u/f_l	Z_1/Z_0	Z_2/Z_0	Z_3/Z_0	R_1/Z_0	R_2/Z_0	R_3/Z_0	VSWR
1.00	1.8374	1.4133	1.0753	1.8734	3.6782	16.7340	1.0004
1.02	1.8364	1.4133	1.0760	1.8734	3.6801	16.5100	1.0004
1.04	1.8354	1.4133	1.0767	1.8734	3.6821	16.2910	1.0005
1.06	1.8344	1.4133	1.0774	1.8736	3.6840	16.0760	1.0006
1.08	1.8334	1.4133	1.0781	1.8737	3.6860	15.8670	1.0006
1.10	1.8324	1.4133	1.0788	1.8740	3.6880	15.6620	1.0008
1.12	1.8314	1.4133	1.0795	1.8743	3.6901	15.4610	1.0009
1.14	1.8304	1.4134	1.0802	1.8747	3.6921	15.2650	1.0011
1.16	1.8295	1.4134	1.0809	1.8752	3.6942	15.0730	1.0013
1.18	1.8285	1.4134	1.0816	1.8758	3.6963	14.8860	1.0015
1.20	1.8275	1.4134	1.0823	1.8764	3.6985	14.7030	1.0018
1.22	1.8265	1.4134	1.0830	1.8770	3.7007	14.5240	1.0020
1.24	1.8255	1.4134	1.0837	1.8778	3.7029	14.3490	1.0023
1.26	1.8245	1.4134	1.0844	1.8786	3.7051	14.1780	1.0026
1.28	1.8235	1.4134	1.0851	1.8794	3.7074	14.0110	1.0030
1.30	1.8225	1.4134	1.0858	1.8803	3.7097	13.8480	1.0034
1.32	1.8215	1.4134	1.0865	1.8813	3.7121	13.6880	1.0037
1.34	1.8205	1.4134	1.0872	1.8824	3.7145	13.5320	1.0042
1.36	1.8195	1.4134	1.0879	1.8835	3.7170	13.3800	1.0046
1.38	1.8185	1.4134	1.0886	1.8847	3.7195	13.2310	1.0051
1.40	1.8175	1.4134	1.0893	1.8859	3.7220	13.0850	1.0055
1.42	1.8166	1.4134	1.0900	1.8872	3.7246	12.9430	1.0060
1.44	1.8156	1.4134	1.0907	1.8885	3.7272	12.8040	1.0066
1.46	1.8146	1.4134	1.0914	1.8899	3.7299	12.6680	1.0071
1.48	1.8136	1.4134	1.0921	1.8914	3.7327	12.5350	1.0077
1.50	1.8126	1.4134	1.0928	1.8929	3.7355	12.4060	1.0083
1.52	1.8116	1.4134	1.0935	1.8945	3.7383	12.2790	1.0089
1.54	1.8106	1.4134	1.0942	1.8961	3.7412	12.1560	1.0096
1.56	1.8096	1.4134	1.0949	1.8978	3.7441	12.0350	1.0102
1.58	1.8086	1.4134	1.0956	1.8995	3.7471	11.9170	1.0109
1.60	1.8076	1.4135	1.0963	1.9013	3.7502	11.8010	1.0116
1.62	1.8066	1.4135	1.0970	1.9032	3.7533	11.6890	1.0123
1.64	1.8056	1.4135	1.0977	1.9051	3.7565	11.5780	1.0131
1.66	1.8046	1.4135	1.0984	1.9070	3.7597	11.4710	1.0138
1.68	1.8037	1.4135	1.0991	1.9090	3.7630	11.3660	1.0146
1.70	1.8027	1.4135	1.0998	1.9111	3.7663	11.2630	1.0154
1.72	1.8017	1.4135	1.1005	1.9132	3.7698	11.1630	1.0162
1.74	1.8007	1.4135	1.1012	1.9153	3.7732	11.0650	1.0171
1.76	1.7997	1.4135	1.1019	1.9175	3.7767	10.9690	1.0180
1.78	1.7987	1.4135	1.1026	1.9198	3.7803	10.8750	1.0188
1.80	1.7977	1.4135	1.1033	1.9220	3.7840	10.7840	1.0197
1.82	1.7967	1.4135	1.1040	1.9244	3.7877	10.6940	1.0207
1.84	1.7957	1.4135	1.1047	1.9268	3.7914	10.6070	1.0216
1.86	1.7947	1.4135	1.1054	1.9292	3.7953	10.5210	1.0226

(continued on next page)

TABLE 11.5 Training data for the design of third-order Wilkinson power dividers/combiners—cont'd

$N=3$							
f_u/f_l	Z_1/Z_0	Z_2/Z_0	Z_3/Z_0	R_1/Z_0	R_2/Z_0	R_3/Z_0	VSWR
1.88	1.7937	1.4135	1.1061	1.9317	3.7992	10.4380	1.0236
1.90	1.7927	1.4135	1.1068	1.9342	3.8031	10.3560	1.0246
1.92	1.7918	1.4135	1.1075	1.9367	3.8071	10.2760	1.0256
1.94	1.7908	1.4135	1.1082	1.9393	3.8112	10.1980	1.0266
1.96	1.7898	1.4135	1.1089	1.9420	3.8153	10.1220	1.0277
1.98	1.7888	1.4135	1.1096	1.9447	3.8195	10.0470	1.0287
2.00	1.7878	1.4135	1.1103	1.9474	3.8238	9.9744	1.0298
2.02	1.7868	1.4135	1.1110	1.9502	3.8281	9.9029	1.0309
2.04	1.7858	1.4136	1.1117	1.9530	3.8325	9.8331	1.0321
2.06	1.7848	1.4136	1.1124	1.9558	3.8370	9.7647	1.0332
2.08	1.7838	1.4136	1.1131	1.9587	3.8415	9.6978	1.0344
2.10	1.7828	1.4136	1.1138	1.9617	3.8461	9.6323	1.0355
2.12	1.7818	1.4136	1.1145	1.9646	3.8507	9.5682	1.0367
2.14	1.7808	1.4136	1.1152	1.9676	3.8554	9.5054	1.0379
2.16	1.7798	1.4136	1.1159	1.9707	3.8602	9.4439	1.0392
2.18	1.7789	1.4136	1.1166	1.9738	3.8650	9.3837	1.0404
2.20	1.7779	1.4136	1.1173	1.9769	3.8699	9.3247	1.0417
2.22	1.7769	1.4136	1.1180	1.9800	3.8748	9.2669	1.0430
2.24	1.7759	1.4136	1.1187	1.9832	3.8798	9.2103	1.0442
2.26	1.7749	1.4136	1.1194	1.9865	3.8849	9.1548	1.0456
2.28	1.7739	1.4136	1.1201	1.9897	3.8900	9.1004	1.0469
2.30	1.7729	1.4136	1.1208	1.9930	3.8952	9.0470	1.0482
2.32	1.7719	1.4136	1.1215	1.9963	3.9004	8.9946	1.0496
2.34	1.7709	1.4136	1.1222	1.9997	3.9057	8.9433	1.0509
2.36	1.7699	1.4136	1.1229	2.0031	3.9111	8.8929	1.0523
2.38	1.7689	1.4136	1.1236	2.0065	3.9165	8.8434	1.0537
2.40	1.7679	1.4136	1.1243	2.0100	3.9220	8.7948	1.0552
2.42	1.7669	1.4136	1.1250	2.0134	3.9275	8.7471	1.0566
2.44	1.7660	1.4136	1.1257	2.0170	3.9331	8.7003	1.0580
2.46	1.7650	1.4136	1.1264	2.0205	3.9387	8.6542	1.0595
2.48	1.7640	1.4136	1.1271	2.0241	3.9444	8.6089	1.0610
2.50	1.7630	1.4137	1.1278	2.0277	3.9502	8.5644	1.0624
2.52	1.7620	1.4137	1.1285	2.0313	3.9560	8.5207	1.0640
2.54	1.7610	1.4137	1.1292	2.0350	3.9619	8.4776	1.0655
2.56	1.7600	1.4137	1.1299	2.0387	3.9678	8.4353	1.0670
2.58	1.7590	1.4137	1.1306	2.0424	3.9738	8.3936	1.0685
2.60	1.7580	1.4137	1.1313	2.0461	3.9798	8.3525	1.0701
2.62	1.7570	1.4137	1.1320	2.0499	3.9858	8.3121	1.0717
2.64	1.7560	1.4137	1.1327	2.0537	3.9920	8.2722	1.0733
2.66	1.7550	1.4137	1.1334	2.0575	3.9981	8.2330	1.0748
2.68	1.7540	1.4137	1.1341	2.0614	4.0044	8.1942	1.0765

(*continued on next page*)

TABLE 11.5 Training data for the design of third-order Wilkinson power dividers/combiners—cont'd

$N = 3$							
f_u/f_l	Z_1/Z_0	Z_2/Z_0	Z_3/Z_0	R_1/Z_0	R_2/Z_0	R_3/Z_0	VSWR
2.70	1.7531	1.4137	1.1348	2.0653	4.0106	8.1561	1.0781
2.72	1.7521	1.4137	1.1355	2.0692	4.0169	8.1184	1.0797
2.74	1.7511	1.4137	1.1362	2.0731	4.0233	8.0813	1.0814
2.76	1.7501	1.4137	1.1369	2.0770	4.0297	8.0446	1.0830
2.78	1.7491	1.4137	1.1376	2.0810	4.0362	8.0084	1.0847
2.80	1.7481	1.4137	1.1383	2.0850	4.0427	7.9726	1.0864
2.82	1.7471	1.4137	1.1390	2.0890	4.0493	7.9373	1.0881
2.84	1.7461	1.4137	1.1397	2.0931	4.0558	7.9024	1.0898
2.86	1.7451	1.4137	1.1404	2.0971	4.0625	7.8679	1.0915
2.88	1.7441	1.4137	1.1411	2.1012	4.0692	7.8337	1.0932
2.90	1.7431	1.4137	1.1418	2.1053	4.0759	7.8000	1.0950
2.92	1.7421	1.4137	1.1425	2.1095	4.0827	7.7666	1.0967
2.94	1.7412	1.4138	1.1432	2.1136	4.0895	7.7335	1.0985
2.96	1.7402	1.4138	1.1439	2.1178	4.0963	7.7008	1.1003
2.98	1.7392	1.4138	1.1446	2.1220	4.1032	7.6684	1.1021
3.00	1.7382	1.4138	1.1453	2.1262	4.1101	7.6363	1.1039
3.02	1.7372	1.4138	1.1460	2.1304	4.1171	7.6045	1.1057
3.04	1.7362	1.4138	1.1467	2.1347	4.1241	7.5730	1.1075
3.06	1.7352	1.4138	1.1474	2.1389	4.1312	7.5418	1.1093
3.08	1.7342	1.4138	1.1481	2.1432	4.1382	7.5108	1.1111
3.10	1.7332	1.4138	1.1488	2.1475	4.1454	7.4801	1.1130
3.12	1.7322	1.4138	1.1495	2.1519	4.1525	7.4496	1.1149
3.14	1.7312	1.4138	1.1502	2.1562	4.1597	7.4194	1.1167
3.16	1.7302	1.4138	1.1509	2.1605	4.1669	7.3894	1.1186
3.18	1.7292	1.4138	1.1516	2.1649	4.1742	7.3597	1.1205
3.20	1.7283	1.4138	1.1523	2.1693	4.1815	7.3301	1.1224
3.22	1.7273	1.4138	1.1530	2.1737	4.1888	7.3008	1.1243
3.24	1.7263	1.4138	1.1537	2.1781	4.1961	7.2717	1.1262
3.26	1.7253	1.4138	1.1544	2.1826	4.2035	7.2428	1.1281
3.28	1.7243	1.4138	1.1551	2.1870	4.2109	7.2140	1.1301
3.30	1.7233	1.4138	1.1558	2.1915	4.2184	7.1855	1.1320
3.32	1.7223	1.4138	1.1565	2.1960	4.2259	7.1571	1.1340
3.34	1.7213	1.4138	1.1572	2.2005	4.2334	7.1289	1.1359
3.36	1.7203	1.4138	1.1579	2.2050	4.2409	7.1009	1.1379
3.38	1.7193	1.4138	1.1586	2.2095	4.2485	7.0731	1.1399
3.40	1.7183	1.4139	1.1593	2.2141	4.2561	7.0454	1.1419
3.42	1.7173	1.4139	1.1600	2.2186	4.2637	7.0179	1.1439
3.44	1.7163	1.4139	1.1607	2.2232	4.2713	6.9905	1.1459
3.46	1.7154	1.4139	1.1614	2.2278	4.2790	6.9634	1.1479
3.48	1.7144	1.4139	1.1621	2.2324	4.2867	6.9363	1.1499
3.50	1.7134	1.4139	1.1628	2.2370	4.2944	6.9094	1.1519

(continued on next page)

TABLE 11.5 Training data for the design of third-order Wilkinson power dividers/combiners—cont'd

$N=3$							
f_u/f_l	Z_1/Z_0	Z_2/Z_0	Z_3/Z_0	R_1/Z_0	R_2/Z_0	R_3/Z_0	VSWR
3.52	1.7124	1.4139	1.1635	2.2416	4.3022	6.8827	1.1540
3.54	1.7114	1.4139	1.1642	2.2462	4.3100	6.8561	1.1560
3.56	1.7104	1.4139	1.1649	2.2509	4.3178	6.8297	1.1581
3.58	1.7094	1.4139	1.1656	2.2555	4.3256	6.8034	1.1601
3.60	1.7084	1.4139	1.1663	2.2602	4.3335	6.7773	1.1622
3.62	1.7074	1.4139	1.1670	2.2649	4.3414	6.7513	1.1643
3.64	1.7064	1.4139	1.1677	2.2696	4.3493	6.7255	1.1663
3.66	1.7054	1.4139	1.1684	2.2743	4.3572	6.6998	1.1684
3.68	1.7044	1.4139	1.1691	2.2790	4.3651	6.6742	1.1705
3.70	1.7034	1.4139	1.1698	2.2837	4.3731	6.6488	1.1726
3.72	1.7025	1.4139	1.1705	2.2884	4.3811	6.6236	1.1747
3.74	1.7015	1.4139	1.1712	2.2932	4.3892	6.5985	1.1768
3.76	1.7005	1.4139	1.1719	2.2979	4.3972	6.5735	1.1790
3.78	1.6995	1.4139	1.1726	2.3027	4.4053	6.5487	1.1811
3.80	1.6985	1.4139	1.1733	2.3075	4.4134	6.5241	1.1832
3.82	1.6975	1.4139	1.1740	2.3123	4.4215	6.4996	1.1853
3.84	1.6965	1.4140	1.1747	2.3170	4.4297	6.4752	1.1875
3.86	1.6955	1.4140	1.1754	2.3218	4.4378	6.4511	1.1896
3.88	1.6945	1.4140	1.1761	2.3267	4.4460	6.4270	1.1918
3.90	1.6935	1.4140	1.1768	2.3315	4.4543	6.4032	1.1940
3.92	1.6925	1.4140	1.1775	2.3363	4.4625	6.3795	1.1961
3.94	1.6915	1.4140	1.1782	2.3411	4.4708	6.3560	1.1983
3.96	1.6906	1.4140	1.1789	2.3460	4.4791	6.3326	1.2005
3.98	1.6896	1.4140	1.1796	2.3508	4.4874	6.3094	1.2026
4.00	1.6886	1.4140	1.1803	2.3557	4.4958	6.2864	1.2048
4.02	1.6876	1.4140	1.1810	2.3606	4.5041	6.2636	1.2070
4.04	1.6866	1.4140	1.1817	2.3655	4.5125	6.2410	1.2092
4.06	1.6856	1.4140	1.1824	2.3703	4.5210	6.2185	1.2114
4.08	1.6846	1.4140	1.1831	2.3752	4.5294	6.1962	1.2136
4.10	1.6836	1.4140	1.1838	2.3801	4.5379	6.1741	1.2158
4.12	1.6826	1.4140	1.1845	2.3850	4.5465	6.1522	1.2181
4.14	1.6816	1.4140	1.1852	2.3900	4.5550	6.1305	1.2203
4.16	1.6806	1.4140	1.1859	2.3949	4.5636	6.1090	1.2225
4.18	1.6796	1.4140	1.1866	2.3998	4.5722	6.0877	1.2247
4.20	1.6786	1.4140	1.1873	2.4048	4.5809	6.0666	1.2270
4.22	1.6777	1.4140	1.1880	2.4097	4.5896	6.0457	1.2292
4.24	1.6767	1.4140	1.1887	2.4146	4.5983	6.0250	1.2314
4.26	1.6757	1.4140	1.1894	2.4196	4.6070	6.0046	1.2337
4.28	1.6747	1.4141	1.1901	2.4246	4.6158	5.9843	1.2359
4.30	1.6737	1.4141	1.1908	2.4295	4.6247	5.9643	1.2382
4.32	1.6727	1.4141	1.1915	2.4345	4.6335	5.9445	1.2404

(continued on next page)

TABLE 11.5 Training data for the design of third-order Wilkinson power dividers/combiners—cont'd

$N = 3$							
f_u/f_l	Z_1/Z_0	Z_2/Z_0	Z_3/Z_0	R_1/Z_0	R_2/Z_0	R_3/Z_0	VSWR
4.34	1.6717	1.4141	1.1922	2.4395	4.6424	5.9249	1.2427
4.36	1.6707	1.4141	1.1929	2.4445	4.6514	5.9056	1.2450
4.38	1.6697	1.4141	1.1936	2.4495	4.6604	5.8864	1.2472
4.40	1.6687	1.4141	1.1943	2.4545	4.6694	5.8676	1.2495
4.42	1.6677	1.4141	1.1950	2.4595	4.6785	5.8489	1.2518
4.44	1.6667	1.4141	1.1957	2.4645	4.6876	5.8305	1.2540
4.46	1.6657	1.4141	1.1964	2.4696	4.6968	5.8124	1.2563
4.48	1.6648	1.4141	1.1971	2.4746	4.7060	5.7945	1.2586
4.50	1.6638	1.4141	1.1978	2.4796	4.7153	5.7768	1.2609
4.52	1.6628	1.4141	1.1985	2.4847	4.7246	5.7594	1.2632
4.54	1.6618	1.4141	1.1992	2.4897	4.7340	5.7423	1.2655
4.56	1.6608	1.4141	1.1999	2.4948	4.7434	5.7254	1.2678
4.58	1.6598	1.4141	1.2006	2.4999	4.7529	5.7087	1.2701
4.60	1.6588	1.4141	1.2013	2.5049	4.7625	5.6923	1.2724
4.62	1.6578	1.4141	1.2020	2.5100	4.7721	5.6762	1.2747
4.64	1.6568	1.4141	1.2027	2.5151	4.7818	5.6604	1.2770
4.66	1.6558	1.4141	1.2034	2.5202	4.7915	5.6448	1.2793
4.68	1.6548	1.4141	1.2041	2.5253	4.8013	5.6295	1.2816
4.70	1.6538	1.4141	1.2048	2.5304	4.8112	5.6144	1.2839
4.72	1.6529	1.4141	1.2055	2.5355	4.8211	5.5996	1.2862
4.74	1.6519	1.4142	1.2062	2.5406	4.8312	5.5851	1.2885
4.76	1.6509	1.4142	1.2069	2.5457	4.8413	5.5709	1.2908
4.78	1.6499	1.4142	1.2076	2.5508	4.8515	5.5569	1.2931
4.80	1.6489	1.4142	1.2083	2.5560	4.8617	5.5432	1.2955
4.82	1.6479	1.4142	1.2090	2.5611	4.8721	5.5297	1.2978
4.84	1.6469	1.4142	1.2097	2.5663	4.8825	5.5166	1.3001
4.86	1.6459	1.4142	1.2104	2.5714	4.8930	5.5037	1.3024
4.88	1.6449	1.4142	1.2111	2.5766	4.9036	5.4910	1.3047
4.90	1.6439	1.4142	1.2118	2.5817	4.9143	5.4787	1.3071
4.92	1.6429	1.4142	1.2125	2.5869	4.9251	5.4666	1.3094
4.94	1.6419	1.4142	1.2132	2.5921	4.9360	5.4547	1.3117
4.96	1.6409	1.4142	1.2139	2.5973	4.9470	5.4432	1.3140
4.98	1.6400	1.4142	1.2146	2.6025	4.9582	5.4318	1.3164
5.00	1.6390	1.4142	1.2153	2.6077	4.9694	5.4208	1.3187
5.02	1.6380	1.4142	1.2160	2.6129	4.9807	5.4100	1.3210
5.04	1.6370	1.4142	1.2167	2.6181	4.9922	5.3994	1.3234
5.06	1.6360	1.4142	1.2174	2.6233	5.0037	5.3891	1.3257
5.08	1.6350	1.4142	1.2181	2.6285	5.0154	5.3790	1.3280
5.10	1.6340	1.4142	1.2188	2.6338	5.0272	5.3692	1.3304
5.12	1.6330	1.4142	1.2195	2.6390	5.0392	5.3596	1.3327
5.14	1.6320	1.4142	1.2202	2.6443	5.0513	5.3502	1.3350

(continued on next page)

TABLE 11.5 Training data for the design of third-order Wilkinson power dividers/combiners—cont'd

$N=3$							
f_u/f_l	Z_1/Z_0	Z_2/Z_0	Z_3/Z_0	R_1/Z_0	R_2/Z_0	R_3/Z_0	VSWR
5.16	1.6310	1.4142	1.2209	2.6495	5.0635	5.3411	1.3374
5.18	1.6300	1.4143	1.2216	2.6548	5.0759	5.3322	1.3397
5.20	1.6290	1.4143	1.2223	2.6601	5.0884	5.3234	1.3420
5.22	1.6280	1.4143	1.2230	2.6654	5.1010	5.3149	1.3444
5.24	1.6271	1.4143	1.2237	2.6707	5.1138	5.3066	1.3467
5.26	1.6261	1.4143	1.2244	2.6760	5.1268	5.2985	1.3490
5.28	1.6251	1.4143	1.2251	2.6813	5.1399	5.2906	1.3514
5.30	1.6241	1.4143	1.2258	2.6866	5.1532	5.2828	1.3537
5.32	1.6231	1.4143	1.2265	2.6919	5.1667	5.2753	1.3561
5.34	1.6221	1.4143	1.2272	2.6973	5.1803	5.2678	1.3584
5.36	1.6211	1.4143	1.2279	2.7026	5.1941	5.2606	1.3607
5.38	1.6201	1.4143	1.2286	2.7080	5.2081	5.2534	1.3631
5.40	1.6191	1.4143	1.2293	2.7133	5.2223	5.2464	1.3654
5.42	1.6181	1.4143	1.2300	2.7187	5.2367	5.2396	1.3677
5.44	1.6171	1.4143	1.2307	2.7241	5.2512	5.2328	1.3701
5.46	1.6161	1.4143	1.2314	2.7295	5.2660	5.2261	1.3724
5.48	1.6151	1.4143	1.2321	2.7349	5.2810	5.2195	1.3747
5.50	1.6142	1.4143	1.2328	2.7403	5.2962	5.2130	1.3771
5.52	1.6132	1.4143	1.2335	2.7457	5.3116	5.2066	1.3794
5.54	1.6122	1.4143	1.2342	2.7511	5.3273	5.2002	1.3817
5.56	1.6112	1.4143	1.2349	2.7566	5.3431	5.1938	1.3841
5.58	1.6102	1.4143	1.2356	2.7621	5.3593	5.1875	1.3864
5.60	1.6092	1.4143	1.2363	2.7675	5.3756	5.1811	1.3887
5.62	1.6082	1.4143	1.2370	2.7730	5.3922	5.1747	1.3910
5.64	1.6072	1.4144	1.2377	2.7785	5.4090	5.1684	1.3934
5.66	1.6062	1.4144	1.2384	2.7840	5.4261	5.1619	1.3957
5.68	1.6052	1.4144	1.2391	2.7895	5.4435	5.1554	1.3980
5.70	1.6042	1.4144	1.2398	2.7950	5.4611	5.1489	1.4003
5.72	1.6032	1.4144	1.2405	2.8006	5.4790	5.1422	1.4027
5.74	1.6023	1.4144	1.2412	2.8061	5.4972	5.1354	1.4050
5.76	1.6013	1.4144	1.2419	2.8117	5.5157	5.1285	1.4073
5.78	1.6003	1.4144	1.2426	2.8173	5.5345	5.1214	1.4096
5.80	1.5993	1.4144	1.2433	2.8229	5.5536	5.1141	1.4119
5.82	1.5983	1.4144	1.2440	2.8285	5.5729	5.1066	1.4142
5.84	1.5973	1.4144	1.2447	2.8341	5.5926	5.0990	1.4166
5.86	1.5963	1.4144	1.2454	2.8397	5.6127	5.0910	1.4189
5.88	1.5953	1.4144	1.2461	2.8454	5.6330	5.0828	1.4212
5.90	1.5943	1.4144	1.2468	2.8510	5.6537	5.0743	1.4235
5.92	1.5933	1.4144	1.2475	2.8567	5.6747	5.0655	1.4258
5.94	1.5923	1.4144	1.2482	2.8624	5.6961	5.0564	1.4281
5.96	1.5913	1.4144	1.2489	2.8681	5.7178	5.0469	1.4304
5.98	1.5903	1.4144	1.2496	2.8739	5.7399	5.0370	1.4327
6.00	1.5894	1.4144	1.2503	2.8796	5.7624	5.0266	1.4350
6.02	1.5884	1.4144	1.2510	2.8854	5.7852	5.0158	1.4373

TABLE 11.6 Training data for the design of fourth-order Wilkinson power dividers/combiners.

$N = 4$									
f_u/f_l	Z_1/Z_0	Z_2/Z_0	Z_3/Z_0	Z_4/Z_0	R_1/Z_0	R_2/Z_0	R_3/Z_0	R_4/Z_0	VSWR
1.70	1.8971	1.6035	1.2571	1.0638	1.4546	3.0082	5.7099	20.7970	1.0073
1.73	1.8961	1.6030	1.2575	1.0644	1.4533	3.0071	5.7077	20.5780	1.0076
1.75	1.8951	1.6025	1.2579	1.0650	1.4527	3.0060	5.7056	20.3630	1.0080
1.78	1.8942	1.6020	1.2582	1.0656	1.4525	3.0051	5.7036	20.1500	1.0083
1.80	1.8932	1.6015	1.2586	1.0662	1.4530	3.0042	5.7016	19.9410	1.0087
1.83	1.8922	1.6010	1.2590	1.0668	1.4539	3.0034	5.6998	19.7340	1.0091
1.85	1.8912	1.6005	1.2594	1.0675	1.4553	3.0027	5.6981	19.5300	1.0095
1.88	1.8902	1.6000	1.2598	1.0681	1.4572	3.0020	5.6965	19.3290	1.0098
1.90	1.8892	1.5995	1.2601	1.0687	1.4595	3.0015	5.6951	19.1310	1.0102
1.93	1.8882	1.5990	1.2605	1.0693	1.4622	3.0010	5.6937	18.9360	1.0106
1.95	1.8873	1.5985	1.2609	1.0699	1.4654	3.0006	5.6924	18.7430	1.0111
1.98	1.8863	1.5980	1.2613	1.0705	1.4690	3.0003	5.6912	18.5530	1.0115
2.00	1.8853	1.5975	1.2616	1.0711	1.4729	3.0001	5.6901	18.3660	1.0119
2.03	1.8843	1.5971	1.2620	1.0718	1.4772	3.0000	5.6891	18.1810	1.0124
2.05	1.8833	1.5966	1.2624	1.0724	1.4818	2.9999	5.6883	17.9990	1.0128
2.08	1.8823	1.5961	1.2628	1.0730	1.4867	2.9999	5.6875	17.8200	1.0133
2.10	1.8813	1.5956	1.2632	1.0736	1.4920	3.0000	5.6868	17.6430	1.0138
2.13	1.8804	1.5951	1.2635	1.0742	1.4975	3.0002	5.6862	17.4680	1.0143
2.15	1.8794	1.5946	1.2639	1.0748	1.5033	3.0005	5.6857	17.2960	1.0148
2.18	1.8784	1.5941	1.2643	1.0754	1.5094	3.0008	5.6853	17.1270	1.0153
2.20	1.8774	1.5937	1.2647	1.0761	1.5157	3.0013	5.6850	16.9600	1.0159
2.23	1.8764	1.5932	1.2650	1.0767	1.5222	3.0018	5.6848	16.7950	1.0164
2.25	1.8754	1.5927	1.2654	1.0773	1.5290	3.0024	5.6847	16.6330	1.0170
2.28	1.8744	1.5922	1.2658	1.0779	1.5359	3.0031	5.6846	16.4730	1.0176
2.30	1.8735	1.5917	1.2662	1.0785	1.5431	3.0038	5.6847	16.3160	1.0182
2.33	1.8725	1.5912	1.2666	1.0791	1.5504	3.0047	5.6848	16.1600	1.0188
2.35	1.8715	1.5908	1.2669	1.0797	1.5579	3.0056	5.6851	16.0070	1.0194
2.38	1.8705	1.5903	1.2673	1.0804	1.5655	3.0066	5.6854	15.8570	1.0200
2.40	1.8695	1.5898	1.2677	1.0810	1.5733	3.0077	5.6858	15.7080	1.0207
2.43	1.8685	1.5893	1.2681	1.0816	1.5812	3.0089	5.6863	15.5620	1.0213
2.45	1.8675	1.5889	1.2684	1.0822	1.5892	3.0101	5.6869	15.4170	1.0220
2.48	1.8665	1.5884	1.2688	1.0828	1.5973	3.0115	5.6876	15.2750	1.0227
2.50	1.8656	1.5879	1.2692	1.0834	1.6055	3.0129	5.6883	15.1350	1.0234
2.53	1.8646	1.5874	1.2696	1.0840	1.6138	3.0144	5.6892	14.9970	1.0241
2.55	1.8636	1.5870	1.2700	1.0847	1.6222	3.0160	5.6901	14.8610	1.0249
2.58	1.8626	1.5865	1.2703	1.0853	1.6306	3.0176	5.6911	14.7270	1.0257
2.60	1.8616	1.5860	1.2707	1.0859	1.6391	3.0194	5.6922	14.5950	1.0264
2.63	1.8606	1.5855	1.2711	1.0865	1.6477	3.0212	5.6934	14.4650	1.0272
2.65	1.8596	1.5851	1.2715	1.0871	1.6563	3.0231	5.6946	14.3370	1.0280
2.68	1.8587	1.5846	1.2718	1.0877	1.6649	3.0251	5.6960	14.2100	1.0289
2.70	1.8577	1.5841	1.2722	1.0883	1.6735	3.0271	5.6974	14.0860	1.0297
2.73	1.8567	1.5837	1.2726	1.0890	1.6822	3.0293	5.6989	13.9630	1.0306
2.75	1.8557	1.5832	1.2730	1.0896	1.6908	3.0315	5.7004	13.8430	1.0315
2.78	1.8547	1.5827	1.2734	1.0902	1.6995	3.0338	5.7021	13.7240	1.0324

(continued on next page)

TABLE 11.6 Training data for the design of fourth-order Wilkinson power dividers/combiners—cont'd

$N = 4$									
f_u/f_l	Z_1/Z_0	Z_2/Z_0	Z_3/Z_0	Z_4/Z_0	R_1/Z_0	R_2/Z_0	R_3/Z_0	R_4/Z_0	VSWR
2.80	1.8537	1.5822	1.2737	1.0908	1.7081	3.0362	5.7038	13.6070	1.0333
2.83	1.8527	1.5818	1.2741	1.0914	1.7168	3.0387	5.7056	13.4910	1.0342
2.85	1.8518	1.5813	1.2745	1.0920	1.7254	3.0413	5.7075	13.3780	1.0352
2.88	1.8508	1.5809	1.2749	1.0926	1.7340	3.0439	5.7094	13.2660	1.0362
2.90	1.8498	1.5804	1.2752	1.0932	1.7425	3.0466	5.7115	13.1560	1.0372
2.93	1.8488	1.5799	1.2756	1.0939	1.7511	3.0494	5.7136	13.0470	1.0382
2.95	1.8478	1.5795	1.2760	1.0945	1.7595	3.0523	5.7157	12.9400	1.0392
2.98	1.8468	1.5790	1.2764	1.0951	1.7680	3.0552	5.7180	12.8350	1.0403
3.00	1.8458	1.5785	1.2768	1.0957	1.7763	3.0583	5.7203	12.7310	1.0414
3.03	1.8449	1.5781	1.2771	1.0963	1.7847	3.0614	5.7227	12.6290	1.0425
3.05	1.8439	1.5776	1.2775	1.0969	1.7929	3.0646	5.7251	12.5280	1.0436
3.08	1.8429	1.5771	1.2779	1.0975	1.8011	3.0679	5.7277	12.4290	1.0447
3.10	1.8419	1.5767	1.2783	1.0982	1.8092	3.0713	5.7303	12.3320	1.0459
3.13	1.8409	1.5762	1.2787	1.0988	1.8173	3.0747	5.7329	12.2360	1.0470
3.15	1.8399	1.5758	1.2790	1.0994	1.8252	3.0782	5.7357	12.1410	1.0482
3.18	1.8389	1.5753	1.2794	1.1000	1.8331	3.0818	5.7385	12.0480	1.0494
3.20	1.8380	1.5749	1.2798	1.1006	1.8409	3.0855	5.7413	11.9560	1.0507
3.23	1.8370	1.5744	1.2802	1.1012	1.8486	3.0893	5.7443	11.8660	1.0519
3.25	1.8360	1.5739	1.2805	1.1018	1.8563	3.0931	5.7473	11.7770	1.0532
1.70	1.8971	1.6035	1.2571	1.0638	1.4546	3.0082	5.7099	20.7970	1.0073
1.73	1.8961	1.6030	1.2575	1.0644	1.4533	3.0071	5.7077	20.5780	1.0076
1.75	1.8951	1.6025	1.2579	1.0650	1.4527	3.0060	5.7056	20.3630	1.0080
1.78	1.8942	1.6020	1.2582	1.0656	1.4525	3.0051	5.7036	20.1500	1.0083
1.80	1.8932	1.6015	1.2586	1.0662	1.4530	3.0042	5.7016	19.9410	1.0087
1.83	1.8922	1.6010	1.2590	1.0668	1.4539	3.0034	5.6998	19.7340	1.0091
1.85	1.8912	1.6005	1.2594	1.0675	1.4553	3.0027	5.6981	19.5300	1.0095
1.88	1.8902	1.6000	1.2598	1.0681	1.4572	3.0020	5.6965	19.3290	1.0098
1.90	1.8892	1.5995	1.2601	1.0687	1.4595	3.0015	5.6951	19.1310	1.0102
1.93	1.8882	1.5990	1.2605	1.0693	1.4622	3.0010	5.6937	18.9360	1.0106
1.95	1.8873	1.5985	1.2609	1.0699	1.4654	3.0006	5.6924	18.7430	1.0111
1.98	1.8863	1.5980	1.2613	1.0705	1.4690	3.0003	5.6912	18.5530	1.0115
2.00	1.8853	1.5975	1.2616	1.0711	1.4729	3.0001	5.6901	18.3660	1.0119
2.03	1.8843	1.5971	1.2620	1.0718	1.4772	3.0000	5.6891	18.1810	1.0124
2.05	1.8833	1.5966	1.2624	1.0724	1.4818	2.9999	5.6883	17.9990	1.0128
2.08	1.8823	1.5961	1.2628	1.0730	1.4867	2.9999	5.6875	17.8200	1.0133
2.10	1.8813	1.5956	1.2632	1.0736	1.4920	3.0000	5.6868	17.6430	1.0138
2.13	1.8804	1.5951	1.2635	1.0742	1.4975	3.0002	5.6862	17.4680	1.0143
2.15	1.8794	1.5946	1.2639	1.0748	1.5033	3.0005	5.6857	17.2960	1.0148
2.18	1.8784	1.5941	1.2643	1.0754	1.5094	3.0008	5.6853	17.1270	1.0153
2.20	1.8774	1.5937	1.2647	1.0761	1.5157	3.0013	5.6850	16.9600	1.0159
2.23	1.8764	1.5932	1.2650	1.0767	1.5222	3.0018	5.6848	16.7950	1.0164
2.25	1.8754	1.5927	1.2654	1.0773	1.5290	3.0024	5.6847	16.6330	1.0170

(continued on next page)

TABLE 11.6 Training data for the design of fourth-order Wilkinson power dividers/combiners—cont'd

$N = 4$									
f_u/f_l	Z_1/Z_0	Z_2/Z_0	Z_3/Z_0	Z_4/Z_0	R_1/Z_0	R_2/Z_0	R_3/Z_0	R_4/Z_0	VSWR
2.28	1.8744	1.5922	1.2658	1.0779	1.5359	3.0031	5.6846	16.4730	1.0176
2.30	1.8735	1.5917	1.2662	1.0785	1.5431	3.0038	5.6847	16.3160	1.0182
2.33	1.8725	1.5912	1.2666	1.0791	1.5504	3.0047	5.6848	16.1600	1.0188
2.35	1.8715	1.5908	1.2669	1.0797	1.5579	3.0056	5.6851	16.0070	1.0194
2.38	1.8705	1.5903	1.2673	1.0804	1.5655	3.0066	5.6854	15.8570	1.0200
2.40	1.8695	1.5898	1.2677	1.0810	1.5733	3.0077	5.6858	15.7080	1.0207
2.43	1.8685	1.5893	1.2681	1.0816	1.5812	3.0089	5.6863	15.5620	1.0213
2.45	1.8675	1.5889	1.2684	1.0822	1.5892	3.0101	5.6869	15.4170	1.0220
2.48	1.8665	1.5884	1.2688	1.0828	1.5973	3.0115	5.6876	15.2750	1.0227
2.50	1.8656	1.5879	1.2692	1.0834	1.6055	3.0129	5.6883	15.1350	1.0234
2.53	1.8646	1.5874	1.2696	1.0840	1.6138	3.0144	5.6892	14.9970	1.0241
2.55	1.8636	1.5870	1.2700	1.0847	1.6222	3.0160	5.6901	14.8610	1.0249
2.58	1.8626	1.5865	1.2703	1.0853	1.6306	3.0176	5.6911	14.7270	1.0257
2.60	1.8616	1.5860	1.2707	1.0859	1.6391	3.0194	5.6922	14.5950	1.0264
2.63	1.8606	1.5855	1.2711	1.0865	1.6477	3.0212	5.6934	14.4650	1.0272
2.65	1.8596	1.5851	1.2715	1.0871	1.6563	3.0231	5.6946	14.3370	1.0280
2.68	1.8587	1.5846	1.2718	1.0877	1.6649	3.0251	5.6960	14.2100	1.0289
2.70	1.8577	1.5841	1.2722	1.0883	1.6735	3.0271	5.6974	14.0860	1.0297
2.73	1.8567	1.5837	1.2726	1.0890	1.6822	3.0293	5.6989	13.9630	1.0306
2.75	1.8557	1.5832	1.2730	1.0896	1.6908	3.0315	5.7004	13.8430	1.0315
2.78	1.8547	1.5827	1.2734	1.0902	1.6995	3.0338	5.7021	13.7240	1.0324
2.80	1.8537	1.5822	1.2737	1.0908	1.7081	3.0362	5.7038	13.6070	1.0333
2.83	1.8527	1.5818	1.2741	1.0914	1.7168	3.0387	5.7056	13.4910	1.0342
2.85	1.8518	1.5813	1.2745	1.0920	1.7254	3.0413	5.7075	13.3780	1.0352
2.88	1.8508	1.5809	1.2749	1.0926	1.7340	3.0439	5.7094	13.2660	1.0362
2.90	1.8498	1.5804	1.2752	1.0932	1.7425	3.0466	5.7115	13.1560	1.0372
2.93	1.8488	1.5799	1.2756	1.0939	1.7511	3.0494	5.7136	13.0470	1.0382
2.95	1.8478	1.5795	1.2760	1.0945	1.7595	3.0523	5.7157	12.9400	1.0392
2.98	1.8468	1.5790	1.2764	1.0951	1.7680	3.0552	5.7180	12.8350	1.0403
3.00	1.8458	1.5785	1.2768	1.0957	1.7763	3.0583	5.7203	12.7310	1.0414
3.03	1.8449	1.5781	1.2771	1.0963	1.7847	3.0614	5.7227	12.6290	1.0425
3.05	1.8439	1.5776	1.2775	1.0969	1.7929	3.0646	5.7251	12.5280	1.0436
3.08	1.8429	1.5771	1.2779	1.0975	1.8011	3.0679	5.7277	12.4290	1.0447
3.10	1.8419	1.5767	1.2783	1.0982	1.8092	3.0713	5.7303	12.3320	1.0459
3.13	1.8409	1.5762	1.2787	1.0988	1.8173	3.0747	5.7329	12.2360	1.0470
3.15	1.8399	1.5758	1.2790	1.0994	1.8252	3.0782	5.7357	12.1410	1.0482
3.18	1.8389	1.5753	1.2794	1.1000	1.8331	3.0818	5.7385	12.0480	1.0494
3.20	1.8380	1.5749	1.2798	1.1006	1.8409	3.0855	5.7413	11.9560	1.0507
3.23	1.8370	1.5744	1.2802	1.1012	1.8486	3.0893	5.7443	11.8660	1.0519
3.25	1.8360	1.5739	1.2805	1.1018	1.8563	3.0931	5.7473	11.7770	1.0532
3.28	1.8350	1.5735	1.2809	1.1025	1.8638	3.0971	5.7504	11.6890	1.0545
3.30	1.8340	1.5730	1.2813	1.1031	1.8712	3.1011	5.7535	11.6030	1.0558

(continued on next page)

TABLE 11.6 Training data for the design of fourth-order Wilkinson power dividers/combiners—cont'd

$N = 4$									
f_u/f_l	Z_1/Z_0	Z_2/Z_0	Z_3/Z_0	Z_4/Z_0	R_1/Z_0	R_2/Z_0	R_3/Z_0	R_4/Z_0	VSWR
3.33	1.8330	1.5726	1.2817	1.1037	1.8785	3.1052	5.7567	11.5180	1.0572
3.35	1.8320	1.5721	1.2821	1.1043	1.8858	3.1093	5.7600	11.4340	1.0585
3.38	1.8311	1.5717	1.2824	1.1049	1.8929	3.1136	5.7633	11.3520	1.0599
3.40	1.8301	1.5712	1.2828	1.1055	1.8999	3.1179	5.7667	11.2700	1.0613
3.43	1.8291	1.5708	1.2832	1.1061	1.9068	3.1223	5.7702	11.1910	1.0627
3.45	1.8281	1.5703	1.2836	1.1068	1.9136	3.1268	5.7737	11.1120	1.0642
3.48	1.8271	1.5699	1.2839	1.1074	1.9203	3.1314	5.7773	11.0340	1.0656
3.50	1.8261	1.5694	1.2843	1.1080	1.9269	3.1361	5.7810	10.9580	1.0671
3.53	1.8251	1.5690	1.2847	1.1086	1.9334	3.1408	5.7847	10.8830	1.0686
3.55	1.8241	1.5685	1.2851	1.1092	1.9397	3.1456	5.7885	10.8090	1.0702
3.58	1.8232	1.5681	1.2855	1.1098	1.9460	3.1505	5.7923	10.7360	1.0717
3.60	1.8222	1.5676	1.2858	1.1104	1.9521	3.1555	5.7962	10.6640	1.0733
3.63	1.8212	1.5672	1.2862	1.1111	1.9581	3.1605	5.8002	10.5940	1.0749
3.65	1.8202	1.5667	1.2866	1.1117	1.9640	3.1657	5.8042	10.5240	1.0765
3.68	1.8192	1.5663	1.2870	1.1123	1.9698	3.1709	5.8083	10.4550	1.0782
3.70	1.8182	1.5659	1.2873	1.1129	1.9754	3.1762	5.8125	10.3880	1.0798
3.73	1.8172	1.5654	1.2877	1.1135	1.9810	3.1816	5.8167	10.3210	1.0815
3.75	1.8163	1.5650	1.2881	1.1141	1.9864	3.1870	5.8210	10.2560	1.0832
3.78	1.8153	1.5645	1.2885	1.1147	1.9917	3.1926	5.8253	10.1920	1.0849
3.80	1.8143	1.5641	1.2889	1.1154	1.9969	3.1982	5.8297	10.1280	1.0867
3.83	1.8133	1.5636	1.2892	1.1160	2.0020	3.2039	5.8341	10.0650	1.0884
3.85	1.8123	1.5632	1.2896	1.1166	2.0069	3.2096	5.8386	10.0040	1.0902
3.88	1.8113	1.5628	1.2900	1.1172	2.0118	3.2155	5.8432	9.9432	1.0920
3.90	1.8103	1.5623	1.2904	1.1178	2.0165	3.2214	5.8478	9.8835	1.0939
3.93	1.8094	1.5619	1.2907	1.1184	2.0211	3.2275	5.8525	9.8246	1.0957
3.95	1.8084	1.5615	1.2911	1.1190	2.0257	3.2335	5.8572	9.7667	1.0976
3.98	1.8074	1.5610	1.2915	1.1197	2.0300	3.2397	5.8620	9.7096	1.0995
4.00	1.8064	1.5606	1.2919	1.1203	2.0343	3.2460	5.8669	9.6533	1.1014
4.03	1.8054	1.5601	1.2923	1.1209	2.0385	3.2523	5.8718	9.5979	1.1034
4.05	1.8044	1.5597	1.2926	1.1215	2.0426	3.2587	5.8768	9.5433	1.1053
4.08	1.8034	1.5593	1.2930	1.1221	2.0465	3.2652	5.8818	9.4895	1.1073
4.10	1.8025	1.5588	1.2934	1.1227	2.0504	3.2718	5.8869	9.4365	1.1093
4.13	1.8015	1.5584	1.2938	1.1233	2.0541	3.2784	5.8921	9.3843	1.1113
4.15	1.8005	1.5580	1.2941	1.1240	2.0577	3.2852	5.8973	9.3329	1.1134
4.18	1.7995	1.5576	1.2945	1.1246	2.0613	3.2920	5.9025	9.2822	1.1154
4.20	1.7985	1.5571	1.2949	1.1252	2.0647	3.2989	5.9078	9.2322	1.1175
4.23	1.7975	1.5567	1.2953	1.1258	2.0680	3.3059	5.9132	9.1830	1.1196
4.25	1.7965	1.5563	1.2957	1.1264	2.0713	3.3129	5.9186	9.1345	1.1217
4.28	1.7956	1.5558	1.2960	1.1270	2.0744	3.3200	5.9241	9.0867	1.1239
4.30	1.7946	1.5554	1.2964	1.1276	2.0774	3.3272	5.9296	9.0396	1.1261
4.33	1.7936	1.5550	1.2968	1.1283	2.0804	3.3345	5.9352	8.9932	1.1283
4.35	1.7926	1.5546	1.2972	1.1289	2.0833	3.3419	5.9409	8.9474	1.1305
4.38	1.7916	1.5541	1.2975	1.1295	2.0860	3.3493	5.9466	8.9023	1.1327

(continued on next page)

TABLE 11.6 Training data for the design of fourth-order Wilkinson power dividers/combiners—cont'd

$N = 4$									
f_u/f_l	Z_1/Z_0	Z_2/Z_0	Z_3/Z_0	Z_4/Z_0	R_1/Z_0	R_2/Z_0	R_3/Z_0	R_4/Z_0	VSWR
4.40	1.7906	1.5537	1.2979	1.1301	2.0887	3.3569	5.9524	8.8578	1.1349
4.43	1.7896	1.5533	1.2983	1.1307	2.0913	3.3645	5.9582	8.8139	1.1372
4.45	1.7886	1.5529	1.2987	1.1313	2.0938	3.3722	5.9641	8.7707	1.1395
4.48	1.7877	1.5524	1.2991	1.1319	2.0962	3.3799	5.9700	8.7280	1.1418
4.50	1.7867	1.5520	1.2994	1.1326	2.0986	3.3878	5.9760	8.6860	1.1441
4.53	1.7857	1.5516	1.2998	1.1332	2.1008	3.3957	5.9820	8.6445	1.1465
4.55	1.7847	1.5512	1.3002	1.1338	2.1030	3.4037	5.9881	8.6036	1.1488
4.58	1.7837	1.5507	1.3006	1.1344	2.1051	3.4118	5.9943	8.5633	1.1512
4.60	1.7827	1.5503	1.3009	1.1350	2.1072	3.4199	6.0005	8.5235	1.1536
4.63	1.7817	1.5499	1.3013	1.1356	2.1091	3.4282	6.0067	8.4843	1.1560
4.65	1.7808	1.5495	1.3017	1.1362	2.1110	3.4365	6.0131	8.4456	1.1585
4.68	1.7798	1.5491	1.3021	1.1369	2.1129	3.4449	6.0194	8.4074	1.1609
4.70	1.7788	1.5487	1.3025	1.1375	2.1146	3.4534	6.0259	8.3697	1.1634
4.73	1.7778	1.5482	1.3028	1.1381	2.1163	3.4619	6.0324	8.3325	1.1659
4.75	1.7768	1.5478	1.3032	1.1387	2.1180	3.4705	6.0389	8.2958	1.1684
4.78	1.7758	1.5474	1.3036	1.1393	2.1196	3.4793	6.0455	8.2595	1.1709
4.80	1.7748	1.5470	1.3040	1.1399	2.1211	3.4880	6.0522	8.2238	1.1735
4.83	1.7739	1.5466	1.3043	1.1405	2.1226	3.4969	6.0589	8.1884	1.1760
4.85	1.7729	1.5462	1.3047	1.1411	2.1240	3.5059	6.0657	8.1536	1.1786
4.88	1.7719	1.5458	1.3051	1.1418	2.1254	3.5149	6.0725	8.1192	1.1812
4.90	1.7709	1.5453	1.3055	1.1424	2.1267	3.5240	6.0794	8.0852	1.1838
4.93	1.7699	1.5449	1.3059	1.1430	2.1280	3.5332	6.0863	8.0516	1.1865
4.95	1.7689	1.5445	1.3062	1.1436	2.1292	3.5424	6.0933	8.0185	1.1891
4.98	1.7679	1.5441	1.3066	1.1442	2.1304	3.5518	6.1004	7.9857	1.1918
5.00	1.7670	1.5437	1.3070	1.1448	2.1316	3.5612	6.1075	7.9534	1.1944
5.03	1.7660	1.5433	1.3074	1.1454	2.1327	3.5707	6.1147	7.9214	1.1971
5.05	1.7650	1.5429	1.3078	1.1461	2.1338	3.5803	6.1219	7.8898	1.1998
5.08	1.7640	1.5425	1.3081	1.1467	2.1348	3.5899	6.1292	7.8586	1.2026
5.10	1.7630	1.5421	1.3085	1.1473	2.1358	3.5997	6.1365	7.8277	1.2053
5.13	1.7620	1.5417	1.3089	1.1479	2.1368	3.6095	6.1439	7.7972	1.2081
5.15	1.7610	1.5413	1.3093	1.1485	2.1378	3.6194	6.1514	7.7671	1.2108
5.18	1.7601	1.5409	1.3096	1.1491	2.1387	3.6293	6.1589	7.7373	1.2136
5.20	1.7591	1.5404	1.3100	1.1497	2.1396	3.6394	6.1665	7.7078	1.2164
5.23	1.7581	1.5400	1.3104	1.1504	2.1405	3.6495	6.1741	7.6787	1.2192
5.25	1.7571	1.5396	1.3108	1.1510	2.1413	3.6597	6.1818	7.6499	1.2220
5.28	1.7561	1.5392	1.3112	1.1516	2.1422	3.6700	6.1896	7.6213	1.2248
5.30	1.7551	1.5388	1.3115	1.1522	2.1430	3.6803	6.1974	7.5931	1.2277
5.33	1.7541	1.5384	1.3119	1.1528	2.1438	3.6908	6.2053	7.5652	1.2305
5.35	1.7531	1.5380	1.3123	1.1534	2.1446	3.7013	6.2132	7.5376	1.2334
5.38	1.7522	1.5376	1.3127	1.1540	2.1454	3.7119	6.2212	7.5103	1.2363
5.40	1.7512	1.5372	1.3130	1.1547	2.1462	3.7226	6.2293	7.4832	1.2392
5.43	1.7502	1.5368	1.3134	1.1553	2.1469	3.7333	6.2374	7.4564	1.2421

(continued on next page)

TABLE 11.6 Training data for the design of fourth-order Wilkinson power dividers/combiners—cont'd

$N = 4$									
f_u/f_l	Z_1/Z_0	Z_2/Z_0	Z_3/Z_0	Z_4/Z_0	R_1/Z_0	R_2/Z_0	R_3/Z_0	R_4/Z_0	VSWR
5.45	1.7492	1.5364	1.3138	1.1559	2.1477	3.7441	6.2456	7.4299	1.2450
5.48	1.7482	1.5361	1.3142	1.1565	2.1484	3.7550	6.2539	7.4037	1.2479
5.50	1.7472	1.5357	1.3146	1.1571	2.1492	3.7660	6.2622	7.3777	1.2509
5.53	1.7462	1.5353	1.3149	1.1577	2.1499	3.7771	6.2706	7.3519	1.2538
5.55	1.7453	1.5349	1.3153	1.1583	2.1506	3.7882	6.2790	7.3264	1.2567
5.58	1.7443	1.5345	1.3157	1.1590	2.1514	3.7994	6.2875	7.3012	1.2597
5.60	1.7433	1.5341	1.3161	1.1596	2.1521	3.8107	6.2961	7.2761	1.2627
5.63	1.7423	1.5337	1.3164	1.1602	2.1529	3.8221	6.3047	7.2513	1.2657
5.65	1.7413	1.5333	1.3168	1.1608	2.1536	3.8336	6.3134	7.2267	1.2686
5.68	1.7403	1.5329	1.3172	1.1614	2.1544	3.8451	6.3222	7.2024	1.2716
5.70	1.7393	1.5325	1.3176	1.1620	2.1551	3.8567	6.3310	7.1782	1.2746
5.73	1.7384	1.5321	1.3180	1.1626	2.1559	3.8684	6.3399	7.1543	1.2777
5.75	1.7374	1.5317	1.3183	1.1633	2.1566	3.8801	6.3488	7.1305	1.2807
5.78	1.7364	1.5313	1.3187	1.1639	2.1574	3.8920	6.3579	7.1070	1.2837
5.80	1.7354	1.5310	1.3191	1.1645	2.1582	3.9039	6.3670	7.0836	1.2867
5.83	1.7344	1.5306	1.3195	1.1651	2.1590	3.9159	6.3761	7.0605	1.2897
5.85	1.7334	1.5302	1.3198	1.1657	2.1598	3.9279	6.3854	7.0375	1.2928
5.88	1.7324	1.5298	1.3202	1.1663	2.1607	3.9401	6.3947	7.0147	1.2958
5.90	1.7315	1.5294	1.3206	1.1669	2.1615	3.9523	6.4040	6.9921	1.2989
5.93	1.7305	1.5290	1.3210	1.1676	2.1624	3.9646	6.4135	6.9697	1.3019
5.95	1.7295	1.5286	1.3214	1.1682	2.1632	3.9770	6.4230	6.9474	1.3050
5.98	1.7285	1.5283	1.3217	1.1688	2.1641	3.9895	6.4326	6.9253	1.3080
6.00	1.7275	1.5279	1.3221	1.1694	2.1650	4.0020	6.4422	6.9033	1.3111
6.03	1.7265	1.5275	1.3225	1.1700	2.1659	4.0146	6.4519	6.8815	1.3142
6.05	1.7255	1.5271	1.3229	1.1706	2.1669	4.0273	6.4617	6.8599	1.3172
6.08	1.7246	1.5267	1.3232	1.1712	2.1678	4.0401	6.4716	6.8384	1.3203
6.10	1.7236	1.5264	1.3236	1.1719	2.1688	4.0529	6.4816	6.8171	1.3234
6.13	1.7226	1.5260	1.3240	1.1725	2.1698	4.0658	6.4916	6.7959	1.3264
6.15	1.7216	1.5256	1.3244	1.1731	2.1708	4.0788	6.5017	6.7749	1.3295
6.18	1.7206	1.5252	1.3248	1.1737	2.1719	4.0919	6.5118	6.7540	1.3326
6.20	1.7196	1.5248	1.3251	1.1743	2.1729	4.1050	6.5221	6.7332	1.3356
6.23	1.7186	1.5245	1.3255	1.1749	2.1740	4.1183	6.5324	6.7126	1.3387
6.25	1.7176	1.5241	1.3259	1.1755	2.1751	4.1316	6.5428	6.6921	1.3417
6.28	1.7167	1.5237	1.3263	1.1762	2.1762	4.1450	6.5533	6.6717	1.3448
6.30	1.7157	1.5233	1.3266	1.1768	2.1773	4.1584	6.5638	6.6514	1.3479
6.33	1.7147	1.5230	1.3270	1.1774	2.1785	4.1720	6.5745	6.6313	1.3509
6.35	1.7137	1.5226	1.3274	1.1780	2.1797	4.1856	6.5852	6.6113	1.3540
6.38	1.7127	1.5222	1.3278	1.1786	2.1809	4.1993	6.5960	6.5914	1.3570
6.40	1.7117	1.5218	1.3282	1.1792	2.1821	4.2130	6.6069	6.5717	1.3601
6.43	1.7107	1.5215	1.3285	1.1798	2.1833	4.2269	6.6178	6.5520	1.3631

(continued on next page)

TABLE 11.6 Training data for the design of fourth-order Wilkinson power dividers/combiners—cont'd

$N = 4$										
f_u/f_l	Z_1/Z_0	Z_2/Z_0	Z_3/Z_0	Z_4/Z_0	R_1/Z_0	R_2/Z_0	R_3/Z_0	R_4/Z_0	VSWR
6.45	1.7098	1.5211	1.3289	1.1805	2.1846	4.2408	6.6289	6.5325	1.3661
6.48	1.7088	1.5207	1.3293	1.1811	2.1859	4.2548	6.6400	6.5130	1.3692
6.50	1.7078	1.5204	1.3297	1.1817	2.1872	4.2689	6.6512	6.4937	1.3722
6.53	1.7068	1.5200	1.3300	1.1823	2.1885	4.2830	6.6625	6.4745	1.3755
6.55	1.7058	1.5196	1.3304	1.1829	2.1899	4.2972	6.6739	6.4554	1.3786
6.58	1.7048	1.5193	1.3308	1.1835	2.1912	4.3115	6.6853	6.4364	1.3817
6.60	1.7038	1.5189	1.3312	1.1841	2.1926	4.3259	6.6969	6.4175	1.3848
6.63	1.7029	1.5185	1.3316	1.1848	2.1940	4.3404	6.7085	6.3987	1.3879
6.65	1.7019	1.5182	1.3319	1.1854	2.1954	4.3549	6.7202	6.3800	1.3910
6.68	1.7009	1.5178	1.3323	1.1860	2.1969	4.3695	6.7320	6.3614	1.3941
6.70	1.6999	1.5174	1.3327	1.1866	2.1983	4.3842	6.7439	6.3429	1.3972
6.73	1.6989	1.5171	1.3331	1.1872	2.1998	4.3990	6.7559	6.3245	1.4004
6.75	1.6979	1.5167	1.3334	1.1878	2.2013	4.4138	6.7680	6.3062	1.4035
6.78	1.6969	1.5163	1.3338	1.1884	2.2028	4.4287	6.7802	6.2880	1.4066
6.80	1.6960	1.5160	1.3342	1.1890	2.2043	4.4437	6.7925	6.2698	1.4097
6.83	1.6950	1.5156	1.3346	1.1897	2.2058	4.4588	6.8048	6.2518	1.4128
6.85	1.6940	1.5153	1.3350	1.1903	2.2074	4.4739	6.8173	6.2339	1.4159
6.88	1.6930	1.5149	1.3353	1.1909	2.2089	4.4891	6.8298	6.2160	1.4190
6.90	1.6920	1.5145	1.3357	1.1915	2.2105	4.5044	6.8425	6.1983	1.4221
6.93	1.6910	1.5142	1.3361	1.1921	2.2121	4.5198	6.8552	6.1806	1.4252
6.95	1.6900	1.5138	1.3365	1.1927	2.2137	4.5352	6.8681	6.1630	1.4283
6.98	1.6891	1.5135	1.3369	1.1933	2.2153	4.5508	6.8810	6.1455	1.4314
7.00	1.6881	1.5131	1.3372	1.1940	2.2169	4.5664	6.8941	6.1281	1.4345
7.03	1.6871	1.5128	1.3376	1.1946	2.2185	4.5820	6.9072	6.1108	1.4376
7.05	1.6861	1.5124	1.3380	1.1952	2.2201	4.5978	6.9205	6.0936	1.4407
7.08	1.6851	1.5120	1.3384	1.1958	2.2218	4.6136	6.9338	6.0765	1.4438
7.10	1.6841	1.5117	1.3387	1.1964	2.2234	4.6295	6.9473	6.0594	1.4469
7.13	1.6831	1.5113	1.3391	1.1970	2.2251	4.6455	6.9608	6.0425	1.4500
7.15	1.6821	1.5110	1.3395	1.1976	2.2267	4.6616	6.9745	6.0256	1.4531
7.18	1.6812	1.5106	1.3399	1.1983	2.2283	4.6777	6.9882	6.0088	1.4562
7.20	1.6802	1.5103	1.3403	1.1989	2.2300	4.6939	7.0021	5.9921	1.4593
7.23	1.6792	1.5099	1.3406	1.1995	2.2317	4.7102	7.0161	5.9755	1.4624
7.25	1.6782	1.5096	1.3410	1.2001	2.2333	4.7266	7.0302	5.9590	1.4655
7.28	1.6772	1.5092	1.3414	1.2007	2.2350	4.7430	7.0444	5.9426	1.4686
7.30	1.6762	1.5089	1.3418	1.2013	2.2366	4.7595	7.0587	5.9262	1.4716
7.33	1.6752	1.5085	1.3421	1.2019	2.2382	4.7761	7.0731	5.9100	1.4747
7.35	1.6743	1.5082	1.3425	1.2026	2.2399	4.7927	7.0877	5.8938	1.4778
7.38	1.6733	1.5079	1.3429	1.2032	2.2415	4.8095	7.1023	5.8778	1.4809
7.40	1.6723	1.5075	1.3433	1.2038	2.2432	4.8263	7.1171	5.8618	1.4840
7.43	1.6713	1.5072	1.3437	1.2044	2.2448	4.8432	7.1320	5.8459	1.4871
7.45	1.6703	1.5068	1.3440	1.2050	2.2464	4.8601	7.1470	5.8301	1.4901
7.48	1.6693	1.5065	1.3444	1.2056	2.2480	4.8772	7.1621	5.8144	1.4932
7.50	1.6683	1.5061	1.3448	1.2062	2.2496	4.8943	7.1773	5.7988	1.4963

(continued on next page)

TABLE 11.6 Training data for the design of fourth-order Wilkinson power dividers/combiners—cont'd

$N = 4$									
f_u/f_l	Z_1/Z_0	Z_2/Z_0	Z_3/Z_0	Z_4/Z_0	R_1/Z_0	R_2/Z_0	R_3/Z_0	R_4/Z_0	VSWR
7.53	1.6674	1.5058	1.3452	1.2069	2.2512	4.9115	7.1927	5.7833	1.4994
7.55	1.6664	1.5054	1.3455	1.2075	2.2528	4.9288	7.2081	5.7679	1.5024
7.58	1.6654	1.5051	1.3459	1.2081	2.2544	4.9461	7.2237	5.7526	1.5055
7.60	1.6644	1.5048	1.3463	1.2087	2.2559	4.9635	7.2395	5.7373	1.5086
7.63	1.6634	1.5044	1.3467	1.2093	2.2575	4.9810	7.2553	5.7222	1.5117
7.65	1.6624	1.5041	1.3471	1.2099	2.2590	4.9986	7.2713	5.7072	1.5147
7.68	1.6614	1.5037	1.3474	1.2105	2.2605	5.0162	7.2874	5.6923	1.5178
7.70	1.6605	1.5034	1.3478	1.2112	2.2620	5.0339	7.3036	5.6774	1.5209
7.73	1.6595	1.5031	1.3482	1.2118	2.2635	5.0517	7.3200	5.6627	1.5239
7.75	1.6585	1.5027	1.3486	1.2124	2.2650	5.0696	7.3364	5.6481	1.5270
7.78	1.6575	1.5024	1.3489	1.2130	2.2664	5.0875	7.3531	5.6336	1.5301
7.80	1.6565	1.5021	1.3493	1.2136	2.2679	5.1055	7.3698	5.6192	1.5332
7.83	1.6555	1.5017	1.3497	1.2142	2.2693	5.1236	7.3867	5.6049	1.5362
7.85	1.6545	1.5014	1.3501	1.2148	2.2707	5.1418	7.4037	5.5907	1.5393
7.88	1.6536	1.5011	1.3505	1.2155	2.2721	5.1601	7.4209	5.5766	1.5424
7.90	1.6526	1.5007	1.3508	1.2161	2.2734	5.1784	7.4382	5.5626	1.5455
7.93	1.6516	1.5004	1.3512	1.2167	2.2748	5.1968	7.4556	5.5487	1.5485
7.95	1.6506	1.5001	1.3516	1.2173	2.2761	5.2152	7.4732	5.5350	1.5516
7.98	1.6496	1.4997	1.3520	1.2179	2.2774	5.2338	7.4909	5.5213	1.5547
8.00	1.6486	1.4994	1.3523	1.2185	2.2787	5.2524	7.5087	5.5078	1.5578
8.03	1.6476	1.4991	1.3527	1.2191	2.2800	5.2711	7.5267	5.4944	1.5609
8.05	1.6466	1.4988	1.3531	1.2198	2.2812	5.2898	7.5449	5.4811	1.5640
8.08	1.6457	1.4984	1.3535	1.2204	2.2825	5.3087	7.5631	5.4680	1.5670
8.10	1.6447	1.4981	1.3539	1.2210	2.2837	5.3276	7.5816	5.4549	1.5701
8.13	1.6437	1.4978	1.3542	1.2216	2.2849	5.3466	7.6002	5.4420	1.5732
8.15	1.6427	1.4974	1.3546	1.2222	2.2860	5.3657	7.6189	5.4292	1.5763
8.18	1.6417	1.4971	1.3550	1.2228	2.2872	5.3848	7.6378	5.4165	1.5794
8.20	1.6407	1.4968	1.3554	1.2234	2.2883	5.4040	7.6568	5.4040	1.5825
8.23	1.6397	1.4965	1.3557	1.2241	2.2895	5.4233	7.6760	5.3916	1.5857
8.25	1.6388	1.4962	1.3561	1.2247	2.2906	5.4427	7.6953	5.3793	1.5888
8.28	1.6378	1.4958	1.3565	1.2253	2.2917	5.4621	7.7148	5.3671	1.5919
8.30	1.6368	1.4955	1.3569	1.2259	2.2928	5.4816	7.7345	5.3551	1.5950
8.33	1.6358	1.4952	1.3573	1.2265	2.2938	5.5012	7.7543	5.3432	1.5981
8.35	1.6348	1.4949	1.3576	1.2271	2.2949	5.5209	7.7743	5.3315	1.6013
8.38	1.6338	1.4945	1.3580	1.2277	2.2960	5.5406	7.7944	5.3199	1.6044
8.40	1.6328	1.4942	1.3584	1.2284	2.2970	5.5604	7.8147	5.3084	1.6076
8.43	1.6319	1.4939	1.3588	1.2290	2.2981	5.5803	7.8352	5.2971	1.6107
8.45	1.6309	1.4936	1.3591	1.2296	2.2991	5.6003	7.8558	5.2859	1.6139
8.48	1.6299	1.4933	1.3595	1.2302	2.3001	5.6203	7.8766	5.2749	1.6171
8.50	1.6289	1.4930	1.3599	1.2308	2.3012	5.6404	7.8976	5.2640	1.6203
8.53	1.6279	1.4926	1.3603	1.2314	2.3022	5.6606	7.9187	5.2532	1.6235

(continued on next page)

TABLE 11.6 Training data for the design of fourth-order Wilkinson power dividers/combiners—cont'd

$N = 4$									
f_u/f_l	Z_1/Z_0	Z_2/Z_0	Z_3/Z_0	Z_4/Z_0	R_1/Z_0	R_2/Z_0	R_3/Z_0	R_4/Z_0	VSWR
8.55	1.6269	1.4923	1.3607	1.2320	2.3032	5.6808	7.9400	5.2427	1.6267
8.58	1.6259	1.4920	1.3610	1.2327	2.3043	5.7012	7.9615	5.2322	1.6299
8.60	1.6250	1.4917	1.3614	1.2333	2.3053	5.7216	7.9832	5.2219	1.6331
8.63	1.6240	1.4914	1.3618	1.2339	2.3064	5.7420	8.0050	5.2118	1.6363
8.65	1.6230	1.4911	1.3622	1.2345	2.3074	5.7626	8.0270	5.2019	1.6396
8.68	1.6220	1.4908	1.3626	1.2351	2.3085	5.7832	8.0492	5.1921	1.6428
8.70	1.6210	1.4905	1.3629	1.2357	2.3096	5.8039	8.0715	5.1824	1.6461
8.73	1.6200	1.4901	1.3633	1.2363	2.3108	5.8247	8.0941	5.1729	1.6494
8.75	1.6190	1.4898	1.3637	1.2369	2.3119	5.8455	8.1168	5.1636	1.6527
8.78	1.6181	1.4895	1.3641	1.2376	2.3131	5.8665	8.1397	5.1545	1.6560
8.80	1.6171	1.4892	1.3644	1.2382	2.3144	5.8874	8.1628	5.1455	1.6593
8.83	1.6161	1.4889	1.3648	1.2388	2.3156	5.9085	8.1861	5.1367	1.6626
8.85	1.6151	1.4886	1.3652	1.2394	2.3169	5.9297	8.2096	5.1280	1.6660
8.88	1.6141	1.4883	1.3656	1.2400	2.3183	5.9509	8.2333	5.1196	1.6693
8.90	1.6131	1.4880	1.3660	1.2406	2.3197	5.9722	8.2572	5.1113	1.6727
8.93	1.6121	1.4877	1.3663	1.2412	2.3211	5.9935	8.2812	5.1031	1.6761
8.95	1.6111	1.4874	1.3667	1.2419	2.3227	6.0150	8.3055	5.0952	1.6795
8.98	1.6102	1.4871	1.3671	1.2425	2.3243	6.0365	8.3299	5.0875	1.6830
9.00	1.6092	1.4868	1.3675	1.2431	2.3259	6.0581	8.3546	5.0799	1.6864
9.03	1.6082	1.4865	1.3678	1.2437	2.3276	6.0797	8.3794	5.0725	1.6899
9.05	1.6072	1.4862	1.3682	1.2443	2.3295	6.1015	8.4045	5.0653	1.6934
9.08	1.6062	1.4859	1.3686	1.2449	2.3314	6.1233	8.4297	5.0582	1.6969
9.10	1.6052	1.4856	1.3690	1.2455	2.3334	6.1451	8.4552	5.0514	1.7005
9.13	1.6042	1.4853	1.3694	1.2462	2.3355	6.1671	8.4808	5.0448	1.7040
9.15	1.6033	1.4850	1.3697	1.2468	2.3377	6.1891	8.5067	5.0383	1.7076
9.18	1.6023	1.4847	1.3701	1.2474	2.3400	6.2112	8.5328	5.0320	1.7112
9.20	1.6013	1.4844	1.3705	1.2480	2.3425	6.2334	8.5591	5.0260	1.7149
9.23	1.6003	1.4841	1.3709	1.2486	2.3450	6.2556	8.5856	5.0201	1.7185
9.25	1.5993	1.4838	1.3712	1.2492	2.3478	6.2779	8.6123	5.0144	1.7222
9.28	1.5983	1.4835	1.3716	1.2498	2.3506	6.3003	8.6393	5.0089	1.7260
9.30	1.5973	1.4832	1.3720	1.2505	2.3536	6.3228	8.6664	5.0037	1.7297
9.33	1.5964	1.4829	1.3724	1.2511	2.3568	6.3453	8.6938	4.9986	1.7335
9.35	1.5954	1.4826	1.3728	1.2517	2.3602	6.3680	8.7214	4.9937	1.7373
9.38	1.5944	1.4823	1.3731	1.2523	2.3637	6.3906	8.7492	4.9890	1.7411
9.40	1.5934	1.4820	1.3735	1.2529	2.3674	6.4134	8.7773	4.9845	1.7450

TABLE 11.7 The predicted impedances Z_2 and Z_4 of the vertical branch lines, as well as the bandwidth (BW), are derived from input values C, Z_1, and Z_3 using the ANN–GA–PSO and ANN–GA–PSO–SM, respectively.

Design	Input			Output					
				Z_2 (Ω)		Z_4 (Ω)		BW (%)	
	C (dB)	Z_1 (Ω)	Z_3 (Ω)	ANN–GA–PSO	ANN–GA–PSO–SM	ANN–GA–PSO	ANN–GA–PSO–SM	ANN–GA–PSO	ANN–GA–PSO–SM
1	1.5	140	90	50.67	49.48	48.10	48.81	50.44	50.02
2	2	140	140	52.73	53.85	59.08	58.34	41.91	42.95
3	3	140	120	49.29	49.59	48.11	48.55	42.63	42.98
4	3	130	130	53.62	57.01	63.21	62.83	25.61	28.86
5	4	160	90	46.18	46.92	44.49	44.23	52.46	50.12

ANN, artificial neural network; GA, genetic algorithms; PSO, particle swarm optimization; SM, space mapping.

TABLE 11.8 Circuit's dimensions of the three-section branch-line couplers, designed for a center frequency of 3.5 GHz (printed on FR-4, $\varepsilon_r = 4.4$, $h = 1.58$ mm).

Design	width, w (mm)					length, l (mm)			
	w_0	w_1	w_2	w_3	w_4	l_1	l_2	l_3	l_4
1	3.021	0.233	3.073	0.925	3.143	12.61	11.73	12.24	11.72
2	3.021	0.233	2.660	0.233	2.316	12.61	11.80	12.61	11.87
3	3.021	0.233	3.062	0.403	3.174	12.61	11.74	12.49	11.72
4	3.021	0.306	2.413	0.306	2.020	12.55	11.85	12.55	11.93
5	3.021	0.135	3.356	0.925	3.687	12.71	11.69	12.24	11.65

TABLE 11.9 The predicted impedances Z_1, Z_2, and the ratio f_{upper}/f_{low} derived from input values R_1 and R_2 using the ANN–GA–PSO and ANN–GA–PSO–SM, respectively.

Design	Input		Output					
			Z_1 (Ω)		Z_2 (Ω)		f_{upper}/f_{low}	
	R_1 (Ω)	R_2 (Ω)	ANN–GA–PSO	ANN–GA–PSO–SM	ANN–GA–PSO	ANN–GA–PSO–SM	ANN–GA–PSO	ANN–GA–PSO–SM
1	95	260	71.63	82.15	52.70	59.91	1.610	1.629
2	107	210	92.84	80.59	70.81	62.72	2.708	2.700
3	138	165	77.52	78.02	63.69	64.00	3.538	3.530

ANN, artificial neural network; *GA*, genetic algorithms; *PSO*, particle swarm optimization; *SM*, space mapping.

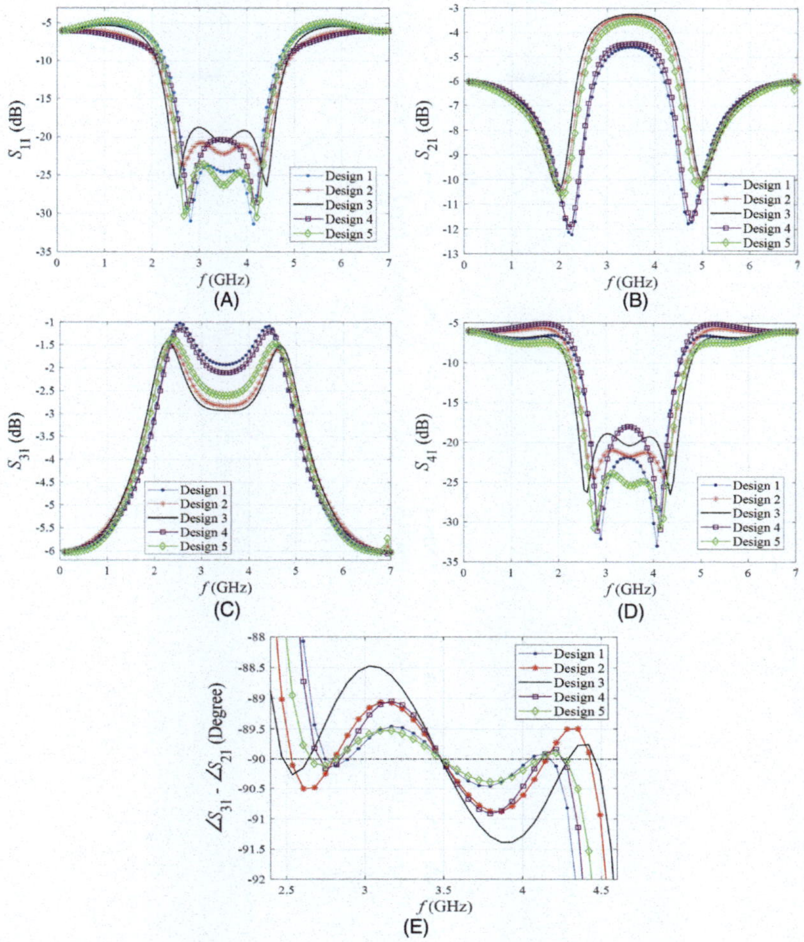

FIGURE 11.10 Simulated S-parameters for the three-section branch-line coupler: (A) Input port return loss (S_{11}), (B) output ports insertion loss (S_{21}), (C) coupling (S_{31}), (D) isolation (S_{41}), and (E) phase differences ($\angle S_{31} - \angle S_{21}$). Simulated performance of the three-section branch-line coupler using dimensions and characteristic impedances predicted by the ANN. *ANN*, Artificial neural network.

TABLE 11.10 The predicted impedances Z_1, Z_2, Z_3, and the ratio f_{upper}/f_{low} derived from input values R_1, R_2, and R_3 using the ANN–GA–PSO and ANN–GA–PSO–SM, respectively.

Design	Input			Output							
				Z_1 (Ω)		Z_2 (Ω)		Z_3 (Ω)		f_{upper}/f_{low}	
	R_1 (Ω)	R_2 (Ω)	R_3 (Ω)	ANN–GA–PSO	ANN–GA–PSO–SM	ANN–GA–PSO	ANN–GA–PSO–SM	ANN–GA–PSO	ANN–GA–PSO–SM	ANN–GA–PSO	ANN–GA–PSO–SM
1	95	185	735	91.86	91.44	70.92	70.68	54.06	54.08	1.1705	1.1793
2	100	200	430	73.51	88.00	59.80	70.22	47.56	55.68	2.2856	2.6201
3	110	215	345	111.57	84.11	92.04	71.61	75.75	60.76	4.6431	4.6210

ANN, artificial neural network; *GA*, genetic algorithms; *PSO*, particle swarm optimization; *SM*, space mapping.

TABLE 11.11 The predicted impedances Z_1, Z_2, Z_3, and the ratio f_{upper}/f_{low} derived from input values R_1, R_2, and R_3 using the ANN–GA–PSO and ANN–GA–PSO–SM, respectively.

	Input				Output									
					Z_1 (Ω)		Z_2 (Ω)		Z_3 (Ω)		Z_4 (Ω)		f_{upper}/f_{low}	
	R_1 (Ω)	R_2 (Ω)	R_3 (Ω)	R_4 (Ω)	ANN–GA–PSO	ANN–GA–PSO–SM	ANN–GA–PSO	ANN–GA–PSO–SM	ANN–GA–PSO	ANN–GA–PSO–SM	ANN–GA–PSO	ANN–GA–PSO–SM	ANN–GA–PSO	ANN–GA–PSO–SM
1	80	150	285	760	92.71	93.63	78.68	79.55	62.71	63.28	53.4	53.90	2.31	2.30
2	105	180	300	400	85.48	89.36	74.8	77.56	62.75	64.80	55.6	56.47	4.41	4.40
3	115	260	375	275	80.04	81.48	78.92	74.86	70.00	68.04	65.41	61.65	8.46	8.45

ANN, artificial neural network; *GA*, genetic algorithms; *PSO*, particle swarm optimization; *SM*, space mapping.

TABLE 11.12 Circuit's dimensions of the two-section Wilkinson power divider/combiner, designed for a center frequency of 3.5 GHz (printed on FR-4, $\varepsilon_r = 4.4$, $h = 1.58$ mm).

Design	width, w (mm)			length, l (mm)	
	w_0	w_1	w_2	l_1	l_2
1	3.02	1.15	2.21	12.16	11.89
2	3.02	1.21	2.03	12.14	11.93
3	3.02	1.30	1.95	12.11	11.94

TABLE 11.13 Circuit's dimensions of the three-section Wilkinson power divider/combiner, designed for a center frequency of 3.5 GHz (printed on FR-4, $\varepsilon_r = 4.4$, $h = 1.58$ mm).

Design	width, w (mm)				length, l (mm)		
	w_0	w_1	w_2	w_3	l_1	l_2	l_3
1	3.02	0.89	1.60	2.65	12.25	12.03	11.81
2	3.02	0.98	1.62	2.52	12.22	12.02	11.83
3	3.02	1.09	1.56	2.15	12.18	12.04	11.90

TABLE 11.14 Circuit's dimensions of the four-section Wilkinson power divider/combiner, designed for a center frequency of 3.5 GHz (printed on FR-4, $\varepsilon_r = 4.4$, $h = 1.58$ mm).

Design	width, w (mm)					length, l (mm)			
	w_0	w_1	w_2	w_3	w_4	l_1	l_2	l_3	l_4
1	3.02	0.836	2.662	1.993	1.242	12.27	12.13	11.93	11.80
2	3.02	0.942	2.455	1.904	1.314	12.23	12.11	11.95	11.84
3	3.02	1.176	2.094	1.731	1.42	12.15	12.08	11.99	11.91

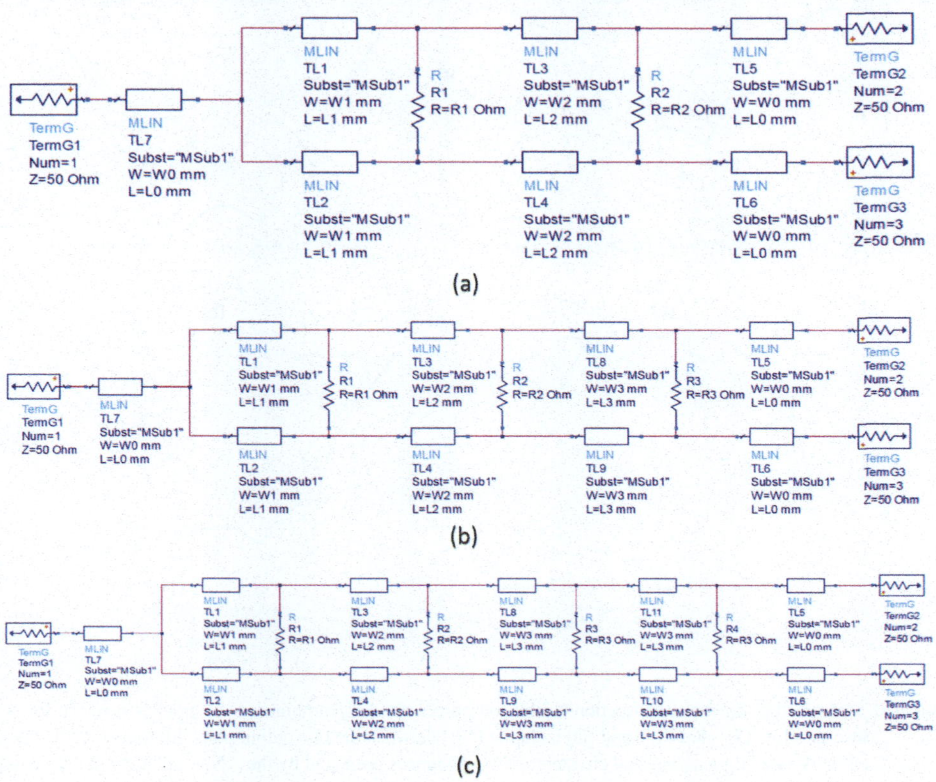

FIGURE 11.11 Simulated circuit schematics for the Wilkinson power dividers/combiners: (A) two-section, (B) three-section, and (C) four-section configurations. 2D circuit simulation for the Wilkinson power dividers/combiners using ADS simulator. *ADS*, Advanced Design System.

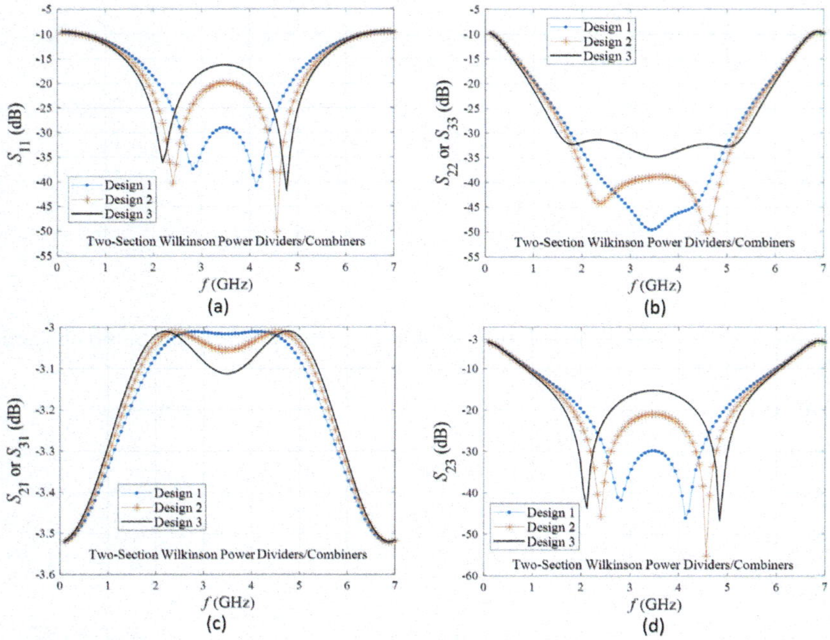

FIGURE 11.12 Simulated S-parameters for the two-section Wilkinson power dividers/combiners: (A) input port return loss (S_{11}), (B) output ports return loss (S_{22} and S_{33}), (C) insertion loss (S_{21} and S_{31}), and (D) isolation (S_{23}) between the two output ports. Simulated performance of the two-section Wilkinson power dividers/combiners using dimensions and characteristic impedances predicted by the ANN. *ANN*, Artificial neural network.

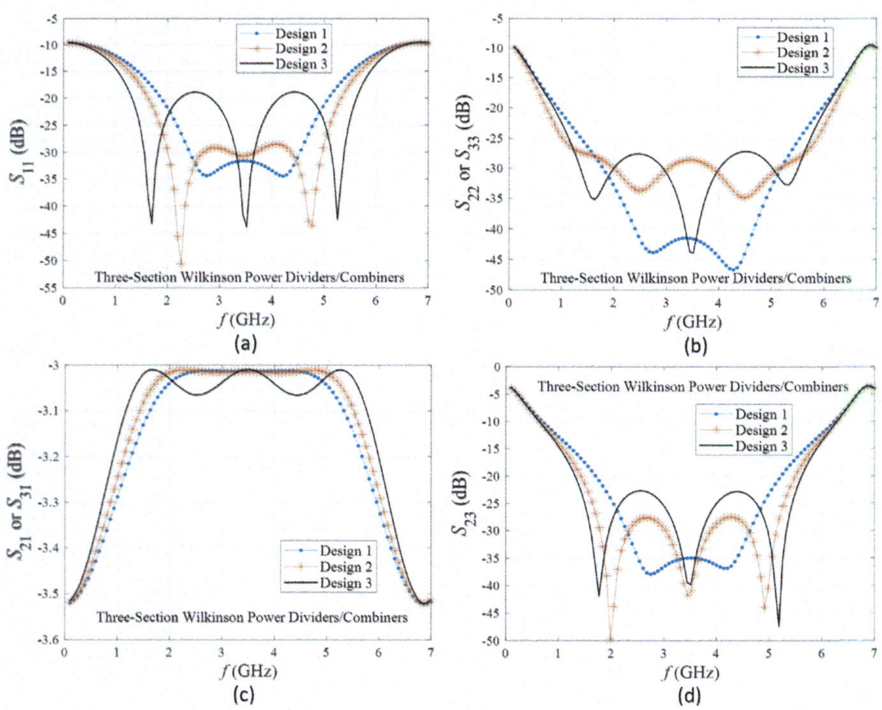

FIGURE 11.13 Simulated S-parameters for the three-section Wilkinson power dividers/combiners: (A) input port return loss (S_{11}), (B) output ports return loss (S_{22} and S_{33}), (C) insertion loss (S_{21} and S_{31}), and (D) isolation (S_{23}) between the two output ports. Simulated performance of the three-section Wilkinson power dividers/combiners using dimensions and characteristic impedances predicted by the ANN. *ANN*, Artificial neural network.

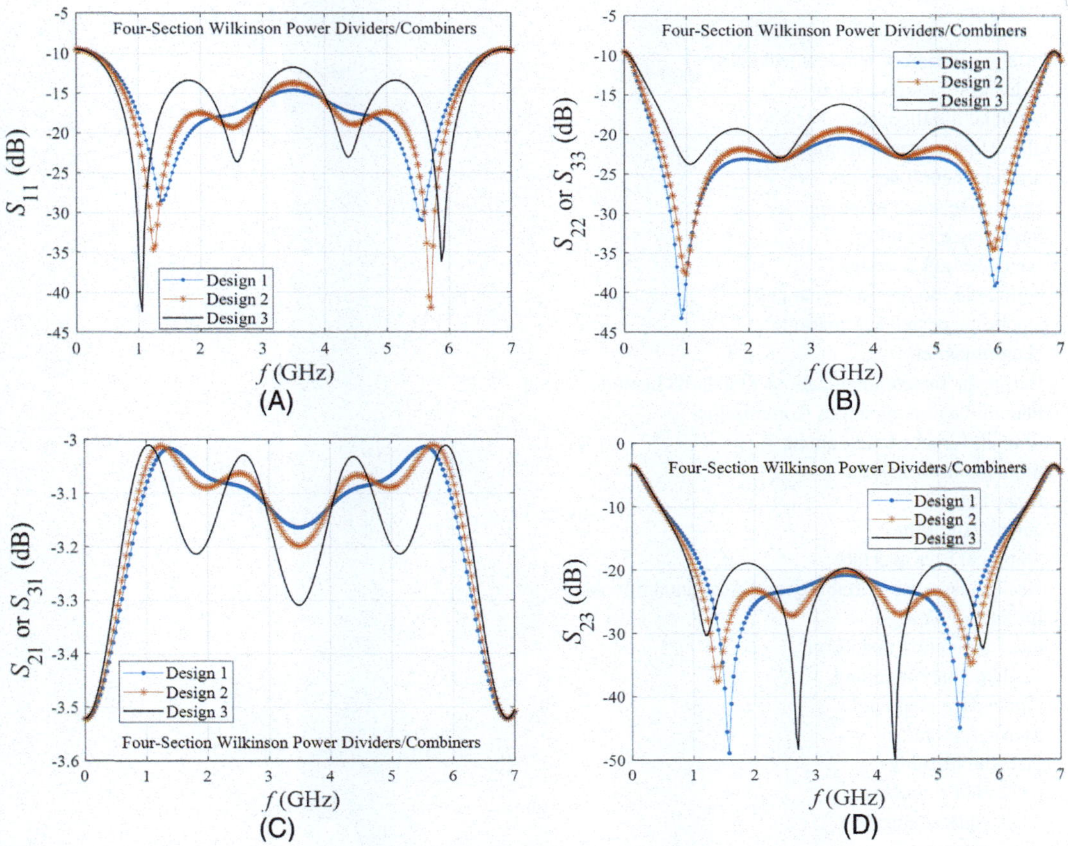

FIGURE 11.14 Simulated *S*-parameters for the four-section Wilkinson power dividers/combiners: (A) input port return loss (S_{11}), (B) output ports return loss (S_{22} and S_{33}), (C) insertion loss (S_{21} and S_{31}), and (D) isolation (S_{23}) between the two output ports. Simulated performance of the four-section Wilkinson power dividers/combiners using dimensions and characteristic impedances predicted by the ANN. *ANN*, Artificial neural network.

11.6 Conclusion

This chapter examines hybrid computational methods—integrating ANN, GA, PSO, and SM—for high-speed circuit design. The ANN–GA–PSO–SM framework streamlines the development of hybrid branch-line couplers and Wilkinson power dividers/combiners, offering significant efficiency and precision improvements by reducing reliance on traditional trial-and-error methods. Leveraging AI enhances both the speed and accuracy of the design process, advancing the performance of high-speed circuit devices and providing a powerful toolset for mechatronic applications.

Further optimization of the ANN–GA–PSO parameters, such as hidden layers, neuron counts, and GA/PSO settings, is recommended to improve prediction accuracy. The current training dataset, sourced from the literature, is also limited in scope. To enhance model accuracy, expanding the dataset with simulations from rigorous EM simulators is recommended. Future research may explore additional AI techniques for complex circuit designs, with AI advancements promising to further revolutionize high-speed circuit design and the development of superior high-frequency components.

In this study the reliability of ANN–GA–PSO–SM techniques for designing high-frequency branch-line couplers and Wilkinson power dividers/combiners is assessed and validated through simulations using the commercial ADS circuit simulator. Simulated results generated from the ANN–GA–PSO–SM-driven specifications align closely with expected circuit performance metrics, demonstrating an acceptable degree of accuracy and effectiveness in meeting design requirements. This validation supports the viability of ANN–GA–PSO–SM as a powerful methodology for high-frequency circuit design optimization.

Acknowledgment

We thank Universiti Teknologi Malaysia (UTM) for providing and supporting the use of MATLAB software for this research.

Abbreviations

5G	fifth generation of wireless technology
ADS	Advanced Design System
AI	artificial intelligence
ALSM	affine linear space mapping
ANN	artificial neural network
ASM	aggressive space mapping
BSM	basic space mapping
CAD	computer-aided design
CRMU	complex ratio measurement units
EDA	electronic design automation
EM	electromagnetic
ETSI	European Telecommunications Standards Institute
FCC	Federal Communications Commission
FDTD	finite difference time domain
FEM	finite element method
FL	fuzzy logic
GA	genetic algorithm
GUI	graphical user interface
HIPAA	Health Insurance Portability and Accountability Act
IoT	Internet of Things
IMD	intermodulation distortion
ImSM	implicit space mapping
ImSM	input space mapping
MATLAB	Matrix Laboratory
ML	machine learning
MoM	method of moments
MSE	mean squared error
MW	microwave
OSM	output space mapping
PCs	personal computers
PCB	printed circuit board
PPA	power, performance, and area
PSO	particle swarm optimization
ReLU	rectified linear units
RF	radio frequency
RMSE	root mean squared error
S-parameters	scattering parameters
SM	space mapping
SMD	surface mount device
SSE	Sum of squared errors

References

Altavilla, D. (2024). Nvidia and Synopsys punctuate AI chip design and acceleration leadership. *Forbes*. Retrieved from https://www.forbes.com/sites/davealtavilla/2024/03/27/nvidia-and-synopsys-punctuate-ai-chip-design-and-acceleration-leadership/.

Bandler, J. W., Biernacki, R. M., Chen, S. H., Grobelny, P. A., & Hemmers, R. H. (1994). Space mapping technique for electromagnetic optimization. *IEEE Transactions on Microwave Theory and Techniques, 42*(12), 2536–2544. doi:10.1109/22.339794.

Bandler, J. W., & Rayas-Sanchez, J. E. (2023). An early history of optimization technology for automated design of microwave circuits. *IEEE Journal of Microwaves, 3*(1), 319–337. https://www.ieee.org/membership-catalog/productdetail/showProductDetailPage.html?product=PER421-ELE. doi:10.1109/JMW.2022.3225012.

Bimana, A., & Sinha, S. (2023). Genetic algorithm for microwave computer-aided design: The state of the art. *IEEE AFRICON Conference*. Institute of Electrical and Electronics Engineers Inc. http://ieeexplore.ieee.org/xpl/conhome.jsp?punumber=1000025. doi:10.1109/AFRICON55910.2023.10293468.

Cadence Design Systems. (2024). Cadence Design Systems: Leading the charge in AI-driven semiconductor design. Medium. Retrieved from https://medium.com/@247newsbrew/cadence-design-systems-leading-the-charge-in-ai-driven-semiconductor-design-bc8f92c49cba

Chen, Y. M. (2003). Passive filter design using genetic algorithms. *IEEE Transactions on Industrial Electronics, 50*(1), 202–207. doi:10.1109/TIE.2002.807664.

Cheng, Q. S., Bandler, J. W., & Rayas-Sánchez, J. E. (2008). Tuning-aided implicit space mapping. *International Journal of RF and Microwave Computer-Aided Engineering, 18*(5), 445–453. doi:10.1002/mmce.20303.

Deb, A., Roy, J. S., & Gupta, B. (2014). Performance comparison of differential evolution, particle swarm optimization and genetic algorithm in the design of circularly polarized microstrip antennas. *IEEE Transactions on Antennas and Propagation, 62*(8), 3920–3928. doi:10.1109/TAP.2014.2322880.

European Telecommunications Standards Institute. (2024). *Standards*. Retrieved from https://www.etsi.org/standards.

Federal Communications Commission. (2024). *Radio frequency safety*. Retrieved from https://www.fcc.gov/general/radio-frequency-safety-0.

Gholami, A., & Nezamoddini, N. (2019). Integrated genetic algorithm and artificial neural network. In *IEEE International Conference on Computational Science and Engineering (CSE) and IEEE International Conference on Embedded and Ubiquitous Computing (EUC)* (pp. 260–262). IEEE.

Howe, H. (1974). *Stripline circuit design* (pp. 94–110). Artech House. ISBN: 9780890060207.

Hung, R., & Tong, S. (2009). Evolving product unit neural networks with particle swarm optimization. In *2009 Fifth International Conference on Image and Graphics* (pp. 624–628). IEEE.

Janson, D. J., & Frenzel, J. F. (1993). Training product unit neural networks with genetic algorithms. *IEEE Expert, 8*(5), 26–33. doi:10.1109/64.236478.

Kennedy, J., & Eberhart, R. (1995). Particle swarm optimization. In *Proceedings International Conference on Neural Networks (ICNN'95)* (pp. 1942–1948). Perth, WA, Australia: IEEE. doi:10.1109/ICNN.1995.488968.

Keysight. (2024a). *Differences in application between power dividers and power splitters. Application note, 1–7.* Keysight Technologies.

Keysight. (2024b). *Keysight EEs of EDA Advanced Design System circuit design cookbook 2.0.* (pp. 5-10) Keysight Technologies.

Levy, R., & Lind, L. F. (1968). Synthesis of symmetrical branch-guide directional couplers. *IEEE Transactions on Microwave Theory and Techniques, 16*(2), 80–89. doi:10.1109/TMTT.1968.1126612.

Li, Y., Feng, F., Zhang, J., Liu, W., Ma, K., & Zhang, Q. J. (2024). Design of fifth-order Chebyshev bandpass filter using group delay of the reflection coefficient based on space mapping. In *IEEE International Conference on Computational Electromagnetics, ICCEM 2024 – Proceedings*. China: Institute of Electrical and Electronics Engineers Inc http://ieeexplore.ieee.org/xpl/mostRecentIssue.jsp?punumber=10558851. doi:10.1109/ICCEM60619.2024.10559187.

Ma, J., Dang, S., Li, P., Watkins, G. T., Morris, K., & Beach, M. (2023). A learning-based methodology for microwave passive component design. *IEEE Transactions on Microwave Theory and Techniques, 71*(7), 3037–3050. http://ieeexplore.ieee.org/xpl/tocresult.jsp?isYear=2009&isnumber=4747395&Submit32=View+Contents. doi:10.1109/TMTT.2023.3238418.

Marks, M., Haupt, C. E., & Chatbots, A. I. (2023). Health privacy, and challenges to HIPAA compliance. *JAMA, 330*(4), 309–310. http://jama.jamanetwork.com/journal.aspx. doi:10.1001/jama.2023.9458.

McQueen, D. (2024). Integrating an AI-powered radio for supercharging device performance. *RCR Wireless News*. Retrieved from https://www.rcrwireless.com/20241009/ai-ml/integrating-an-ai-powered-radio-for-supercharging-device-performance.

Mehradfar, A., Zhao, X., Niu, Y., Babakniya, S., Alesheikh, M., Aghasi, H., & Avestimehr, S. (2024). AICircuit: A multi-level dataset and benchmark for AI-driven analog integrated circuit design. *arXiv*. https://arxiv.org. doi:10.48550/arXiv.2407.18272.

Melgarejo, J. C., Ossorio, J., Rubio, D., Cogollos, S., Guglielmi, M., & Boria, V. E. (2022). Recent advances on aggressive space mapping techniques for waveguide filters design and tuning. In *IEEE MTT-S International Microwave Symposium Digest* (pp. 121–124). Institute of Electrical and Electronics Engineers Inc. doi:10.1109/IMS37962.2022.9865299.

Mitchell, M., & Forrest, S. (1994). Genetic algorithms and artificial life. *Artificial Life, 1*(3), 267–289. doi:10.1162/artl.1994.1.267.

Mohammed, H. J., Abdulsalam, F., Abdulla, A. S., Ali, R. S., Abd-Alhameed, R. A., Noras, J. M., Abdulraheem, Y. I., Ali, A., Rodriguez, J., & Abdalla, A. M. (2016). Evaluation of genetic algorithms, particle swarm optimisation, and firefly algorithms in antenna design. *2016 13th International Conference on Synthesis, Modeling, Analysis and Simulation Methods and Applications to Circuit Design, SMACD 2016*. Institute of Electrical and Electronics Engineers Inc. doi:10.1109/SMACD.2016.7520747.

Muraguchi, M., Yukitake, T., & Naito, Y. (1983). Optimum design of 3-Db branch-line couplers using microstrip lines. *IEEE Transactions on Microwave Theory and Techniques, 31*(8), 674–678. doi:10.1109/TMTT.1983.1131568.

Ohira, M., Yamashita, A., Ma, Z., & Wang, X. (2018). Automated microstrip bandpass filter design using feedforward and inverse models of neural network. In *Asia-Pacific Microwave Conference Proceedings, APMC* (pp. 1292–1294). Institute of Electrical and Electronics Engineers Inc. https://ieeexplore.ieee.org/xpl/conhome/1000455/all-proceedings. doi:10.23919/APMC.2018.8617627.

Ohira, M., Asai, Y., & Ma, Z. (2023). A deep-reinforcement-learning assisted microstrip BPF design approach for multiple specifications. In *Asia-Pacific Microwave Conference Proceedings, APMC* (pp. 288–290). Institute of Electrical and Electronics Engineers Inc. https://ieeexplore.ieee.org/xpl/conhome/1000455/all-proceedings. doi:10.1109/APMC57107.2023.10439910.

Oluyemi, O., Laforge, P., & Bais, A. (2023). Space mapping technique on microwave filter circuit model using artificial neural network for parameter extraction. In *Canadian Conference on Electrical and Computer Engineering* (pp. 337–341). Institute of Electrical and Electronics Engineers Inc. http://ieeexplore.ieee.org/xpl/conhome.jsp?punumber=1000225. doi:10.1109/CCECE58730.2023.10288716.

Ossorio, J., Vague, J., Boria, V. E., & Guglielmi, M. (2017). Efficient implementation of the aggressive space mapping technique for microwave filter design. In *European Microwave Week 2017: "A Prime Year for a Prime Event", EuMW 2017 – Conference Proceedings; 47th European Microwave Conference, EuMC 2017* (pp. 644–647). Institute of Electrical and Electronics Engineers Inc. doi:10.23919/EuMC.2017.8230930.

Palmes, P. P., Hayasaka, T., & Usui, S. (2005). Mutation-based genetic neural network. *IEEE Transactions on Neural Networks, 16*(3), 587–600. doi:10.1109/tnn.2005.844858.

Reed, J. (1958). The multiple branch waveguide coupler. *IEEE Transactions on Microwave Theory and Techniques, 6*(4), 398–403. doi:10.1109/TMTT.1958.1125213.

Rosa, S. (2024). *Keysight introduces Quantum Circuit Simulation the first circuit environment with frequency-domain flux quantization.* Keysight Technologies Retrieved from https://www.keysight.com/my/en/about/newsroom/news-releases/2024/0917-pr24-116-keysight-introduces-quantum-circuit-simulation-the-first-circuit-environment-with-frequency-domain-flux-quantization.html.

Sans, M., Selga, J., Rodriguez, A., Bonache, J., Boria, V. E., & Martin, F. (2014). Design of planar wideband bandpass filters from specifications using a two-step aggressive space mapping (ASM) optimization algorithm. *IEEE Transactions on Microwave Theory and Techniques, 62*(12), 3341–3350. http://ieeexplore.ieee.org/xpl/tocresult.jsp?isYear=2009&isnumber=4747395&Submit32=View+Contents. doi:10.1109/TMTT.2014.2365477.

Sythoff, J. (2023). *The Total Economic Impact™ of Keysight Eggplant test automation software: Cost savings and business benefits enabled by Eggplant test* (pp. 1–19). Forrester Research, Inc.

Takahashi, D. (2023). *Synopsys.ai Copilot accelerates AI-driven chip design.* VentureBeat. Retrieved from https://venturebeat.com/ai/synopsys-ai-copilot-accelerates-ai-driven-chip-design/.

Tang, W., Zhou, P., Chen, X., & Chow, Y. L. (2006). Design and optimization of microstrip passive circuits by field-based equivalent circuit model with aggressive space mapping. In *IEEE 2007 International Symposium on Microwave, Antenna, Propagation and EMC Technologies for Wireless Communications, MAPE* (pp. 909–912). https://ieeexplore.ieee.org/document/4393776. doi:10.1109/MAPE.2007.4393776.

Temes, G. C., & Calahan, D. A. (1967). Computer-aided network optimization the state-of-the-art. *Proceedings of the IEEE, 55*(11), 1832–1863. doi:10.1109/PROC.1967.6015.

Weile, D. S., & Michielssen, E. (1997). Genetic algorithm optimization applied to electromagnetics: A review. *IEEE Transactions on Antennas and Propagation, 45*(3), 343–353. doi:10.1109/8.558650.

Yeung, S. H., Chan, W. S., Ng, K. T., & Man, K. F. (2012). Computational optimization algorithms for antennas and RF/microwave circuit designs: An overview. *IEEE Transactions on Industrial Informatics, 8*(2), 216–227. doi:10.1109/TII.2012.2186821.

You, K. Y., Al-Areqi, N., Chong, J. C., Lee, K. Y., Cheng, E. M., & Lee, Y. S. (2018). Analytical modeling of conventional and miniaturization three-section branch-line couplers. *Journal of Electrical Engineering and Technology, 13*(2), 858–867. http://www.jeet.or.kr/ltkPSWeb/pub/pubfpfile.aspx?ppseq=2026. doi:10.5370/JEET.2018.13.2.858.

You, K. Y., & Sim, M. S. (2025). Integrating artificial neural networks in microwave electronic design automation. In *Expert artificial neural network applications for science and engineering* (pp. 51–146). IGI Global. https://www.igi-global.com/book/expert-artificial-neural-network-applications/347395. doi:10.4018/979-8-3693-7250-0.ch003.

You, K. Y. (2016). *RF coaxial slot radiators: Modeling, measurements, and applications*: 285. Artech House ISBN: 9781608078226.

You, K. Y., Al-Areqi, N. N., Lee, C. Y., & Lee, Y. S. (2020). Computer-aided design and applications of planar branch-line coupler circuits. In J. Zbitou, C. Pruncu, & A. Errkik (Eds.), *Handbook of research on recent developments in electrical and mechanical engineering* (pp. 1–63). IGI Global. doi:10.4018/978-1-7998-0117-7.ch001.

Zhang, Q. J., Bandler, J., Koziel, S., Kabir, H., & Zhang, L. (2010). ANN and space mapping for microwave modelling and optimization. In *IEEE MTT-S International Microwave Symposium Digest* (pp. 980–983). Canada. doi:10.1109/MWSYM.2010.5515941.

Zhang, W., Feng, F., Liu, W., Yan, S., Zhang, J., Jin, J., & Zhang, Q. J. (2021). Advanced parallel space-mapping-based multiphysics optimization for high-power microwave filters. *IEEE Transactions on Microwave Theory and Techniques, 69*(5), 2470–2484. http://ieeexplore.ieee.org/xpl/tocresult.jsp?isYear=2009&isnumber=4747395&Submit32=View+Contents. doi:10.1109/TMTT.2021.3065972.

Chapter 12

Bearing fault diagnosis via QPSO algorithm under strong noise background and complex variable speed conditions

Tao Gong[a], Zhongqiu Wang[2], Jianhua Yang[a], Shangyuan Li[a] and Zhen Shan[b]
[a] *Jiangsu Key Laboratory of Mine Mechanical and Electrical Equipment, School of Mechatronic Engineering, China University of Mining and Technology, Xuzhou, China,* [b] *School of Computer Science and Technology, China University of Mining and Technology, Xuzhou, China*

12.1 Introduction

Bearings are key components in rotating machinery, extensively utilized across various mechanical equipment. Because of their complex operating environment and harsh operating conditions, bearings are prone to failure, and their operating status directly affects the reliability and safety of the entire equipment. Early-stage fault signals are often weak and easily corrupted with strong background noise. Therefore it is of great engineering value and practical significance to extract weak fault features of bearings under strong noise background.

During the mechanical equipment fault diagnosis the variable speed condition is the most typical speed condition. Strictly speaking, all mechanical equipment operates under variable speed conditions, where rotational speed variations make traditional signal processing methods ineffective, making it difficult to accurately identify fault features. The fault feature extraction at an early stage under such conditions holds significant importance. Currently the majority of equipment primarily operates under variable speed conditions, where frequent speed variations result in nonstationary vibration signals. For instance the mine hoist frequently experiences acceleration, steady speed, and deceleration conditions during the lifting process, causing frequent impacts on bearings (Zhao et al., 2024). Industrial robots, widely employed in manufacturing assembly and welding because of their high efficiency and capability to perform repetitive and hazardous tasks, also experience substantial joint impacts, making bearings prone to damage (Chen et al., 2024a; Kim et al., 2020; Yin et al., 2023). Consequently, research on fault diagnosis under variable speed conditions has practical significance.

There are two types of main methods for fault diagnosis under variable speed conditions: time-frequency analysis method (Feng et al., 2013; Li et al., 2023) and order analysis method (Lu et al., 2019a; Sapena-Bano et al., 2016). The former mainly utilizes the advantage of time-frequency representations in displaying both temporal and spectral characteristics of signals, enabling the observation of frequency variations over time in the signal. The basic principle is to arrange the vibration signal in both time and frequency domains. Through time-frequency diagram the rotational frequency, harmonic components, and fault characteristics frequency of the signal can be visually observed, thereby enabling the identification of bearing fault features. However, time-frequency analysis methods are computationally intensive and complex in implementation. The methods based on time-frequency analysis mainly include Wigner–Ville distribution (Chen & Li, 2024), short-time Fourier transform (Li et al., 2020), wavelet time-frequency analysis (Arts & van den Broek, 2022), and generalized S-transform (Wang et al., 2023). The latter mainly utilizes the idea of signal conversion to collect speed pulses while collecting vibration signals. Through the angular resampling technique, nonstationary time-domain signals are transformed into stationary

Computational Intelligence in Mechatronics: Solving Real-World Problems in Electronic Systems Design. DOI: https://doi.org/10.1016/B978-0-443-27580-7.00030-5

angular-domain signals, which is similar to vibration signals under constant speed conditions. Traditional analysis methods can be used to identify and judge faults.

Compared with the time-frequency analysis method the order analysis method is more convenient, accurate, and fast in calculation, which is widely used in fault diagnosis. Order analysis methods are primarily categorized into hardware order analysis (Berry, 1991), computed order analysis (COA) (Cheng et al., 2014; Fyfe & Munck, 1997; Gong et al., 2022), and tacholess order analysis (Choudhury et al., 2021; Schmidt et al., 2018). The hardware order analysis is achieved by sampling the pulse signal of the tachometer via hardware circuits. However, this method is limited by hardware in extracting fault characteristics during rapid speed variations and incurs higher costs. Tacholess order analysis, which estimates rotational speed by extracting instantaneous frequency components from vibration signals, offers the advantage of not needing the tachometer and low cost. Nevertheless, this method demonstrates poor resistance to noise. In summary, COA technology overcomes the limitations and complexity associated with hardware order analysis. Compared to tacholess order analysis, it exhibits strong resistance to environmental noise and can effectively extract fault characteristic information during speed fluctuations.

Stochastic resonance (SR) (Lu et al., 2019b; Yang et al., 2024) is a dynamic phenomenon in which the system output is significantly enhanced under the synergism of the weak low-frequency signal, noise, and nonlinear system, manifesting as the enhancement of the weak signal. Traditional weak signal detection methods (Buzzoni et al., 2018; Cheng et al., 2024; Hou et al., 2023; Martins et al., 2023) aim at using various principles to suppress noise as much as possible and then extracting useful signals. However, SR is different from these methods in that it enhances the weak feature signal by converting part of the noise energy into signal energy through a nonlinear system and converts the noise energy into the energy of the required characteristic signal. This type of method can not only enhance the signal but also effectively suppress interference noise. The fault characteristic signals related to mechanical equipment failures are often submerged in strong background noise. Traditional filtering and denoising techniques may filter out useful feature information, whereas SR not only does not remove useful information but also enhances the weak signal. Combined with other signal analysis methods, it can achieve the enhancement of fault signals under strong noise backgrounds and the identification of unknown fault signals. Many scholars have developed the SR method in fault diagnosis, He et al. (2024a) proposed a cascaded segmented unsaturated asymmetric underdamped tri-stable SR system and derived the theoretical formula for the spectral amplification factor under this system based on the adiabatic approximation theory. The output results showed good performance in signal detection of bearing inner raceway faults and outer raceway faults. Furthermore, they (He et al., 2024b) constructed an unsaturated piecewise linear quad-stable SR system, which exhibited higher output signal-to-noise ratio (SNR) compared to conventional quad-stable systems, thereby improving the detection of weak fault characteristic signals. Li et al. (2024) proposed a novel feature extraction method based on a coupled resonance system, which enhances SR through vibrational resonance. This system combines the advantages of both vibrational resonance and SR, enabling effective detection of weak characteristic signals. Chen et al. (2024b) discovered a bistable SR system based on fractional exponential power, which effectively overcomes the unsaturated effect of the system. They also defined a new SNR to address the deviation between actual and theoretical characteristic frequencies. Hou et al. (2025) introduced an innovative frequency modulation signal detection method for rub-impact faults of the structural rotor, and the effectiveness of the proposed diagnosis method is verified through simulation and experiments, demonstrating high noise resistance. We (Gong et al., 2024b) proposed a hybrid diagnosis strategy for motor bearings under complex variable speed conditions; two types of typical speed conditions are considered, which are representative of the joint in the industrial robot. Therefore utilizing adaptive SR has great potential in extracting weak fault characteristic signals.

Current research on weak fault feature extraction under complex variable speed conditions remains limited, particularly for wide-range and rapid speed regulation scenarios. Such complex speed conditions pose significant challenges for extracting weak fault features, and strong noise interference often effects the fault feature identification. Moreover, there is a lack of effective signal processing methods for feature extraction. Inspired by previous studies, this research proposes a hybrid fault diagnosis strategy for weak signal extraction, which includes an adaptive SR method based on the optimization algorithm and nonstationary signal processing techniques, to achieve effective extraction and enhancement of weak fault features.

The remainder of this study is organized as follows. Section 12.2 introduces the nonstationary signal extraction and enhancement methods. Section 12.3 presents numerical simulation results demonstrating the extraction of nonstationary weak fault features under complex variable speed conditions. Section 12.4 validates the effectiveness of the proposed method through experimental data analysis for three different fault types of bearings. Finally, relevant conclusions are presented in Section 12.5.

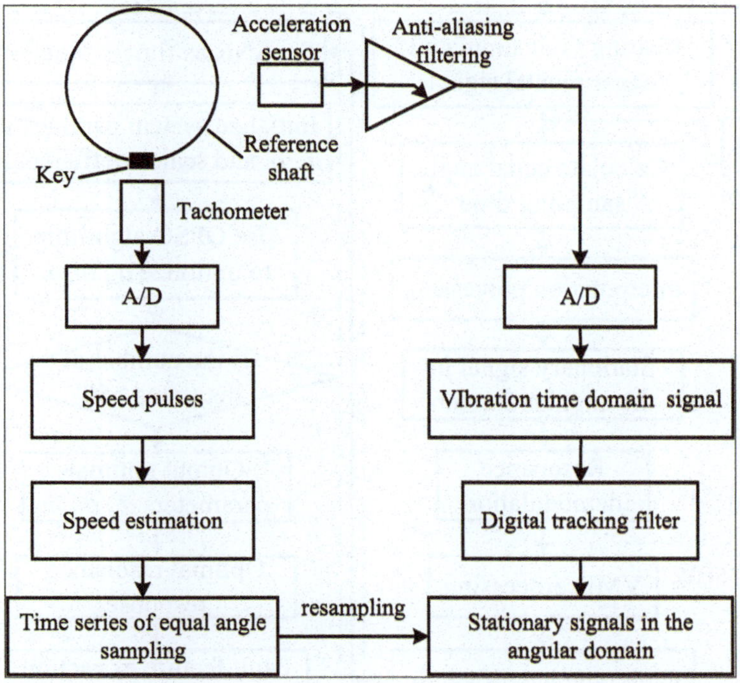

FIGURE 12.1 Schematic diagram of order analysis. *A/D*, Analog-to-digital.

TABLE 12.1 The physical parameters of the 6204-type bearing.

Type	6204
External diameter	47 mm
Bore diameter	20 mm
Pitch diameter	33.5 mm
Width	17 mm
Rolling element diameter	8 mm
Number of rolling elements	8
Contact angle	0

12.2 Nonstationary signal extraction and enhancement methods

12.2.1 Order analysis method for nonstationary signals

COA method is a convenient and effective approach for processing nonstationary vibration signals. In this technique a tachometer is installed on the front side of the rotating shaft to collect the speed signal, whereas an acceleration sensor placed at the fixed end of the shaft collects the vibration signal via consistent angular intervals. This setup enables the transformation of the nonstationary vibration characteristics from the time domain into stationary vibration information in the angular domain. The fundamental principle is illustrated in Fig. 12.1.

The main steps of the algorithm are as follows:

1. Determine the cut-off frequency and sampling order of antialiasing filter.

   To prevent frequency aliasing in the analog-to-digital (A/D) conversion process the sampling frequency must satisfy the Nyquist sampling criterion. Specifically, f_s must be greater than twice the maximum frequency f_{max} of the signal being analyzed. The maximum sampling order, O_{max}, is determined by the maximum frequency, f_{max}, of the signal under analysis, satisfying the following conditions: $f_{max} = R_{max}/60 \times O_{max}$, where R_{max} is the maximum speed of the reference shaft.

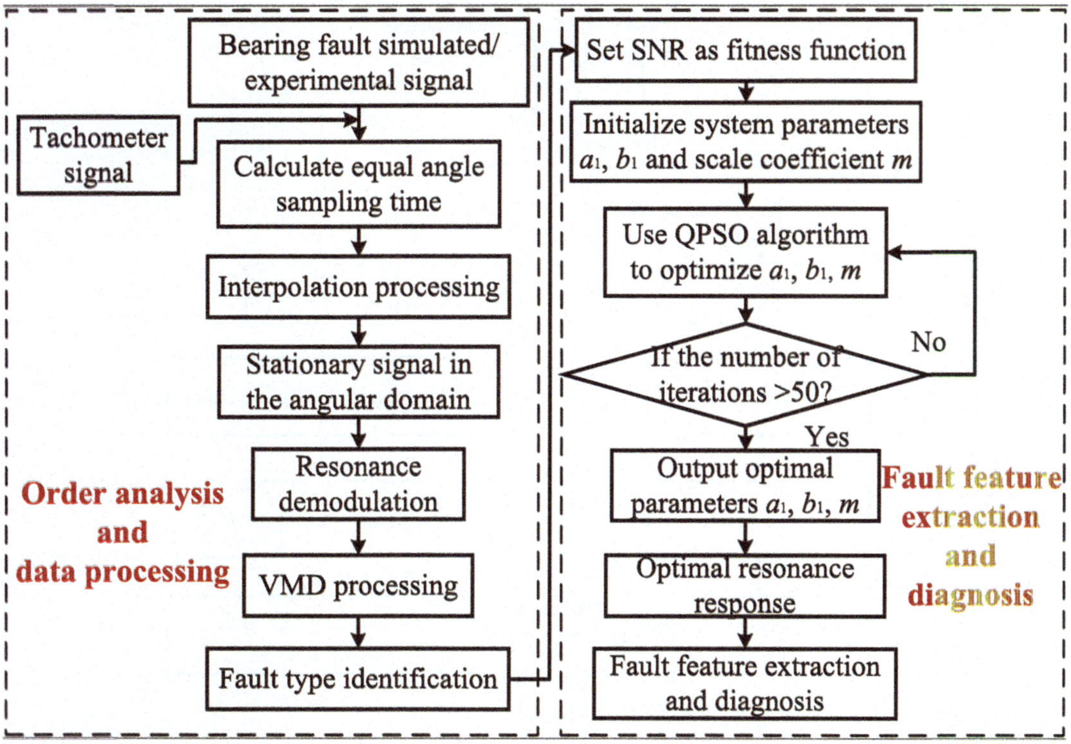

FIGURE 12.2 Technology roadmap for the proposed method. *QPSO*, Quantum particle swarm optimization; *SNR*, signal-to-noise ratio; *VMD*, variational mode decomposition.

2. Determine the resampling time

Assuming that the motion of the rotation shaft in a short time conforms to the law of constant angular acceleration, the cumulative rotation angle $\theta(t)$ of the reference shaft can be expressed as a quadratic polynomial with

$$\theta(t) = b_0 + b_1 t + b_2 t^2, \tag{12.1}$$

where t is the resampling time, and b_0, b_1, and b_2 are unknown parameters.

For the three consecutive pulse time points t_1, t_2, and t_3, the corresponding rotation angles are 0, 2π, and 4π, respectively. This indicates that the pulse is collected once each complete revolution of the reference shaft, which is expressed as

$$\begin{cases} \theta(t_1) = 0 \\ \theta(t_2) = 2\pi . \\ \theta(t_3) = 4\pi \end{cases} \tag{12.2}$$

Substituting Eq. (12.2) into Eq. (12.1), the matrix form is

$$\begin{pmatrix} 0 \\ 2\pi \\ 4\pi \end{pmatrix} = \begin{pmatrix} 1 & t_1 & t_1^2 \\ 1 & t_2 & t_2^2 \\ 1 & t_3 & t_3^2 \end{pmatrix} \begin{pmatrix} b_0 \\ b_1 \\ b_2 \end{pmatrix}. \tag{12.3}$$

Solving Eq. (12.3) to obtain the values of b_0, b_1, and b_2, and substituting them into Eq. (12.1), one obtains

$$t = \frac{1}{2b_2}\left[\sqrt{4b_2(k\Delta\theta - b_0) + b_1^2} - b_1\right], \tag{12.4}$$

where k is the interpolation coefficient, and $\Delta\theta$ is the angle increment.

3. Interpolation filtering for nonstationary signals

Interpolation algorithm is used for amplitude interpolation filtering during resampling, which can realize the transformation from nonstationary signals to stationary signals. The signal is transformed from the time domain to the angular domain, and then the signal component is identified by spectrum analysis.

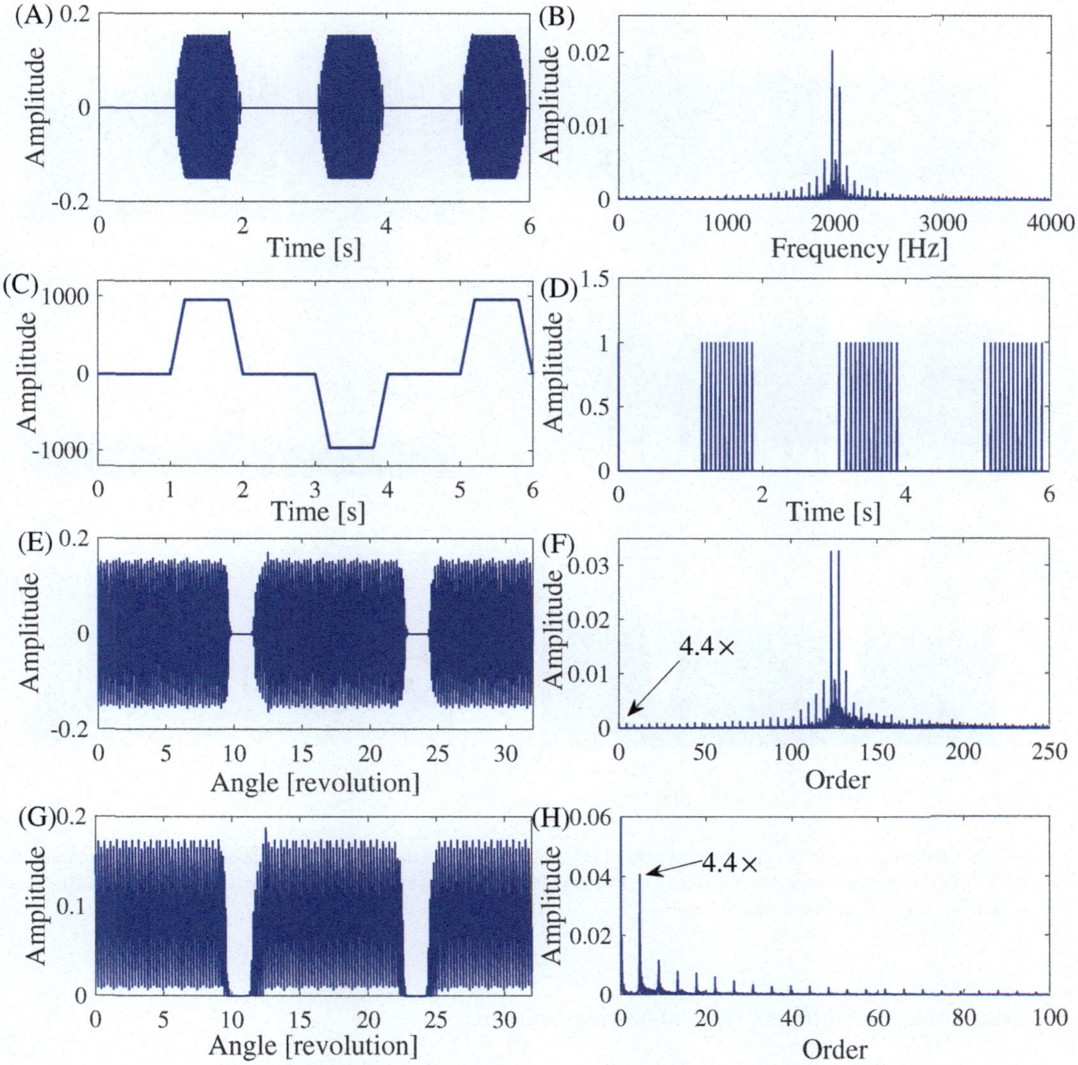

FIGURE 12.3 The simulated signal of the bearing outer raceway fault: (A) the time-domain waveform, (B) the spectrum, (C) the speed curve, (D) the speed pulse signal, (E) the angular-domain waveform, (F) the order spectrum, (G) the angular-domain waveform (by resonance demodulation), and (H) the order spectrum (by resonance demodulation).

COA is an effective method for nonstationary signal processing in fault diagnosis. During this process the acceleration sensor and current sensor synchronously collect the vibration signal and speed signal of the reference shaft. These two kinds of signals are preprocessed, respectively, where the pulse time of angular sampling is determined by the resampling algorithm, and the signal resampling in the angular domain is realized by interpolation algorithm. The nonstationary signal is converted into a stationary signal, and then it can be processed using stationary signal processing method. This method has been widely applied in fault diagnosis of mechanical equipment with its advantages of convenience, low cost, and high reliability. During the analysis process a shaft is selected as the reference axis, and its rotational frequency is used as the reference. The order is defined as the multiple of the rotational frequency, as

$$O_c = \frac{f_c}{f_r} = \frac{f_c}{n_r/60}, \tag{12.5}$$

where O_c and f_c denote the characteristic order and characteristic frequency, respectively. f_r and n_r represent the rotational frequency and speed of the reference shaft. The characteristic order reflects vibration characteristics unrelated to the rotational frequency, serving as an indicator of the motion condition of the shaft.

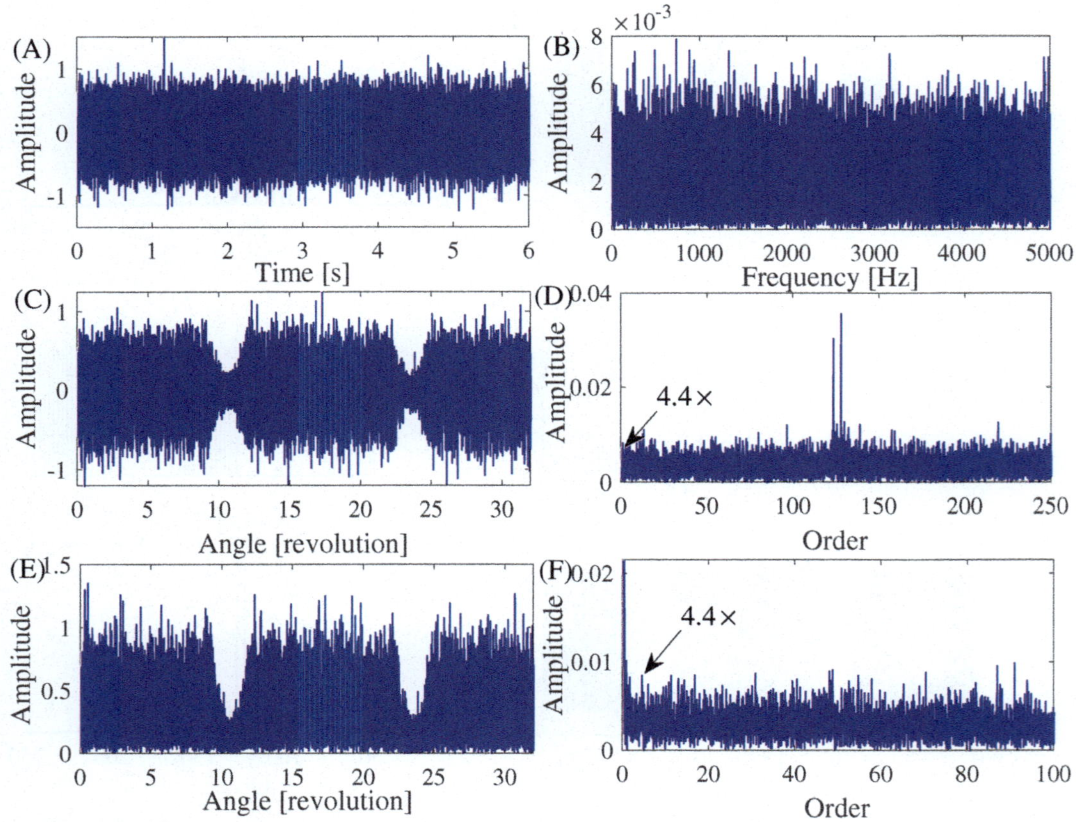

FIGURE 12.4 The simulated signal of the bearing outer raceway fault (with noise): (A) the time-domain waveform, (B) the spectrum, (C) the angular-domain waveform (by COA), (D) the order spectrum (by COA), (E) the angular-domain waveform (by resonance demodulation), and (F) the order spectrum (by resonance demodulation). *COA*, Computed order analysis.

The characteristic orders for different types of bearing faults are

$$\begin{cases} O_i = \frac{f_i}{f_r} = \frac{1}{2}\left(1 + \frac{D_b}{D_c}\cos\beta\right)z \\ O_o = \frac{f_o}{f_r} = \frac{1}{2}\left(1 - \frac{D_b}{D_c}\cos\beta\right)z \\ O_b = \frac{f_b}{f_r} = \frac{1}{2}\frac{D_c}{D_b}\left(1 - \frac{D_b^2}{D_c^2}\cos^2\beta\right) \end{cases}, \tag{12.6}$$

where O_i, O_o, and O_b denote the theoretical characteristic orders of inner raceway fault, outer raceway fault, and rolling element fault, respectively. As shown in Eq. (12.6), these fault characteristic orders for different fault types are independent of the speed conditions for the reference shaft, depending on the intrinsic parameters of the bearing. This method is extensively applied in bearing fault diagnosis under variable speed conditions.

12.2.2 QPSO algorithm for parameter optimization

Quantum particle swarm optimization (QPSO) (Flori et al., 2022; Gong et al., 2024a) algorithm is an emerging intelligent optimization algorithm that utilizes the quantum mechanical principles to enhance the parameter optimization process. This algorithm enables simultaneous optimization of multiple parameters by modeling particle behavior within a quantum framework, effectively overcoming the limitations of the traditional PSO algorithm, particularly their parameter dependency and susceptibility to local optima. Consequently, QPSO has gained significant performance in parameter optimization applications. In this study the SNR is used as the fitness function in the QPSO algorithm. By maximizing the SNR as the optimization objective the algorithm achieves optimal SR response, thereby facilitating optimal resonance enhancement of characteristic signals.

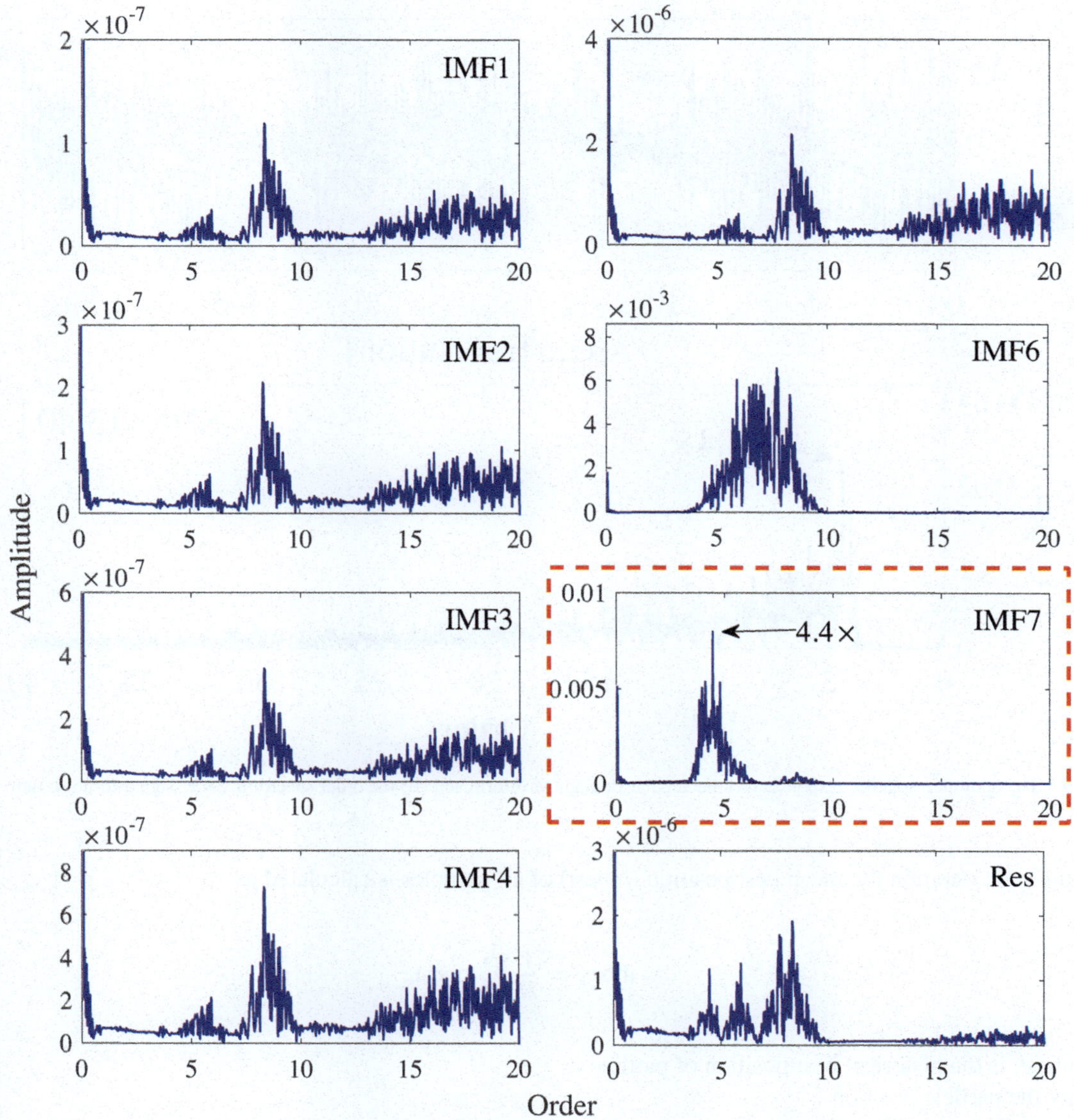

FIGURE 12.5 The order spectra of IMFs decomposed by VMD. *IMFs*, Intrinsic mode functions; *VMD*, variational mode decomposition.

The flowchart of the algorithm is summarized as follows:

1. Parameter initialization

 Population size (N): The number of particles in the swarm.

 Iterations (T): The maximum number of iterations.

 Position (x_i) and velocity (v_i) of individual particles.

 Inertia factor (w) and contraction factor (α): Typically used to control the convergence of particles.

 Individual optimal position ($pbest_i$) and global optimal position ($gbest_i$): The historical best positions of each particle and the entire particle.

2. Introduction of quantum behavior mechanism

 QPSO uses probability distributions in quantum mechanics to describe the state of particles, rather than deterministic velocities and positions. In QPSO the state of each particle can be represented as a probability cloud, which contains positional uncertainty.

3. Calculate the average gravitational center of an individual

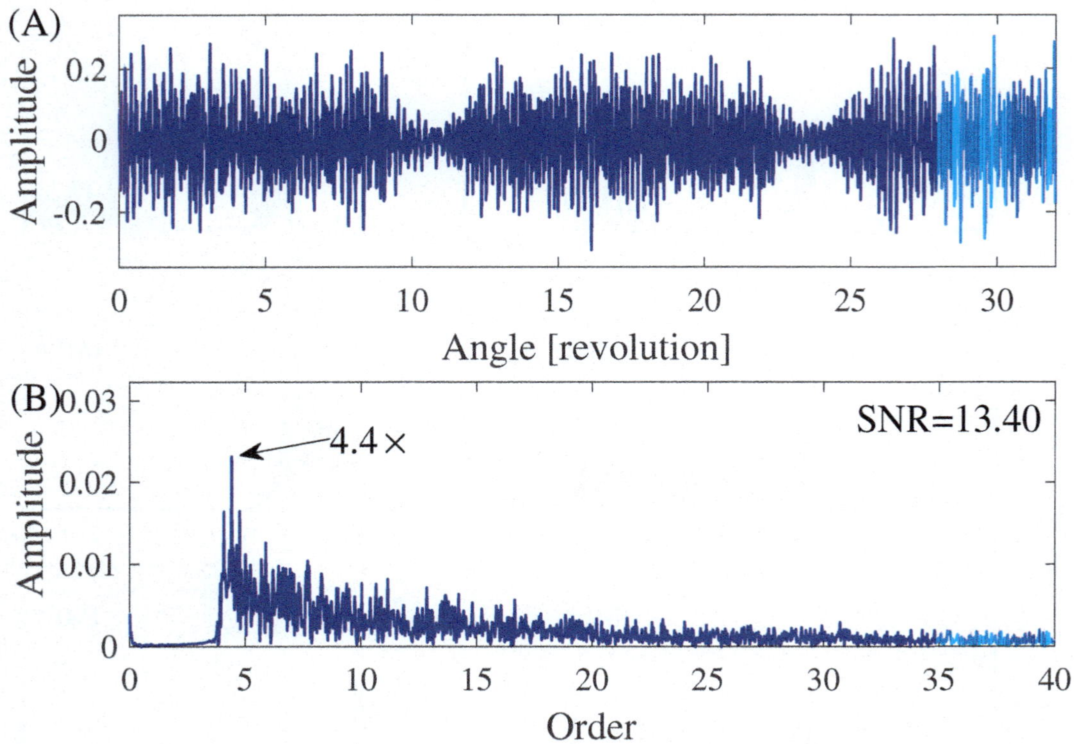

FIGURE 12.6 The optimal resonance response: (A) the angular-domain waveform and (B) the order spectrum. *SNR*, Signal-to-noise ratio.

During each iteration the mean best position (mbest) of all particles is calculated as

$$\text{mbest} = \frac{1}{N}\sum_{i=1}^{N} \text{pbest}_i, \tag{12.7}$$

where pbest_i is the historical best position of particle i.

4. Updating the particle position

The particle position is updated according to the quantum update rule, as

$$x_i^{t+1} = p_i \pm \gamma \cdot \left|\text{mbest} - x_i^t\right| \cdot \ln\left(\frac{1}{r}\right), \tag{12.8}$$

where p_i represents the individual optimal position (pbest_i), γ denotes the contraction factor that controls the search scope, and r is a random number uniformly distributed in the interval [0, 1].

5. Fitness evaluation

For each updated particle position x_i the corresponding fitness value is evaluated. In the optimization problems, this value typically represents the objective function value.

6. Updating the individual and global best position

If the current fitness value of particle i is better than its historical best fitness value, update the pbest of the particle. If the current fitness value of particle i is better than the gbest of the entire swarm, update the gbest.

7. Determining the termination condition

The algorithm terminates if the maximum number of iterations T is reached or if the change in fitness value satisfies a specified precision requirement. Otherwise, return to step 3 and proceed to the next iteration.

8. Outputting the results

When the termination condition is satisfied, output the gbest and its corresponding fitness value, which represent the optimal solution of the algorithm.

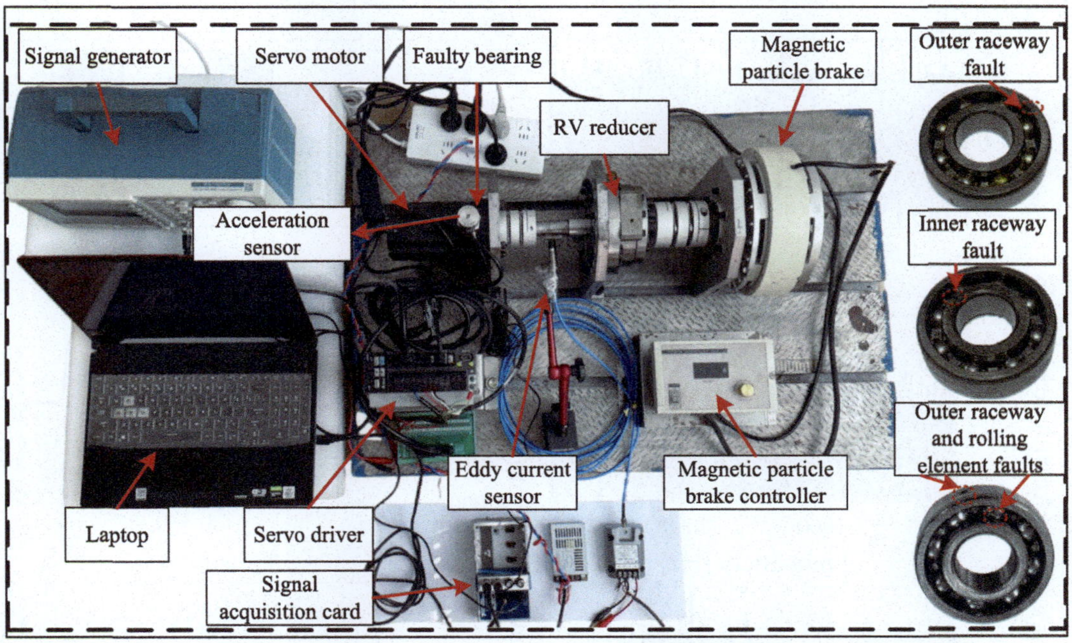

FIGURE 12.7 Experimental rig.

12.2.3 Variational mode decomposition (VMD)

Variational mode decomposition (VMD) constructs and solves constrained variational problems, using the nonrecursive processing strategy to decompose the target signal into several intrinsic mode functions (IMFs). By selecting and identifying IMF components, it becomes possible to identify and extract bearing fault characteristic signals under noise background. Compared with empirical mode decomposition, VMD has better accuracy and noise resistance in the decomposition of complex data.

Assuming that the original signal F_1 can be decomposed into l IMF components, the constrained model is shown as

$$\begin{cases} \min\limits_{\{u_k\},\{\omega_k\}} \left\{ \sum_l \left\| \partial t\left[\left(\delta(t) + \frac{j}{\pi t} \right) \times u_l(t) \right] e^{-j\omega_l t} \right\|_2^2 \right\}, \\ \text{s.t.} \sum_l u_l = F_1 \end{cases} \tag{12.9}$$

where l, $\{u_l\}$, and $\{\omega_l\}$ are the number of decomposed modes, the lth component obtained through decomposition, and the center frequency of that component, respectively.

Introducing the quadratic penalty factor α and Lagrange multiplication operator λ to transform the constrained problem into an unconstrained variational problem, one gets

$$L(\{u_l\}, \{\omega_l\}, \lambda) = \alpha \sum_l \left\| \partial t\left[\left(\delta(t) + \frac{j}{\pi t} \right) \times u_l(t) \right] e^{-j\omega_l t} \right\|_2^2 + \left\| F_1(t) - \sum_l u_l(t) \right\|_2^2 + \left\langle \lambda(t), F(t) - \sum_l u_l(t) \right\rangle. \tag{12.10}$$

The specific process of this algorithm is as follows:

1. Initialize $\{u_l^1\}$, $\{\omega_l^1\}$, λ^1, $h = 0$, where $\{u_l^1\}$ and $\{\omega_l^1\}$ are the initial value and the center frequency of the lth IMF component, respectively. λ^1 is the initial value of the Lagrange multiplication operator, and h is the number of iterations.
2. Execute $h = h + 1$ and loop.
3. Update u_l and ω_l the updated equation is

$$\hat{u}_l^{h+1}(\omega) \leftarrow \frac{\widehat{F_1}(\omega) - \sum_{q<l} \hat{u}_q^{h+1}(\omega) + \sum_{q>1} \hat{u}_q^h(\omega) + \hat{\lambda}^h(\omega)/2}{1 + 2\alpha\left(\omega - \omega_l^h\right)^2}, \tag{12.11}$$

and

$$\omega_l^{h+1} \leftarrow \frac{\int_0^\infty \omega \left| \hat{u}_l^{h+1}(\omega) \right|^2 d\omega}{\int_0^\infty \left| \hat{u}_l^{h+1}(\omega) \right|^2 d\omega}, \tag{12.12}$$

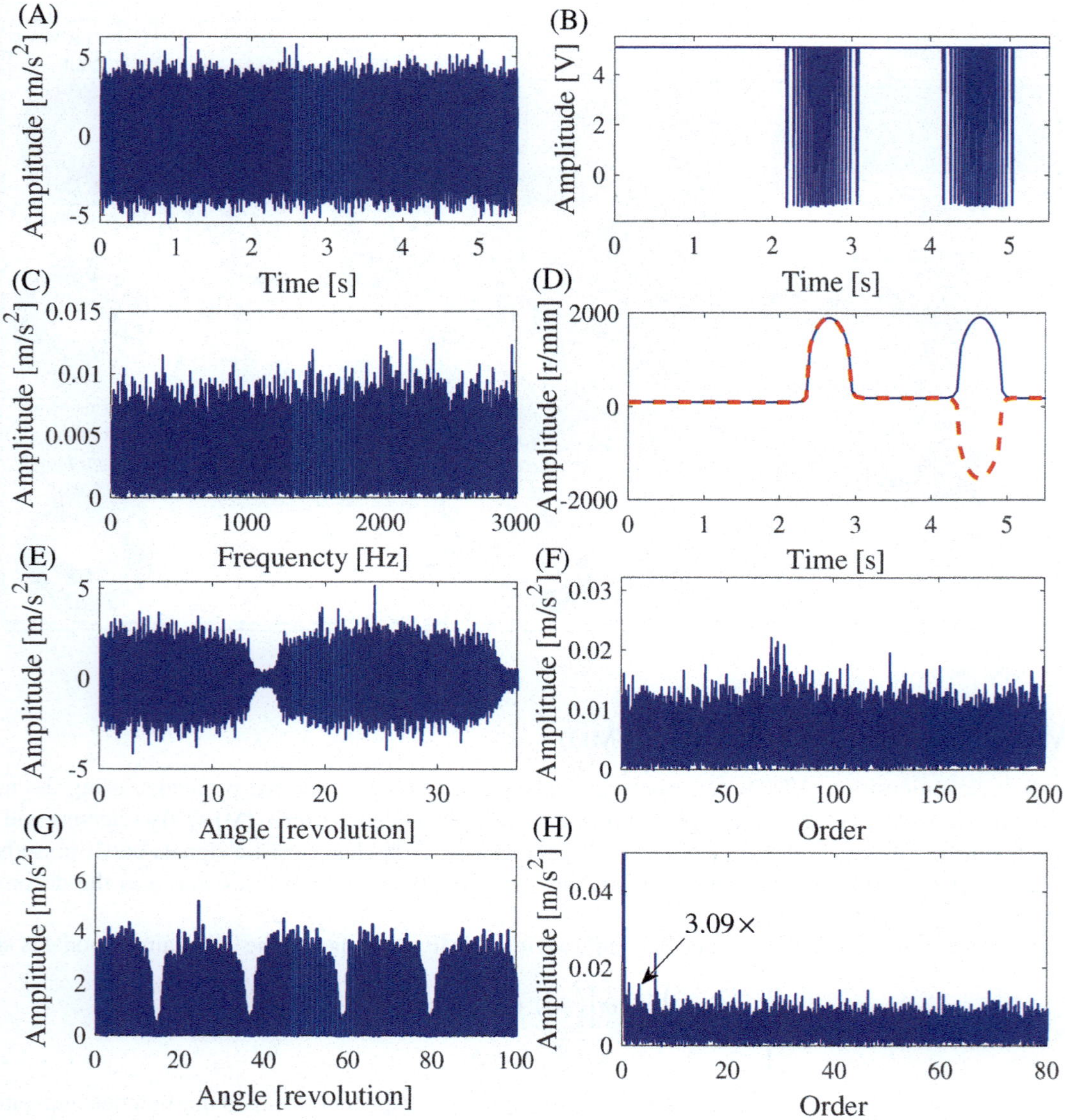

FIGURE 12.8 Experimental signal with its fault on the bearing outer raceway: (A) the time-domain waveform, (B) the speed pulse signal, (C) the spectrum, (D) the speed signal, (E) the angular-domain waveform (by COA), (F) the order spectrum (by COA), (G) the angular-domain waveform (by resonance demodulation), and (H) the order spectrum (by resonance demodulation). *COA*, Computed order analysis.

where $\hat{u}_l^{h+1}(\omega)$, $\hat{F}_1(\omega)$, $\hat{u}_q^{h+1}(\omega)$, $\hat{u}_q^{h}(\omega)$, and $\hat{\lambda}^h(\omega)$ are the Fourier transforms of $u_l^{h+1}(t)$, $F(t)$, $u_q^{h+1}(t)$, $u_q^{h}(t)$, and $\hat{\lambda}^h(t)$. $u_l^{h+1}(t)$ and $u_q^{h+1}(t)$ are the $(h+1)$ updated results of the lth and qth IMF components. $u_q^h(t)$ and $\hat{\lambda}^h(t)$ are the hth updated result of the qth IMF component and the hth updated result of λ, respectively. ω_l^h and ω_l^{h+1} are the hth and $(h+1)$th updated results of the center frequency for the lth IMF component, respectively. When the number of decompositions reaches l, the loop is exited and the process proceeds to the next step.

4. Update λ and update the equation to

$$\hat{\lambda}^{h+1}(\omega) \leftarrow \hat{\lambda}^h(\omega) + \nu\left(\widehat{F_1}(\omega) - \sum_l \hat{u}_l^{h+1}(\omega)\right), \tag{12.13}$$

where ν represents the noise margin, which satisfies the fidelity requirements for signal decomposition.

5. Give the precision and when the loop will terminate once the desired precision is achieved; otherwise, it will reiterate step 2.

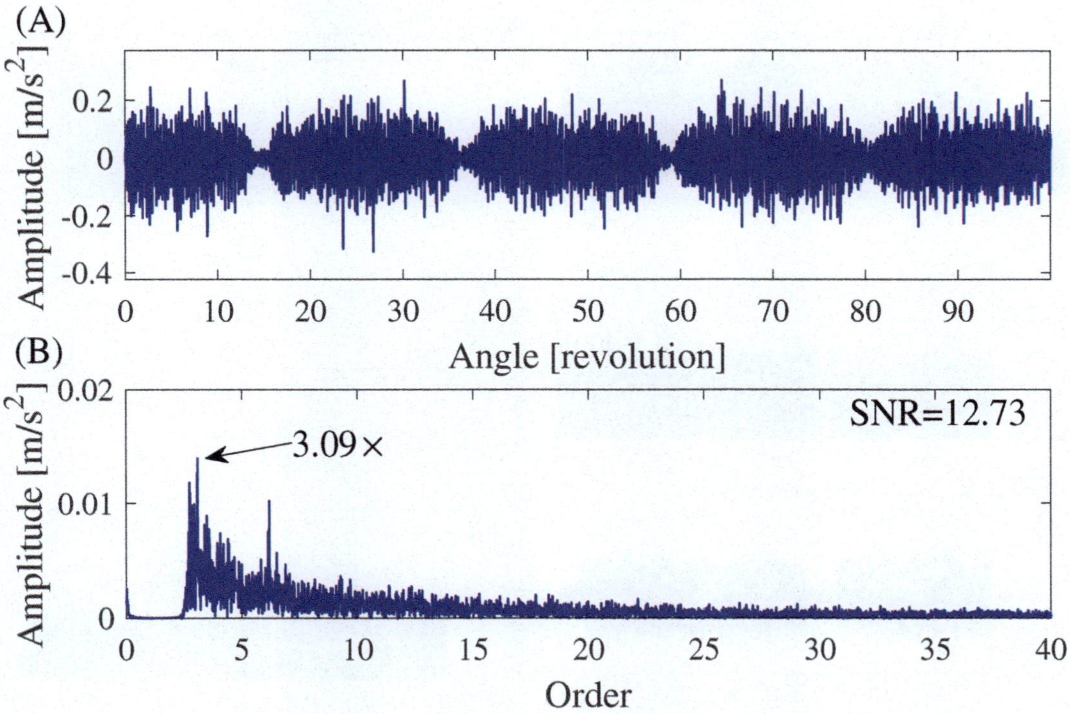

FIGURE 12.9 The optimal resonance response: (A) the angular-domain waveform and (B) the order spectrum. *SNR*, Signal-to-noise ratio.

12.2.4 Adaptive SR method for weak characteristic signals

The SR method is an effective approach for extracting weak characteristic signals under strong noise background. This phenomenon is driven by the synergism of the periodic force, random noise, and nonlinear system. A classical dynamical equation for the bistable system can be expressed as

$$\begin{cases} \frac{dx_0(t)}{dt} = -\frac{\partial U_0(x)}{\partial x} + N_0(t) + s_0(t) \\ U_0(x) = -\frac{a_0}{2}{x_0}^2 + \frac{b_0}{4}{x_0}^4 \\ s_0(t) = A_0 \cos(2\pi f_0 t) \end{cases} , \tag{12.14}$$

where $x_0(t)$ represents the system response, and $U_0(x)$ is the potential function of the bistable system. $a_0 > 0$ and $b_0 > 0$ both are system parameters. $N_0(t) = \sqrt{2D_0}\xi(t)$ denotes Gaussian white noise and D_0 represents the noise intensity. $\xi(t)$ is standard Gaussian white noise with a mean of 0 and a variance of 1. $s_0(t)$ is a low-frequency periodic signal with amplitude A_0 and frequency f_0. The Gaussian white noise satisfies the following statistical characteristics, as

$$\langle N_0(t)\rangle = 0,\ \langle N_0(t), N_0(0)\rangle = 2D\delta(t), \tag{12.15}$$

where $\langle N_0(t)\rangle$ represents the first moment, which is the expectation of the noise. $\langle N_0(t),N_0(0)\rangle$ denotes the second moment, which is the mean square value of the noise. $\delta(t)$ denotes the Dirac delta function.

In practical engineering, signals generally contain high frequencies. The conditions for the occurrence of SR require that the input signal of the system satisfies the small-parameter condition, meaning the signal frequency is less than 1 Hz and signal amplitude is less than 1. Therefore real-world signals often do not satisfy the conditions for SR. To address this issue the scale transformation is introduced.

The high-frequency periodic signal $A\cos(2\pi ft)$ is used to simulate actual periodic signal, and the corresponding dynamics equation in the bistable system is expressed as

$$\frac{dx(t)}{dt} = ax - bx^3 + A\cos(2\pi ft) + \sqrt{2D}\xi(t), \tag{12.16}$$

where parameters a and b represent the large parameters of system ($a \gg 1$, $b \gg 1$). A and f denote the amplitude and frequency of the high-frequency signal. $x(t)$ is the system response.

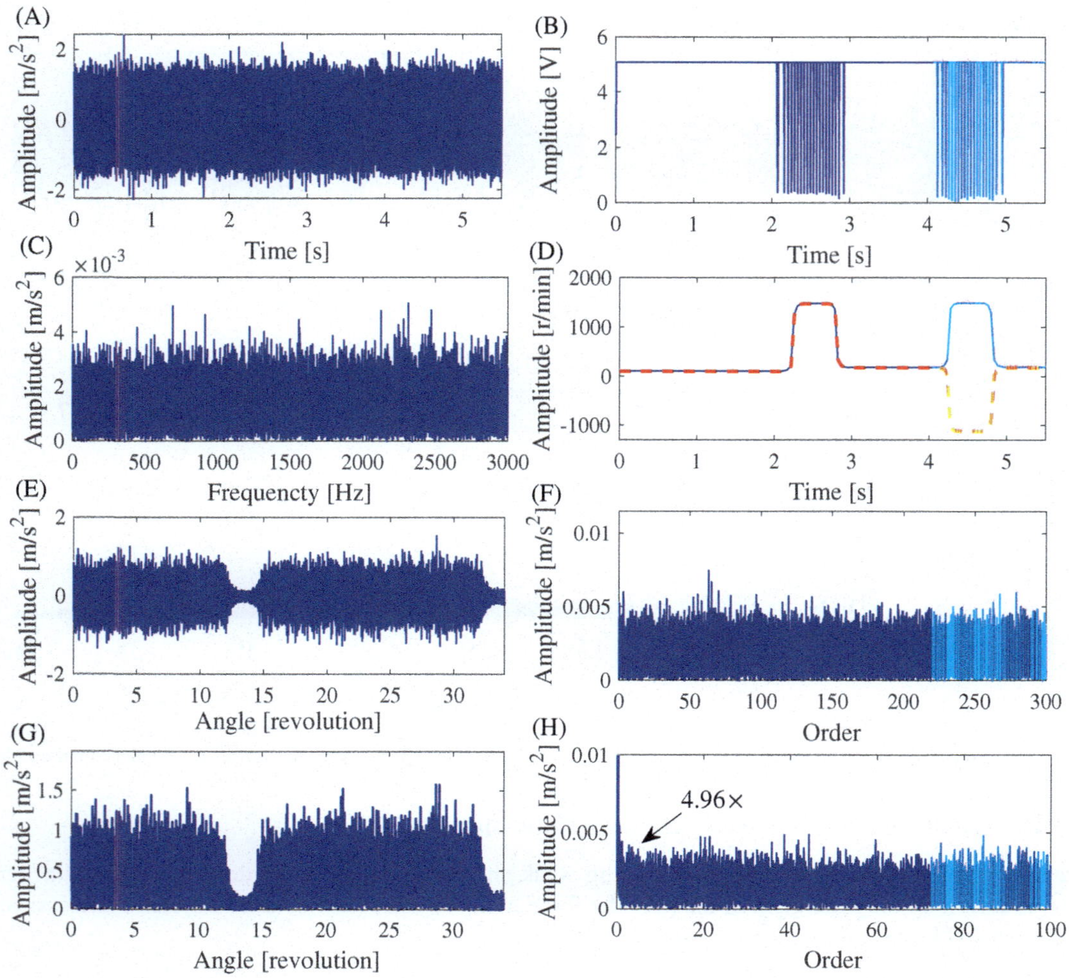

FIGURE 12.10 Experimental signal with its fault on the bearing inner raceway: (A) the time-domain waveform, (B) the speed pulse signal, (C) the spectrum, (D) the speed signal, (E) the angular-domain waveform (by COA), (F) the order spectrum (by COA), (G) the angular-domain waveform (by resonance demodulation), and (H) the order spectrum (by resonance demodulation). *COA*, Computed order analysis.

Introducing the variable substitution $x(t) = z(\tau)$, $\tau = mt$, Eq. (12.16) is expressed as

$$\frac{dz(\tau)}{d\tau} = \frac{a}{m}z(\tau) - \frac{b}{m}z^3(\tau) + \frac{A}{m}\cos\left(\frac{2\pi f \tau}{m}\right) + \sqrt{\frac{2D}{m}}\xi(\tau), \tag{12.17}$$

where τ represents the large time scale, and m denotes the large-scale coefficient ($m \gg 1$).

Letting $a_1 = a/m$, $b_1 = b/m$, and $f_1 = f/m$, Eq. (12.17) is expressed as

$$\frac{dz(\tau)}{d\tau} = a_1 z(\tau) - b_1 z^3(\tau) + \frac{A}{m}\cos(2\pi f_1 \tau) + \sqrt{\frac{2D}{m}}\xi(\tau), \tag{12.18}$$

where $a_1 \ll 1$ and $b_1 \ll 1$ both represent small system parameters. $f_1 \ll 1$ Hz is the characteristic frequency of the low-frequency signal. In Eq. (12.18) the system parameters are significantly less than 1, and the equivalent frequency is less than 1 Hz, thereby satisfying the small-parameter conditions for classical SR. However, compared to Eq. (12.14), the amplitude of the original signal A/m and the noise intensity D/m become $1/m$ times that of the original signal. To satisfy the resonance requirements in terms of amplitude, both the signal amplitude and noise intensity are simultaneously multiplied by m, one obtains

$$\frac{dz(\tau)}{dt} = a_1 z(\tau) - b_1 z^3(\tau) + A\cos(2\pi f_1 \tau) + \sqrt{2D}\xi(\tau). \tag{12.19}$$

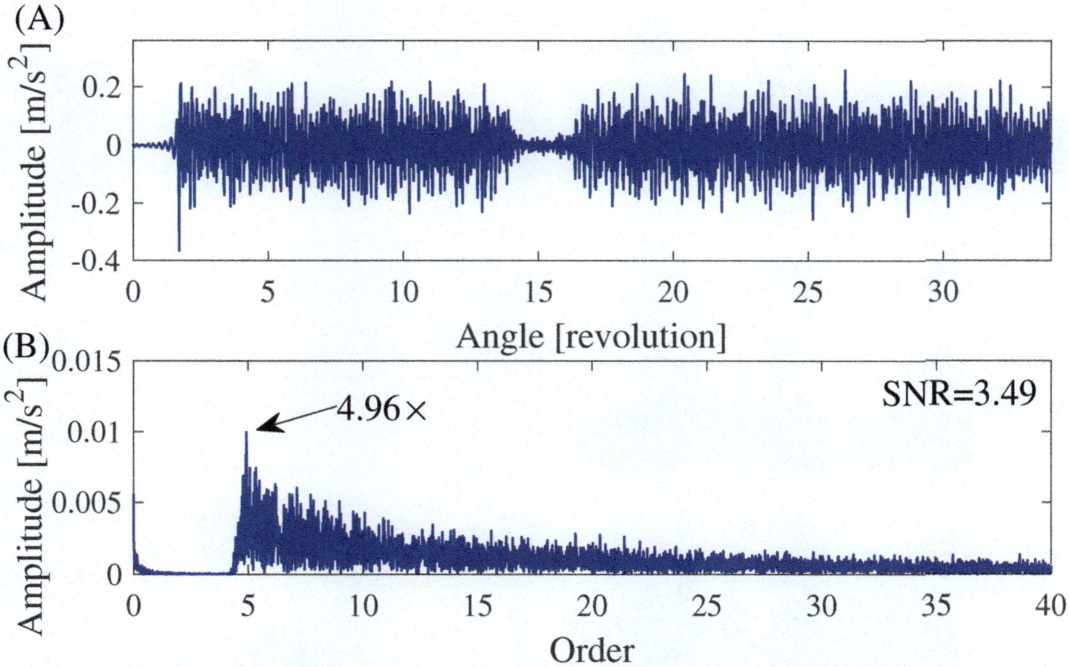

FIGURE 12.11 The optimal resonance response: (A) the angular-domain waveform and (B) the order spectrum. *SNR*, Signal-to-noise ratio.

Eq. (12.19) satisfies the small-parameter conditions of classical SR (signal amplitude and frequency are both far less than 1), and for it to be restored to the time scale t, there is

$$\frac{\mathrm{d}x(t)}{\mathrm{d}t} = ax(t) - bx^3(t) + mA\cos(2\pi ft) + \sqrt{2Dm}\xi(t). \tag{12.20}$$

Eq. (12.20) represents a large-parameter SR model, whose dynamic essence is identical to that of Eq. (12.14), and can achieve SR under high-frequency signal excitation.

As the signal processed by COA is transformed into the angular domain signal, the SNR is defined in the angular domain as the fitness function for optimization, which is similar to the SNR in the time domain, as shown below.

$$\begin{cases} \mathrm{SNR} = 10\log\frac{S(O_c)}{P(O_c)} \\ S(O_c) = |X(p)|^2 \\ P(O_c) = \frac{1}{2M}\sum_{j=1}^{M}\left(|X(p-j)|^2 + |X(p+j)|^2\right) \end{cases}, \tag{12.21}$$

where $S(O_c)$ denotes the power value at the characteristic order O_c, and $X(p)$ represents the amplitude at the pth data point in the spectrum. $P(O_c)$ is the average power within a specified range centered at the characteristic order O_c; p indicates the position index of the characteristic order, and M is the number of data points with left–right symmetry of the characteristic order O_c.

12.2.5 Proposed fault diagnosis method

The overall technology roadmap is shown in Fig. 12.2, mainly divided into the following steps:

Step 1: Use the vibration acceleration sensor to collect the vibration signal of the faulty bearing under variable speed conditions and use an eddy current sensor to synchronously collect the speed pulse signal of the reference shaft.

Step 2: Perform COA on the vibration signal in the time domain. Determine the time of angular resampling through the tachometer signal, select an appropriate resampling frequency, and perform differential processing on the vibration signal to obtain the stationary vibration signal in the angular domain.

Step 3: Resonance demodulation and VMD processing are sequentially applied to the processed angular signal, and then the fault type of the bearing is identified through the IMF components.

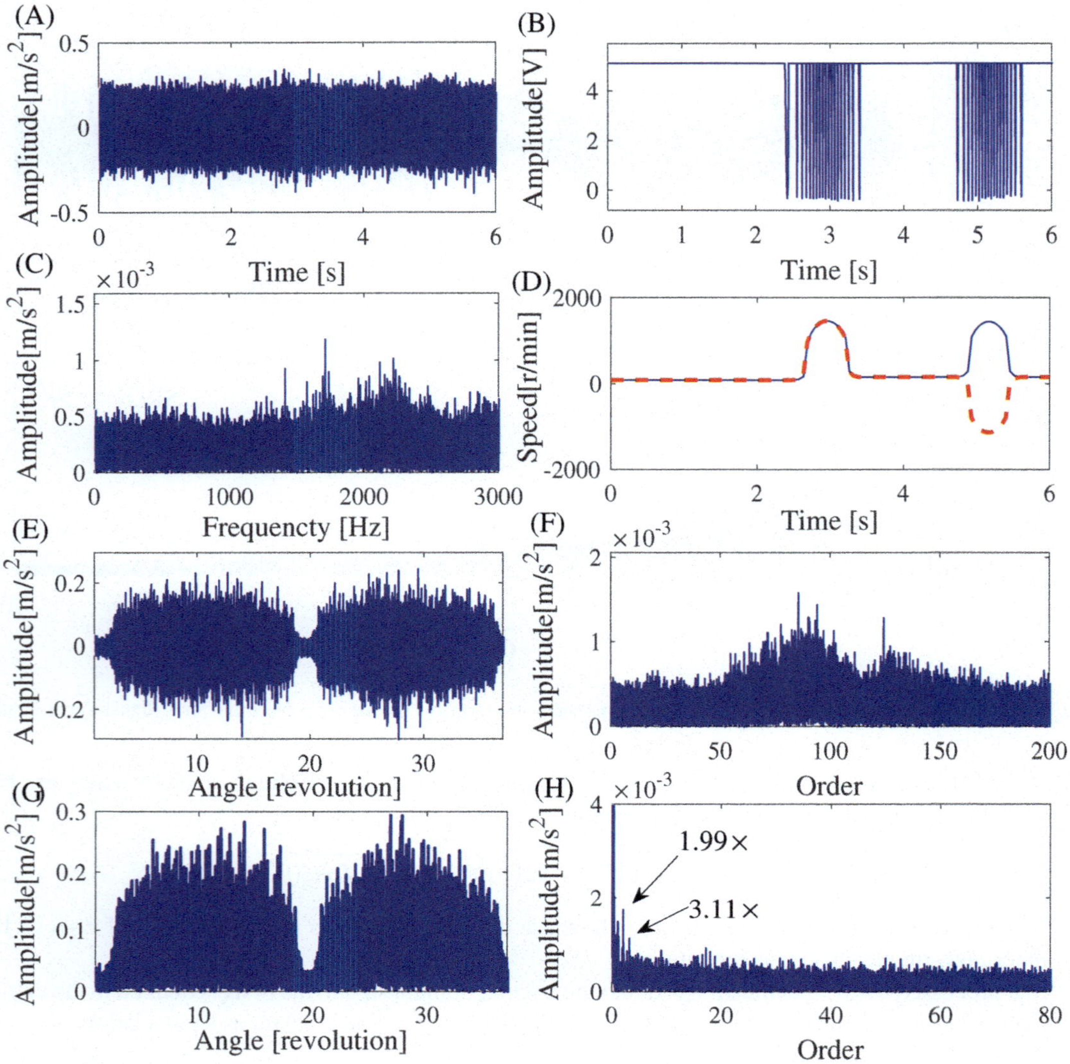

FIGURE 12.12 Experimental signal with its fault on the bearing outer raceway and rolling element: (A) the time-domain waveform, (B) the speed pulse signal, (C) the spectrum, (D) the speed signal, (E) the angular-domain waveform (by COA), (F) the order spectrum (by COA), (G) the angular-domain waveform (by resonance demodulation), and (H) the order spectrum (by resonance demodulation). *COA*, Computed order analysis.

Step 4: Use the SR method based on QPSO algorithm to process the angular signal in step 3, set SNR as the fitness function, initialize system parameters and scale coefficient, optimize system parameters and scale coefficient with the goal of maximizing SNR through QPSO algorithm, and finally obtain the optimal resonance response to extract fault features.

12.3 Numerical simulation analysis

The simulated signal for bearing outer raceway fault under variable speed conditions is defined as

$$s(t) = \sum_{i=0}^{H} A_h \exp\{-2\pi\delta f_n[t - v_h(t)]\}\sin\{2\pi f_n[t - v_h(t)] + \phi\}u[t - v_h(t)], \tag{12.22}$$

where H and A_h represent the number of bearing fault pulses and the amplitude of each pulse, respectively. δ and f_n are the damping characteristic factor and the natural frequency of the bearing, respectively. $v_h(t) = \sum_{i=0}^{H} T_i$ is the total pulse time, where T_i represents the ith pulse cycle at the occurrence time t_i. $u(t)$ represents the unit step function.

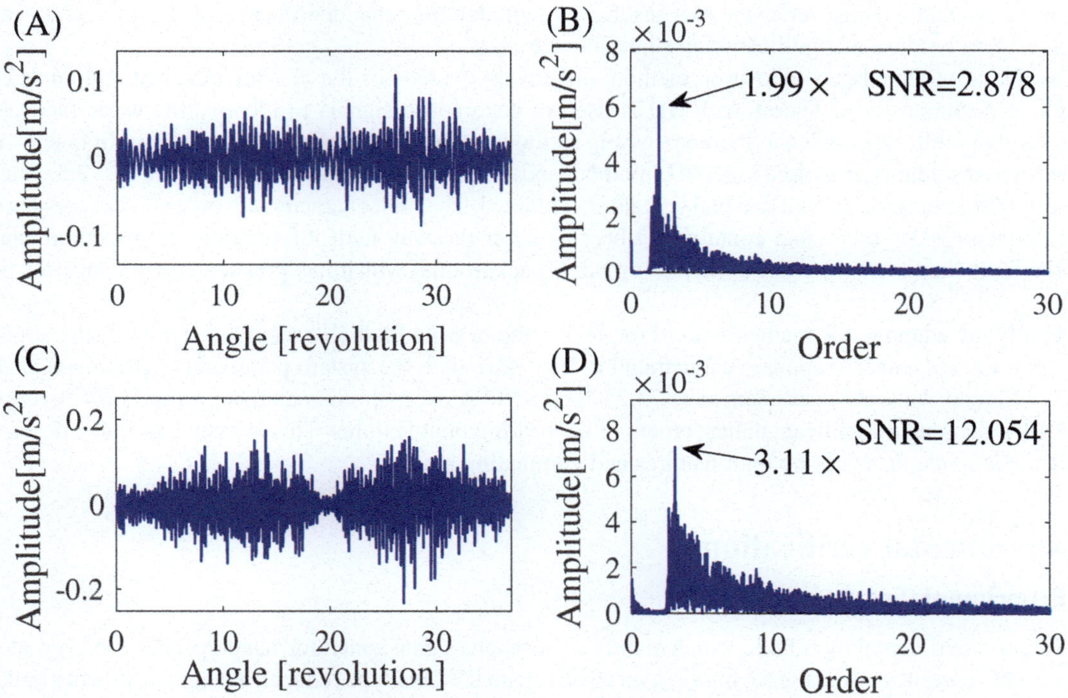

FIGURE 12.13 The optimal resonance response: (A) and (B) the angular-domain waveform and order spectrum at 1.99×, respectively. (C) and (D) the angular-domain waveform and order spectrum at 3.11×, respectively. *SNR*, Signal-to-noise ratio.

Assuming the rotational speed of the shaft is $f(t)$, the fault characteristic frequency of the bearing is expressed as FCO $\times f(t)$. The starting time of the fault characteristic signal is t_0, and the occurrence time of the ith pulse is expressed as

$$t_i = \frac{1}{f(t_i - 1)\mathrm{FCO}}, \tag{12.23}$$

According to Eq. (12.6) the fault characteristic order is related to the inherent parameters of the bearing. Without loss of generality the fault characteristic order of the simulated signal is set to 4.4×. The other simulated parameters are $f_0 = 2000$ Hz, $\delta = 120{,}000$, and $f_s = 10{,}240$ Hz.

Fig. 12.3 depicts the simulated signal of the bearing outer raceway fault under forward and reverse rotational movement. The speed condition follows the rule of linear increase–stationary–linear decrease, then reverse linear increase–stationary–linear decrease, and then linear increase–stationary–linear decrease. Fig. 12.3A and B shows the vibration signals of the bearing outer raceway fault with the speed changes. The time-domain waveform in Fig. 12.3A simulates the amplitude of the fault characteristic signal that varies with the speed, which has practical significance. As the speed of the shaft increases, it will inevitably lead to strong vibration, resulting in an increase in amplitude. When the speed of the shaft decreases, the vibration is relatively small and the amplitude decreases. Fig. 12.3B shows the corresponding spectrum. Because of the frequency ambiguity caused by directly Fourier transform analysis of the nonstationary signal, no useful information reflecting the fault feature will be obtained. The subsequent steps require COA processing. The speed curve of the reference shaft is shown in Fig. 12.3C, which represents the speed characteristics of the reference shaft, and Fig. 12.3D shows the corresponding speed pulse signal. Fig. 12.3E and F show the angular-domain signal and order spectrum of bearing fault characteristics after COA processing, respectively. The fault characteristics order [4.4×] can be observed in the order spectrum. To highlight the order fault characteristic, resonance demodulation technology is used to process the angular-domain signal in Fig. 12.3E. The processed results are shown in Fig. 12.3G and H, which are the angular-domain signal and order spectrum after demodulation, respectively. The obvious fault characteristic order [4.4×] and its multiples appear in the order spectrum.

The simulated signal is added to Gaussian white noise with an intensity of 0.045 to simulate the strong noise background. The fault characteristic signals with noise are shown in Fig. 12.4A and B. Subsequently the signal is processed via COA, with the processed results shown in Fig. 12.4C and D. The characteristic order related to bearing fault is unable to be found in the order spectrum, which needs further analysis. Then the signal is processed by resonance demodulation, with the results shown in Fig. 12.4E and F. However, even after resonance demodulation, the identification of the fault characteristic order remains challenging because of significant noise interference.

VMD, as a kind of signal decomposition method, effectively overcomes the aliasing effect of endpoints compared to empirical mode decomposition. Therefore VMD is used to decompose signals to identify the weak fault characteristic order. First, bandpass filtering is used as a preprocessing method to filter out high-frequency and low-frequency interference signals. The filtered signal is processed via VMD method, and the processed results are shown in Fig. 12.5, where the order spectra of seven IMF components and residual signals are obtained. The fault characteristic order [4.4×] appears in the IMF7 component, indicating that the signal contains the bearing outer raceway fault information. This method can effectively determine the fault characteristic order under strong noise background, which has practical significance for bearing fault diagnosis.

Subsequently the adaptive SR method is used to process the bearing fault characteristic signals in the angular domain. The optimal system resonance response is illustrated in Fig. 12.6, with the system parameters determined as $a_1 = 0.0073$ and $b_1 = 0.1372$, and the scale coefficient m is 73.25. The SNR at the fault characteristic order [4.4×] is improved from 10.88 to 13.40, indicating a significant enhancement of the bearing fault features. This demonstrates the effectiveness of the adaptive SR method in enhancing the fault features and eliminating noise.

12.4 Experimental verifications

12.4.1 Experimental setup

The rotor system experimental rig is built, which mainly consists of a signal generator, a laptop, a 1A206-type accelerometer, a 5E103-type eddy current sensor, a servo motor, a servo driver, an RV-20E-type reducer, a magnetic powder brake, a NI9234 signal acquisition card, a magnetic powder brake controller, and three 6204-type faulty bearings, as shown in Fig. 12.7. Because of the flexibility for servo motor control the motion of any speed characteristic can be simulated.

In this experiment the signal generator is used to produce voltage analog signals that are input into the servo drive to control the forward and reverse rotation of the motor with variable speed conditions. The servo drive is configured in analog control mode, with the signal generator capable of producing any analog voltage within the range of −5 to +5V. Positive and negative values of the voltage corresponded to the forward and reverse rotation of the servo motor, respectively. First the signal generator generates analog signals corresponding to different kinds of rotational speed, which are then input into the servo drive to control the servo motor to produce the corresponding rotational motion. Subsequently a vibration acceleration sensor is employed to collect the vibration signals during the operation of the bearing. These signals are transmitted through a signal acquisition card to the laptop and ultimately saved on the computer for analysis and processing using MATLAB software. The sampling frequency is set as 51,200 Hz. The test bearings, located at the end of the servo motor, are three 6204-type deep groove ball bearings, and their physical parameters are listed in Table 12.1. According to Eq. (12.6) the theoretical characteristic orders of the bearing outer raceway fault, inner raceway fault, and rolling element fault are calculated to be 3.045×, 4.955×, and 1.974×, respectively.

12.4.2 Bearing outer raceway fault feature extraction

The control mode of analog input is employed to regulate the servo motor for complex speed conditions, utilizing the voltage from the signal generator to control the forward and reverse motion of the motor. The acquired vibration signals, as shown in Fig. 12.8A and C, are entirely corrupted with strong noise. Correspondingly the speed pulse signal is illustrated in Fig. 12.8B, with Fig. 12.8D presenting the speed signal calculated from the pulse signal. Herein the dotted line represents the actual speed signal, whereas the solid line is the measured speed signal. Because of the limitation of the eddy current sensor to only recognize the pulse per revolution of the reference shaft and not its direction of rotation the calculated signal reflects the rotational speed characteristics in one direction. For the convenience of understanding the actual speed signal is provided. The time-domain vibration signal is processed via the COA method, with Fig. 12.8E showing the angular-domain signal after resampling. Fig. 12.8F is the order spectrum, where the fault characteristic order [3.09×] is corrupted with noise. The signal is processed by the resonance demodulation technology to highlight the fault characteristic order, as shown in Fig. 12.8G and H, where the fault characteristic order of [3.09×] is too weak to identify the fault feature. However, the strong noise interference makes it difficult to judge whether the fault characteristic order is induced by noise or the actual fault, potentially leading to misjudgment. Consequently the adaptive SR method is implemented to process the signal.

The adaptive SR method is used to extract and enhance the weak fault characteristic signal; the optimized system parameters are $a_1 = 0.0001$ and $b_1 = 0.2886$, and the scale coefficient m is 16.7573. The SNR at the fault order [3.09×]

is enhanced from 10.63 to 12.73. The final resonance response is shown in Fig. 12.9, where the fault characteristic order [3.09×] is significantly enhanced and the surrounding interference is significantly suppressed. Therefore it can be known that the fault is on the bearing outer raceway.

12.4.3 Bearing inner raceway fault feature extraction

The processing method for the characteristic signal of the bearing inner raceway fault is consistent with the processing method for the vibration signal of the bearing outer raceway fault. The collected signal is shown in Fig. 12.10A and C, where the signal is corrupted with strong noise. Fig. 12.10B is the speed pulse signal, and Fig. 12.10D is the calculated speed signal, whose speed condition corresponds to the motion process of one cycle of forward and backward motion. Fig. 12.10E and F shows the angular-domain signal and order spectrum of the signal processed by COA. To highlight the fault characteristic order the resonance demodulation technology is used to process the angular-domain signal. In Fig. 12.10H the fault characteristic order in the order spectrum is hard to identify. Subsequently the adaptive SR method is used for processing this signal.

The angular signal is processed by the adaptive SR method; the result is shown in Fig. 12.11. The SNR at the fault characteristic order [4.96×] increases from 0.8 to 3.49 in Fig. 12.11B, and the noise is significantly eliminated with the optimized system parameters $a_1 = 0.0001$ and $b_1 = 0.666$ and scale coefficient $m = 73.1983$. So, it can be known that the fault is on the bearing inner raceway.

12.4.4 Bearing compound fault feature extraction

To demonstrate the generality of the proposed method in fault diagnosis, we validate it using data of bearing compound faults. Fig. 12.12A and C shows the complex vibration signal and its spectrum under the pitting fault of the bearing's outer raceway and rolling element, respectively. The speed pulse signal and calculated speed signal are shown in Fig. 12.12B and D, respectively. Fig. 12.12E and F is the angular-domain signal and its order spectrum via COA, respectively. To highlight the fault characteristic order, resonance demodulation technology is used for processing. The results are shown in Fig. 12.12G and H, where the fault characteristic orders of 1.99× and 3.11× can be weakly observed, but the surrounding interference is significant, making it difficult to distinguish them. Because of factors such as machining errors in the bearing, there may be slight errors between the actual fault characteristic order and the theoretical fault characteristic order. Therefore 1.99× and 3.11× are considered the fault characteristic orders of the bearing rolling element and outer raceway.

The adaptive SR method is used to extract and enhance the fault characteristic orders of the bearing compound fault. As shown in Fig. 12.13 the SNR at the fault characteristic order of 1.99× is enhanced from 0.803 to 2.878, and that at the order of 3.11× is enhanced from 10.643 to 12.054. The feature information of bearing rolling element faults and outer raceway faults is effectively extracted and enhanced.

12.5 Conclusion

In this study a bearing fault diagnosis algorithm based on QPSO algorithm is proposed to extract weak feature information under strong noise background and complex variable speed conditions. The complex speed conditions include continuous forward and reverse movement and intermittent movement, which basically cover most complex speed conditions. First a nonstationary simulated signal for bearing outer raceway fault is constructed, and it is analyzed by transforming the nonstationary signal into a stationary signal via COA. Then, resonance demodulation technology is used to highlight the fault characteristic order; VMD method is used to separate the fault characteristic components and determine the fault type of the bearing. Finally the adaptive SR method based on QPSO algorithm is utilized to effectively extract weak fault feature information. The proposed method is validated through multiple sets of experimental data, all of which indicate its good applicability in diagnosing single and compound fault modes of the bearing under strong noise background and complex variable speed conditions.

Data availability

The datasets generated during and analyzed during the current study are available from the corresponding author on reasonable request.

Acknowledgments

The work is supported by the Fundamental Research Funds for the Central Universities (Grant Nos. 2024-10940), the Postgraduate Research & Practice & Innovation Program of Jiangsu Province (Grant Nos. KYCX24_2711), the Graduate Innovation Program of China University of Mining and Technology (Grant Nos. 2024WLKXJ067), and the Priority Academic Program Development of Jiangsu Higher Education Institutions.

References

Arts, L. P. A., & van den Broek, E. L. (2022). The fast continuous wavelet transformation (fCWT) for real-time, high-quality, noise-resistant time–frequency analysis. *Nature Computational Science, 2*(1), 47–58. https://www.nature.com/natcomputsci/journal-information. doi:10.1038/s43588-021-00183-z.

Berry, J. E. (1991). How to track rolling element bearing health with vibration signature analysis. *Sound and Vibration, 25*(11), 24–35.

Buzzoni, M., Antoni, J., & D'Elia, G. (2018). Blind deconvolution based on cyclostationarity maximization and its application to fault identification. *Journal of Sound and Vibration, 432*, 569–601. doi:10.1016/j.jsv.2018.06.055.

Chen, C., Wang, T., Lu, K., Liu, Y., & Cheng, L. (2024). Compact convolutional transformers-generative adversarial network for compound fault diagnosis of industrial robot. *Engineering Applications of Artificial Intelligence, 138*, 109315.

Chen, J., Zhang, X., Chen, Z., Zi, Y., Chen, Y., & Shi, Z. (2024). A fractional exponential power bistable stochastic resonance method for rolling bearing weak features extraction. *IEEE Transactions on Instrumentation and Measurement, 73*, 1–14. doi:10.1109/tim.2024.3369151.

Chen, J. Y., & Li, B. Z. (2024). The short-time Wigner-Ville distribution associated with linear canonical transform. In *Proceedings of SPIE—The International Society for Optical Engineering* (p. 13091). SPIE. https://www.spiedigitallibrary.org/conference-proceedings-of-spie. doi:10.1117/12.3023189.

Cheng, J., Yang, Y., Shao, H., Hu, N., Cheng, Z., & Cheng, J. (2024). Empirical mixed Ramanujan Fourier decomposition and its application to early fault diagnosis of planetary gears. *Journal of Vibration and Control, 30*(17–18), 3934–3946. doi:10.1177/10775463231204478.

Cheng, W., Gao, R. X., Wang, J., Wang, T., Wen, W., & Li, J. (2014). Envelope deformation in computed order tracking and error in order analysis. *Mechanical Systems and Signal Processing, 48*(1–2), 92–102. http://www.elsevier.com/inca/publications/store/6/2/2/9/1/2/index.htt. doi:10.1016/j.ymssp.2014.03.004.

Choudhury, M. D., Hong, L., & Dhupia, J. S. (2021). A novel tacholess order analysis method for bearings operating under time-varying speed conditions. *Measurement, 186*, 110127. doi:10.1016/j.measurement.2021.110127.

Feng, Z., Liang, M., & Chu, F. (2013). Recent advances in time–frequency analysis methods for machinery fault diagnosis: A review with application examples. *Mechanical Systems and Signal Processing, 38*(1), 165–205. doi:10.1016/j.ymssp.2013.01.017.

Flori, A., Oulhadj, H., & Siarry, P. (2022). Quantum particle swarm optimization: An auto-adaptive PSO for local and global optimization. *Computational Optimization and Applications, 82*(2), 525–559. doi:10.1007/s10589-022-00362-2.

Fyfe, K. R., & Munck, E. D. S. (1997). Analysis of computed order tracking. *Mechanical Systems and Signal Processing, 11*(2), 187–205. doi:10.1006/mssp.1996.0056.

Gong, C., Zhou, N., Xia, S., & Huang, S. (2024). Quantum particle swarm optimization algorithm based on diversity migration strategy. *Future Generation Computer Systems, 157*, 445–458. doi:10.1016/j.future.2024.04.008.

Gong, T., Wang, Z., Ma, Q., & Yang, J. (2024). Motor bearing fault diagnosis in an industrial robot under complex variable speed conditions. *Journal of Computational and Nonlinear Dynamics, 19*(2). doi:10.1115/1.4064250.

Gong, T., Yang, J., Liu, S., & Liu, H. (2022). Non-stationary feature extraction by the stochastic response of coupled oscillators and its application in bearing fault diagnosis under variable speed condition. *Nonlinear Dynamics, 108*(4), 3839–3857. doi:10.1007/s11071-022-07373-y.

He, L., Bi, L., & Xiong, Q. (2024). A cascaded piecewise unsaturated asymmetric under-damped tri-stable stochastic resonance system and its application in bearing fault diagnosis. *Nonlinear Dynamics, 112*(2), 925–943. doi:10.1007/s11071-023-09129-8.

He, L., Jiang, Z., & Chen, Y. (2024). Unveiling the principles of stochastic resonance and complex potential functions for bearing fault diagnosis. *Chaos, Solitons & Fractals, 180*, 114484. doi:10.1016/j.chaos.2024.114484.

Hou, B., Wang, D., Xia, T., Peng, Z., & Tsui, K.-L. (2023). Difference mode decomposition for adaptive signal decomposition. *Mechanical Systems and Signal Processing, 191*, 110203. doi:10.1016/j.ymssp.2023.110203.

Hou, Y., Wang, H., Wang, Y., Wu, P., Huang, W., & Wu, D. (2025). Structural rotor rub-impact diagnosis under intricate noise interferences based on targeted component extraction and stochastic resonance enhancement. *Structural Health Monitoring, 24*(1), 255–294. doi:10.1177/14759217241231897.

Kim, Y., Park, J., Na, K., Yuan, H., Youn, B. D., & Kang, C.-S. (2020). Phase-based time domain averaging (PTDA) for fault detection of a gearbox in an industrial robot using vibration signals. *Mechanical Systems and Signal Processing, 138*, 106544. doi:10.1016/j.ymssp.2019.106544.

Li, J., Cheng, X., Zhang, S., Meng, Z., & Cao, L. (2024). Fault feature extraction method of rolling bearings based on coupled resonance system with vibrational resonance-assisted enhanced stochastic resonance. *Mechanical Systems and Signal Processing, 208*, 111069. doi:10.1016/j.ymssp.2023.111069.

Li, L., Cai, H., Han, H., Jiang, Q., & Ji, H. (2020). Adaptive short-time Fourier transform and synchrosqueezing transform for non-stationary signal separation. *Signal Processing, 166*, 107231. doi:10.1016/j.sigpro.2019.07.024.

Li, Y., Zhang, X., Chen, Z., Yang, Y., Geng, C., & Zuo, M. J. (2023). Time-frequency ridge estimation: An effective tool for gear and bearing fault diagnosis at time-varying speeds. *Mechanical Systems and Signal Processing, 189*, 110108. doi:10.1016/j.ymssp.2023.110108.

Lu, S., Yan, R., Liu, Y., & Wang, Q. (2019). Tacholess speed estimation in order tracking: A review with application to rotating machine fault diagnosis. *IEEE Transactions on Instrumentation and Measurement, 68*(7), 2315–2332. doi:10.1109/tim.2019.2902806.

Lu, S., He, Q., & Wang, J. (2019). A review of stochastic resonance in rotating machine fault detection. *Mechanical Systems and Signal Processing, 116*, 230–260. doi:10.1016/j.ymssp.2018.06.032.

Martins, D. H. C. S. S., de Lima, A. A., Gutiérrez, R. H. R., Pestana-Viana, D., Netto, S. L., Vaz, L. A. P., da Silva, E. A. B., & Haddad, D. B. (2023). Improved variational mode decomposition for combined imbalance-and-misalignment fault recognition and severity quantification. *Engineering Applications of Artificial Intelligence, 124*. https://www.journals.elsevier.com/engineering-applications-of-artificial-intelligence. doi:10.1016/j.engappai.2023.106516.

Sapena-Bano, A, Burriel-Valencia, J., & Pineda-Sanchez, M. (2016). The harmonic order tracking analysis method for the fault diagnosis in induction motors under time-varying conditions. *IEEE Transactions on Energy Conversion, 32*(1), 244–256.

Schmidt, S., Heyns, P. S., & de Villiers, J. P. (2018). A tacholess order tracking methodology based on a probabilistic approach to incorporate angular acceleration information into the maxima tracking process. *Mechanical Systems and Signal Processing, 100*, 630–646. doi:10.1016/j.ymssp.2017.07.053.

Wang, H., Fang, Z., Wang, H., Li, Y., Geng, Y., Chen, L., & Chang, X. (2023). A novel time-frequency analysis method for fault diagnosis based on generalized S-transform and synchroextracting transform. *Measurement Science and Technology, 35*(3), 036101. doi:10.1088/1361-6501/ad0e59.

Yang, J., Wang, Z., Guo, Y., Gong, T., & Shan, Z. (2024). A novel noise-aided fault feature extraction using stochastic resonance in a nonlinear system and its application. *IEEE Sensors Journal, 24*(7), 11856–11866. doi:10.1109/jsen.2024.3365105.

Yin, T., Lu, N., Guo, G., Lei, Y., Wang, S., & Guan, X. (2023). Knowledge and data dual-driven transfer network for industrial robot fault diagnosis. *Mechanical Systems and Signal Processing, 182*, 109597. doi:10.1016/j.ymssp.2022.109597.

Zhao, J., Han, J., Ma, C., Xiao, X., Jiang, Y., & Zhang, Y. (2024). Research on operation characteristics of hoisting conveyance under mining deformation action in mine vertical shaft. *Nonlinear Dynamics, 112*(4), 2629–2659. doi:10.1007/s11071-023-09172-5.

Chapter 13

Classification of gray zone on PCB using Bayesian optimization-random forest method

Yılmaz Seryar Arikuşu[a] **and Göksu Taş**[b]

[a] *Department of Electrical Engineering, Manisa Celal Bayar University, Manisa, Turkey,* [b] *Faculty of Agricultural Sciences and Technology, Department of Precision Agriculture and Agricultural Robots, Sivas University of Science and Technology, Sivas, Türkiye*

13.1 Introduction

Applications for the Internet of Things (IoT) have fundamentally altered our way of life and greatly enhanced the productivity of both individuals and businesses. Billions of common items are now outfitted with cutting-edge sensors, wireless networks, and cutting-edge computer power. Wearable technology, smart home applications, sophisticated healthcare systems, "smart cities," and industrial automation are the results of this abundance. IoT is ready to transition into commonplace commercial application after years of uncertainty: Businesses are embracing IoT technology at an increasing rate; by 2023, it is expected that there will be 43 billion IoT-connected devices globally. The IoT is a reflection of this growing trend toward tangible objects with the computational and communication power to pool resources and obtain data in real time (Sestino et al., 2020). IoT uses communication technology and sensors to sense and communicate data in real-time, allowing for quick calculations and the best possible decision-making. IoT can also assist the energy industry in moving from a centralized to a distributed, intelligent, and integrated energy system. Deploying local, distributed renewable energy sources (RESs) such as wind and solar power is crucial, as is enabling numerous small-scale energy consumers to become prosumers by pooling their supply and optimizing their demand whenever it benefits the grid. With the use of sensors and communication technologies, IoT-based systems automate, integrate, and control activities. Large-scale data gathering and the application of clever algorithms for real-time data analysis can aid in the monitoring of trends in energy usage across many users and devices across a range of time periods and in the more effective management of that consumption (Hossein Motlagh et al., 2020). One of the most critical objects of a system for IoT devices is printed circuit boards (PCBs). PCB is of critical importance because it is the junction and distribution point of electrical and electronic connections in all devices containing electronic components and because it organizes and supports them.

Consumer gadgets, such as tablets, laptops, smart TVs, and smartphones, are becoming more and more common, which has increased demand for high-quality PCBs made in bulk. However, there are significant obstacles in maintaining excellent PCB quality at such scales (Fung et al., 2024). PCBs must have strong stability, a high level of interference resistance, and cutting-edge characteristics, including improved integration, compactness, and high-speed transmission capabilities because they are essential parts of electronic devices (Chen et al., 2023). PCB quality control that works well is essential. Even small flaws can cause big problems, such as lower manufacturing quality from hidden defects; higher production costs from fixing or discarding defective products; longer time to market while spending more time on quality control; and possible user-safety hazards such as short circuits, component failure, or overheating (Jian et al., 2025). For this reason, many studies have been conducted in the literature on improving PCB production and error detection. In place of manual inspection, Zhou et al. (2021) suggested a multiview template matching technique for PCB component identification. The

Computational Intelligence in Mechatronics: Solving Real-World Problems in Electronic Systems Design. DOI: https://doi.org/10.1016/B978-0-443-27580-7.00020-2

experiment showed a considerable improvement in performance using the multiview strategy when comparing it with the single-view approach. Machine learning (ML) was suggested by Bahrebar et al. (2022) as a way to forecast PCB corrosion conditions. The degrees of electrochemical migration (ECM) and leakage current (LC) in electronic devices exposed to corrosive environments are predicted using predictive models. The ECM and LC models' predictions are used to direct the training process in supervised ML techniques (Bahrebar et al., 2022). For the purpose of identifying hardware Trojan components on PCBs, Piliposyan and Khursheed (2022) developed a hybrid object identification technique that blends conventional computer vision with the siamese neural network (SNN). A multiscale feature extraction modification of the vision transformer (ViT) was proposed by Lou et al. (2023). Each level of the multistage ViT architecture used by the approach focuses on collecting characteristics at a certain size. A tiny defect detection network (TDD-Net) was developed especially to find minute flaws on PCB surfaces by Ding et al. (2019). TDD-Net is based on the Faster Region-based Convolutional Neural Network (R-CNN) architecture, but it makes three key changes: It applies a multiscale feature fusion method for the model, designs appropriate anchors, and applies a data augmentation technique (Ding et al., 2019). An expanded Feature Pyramid Network (FPN) model for hard-sample PCB identification was presented by Li et al. (2021). In addition, the model incorporates the focused loss function to lower sample loss and raise the weight of hard samples (Li et al., 2021). As seen in the literature research, problems occur on PCB boards, which perform very important tasks for IoT systems, after they are produced or after they have been operated for a certain period of time.

As PCBs contain many types of electrical and electronic materials and the production process is complex and detailed, many problems can occur on the board. Although many solutions to this problem have been presented in the literature, a solution to the problem was presented in this study using the optimization-supported ML method based on ML. In this study, the gray areas on PCBs which are critical to maintain the production quality and avoid connection related problems were identified using ML-based classification technique. In addition, the most important problem, when classifying with ML methods, is the appropriate determination of hyperparameters. Therefore the minimum leaf number value of the RF algorithm was determined with the Bayesian optimization (BO) algorithm (BOA), and classification was made faster. Although ML experiments were performed, the performance comparison of RF, BO-RF, K-Nearest Neighbors (KNN), and support vector machine (SVM) methods was also presented.

13.2 Material method

PCBs make it easy to create complex systems by combining various circuit components. Manually finding and testing test points is an error-prone process, especially on old PCBs or PCBs without schematic data. In this chapter, Tan et al. created a dataset to estimate test points on PCBs. This dataset consists of images of an electronic circuit board consisting of 71,040-pixel records, to which a two-stage clustering procedure was applied. The digital image consists of *X* and *Y* coordinates (physical location) and color values [red, green, blue (RGB)] of each pixel (Tan & Kit, 2016; Tan & Tan, 2016). As the test points are usually gray, the RGB values of gray pixels are quite close to each other. The value to be predicted consists of the class values of gray points. The representative PCB surface image is shown in Fig. 13.1, demonstrating the numerous gray circular dots on the board that correspond to test pads. As the class sizes in the dataset to be predicted are not equal, the data sizes are equalized. Thus 23.818 data were used in this study. In the testing phase, 33% of the total data was used to evaluate the gray zone prediction. In this study, all calculations were obtained using the MATLAB software program. All ML methods' calculations were performed on a single computer with a Windows 10 Intel Core I7-8565U 1.80 GHz CPU and 24 GB RAM.

13.2.1 Bayesian optimization

A variety of options need to be investigated to create the best ML model. Hyperparameter tuning is the process of creating the perfect model architecture with the best possible hyperparameter configuration. Building an effective ML model is thought to require carefully adjusting hyperparameters, particularly for deep neural networks and tree-based models that have a large number of hyperparameters (Yang & Shami, 2020). Because of the various types of hyperparameters used by ML algorithms—categorical, discrete, and continuous—the procedure of fine-tuning them varies across them. In this study, hyperparameter tuning was performed with BOA. The BOA models potential solutions and generates new candidate solutions using Bayesian networks to develop a population of candidate solutions (strings) to the given issue. A specific area of the search space may be skewed in the first population of candidate solutions produced by BOA, which is generated at random with a uniform distribution (Pelikan et al., 2002). Using the Bayes formula, it integrates previous knowledge about the unknown function with sample data to get posterior information about the function distribution. This approach seeks to determine the greatest value for an unknown objective function $f(x)$ at the sampling point. The objective unknown function

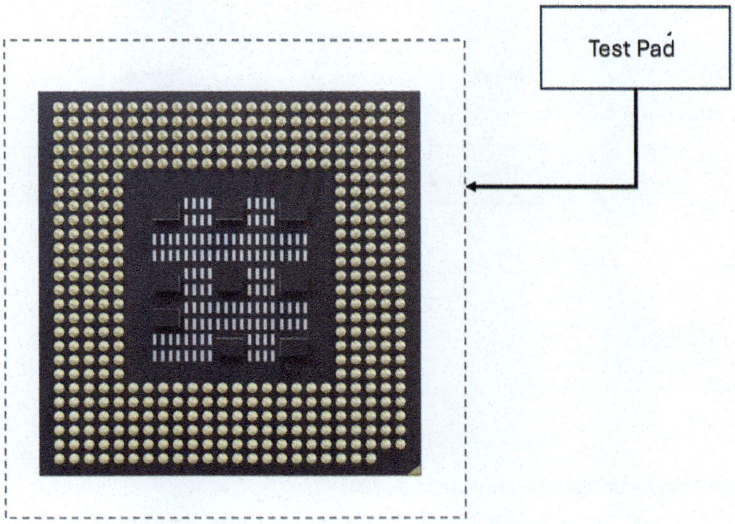

FIGURE 13.1 A representative test pad: an example of PCB surface. *PCB*, Printed circuit board.

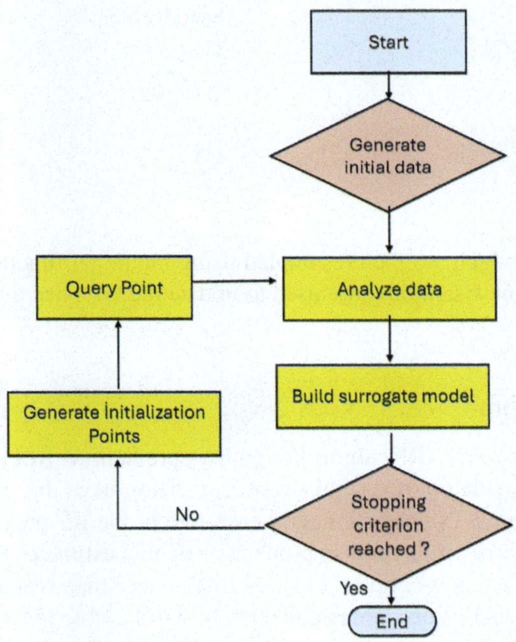

FIGURE 13.2 The Bayesian algorithm flowchart is about the flowchart of Bayesian optimization.

is given as

$$x^* = argmax f(x)$$

$$x \in X$$

where X is a design space of interest x (Shahriari et al., 2016).

The Bayes algorithm flowchart is illustrated in Fig. 13.2.

The initial data points are produced by repeatedly sampling the target function with a specified set of hyperparameters. These data points include the hyperparameters and the target function value. The unknown goal function is represented a priori in BO. Gaussian process (GP), a probabilistic model, is typically used to generate this a priori model. The acquisition function determines which hyperparameters should be assessed next in the BO step. When there is ambiguity regarding the target function, this function directs the learning process by conducting more sampling. The acquisition function determines

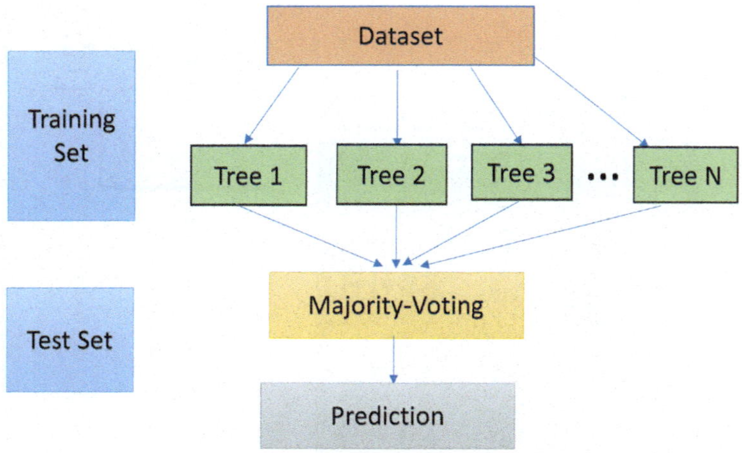

FIGURE 13.3 Flowchart of the random forest algorithm (Arıkuşu et al., 2023).

TABLE 13.1 Hyperparameter values found by Bayesian optimization.

Hyperparameter	Search area	Find
numTrees	1–20	2
MinLeafSize	1–20	13
MaxNumSplit	1–20	2
NumVariablesToSample	1–4	1

the optimal set of hyperparameters, which are then resampled using the target function. The target function value and the hyperparameters that came from the new sampling are used to update the GP once more. When the conditions are met, the stopping criterion is activated.

13.2.2 Random forest algorithm

An ML technique called the random forest (RF) algorithm gathers predictions from numerous decision trees and groups them into various data subsets. It is made up of several separate decision trees that cooperate with one another. The class with the highest votes becomes the model's prediction, and each tree in the RF provides the class prediction. It is crucial to have low model correlation. Low correlation models produce ensemble estimates that are more accurate than any of the individual estimations. The reason for this is simple, provided that no tree makes faults in the same direction all the time, trees shield one another from individual errors. Trees collectively will be able to move in the appropriate direction even though many of them will be correct and some will be incorrect. The RF algorithm flowchart is shown in Fig. 13.3. The steps of the RF algorithm shown in are explained in order. First, bootstrapped sampling is used by RF to generate distinct data samples for every tree. By repeatedly and randomly selecting samples from the data collection, this enhances variety. This bootstrapped sample is used to generate each decision tree in the RF. The dataset's attributes, these trees continuously segment it and make judgments based on those divisions by analyzing. Every node in a tree operates with a randomly chosen subset. This lowers the possibility of overfitting and enables each tree to concentrate on distinct properties. Each tree provides predictions in classification tasks, and the class with the most votes is selected as the final prediction. The final prediction in regression tasks is derived by averaging the predictions made by the trees. The final predictions are produced by RF by adding the outcomes of each tree (Arıkuşu et al., 2023).

13.3 Results and discussions

In this study, optimization-based ML method was used by following the classification process to detect the gray area on the PCB the search area and the found value of the hyperparameter values of the RF ML method are given in Table 13.1. The hyperparameter values were obtained using the Bayesian algorithm. The new values obtained were applied to the RF

TABLE 13.2 Machine learning prediction results.

Hyperparameter	Accuracy
K-means (Tan & Kit, 2016)	0.9375
SVM	0.9098
RF	0.8829
BO-RF	**0.9526**

BO-RF, Bayesian optimization-random forest; *RF*, random forest; *SVM*, support vector machine.

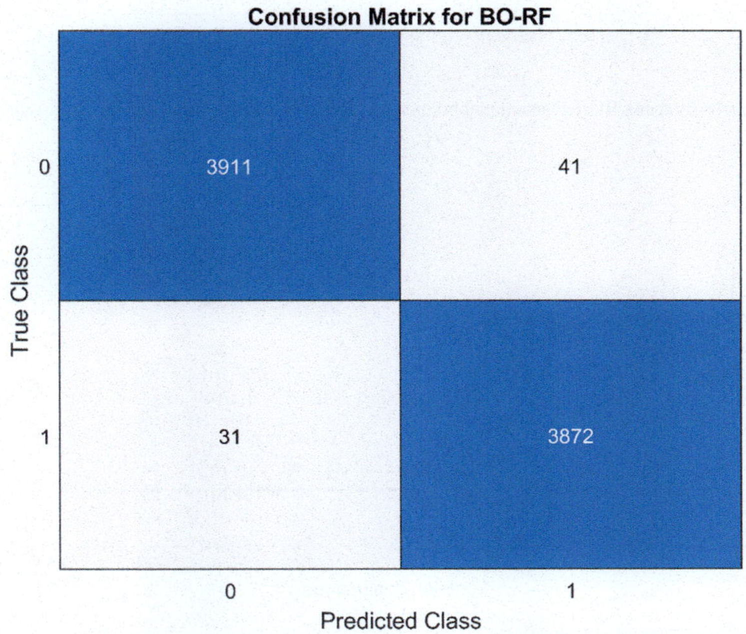

FIGURE 13.4 Confusion matrix for BO-RF method: classification results of BO-RF. *BO-RF*, Bayesian optimization-random forest.

algorithm and the BO-RF method was obtained. The RF method consists of hyperparameter values obtained through trial and error.

In this study, min–max normalization method was used. The purpose of normalization is to provide comparisons between data and to avoid over- and underfitting. Variables in the normalizing method can only range from 0 to 1. Min–max normalization method is the following term:

$$a = \frac{a_i - a_{\min}}{a_{\max} - a_{\min}}$$

where a, a_i, $a_{\min}$, and $a_{\max}$ described normalized value, value to normalize, minimum value, and maximum value, respectively. The ML experiment results performed on the PCB dataset are given in Table 13.2. These ML experiments were repeated 10 times, and the results were averaged. Thus it was aimed to obtain more reliable results. The BO-RF method achieved 95.26% more successful results than other methods. In addition, all models successfully detected the gray area on the PCB.

Confusion matrix is the most preferred metric among the prediction metrics in solving ML-based problems in classification problems. In this study the aim was to make a more accurate classification by reorganizing the dataset. The confusion matrix is among the metrics used to measure the success of the classification result. Confusion matrix is in a table format and shows the relationship between the estimated and actual data by comparing the actual value of the model with the estimated value. Fig. 13.3 shows the confusion matrix resulting from the estimation obtained with the BO-RF method. The classes of the gray area consist of 1 and 0. When Fig. 13.4 is examined, the part shown in *blue color* shows the correct values, whereas the other color shows the wrong prediction values. It successfully predicted 3872 test data but incorrectly forecasted 31 data from class 1. Furthermore, it forecasted 41 data erroneously while correctly predicting 3911 data belonging to class 0.

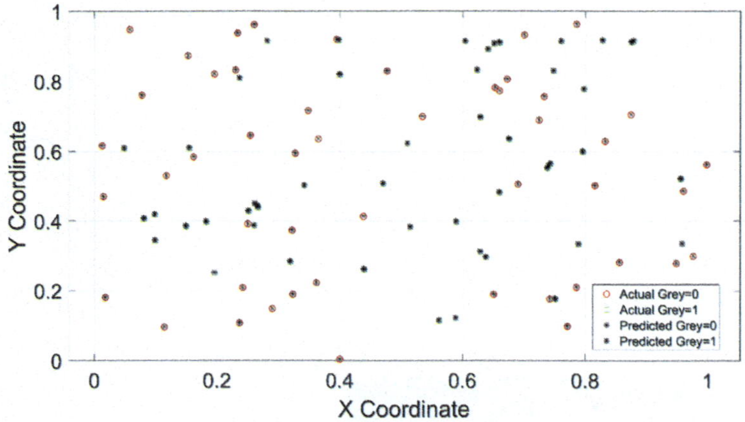

FIGURE 13.5 Classification results by using BO-RF actual and predicted values. *BO-RF*, Bayesian optimization-random forest.

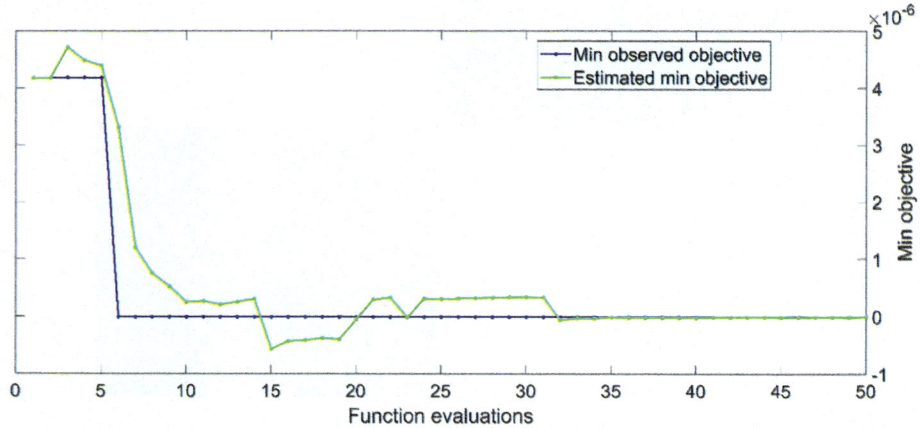

FIGURE 13.6 Optimization results process of optimization.

The estimated results of the BO-RF method and the actual values are given in Fig. 13.5. Actual and estimated values are shown in different shapes and colors in Fig. 13.5. *Star* and *circle symbols* are given for two separate classes. Although the *circle symbol* is used for real data, the values resulting from the prediction are shown with a *star*. When Fig. 13.5 is examined, the correct predicted results are seen with the overlap of the real and predicted values.

Fig. 13.6 shows the process of searching for suitable values obtained when trying to find the hyperparameter values of the RF with the BO method used in this study. In Fig. 13.6 the *x*-axis gives the number of trials performed during the optimization, whereas the values given on the *y*-axis give the minimum error value that occurs during each trial.

13.4 Conclusion

In this study, classification was performed to detect the gray area on the PCB using ML and optimization methods. The PCB dataset was arranged, and the number of samples in the classes was equalized to provide a more appropriate and regular class distribution for classification. In addition, the question of how effective the BO method will be in ML management was sought. As a result of the experiments the RF method created with the hyperparameter values determined using BO was the method that made the most successful prediction. The RF method, which consists of hyperparameter values obtained by the traditional trial and error method, was the least successful method. For this reason the hyperparameter search process with the BO method was faster and more efficient than the long-lasting hyperparameter determination process with the trial-and-error method. It is very important to determine hyperparameters quickly, especially in solving critical problems that require speed. Choosing the appropriate hyperparameter helps to shorten the training time and also helps to increase the overall performance of the model. The BO-RF method was the most successful method, reaching an accuracy value of 95.26% according to the estimation results that were run 10 times and averaged compared to other methods. It has been concluded that the BO method can be used successfully in ML methods. The authors plan to perform error detection on the image dataset using deep learning and ML methods in future studies.

References

Arıkuşu, Y. S., Bayhan, N., & Tiryaki, H. (2023). Determination of energy savings via fuel consumption estimation with machine learning methods and rule-based control methods developed for experimental data of hybrid electric vehicles. *Energies, 16*(24), 7970. doi:10.3390/en16247970.

Bahrebar, S., Homayoun, S., & Ambat, R. (2022). Using machine learning algorithms to predict failure on the PCB surface under corrosive conditions. *Corrosion Science, 206*, 110500. doi:10.1016/j.corsci.2022.110500.

Chen, I. C., Hwang, R. C., & Huang, H. C. (2023). PCB defect detection based on deep learning algorithm. *Processes, 11*(3). http://www.mdpi.com/journal/processes. doi:10.3390/pr11030775.

Ding, R., Dai, L., Li, G., & Liu, H. (2019). TDD-net: A tiny defect detection network for printed circuit boards. *CAAI Transactions on Intelligence Technology, 4*(2), 110–116. doi:10.1049/trit.2019.0019.

Fung, K. C., Xue, K. W., Lai, C. M., Lin, K. H., & Lam, K. M. (2024). Improving PCB defect detection using selective feature attention and pixel shuffle pyramid. *Results in Engineering, 21*. https://www.sciencedirect.com/science/journal/25901230. doi:10.1016/j.rineng.2024.101992.

Hossein Motlagh, N., Mohammadrezaei, M., Hunt, J., & Zakeri, B. (2020). Internet of Things (IoT) and the energy sector. *Energies, 13*(2), 494. doi:10.3390/en13020494.

Jian, T. S., Fauadi, M. H. F. M., Yahaya, S. H., Noor, A. Z. M., & Saptari, A. (2025). A deep learning approach for automated PCB defect detection: A comprehensive review. *Multidisciplinary Reviews, 8*(1). https://malque.pub/ojs/index.php/mr/article/download/5118/2443. doi:10.31893/multirev.2025011.

Li, C. J., Qu, Z., Wang, S. Y., Bao, K. H., & Wang, S. Y. (2021). A method of defect detection for focal hard samples PCB based on extended FPN model. *IEEE Transactions on Components, Packaging and Manufacturing Technology, 12*(2), 217–227.

Lou, L., Lu, K., & Xue, J. (2023). Multi-scale vision transformer for defect object detection. *Procedia Computer Science, 222*, 397–406. doi:10.1016/j.procs.2023.08.178.

Pelikan, M., Sastry, K., & Goldberg, D. E. (2002). Scalability of the Bayesian optimization algorithm. *International Journal of Approximate Reasoning, 31*(3), 221–258. doi:10.1016/S0888-613X(02)00095-6.

Piliposyan, G., & Khursheed, S. (2022). Computer vision for hardware Trojan detection on a PCB using Siamese neural network. *IEEE Physical Assurance and Inspection of Electronics, PAINE 2022*. Institute of Electrical and Electronics Engineers Inc. http://ieeexplore.ieee.org/xpl/mostRecentIssue.jsp?punumber=10014680. doi:10.1109/PAINE56030.2022.10014967.

Sestino, A., Prete, M. I., Piper, L., & Guido, G. (2020). Internet of Things and big data as enablers for business digitalization strategies. *Technovation, 98*, 102173. doi:10.1016/j.technovation.2020.102173.

Shahriari, B., Swersky, K., Wang, Z., Adams, R. P., & De Freitas, N. (2016). Taking the human out of the loop: A review of Bayesian optimization. *Proceedings of the IEEE, 104*(1), 148–175. http://ieeexplore.ieee.org/xpl/RecentIssue.jsp?punumber=5. doi:10.1109/JPROC.2015.2494218.

Tan S., & Tan S. (2016). *Printed circuit board processed image*. UCI Machine Learning Repository. doi: 10.24432/C5DK8H.

Tan, S. C., & Kit, S. T. W. (2016). Fast retrievals of test-pad coordinates from photo images of printed circuit boards. In *International Conference on Advanced Mechatronic Systems, ICAMechS* (pp. 464–467). IEEE Computer Society. http://ieeexplore.ieee.org/xpl/conhome.jsp?punumber=1800493. doi:10.1109/ICAMechS.2016.7813492.

Yang, L., & Shami, A. (2020). On hyperparameter optimization of machine learning algorithms: Theory and practice. *Neurocomputing, 415*, 295–316. doi:10.1016/j.neucom.2020.07.061.

Zhou, J., Liu, Y., Zhang, X., & Yang, Z. (2021). Multi-view based template matching method for surface defect detection of circuit board. *Journal of Physics: Conference Series, 1983*(1), 012063. doi:10.1088/1742-6596/1983/1/012063.

Chapter 14

Estimation of PID parameters for vertical motion of a robot arm using by GTO-LSTM method

Göksu Taş

Faculty of Agricultural Sciences and Technology, Department of Precision Agriculture and Agricultural Robots, Sivas University of Science and Technology, Sivas, Türkiye

14.1 Introduction

In today's shift to human-centered, sustainable, and resilient production under Industry 5.0, industrial companies are working toward the following goals: flexible production of intricate and precise mechanical parts that reduces reliance on manual operation; transformable production with minimal changeover times when manufacturers introduce new products; and occupational health of employees that prevents musculoskeletal disorders caused by awkward postures, excessive effort, and repetitive movements. Human–robot collaboration (HRC) is a well-known production architecture that combines the superior strength and precision of robots with the sophisticated intelligence and adaptability of people to lead the way in this human-centric smart manufacturing paradigm. HRC systems can increase productivity by switching production processes to flexible automation (Li et al., 2023). The degree of contact between the human and robotic agent or agents changes and improves as the area develops. The subfield of human–robot cooperation captures the ultimate goal of achieving the smooth team dynamics and communication fluency of an all-human team. Depending on their intended use, robots in that sector can be either industrial or social. Although social robots do not aim to replace humans, industrial robots have historically been fixed and programmed to work at high speeds and torques and carry out manufacturing (Taş & Bal, 2023) and related tasks, including welding, painting, assembling, pick-and-place, and material handling. Rather, they are designed to help people in service-related businesses. Generally speaking, they have been lightweight, portable, and designed to function in a range of settings. In the industrial context, HRC combines the needs of these two previously distinct fields to create intelligent, mobile robots that are conscious of their environment and their human companion (Mukherjee et al., 2022).

When humans see something new, they can nearly instantly figure out how to pick it up. Robotics skills are far behind. One of the most important challenges is robotic grasping and manipulation. The goal of building cognitive robots with human-level dexterity has been pursued for many years. It is still an unresolved issue in spite of corporate and scientific attention (Kumra & Kanan, 2017; Zeng et al., 2022). Robot systems must be automatically configured (plug and produce) because of shorter product lifecycles and the continually increasing need for customization, which calls for more adaptable and changeable production systems. There is a lot of interest in creating robots that can function in dynamic and unstructured settings, such as bin-picking, daily life in the home, or professional services. Robotic grasping techniques use learning-based techniques to automatically set up for the job at hand without the need for human interaction, allowing for a large reduction in programming work. Because machine learning can generalize to new objects, it is a viable method for robotic grasping (Kleeberger et al., 2020).

To improve the robot tracking system, Ma et al. (2017) used the fuzzy BP neural network PID control method, which is based on the conventional PID control algorithm. To identify, create, and verify the control strategy, Angel and Viola developed a collaborative simulation model of a delta robot and presented a delta manipulator tracking control based on a fractional-order PID controller and computed torque control method. Performance metrics, such as trajectory tracking, joint torque, and joint and space errors, were used to assess how resilient the controller was to outside disruptions (Angel & Viola, 2018). A novel symmetric, adaptive, variable-admittance control technique for dual-arm cooperative robot position

Computational Intelligence in Mechatronics: Solving Real-World Problems in Electronic Systems Design. DOI: https://doi.org/10.1016/B978-0-443-27580-7.00021-4

tracking was presented by Jinjun et al. For the first time, symmetrical adaptive variable-admittance control with two-arm coordination was achieved by adjusting the admittance parameters online to correct for the unknown trajectory deviation to monitor the intended location and force depending on the tracking error. The approach could accomplish high tracking performance of location and force, according to the simulation experiment (Jinjun et al., 2019). An optimum state-dependent Riccati equation based on the secure digital input/output (SDIO) controller was suggested by Kalamian et al. This method's two primary benefits are its capacity to develop controllers for a larger variety of nonlinear systems while maintaining system stability and the observer's ability to constantly estimate the output signal's state at a time-varying rate at each sample (Kalamian et al., 2019).

In this study an optimization-based deep learning model is proposed to control a robot arm during its vertical movement. It is shown that the mobility of autonomous vehicles and devices can be increased along with automated devices by estimating PID parameters in a suitable, fast, and reliable way. In addition, the performance of deep learning algorithms is compared, and it is shown that they can be used in robotic applications. The success of the proposed method is given using the most commonly used regression error metrics in the literature. The optimization-based deep learning solution for the vertical movement of the robot arm is a very new study that leads to new studies.

14.2 Materials and methods

In this study a robot dataset was used to obtain its mathematical model, and PID parameters were estimated by deep learning methods. In addition, the giant trevally optimizer (GTO) algorithm was used to determine hyperparameters in skin learning. In this study, the Python programming language was used, and libraries such as Matplotlib, Control, Pandas, and Scikit-Learn were used.

14.2.1 Robot arm

In this study, the NAO robot dataset was used (Nugroho et al., 2023). In addition, the mathematical model of the NAO robot was obtained in another study. According to the model, there are six connections on the right arm of the NAO robot. The Euler–Lagrange equation procedure was applied for the robot model. The Euler–Lagrange equation is given in Eq. (14.1) (Jimenez-Vazquez et al., 2021).

$$F_1(s) = \frac{d}{dt}\frac{\partial L}{\partial \dot{\theta}_t} - \frac{\partial L}{\partial \theta_t} = \tau_t \tag{14.1}$$

The output transfer function is given by Eqs. (14.2) and (14.3) according to the reference position according to the vertical and horizontal positions, respectively (Jimenez-Vazquez et al., 2021).

$$F_1(s) = \frac{Q_1}{Q_{1r}} = \frac{5.1887}{s + 5.1576} \tag{14.2}$$

$$F_2(s) = \frac{Q_2}{Q_{2r}} = \frac{6.2668}{s + 6.4456} \tag{14.3}$$

14.2.2 Giant trevally optimization

Within the jack family of marine predators is the massive giant trevally, or *Caranx ignobilis*. Another name for it is the giant kingfish. They are widespread throughout the Pacific and Indian seas, particularly in the regions surrounding Australia and New Zealand. They can also be found around the Hawaiian Islands and off the coast of East Africa (Sadeeq & Abdulazeez, 2022). The gigantic trevally is a source of inspiration for the GTO approach, which is a metaheuristic algorithm. The giant trevally employs a variety of strategies, including choosing an ideal hunting area, engaging in systematic foraging activities, and jumping out of the water to seize food. The GTO algorithm duplicates these tactics into a three-phase procedure: extensive search, selection of area, and attack (Hashish et al., 2023).

14.2.2.1 Extensive search

The GTO method uses a mathematical model based on Levy flights, a type of random walk, to replicate the great distances huge trevallies traverse in search of food. This stage helps the algorithm avoid local optima and enhances its exploration capabilities. The following illustration provides an example of the equation utilized in this phase. This expression is given

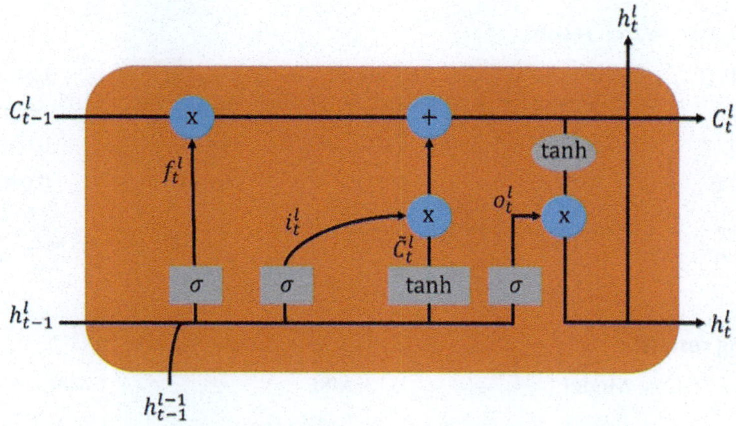

FIGURE 14.1 LSTM block. *LSTM*, Long-short term memory.

by Eq. (14.4).

$$X(t+1) = \text{Bestpx}R + (\text{Maximum} - \text{Minimum})XR + \text{MinimumXLevy(Dim)} \tag{14.4}$$

where the gigantic trevally's position vector in the subsequent iteration is represented by $X(t+1)$, the best position gained is represented by Bestp, a random number between 0 and 1 is represented by R, and the Levy flight is represented by Levy(Dim).

14.2.2.2 Choosing area

During this stage the system determines the best hunting area inside the search space by looking at the availability of food. This phenomenon may be quantitatively replicated using the following equation. This expression is given by Eq. (14.5).

$$X(t+1) = \text{BestpxAx}R + \text{Mean}_{\text{Info}} - \text{Xi}(t)xR \tag{14.5}$$

where Xi(t) indicates the current position, R is a random integer, and A is a parameter that regulates position change. The title Mean_Info for these huge trevallies implies that all of the data that was collected from earlier sites was effectively used.

14.2.2.3 Attacking

The last phase of the algorithm replicates the trevally's assault on its victim. The trevally's behavior is affected by light refraction, which also affects its vision. The method uses Snell's equation to compute the visual distortion ν to mimic this behavior. Next, it uses the following to replicate the trevally's attack. This expression is given by Eq. (14.6).

$$X(t+1) = \mathcal{L} + \mathcal{V} + \mathcal{H} \tag{14.6}$$

Thus the algorithm may transition from the exploration phase to the exploitation phase by using the variables $X(t+1)$ for the next position, $\mathcal{L}$ for the launch speed, ν for the visual distortion, and $\mathcal{H}$ for the jumping slope function (Hashish et al., 2023).

14.2.3 Long-short term memory

Even in cases where the minimal time lags are extremely long, the long-short term memory (LSTM) model is a potent recurrent neural system specifically engineered to overcome the exploding/vanishing gradient problems that generally arise when learning long-term dependencies. In general, this may be avoided by employing a constant error carousel (CEC), which keeps the error signal contained in the cell of each unit. In actuality, these cells are recurrent networks in and of themselves. Their intriguing design comes from the way the input and output gates, which together create the memory cell, are added to the CEC. One time step lag is indicated by the self-recurrent connections. The design of a vanilla LSTM block is shown and includes peephole connections, activation functions, gates, input signal $x(t)$, and output $y(t)$. The block's output is repeatedly linked to both the block's input and every gate. Fig. 14.1 is given to represent an LSTM block.

To provide light on the operation of the LSTM model, let us consider a network with N processing blocks and M inputs. Below is a description of this recurrent neural network system's forward model (Van Houdt et al., 2020).

TABLE 14.1 GTO optimization search results.

Parameter	Units_1		Units_2		Learning_rate	
	Area	Found	Area	Found	Area	Found
K_p	4–512	132.66	4–512	24.97	0.000001–0.1	0.014
K_i	4–512	74.330	4–512	66.10	0.000001–0.1	0.011

GTO, giant trevally optimizer.

TABLE 14.2 Deep learning results.

Parameter	Model	MSE	MAE	R^2
K_p	BiGRU	0.001441	0.034015	0.858131
	LSTM	0.000432	0.015738	0.957494
	GRU	0.001343	0.031351	0.867774
	GTO-LSTM	0.000281	0.013646	0.972302
K_i	BiGRU	0.000595	0.019033	0.993153
	LSTM	0.000449	0.015858	0.994839
	GRU	0.000341	0.015327	0.996082
	GTO-LSTM	0.000146	0.009088	0.998321
Average	BiGRU	0.001018	0.026524	0.925642
	LSTM	0.000441	0.015798	0.976167
	GRU	0.000842	0.023339	0.931928
	GTO-LSTM	0.000214	0.011367	**0.985312**

GTO, giant trevally optimizer; *LSTM*, long-short term memory.

The first step necessitates a choice on what data should be eliminated from the cell state. The "forget gate layer," one of the sigmoid layers, makes these judgments. When making decisions, x_t and h_{t-1} are taken into account, and the results for each integer in cell C_{t-1} might be any value between 0 and 1. If the result is a "0," it means that the data must be deleted; if it is a "1," it means that the data must be kept. This expression is given by Eq. (14.7).

$$f_t = \sigma\left(W_f \cdot [h_{t-1}, x_t] + b_f\right) \tag{14.7}$$

Followed by that, one must now arrange what data should be kept in the cells. There are two components to this technique. First the values that need to be updated are resolved by the input gate layer, which is also a sigmoid coat layer. Second, in this condition, a tan$h$ layer—which is meant to be used for addition—generates the vector of a new character, t. This expression is given by Eqs. (14.8) and (14.9).

$$i_t = \sigma\left(W_f \cdot [h_{t-1}, x_t] + b_t\right) \tag{14.8}$$

$$\tilde{c}_t = \tan h\left(W_f[h_{t-1}, x_t] + b_c\right) \tag{14.9}$$

This stage involves execution, which comes after all the preparation and decision-making. C_t is updated from cell state C_{t-1}. Now, as previously anticipated, f_t is multiplied by the previous state, and then $(i_t \times \widetilde{C}_t)$ is added to make the information disappear. The new character's value, which is obtained in this way, is scaled by a limit in line with the choice to update the cell state value. This expression is given by Eq. (14.10).

$$C_t = f_t\tilde{c}^*_{t-1} + i_tC_t \tag{14.10}$$

In conclusion, planning the output is now crucial. This is determined by the cell state, but the output will be filtered. To show the necessary component of the cell state as the output a sigmoid layer selects it first. The state of this cell is then transmitted along tan$h$ (to limit the results from −1 to 1), and it may subsequently be enhanced by multiplying it with the sigmoid gate layer result to get the precise output that was intended (Bansal et al., 2022). This expression is given by

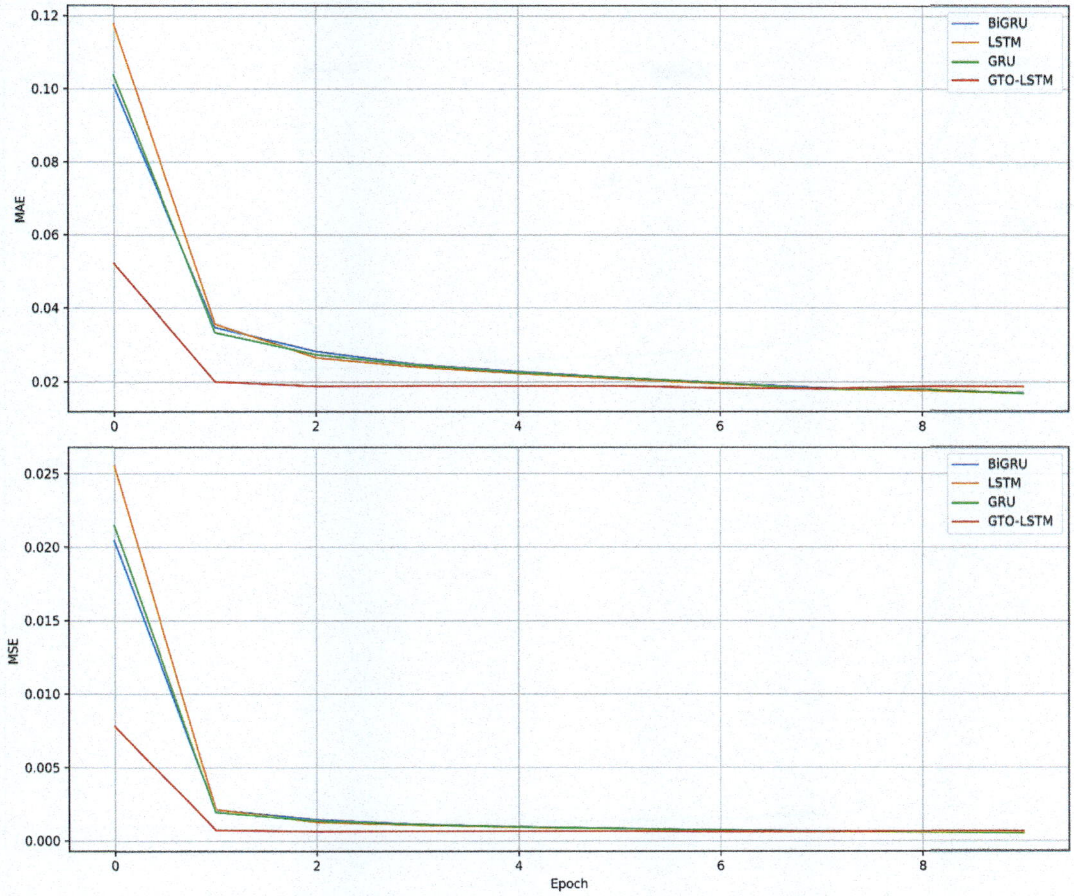

FIGURE 14.2 Result of estimation of parameter K_p. *GTO*, Giant trevally optimizer; *LSTM*, long-short term memory.

Eqs. (14.11) and (14.12).

$$o_t = \sigma(W_o[h_{t-1}, x_t] + b_0) \tag{14.11}$$

$$h_t = \sigma_t \times \tan h(C_t) \tag{14.12}$$

14.3 Results

In this study the PID parameters obtained based on the angle and joint data of a robot arm were successfully estimated with the metaheuristic optimization-based deep learning method. Among the PID parameters the K_d value was kept constant, and the K_p and K_i values were estimated. The deep learning hyperparameter values were determined separately for both control parameters. The hyperparameter values and search area obtained separately for the K_p and K_i values are given in Table 14.1.

Table 14.1 shows the GTO optimization search results. When searching with this optimization method the population number was taken as 10 and the epoch value was taken as 5. Table 14.2 shows the estimation results of deep learning methods. For all deep learning methods the epoch number was determined as 10 and the batch size value was determined as 64.

According to the results in Table 14.2 the GTO-LSTM method became the most successful method by reaching a value of 98.5312% according to the average R^2 metric. The LSTM method became the second most successful method according to the average R^2 metric and reached a value of 97.6167%. The GRU method became the third most successful method by reaching a value of 93.1928% according to the average R^2 metric. The BiGRU method became the least successful method by reaching a value of 92.5642% according to the average R^2 metric.

Fig. 14.2 shows the changes in the MSE and MAE metrics that emerged during deep learning training while estimating the K_p value. As can be seen from the figure, the GTO-LSTM method made the least estimation error by making a prediction

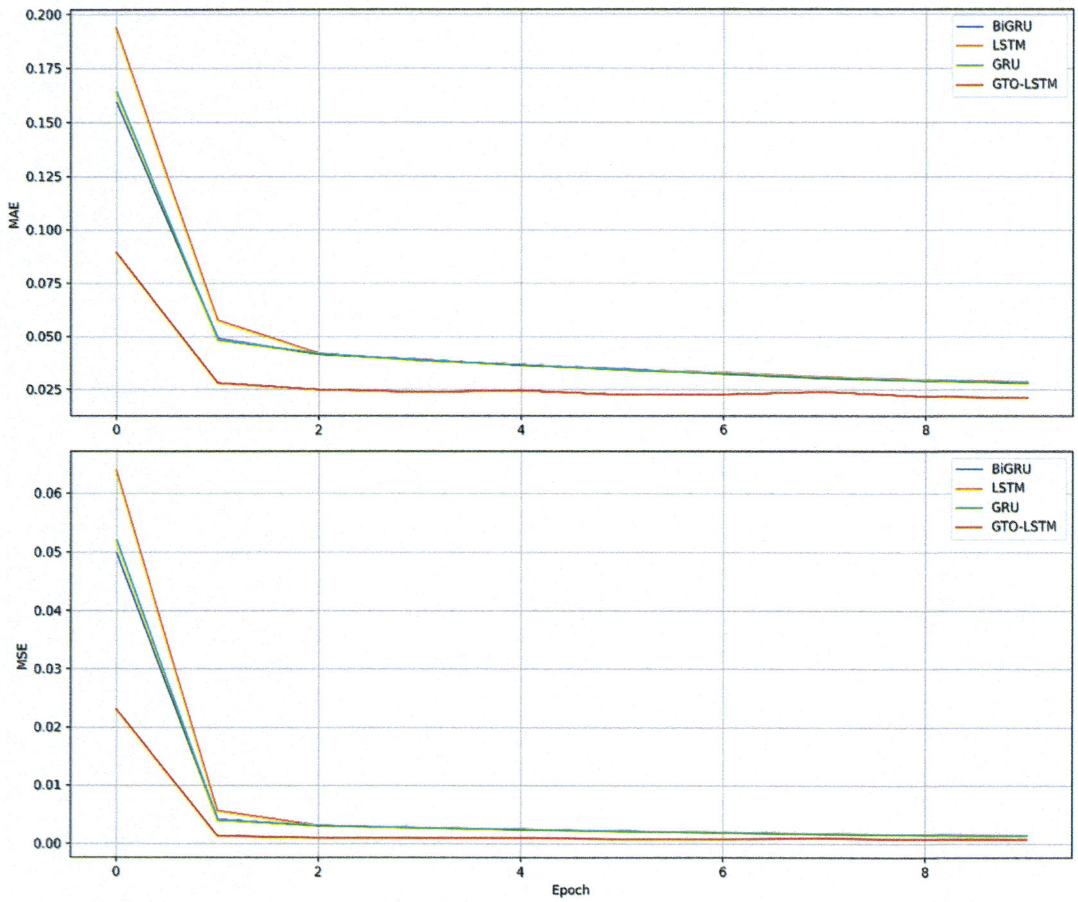

FIGURE 14.3 Estimation result of the parameter K_i about K_p estimation. *GTO*, Giant trevally optimizer; *LSTM*, long-short term memory.

of 0.013646 according to the MAE metric and 0.000281 according to the MSE metric. The LSTM method was the second method with the least error in prediction, making errors of 0.000432 and 0.015738 according to the MSE and MAE metrics, respectively. The GRU method was the third method with the least error in prediction, making errors of 0.001343 and 0.031351 according to the MSE and MAE metrics, respectively. The BiGRU method was the method with the most error in prediction, making errors of 0.001441 and 0.034015 according to the MSE and MAE metrics, respectively, and was the least successful method.

Fig. 14.3 shows the changes in the MSE and MAE estimation error metrics that occur while estimating the K_i parameter of the models during the deep learning training conducted in this study. The GTO-LSTM method made the least estimation error while estimating the K_i parameter, whereas the changes in the MSE and MAE metrics that occur during the deep learning training were 0.000146 and 0.009088, respectively. The LSTM method became the third method that made the least error by making 0.000449 and 0.015858 estimation errors according to the MSE and MAE metrics, respectively. In addition, the GRU method became the second method that made the least estimation error by making 0.000341 and 0.015327 estimation errors according to the MAE and MSE metrics, respectively. The BiGRU method became the method that made the most estimation errors according to the MSE and MAE metrics, respectively.

14.4 Conclusion

Thanks to the healthy, fast, and safe control of robots, efficiency in production and user safety are very important in many aspects. For this reason, in this study, a new dataset was obtained from the PID parameters that emerged by applying the modeling for the vertical movement of a robot arm using a publicly available dataset. The obtained dataset includes K_p, K_i, K_d, vertical error, and horizontal error parameters. In this study, estimation was made only using K_p, K_i, K_d, and vertical error data. The K_d parameter value was kept constant, and K_p and K_i values were estimated according to K_i and error vertical parameters. Hyperparameter values were obtained for K_p and K_i using separate GTO optimization. The obtained hyperparameter values were applied to the LSTM method, and the GTO-LSTM method was obtained. The performance of four separate deep learning models was compared in the deep learning experiments. The GTO-LSTM method became the most successful method by achieving 98.5312% estimation success according to the R^2 metric according to the average of all experiments. The author plans to work on new robot models for future studies.

References

Angel, L., & Viola, J. (2018). Fractional order PID for tracking control of a parallel robotic manipulator type delta. *ISA Transactions, 79*, 172–188. doi:10.1016/j.isatra.2018.04.010.

Bansal, M., Goyal, A., & Choudhary, A. (2022). A comparative analysis of K-nearest neighbor, genetic, support vector machine, decision tree, and long short term memory algorithms in machine learning. *Decision Analytics Journal, 3*, 100071. doi:10.1016/j.dajour.2022.100071.

Hashish, M. S., Hasanien, H. M., Ullah, Z., Alkuhayli, A., & Badr, A. O. (2023). Giant trevally optimization approach for probabilistic optimal power flow of power systems including renewable energy systems uncertainty. *Sustainability, 15*(18), 13283. doi:10.3390/su151813283.

Jimenez-Vazquez, E., Ayala-Rodriguez, J., Navarro-Duran, D., & Lopez-Caudana, E. (2021). Model approximation of an arm of the NAO™ robot using system identification. *2nd International Symposium on Instrumentation, Control, Artificial Intelligence, and Robotics, ICA-SYMP 2021*. Institute of Electrical and Electronics Engineers Inc. doi:10.1109/ICA-SYMP50206.2021.9358430.

Jinjun, D., Yahui, G., Ming, C., & Xianzhong, D. (2019). Symmetrical adaptive variable admittance control for position/force tracking of dual-arm cooperative manipulators with unknown trajectory deviations. *Robotics and Computer-Integrated Manufacturing, 57*, 357–369. doi:10.1016/j.rcim.2018.12.012.

Kalamian, N., Niri, M. F., & Mehrabizadeh, H. (2019). Design of a suboptimal controller based on Riccati equation and state-dependent impulsive observer for a robotic manipulator. In *Proceedings—2019 6th International Conference on Control, Instrumentation and Automation, ICCIA 2019*. Institute of Electrical and Electronics Engineers Inc. doi:10.1109/ICCIA49288.2019.9030988.

Kleeberger, K., Bormann, R., Kraus, W., & Huber, M. F. (2020). A survey on learning-based robotic grasping. *Current Robotics Reports, 1*(4), 239–249. doi:10.1007/s43154-020-00021-6.

Kumra, S., & Kanan, C. (2017). Robotic grasp detection using deep convolutional neural networks. In *IEEE International Conference on Intelligent Robots and Systems* (pp. 769–776). Institute of Electrical and Electronics Engineers Inc. doi:10.1109/IROS.2017.8202237.

Li, S., Zheng, P., Liu, S., Wang, Z., Vincent Wang, X., Zheng, L., & Wang, L. (2023). Proactive human–robot collaboration: Mutual-cognitive, predictable, and self-organising perspectives. *Robotics and Computer-Integrated Manufacturing, 81*, 102510. doi:10.1016/j.rcim.2022.102510.

Ma, L., Yao, Y., & Wang, M. (2017). The optimizing design of wheeled robot tracking system by PID control algorithm based on BP neural network. In *Proceedings—2016 International Conference on Industrial Informatics—Computing Technology, Intelligent Technology, Industrial Information Integration, ICIICII 2016* (pp. 34–39). Institute of Electrical and Electronics Engineers Inc. doi:10.1109/ICIICII.2016.0020.

Mukherjee, D., Gupta, K., Chang, L. H., & Najjaran, H. (2022). A survey of robot learning strategies for human-robot collaboration in industrial settings. *Robotics and Computer-Integrated Manufacturing, 73*. doi:10.1016/j.rcim.2021.102231.

Nugroho, A., Yuniarno, E. M., & Purnomo, M. H. (2023). ARKOMA dataset: An open-source dataset to develop neural networks-based inverse kinematics model for NAO robot arms. *Data in Brief, 51*. doi:10.1016/j.dib.2023.109727.

Sadeeq, H. T., & Abdulazeez, A. M. (2022). Giant trevally optimizer (GTO): A novel Metaheuristic algorithm for global optimization and challenging engineering problems. *IEEE Access, 10*, 121615–121640. doi:10.1109/ACCESS.2022.3223388.

Taş, G., & Bal, C. (2023). Bloomed or non-bloomed fruit tree classification with transfer learning. *Remote Sensing Letters, 14*(9), 981–992. doi:10.1080/2150704X.2023.2258457.

Van Houdt, G., Mosquera, C., & Nápoles, G. (2020). A review on the long short-term memory model. *Artificial Intelligence Review, 53*(8), 5929–5955. doi:10.1007/s10462-020-09838-1.

Zeng, A., Song, S., Yu, K. T., Donlon, E., Hogan, F. R., Bauza, M., … Rodriguez, A. (2022). Robotic pick-and-place of novel objects in clutter with multi-affordance grasping and cross-domain image matching. *International Journal of Robotics Research, 41*(7), 690–705. doi:10.1177/0278364919868017.

Chapter 15

Mechatronic design and implementation of a modular multiagent robotic system with a coupling mechanism for collaborative object transportation

Alberto Luviano-Juárez[a], Jaime Guzmán-López[b], David Bautista-Lázaro[c] and Mario Ramirez-Neria[d]

[a]*National Polytechnic Institute UPIITA, Mexico City, Mexico,* [b]*Intelligent Computing Laboratory, National Polytechnic Institute—CIDETEC, Mexico City, Mexico,* [c]*Advanced Technologies, National Polytechnic Institute—UPIITA, Mexico City, Mexico,* [d]*Institute for Applied Research and Technology (INIAT), Ibero-American University, Mexico City, Mexico*

15.1 Introduction

Modern manufacturing systems must be flexible, reactive, integrated, and cost-effective simultaneously to enable industrial companies to stay competitive in an international competition (Mohamed, 2018).

An automated guided vehicle (AGV) is an industrial transport system, which is associated with the concept of flexible manufacturing systems (FMS) by automating the transportation of materials and replacing traditional production lines (Espinosa et al., 2020).

AGV systems have been widely implemented across various sectors, including logistics warehouses, the agronomy industry, and military operations, among others (Espinosa et al., 2020; Moreno & Ceballos, 2019). However, most available commercial AGV solutions are designed to handle loads of at least 1000 kg, making them unsuitable for small and medium-sized enterprises (SMEs). These businesses often lack the necessary infrastructure to support such systems, and the cost-benefit ratio is not favorable enough to justify their implementation.

As a result, AGV systems with smaller dimensions and reduced load capacity are being developed (Tuci et al., 2018). A multiagent system (MAS) is a network of problem-solving entities that work together to find the answer to problems that are beyond the individual capabilities or knowledge of each entity (Farahvash & Boucher, 2004). This approach enhances operational flexibility while addressing the specific needs of smaller-scale operations.

Although this may be true, deploying multiple robots designed to work together simultaneously generates a substantial amount of data. As the number of agents increases, so does the volume of data, leading to significantly higher computational costs required to manage and control each individual unit. This can pose a challenge in terms of system efficiency and resource allocation.

The integration of *Industry 4.0* technologies, such as Internet of Things (IoT) protocols, allows high-volume communication between machines, devices, and sensors to make production processes more efficient, flexible, and automated (Soori et al., 2023).

AGVs needs to plan a collision-free, safe, and feasible path to avoid obstacles and accurately reach desired location in the working environment while performing transportation of goods or reaching the charging station (Tang et al., 2021). However,

Computational Intelligence in Mechatronics: Solving Real-World Problems in Electronic Systems Design. DOI: https://doi.org/10.1016/B978-0-443-27580-7.00002-0

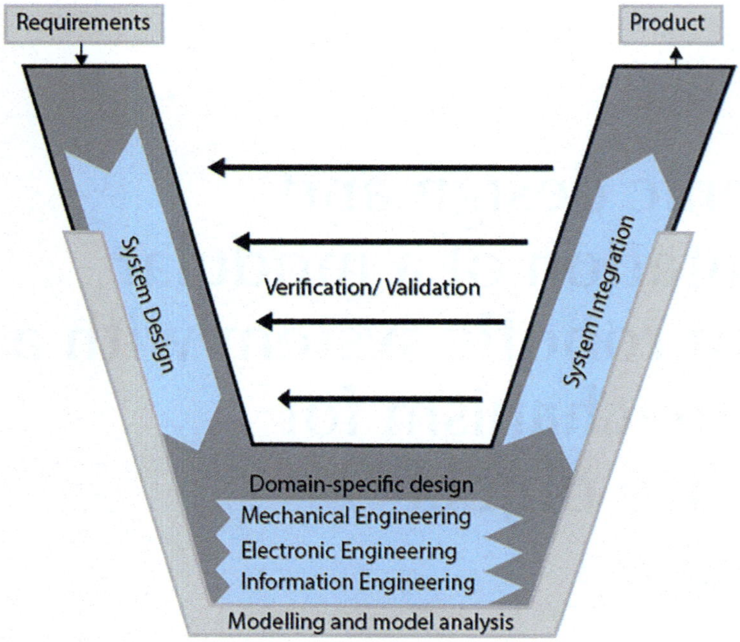

FIGURE 15.1 The V-Model. The V-Model according to guideline VDI 2206.

mobile robots that move over the ground have limited perception range and are less comprehensive of their environment because of occlusion of their sensors (Borges et al., 2019).

This limitation can be addressed by implementing an external computer vision system capable of detecting obstacles in the environment and performing path planning operations. In addition, the system can be adapted to execute visual servoing of the robotic agents, allowing for continuous tracking of each agent's global position and ensuring more precise coordination and navigation.

On the other hand, control algorithms for multiagent robotic systems require a significant amount of coordination of actions to sustain the transport (Tuci et al., 2018).

The purpose of this chapter is to present the design and implementation of a multiagent robotic system capable of collaborative payload transport through mechanical coupling between agents. This approach aims to reduce computational costs and simplify the algorithms required for coordination, making the system more efficient and easier to implement. Moreover, the system implements an IoT network architecture for communication and incorporates an external computer vision system for path planning and visual servoing.

15.1.1 Mechatronic design methodology

The design was based on the VDI-2206 methodology for mechatronic design (VDI-Fachbereich Produktentwicklung, 2004) for function and system architecture modeling and design; the V model of this methodology is shown in Fig. 15.1. This methodology allows the designer to have multiple integration and validation stages along the project, allowing the engineers to validate the requirements without compromising the functionality and synergy of the system. The methodology defines five stages for the development of a project.

- **System design:** The first stage defines the overall architecture and structure of the system. It facilitates the identification of subsystems and their interactions by establishing requirements, specifications, and performance criteria.
- **Domain-specific design:** The second stage focuses on designing components related to a specific application domain, incorporating domain knowledge and best practices, such as ensuring compatibility between relevant standards and technologies.
- **System Integration:** This stage integrates individual components to ensure compatibility across the system, addressing integration challenges and synergies among multidomain subsystems.

- **Modeling and model analysis:** As a parallel process during design and integration, this stage emphasizes project modeling to represent system behavior and interactions, enabling the generation of models that enhance design decisions through simulations and multimodel analysis.
- **Assurance of properties:** This parallel process verifies that properties and requirements are met at every stage of design, helping to identify metric goals and guiding the design toward a requirements-based solution.

15.2 Design of the system

15.2.1 Requirements and function definition

The first step in the VDI-2206 methodology is to define the system's requirements and functions. For the proposed system the agents were designed to meet the standard and safety needs of a convenience store. Their primary functions include transporting objects both individually and collaboratively, with the added innovation of a coupling mechanism to enhance payload transport capabilities.

To build the systems and modules that constitute the entire system, known as the system functional architecture (SFA), the functional breakdown structure (FBS) technique was used to segment the system's functions effectively.

FBS is an indentured listing of the system's functions, starting from a generalized concept and breaking them down into specific functions that cannot be further decomposed (Dehoff et al., 2009). This hierarchical structure helps to clearly define each function within the system architecture.

Using this technique the systems and modules were identified for the proposed system. Each system and module is outlined as follows:

- **S1. Mechanical and structural system.** This system is responsible for integrating and supporting the mechanical and electronic components of each agent.
- **M1.1. Displacement module.** This module is responsible for generating, controlling, and monitoring the locomotion of each agent.
- **M1.2. Coupling module.** This module is responsible for the coupling between agents.
- **M1.3. Structural module.** This module outlines the main structural framework of each agent, designed to house and integrate all internal modules and systems within the robotic agents. In addition, it provides the necessary support for carrying the load of the objects being transported.
- **S2. Agent processing system.** This system is responsible for data processing and coordinating the actuators within the robotic agent.
- **M2.1. Agent processing module.** This module is responsible for processing, management, and validation of the information acquired by external systems or agents.
- **M2.2. Agent monitoring module.** This module is responsible for monitoring the internal information of each agent.
- **S3. Communication and IoT management system.** This system manages data transmitted via IoT protocols.
- **M3.1. Communication module.** This module is responsible for transmitting data from the robotic agents and external systems to the IoT management module.
- **M3.2. IoT management module.** This module manages the exchange of information through IoT channels.
- **S4. Battery management system.** This system is responsible for monitoring, distributing, and protecting the electrical supply for the robotic agents.
- **M4.1. Power distribution module.** This module is responsible for distributing the appropriate voltage levels and the physical layer of the data distribution bus.
- **M4.2. Power regulation module.** This module is responsible for regulating and managing the various voltage levels required by the modules and systems within the robotic agent.
- **M4.3. Battery recharge module.** This module monitors the charging and discharging cycles of the robotic agent's battery.
- **S5. Computer vision system.** This system is responsible for monitoring and recognition of the elements inside the working space.
- **M5.1. Recognition module.** This module is responsible for recognizing and differentiating between the robotic agents and the objects to be transported. In addition, it processes data received from the environment monitoring module.
- **M5.2. Environment monitoring module.** This module is responsible for capturing and monitoring the working space.
- **S6. Human–machine interface (HMI).** This system is responsible for data visualization and facilitating interaction between the system and the operator.

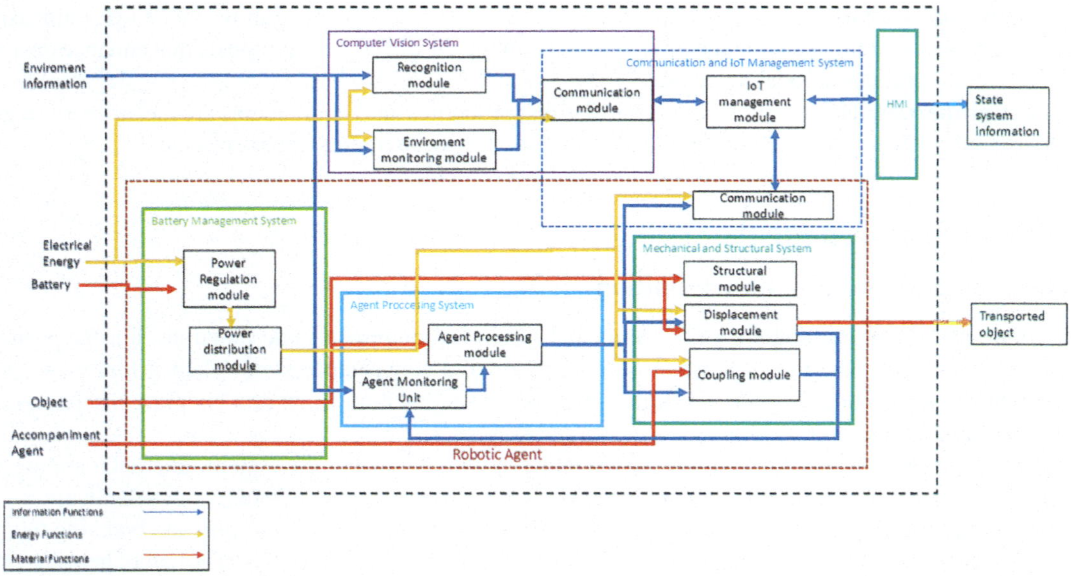

FIGURE 15.2 System physical architecture. Representation of the systems and their corresponding modules, highlighting their integration and relationships. *HMI*, Human–machine interface; *IOT*, internet of things.

The relationships between systems and modules are illustrated in Fig. 15.2, which represents the system physical architecture (SPA). The diagram details the flow of energy, information, and materials throughout the system.

In the following sections the development of the relevant modules and systems in the specific domain is described.

15.2.2 Design of M1.1: Displacement module

15.2.2.1 Motor torque and power analysis

The selection of mechanical displacement components involved calculating the design load using Eq. (15.1), which accounts for load distribution in a four-wheel mobile robot with an expected payload of 5 kg and an agent weight of 2 kg.

$$\text{Load}_{\text{Design}} = \frac{m_R + \left(m_{\text{Obj}} \times \text{FS}\right)}{\text{No. Wheels}} \tag{15.1}$$

where $\text{Load}_{\text{Design}}$ is the design load in kg; m_R is the weight of robot; m_{Obj} is the weight of object; FS is the factor of security; and No. Wheels is the number of wheels per agent.

To fulfill safety requirements for in-house warehouse robots the maximum linear speed of the robot should not exceed 0.5 m/s.

The component selection was realized, considering the restrains and requirements for motor torque and linear speed, by comparing the model combination for commercial DC motors and omnidirectional wheels of 45 degrees (the kinematics of 45-degree wheels is expanded in the following section).

The motor parameters were taken from characteristics curves in Pololu Corporation (2019) to ensure the maximum efficiency between mechanical parameters and electrical power as indicated in the datasheet of the motor.

The driver was selected to maximize efficiency during continuous operation and to safely support the motor's maximum power under peak torque conditions.

15.2.2.2 Omnidirectional four-wheeled mobile robot inverse kinematic

Given the limited space in which the robotic agent will operate, the ability to move in all directions within the *XY* plane without needing to rotate beforehand is essential. To achieve this type of movement, omnidirectional robots utilize wheels equipped with rollers arranged in two angular configurations: 45 and 90 degrees, these kinds of wheels are also referred to as Mecanum wheels.

However, equipping a mobile robot with these types of wheels does not necessarily guarantee omnidirectional movement. In Li et al. (2019), various configurations of four-wheeled robots were analyzed, as illustrated in Fig. 15.3. These

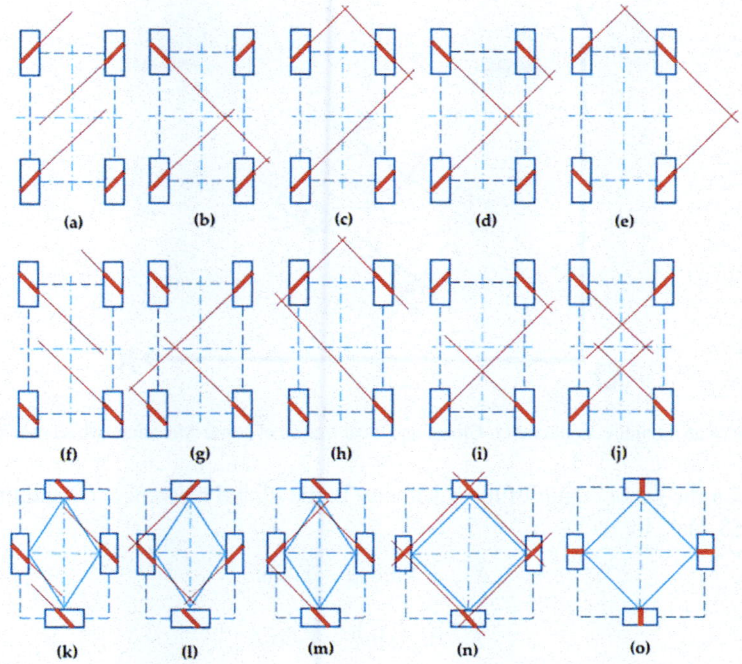

FIGURE 15.3 Configurations of four-Mecanum-wheel mobile robot: (a)-(j) corner-mounted rectangular configurations with axles parallel to longitudinal axis; (k)-(n) centripetal; (o) circular centripetal array with orthogonal Mecanum wheels. *(From Li, Y., Dai, S., Zhao, L., Yan, X., & Shi, Y. (2019). Topological design methods for Mecanum wheel configurations of an omnidirectional mobile robot.* Symmetry, 11*(10). https://doi.org/10.3390/sym11101268.)*

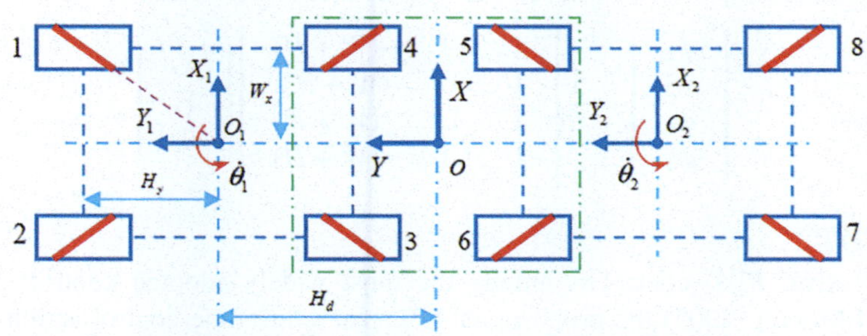

FIGURE 15.4 Tandem configuration of two omnidirectional mobile robots. A tandem configuration composed of two symmetrical four-Mecanum-wheel subconfigurations. *(From Li, Y., Dai, S., Zhao, L., Yan, X., & Shi, Y. (2019). Topological design methods for Mecanum wheel configurations of an omnidirectional mobile robot.* Symmetry, 11*(10). https://doi.org/10.3390/sym11101268.)*

configurations were evaluated based on two key indicators: whether omnidirectional movement was achieved and how well the system performed in terms of maneuverability.

When operating collaboratively, it is essential to ensure that the tandem configurations formed by the coupling of two or more agents also achieve omnidirectional movement. Configuration *e* from Fig. 15.3 was selected because it not only enables omnidirectional movement but also offers strong performance. Furthermore, when two agents with this configuration are coupled, the resulting tandem configuration corresponds to *j* in Fig. 15.3, as depicted in Fig. 15.4.

As a key design consideration, it is crucial to ensure that the tandem configuration avoids singularities. Otherwise the system will fail to achieve omnidirectional movement when coupling two or more agents.

Once the configuration type has been selected, and given the implementation of path-following functions, it is essential to calculate the speed assigned to each wheel to allow the robot to move in the *XOY* plane according to the specified velocities $\dot{X}$, $\dot{Y}$, and $\dot{\theta}$. Fig. 15.5 illustrates the global frame of reference *XOY*, which contains the mobile robot's frame of reference $X_rO_rY_r$, located at a distance *X, Y* from the global origin. The frame $X_{r_i}O_{r_i}Y_{r_i}$ corresponds to the *i*th wheel, where $i = 1, 2, 3, 4$. Each wheel rotates around its X_{r_i} axis with an angular displacement of ϕ_i, the roller angle is defined as γ_i, and *H* and *W* represent the vertical and horizontal distances from the robot's center of mass to the center of mass of the wheels.

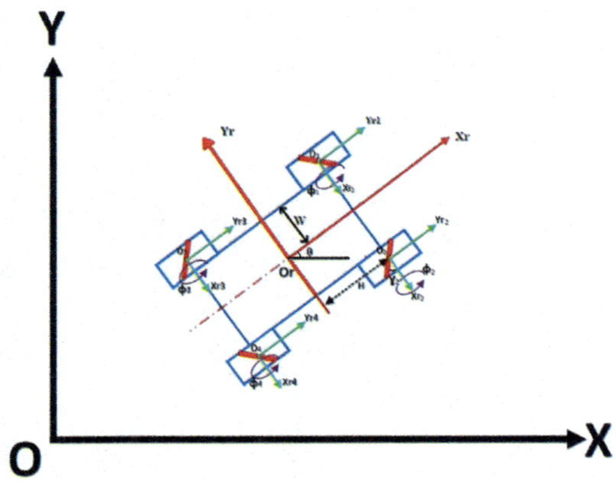

FIGURE 15.5 Reference frame of an omnidirectional robot. Global and local frames of a four-Mecanum-wheel mobile robot.

The relationship between the robot center of mass velocity in the global reference, O, and the robot local reference, O_r are defined in Eqs. (15.2–15.4).

$$\dot{X}_{r_O} = \dot{X}_r \cos\theta - \dot{Y}_r \sin\theta \tag{15.2}$$

$$\dot{Y}_{r_O} = \dot{X}_r \sin\theta + \dot{Y}_r \cos\theta \tag{15.3}$$

$$\dot{\theta}_{r_O} = \dot{\theta} \tag{15.4}$$

Therefore given linear and angular velocities, the angular speed of each wheel can be calculated. For the configuration implemented the angles of the rollers are defined ad $\gamma_1 = -\gamma_2 = \gamma_3 = -\gamma_4$, where $|\gamma_i| = 45$ degrees, where $i = 1, 2, 3, 4$, and all wheels have the same radius R. Eq. (15.5) was used to calculate the inverse kinematic.

$$\begin{bmatrix} \dot{\phi}_1 \\ \dot{\phi}_2 \\ \dot{\phi}_3 \\ \dot{\phi}_4 \end{bmatrix} = \frac{1}{R} \begin{bmatrix} 1 & -1 & -(W+H) \\ 1 & 1 & (W+H) \\ 1 & 1 & -(W+H) \\ 1 & -1 & -(W+H) \end{bmatrix} \begin{bmatrix} \dot{X}_r \\ \dot{Y}_r \\ \dot{\theta}_r \end{bmatrix} \tag{15.5}$$

15.2.2.3 *Control strategy*

Proportional–integral–derivative (PID) control is among the most widely adopted control strategies (Knospe, 2006). According to Johnson and Moradi (2005) the proportional (P) term adjusts the control action in proportion to the error magnitude between the output signal and the reference. The integral (I) term addresses steady-state errors, enhancing accuracy by compensating for any residual discrepancies. Finally the derivative (D) term incorporates a predictive element, anticipating future errors to improve control stability.

This kind of control can be implemented in DC motors to reach the desired velocity by controlling the Pulse Width Modulation (PWM) sent into the actuators to control the speed of the motor's shaft.

15.2.2.3.1 Path-following control strategy

For the path-following task a proportional (P) control approach was implemented because of the high precision of the computer vision system, which can measure the position of the robotic agents with an accuracy of up to 3 mm.

The proposed controller is presented in Eq. (15.6), where the output corresponds to the velocities in the global frame along the x-axis, y-axis, and rotational axis. The velocities marked with an asterisk indicate the desired velocity and position for the target trajectory.

$$\begin{bmatrix} \dot{x} \\ \dot{y} \\ \dot{\theta} \end{bmatrix} = \begin{bmatrix} \dot{x}^* \\ \dot{y}^* \\ \dot{\theta}^* \end{bmatrix} - \begin{bmatrix} k_x & 0 & 0 \\ 0 & k_y & 0 \\ 0 & 0 & k_\theta \end{bmatrix} \begin{bmatrix} x - x^* \\ y - y^* \\ \theta - \theta^* \end{bmatrix} \tag{15.6}$$

15.2.2.3.2 Validation of path-following control strategy

The control strategy was first validated through numerical simulation prior to physical implementation. Fig. 15.6 provides

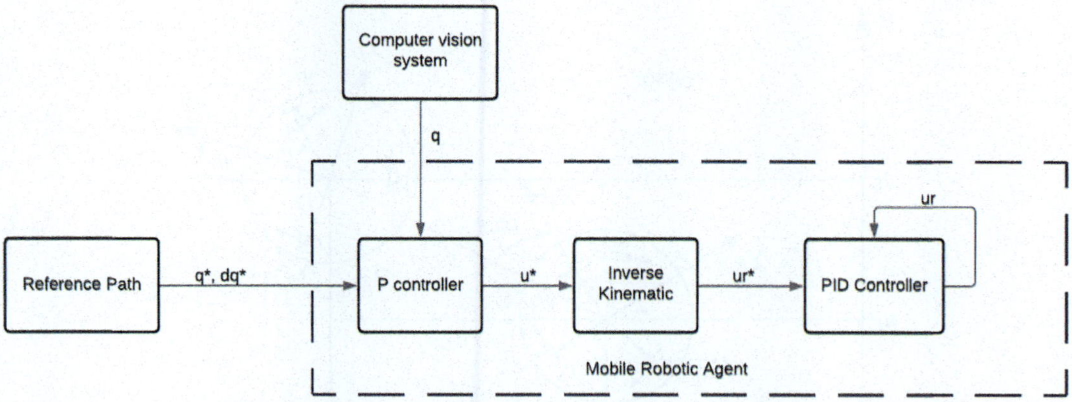

FIGURE 15.6 Control strategy for path-following task. The diagram shows the flow of the information to achieve the control of velocity in each wheel. *PID*, Proportional-integral-derivative.

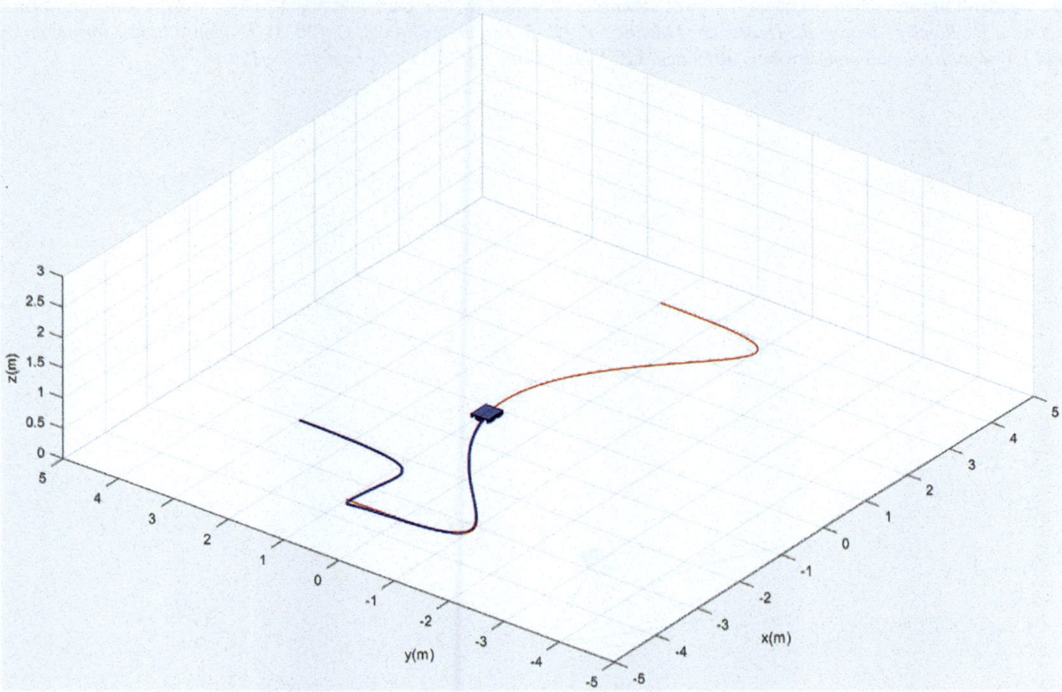

FIGURE 15.7 Simulation of path following for a single robotic agent. A simulated robot performed a proposed trajectory.

an overview of the simulated strategy, where q denotes the position in the global frame, q^* and dq^* represent the desired position and velocities, respectively, also in the global frame. The term u^* denotes the velocities in the mobile robot frame, whereas ur and ur^* represent the actual and desired speeds of each wheel, respectively.

The strategy was then simulated for a desired trajectory in which the mobile robotic agent moved between two points within the simulated working environment, with motion along the X and Y axes and no desired rotation, as illustrated in Fig. 15.7.

15.2.2.3.3 Leader–follower control strategy

Collaborative work is required to transport larger or heavier objects. The leader–follower scheme is widely used in autonomous mobile robotics, where an agent, designated as the *leader*, serves as the reference for other agents, termed *followers*, during the path-following task, as shown in Fig. 15.8 from Ramírez-Neria et al. (2023).

In this system the robotic agents are coupled to the leader, maintaining a fixed distance between agents with minimal perturbations, which provides an advantage when implementing this scheme.

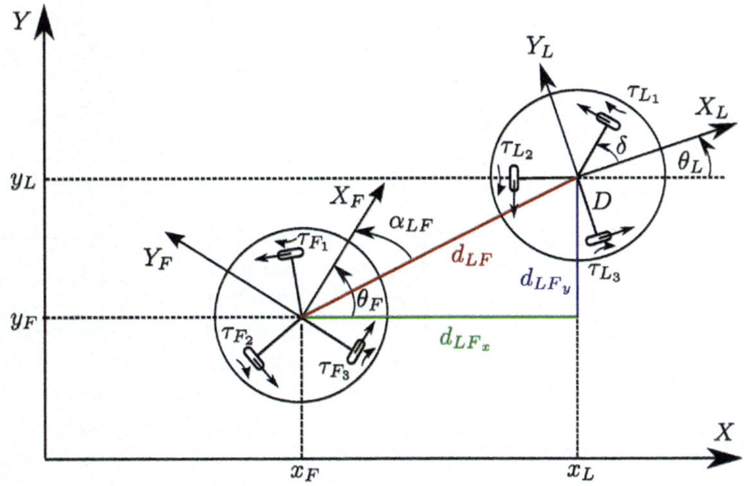

FIGURE 15.8 Leader–follower scheme. Leader–follower scheme represented by two omnidirectional robots. (*From Ramírez-Neria, M., González-Sierra, J., Madonski, R., Ramírez-Juárez, R., Hernandez-Martinez, E. G., & Fernández-Anaya, G. (2023). Leader–follower formation and disturbance rejection control for omnidirectional mobile robots.* Robotics, 12*(5). https://doi.org/10.3390/robotics12050122.*)

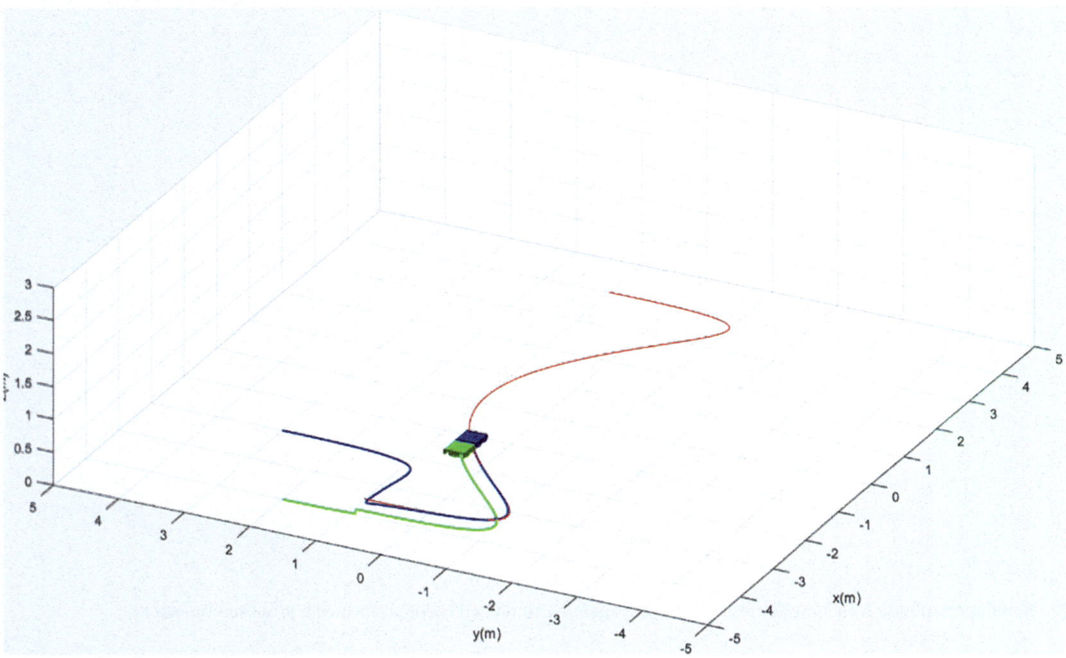

FIGURE 15.9 Simulation of a collaborative path-following task. Two simulated robots performed a proposed trajectory in leader–follower scheme.

15.2.2.3.4 Validation of collaborative path-following control strategy

By maintaining the same control scheme for the leader agent and using the leader's actual position as a reference for the follower unit a simulation was performed under these conditions and along the same trajectory, as shown in Fig. 15.9.

15.2.3 Design of M1.2: Coupling module

15.2.3.1 Design of coupling mechanism

The selected coupling mechanism between robotic agents is a four-finger gripper. The design process begins by calculating the required force for the gripper to maintain coupling between units under the worst case scenario, which occurs when two units move in opposite directions at maximum acceleration.

The estimated mass of each robotic agent is 12 kg, with a maximum acceleration limited to 0.15 m/s^2, resulting in a required force of 1.8 N. In the worst case scenario, this force is exerted by each unit in opposite directions, doubling the force required by the coupling mechanism. To ensure reliability a safety factor of 4 was applied based on Mott (1995), resulting in a total force requirement for the mechanism as described in the following equation:

$$F_f = 2 \cdot \text{F.S.} \cdot F = (2)(4)(1.8\ \text{N}) = 14.4\ \text{N} \tag{15.7}$$

The relationship between the total force F_T calculated and the force that a single finger must exert is described in Nagel & Odrey, 2012,as shown in Eq. (15.8), where F_f is the force on each finger, μ represents the static friction coefficient, and x_f denotes the number of fingers in the mechanism.

$$F_f = \frac{F_T}{\mu x_f} \tag{15.8}$$

The tips of the fingers, as well as the female receptacle of the mechanism, are coated with neoprene to enhance adherence. This results in a static friction coefficient of 0.947, which corresponds to a force of 3.8 N exerted by each finger, as calculated in the following equation:

$$F_f = \frac{7.2\ \text{N}}{0.947 \cdot 4} = 3.8\ \text{N} \tag{15.9}$$

Given the dimensions of the robotic agent, two mechanisms had to fit within its largest dimension of 370 mm. This requirement established a maximum elongation of 187 mm for the mechanism, resulting in a maximum allowable finger length of 93.5 mm. However, considering that the length of the finger affects the actuator's moment arm, a finger length of 70 mm was proposed.

To select the material of the mechanism the Ashby methodology (Ashby, 2011) for material selection for engineering was implemented. The methodology results in a performance index P, which is a function that relates the material properties, as described in Eq. (15.10). The objective is to minimize mass, which requires maximizing P. This is achieved by selecting a material with a modulus of elasticity significantly greater than its density.

$$P = \frac{E^{\frac{1}{2}}}{\rho} \tag{15.10}$$

Subsequently the calculation of the minimum Young's modulus required for the link is performed in Eq. (15.11) by proposing a rectangular cross-section of 5 mm and a deflection of 1 mm.

$$E_{\min} = \frac{\text{ML}^2}{4\delta b^4} = \frac{(266\ \text{N} \cdot \text{mm})(70\ \text{mm})^2}{(4)(1\ \text{mm})(5\ \text{mm})^4} = 0.52\ \text{GPa} \tag{15.11}$$

Using this factor, along with the graph presented in Ashby (2011), which compares the modulus of elasticity to material density, Fig. 15.10 illustrates the materials available for constructing the gripper mechanism.

Fig. 15.10 indicates that the link can be constructed from wood, various polymers, composite materials, metals, and ceramics. However, based on the criterion of ease of manufacturing, ceramics are excluded. In addition, as the structural system primarily consists of metallic or printed elements, wood is also omitted. Metals, which possess a higher density than the other alternatives, are discarded in favor of lightweight materials. Ultimately, materials suitable for 3D printing are prioritized; thus examples of materials that meet the proposed requirements include acrylonitrile butadiene styrene (ABS), certain 3D printing resins, and carbon fiber reinforced polymers (CFRP).

Regarding the gripper mechanism the design shown in Fig. 15.11 used the geometrical parameters: r = 20 mm, L = 20 mm, L' = 15 mm, L_1 = 15 mm, M = 20 mm, and K = 30 mm.

By analyzing the kinematics of the mechanism, it is determined that an actuator with a torque rating of 0.671 kg • cm is required to exert a force of 3.8 N at the tip of the finger. Because of the fact that the frictional forces between the links and the rotation joints, as well as the weight of the links, were not considered in the analysis, a factor of two is applied to the calculated moment. Therefore the servomotor must provide at least 1.34 kg • cm. Based on this the MG90S servomotor has the necessary characteristics as an actuator to perform the gripping; however, in practice, this actuator has drawbacks such as overheating when used for more than an hour. Therefore the alternative with superior immediate performance is the MG995.

The coupling mechanism was designed in CAD software based on the results from the previous analysis; male and female modules are shown in Fig. 15.12.

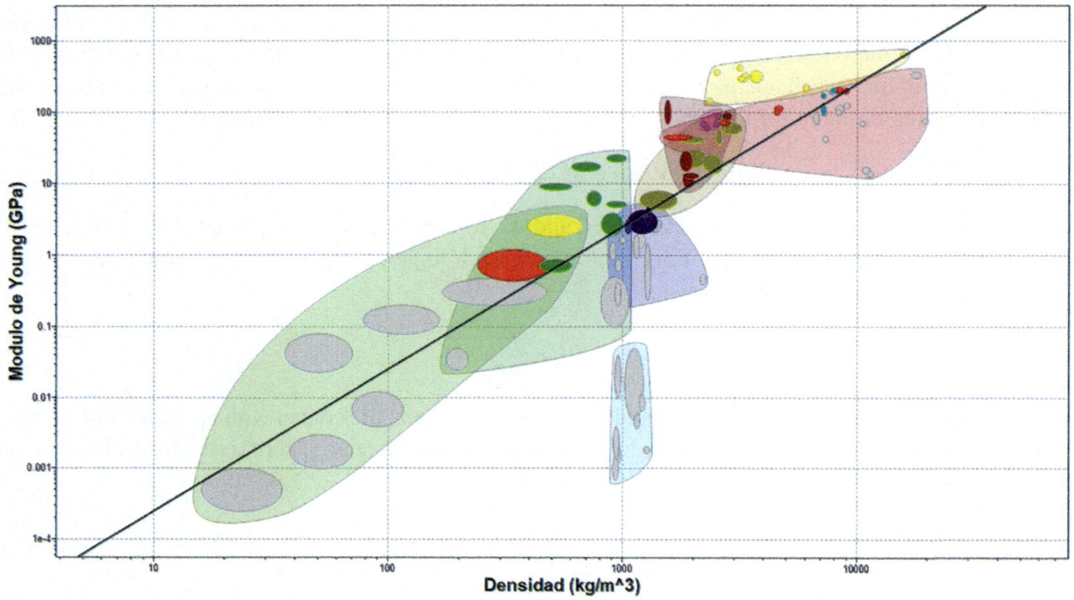

FIGURE 15.10 Ashby plot of density versus Young's modulus with a lower limit. Adecuated materials for the gripper manufacturing.

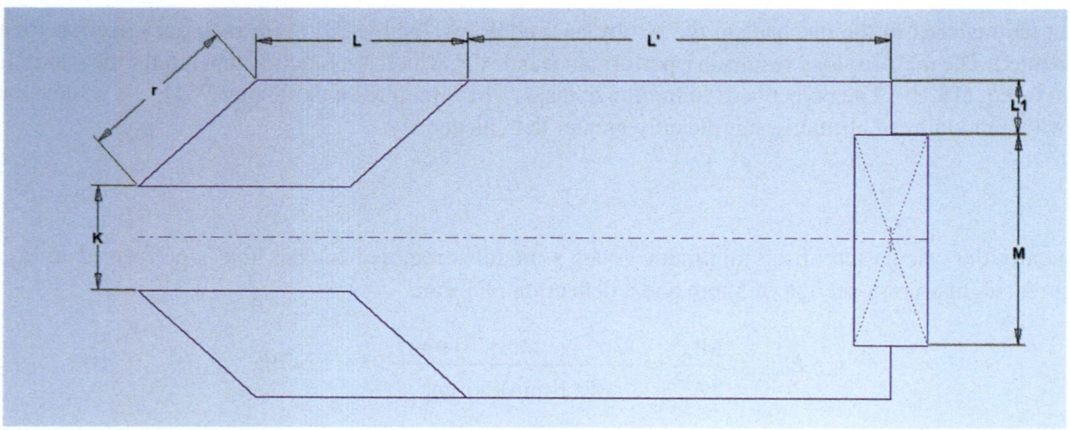

FIGURE 15.11 Free body diagram of half of the coupling mechanism. Free body diagram of a two-finger gripper obtained.

15.2.3.2 Numerical validation of the coupling mechanism

The numerical validation of this module was performed using structural finite element analysis (FEA) in Ansys, with ABS selected as the material for the coupling mechanism, characterized by an elastic modulus of 1.628 GPa. The results of the analysis are presented in Fig. 15.13, which illustrates the response when a torque of 11 kg • cm is applied to the driving gear.

Fig. 15.13 illustrates that the highest concentration of stresses is found between the contacting teeth of the gears, with a maximum value of 0.118 GPa. Given that the modulus of elasticity of ABS is 1.628 GPa, this indicates that significant deformations or failure of the part are unlikely to occur in the critical areas.

The second study analyzes the stresses and deformations in the 4-bar mechanism when the driving gear is positioned at an angle of 30 degrees with respect to the horizontal. This angle corresponds to the moment where the finger is in contact with the female module. To simulate the attachment a fixed joint is applied at the end of the flange, which prevents movement of this component. Fig. 15.14 illustrates the deformation of the links when a driving torque of 11 kg • cm is applied to the driving gear.

As indicated by the scale in Fig. 15.14 the largest deformations occur at the teeth of the gears, remaining below the maximum allowable deformation of 1 mm established during material selection.

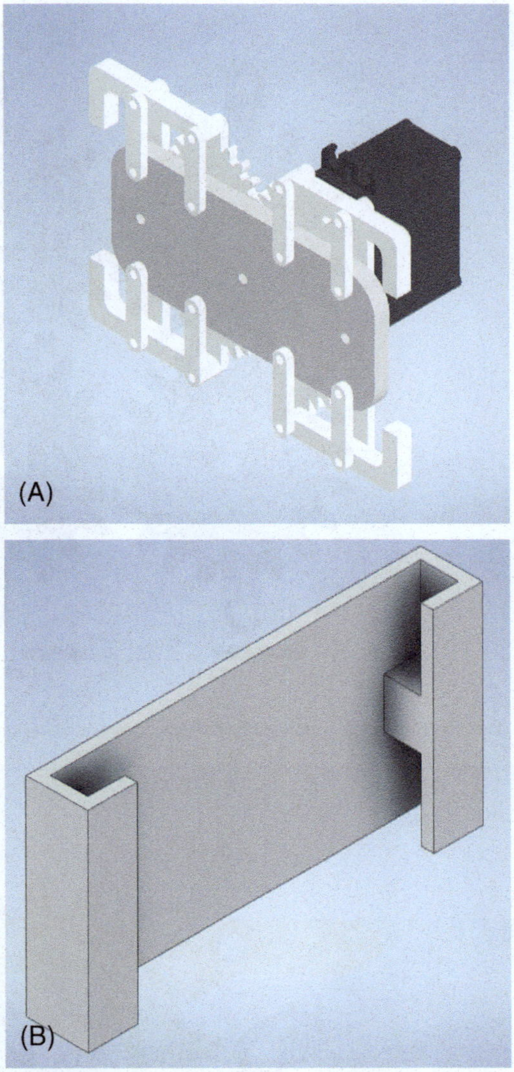

FIGURE 15.12 CAD model of the coupling mechanism: (A) male module and (B) female module.

15.2.4 Design of M1.3: Structural module

To enable the modularity of the agents the structure was designed using an aluminum structural profile. The deflection and stress of the elements were evaluated through CAE analysis in Autodesk Inventor, as shown in Fig. 15.15. The structure was assessed by considering the maximum payload as a distributed force across the elements. The deflection obtained from the analysis is a negligible value compared to the material's elasticity, ensuring the structural and shape stability of the system, as well as identifying the points of highest stress and deflection.

15.2.5 Design of S2: Robotic processing unit

Each agent implements a hierarchical distributed processing of three stages. Two stages are embedded in each agent and an external stage for agent distribution, logistics, and local positioning.

The embedded stages communicate between them to ensure the agent's correct behavior:

- **Interface processing unit (IPU):** Embedded unit in charge of processing at high speed the measurement of the external sensors (motor speed, current consumption, and battery voltage).
- **Central processing unit (CPU):** Processing unit in charge of the following of trajectories, motor speed control, and state machines.

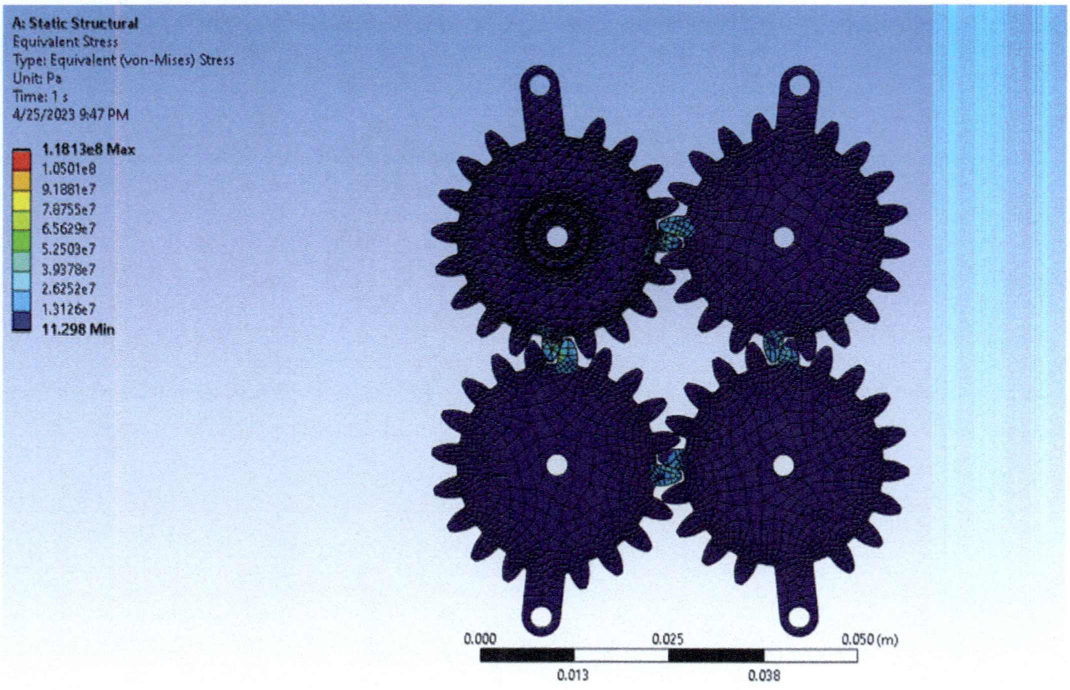

FIGURE 15.13 Stress analysis on the gears. Deformation analysis applied on the power transmission stage of the coupling mechanism.

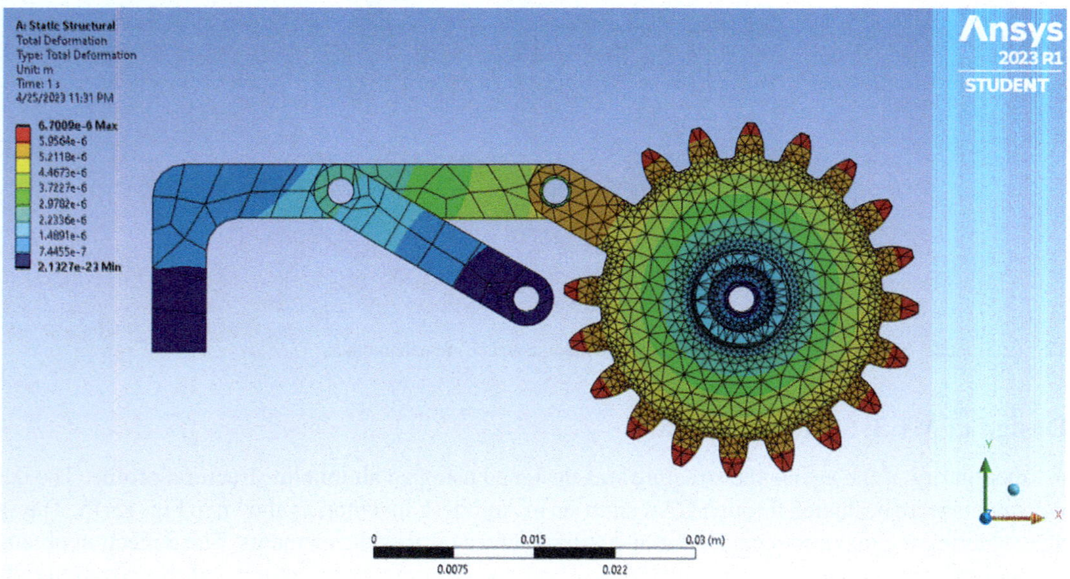

FIGURE 15.14 Stress analysis on the 4-bar mechanism. Deformation analysis applied on the finger structure of the coupling mechanism.

- **Logistics and positioning processing unit (LPPU):** The external stage allows identifying the agents in a local coordinated frame for path generation and mapping.

Every agent is autonomous and can operate as a single unit without requiring other agent information.

15.2.6 Design of S3: Communication system and Internet of Things interface

To ensure communication between the LPPU system and the agents, each agent of the system is fitted with a communication module that allows interagent communication for cooperative behavior and communication with LPPU system for local positioning and path generation.

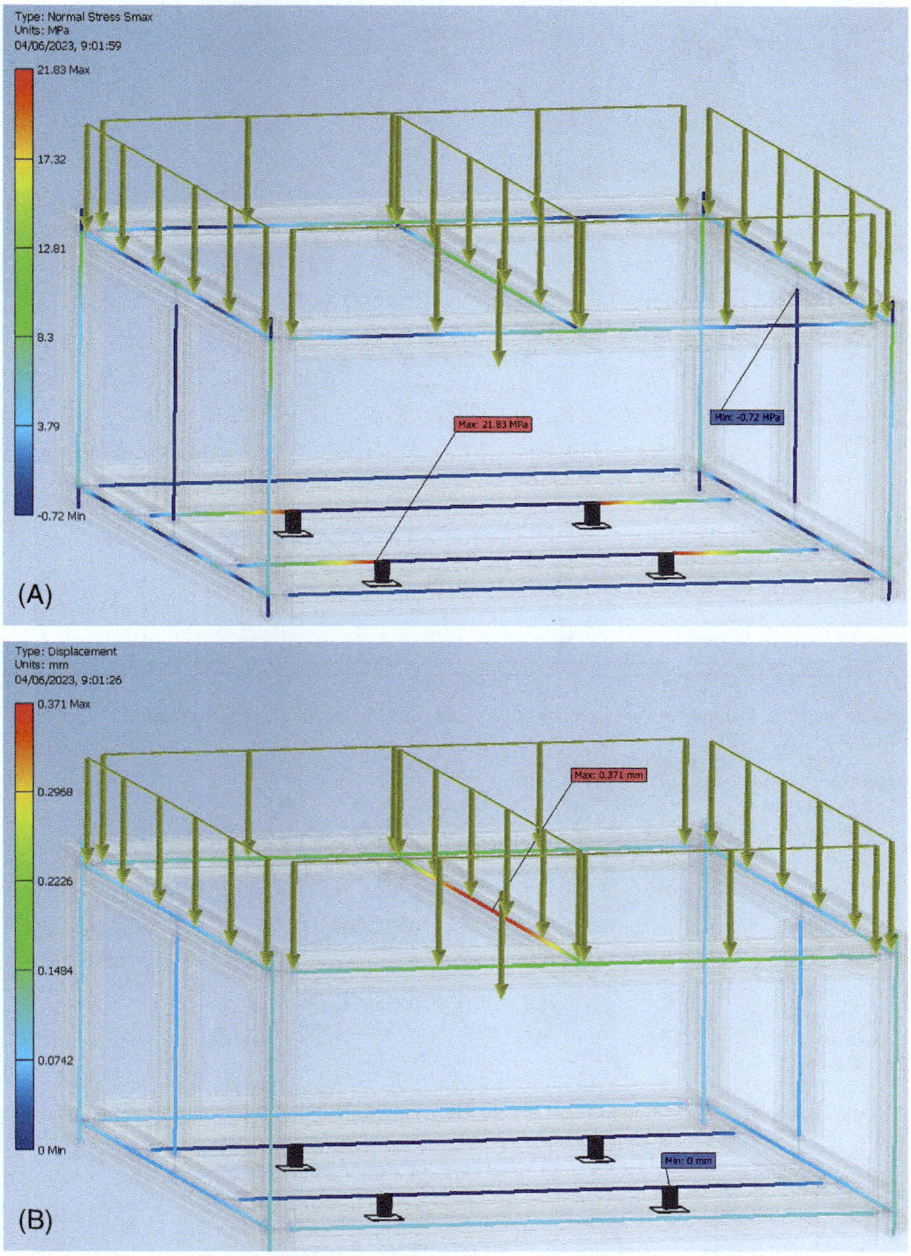

FIGURE 15.15 FEA conducted on the structural module: (A) deformation analysis and (B) stress analysis.

For the implementation, it was required to implement a Wi-Fi network for IP addressing of all the agents and external systems as shown in Fig. 15.16.

15.2.6.1 Communication between LPPU and agent

The Message Queuing Telemetry Transport (MQTT) protocol was implemented using a Mosquitto Broker to facilitate data exchange between the agents and the LPPU system. The subscription and publishing channels, including the IP addresses of the broker and the Flask server, are detailed in Fig. 15.17.

15.2.7 Design of S4: Power management system

To achieve autonomy in device operation the power distribution and regulation modules were designed based on the power requirements for each selected module and component. The power distribution scheme is illustrated in Fig. 15.18.

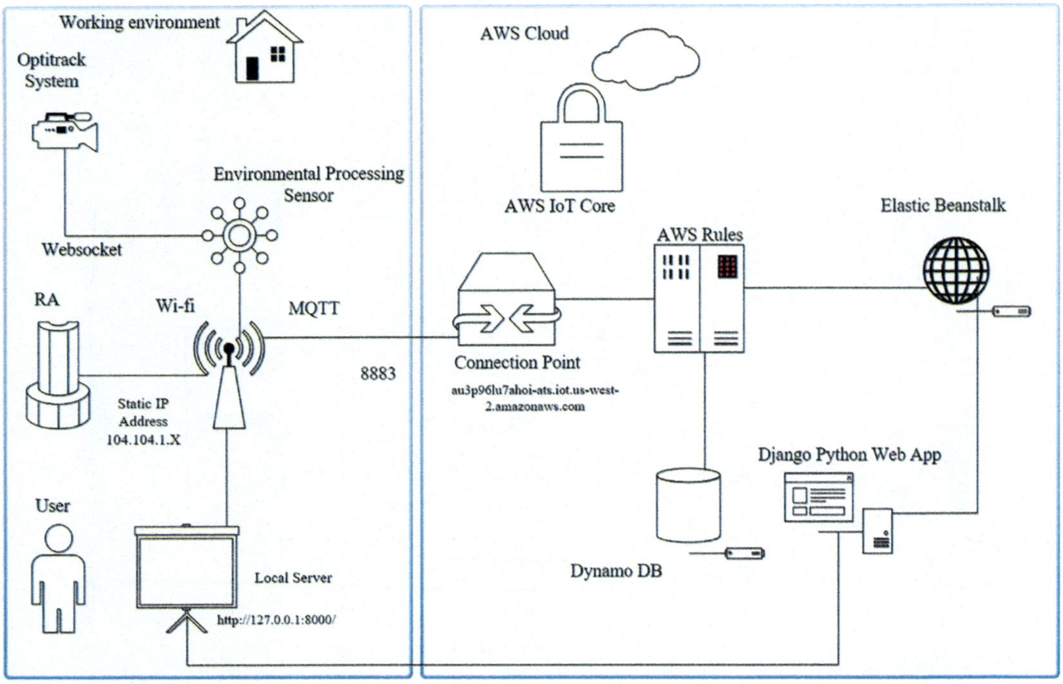

FIGURE 15.16 Network architecture. Diagram of the communication system and Internet of Things interface.

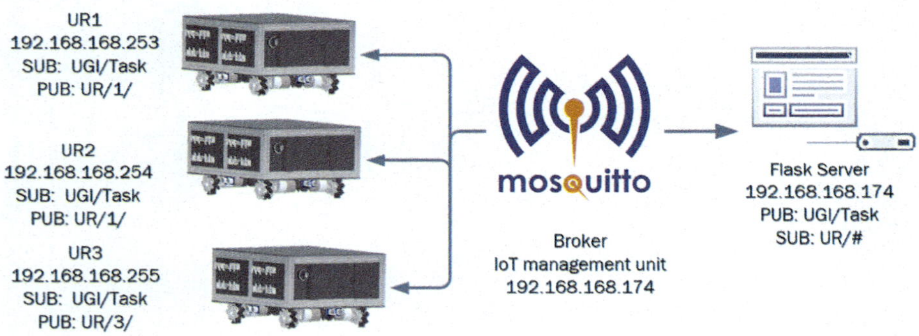

FIGURE 15.17 MQTT topics architecture. MQTT architecture and channel configuration for IoT communication. *IOT*, Internet of things.

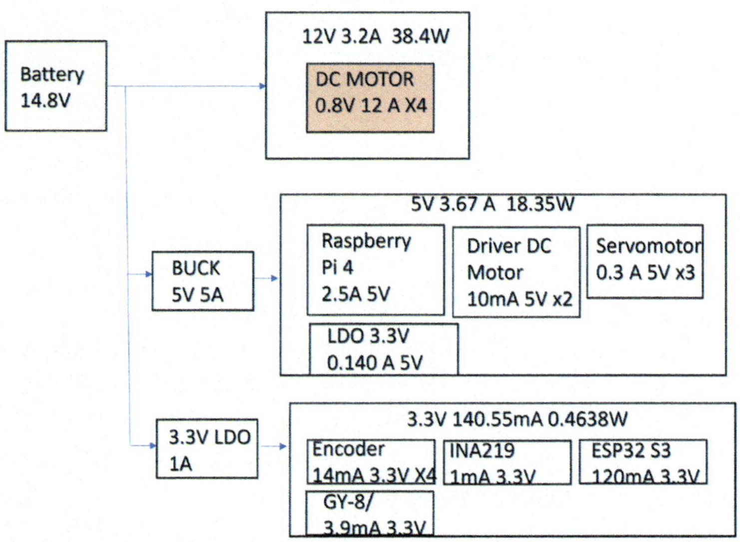

FIGURE 15.18 Power distribution. Voltage levels and distribution diagram.

FIGURE 15.19 ArUco marker. ArUco marker example.

FIGURE 15.20 Validation of ArUco marker detection. Identification of multiple objects by implementing ArUco markers.

15.2.8 Design of S5: Computer vision system

15.2.8.1 M5.1: Recognition module

ArUco markers were used to identify the boundaries of the working area and designate zones for agent movement. Each marker consists of a black border containing a binary matrix that serves as a unique identifier (ID), an example of which is shown in Fig. 15.19. This ID enables the camera system to determine the original orientation of the marker, even when it appears rotated in the captured image.

Measurements were taken of three objects of different sizes and weights, to which an ArUco marker was assigned to validate the real-time detection of these objects within a controlled environment. Using a Logitech C992 Pro camera at a resolution of 1280 × 720 pixels and 60 frames per second the program detects markers in the environment and displays the previously recorded information about the object according to its corresponding identifier. Fig. 15.20 shows the result of detecting multiple objects in the scene.

15.2.8.2 M5.2: Environment monitoring module

The A^* algorithm was selected for path planning based on the results of Costa and Silva (2019) because of its notable performance in static environments and low computational cost, making it suitable for implementation on an embedded system.

This intelligent search algorithm finds the path of least cost between two points by considering both the heuristic distance between the current node and the goal node, $h(n)$, and the distance covered from the origin node to the current node, $g(n)$. This approach enables finding the lowest cost trajectory without needing to explore all possible routes.

The algorithm maintains two data lists while evaluating each of the nodes. The first is a list where open nodes are stored, which will be evaluated using the evaluation function shown in Eq. (15.12) and are sorted from lowest to highest value, thus prioritizing nodes with the lowest cost. The second stores the values of nodes that have already been evaluated, also called the closed set, along with their cost and the node from which they originated.

$$f(n) = g(n) + h(n) \tag{15.12}$$

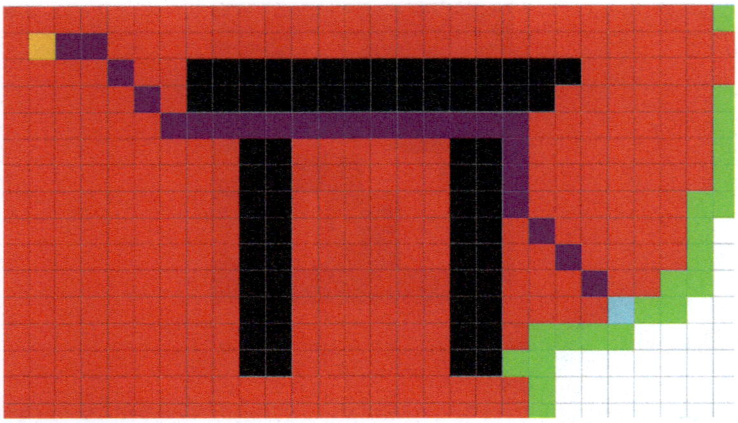

FIGURE 15.21 *A** algorithm example. Simulation of the *A** algorithm for path planning in a virtual environment.

The heuristic distance was calculated using Eq. (15.13). However, in Costa and Silva (2019), other functions can be found.

$$g(n) = \sqrt{(x_1 - x_2)^2 + (y_1 - y_2)^2} \tag{15.13}$$

To implement navigation with computer vision the environment is captured via a camera and processed to detect obstacles and restricted zones. The map is then digitized, and restricted areas are dilated by the robot's radius to avoid collisions. The environment is gridded, defining nodes within the space: smaller grid squares increase resolution but raise computational costs. In practice, grid cells should be at least the diameter of the robot.

The initial and target positions are identified, and an algorithm is applied, allowing eight possible moves (up, down, left, right, and four diagonals). Movement costs are typically unitary for single-axis moves and $\sqrt{2}$ for diagonals.

With these considerations the algorithm can map a low-cost path, enabling the robot to navigate a static environment without collisions. Fig. 15.21 presents an example of a path-planning exercise using the A^* algorithm, where red squares represent nodes in the closed list, green squares represent nodes in the open list, white squares were not analyzed, black squares indicate obstacles, and purple squares trace the lowest cost path from the start node to the goal node.

Once the lowest cost path is obtained, key points along the trajectory are identified, which indicate changes in movement. These points are used to define smooth point-to-point trajectories using fifth-order Bézier polynomials. An iterative process was employed to obtain trajectories whose speed does not exceed 200 RPM, which is the limit provided by the motors of the robotic agents.

15.2.9 Design of S6: Human–machine interface

To meet the multithreading requirements for image processing and IoT brokering the HMI was implemented on a web server using Python with the Flask framework. This interface allowed for real-time validation of messages sent and received via the broker and displayed operational data graphically, as shown in Fig. 15.22.

15.3 Experimental results

15.3.1 Validation of coupling mechanism

The coupling mechanism was manufactured using 3D printing. Active components were printed in Phrozen TR250 LV resin, which meets the mechanical property requirements identified in the Ashby analysis. Fig. 15.23 presents a set of elements comprising the active parts of a single coupling mechanism.

With the same technique the wall that supports the male module and the servomotor was manufactured (Fig. 15.24).

The final component of the mechanism was also 3D-printed, using PLA. After obtaining all parts, including screws and the servomotor, the assembly was completed and integrated into the robotic agent, as shown in Fig. 15.25.

The female module was also 3D-printed using PLA and subsequently assembled into the robot's structure, as shown in Fig. 15.26.

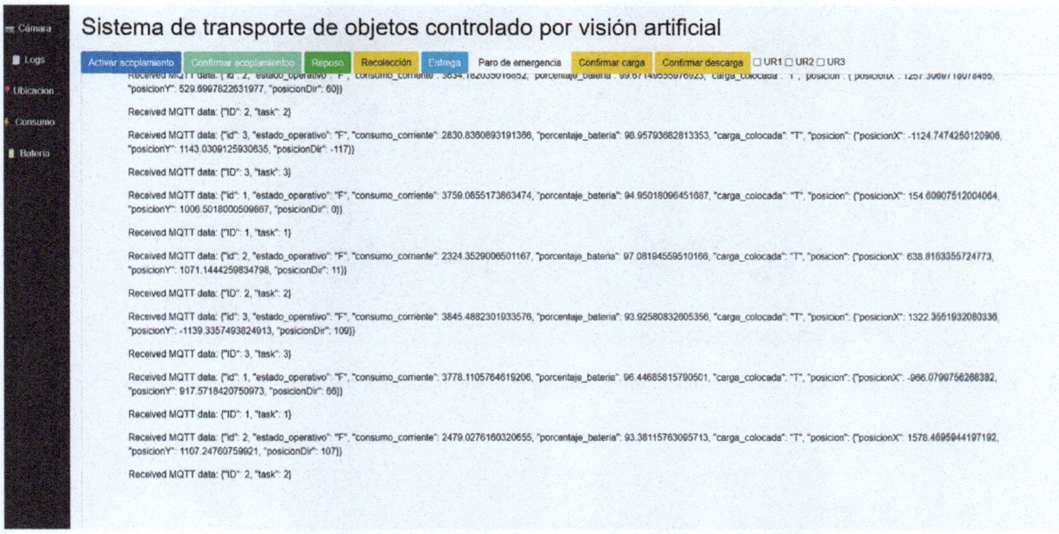

FIGURE 15.22 HMI system. Data transmitted and received through the MQTT broker is shown in the HMI. *HMI*, Human–machine interface.

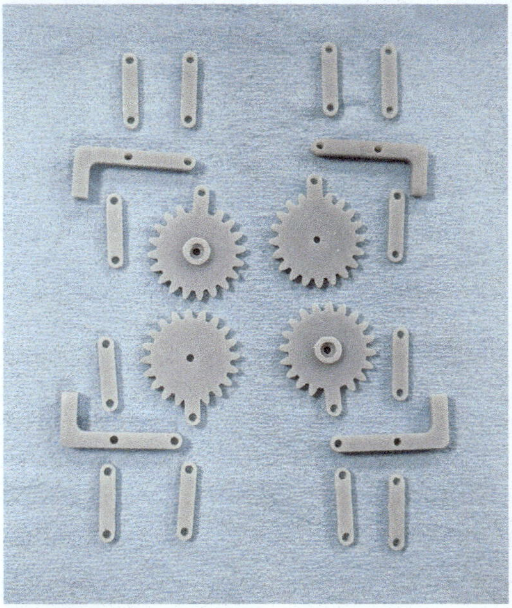

FIGURE 15.23 Manufacturing of the male module of the coupling mechanism. 3D printed pieces in resin of male coupling mechanism.

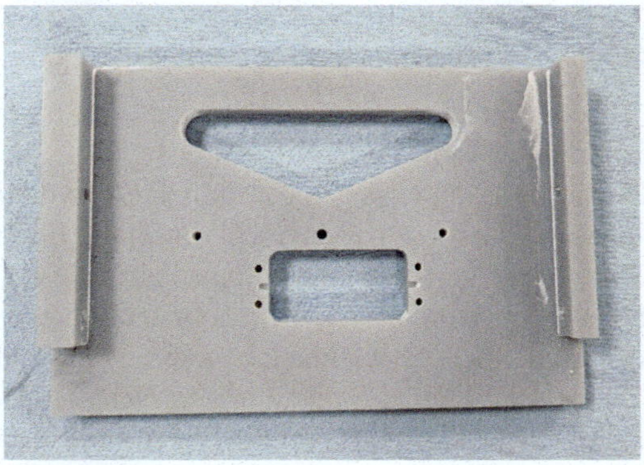

FIGURE 15.24 Manufacturing of the female module of the coupling mechanism. 3D printed wall in resin of female coupling mechanism.

FIGURE 15.25 Assembly of the male coupling mechanism. Integration of the male module of the coupling mechanism into the structure of the mobile robot.

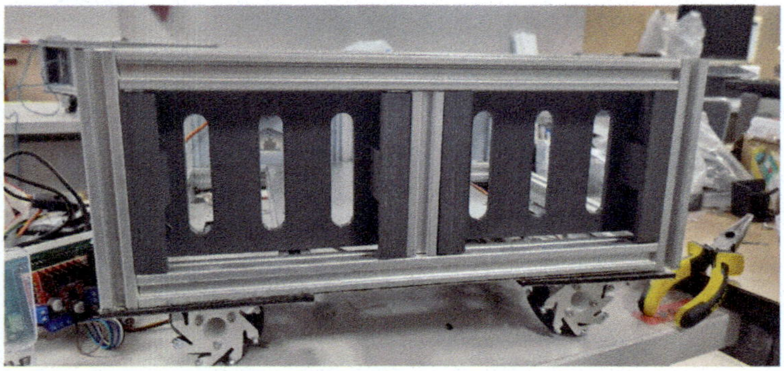

FIGURE 15.26 Assembly of the female coupling mechanism. Integration of the female module of the coupling mechanism into the structure of the mobile robot.

Neoprene was added to the tip of each finger and to the female module to enhance grip between agents. Once both the female and male modules were operational, an experiment was conducted to validate the gripping force, ensuring a secure connection between robotic agents. A video of the experiment can be viewed at https://youtube/IOdmf_8kgtA?si=6LNYgY9e5WDU2u-B, and a summary is shown in Fig. 15.27, where the structure of an agent is lifted using only one coupling mechanism.

15.3.2 Validation of structural module

Once the robotic agents were assembled, the first experiment conducted involved applying various payloads in both single and collaborative schemes. Fig. 15.28 illustrates a payload of 4.4 kg applied to a single unit.

Meanwhile, Fig. 15.29 shows a payload of 16 kg applied to a configuration involving three agents.

15.3.3 Validation of computer vision system

Using a Logitech C992 Pro camera at a resolution of 1280 × 720 pixels and 60 frames per second the calibration was performed with MATLAB Camera Calibrator as shown in Fig. 15.30.

ArUco markers were then placed in the working environment, utilizing two dictionaries: one to define the boundaries and the other to designate the locations where the robotic agents will transition. Fig. 15.31 illustrates the distribution of the markers in the working environment for a single trajectory, including an obstacle positioned in the middle.

Distortion in the camera was eliminated by incorporating the parameters calculated during calibration. A homography algorithm was then applied to the undistorted image to obtain a top view, as shown in Fig. 15.32.

The obstacle is then filtered by adjusting the Hue, Saturation and Value (HSV) parameters and applying standard morphological transformations to the image, which can be controlled by the user through the HMI, as shown in Fig. 15.33.

Once the system has identified the working environment boundaries, target positions, and obstacles, the A^* algorithm is executed. Obstacles are dilated by a factor equal to the agent's radius, and a grid size at least as large as the robot's radius

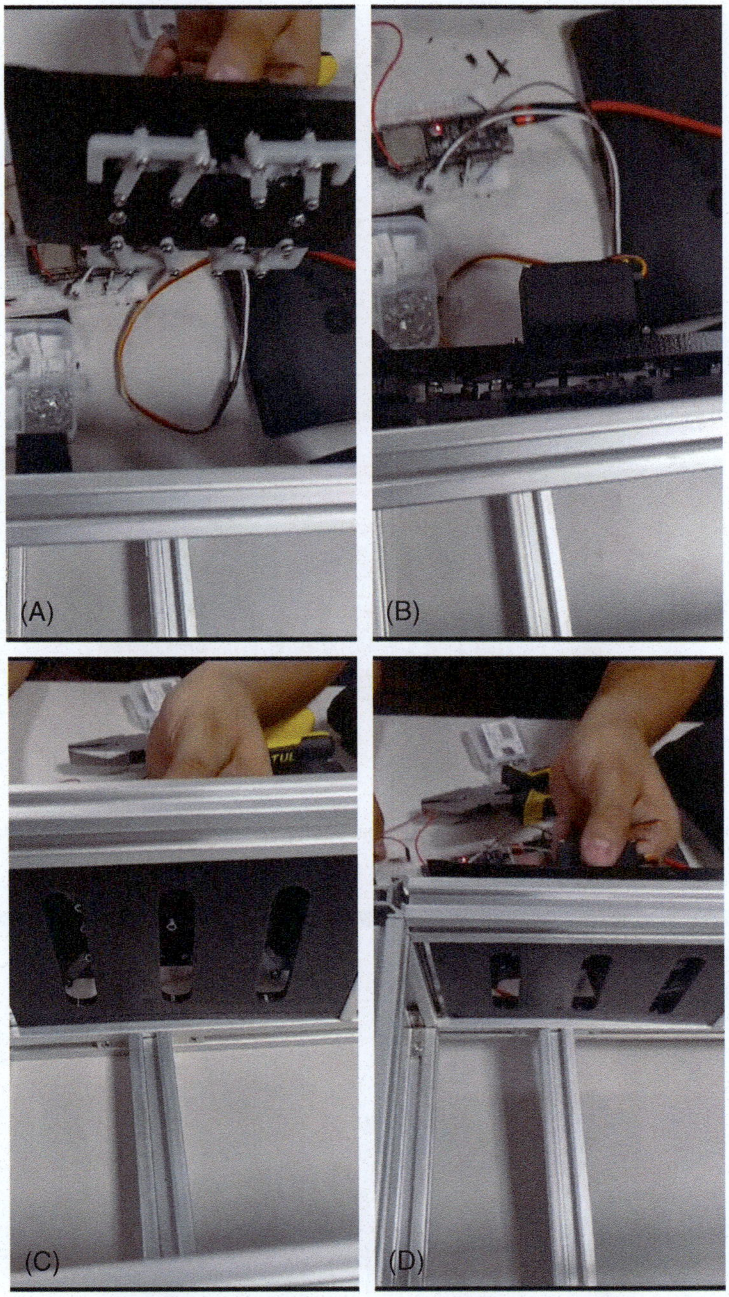

FIGURE 15.27 Validation of the coupling process: (A) gripper in open position, (B) module alignment, (C) gripper in closed position, and (D) coupling validation.

is applied. Based on these parameters the system generates a set of waypoints for the user, as shown in Fig. 15.34. In this figure, *pink points* indicate the lowest cost trajectory, *blue crosses* mark nodes in the closed set, and the *line* represents a safe path for the agent to move without collisions.

The algorithm identifies the lowest cost trajectory without analyzing the entire environment. Once the waypoints are acquired, a smooth trajectory is generated using Bezier polynomials. Through inverse kinematics the resulting velocities are translated to wheel speeds, and the system iterates until achieving time and position vectors that keep the robotic agent's wheels below 200 RPM. Fig. 15.35 shows the resulting trajectory calculated for the current example.

Then the algorithm generates three CSV files containing the desired position, velocity, and acceleration along the x and y axes, as well as rotational data. These files are sent to each robotic agent for execution during the path-following task.

FIGURE 15.28 Validation of payload for single agent. A 4.4 kg payload applied to a robotic agent for validation of the structural module.

FIGURE 15.29 Validation of payload for multiple agents. A 16 kg payload applied to three robotic agents for validation of the structural module in a collaborative scheme.

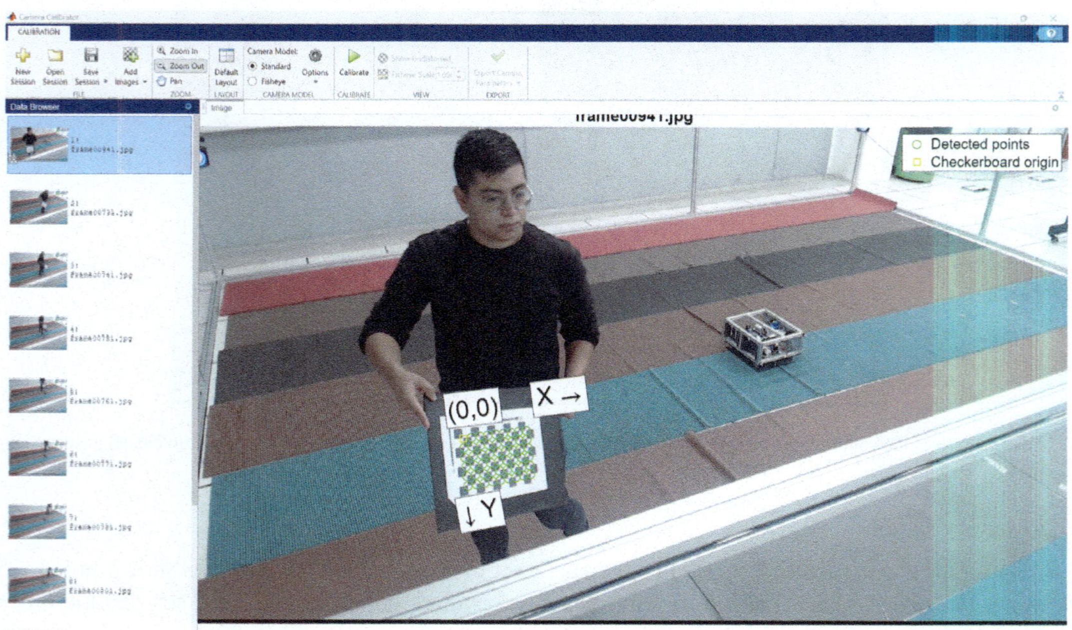

FIGURE 15.30 Camera calibration. Camera calibration using a chess pattern.

Along with the computer vision system a motion tracking system was implemented to enable visual servoing of the robotic agents. Six reflective markers were placed on the robotic structure. Fig. 15.36 shows a robotic agent in the working environment after camera calibration. Once calibrated the software detects the markers and groups them to create a rigid body in the tracking software, assigning it an identifier and monitoring the position of its center of mass.

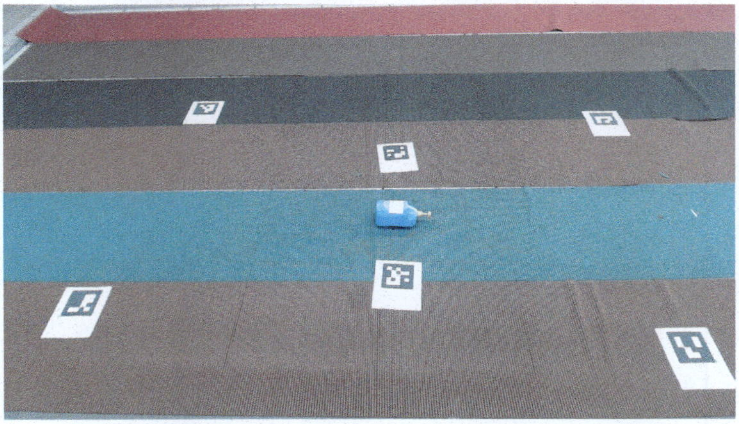

FIGURE 15.31 Distribution of markers in the working environment. An experimental disposition of markers and obstacles for validation of the path planning algorithm.

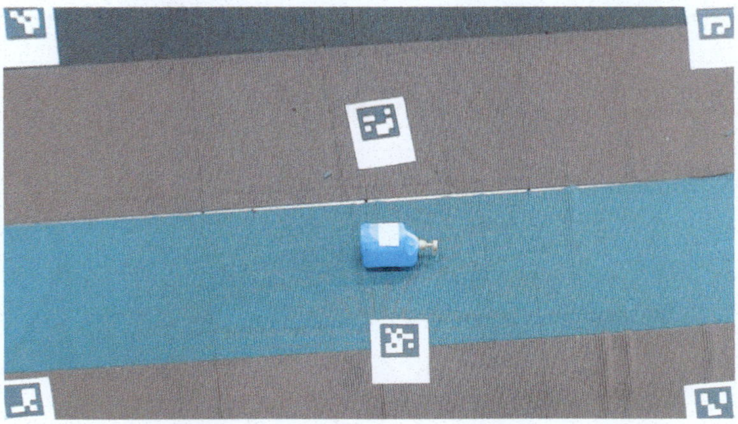

FIGURE 15.32 Artificial top view of the environment. Top view of the environment obtained by applying a homography algorithm.

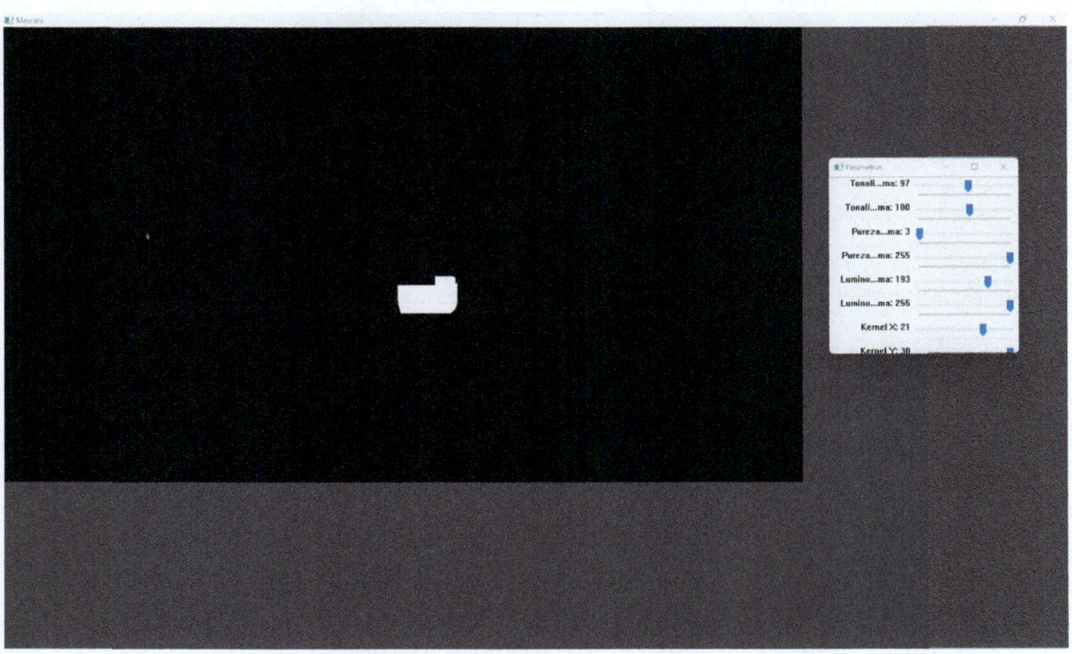

FIGURE 15.33 Obstacle identification with computer vision. Filtering in HSV of the obstacle using the HMI. *HMI*, Human–machine interface.

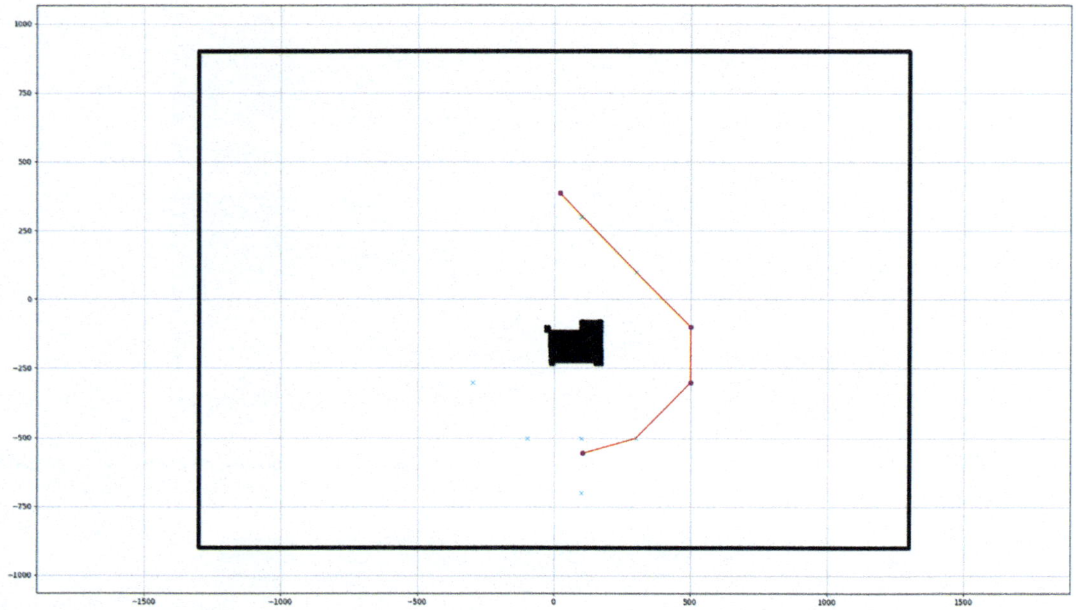

FIGURE 15.34 Path planning. Lowest cost trajectory obtained with A^* algorithm.

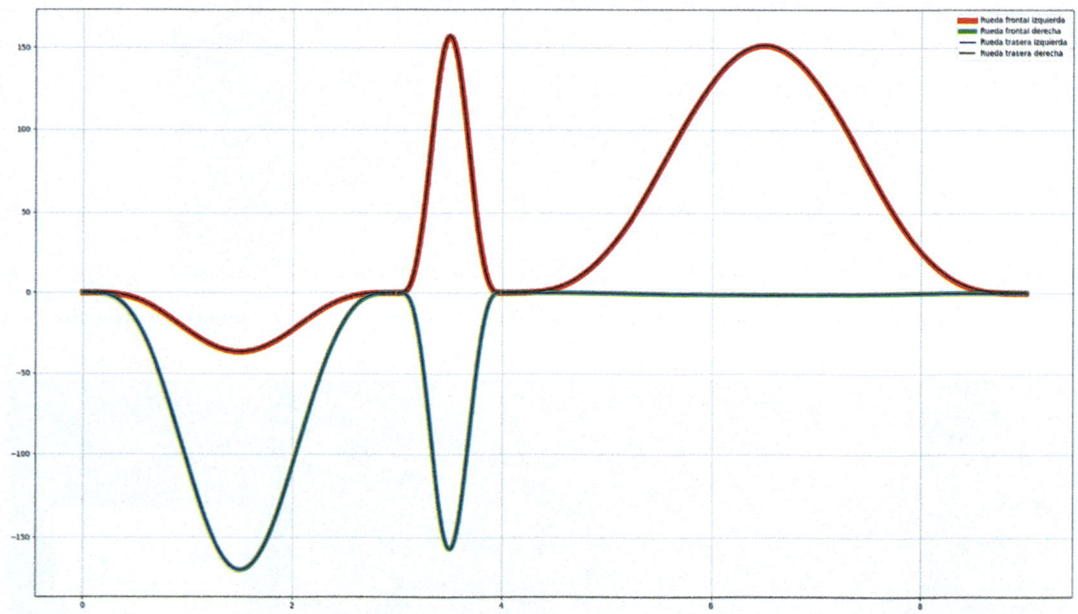

FIGURE 15.35 Robotic agent's wheel speed profiles. Wheel speed profiles generated to achieve the desired motion of the mobile robot.

15.3.4 System validation

15.3.4.1 Single robotic agent

System validation began with a single robotic agent, tasked with navigating between two target points identified by the computer vision system while avoiding a centrally placed obstacle. The full operation is available at https://youtu.be/mE3aXwP46BU?si=oi2PmVU2JhbwDZJ5, with key results summarized in Fig. 15.37.

15.3.4.2 Automatic coupling

Following successful control of individual robotic agents within the working environment, the coupling task was tested to enable collaborative behavior. A video demonstration is available at https://youtu.be/ED3-UIFJt18?si=qxv_JPUtB8FH00Ir, and Fig. 15.38 illustrates the successful automatic coupling between two robotic agents.

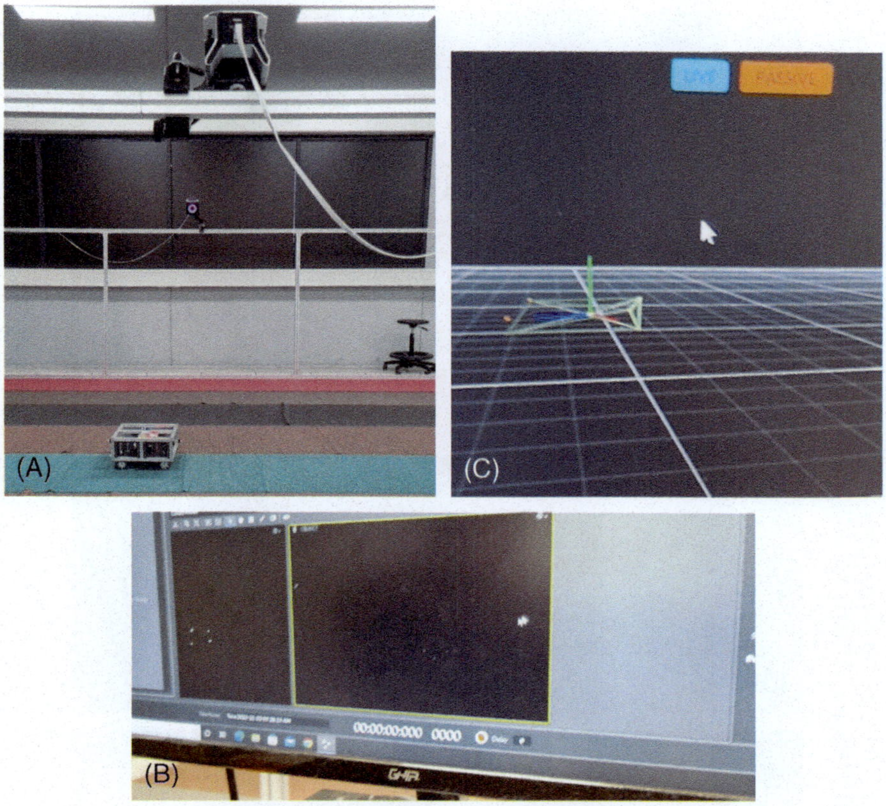

FIGURE 15.36 Validation of visual servoing: (A) robotic agent positioned in the environment for virtual model creation, (B) camera calibration, and (C) rigid body model of the robotic agent.

15.3.4.3 Multiagent validation

Finally, three robotic agents were placed in the working environment to assess their simultaneous performance. A series of experiments, viewable at https://youtu.be/Gx4PRrMTiYY?si=TdZ6bRePYuO2acTp, were conducted in both single-agent and multiagent modes. In all cases the agents successfully achieved the desired displacement.

15.4 Conclusions

This chapter presented the design and implementation of a modular multiagent robotic system with a coupling mechanism for collaborative object transportation. Multiple experiments were conducted in both single- and multi-agent modes. A detailed overview of the entire implementation process, including user interaction, is available at https://youtu.be/yxg_vZpbbh8?si=OrPV4UTTufFOEn_B.

Experimental results demonstrate that the coupling mechanism effectively maintained a stable connection between robotic agents, facilitating the use of a simplified control algorithm. The design enables autonomous coupling and decoupling without requiring precise alignment, and its high replicability and ease of manufacture make it suitable for potential high-volume commercial applications.

The artificial vision system provides the flexibility essential for compliance with Industry 4.0 standards. Through the implementation of IoT solutions the system becomes scalable, whereas the motion tracking system accurately monitors agent positions, even during instances of occlusion. The generated trajectories ensure safe navigation for the robotic agents.

To enhance the design, proximity sensors could be implemented to ensure safe navigation in dynamic environments. Multiagent performance could also improve through a collaborative scheme where follower agents use the leader's velocity as input. However, thanks to the coupling mechanism's robust performance, this enhancement would not significantly increase algorithmic complexity.

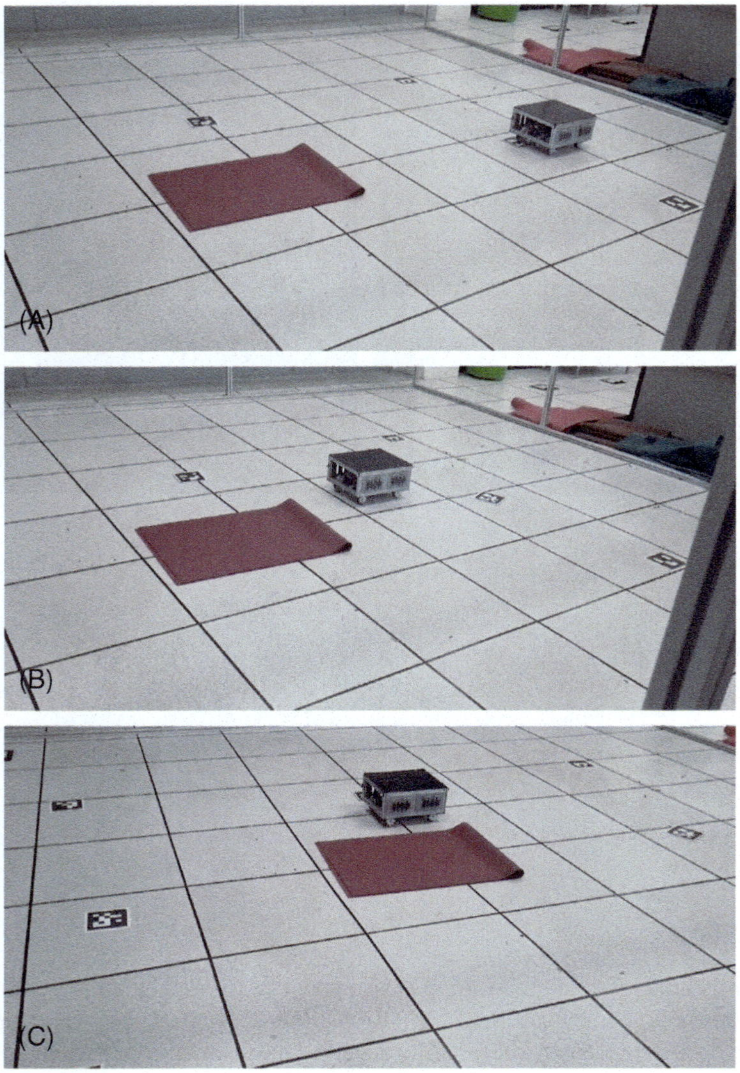

FIGURE 15.37 Autonomous navigation of a single robotic agent with collision avoidance of static obstacle: (A) Departure from initial position; (B) avoidance of static obstacle; (C) approach to the goal position.

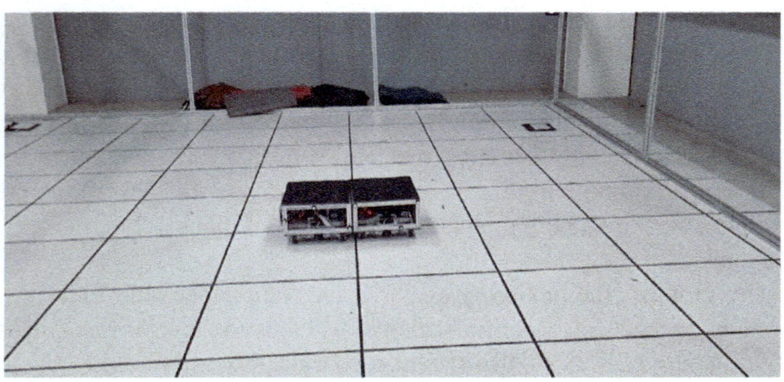

FIGURE 15.38 Autonomous coupling between two robotic agents. Validation of automatic coupling between two robotic agents.

References

Ashby, M. F. (2011). Chapter 5—Materials selection—The basics. In *Materials selection in mechanical design* (pp. 97–124). Butterworth-Heinemann. doi:10.1016/B978-1-85617-663-7.00005-9.

Borges, C. D. B., Almeida, A. M. A., Paula Júnior, I. C., & de Mesquita Sá Junior, J. J. (2019). A strategy and evaluation method for ground global path planning based on aerial images. *Expert Systems With Applications, 137*, 232–252. doi:10.1016/j.eswa.2019.06.067.

Costa, M. M., & Silva, M. F. (2019). A survey on path planning algorithms for mobile robots. *2019 IEEE International Conference on Autonomous Robot Systems and Competitions (ICARSC)*. doi:10.1109/ICARSC.2019.8733623.

Dehoff, B., Levack, D., & Rhodes, R. (2009). The functional breakdown structure (FBS) and its relationship to life cycle cost. *45th AIAA/ASME/SAE/ASEE Joint Propulsion Conference & Exhibit, 2–5 August 2009, Denver, Colorado*. doi:10.2514/6.2009-5344.

Espinosa, F., Santos, C., & Sierra-García, J. (2020). Transporte multi-AGV de una carga: estado del arte y propuesta. *Revista Iberoamericana de Automática e Informática industrial, 18*, 82. doi:10.4995/riai.2020.12846.

Farahvash, P., & Boucher, T. O. (2004). A multi-agent architecture for control of AGV systems. *Robotics and Computer-Integrated Manufacturing, 20*(6), 473–483. doi:10.1016/j.rcim.2004.07.005.

Johnson, M. A., & Moradi, M. H. (2005). *PID control technology PID control: New identification and design methods* (pp. 1–46). Springer London. doi:10.1007/1-84628-148-2-1.

Knospe, C. (2006). PID control. *IEEE Control Systems Magazine, 26*(1), 30–31. doi:10.1109/MCS.2006.1580151.

Li, Y., Dai, S., Zhao, L., Yan, X., & Shi, Y. (2019). Topological design methods for Mecanum wheel configurations of an omnidirectional mobile robot. *Symmetry, 11*(10). doi:10.3390/sym11101268.

Mohamed, M. (2018). Challenges and benefits of industry 4.0: An overview. *International Journal of Supply and Operations Management, 5*(3), 256–265. doi:10.22034/2018.3.7.

Moreno, S. A. M., & Ceballos, N. D. M. (2019). Vehículos de guiado autónomo (AGV) en aplicaciones industriales: una. *Revista Politécnica, 15*(28), 117–137.

Mott, R. L. (1995). *Diseño De Elementos De Maquinas*. Prentice Hall Hispanoamericana.

Nagel, R. N., & Odrey, N. G. (2012). Robot end effectors. In *Industrial robotics: Technology, programming, and applications* (p. 118). Tata McGraw-Hill Education.

Pololu Corporation. (2019). *37D metal gearmotors, revision 1.2*. Pololu Corporation https://www.pololu.com/file/0J1736/pololu-37d-metal-gearmotors-rev-1-2.pdf.

Ramírez-Neria, M., González-Sierra, J., Madonski, R., Ramírez-Juárez, R., Hernandez-Martinez, E. G., & Fernández-Anaya, G. (2023). Leader–follower formation and disturbance rejection control for omnidirectional mobile robots. *Robotics, 12*(5), 122. doi:10.3390/robotics12050122.

Soori, M., Arezoo, B., & Dastres, R. (2023). Internet of things for smart factories in industry 4.0, a review. *Internet of Things and Cyber-Physical Systems, 3*, 192–204. doi:10.1016/j.iotcps.2023.04.006.

Tang, G., Tang, C., Claramunt, C., Hu, X., & Zhou, P. (2021). Geometric A-star algorithm: An improved A-star algorithm for AGV path planning in a port environment. *IEEE Access, 9*, 59196–59210. doi:10.1109/ACCESS.2021.3070054.

Tuci, E., Alkilabi, M., & Akanyeti, O. (2018). Cooperative object transport in multi-robot systems: A review of the state-of-the-art. *Frontier in Robotics and AI, 5*. Available from: http://dx.doi.org/10.3389/frobt.2018.00059 . doi:10.3389/frobt.2018.00059.

VDI-Fachbereich Produktentwicklung. (2004). *Design methodology for mechatronic systems (Norme VDI. 2206)*.

Chapter 16

The role of UAVs in wildlife monitoring: A case study of conservation efforts

Harpreet Kaur Channi
Eudoxia Research Centre (Affiliated to Eudoxia Research University, USA), Mumbai, Maharashtra, India

16.1 Introduction

Unmanned aerial vehicles (UAVs), commonly referred to as drones, have become a transformative tool in wildlife conservation, significantly enhancing the ability of conservationists and researchers to monitor ecosystems and protect endangered species. Traditionally, wildlife monitoring involved manual fieldwork, which could be costly, time-consuming, and potentially disruptive to animal habitats. The introduction of UAVs in conservation has offered a noninvasive, efficient, and often safer alternative, allowing researchers to capture comprehensive data from diverse and inaccessible areas with minimal human interference. This chapter examines the transformative impact of UAVs in wildlife conservation, focusing on their role in data collection, species monitoring, and environmental assessment (Zhu, 2024). One of the most significant advantages of UAV technology in wildlife conservation is its ability to collect high-resolution data with speed and precision. Equipped with advanced cameras and sensors, including thermal imaging, multispectral, and hyperspectral cameras, drones can capture detailed images and videos that reveal important information about species populations, behaviors, and habitats. This data can be especially valuable for tracking endangered species, whose dwindling populations often make them difficult to observe directly. UAVs can monitor these animals from a distance, allowing researchers to gather insights into population dynamics, health status, and even social behaviors without disturbing the animals. For example, UAVs are now frequently used to monitor African elephants, tracking their movements, studying their social groups, and identifying poaching threats across vast landscapes (Millner et al., 2024). In addition, drones offer unparalleled access to remote or hazardous terrains, enabling conservationists to collect data in areas that would otherwise be challenging or dangerous to reach. In regions such as dense rainforests, high mountains, or vast oceans, UAVs allow researchers to survey wildlife populations without the need for extensive and potentially invasive field operations. For instance, UAVs have been employed to track elusive species such as orangutans in the thick jungles of Southeast Asia, a task that would otherwise be daunting because of dense foliage and harsh environmental conditions. Similarly, drones have been deployed over the Pacific Ocean to monitor marine wildlife like whales and sharks, enabling researchers to study migration patterns, population health, and environmental impacts without requiring intrusive tagging or close-proximity observation (Buchelt et al., 2024).

UAV technology has also proven valuable in reducing human-wildlife conflicts and supporting antipoaching efforts. Drones can provide real-time monitoring of critical areas, enabling conservationists and park rangers to spot illegal activities and swiftly respond to potential threats. In this way, UAVs act as a deterrent to poachers and can even capture evidence to support legal action against illegal hunting. In Africa, for instance, UAVs have become an essential tool in antipoaching operations to protect rhinos, elephants, and other threatened species from illegal hunting, adding a technological layer of protection that can operate day or night, in nearly any weather condition. Despite these advantages the integration of UAVs in wildlife conservation still faces some challenges. Regulatory restrictions on drone flights, especially in protected areas or near borders, can complicate their deployment. Moreover, the limited battery life and range of many drones can restrict their ability to cover expansive areas continuously. Nonetheless, the ongoing advancement of UAV technology, coupled with machine learning (ML) and artificial intelligence (AI), promises to overcome these limitations and expand the capabilities of drones in wildlife conservation. In the future, UAVs will likely play an even more crucial role in biodiversity protection,

Computational Intelligence in Mechatronics: Solving Real-World Problems in Electronic Systems Design. DOI: https://doi.org/10.1016/B978-0-443-27580-7.00007-X

ecosystem monitoring, and environmental sustainability, making them an indispensable asset for conservationists globally (Buchelt et al., 2024).

16.1.1 Overview of UAV technology in ecological research

UAVs, or drones, have rapidly advanced and diversified over the past two decades, evolving into powerful tools for ecological research. Initially developed for military applications, UAVs are now extensively used in environmental sciences, revolutionizing how researchers study and manage ecosystems. UAV technology has matured to include a range of models, from small, lightweight drones capable of short flights to large, sophisticated drones designed for extended monitoring missions across vast landscapes. Equipped with high-resolution cameras, multispectral and hyperspectral sensors, thermal imaging, and GPS systems, drones offer diverse data collection capabilities that are adaptable to various ecological needs. They enable the rapid, detailed, and often automated collection of data on wildlife populations, vegetation health, environmental changes, and habitat quality. The flexibility, mobility, and data accuracy that UAVs provide have led to a surge in ecological applications, from tracking species populations to assessing the impacts of climate change on ecosystems (Chetverikov et al., 2024).

16.1.2 Importance of noninvasive monitoring in wildlife conservation

Noninvasive monitoring is essential in wildlife conservation, as traditional monitoring methods can disturb animal behavior, cause stress, and even impact population health. UAVs provide a nondisruptive means of observing and gathering data on wildlife, ensuring that conservation activities do not negatively affect the very species they aim to protect. By flying above or around animals, drones can collect data without physically interacting with them or their habitats, reducing the risk of human-wildlife conflicts and maintaining natural behaviors. This method is especially beneficial for monitoring sensitive or endangered species, where minimal disturbance is crucial. In addition, noninvasive monitoring with drones reduces the need for physical tagging or tracking, which can sometimes harm animals or alter their natural behavior. UAVs, thus, allow conservationists to conduct more ethical, sustainable, and respectful research, aligning with conservation goals of protecting wildlife and promoting biodiversity (Kayaoglu et al., 2026).

To illustrate the role of UAVs in wildlife conservation, this chapter presents a focused case study on monitoring elephant populations in Africa. African elephants face numerous threats, including habitat loss, poaching, and human-wildlife conflicts, making them a priority for conservation efforts. The vast landscapes they inhabit, combined with the need for frequent population assessments, make UAVs an ideal tool for monitoring these majestic animals. The selected case study will explore how drones equipped with high-resolution cameras and thermal imaging are used to locate, count, and track elephants across diverse terrains, providing crucial data for conservation strategies. This case study demonstrates the practical applications of UAV technology in monitoring large mammals, the benefits of noninvasive data collection, and the challenges faced in deploying UAVs in complex, expansive habitats. By focusing on African elephants, this chapter highlights the potential of UAV technology to enhance species protection and contribute to broader conservation goals (Gonzalez Nunez et al., 2024).

16.2 Capabilities of UAVs in wildlife monitoring

UAVs, or drones, have brought transformative capabilities to wildlife monitoring by allowing conservationists to collect detailed and real-time data across various ecosystems, as shown in Fig. 16.1. Equipped with cutting-edge sensors, high-resolution cameras, and advanced GPS systems, UAVs enable researchers to observe animal populations, study habitat dynamics, and detect potential threats without disturbing wildlife. Their versatility and adaptability across diverse terrains make UAVs essential for monitoring hard-to-reach locations such as dense forests, open savannas, and extensive marine environments (Mitra & Das, 2024).

One of the core capabilities of UAVs in wildlife monitoring is high-resolution imaging, which captures detailed visuals of animals and their habitats from a distance. These drones can provide data on animal populations, behavioral patterns, and habitat conditions, giving researchers valuable insights that would be challenging to obtain through traditional, on-the-ground methods. Advanced imaging technologies, such as thermal and infrared sensors, further enhance UAVs' utility by enabling nighttime monitoring and detecting heat signatures of animals hidden within dense vegetation. This capability is particularly beneficial for tracking elusive or nocturnal species, allowing for continuous data collection across both day and night (Dokoro et al., 2024).

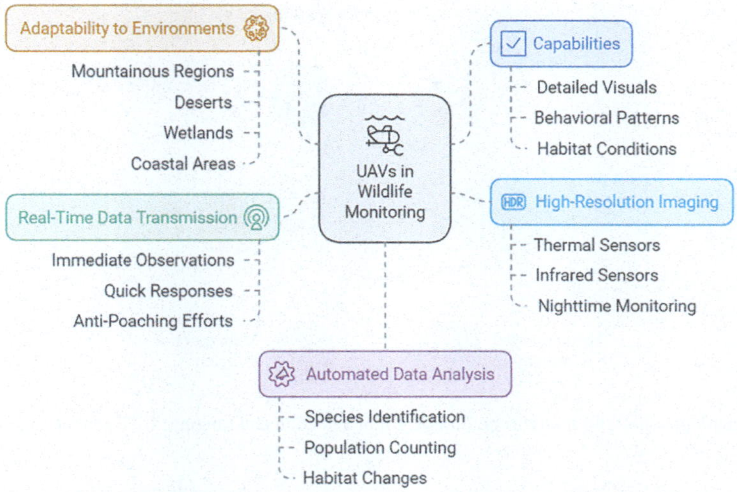

FIGURE 16.1 Capabilities of UAVs in wildlife monitoring. *UAVs*, Unmanned aerial vehicles.

Another key capability of UAVs is real-time data transmission, which enables conservationists to make immediate observations and respond to potential threats, such as poaching or environmental hazards. By deploying drones with real-time monitoring systems, researchers and conservation teams can cover large areas efficiently, providing timely updates and quick responses to incidents. This capacity is invaluable for managing wildlife reserves, where real-time monitoring allows for rapid detection of illegal activities, potentially reducing human-wildlife conflicts and supporting antipoaching efforts. For example, UAVs used in antipoaching initiatives can detect unusual activities in restricted areas and alert park authorities, allowing for immediate interventions (He et al., 2024).

UAVs also support automated data analysis through ML and AI, which is essential for handling the vast amounts of data generated from extensive monitoring. With AI-enabled processing, drones can identify specific species, count population numbers, and even differentiate between individual animals based on size, markings, or movement patterns. These automated features save time and increase the accuracy of data analysis, empowering conservationists to make data-driven decisions. ML algorithms trained on UAV-collected imagery can also detect subtle changes in habitats over time, helping identify early signs of environmental stress or degradation that may impact wildlife populations (Afridi et al., 2024).

The adaptability of UAVs to various habitats and environmental conditions further highlights their value in wildlife monitoring, as shown in Fig. 16.1. From mountainous regions and deserts to wetlands and coastal areas, drones can navigate diverse landscapes and challenging weather conditions to gather comprehensive data. This capability allows for long-term ecological studies across different ecosystems, aiding in comparative analysis and understanding of species adaptations, migrations, and the impacts of climate change. In summary, UAVs have become essential tools in wildlife monitoring because of their high-resolution imaging, real-time data collection, automated analysis, and versatility in covering diverse terrains. By enhancing monitoring efficiency, improving data accuracy, and minimizing human impact on natural environments, UAVs play an increasingly critical role in supporting sustainable wildlife conservation and ecosystem protection.

16.2.1 Imaging and sensor technologies

The integration of advanced imaging and sensor technologies has significantly enhanced the capabilities of UAVs for wildlife monitoring, allowing for detailed and noninvasive data collection as shown in Fig. 16.2. High-resolution imaging, along with specialized sensors such as thermal, multispectral, and hyperspectral cameras, enables drones to capture detailed visuals and detect subtle environmental changes, all of which contribute valuable information to conservation efforts (Afridi et al., 2024; Anandaraj et al., 2024; Novković et al., 2024).

16.2.1.1 High-resolution cameras

One of the most commonly used technologies in UAV wildlife monitoring is high-resolution photography and videography, which allows drones to capture clear, detailed images of animals and their habitats. High-resolution cameras provide researchers with visual data on population densities, animal behaviors, and physical characteristics such as age or health

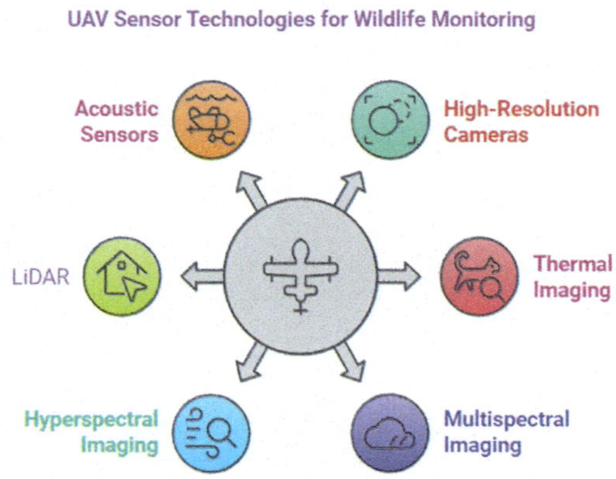

FIGURE 16.2 UAV sensor technologies for wild monitoring. *LiDAR*, Light detection and ranging; *UAV*, unmanned aerial vehicle.

indicators. This technology allows conservationists to monitor animals without direct contact, minimizing disturbance while still gathering essential information. High-resolution imagery can also be used to study vegetation cover and landscape features, supporting habitat assessments.

16.2.1.2 Thermal imaging

Thermal imaging technology has proven particularly beneficial for monitoring nocturnal or camouflaged species. Thermal sensors detect heat signatures, allowing drones to locate animals even in low-visibility conditions such as dense forests, underbrush, or nighttime hours. For example, drones equipped with thermal cameras are frequently used to track animals like elephants, orangutans, and nocturnal predators, offering conservationists valuable insights into species that are challenging to observe through standard imaging. Thermal imaging also plays a crucial role in detecting illegal activities, such as poaching, as human presence in remote areas can be identified by heat signatures.

16.2.1.3 Multispectral and hyperspectral imaging

Multispectral and hyperspectral imaging technologies are particularly useful for monitoring vegetation health, land cover changes, and water quality—factors that directly impact wildlife habitats. Multispectral cameras capture data across multiple wavelengths of light, including visible and near-infrared bands, to reveal environmental details invisible to the human eye. Hyperspectral imaging goes further, collecting hundreds of wavelengths, which allows for detailed analysis of habitat quality, vegetation health, and environmental stressors. These technologies help conservationists assess the health and sustainability of ecosystems, as well as the availability of resources that animals rely on for survival.

16.2.1.4 LiDAR (light detection and ranging)

LiDAR technology is a remote sensing method that uses laser light to create detailed, 3D maps of landscapes. When deployed via UAVs, LiDAR can penetrate forest canopies to provide detailed maps of forest structure, ground cover, and vegetation density. This is particularly useful in dense forests where traditional imaging may be obscured. In wildlife conservation, LiDAR is often used to study habitat structure and monitor changes in landscape features over time, helping conservationists understand how these changes may impact species and their habitats.

16.2.1.5 Acoustic sensors

In addition to visual sensors, UAVs can also carry acoustic sensors to detect and record sounds. This is especially beneficial for studying vocal species, such as birds, primates, and marine mammals. Acoustic sensors capture sounds in remote or underwater environments where visual tracking is challenging, providing researchers with data on species presence, vocalizations, and social interactions. These sensors also allow conservationists to monitor areas for illegal activities, as certain sounds—such as gunshots—can indicate poaching activities. In summary, imaging and sensor technologies significantly expand the range and depth of data that UAVs can collect for wildlife monitoring. By utilizing a combination of high-resolution cameras, thermal imaging, multispectral and hyperspectral cameras, LiDAR, and acoustic sensors, drones provide conservationists with a comprehensive, noninvasive toolset for understanding and protecting wildlife and their ecosystems.

16.2.2 Real-time data collection and analysis

Real-time data collection and analysis are among the most transformative capabilities that UAVs bring to wildlife monitoring, enabling conservationists to quickly assess and respond to dynamic situations in the field. Through advancements in imaging, sensor technology, and data processing algorithms, drones can now transmit data instantly from remote environments, providing actionable insights in real time. This capability is invaluable for monitoring animal movements, assessing environmental conditions, and quickly identifying threats such as poaching or habitat degradation (Stone et al., 2024).

16.2.2.1 Real-time data transmission

UAVs equipped with satellite or cellular connections can transmit data directly to ground stations, enabling conservation teams to access live footage and data from even the most remote or challenging locations. This real-time access allows for continuous monitoring without requiring personnel to be physically present, reducing disturbance to wildlife and enabling rapid interventions when necessary. In antipoaching operations, for example, drones can detect suspicious activity and immediately relay the information to authorities, allowing for a quick response that can prevent illegal hunting activities.

16.2.2.2 Immediate threat detection

Real-time data collection allows UAVs to act as immediate surveillance tools, particularly in areas prone to poaching, illegal logging, or other human-wildlife conflicts. Equipped with infrared or thermal sensors, UAVs can detect human presence or unauthorized movement, even in low-light conditions. By instantly analyzing patterns or detecting anomalies, drones provide conservationists with timely alerts, enabling rapid responses to threats. This capability has been successfully implemented in wildlife reserves where real-time monitoring has helped to reduce poaching incidents by quickly notifying authorities of intrusions.

16.2.2.3 Automated species identification and counting

Real-time data analysis also enables UAVs to automate the identification and counting of animals through ML algorithms. Advanced algorithms, often trained on large datasets of images, can detect specific animal species, estimate population counts, and even recognize individual animals based on unique features. These capabilities allow drones to analyze data while in flight, providing conservationists with up-to-date information on animal populations and movement patterns without requiring extensive postflight data processing. This automation reduces time, enhances data accuracy, and allows for more frequent and extensive surveys.

16.2.1.4 Habitat assessment and environmental monitoring

In addition to tracking wildlife, real-time data from UAVs can assess environmental variables, including vegetation health, water availability, and land use changes, which are critical to habitat stability. Equipped with multispectral or hyperspectral cameras, drones can analyze habitat conditions and detect signs of environmental stress or degradation that could affect wildlife. Real-time data on factors, such as vegetation cover, soil moisture, and water quality, supports proactive conservation strategies, allowing for early interventions to preserve habitats essential for species survival.

16.2.2.5 Machine learning for real-time pattern recognition

ML models integrated into UAV systems enable pattern recognition and anomaly detection in real time. These models are especially useful in identifying animal movement patterns, seasonal migration routes, and potential disease outbreaks based on observed behaviors or symptoms. By continuously analyzing data during flight, ML algorithms can identify unusual behaviors that may indicate threats to animal health or habitat stability. For instance, rapid changes in migration patterns could signal environmental disturbances, allowing conservationists to investigate and address the underlying causes.

16.2.2.6 Reduced data processing time and faster decision-making

Real-time analysis reduces the time typically required for data processing, which is crucial in scenarios requiring prompt decision-making. Conservation efforts often depend on up-to-date information to implement effective interventions, such as relocating animals from a threatened area or reinforcing surveillance in response to illegal activities. The ability to quickly analyze and interpret drone-collected data significantly enhances the capacity for fast, informed decisions, leading to more effective conservation management. In summary the real-time data collection and analysis capabilities of UAVs have revolutionized wildlife conservation by enhancing situational awareness, improving response times, and supporting data-driven decisions. By integrating live data transmission, automated species identification, and ML-based pattern recognition, drones provide a comprehensive and responsive toolset for addressing the complex challenges of wildlife monitoring and habitat protection (Bhute & Bhute, 2025; Laghari et al., 2024).

16.2.3 Applications for population assessment and habitat monitoring

UAVs play a crucial role in population assessment and habitat monitoring by offering a noninvasive, efficient, and scalable approach to gather extensive data on wildlife populations and their environments. Equipped with high-resolution imaging, multispectral sensors, and real-time data processing capabilities, drones provide conservationists with valuable insights into population densities, migration patterns, habitat health, and ecosystem changes. These applications are essential for developing effective conservation strategies and ensuring the sustainable management of natural habitats (Bhute & Bhute, 2025).

16.2.3.1 Population counting and density estimation

One of the primary applications of UAVs in wildlife monitoring is the accurate assessment of population numbers and density distribution. Traditional counting methods, which involve field surveys or aerial photography, can be labor-intensive, costly, and potentially disruptive to wildlife. UAVs, however, provide an efficient and less intrusive alternative by capturing high-resolution images that can be analyzed for precise counts. Automated ML algorithms can further enhance this process by recognizing and counting individual animals from images, even in large herds or dense forested areas. Drones have proven particularly useful in monitoring populations of large mammals, birds, and marine species, providing valuable data on population dynamics that inform conservation policies.

16.2.3.2 Tracking movement patterns and migration routes

UAVs equipped with GPS tracking and real-time data transmission capabilities enable researchers to monitor animal movements over time. By observing animals' movement patterns, conservationists can gain insights into migration routes, seasonal behaviors, and habitat preferences. These data are critical for understanding how species adapt to changes in their environments, such as habitat fragmentation or climate shifts. Tracking movement patterns helps in designing wildlife corridors that facilitate safe migration and can aid in preventing human-wildlife conflicts. For example, UAVs have been effectively used to monitor elephant migration routes in Africa, allowing conservationists to protect critical migratory pathways and reduce encounters with human communities.

16.2.3.3 Health assessment of populations

High-resolution imaging and thermal sensors on UAVs allow researchers to monitor individual animals for signs of disease, malnutrition, or injury, providing valuable information on population health. Thermal imaging, for instance, can detect body temperature anomalies in animals, which may indicate fever or stress. Multispectral imaging can identify changes in body condition or behavior that signal potential health issues. These capabilities enable conservationists to proactively address emerging health concerns within populations and implement preventative measures, particularly during disease outbreaks. This early detection reduces the risk of widespread disease transmission and supports healthier, more resilient populations.

16.2.3.4 Habitat condition monitoring

UAVs equipped with multispectral and hyperspectral sensors provide detailed information on habitat conditions by measuring vegetation health, water quality, and soil properties. These sensors can capture data across various wavelengths, revealing insights into plant health, canopy cover, and land use changes that may impact wildlife. For instance, multispectral imaging can detect chlorophyll levels in vegetation, indicating overall ecosystem health. UAVs are also used to assess the availability and quality of water resources, which are critical for many species, especially in arid regions. Regular UAV-based habitat assessments allow conservationists to track environmental changes, such as deforestation or desertification, and develop strategies to mitigate these threats.

16.2.3.5 Assessing the impact of environmental changes

As climate change and human activities continue to alter natural landscapes, UAVs play a crucial role in monitoring the impacts on wildlife and their habitats. By conducting periodic assessments of ecosystems, UAVs can document changes in land cover, vegetation density, and the expansion of urban or agricultural areas into wildlife habitats. These assessments are essential for understanding how environmental changes influence species distribution and behavior. For instance, UAVs can monitor coastal erosion and rising water levels in marine ecosystems, helping conservationists track the effects on local marine populations. In terrestrial environments, drones can document deforestation rates, which can significantly affect species that depend on forested habitats for food and shelter.

16.2.3.6 *Biodiversity assessment and species distribution mapping*

Using UAVs to map species distribution and biodiversity across different habitats enables researchers to identify biodiversity hotspots and prioritize areas for conservation. Multispectral and high-resolution imaging can identify different plant species, track their distribution, and assess the variety of habitats within an ecosystem. This mapping capability is particularly useful in ecologically diverse areas, where understanding species distribution is key to maintaining ecosystem balance and protecting vulnerable species. In addition, UAVs can monitor the spread of invasive species, which threaten native biodiversity, by identifying nonnative plant or animal populations.

16.2.3.7 *Long-term ecological monitoring and change detection*

UAVs allow for consistent, repeatable data collection over time, making them ideal for long-term ecological monitoring. By conducting regular UAV surveys, conservationists can establish baselines for wildlife populations and habitat conditions, then track changes over months or years. This long-term monitoring provides essential insights into the impacts of conservation interventions, climate variations, and human activities on ecosystems. ML algorithms can also analyze temporal data, detecting changes in animal populations, vegetation patterns, or habitat use over time. This information is vital for evaluating the effectiveness of conservation strategies and adapting them as needed to ensure sustainable ecosystem management. In summary, UAVs are powerful tools for population assessment and habitat monitoring, offering conservationists a scalable, noninvasive, and data-rich approach to studying wildlife. By facilitating accurate population counts, tracking movement patterns, monitoring health, assessing habitat conditions, and supporting long-term ecological studies, drones play a pivotal role in advancing wildlife conservation efforts and promoting sustainable ecosystem management (Hariharan et al., 2024; Rolland et al., 2024; Zhou & Chibaya, 2024).

16.3 Literature survey

The use of UAVs in wildlife conservation has emerged as a revolutionary development in ecological research, transforming traditional approaches to studying and managing wildlife populations. Originally developed for military and surveillance purposes, UAV technology has evolved to become an indispensable tool in various fields, including agriculture, forestry, environmental monitoring, and now conservation science. As the demand for data-driven and sustainable conservation practices has grown, so too has the adoption of drones for tracking, monitoring, and protecting vulnerable species. Their adaptability and cost-effectiveness allow them to gather valuable data in environments that are difficult to access by foot or other means, such as dense forests, open savannas, and expansive ocean areas. This capability is particularly crucial in the face of pressing global issues such as habitat loss, climate change, and poaching, which threaten many species with extinction.

UAVs have proven to be valuable in ecological research and conservation for their ability to capture high-resolution imagery and real-time data, offering insights into species populations, habitats, and migration patterns. Equipped with advanced sensors, such as multispectral and thermal cameras, drones can gather comprehensive data that researchers can analyze to understand animal health, monitor population dynamics, detect environmental changes, and even predict future trends. This data allows conservationists to develop and implement more effective strategies for wildlife protection. For instance, thermal imaging technology enables drones to locate animals even in low-light or heavily forested environments, making them essential for tracking nocturnal or elusive species that are otherwise hard to observe. In addition, GPS-enabled drones facilitate mapping of landscapes and habitats, contributing valuable information about ecosystem changes and their potential impact on local wildlife. One of the most important benefits of UAVs in conservation is their ability to conduct noninvasive monitoring (Kemarau et al., 2024). Traditional methods, such as ground surveys or animal tagging, can be costly, labor-intensive, and, in some cases, stressful or harmful to the animals being studied. By contrast, drones provide a way to monitor wildlife without direct interference, reducing stress on animals and preserving their natural behaviors. This is especially beneficial when studying endangered or vulnerable species, where disturbance can have adverse effects on population health or breeding success. Drones offer the ability to observe animals from a distance, capturing candid insights into behaviors, social interactions, and ecosystem roles that may be altered by human presence. Moreover, the real-time data captured by drones can alert conservationists to potential threats, such as poaching activities, allowing for immediate intervention to protect at-risk species (Blanton et al., 2024). The increasing adoption of UAVs in conservation also reflects broader advancements in technology and data science, including the integration of ML and AI for automated analysis. ML algorithms can process large volumes of UAV-collected data to identify individual animals, detect species, and analyze habitat changes over time. This capability greatly enhances the speed and accuracy of research findings, providing conservationists with actionable insights to address environmental and ecological challenges more effectively. However, despite these benefits, the use of UAVs in wildlife conservation is not without its challenges. Issues such as regulatory restrictions, limited flight times, high costs for advanced drones, and the need for trained personnel to operate and maintain UAVs can present hurdles for conservation organizations, particularly those with limited funding. Nonetheless, as UAV technology continues to improve, it is expected that drones will play an increasingly critical role in biodiversity protection and the sustainable management of ecosystems worldwide (Tripathi et al., 2024). The use of UAVs in wildlife monitoring has gained significant attention in recent years because of their ability to survey large areas and navigate challenging environments autonomously (Boiteau et al., 2024). UAVs equipped with advanced sensor technologies have been utilized in various applications, including monitoring clandestine graves for forensic investigations (Molina et al., 2024) and tracking Holstein cattle with high precision (Luo et al., 2024). However, challenges such as interference with civil aircrafts because of limited channel capacity have been identified (Liao et al., 2024). Incorporating biological data into autonomous remote sensing systems for in situ imageomics has shown promise in studying wildlife behavior and ecosystems (Kline et al., 2024). The role of hunters as citizen scientists in wildlife health research and monitoring has also been recognized, with significant contributions to sample collection and surveillance activities in Italy (Zanet et al., 2024). Advancements in primate surveillance using image recognition techniques from UAVs have been proposed to enhance monitoring effectiveness (He et al., 2024). Collaborative UAV 3D mapping has been explored as a tool for wildlife research in Kenyan savannas, utilizing visual simultaneous localization and mapping (V-SLAM) and structure-from-motion (SfM) frameworks for habitat mapping and monitoring solutions (Shukla et al., 2024). In addition, the development of a multi-UAV dataset for multiobject tracking and reidentification of wild antelopes has been introduced to address the challenges of monitoring and identifying individual animals in the wild (Yuan et al., 2024). Overall, the integration of UAVs and advanced technologies in wildlife monitoring presents opportunities for improving conservation efforts and understanding animal behavior in diverse ecosystems. Further research and technological advancements are needed to address challenges such as interference with other aircraft and enhancing the automation of image interpretation for more effective monitoring (Boroujeni et al., 2024).

16.3.1 Problem formulation

In the context of African elephant conservation, there is an urgent need to monitor and protect this vulnerable species because of threats such as poaching, habitat loss, and human-wildlife conflict. Traditional monitoring techniques, while effective, are often labor-intensive, costly, and intrusive, potentially disrupting natural behaviors. The application of UAVs equipped with advanced sensors and ML algorithms offers a promising solution to this challenge by enabling efficient, noninvasive, and real-time monitoring of elephant populations and behaviors across vast landscapes. However, despite the advantages of UAV technology, several challenges exist, including processing large amounts of drone-collected data, accurately identifying animals in diverse terrains, and extracting actionable insights to support conservation decisions (Whig et al., 2024). Following are the main objectives:

1. Leverage ML algorithms to automatically identify and count African elephants from UAV-collected imagery, enabling accurate and real-time population estimates to inform conservation strategies.
2. Use ML models to analyze movement patterns and habitat preferences of elephants based on drone-collected data, helping researchers understand seasonal migrations, habitat utilization, and areas at risk from human activity.
3. Implement predictive ML algorithms to identify patterns and potential high-risk areas for poaching based on environmental and UAV-collected data, allowing conservation teams to proactively allocate resources to protect elephants.

These objectives, rooted in ML, aim to enhance the precision and efficiency of elephant conservation efforts through data-driven, proactive monitoring methods.

16.4 Methodology

Monitoring elephant populations and behaviors across vast and often inaccessible landscapes presents significant challenges for conservationists. Fig. 16.3 shows methodology flow process. UAVs equipped with high-resolution cameras, thermal

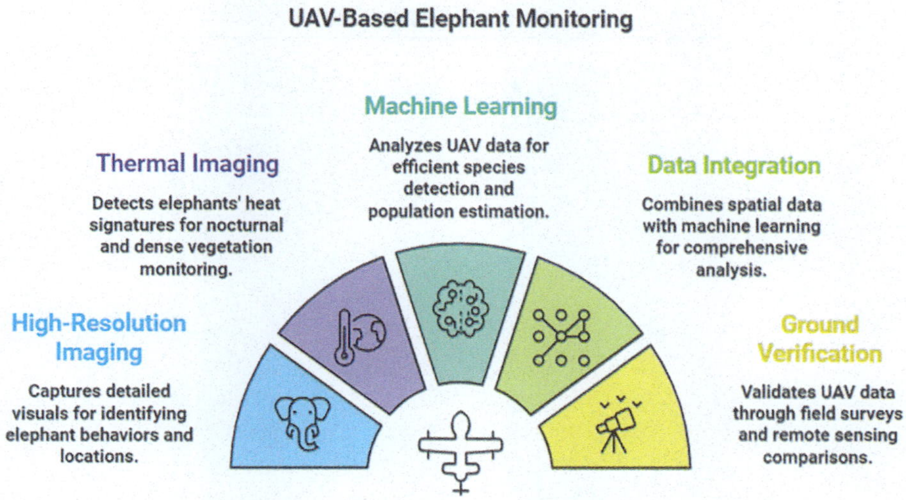

FIGURE 16.3 Methodology flow process. *UAV*, Unmanned aerial vehicle.

imaging, and ML algorithms have become essential tools for overcoming these challenges. High-resolution imaging provides detailed visuals necessary for identifying and tracking elephants, even in dense vegetation or over large areas. UAVs capture images with resolutions of up to 20–30 megapixels, ensuring precise detection of elephants and integration of GPS data for accurate mapping of migration routes and behaviors (Kumar et al., 2024).

Thermal imaging is particularly useful for nocturnal monitoring, as elephants are primarily active at night. Thermal sensors detect heat emitted by elephants, allowing their tracking even in dense forests or after dark when traditional methods would fail. The contrast between the heat signatures of elephants and the cooler surroundings helps conservationists identify their location efficiently. ML algorithms, especially computer vision models, are employed to automatically detect and classify elephants in UAV-captured images by analyzing features such as body size, shape, and movement. These models can differentiate elephants from other wildlife and even estimate populations by counting individuals in large groups, significantly reducing the time and errors associated with manual counts. ML can also analyze behavioral patterns, such as foraging or migration, providing insights into the social dynamics of elephant herds and their responses to environmental or human-induced threats. The data collected through UAVs are processed in real-time, with ML algorithms providing immediate analysis of elephant locations and population estimates. Regular UAV surveys enable long-term monitoring of population trends and habitat use. To ensure the accuracy of these data, field validation is conducted by ground teams who cross-check the UAV-detected locations and behavior of elephants (Stemle et al., 2024). In conclusion the integration of high-resolution imaging, thermal sensors, and ML allows for efficient, noninvasive monitoring of elephant populations. These technologies enhance the accuracy of population assessments, provide insights into elephant behavior, and help conservationists make informed decisions in the management and protection of elephant species across Africa.

16.4.1 Data collection

This dataset represents a collection of images used for monitoring elephants in wildlife conservation efforts, utilizing UAVs. Each entry in the dataset corresponds to a specific image of elephants captured in different resolution formats and types of imaging (Shah et al., 2024). The dataset includes the following columns:

1. **Image_File:** This column lists the filenames of the images taken, each of which depicts elephants in various settings. For instance, the first entry, "elephant_001.jpg," refers to the first captured image in the dataset.
2. **Label:** The label indicates that each image represents an elephant, confirming that the species under consideration for monitoring is the elephant.
3. **Image_Resolution:** This column describes the resolution of the images. Images with higher resolutions like "1920 × 1080" (Full HD) provide fine details, which are crucial for accurately identifying and tracking individual elephants in the field. Lower resolution images, such as "1280 × 720," are used for thermal imaging, which may not require the same level of detail but still offers enough information to track elephants.

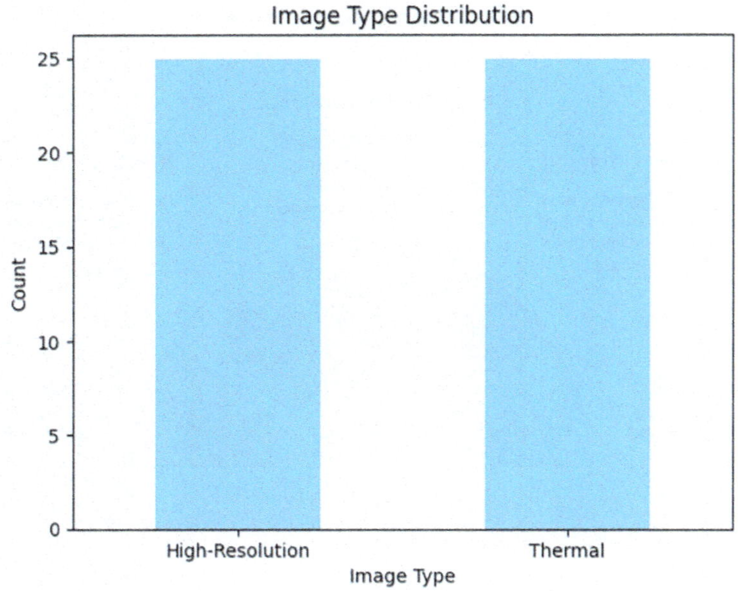

FIGURE 16.4 Image type distribution.

4. **Image_Type:** This column specifies the type of imaging used. "High-Resolution" refers to standard visual images taken with high-definition cameras, whereas "Thermal" images use infrared technology to detect heat signatures, which is particularly useful for monitoring elephants at night or in dense environments where visibility is limited.
5. **Thermal_Heat_Signature:** This column describes the thermal heat signature of the elephants. "Visible" refers to images captured in the visible spectrum, where the elephants are clearly visible. "Strong" and "Moderate" refer to the intensity of the heat signature detected by thermal imaging. Strong heat signatures indicate a clear and distinct temperature difference between the elephants and their surroundings, making them easier to track, whereas moderate signatures suggest a less distinct contrast (Ardiny et al., 2024). Table 16.1 shows dataset that is useful for ML models and image analysis tasks aimed at detecting, tracking, and monitoring elephants across vast landscapes, particularly in difficult-to-reach or remote areas. The combination of high-resolution and thermal images helps conservationists to not only track elephants during the day but also monitor their behavior during the night or in environments with limited visibility.

Fig. 16.4 shows various insights into the dataset related to elephant monitoring. This graph shows the distribution of image types in the dataset, specifically comparing "High-Resolution" and "Thermal" images. It highlights that half of the images are high-resolution, which are typically used for clear daytime imagery, whereas the other half consists of thermal images, which are vital for nighttime or low-visibility conditions, such as when tracking elephants in dense forests or during nocturnal activity.

Fig. 16.5 illustrates the proportion of different thermal heat signatures present in the dataset, categorized into "Visible," "Strong," and "Moderate" signatures. It reflects the varying intensity of thermal imaging used to detect elephants, with the "Strong" and "Moderate" categories covering a larger share. The chart emphasizes the importance of thermal imaging, particularly for detecting elephants in challenging environments where visibility is poor.

Fig. 16.6 visualizes the distribution of image resolutions, specifically comparing "1920 × 1080" (high-resolution) and "1280 × 720" (thermal) image types. It shows that most images are captured at high resolution, useful for daytime monitoring, while a smaller proportion are thermal images, typically captured in lower resolution for night-time surveillance. This highlights the varying requirements for resolution based on the imaging method and the time of day.

Fig. 16.7 combines both image resolution and thermal heat signature data, showing how these two factors correlate in the dataset. It categorizes image resolution against different thermal heat signatures (Visible, Strong, Moderate), demonstrating how both resolution and thermal signature type are distributed across the dataset. This graph provides insights into how different imaging methods (high-resolution and thermal) are applied to various thermal heat signatures, crucial for understanding the use of both types of images in different environmental conditions and times of day.

Together, these graphs allow us to understand how different imaging technologies, such as high-resolution and thermal imaging, are distributed across the dataset, their use cases in detecting elephants under varying conditions, and how thermal signatures are categorized based on the images' characteristics (Davidge et al., 2024). The analysis of the graphs underscores

TABLE 16.1 Dataset.

Image_file	Label	Image_resolution	Image_type	Thermal_heat_signature
	Elephant	1920 × 1080	High-resolution	Visible
	Elephant	1920 × 1080	High-resolution	Visible
	Elephant	1920 × 1080	High-resolution	Visible
	Elephant	1920 × 1080	High-resolution	Visible
	Elephant	1920 × 1080	High-resolution	Visible
	Elephant	1280 × 720	Thermal	Strong
	Elephant	1280 × 720	Thermal	Strong
	Elephant	1280 × 720	Thermal	Moderate

(*continued on next page*)

TABLE 16.1 Dataset—cont'd

Image_file	Label	Image_resolution	Image_type	Thermal_heat_signature
	Elephant	1280 × 720	Thermal	Strong
	Elephant	1280 × 720	Thermal	Moderate

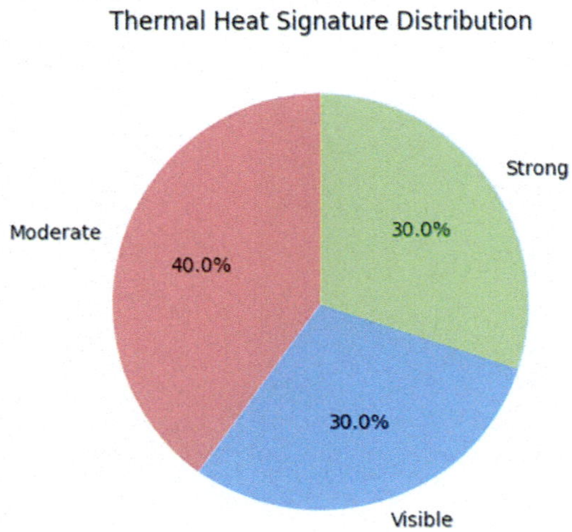

FIGURE 16.5 Thermal heat signature distribution.

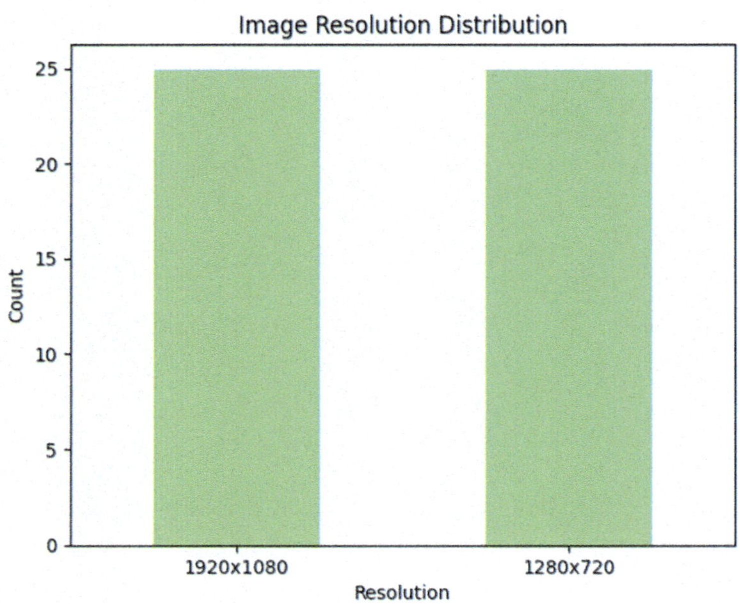

FIGURE 16.6 Image resolution distribution.

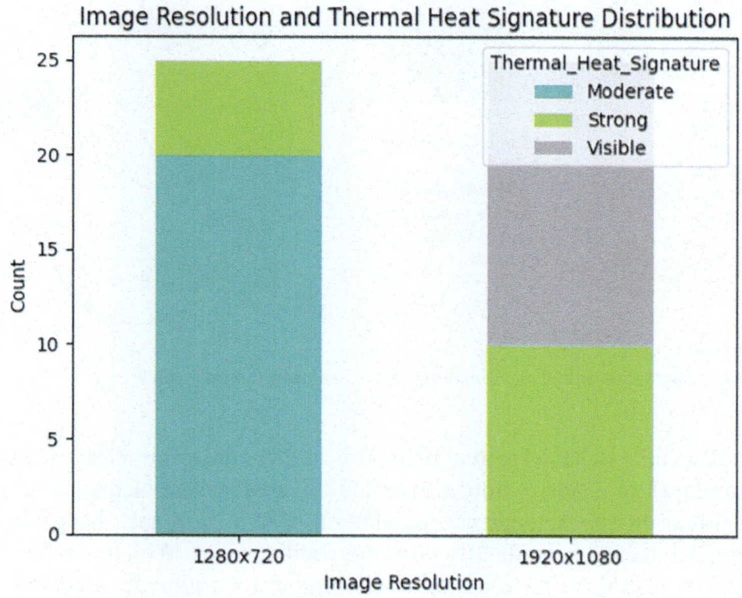

FIGURE 16.7 Image resolution and thermal heat signature distribution.

the crucial role of UAVs equipped with high-resolution and thermal imaging in wildlife conservation. The majority of images captured were high-resolution (1920 × 1080), particularly for visible spectrum images, which are essential for clear identification and monitoring of elephants. Thermal images, particularly with "Strong" and "Moderate" heat signatures, demonstrate the effectiveness of thermal sensors in detecting elephants in low-visibility conditions, such as nighttime or dense vegetation. The correlation between image types and thermal signatures shows the complementary roles of visible and thermal data—whereas high-resolution images provide detailed behavioral analysis, thermal imaging aids in tracking elephants in challenging environments. The combination of both technologies allows for a more comprehensive understanding of elephant behavior and movement patterns. These findings confirm the value of UAVs in improving data accuracy, minimizing human impact, and enhancing wildlife conservation efforts (Bakar et al., 2024).

16.5 Benefits of UAV technology in conservation

UAVs have revolutionized wildlife conservation, providing numerous benefits that enhance research, protect biodiversity, and support ecosystem sustainability. The following sections outline the primary benefits of UAV technology in conservation efforts, particularly focusing on improving research efficiency, minimizing human-wildlife conflicts, and supporting antipoaching operations (Giannetti et al., 2024).

16.5.1 Efficiency in research and data accuracy

One of the most significant advantages of UAV technology in wildlife conservation is its ability to increase the efficiency of research and improve data accuracy. UAVs can cover large, often inaccessible areas quickly and cost-effectively, allowing researchers to gather comprehensive data without the time and expense of traditional methods such as ground surveys or manned aerial surveys. The ability of UAVs to fly at low altitudes provides high-resolution images and real-time data that enhance the accuracy of wildlife population assessments, habitat mapping, and behavior analysis. For example, in the case of elephant monitoring, UAVs equipped with high-definition cameras and thermal sensors can capture detailed images of animals even in dense forests or at night. This ensures that conservationists can accurately track the movements and population dynamics of species without disturbing their natural habitat. Furthermore, ML algorithms can process this data, allowing for quick and reliable analysis of wildlife populations, migration patterns, and behavior, which significantly reduces the time spent on manual data interpretation.

16.5.2 Minimization of human-wildlife conflicts

Human-wildlife conflict is a major challenge in many conservation areas, particularly where human populations and wildlife habitats overlap. UAV technology plays a crucial role in minimizing these conflicts by providing early detection of wildlife

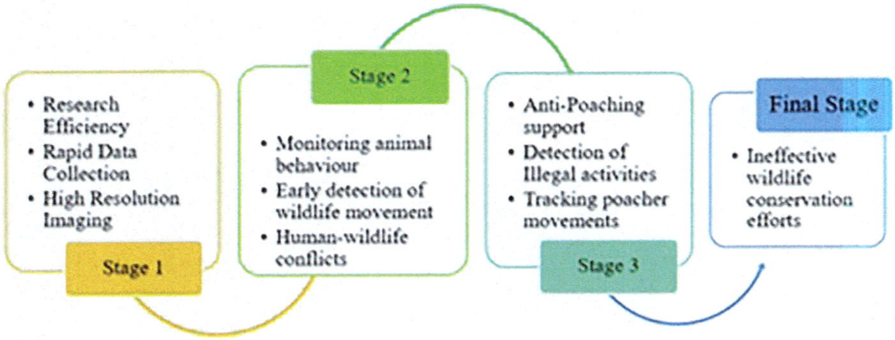

FIGURE 16.8 Enhancing wildlife conservation with UAV technology. *UAVs*, Unmanned aerial vehicles.

movement and monitoring of wildlife behavior in real time. This allows conservationists to implement proactive measures such as alerting local communities of nearby wildlife, especially in areas where elephants or other large animals may be approaching agricultural fields or human settlements. In addition, UAVs can track changes in animal behavior, such as migration patterns or feeding habits, and help identify potential conflict zones. With this information, local authorities can take preventive actions, such as creating buffer zones or modifying infrastructure, to reduce the risk of conflict. UAVs also help monitor endangered species in a way that minimizes human interaction, reducing the likelihood of stress or aggressive behavior in animals. As a result, UAVs are a key tool in fostering peaceful coexistence between humans and wildlife.

16.5.3 UAV support in antipoaching operations

Poaching remains one of the most significant threats to wildlife populations, particularly for species such as elephants, rhinos, and tigers. UAVs provide critical support in antipoaching efforts by offering real-time surveillance of vast and remote areas that are difficult to patrol on foot or by vehicle. Equipped with high-resolution cameras, infrared sensors, and thermal imaging technology, UAVs can detect illegal activities, such as poaching or habitat destruction, even in areas that are otherwise hard to monitor.

By using UAVs, conservation teams can identify suspicious activity, track poachers' movements, and provide timely information to law enforcement authorities, who can respond quickly to prevent illegal hunting or poaching incidents, as shown in Fig. 16.8. UAVs can also be used to monitor known poaching hotspots, making it easier to deploy security teams to high-risk areas. The presence of UAVs in conservation zones can act as a deterrent to poachers, knowing that surveillance is constant and difficult to avoid. In addition, the use of UAVs in conjunction with other technologies, such as GPS tracking collars and automated alert systems, enhances the effectiveness of antipoaching operations, ultimately contributing to the protection of endangered species.

UAV technology offers substantial benefits to conservation efforts by enhancing research efficiency, improving data accuracy, and supporting efforts to minimize human-wildlife conflicts and combat poaching. These advantages make UAVs an invaluable tool in modern wildlife conservation strategies, ensuring that efforts to protect biodiversity are more effective, informed, and sustainable. By integrating UAVs into conservation practices, researchers and conservationists can gain deeper insights into wildlife populations, behaviors, and habitats, leading to better management practices and the long-term preservation of ecosystems (Cecchinato et al., 2024; Hassan, 2024; Osmani & Schulz, 2024; Mayer et al., 2024).

16.6 Technical and regulatory challenges

Although UAV technology has proven to be highly effective in wildlife conservation, its deployment is not without challenges. These challenges span regulatory limitations, technical constraints in wildlife habitats, and ethical considerations. Addressing these issues is crucial for the continued growth and success of UAVs in conservation (Das, 2024).

16.6.1 Navigating regulatory limitations

One of the most significant challenges in deploying UAVs for wildlife conservation is navigating the complex regulatory landscape surrounding drone use. Many countries have strict regulations governing the operation of UAVs, especially in protected or remote areas. These regulations often limit the altitude at which drones can fly, the distances drones can cover, and the areas where they are allowed to operate. In addition, drone operators must obtain special permits to fly UAVs in national parks, wildlife reserves, or protected areas, where conservation work is often focused.

These regulations are in place to ensure safety, privacy, and airspace management, but they can also create significant barriers to the widespread adoption of UAVs in conservation. For instance, conservationists may face delays in securing the necessary permits or may be restricted from flying drones in certain regions during critical research periods, such as mating seasons or migration events. Overcoming these regulatory hurdles requires collaboration between conservation organizations, local authorities, and aviation bodies to develop policies that support both conservation efforts and safe drone operations. In some cases, regulations may need to be updated or adapted to better align with the technological advancements in UAVs (Ali et al., 2024).

16.6.2 Overcoming technical constraints in wildlife habitats

Technical challenges are another key issue when it comes to the use of UAVs in wildlife conservation. Many wildlife habitats are located in remote or rugged environments that present operational challenges for UAVs. Dense forests, mountainous terrain, or areas prone to extreme weather conditions can limit the effectiveness of drones in gathering data. For example, UAVs equipped with sensors or cameras may struggle to capture high-quality images in areas with low visibility, such as dense foliage, fog, or rain.

Another technical constraint is the limited battery life of UAVs, which often restricts the duration and range of flights. In large conservation areas, drones may not have sufficient battery capacity to cover vast distances or conduct long-duration surveys without the need for recharging. Furthermore the reliance on GPS for navigation can be problematic in certain environments, such as areas with dense tree canopies or deep valleys, where GPS signals may be weak or unreliable. To address these technical challenges, researchers are developing more robust UAV systems equipped with better sensors, improved navigation capabilities, and longer battery lives. In addition, advances in drone autonomy, such as automated flight planning and obstacle detection, are helping to reduce the risks posed by challenging habitats. In the future, hybrid UAVs that combine fixed-wing and rotary capabilities may provide greater versatility and the ability to operate in diverse terrains (Fang et al., 2024).

16.6.3 Ethical considerations in UAV deployment

Although UAVs provide significant benefits in wildlife conservation, their deployment must also consider ethical implications. One of the primary concerns is the potential disturbance to wildlife. Despite being noninvasive, drones can still cause stress to animals, particularly species that are sensitive to noise or human presence. The sound of UAVs, even at high altitudes, can disturb animals, leading to changes in behavior, such as abandoning nesting sites, disrupting mating rituals, or altering migration routes.

Furthermore the repeated presence of drones in certain areas may condition animals to associate UAVs with human activity, leading to habituation or, conversely, increased fear. For instance, elephants, which are already stressed by poaching and habitat destruction, may become agitated by drone surveillance, potentially altering their behavior in ways that could impact their health or conservation. To mitigate these concerns, ethical guidelines for UAV deployment are being developed, emphasizing the need for minimal disturbance to wildlife. Conservationists are encouraged to use UAVs during times when animals are least active, such as during the night or when migration patterns suggest lower levels of interaction. In addition, UAVs should be operated at a safe distance to reduce noise pollution, and flight paths should be carefully planned to avoid direct interference with animals' daily routines or sensitive areas, such as breeding grounds or calving sites.

Another ethical consideration involves the protection of local communities. In some cases, UAVs may be used to monitor areas inhabited by indigenous or local communities. These communities may have concerns about privacy or the surveillance of their lands, leading to potential conflicts. In such cases, it is essential to engage with local stakeholders and obtain informed consent before deploying UAVs in these regions (Francis et al., 2024). Although UAV technology offers significant advantages for wildlife conservation, navigating the technical, regulatory, and ethical challenges is essential to ensuring that drones are used effectively and responsibly. Regulatory limitations must be addressed to allow for the widespread use of UAVs in conservation, whereas technological advancements will continue to improve the capability of drones to operate in diverse and challenging environments. Ethical considerations, particularly concerning animal welfare and community engagement, must be carefully weighed to avoid unintended negative consequences. By overcoming these challenges, UAVs can continue to play a vital role in the protection and preservation of wildlife and ecosystems (Gowtham et al., 2024).

16.6.4 Future prospects of UAVs in wildlife conservation

The future prospects of UAVs in wildlife conservation are extremely promising, driven by continuous advancements in sensor technology and AI. The integration of more sophisticated sensors, such as multispectral and hyperspectral imaging, along with AI-powered analysis, will significantly enhance the ability of UAVs to monitor ecosystems with unprecedented precision. These advancements will enable drones to detect subtle changes in animal behavior, track movements in real-time, and analyze vegetation health, thus providing deeper insights into wildlife populations and their habitats. As UAV capabilities evolve, the scope of their applications is expected to expand across diverse ecosystems, from dense forests to vast savannas, enabling more comprehensive and targeted conservation efforts in remote or hard-to-reach areas. This will not only streamline efforts in monitoring species like elephants and other endangered wildlife but also allow for the protection of critical habitats such as coral reefs, wetlands, and grasslands. Moreover, the increased use of UAVs in biodiversity conservation is expected to have a significant impact on ecosystem sustainability. With more accurate data and enhanced monitoring capabilities, conservationists will be better equipped to assess the health of ecosystems, track the effects of climate change, and detect early signs of environmental degradation. As UAVs become more affordable, efficient, and accessible, their integration into conservation strategies will empower stakeholders at all levels to make informed decisions, drive positive change, and significantly bolster the preservation of biodiversity for future generations (Frederiksen et al., 2024).

16.7 Conclusion

In conclusion, UAVs have proven to be a transformative tool in wildlife conservation, offering unprecedented opportunities for noninvasive monitoring and research. Through advanced imaging technologies, including high-resolution cameras and thermal sensors, UAVs provide detailed and accurate data on wildlife populations, migration patterns, and habitat changes, even in remote or difficult-to-reach areas. These capabilities enable conservationists to track species such as elephants, monitor ecosystems, and collect data without disturbing natural habitats or wildlife behavior. ML algorithms further enhance the efficacy of UAVs by automating the detection, classification, and counting of species, allowing for more efficient and scalable conservation efforts. UAVs not only improve data accuracy but also help reduce the risks and costs associated with traditional wildlife monitoring techniques, such as ground surveys and satellite imaging.

In addition, UAVs play a crucial role in minimizing human-wildlife conflicts and supporting antipoaching operations. By providing real-time surveillance, UAVs enable authorities to monitor high-risk areas and identify potential threats, enhancing the responsiveness of conservation teams. This real-time capability is vital in mitigating poaching activities and protecting endangered species. The deployment of UAVs also ensures greater safety for field researchers, reducing the need for human presence in dangerous or inaccessible regions. However, despite their numerous benefits, the use of UAVs in wildlife conservation is not without challenges. Regulatory limitations, technical constraints in complex environments, and ethical considerations surrounding wildlife disturbance must be carefully navigated to ensure responsible and effective deployment. Future advancements in sensor technology, AI, and data analysis will continue to push the boundaries of what UAVs can achieve, enabling broader and more impactful conservation efforts. As UAV technology becomes more accessible and cost-effective, it is poised to make an even greater contribution to biodiversity protection and ecosystem sustainability, helping to secure a healthier planet for future generations.

References

Afridi, S., Hlebowicz, K., Cawthorne, D., & Lundquist, U. P. S. (2024). Unveiling the impact of drone noise on wildlife: A crucial research imperative. *International Conference on Unmanned Aircraft Systems (ICUAS) 2024*. IEEE.

Ali, M. F., Jayakody, D. N. K., & Muthuchidambaranathan, P. (2024). Revolutionizing firefighting: UAV-based optical communication systems for wildfires. *Photonics, 11*(7), 656.

Anandaraj, A. P. S., Dhivya, R., Ateeq, K., & Subramaniam, S. (2024). A survey of AI integration in unmanned aerial vehicles (UAVs) using digital twin technology: Advancements and applications. In *Digital twin technology and AI implementations in future-focused businesses* (pp. 14–26). IGI Global.

Ardiny, H., Beigzadeh, A., & Mahani, H. (2024). Applications of unmanned aerial vehicles in radiological monitoring: A review. *Nuclear Engineering and Design, 422*, 113110.

Bakar, N. A. A., Jaafar, W. S. W. M., Maulud, K. N. A., Kamarulzaman, A. M. M., Saad, S. N. M., & Mohan, M. (2024). Monitoring mangrove-based blue carbon ecosystems using UAVs: A review. *Geocarto International, 39*(1), 2405123.

Bhute, A., & Bhute, H. (2025). A comprehensive review on digital innovations for renewable energy and conservation. *Digital innovations for renewable energy and conservation*, 1–20.

Blanton, A., Ewane, E. B., McTavish, F., Watt, M. S., Rogers, K., Daneil, R., & Mohan, M. (2024). Ecotourism and mangrove conservation in Southeast Asia: Current trends and perspectives. *Journal of Environmental Management, 365*, 121529.

Boiteau, S., Vanegas, F., & Gonzalez, F. (2024). Framework for autonomous UAV navigation and target detection in global-navigation-satellite-system-denied and visually degraded environments. *Remote Sensing, 16*(3), 471.

Boroujeni, S. P. H., Razi, A., Khoshdel, S., Afghah, F., Coen, J. L., O'Neill, L., Fule, P., Watts, A., Kokolakis, N.-M. T., & Vamvoudakis, K. G. (2024). A comprehensive survey of research towards AI-enabled unmanned aerial systems in pre-, active-, and post-wildfire management. *Information Fusion, 108*, 102369.

Buchelt, A., Adrowitzer, A., Kieseberg, P., Gollob, C., Nothdurft, A., Eresheim, S., & Holzinger, A. (2024). Exploring artificial intelligence for applications of drones in forest ecology and management. *Forest Ecology and Management, 551*, 121530.

Cecchinato, N., Scagnetto, I., Toma, A., Drioli, C., & Foresti, G. L. (2024). A broadcast sub-GHz framework for unmanned aerial vehicles clock synchronization. *Integrated Computer-Aided Engineering, 31*(1), 59–75.

Chetverikov, B., Różycki, S., Malitskyy, A., & Babiy, L. (2024). Application of orthophoto maps created from UAV aerial images for monitoring historical and cultural heritage lands. *Journal of Environmental and Earth Sciences, 6*(2), 144–163.

Das, S. (2024). Transforming agriculture: Harnessing robotics and drones for sustainable farming solution. *Journal of Experimental Agriculture International, 46*(7), 219–231.

Davidge, L. R., Knox, C. D., & Monks, J. M. (2024). Flying towards the future: Using drones to detect lizards in remote alpine terrain. *Drones, 8*(3), 79.

Dokoro, H. A., Hassan, I., Yarda, M. Y., & Umar, M. (2024). Exploring the technological advancement in drone technology for surveillance. *Gombe State Polytechnic Bajoga, Journal of Science and Technology, 1*(1), 78–85.

Fang, H., Wu, C., Wang, X., Zhou, F., Chang, Y., & Yan, L. (2024). Online infrared UAV target tracking with enhanced context-awareness and pixel-wise attention modulation. *IEEE Transactions on Geoscience and Remote Sensing, 62*, 1–17.

Francis, R. J., Kingsford, R. T., Moseby, K., Read, J., Pedler, R., Fisher, A., & West, R. (2024). Tracking landscape scale vegetation change in the arid zone by integrating ground, drone and satellite data. *Remote Sensing in Ecology and Conservation, 10*(3), 374–387.

Frederiksen, M. H., Wolf, P., & Klotz, U. (2024). Citizen visions of drone uses and impacts in 2057: Far-future insights for policy decision-makers. *Technological Forecasting and Social Change, 204*, 123438.

Giannetti, D., Patelli, N., Palazzetti, L., Betti Sorbelli, F., Pinotti, C. M., & Maistrello, L. (2024). First use of unmanned aerial vehicles to monitor *Halyomorpha halys* and recognize it using artificial intelligence. *Pest Management Science, 80*, 4074–4084.

Gonzalez Nunez, J. A., Gonzalez Nunez, J. G., Akbas, M. I., Currier, P., & Macchiarella, N. D. (2024). Real-time detection of sea turtles using UAV and neural networks on edge devices. *Journal of Aviation/Aerospace Education & Research, 33*(5), 1.

Gowtham, G., Jaganraj, R., Nithya, S., & Arunkumar, K. (2024). Modelling and numerical analysis of attitude control flaps of a UAV. *International Journal on Interactive Design and Manufacturing (IJIDeM), 18*(5), 3043–3057.

Hariharan, R., Saxena, A., Dhote, V., Srisathirapathy, S., Almusawi, M., & Kumar, J. R. R. (2024). Advancements in solar-powered UAV design leveraging machine learning: A comprehensive review. In *E3S Web of Conferences Vol. 540* (p. 02024). France: EDP Sciences.

Hassan, M. (2024). *Integration of UAVs in society: Opportunities and challenges from public viewpoint on UAVs in Civilian Airspace*.

He, G., Zhang, X., Wang, J., Xu, P., Hou, X., Dong, W., Lei, Y., Jin, X., Wang, W., Tian, W., Huang, Y., Li, D., Qin, T., Wang, J., Pan, R., Li, B., & Guo, S. (2024). Advancing primate surveillance with image recognition techniques from unmanned aerial vehicles. *American Journal of Primatology, 87*, e23676.

Kayaoglu, M., Unal, S., & Biyik, H. (2026). Design and simulation of a fuel cell–battery hybrid power system for extended-endurance quadrotor UAVs. *International Journal of Hydrogen Energy, 204*, 153278.

Kemarau, R. A., Sakawi, Z., Suab, S. A., Eboy, O. V., Jaafar, W. S. W. M., Maulud, K. N. A., & Nor, N. N. F. M. (2024). Global perspectives on unmanned aerial vehicles technology in social sciences: Applications, innovations, and future research directions. *Geocarto International, 39*(1), 2413547.

Kline, J. M., Kholiavchenko, M., Brookes, O., Berger-Wolf, T., Stewart, C. V., & Stewart, C. (2024). Integrating biological data into autonomous remote sensing systems for in situ imageomics: A case study for Kenyan Animal behavior sensing with unmanned aerial vehicles (UAVs). *arXiv*. doi:10.48550/arXiv.2407.16864.

Kumar, A., Tallapaneni, K., Yadav, A., & Gautam, H. (2024). Enhancing home security with AI-driven drone surveillance. *Journal of Sensor Networks and Data Communications, 4*(2), 01–10.

Laghari, A. A., Jumani, A. K., Laghari, R. A., Li, H., Karim, S., & Khan, A. A. (2024). Unmanned aerial vehicles advances in object detection and communication security review. *Cognitive Robotics, 4*, 128–141.

Liao, Y., Jia, Z., Dong, C., Zhang, L., Wu, Q., Hu, H., & Han, Z. (2024). Interference analysis for coexistence of UAVs and civil aircrafts based on automatic dependent surveillance-broadcast. *IEEE Transactions on Vehicular Technology, 73*, 15911–15915.

Luo, W., Zhang, G., Yuan, Q., Zhao, Y., Chen, H., Zhou, J., & Yu, Z. (2024). High-precision tracking and positioning for monitoring Holstein cattle. *PLoS One, 19*(5), e0302277.

Mayer, M., Furuhovde, E., Nordli, K., Myriam Ausilio, G., Wabakken, P., Eriksen, A., & Zimmermann, B. (2024). Monitoring GPS-collared moose by ground versus drone approaches: Efficiency and disturbance effects. *Wildlife Biology*, e01213.

Millner, N., Newport, B., Sandbrook, C., & Simlai, T. (2024). Between monitoring and surveillance: Geographies of emerging drone technologies in contemporary conservation. *Progress in Environmental Geography, 3*(1), 17–39.

Mitra, A., & Das, S. (2024). Leveraging AI-enabled WSNs for environmental monitoring: An in-depth analysis with RF-pose technology. In *Wireless Ad-hoc and sensor networks* (pp. 214–230). CRC Press.

Molina, C. M., Wisniewski, K. D., Salamanca, A., Saumett, M., Rojas, C., Gómez, H., & Pringle, J. K. (2024). Monitoring of simulated clandestine graves of victims using UAVs, GPR, electrical tomography and conductivity over 4-8 years post-burial to aid forensic search investigators in Colombia, South America. *Forensic Science International, 355*, 111919.

Novković, M., Cvijanović, D., Anđelković, A., Mesaroš, M., Drešković, N., & Radulović, S. (2024). Using UAV approach in monitoring of macrophytes and their habitats within small water bodies of pristine riparian wetlands: A case study of the Middle Danube Area (Serbia). *Biologia Serbica, 46*(1), 48–66.

Osmani, K., & Schulz, D. (2024). Comprehensive investigation of unmanned aerial vehicles (UAVs): An in-depth analysis of avionics systems. *Sensors, 24*(10), 3064.

Rolland, E., Laporte-Devylder, L., & Christensen, A. (2024). Advancing wildlife monitoring in gregarious species with drone swarms. *Distributed Computing and Artificial Intelligence, 21st International Conference*. Springer.

Shah, I. A., Jhanjhi, N. Z., & Rajper, S. (2024). Use of deep learning applications for drone technology. In *Cybersecurity issues and challenges in the drone industry* (pp. 128–147). IGI Global.

Shukla, V., Morelli, L., Trybala, P., Remondino, F., Gan, W., Yu, Y., & Wang, X. (2024). Exploring the potential of collaborative UAV 3D mapping in Kenyan savanna for wildlife research. arXiv preprint *arXiv:2409.15914*. doi: 10.48550/arXiv.2409.15914

Stemle, L. R., Sorfleet, J. M., Moore, C. L., Christie, J. T., Searcy, C. A., & Rothermel, B. B. (2024). Growth and survival outcomes for immature gopher tortoises in contrasting habitats: A test of drone-based habitat assessment. *Ecology and Evolution, 14*(11), e70509.

Stone, Z. L., Macdermid, K., Muller, C. G., Armstrong, D. P., & Parker, K. A. (2024). Comparing cost-effectiveness of radio and drone telemetry with playback surveys for assessing translocation outcomes. *Journal of Applied Ecology, 61*, 2525–2537.

Tripathi, R. N., Ramachandran, A., Tripathi, V., Badola, R., & Hussain, S. A. (2024). Optimizing riparian habitat conservation: A spatial approach using aerial and space technologies. *IEEE Journal of Selected Topics in Applied Earth Observations and Remote Sensing, 17*, 16932–16943.

Whig, P., Bhatia, A. B., Nadikatu, R. R., Alkali, Y., & Sharma, P. (2024). GIS and remote sensing application for vegetation mapping. In *Geo-environmental hazards using AI-enabled geospatial techniques and earth observation systems* (pp. 17–39). Switzerland: Springer Nature.

Yuan, S., Yang, Y., Nguyen, T. H., Nguyen, T. M., Yang, J., Liu, F., Li, J., Wang, H., & Xie, L. (2024). MMAUD: A comprehensive multi-modal anti-UAV dataset for modern miniature drone threats. *arXiv* preprint *arXiv:2402.03706*. doi:10.48550/arXiv.2402.03706.

Zanet, S., Benatti, F., Poncina, M., Pasetto, C., Chiari, M., Sorrenti, M., & Ferroglio, E. (2024). The role of hunters in wildlife health research and monitoring: Their contribution as citizen scientists in Italy. *Animals, 14*(15), 2204.

Zhou, Z., & Chibaya, T. (2024). The opportunities and challenges of harnessing digital technologies in wildlife tourism resources conservation in Zimbabwe. Tourism and Hospitality for sustainable development: Volume one: Technological Innovations and Development Realities, 111–128.

Zhu, X. (2024). Integrating genomics and remote sensing technologies for wildlife monitoring. *Animal Molecular Breeding, 14*, 130–140.

Chapter 17

Impact evaluation, modeling, and optimization of machining features of heat-treated and non-heat-treated X40CrMoV51 alloy in PMEDM process by GRA and NSGA II-EAMR algorithms

Van-Tao Le
ATC, Le Quy Don Technical University, Hanoi, Vietnam

17.1 Introduction

Hot work tool steel is primarily used in metal forming processes that take place at temperatures above 600°C. The key processes include forging, extrusion, and casting of nonferrous metals, especially aluminum alloys. During these operations the dies are subjected to both mechanical stress and high temperatures simultaneously. As a result the tool steel must possess crucial properties such as high hot strength, toughness (shock resistance), and resistance to both thermal and mechanical fatigue (Wu et al., 2023). These attributes are vital for ensuring the dies perform effectively and last under the harsh conditions of hot-forming operations (Gapontseva et al., 2021). X40CrMoV51 is an alloy known for its excellent resistance to high temperatures, corrosion, oxidation, and toughness. It is widely used in the manufacturing industry, particularly in applications such as hot working dies, die-casting and extrusion dies, wear-resistant tools, pressure die-casting tools, and pressing tools for both light and heavy metals (Yan et al., 2001).

In the fabrication sector, both heat-treated and nonheat-treated states of the X40CrMoV51 alloy are used to shape components. The choice between these material states depends on the intended use of the final product and the sequence of processing steps needed to meet accuracy and technical requirements established during the design phase. However, there remains limited understanding of the differences in machining properties between these two material states (Kamenskikh et al., 2023; Le et al., 2025b;Pedroso et al., 2024). One widely utilized method is the EDM (electrical-discharge machining) process, which is favored for its precision and lack of required cutting forces. In EDM, material is removed through a series of spark discharges that produce intense heat between the tool electrode and the workpiece (Rezaei Ashtiani & Hojati, 2022). This heat causes the material of both the tool electrode and workpiece to vaporize and melt (Thakur et al., 2023). However, as the process involves heat, it often results in significant surface defects (Srivastava et al., 2023). In addition, the low machining efficiency of EDM presents a major concern, as it not only affects product quality but also increases production costs. To tackle these issues, several solutions have been proposed to improve machining efficiency (Gerçekcioğlu & Albaşkara, 2023). These include optimizing injection flow during spark discharge, incorporating ultrasonic vibration into the workpiece or electrode, using electrode rotation, and adopting powder-mixed electro-discharge machining (PMEDM). These solutions aim to enhance the discharge channel's effectiveness, improving both surface quality and machining efficiency. Of these approaches, PMEDM is particularly notable for its compact machining conditions, superior efficiency, and its ability to produce high-quality surface finishes on products (Sahu & Mandal, 2020).

Computational Intelligence in Mechatronics: Solving Real-World Problems in Electronic Systems Design. DOI: https://doi.org/10.1016/B978-0-443-27580-7.00035-4

Enhancing machining efficiency and product quality is a key goal in machining processes. The primary strategy focuses on machine design, adopting new machining technologies, and advancing intelligent control and device systems. At the same time the relationship between machining conditions and outcomes is studied, with optimization techniques used to identify optimal parameters as a secondary approach (Gao et al., 2024; Le et al., 2025c; Zhong, 2021). Although hardware-based solutions can be costly when updating or replacing existing equipment, improvements within current machinery are limited. Therefore intelligent optimization of machining processes becomes a prudent choice. This method is more economically feasible and aligns with sustainable development goals, avoiding the high costs and potential resistance from manufacturers associated with major hardware upgrades. Consequently, selecting the optimal machining conditions is crucial to meeting machining requirements, including cost reduction, improved quality, and increased productivity (Wong et al., 2012). To achieve this, various statistical techniques in the design of experiments (DOE) are used for planning, analysis, and consideration of factors influencing the machining process. Methods, such as factorial design, the Taguchi method, and response surface methodology (RSM), are commonly applied for experiment design and analysis. RSM is a specific type of experiment design used for optimization and to evaluate how input factors affect output attributes (Matys et al., 2022). Its ability to analyze interactions between variables and minimize downtime makes it particularly useful. The Box–Behnken design in RSM is especially beneficial for analyzing quantitative variables, as it offers a more efficient approach than other three-level design methods by allowing selection of points from a three-level factorial arrangement (Vijayakumar et al., 2023). Multiple optimization algorithms have been applied to solve multiobjective (MO) problems, including techniques such as the technique for order of preference by similarity to the ideal solution (TOPSIS), gray relational analysis (GRA), MO particle swarm optimization (MOPSO), and the nondominated sorted-genetic algorithm II (NSGA II). Although TOPSIS and GRA yield a single optimal solution based on control parameters and experiment responses, they may not fully capture the relationship between control variables and machining attributes. In contrast, MOPSO and NSGA II generate Pareto fronts of feasible solutions, which better reflect the relationship between control variables and machining characteristics. The final solution is then chosen from these Pareto fronts based on the required criteria. Although TOPSIS is often used for final solution selection, it is time-consuming and complex to calculate. Recently the evaluation by an area based on method of ranking (EAMR) has been employed for final solution selection. In this investigation, both GRA and NSGA II-EAMR algorithms were applied, and their optimal results were evaluated.

It can be seen that the evaluation of the machining features, comprising machining performance and surface features of the two material states in the PMEDM process, is very limited. Furthermore a combination of EDM operation with tungsten compound powder mixed in an insulated solvent has a factual sense. As the tungsten compound powder is a metal compound powder, its physical, chemical, and mechanical characteristics are necessary for actual applications of surface components, such as a high melting temperature (about 3422°C); low thermal deformation; good thermal conductivity; excellent thermal strength; corrosion resistance of acids, alkali, and oxidation; and excellent hardness. Meanwhile, the EDM process with the incorporation of tungsten compound powder has not been fully considered in terms of aspects such as process and machining features.

17.2 Mathematical model

A Box–Behnken experimental design was established to institute the mathematical model, as shown in Table 17.1. In this, MRR and tool wear rate (TWR) are calculated using Eqs. (17.1) and (17.2) (Le et al., 2025b; Le, 2025b). Meanwhile, surface roughness (R_a) of machined surfaces was measured using a Mitutoyo SJ-201 tester. Measurements were carried out in an initialed configuration with a cutoff length of 0.8 mm and a track length of 4 mm and three repetitions.

$$\mathrm{MRR}\left(\frac{g}{\min}\right) = \frac{\mathrm{Workpieceweightbeforemachining} - \mathrm{Workpieceweightaftermachining}}{\mathrm{Machiningtime}} \quad (17.1)$$

$$\mathrm{TWR}\left(\frac{g}{\min}\right) = \frac{\mathrm{Electrodeweightbeforemachining} - \mathrm{Electrodeweightaftermachining}}{\mathrm{Machiningtime}} \quad (17.2)$$

The mathematical models for $\mathrm{MRR_{ht}}$, $\mathrm{TWR_{ht}}$, R_{aht}, $\mathrm{MRR_{nht}}$, $\mathrm{TWR_{nht}}$, and R_{anht} were established using Minitab software (version 19), corresponding to Eqs. (17.3)–(17.8).

$$\begin{aligned}\mathrm{MRR_{ht}} &= 0.134596 - 0.00032T_{on} - 0.052096I_p + 0.000496A_p + 0.006283I_p^2 \\ &+0.000038A_p^2 + 0.000015T_{on}^2 - 0.000050I_p * T_{on} - 0.000117I_p * A_p - 0.000012T_{on} * A_p\end{aligned} \quad (17.3)$$

TABLE 17.1 Matrix of variables for the experiment and achieved output data.

No.	Machining parameters			Measured machining responses (avg.)					
	I_p (A)	T_{on} (μs)	A_p (g/L)	MRR_{ht} (g/min)	TWR_{ht}(g/min)	R_{aht} (μm)	MRR_{nht} (g/min)	TWR_{nht}(g/min)	R_{anht} (μm)
Empirical results for elaborating models									
1	6	32	20	0.0467	0.0051	3.18	0.0441	0.0054	3.09
2	6	16	10	0.0426	0.0047	3.12	0.0458	0.0051	2.92
3	6	48	10	0.0496	0.0048	3.43	0.0414	0.0055	3.24
4	5	48	20	0.0416	0.0041	2.73	0.0316	0.0043	2.58
5	5	32	10	0.0275	0.0039	2.59	0.0261	0.00379	2.41
6	5	32	10	0.0276	0.004	2.62	0.0259	0.00385	2.43
7	4	48	10	0.0346	0.0048	2.59	0.0326	0.0051	2.4
8	5	32	10	0.0278	0.0041	2.64	0.0262	0.00378	2.38
9	5	16	0	0.0253	0.0032	2.38	0.0222	0.0041	2.21
10	5	48	0	0.0386	0.0048	2.63	0.0327	0.0063	2.55
11	4	16	10	0.0244	0.0021	2.11	0.0211	0.0025	1.94
12	4	32	20	0.0338	0.0038	2.43	0.0315	0.00395	2.18
13	5	16	20	0.0358	0.0029	2.35	0.0355	0.0031	2.11
14	6	32	0	0.044	0.0051	3.53	0.0402	0.0064	3.08
15	4	32	0	0.0264	0.0045	2.35	0.0245	0.00505	2.29

MRR, material removal rate; *TWR*, tool wear rate.

$$TWR_{ht} = 0.00495 + 0.00035T_{on} - 0.003188I_p - 0.000116A_p + 0.000488I_p^2 + 1.37500E-06A_p^2 - 1.51367E-06T_{on}^2 - 0.000041I_p{*}T_{on} + 0.000017I_p{*}A_p - 6.25000E-07T_{on}{*}A_p \quad (17.4)$$

$$R_{aht} = 5.52417 + 0.041615T_{on} - 2.06417I_p + 0.048167A_p + 0.272917I_p^2 - 0.000171A_p^2 - 0.000301T_{on}^2 - 0.002656I_p * T_{on} - 0.01075I_p * A_p + 0.000203T_{on}{*}A_p \quad (17.5)$$

$$MRR_{nht} = 0.113279 + 0.001001T_{on} - 0.052217I_p + 0.000968A_p + 0.006867I_p^2 + 0.000021A_p^2 + 8.95182E-06T_{on}^2 - 0.000248I_p * T_{on} - 0.000078I_p * A_p - 0.000022T_{on} * A_p \quad (17.6)$$

$$TWR_{nht} = 0.012644 + 0.000238T_{on} - 0.005667I_p - 0.000156A_p + 0.000747I_p^2 + 6.46667E-06A_p^2 - 1.30208E-08T_{on}^2 - 0.000034I_p * T_{on} + 2.50000E-06I_p * A_p - 1.56250E-06T_{on} * A_p \quad (17.7)$$

$$R_{anht} = 5.98042 + 0.031224T_{on} - 2.09917I_p - 0.022708A_p + 0.257917I_p^2 - 0.000046A_p^2 - 0.000155T_{on}^2 - 0.002187I_p * T_{on} + 0.003I_p * A_p + 0.000203T_{on} * A_p \quad (17.8)$$

Analysis of variance (ANOVA) was employed to assess the adequacy of the developed models, with a confidence level of 95% and a significance level of 5%. The ANOVA results for the MRR_{ht}, TWR_{ht}, R_{aht}, MRR_{nht}, TWR_{nht}, and R_{anht} models are presented in Tables 17.2–17.7, respectively. These tables contain detailed information regarding the variance components and the significance of each model term, enabling an intimate evaluation of the models' adequacy in explaining the variability in the responses. The significance of model terms is determined by their corresponding P values and is considered significant when they are smaller than 5%. Therefore the terms comprising (T_{on}, I_p, A_p, $T_{on}{*}A_p$, $I_p{*}A_p$, I_p^2, T_{on}^2, and A_p^2), (T_{on}, I_p, A_p, $I_p{*}T_{on}$, $I_p{*}A_p$, I_p^2, and T_{on}^2), and (T_{on}, I_p, A_p, $I_p{*}A_p$, and I_p^2) are significant for the MRR_{ht}, TWR_{ht}, and R_{aht} models, respectively. Meanwhile, the terms comprising (T_{on}, I_p, A_p, $I_p{*}T_{on}$, $T_{on}{*}A_p$, I_p^2, T_{on}^2, and A_p^2), (T_{on}, I_p, A_p, $I_p{*}T_{on}$, $T_{on}{*}A_p$, I_p^2, and A_p^2), and (T_{on}, I_p, A_p, $I_p{*}T_{on}$, I_p^2, and T_{on}^2) are significant for the MRR_{nht}, TWR_{nht}, and R_{anht} models, respectively. The development models have been validated for precision/adequacy using coefficients such as "R^2," "R^2(adj)," and "R^2(pred)." Specifically the "R^2" values for the MRR_{ht}, TWR_{ht}, R_{aht}, MRR_{nht}, TWR_{nht}, and R_{anht} models are 0.9971, 0.9911, 0.9919, 0.9958, 0.9926, and 0.9985, respectively. This indicates strong consistency between empirical and predictive values. Furthermore the "R^2(pred)" values of these models (MRR_{ht} of 0.9546, TWR_{ht} of 0.8838, R_{aht} of 0.8776, MRR_{nht} of 0.9342, TWR_{nht} of 0.8846, and R_{anht} of 0.9836) align well with the "R^2(adj)" values (MRR_{ht} of 0.9919, TWR_{ht} of 0.977, R_{aht} of 0.9773, MRR_{nht} of 0.9884, TWR_{nht} of 0.9794, and R_{anht} of 0.9957).

Besides, in the normal probability distribution chart of MRR_{ht}, TWR_{ht}, R_{aht}, MRR_{nht}, TWR_{nht}, and R_{anht} (Fig. 17.1A–F), the residuals have been standardized and distributed almost evenly along the straight line. Therefore the models are considered reliable from a statistical perspective (Le et al., 2022, 2023; Chaudhari et al., 2022; Van Tao et al., 2024). Foundationally on these criteria, it can be affirmed that the models developed for MRR_{ht}, TWR_{ht}, R_{aht}, MRR_{nht}, TWR_{nht}, and R_{anht} are validated over the entire design space and can be used to predict optimum responses.

17.3 The influence of control parameters on machining features

17.3.1 The influence of control parameters on MRR_{ht} and MRR_{nht}

The main impact on the MRR is depicted in Fig. 17.2A and B, this proves that T_{on}, I_p, and A_p are influential for MRR in the two material states. Specifically, MRR_{ht} increases when I_p, A_p, and T_{on} increase from 4 to 6 A, from 0 to 20 g/L, and from 16 to 48 μs, respectively (Fig. 17.2A). Meanwhile, a similar influence trend also occurs with MRR_{nht}, as depicted in Fig. 17.2B. However, as I_p, A_p, and T_{on} increase from 5 to 6 A, from 10 to 20 g/L, and from 32 to 48 μs, respectively, a rapid rise is applied in MRR_{ht} (Fig. 17.2A), and a slow increase is for MRR_{nht} (Fig. 17.2B). When T_{on} and I_p increase, the heat energy applied to the workpiece is intensified. Consequently, this heightened heat energy leads to a greater level of material melting on the surface of the workpiece (Le et al., 2019a, 2020; Le, 2021b; Le & Nguyen, 2022; Xie et al., 2024). Furthermore the inclusion of powder in the solvent medium reduces the viscosity of the dielectric fluid. This reduction in viscosity facilitates the expulsion of molten material from the surface of the workpiece at the conclusion of the spark discharge process. These factors collectively contribute to enhancing MRR in two material states (Ekmekci et al., 2016).

TABLE 17.2 ANOVA of MRR_{ht} mathematical model.

	Sum of squares	Mean square	*F* value	*P* value	Remark	Contribution (%)
Model	0.0009879	0.0001098	192.3	<.0001	Significant	
I_p	8.955E − 05	8.955E − 05	156.9	<.0001	Significant	51.51
T_{on}	2.026E − 05	2.026E − 05	35.50	.001905	Significant	15.82
A_p	1.765E − 05	1.765E − 05	30.92	.002587	Significant	7.68
I_p*T_{on}	2.560E − 06	2.560E − 06	4.485	.08776	Not significant	0.26
I_p*A_p	5.522E − 06	5.522E − 06	9.674	.02654	Significant	0.56
$T_{on}*A_p$	1.406E − 05	1.406E − 05	24.64	.004236	Significant	1.07
I_p^2	0.0001458	0.0001458	255.4	<.0001	Significant	13.1
T_{on}^2	5.568E − 05	5.568E − 05	97.54	.0001814	Significant	4.56
A_p^2	5.355E − 05	5.355E − 05	93.81	.0001992	Significant	5.09
Lack of fit	2.807E − 06	9.358E − 07	40.11	.02443	Significant	0.35

"R^2" = 0.9971, "R^2(adj)" = 0.9919, and "R^2(pred)" = 0.9546

ANOVA, analysis of variance; MRR, material removal rate.

TABLE 17.3 ANOVA of TWR_{ht} mathematical model.

	Sum of squares	Mean square	F value	P value	Remark	Contribution (%)
Model	1.028E − 05	1.142E − 06	61.72	.0001369	Significant	
I_p	5.652E − 07	5.652E − 07	30.55	.002657	Significant	23.14
T_{on}	2.957E − 06	2.957E − 06	159.8	<.0001	Significant	38.84
A_p	2.290E − 07	2.290E − 07	12.38	.01695	Significant	3.45
$I_p{}^*T_{on}$	1.690E − 06	1.690E − 06	91.35	.0002123	Significant	16.16
$I_p{}^*A_p$	1.225E − 07	1.225E − 07	6.622	.04984	Significant	1.53
$T_{on}{}^*A_p$	4.000E − 08	4.000E − 08	2.162	.2014	Not significant	0.22
$I_p{}^2$	8.775E − 07	8.775E − 07	47.43	.0009879	Significant	9.05
$T_{on}{}^2$	5.544E − 07	5.544E − 07	29.97	.002771	Significant	5.68
$A_p{}^2$	6.981E − 08	6.981E − 08	3.773	.1097	Not significant	0.93
Lack of fit	7.250E − 08	2.417E − 08	2.417	.3061	Not significant	0.79
"R^2" = 0.9911, "R^2(adj)" = 0.975, and "R^2(pred)" = 0.8838						

ANOVA, analysis of variance; TWR, tool wear rate.

TABLE 17.4 ANOVA of R_{aht} mathematical model.

	Sum of squares	Mean square	F value	P value	Remark	Contribution (%)
Model	2.415	0.2683	68.05	.0001077	Significant	
I_p	0.1374	0.1374	34.85	.001985	Significant	73.35
T_{on}	0.03917	0.03917	9.933	.02533	Significant	10.35
A_p	0.03727	0.03727	9.450	.02765	Significant	0.21
$I_p{}^*T_{on}$	0.007225	0.007225	1.832	.2338	Not significant	0.3
$I_p{}^*A_p$	0.04622	0.04622	11.72	.01876	Significant	1.9
$T_{on}{}^*A_p$	0.004225	0.004225	1.071	.3481	Not significant	0.17
$I_p{}^2$	0.2750	0.2750	69.74	.0004028	Significant	11.99
$T_{on}{}^2$	0.02194	0.02194	5.564	.06485	Not significant	0.88
$A_p{}^2$	0.001078	0.001078	0.2733	.6235	Not significant	0.04
Lack of fit	0.01845	0.00615	9.711	.09480	Not significant	0.76
"R^2" = 0.9919, "R^2(adj)" = 0.9773, and "R^2(pred)" = 0.8776						

ANOVA, analysis of variance.

Observing the comparison of MRR in two material states the impact of control parameters on MRR in these two material states is different; the value of MRR_{ht} is larger than the value of MRR_{nht} when received the same control parameters. This proves that MRR_{ht} has been improved better than MRR_{nht} with the same control parameters. The cause of this phenomenon is the different thermal conductivity of the two material states. Indeed the heat-treated material state has better thermal conductivity than the nonheat-treated material state (Wu et al., 2023). Therefore the molten metal produced from the heat-treated material state converts to the solidified state more quickly, and they are ejected from the surface by the force of the explosion at the end of the discharge (Pervaiz et al., 2020).

Based on the analysis of contour plots as presented in Fig. 17.3A–F the incorporated influence of the control parameters on MRR_{ht} and MRR_{nht} is depicted. In which I_p expresses the most influence on MRR_{ht} with the contribution of 51.51%, followed by T_{on} (15.82%) and A_p (7.68%), as indicated in Table 17.2. Meanwhile, according to Table 17.5 is the ANOVA results of MRR_{nht}, I_p expresses the most contribution for MRR_{nht} with 54.09%, followed by A_p (7.56%) and T_{on} (2.66%). Observing the contour lines (as in Fig. 17.3A–F) between the pairs of control parameters such as I_p and T_{on}, I_p and A_p, and T_{on} and A_p, we notice that the contour lines are different in the two material states. This proves that the two material states have different influences when the control parameter changes. Furthermore, in the combined influence for each pair

TABLE 17.5 ANOVA of MRR_{nht} mathematical model.

	Sum of squares	Mean square	F value	P value	Remark	Contribution (%)
Model	0.0008789	9.766E − 05	133.1	<.0001	significant	
I_p	0.0001135	0.0001135	154.6	<.0001	significant	54.09
T_{on}	7.831E − 05	7.831E − 05	106.7	.0001463	significant	2.66
A_p	1.144E − 05	1.144E − 05	15.59	.01087	significant	7.56
$I_p{*}T_{on}$	6.320E − 05	6.320E − 05	86.13	.0002443	significant	7.16
$I_p{*}A_p$	2.403E − 06	2.403E − 06	3.274	.1302	not significant	0.27
$T_{on}{*}A_p$	5.184E − 05	5.184E − 05	70.64	.0003908	significant	5.87
I_p^2	0.0001741	0.0001741	237.2	<.0001	significant	18.15
T_{on}^2	1.939E − 05	1.939E − 05	26.42	.003644	significant	1.9
A_p^2	1.694E − 05	1.694E − 05	23.08	.004867	significant	1.92
Lack of fit	3.622E − 06	1.207E − 06	51.75	.01902	significant	0.41
"R^2" = 0.9958, "R^2(adj)" = 0.9884, and "R^2(pred)" = 0.9342						

ANOVA, analysis of variance; MRR, material removal rate.

TABLE 17.6 ANOVA of TWR_{nht} mathematical model.

	Sum of squares	Mean square	F value	P value	Remark	Contribution (%)
Model	1.743E − 05	1.936E − 06	74.98	<.0001	Significant	
I_p	1.428E − 06	1.428E − 06	55.29	.0006933	Significant	23.97
T_{on}	2.527E − 06	2.527E − 06	97.87	.0001800	Significant	29.19
A_p	2.137E − 07	2.137E − 07	8.277	.03471	Significant	18.54
$I_p{*}T_{on}$	1.210E − 06	1.210E − 06	46.86	.001016	Significant	6.9
$I_p{*}A_p$	2.500E − 09	2.500E − 09	0.09681	.7682	Not significant	0.01
$T_{on}{*}A_p$	2.500E − 07	2.500E − 07	9.681	.02651	Significant	1.43
I_p^2	2.059E − 06	2.059E − 06	79.71	.0002936	Significant	10.41
T_{on}^2	4.103E − 11	4.103E − 11	0.001589	.9697	Not significant	0.06
A_p^2	1.544E − 06	1.544E − 06	59.79	.0005779	Significant	8.76
Lack of fit	1.263E − 07	4.208E − 08	29.36	.03312	Significant	0.72
"R^2" = 0.9926, "R^2(adj)" = 0.9794, and "R^2(pred)" = 0.8846						

ANOVA, analysis of variance; TWR, tool wear rate.

of control parameters, the area for the highest value of MRR_{ht} is wider than that of MRR_{nht}, as indicated in Fig. 17.3. The analysis outlined above reveals that a simultaneous increase in both T_{on} and I_p results in heightened discharge energy, which leads to an increase in MRR in two material states (Rezaei Ashtiani & Hojati, 2022). However, this may introduce instability in the machining process because of poor debris ejection within the machining gap, potentially resulting in short circuits. The inclusion of powder into the insulating solvent promotes stratified discharge, thereby enhancing MRR in two material states (Ekmekci et al., 2016). Nevertheless, instances of abnormal deterioration in MRR of two material states by short circuits or arcing phenomena, which are caused by the accumulation of debris from previous spark discharge operations (Eyercioglu & Gov, 2021), or improper adjustment of I_p and T_{on} results in low powder density in subsequent discharge processes (Oghazi et al., 2023).

In this study, MRR results in two states of X40CrMoV51 alloy were acquired by experimentalizing with tungsten compound powder that is agreeable to the prior studies (Alam et al., 2021; Bhattacharya et al., 2013; Kansal et al., 2007; Sana et al., 2024; Schubert et al., 2022; Srivastava et al., 2023) with various types of powders such as Si, Gr, Al, TiO_2, SiC, Cr, carbon nanotube, and Ag. Although these investigations into the impact of different process variables were diverse, this study found that the fundamental principle governing MRR in the two material states remained similar. However, compared

TABLE 17.7 ANOVA of R_{anht} mathematical model.

	Sum of squares	Mean square	*F* value	*P* value	Remark	Contribution (%)
Model	2.141	0.2379	361.4	<.0001	Significant	
I_p	0.1242	0.1242	188.7	<.0001	Significant	72.47
T_{on}	0.03885	0.03885	59.01	.0005959	Significant	14.79
A_p	0.005211	0.005211	7.916	.03739	Significant	0.17
$I_p{*}T_{on}$	0.004900	0.004900	7.443	.04137	Significant	0.23
$I_p{*}A_p$	0.003600	0.003600	5.468	.06651	Not significant	0.17
$T_{on}{*}A_p$	0.004225	0.004225	6.418	.05232	Not significant	0.2
I_p^2	0.2456	0.2456	373.1	<.0001	Significant	11.52
T_{on}^2	0.005785	0.005785	8.788	.03136	Significant	0.33
A_p^2	7.756E − 05	7.756E − 05	0.1178	.7454	Not significant	0.22
Lack of fit	0.002025	0.0006750	1.066	.5175	Not significant	0.09

"R^2" = 0.9985, "R^2(adj)" = 0.9957, and "R^2(pred)" = 0.9836

ANOVA, analysis of variance.

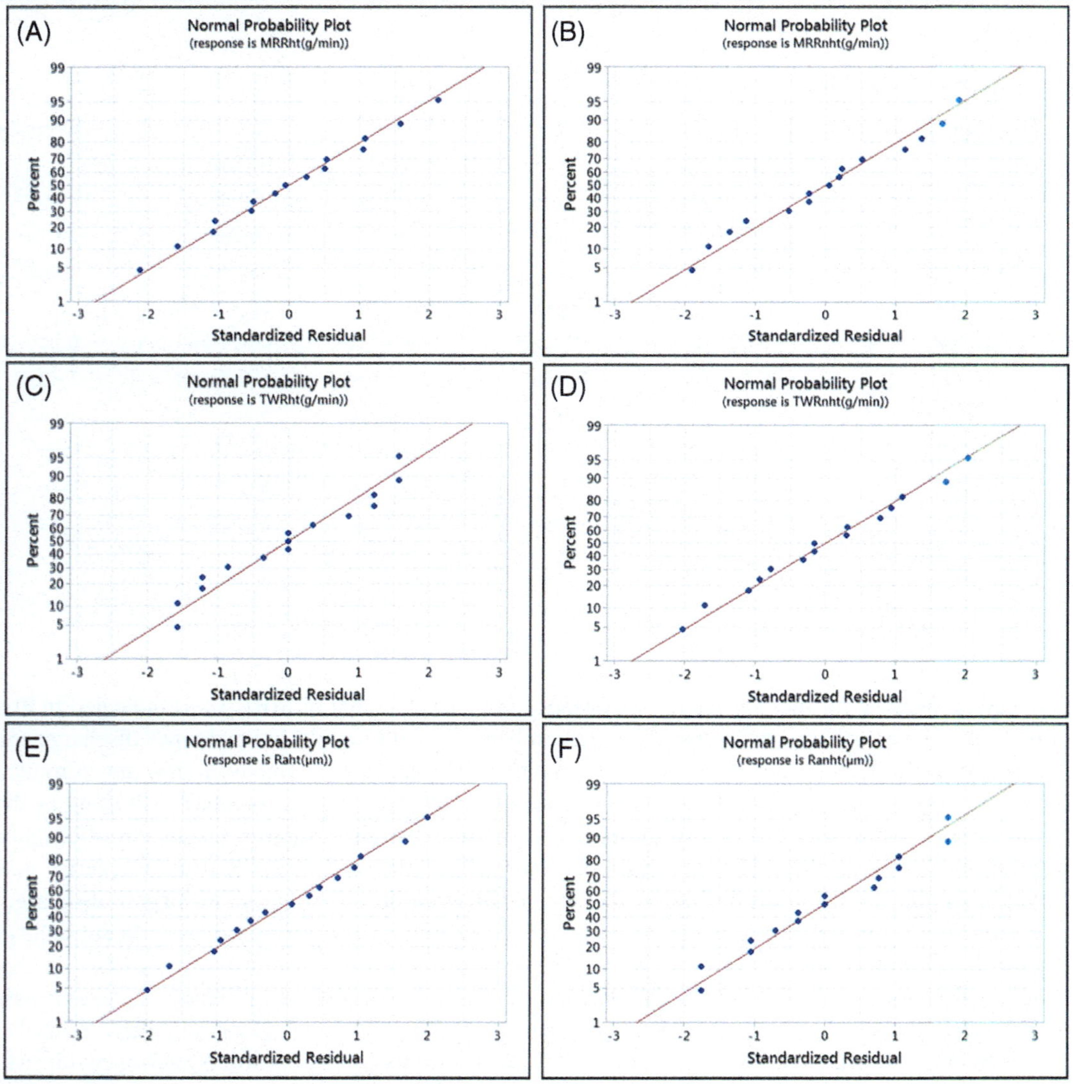

FIGURE 17.1 The normal probability plots. *MRR*, Material removal rate; *TWR*, tool wear rate.

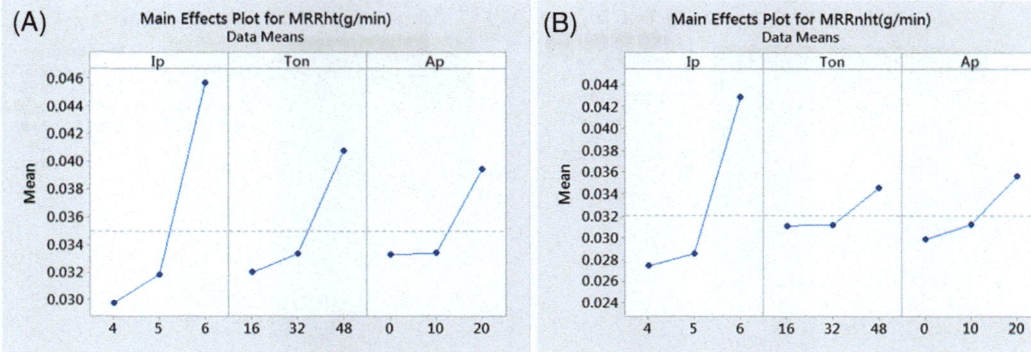

FIGURE 17.2 The main influential plot of control parameters on MRR: (A) MRR_{ht} and (B) MRR_{nht}. *MRR*, Material removal rate.

to previous research, the improvements in MRR observed in this study were comparatively modest. This deviation could be attributed to differences in powder characteristics, particularly their lower melting points in prior studies, as well as variations in selected process electrical variables resulting in higher energy inputs compared to the present investigation.

17.3.2 The influence of control parameters on TWR_{ht} and TWR_{nht}

In Fig. 17.4A and B, they express the main influence of the control parameters on TWR_{ht} and TWR_{nht}. In which, TWR_{ht} and TWR_{nht} augment with an increase in T_{on} and I_p in the entire investigation space, whereas the increase of A_p in the range of 0 to 10 g/L has rapidly reduced TWR_{ht} and TWR_{nht} and continuing to increase A_p in the range of 10–20 g/L caused a slow decline in TWR_{ht} and TWR_{nht}. The reason is that there is a reasonable combination of T_{on}, I_p, and A_p (ranging from 0 to 10 g/L), leading to the powder density in the succeeding discharge being sensible. This results in improved heat transfer to the surrounding environments. As a result the energy of thermal is brought to the electrode with a diminution, and the amount of removed electrode material is decreased. However, A_p in the range from 10 to 20 g/L, it was found that TWR_{ht} and TWR_{nht} have a slow decline. This is owing to an inappropriate powder amount during spark discharge and causes extensive wear of the electrode (Chawla & Mittal, 2023). Observing the comparison of TWR in the two material states the impact of control parameters on TWR in these two material states is different; the value of TWR_{ht} is smaller than the value of TWR_{nht} when received the same control parameters. This proves that TWR_{ht} has been improved better than TWR_{nht} with the same control parameters. The cause of this phenomenon is believed that the heat-treated state has better thermal conductivity than the nonheat-treated state (Wu et al., 2023). As a result, heat transfer to the electrode is reduced, which makes the electrode less worn.

The contour plots in the incorporated influence of the control parameters on TWR_{ht} and TWR_{nht} are depicted in Fig. 17.5A–F. In what, T_{on} expresses the most influence on TWR_{ht} with the contribution of 38.84%, followed by I_p (23.14%) and A_p (3.45%), as indicated in Table 17.3. Meanwhile, according to Table 17.6 is the ANOVA results of TWR_{nht}, T_{on} expresses the most contribution for TWR_{nht} with 29.19%, followed by I_p (23.97%) and A_p (18.54%). Besides, observing in Fig. 17.5, in the combined influence for each pair of control parameters, it is seen that the area for the lowest value of TWR_{ht} is wider than that of TWR_{nht}. It is evident that both I_p and T_{on} are simultaneously increased; there is also an escalation in the heat energy of the discharge channel. This heat energy is absorbed into the larger electrode; there is an increase in TWR of two material states (Shabarinathan et al., 2024; Le, 2021d). Introducing powder into the insulated solvent promotes stratified discharge; this leads to an improvement in TWR (Ekmekci et al., 2016). Nevertheless, abnormal deterioration of TWR in two material states can occur because of phenomena such as short circuits or arcing, caused by a large amount of debris remaining from previous discharge processes in the next discharge area. In addition, improper combinations of T_{on} and I_p can lead to high pressure during gas bubble disintegration in the preceding discharge, resulting in low powder density in the subsequent discharge (Hosni & Lajis, 2020). These factors contribute to a decrease in discharge energy, resulting in reduced thermal energy transmitted to the electrode and an abnormal drop in TWR.

According to the research workings investigated of the TWR in the PMEDM process (including the kinds of powders such as Ag, Cr, Si, carbon nanotube, TiO_2, and SiC) (Chaudhari et al., 2022; Ekmekci et al., 2015; Gill & Kumar, 2018; Jadam et al., 2019; Srivastava et al., 2023). These investigations have pointed out the various impacts on TWR. This relates to the different thermal and electrical conductivity of each powder and workpiece material. Regarding types of powders accompanied by good thermal conductance, such as Ag and carbon nanotubes, the TWR has a greater value versus that of normal EDM. On the contrary, for powders with medium thermal conductivity such as Cr, TiO_2, SiC, and Si, the TWR has

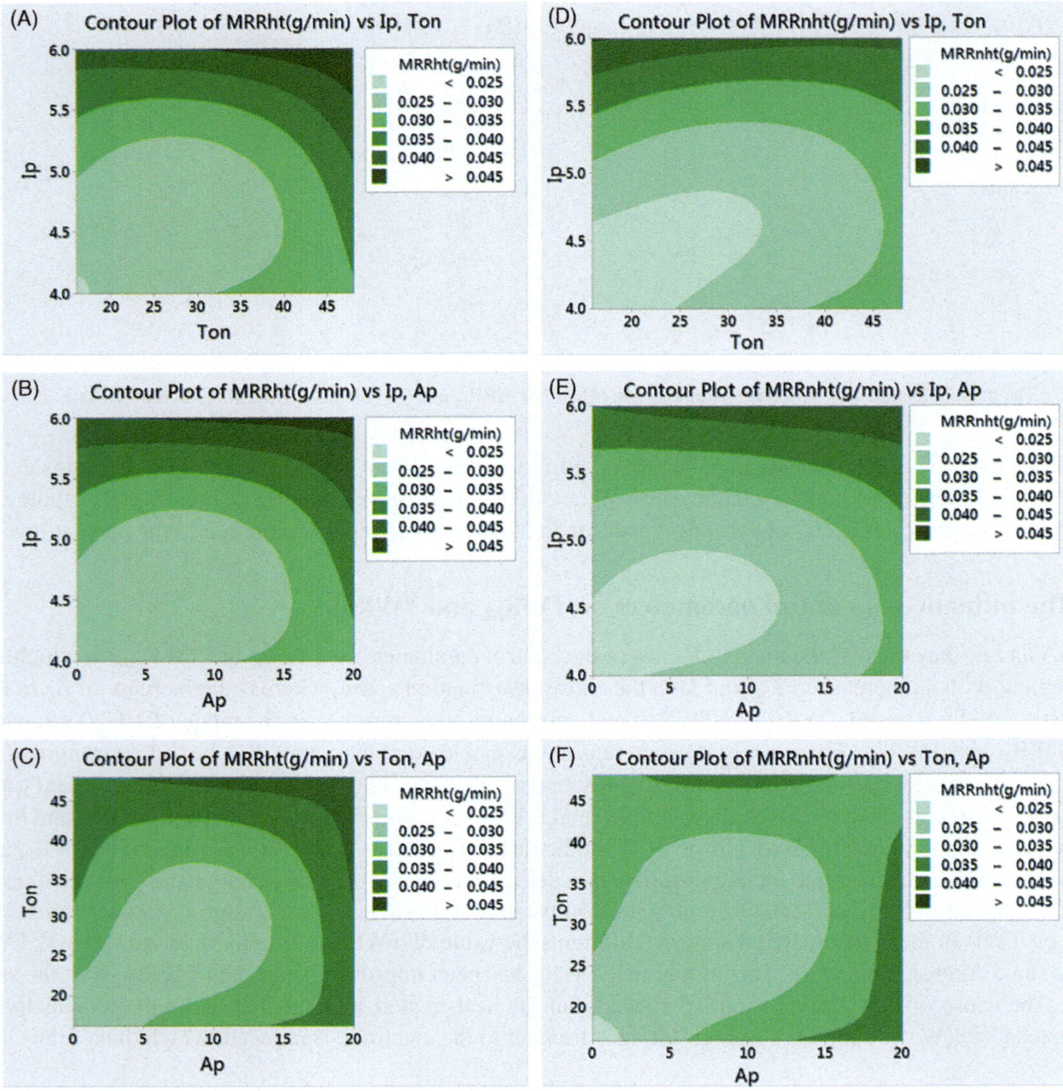

FIGURE 17.3 The incorporated influence of control parameters on MRR_{ht} and MRR_{nht}. *MRR*, Material removal rate.

a smaller value versus that of normal EDM. In this study the powder has a lower thermal conductivity versus that of carbon nanotubes and Ag but a higher thermal conductivity versus Cr, TiO_2, SiC, and Si. The TWR values by PMEDM are higher than EDM's in the entire design space. The variation trend of TWR in this investigation is appropriate with Ag powder and carbon nanotubes (Korgal et al., 2024; Tufail et al., 2024).

17.3.3 The influence of control parameters on R_{aht} and R_{anht}

Fig. 17.6A and B describe the primary influence of control parameters on R_{aht} and R_{anht}. In what an increase in I_p and T_{on} in the entire investigation space results in an augmentation in R_{aht} and R_{anht}, whereas an augmentation in A_p in the whole investigation space leads to a deduction in R_{aht} and R_{anht}. The reason is because of the congruous or incongruous combination of T_{on}, I_p, and A_p; this constitutes the appropriate or inappropriate pressure of the gas bubble decomposition in the preceding discharge and leads to the subsequent discharge having a high or low density of powder (Ishfaq et al., 2022; Le et al., 2019b, 2022; Le, 2021c). These causes change the discharge channel about discharge gap, stability, and uniformity, which results in a decline or an increase in the roughness. When R_a is compared in the two material states, the impact of control parameters on R_a in these two material states is different; the value of R_{aht} is larger than the value of R_{anht} when receiving the same control parameters. This proves that R_{anht} has been improved better than R_{aht} with the same control parameters. The cause of this phenomenon is believed that the heat-treated state has better thermal conductivity than the nonheat-treated state

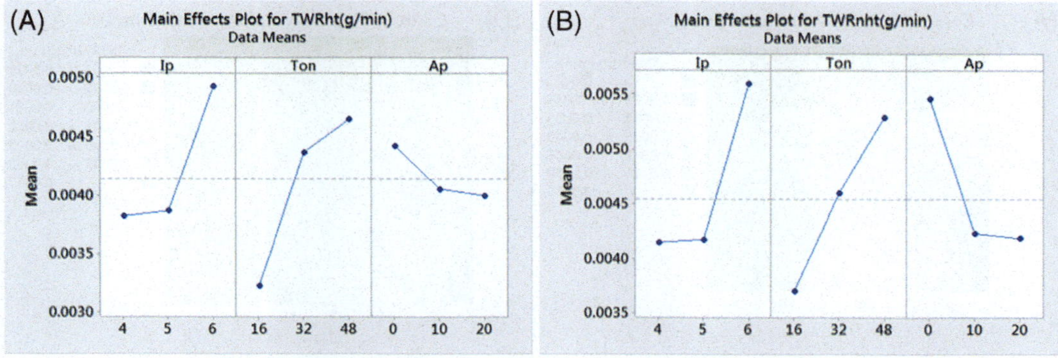

FIGURE 17.4 The main influential plot of control parameters on TWR: (A) TWR_{ht} and (B) TWR_{nht}. *TWR*, Tool wear rate.

(Wu et al., 2023). As a result the molten metal produced from the heat-treated material state converts to the solidified state more quickly and partly adheres to the surface. Another factor contributing to this phenomenon could be the presence of heterogeneous metallurgy phases and the morphology of the intergranular phase on the surface, which vigorously influence polycrystalline composites in mechanical and thermal properties. This results in heterogeneous corrosion when the material is eroded (Yildiz et al., 2017).

In Fig. 17.7A–F, they describe the contour plots of the incorporated influence of the control parameters on R_{aht} and R_{anht}. In which I_p expresses the most influence on R_{aht} with the contribution of 73.35%, followed by T_{on} (10.35%) and A_p (0.21%), as indicated in Table 17.4. Meanwhile, according to Table 17.7 is the ANOVA results of R_{anht}, I_p expresses the most contribution for R_{anht} with 72.47%, followed by T_{on} (14.79%) and A_p (0.17%). Besides, observing in Fig. 17.7, in the combined influence for each pair of control parameters, it is seen that the area for the smallest value of R_{anht} is wider than that of R_{aht}. It was found that when both I_p and T_{on} are increased simultaneously, there is an augmentation in discharge energy, which results in an increase of R_a in two material states (Ribeiro da Silva Junior et al., 2017). Introducing powder into the insulated solvent leads to stratified discharge and expands the machining gap. Consequently, this reduces energy and charge density in the discharge channel, and the craters are created on the surface of the workpiece, reducing the depth and width, which improves R_a (Le & Le, 2021; Wang et al., 2023).

Surface roughness has been investigated in previous works with different types of powder (including Gr, B_4C, Mo, Ag, SiC, Si, carbon nanotube, and Al) (Amorim et al., 2017; Bhattacharya et al., 2012; Chaudhari et al., 2024; Chen et al., 2017; Devgan et al., 2021; Philip et al., 2021; Sahu et al., 2018; Schubert et al., 2022; Shard et al., 2018). The various impacts on surface roughness have been depicted in these investigations, which are associated with the different thermal and electrical conductivity of each powder, solvent, and workpiece material. In addition, the values of processing parameters are dissimilar. However, the trend of roughness change in this study with previous studies is not much different. The variation trend of R_a in this investigation is appropriate with Ag powder and carbon nanotubes (Jadam et al., 2019; Korgal et al., 2024).

17.4 The optimum results

The multicriteria optimization problem of the current study is described as follows:

Finding $x = [T_{on}, I_p, A_p]$ to RMR and {TWR and R_a} simultaneously reaches maximum and minimum, respectively, for each material state (heat-treated and nonheat-treated states).

Process parameters are in conditions: $16 \leq T_{on} \leq 48$ (μs), $4 \leq I_p \leq 6$ (A), and $0 \leq A_p \leq 20$ (g/L).

To solve these goal optimization problems the GRA and NSGA II-EAMR algorithms were implemented, and the optimization results were evaluated and compared.

GRA: Gray is a MO hybrid optimization approach; it was instituted by Deng (Kumar et al., 2022, 2024; Ozay et al., 2021; Tebassi et al., 2023). He developed GRA as a technique based on gray system theory. The procedure of GRA is represented as follows:

All the output values need to be normalized using Eqs. (17.9) and (17.10) for the "lower-the-better" approach (i.e., TWR and R_a) and the "higher-the-better" approach (i.e., MRR), respectively.

$$\text{For lower} - \text{the} - \text{better,}\ Y_{ij} = \frac{y_{ij}^{max} - y_{ij}}{y_j^{max} - y_j^{min}} \tag{17.9}$$

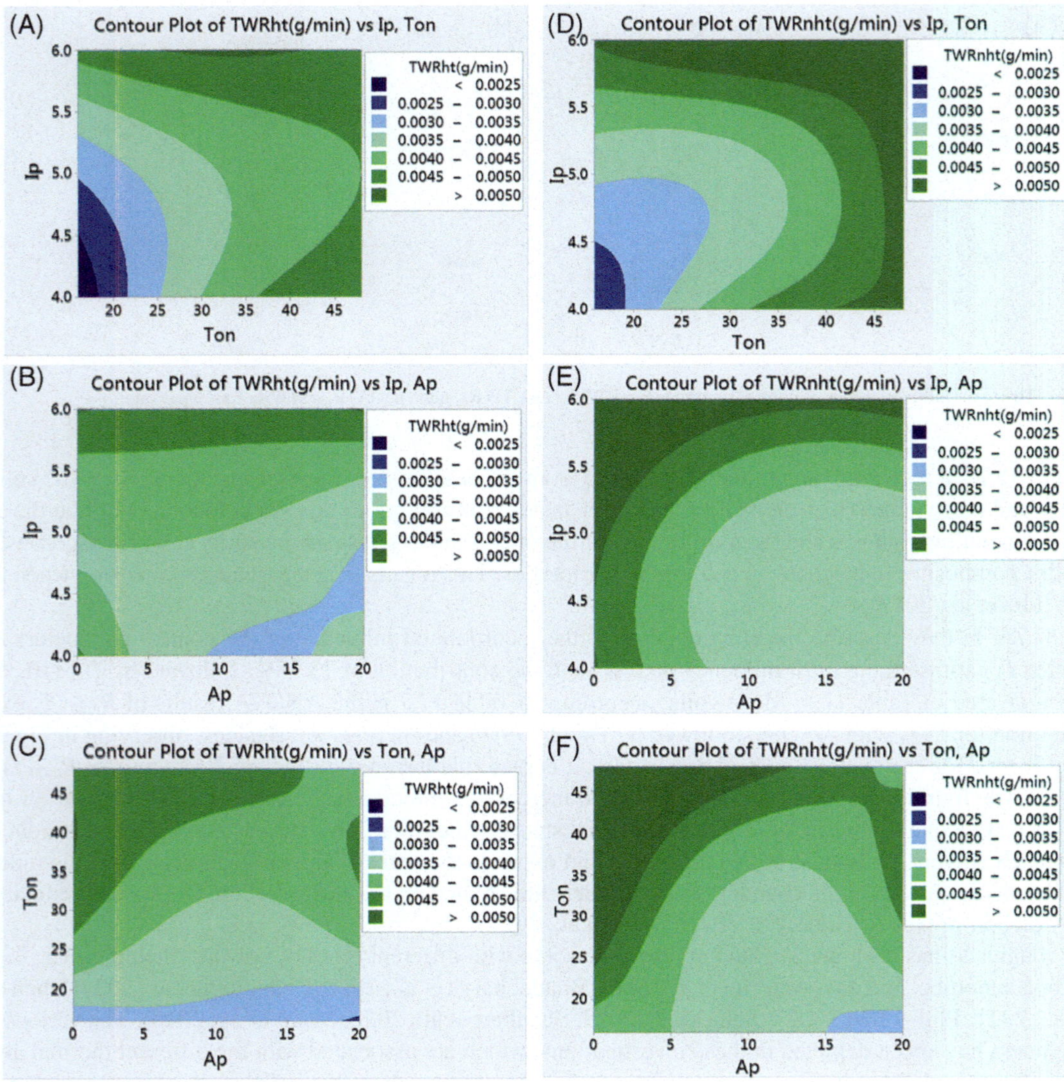

FIGURE 17.5 The incorporated influence of control parameters on TWR_{ht} and TWR_{nht}. *TWR*, Tool wear rate.

$$\text{For higher} - \text{the} - \text{better},\ Y_{\text{ij}} = \frac{y_{ij} - y_{\text{ij}}^{\text{min}}}{y_j^{\text{max}} - y_j^{\text{min}}} \tag{17.10}$$

where y_{ij} is the value for *i*th alternative corresponding to *j*th attribute. y_{ij}^{max} is the largest value of *j*th attribute. y_{ij}^{max} is the smallest value of *j*th attribute.

Computing gray relational coefficients (GRCs):

$$\gamma_{\text{ij}} = \frac{(\Delta_j^{\text{min}_j^{\text{max}}}}{\left(\Delta_{\text{ij}} + \xi \Delta_j^{\text{max}}\right)} \tag{17.11}$$

With $\Delta_{\text{ij}} = 1 - Y_{\text{ij}}$ and called the deviation sequence; $\xi \in [0,1]$ is defined as the distinguishing coefficient and its value is normally selected as 0.5 (Davim, 2012; Tebassi et al., 2023); $\Delta_j^{\text{min}} = \min(\Delta_{1j}, \Delta_{2j}, \ldots, \Delta_{mj})$; $\Delta_j^{\text{max}} = max(\Delta_{1j}, \Delta_{2j}, \ldots, \Delta_{mj})$.

Establishing Gray relational grades (GRGs): GRGs are defined by the mean of GRCs (Do et al., 2025; Wang, 2019) (Eq. 17.12).

$$\text{GRG} = \lambda_i = \frac{1}{n} \sum_{j=1}^{n} w_j \gamma_{\text{ij}} \tag{17.12}$$

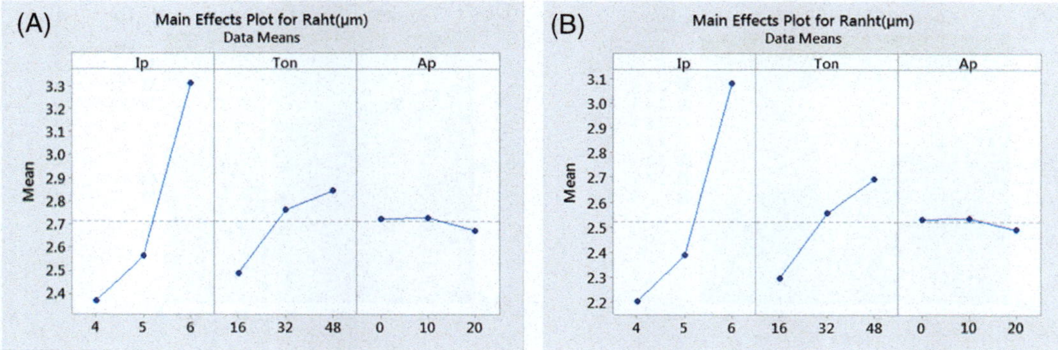

FIGURE 17.6 The main influential plot of control parameters on R_a: (A) R_{aht} and (B) R_{anht}.

The optimum alternative corresponds to the highest value of λ_i.

Weight: The weight of output responses was assumed to be w_j (j = 1, 2..., n), where $w_j \in (0, 1)$ and $\sum_{j=1}^{n} w_j = 1$. The weight is determined by the entropy methodology (Jain & Parashar, 2021; Kuo et al., 2008; Li et al., 2011), and the index weight is calculated as follows:

The normalization for indexes:

$$r'_{ij} = \frac{x_{ij}}{\max_i x_{ij}}, (i = 1, ..., m; j = 1, ..., n) \tag{17.13}$$

or

$$r'_{ij} = \frac{\min_j x_{ij}}{x_{ij}}, (i = 1, ..., m; j = 1, ..., n) \tag{17.14}$$

Calculating entropy for the index:

$$E_j = -\frac{\sum_{i=1}^{m} f_{ij}\ln f_{ij}}{\text{In } m}, (i = 1 \ldots, m; j = 1, \ldots n) \tag{17.15}$$

with

$$f_{ij} = \frac{r'_{ij}}{\sum_{i=1}^{m} r_{ij}}, (i = 1, ..., m; j = 1, ..., n) \tag{17.16}$$

To calculate the entropy weight of the index (Vijayakumar et al., 2023):

$$w_j = \frac{1 - E_j}{n - \sum_{j=1}^{n} E_j}, \sum_{j=1}^{n} w_j = 1, (j = 1, ..., n) \tag{17.17}$$

EAMR: This method is evaluated by an area-based method of ranking (EAMR) (Ghorabaee et al., 2016; Le et al., 2025a). The response (r_{ij}) is normalized as follows:

$$r_{ij} = \frac{y_{ij}}{\text{max}y_{ij}} \tag{17.18}$$

The normalized response with weight (v_{ij}) is figured out as

$$v_{ij} = \omega_j(r_{ij}) \tag{17.19}$$

The normalization solution of the criteria is calculated as

$$V_i^+ = v_{i1}^+ + v_{i2}^+ + \ldots v_{im}^+ \tag{17.20}$$

For larger criteria is better (17.21)

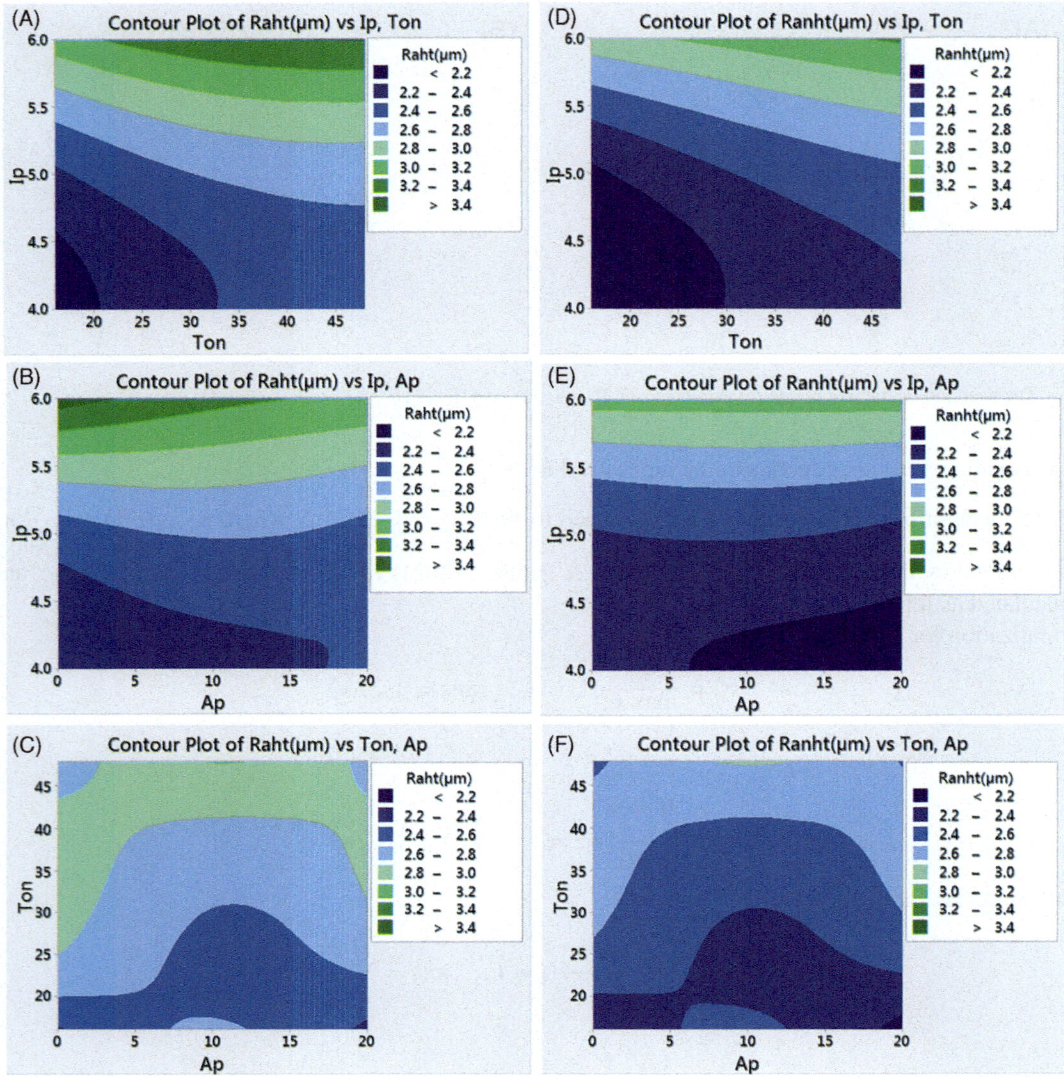

FIGURE 17.7 The incorporated influence of control parameters on R_{aht} and R_{anht}.

For smaller criterion is better (Van Tao et al., 2024).

The performance indicator (G_i) is calculated as

$$G_i = \frac{V_i^+}{V_i^-} \tag{17.22}$$

The largest value of G_i is the best solution.

NSGA II: NSGA II is a popular NSGA for MO optimization (Davim, 2012; Khan et al., 2023; Hoseini Rahdar et al., 2016; Pasandideh et al., 2013; Vinet & Zhedanov, 2011). It is a very effective algorithm that incorporates elitism, and no sharing parameter must be chosen a priori. The strength of NSGA II is to obtain a large number of solutions in both continuous and discrete design spaces with low computation time and produce global results. The block schema of the NSGA II algorithm is shown in Fig. 17.8 and described in the following:

The initialized population: The population size is decided on the optimum issue and parameter constraints.

Nondominated sort procedure: The main/parent population (P_p) is diverged into n subsets, including $P_1, P_2, \ldots, P_n$. The individual P_i dominates P_{i+1}, and is calculated as

$$P_i = \{i/n_i, i \in \{1, 2, ..., N\}\} \tag{17.23}$$

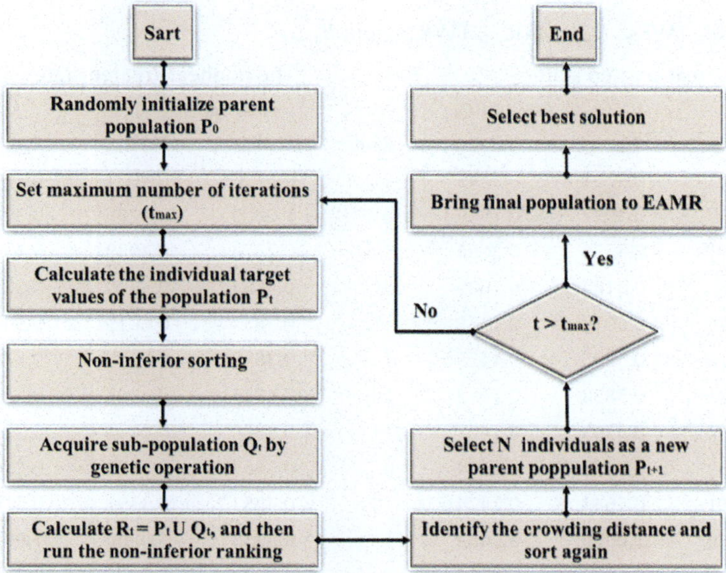

FIGURE 17.8 The schematic operation of NSGA II-EARM. *NSGA*, Nondominated sorted-genetic algorithm.

With individual P_k is resolved as

$$P_k = \{\text{Allindividual}/n_i - k + 1, k = 2, 3, ..., n\} \tag{17.24}$$

Computing the distance of two individuals($P[i]_d$): Owing to the diversity and distribution of each individual, hence the ranking uses the swarm distance techniques.

$$P[i]_d = \left|P[i+1]_{y1} - P[i-1]_{y1}\right| + \left|P[i+1]_{y2} - P[i-1]_{y2}\right| \tag{17.25}$$

In there, y_1 and y_2 correspond to the objectives.

Selection: With the aid of the binary tournament the individuals are sorted based on nondomination and with assigned crowding distance. The finest individuals are designated. Evaluation for two random solutions, in which the lower rank solution is selected.

Genetic operators: The crossover and mutation operations with a minimum probability are employed for the generated offspring.

The best individual is selected from sorting the current population: The worst individuals are removed to form the newest generation, which is repeated till reaching the maximum number of generations.

As previously mentioned, the GRA and NSGA II-EARM algorithms were employed to make the multicriteria decision. The predictive models of MRR_{ht}, TWR_{ht}, R_{aht}, MRR_{nht}, TWR_{nht}, and R_{aht} correspond to Eqs. (17.3)–(17.8) were used as objective functions for each material state in NSGA II-EARM algorithm. The weights for the three responses in each material state were obtained through the entropy weight method according to Eq. (17.17); the results of the heat-treated state are $w_{1ht} = 0.383$ for MRR_{ht}, $w_{2ht} = 0.493$ for TWR_{ht}, and $w_{3ht} = 0.124$ for R_{aht}, whereas the results of the nonheat-treated state are $w_{1nht} = 0.39$ for MRR_{nht}, $w_{2nht} = 0.467$ for TWR_{nht}, and $w_{3nht} = 0.143$ for R_{anht}.

17.4.1 Optimization by GRA algorithm

Table 17.8 illustrates all the GRCs and GRG values corresponding to ranks. The control parameters associated with the highest GRG (Rank 1: run 11th for the nonheat-treated and heat-treated states). The outcome, a set of optimization control parameters attained by the GRA algorithm, is $T_{on} = 16$ μs, $I_p = 4$ A, and $A_p = 10$ g/L for both material states. The output attributes achieved at the optimum control parameters are $R_{aht} = 2.14$ μm, $MRR_{ht} = 0.0244$ g/min, $TWR_{ht} = 0.0021$ g/min, $R_{anht} = 1.94$ μm, $MRR_{nht} = 0.0211$ g/min, and $TWR_{nht} = 0.0025$ g/min.

TABLE 17.8 GRA for MRR_{ht}, TWR_{ht}, R_{aht}, MRR_{nht}, TWR_{nht}, and R_{anht}.

Run	The heat-treated state					The nonheat-treated state				
	$GRCs_{ht}$			GRG_{ht}		$GRCs_{nht}$			GRG_{nht}	
	MRR_{ht}	TWR_{ht}	R_{aht}	Value	Rank	MRR_{nht}	TWR_{nht}	R_{anht}	Value	Rank
1	0.8129	0.3333	0.3791	0.1742	4	0.8790	0.4021	0.3611	0.1941	4
2	0.6429	0.3659	0.3932	0.1584	8	1.0000	0.4286	0.3988	0.2157	3
3	1.0000	0.3571	0.3333	0.2001	2	0.7373	0.3939	0.3333	0.1731	6
4	0.6117	0.4286	0.5184	0.1699	5	0.4652	0.5200	0.5039	0.1654	10
5	0.3631	0.4545	0.6019	0.1459	11	0.3853	0.6019	0.5804	0.1714	8
6	0.3642	0.4412	0.5695	0.1425	12	0.3829	0.5909	0.5702	0.1689	9
7	0.4565	0.3571	0.5853	0.1412	13	0.4834	0.4286	0.5856	0.1574	13
8	0.3663	0.4286	0.5595	0.1403	14	0.3865	0.6037	0.5963	0.1727	**7**
9	0.3415	0.5769	0.7257	0.1684	6	0.3435	0.5493	0.7065	0.1638	11
10	0.5122	0.3571	0.5644	0.1474	10	0.4853	0.3391	0.5159	0.1404	15
11	**0.3333**	**1.0000**	**1.0000**	**0.2482**	**1**	**0.3333**	**1.0000**	**1.0000**	**0.2467**	**1**
12	0.4437	0.4688	0.6720	0.1614	7	0.4634	0.5735	0.7303	0.1843	5
13	0.4773	0.6522	0.7515	0.1992	3	0.5453	0.7647	0.7927	0.2277	2
14	0.6923	0.3333	0.3479	0.1575	9	0.6880	0.3333	0.3631	0.1586	12
15	0.3520	0.3846	0.7515	0.1392	15	0.3670	0.4333	0.6500	0.1461	14

GRA, gray relational analysis; GRCs, gray relational coefficients; GRG, gray relational grade; MRR, material removal rate; TRL, tool wear rate.

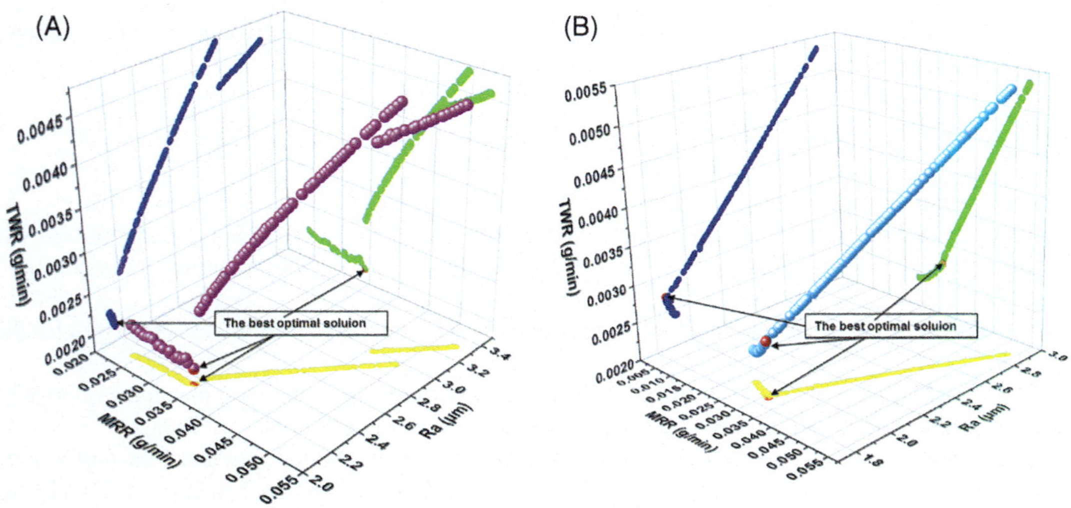

FIGURE 17.9 3D and 2D Pareto of feasible optimal solutions by NSGA II: (A) heat-treated and (B) nonheat-treated material states. *MMR*, Material removal rate; *NSGA*, nondominated sorted-genetic algorithm; *TWR*, tool wear rate.

17.4.2 Optimization by NSGA II incorporated with EARM (NSGA II-EARM)

The NSGA II algorithm was used to optimize for two different material states, set as population size = 100, crossover percentage = 0.8, mutation percentage = 0.2, and maximum number of iterations (t_{max}) = 100. As a result the Pareto of 100 feasible optimum solutions was obtained from the NSGA II algorithm, as represented in Fig. 17.9, where Fig. 17.9A and B is the response for the heat-treated and nonheat-treated states, respectively. However, it can be found that it is not easy to choose one solution that meets the requirements (i.e., maximization for MRR and minimization for TWR and R_a). Hence the EARM method was used to find a single solution, and the best optimum solution with the highest G_i values of 1.678 and 0.693 corresponding to the heat-treated and nonheat-treated states was formed. For the heat-treated state the optimum parameter values of I_p, T_{on}, and A_p are 4 A, 16 μs, and 20 g/L, respectively, and the obtained optimum response values of MRR_{ht}, TWR_{ht}, and R_{aht} are 0.03416 g/min, 0.001984 g/min, and 2.1528 μm, respectively. For the nonheat-treated state the optimum parameter values of I_p, T_{on}, and A_p are 4 A, 16 μs, and 20 g/L, respectively, and the attained optimum response values of MRR_{nht}, TWR_{nht}, and R_{anht} are 0.03114 g/min, 0.002719 g/min, and 1.8626 μm, respectively.

17.4.3 Comparison of various optimization results

Table 17.9 exhibits the compared optimization results attained by GRA and NSGA II-EAMR. Herein the divergence in results between the method of A versus the method of B was specified by Eq. (17.26), with attribute (*A*) and attribute (*B*) being the values of the similar attribute achieved by the method of *A* and *B*, respectively.

$$\text{The divergence of A vs. B} = \frac{\text{Attribute}(A) - \text{Attribute}(B)}{\text{Attribute}(B)} \times 100\% \qquad (17.26)$$

Table 17.9 also shows that for the heat-treated state, the optimum solution gained from NSGA II-EAMR gives the highest value of MRR_{ht} and R_{aht} and the lowest value of TWR_{ht}. Meanwhile, for the nonheat-treated state, the optimum solution attained from NSGA II-EAMR also provides the highest value of MRR_{nht} and TWR_{nht} and the lowest value of R_{aht}. To maximize the MRR for these two material states when compared to the GRA methods, it was found that the NSGA II-EAMR method produced results with the highest MRR value (MRR_{ht} = 0.03416 g/min and MRR_{nht} = 0.03114 g/min). The result of two optimization methods was compared to minimize the TWR and R_a for these two material states. For the heat-treated state the NSGA II-EAMR method offers the result with the lowest TWR_{ht} of 0.001984 g/min, whereas the GRA method provides the lowest R_{aht} of 2.14 μm. For the nonheat-treated state the GRA algorithm suggests the lowest TWR_{nht} of 0.0025 g/min, whereas the NSGA II-EAMR method produces the lowest R_{anht} of 1.8626 μm.

With the mentioned-earlier results, in manufacturing conditions, it is required to reduce the TWR; the set of the optimum control parameters {I_p = 4 A, T_{on} = 16 μs, and A_p = 20 g/L} for the heat-treated state or {I_p = 4 A, T_{on} = 16 μs, and A_p = 10 g/L} for the nonheat-treated state is preferred to select. In contrast, it is required to reduce the R_a, the set of the optimum control parameters {I_p = 4 A, T_{on} = 16 μs, and A_p = 10 g/L} for the heat-treated state or {I_p = 4 A, T_{on} = 16 μs, and A_p = 20 g/L} for the nonheat-treated state is voted for selection. However, to improve the MRR, the set of optimum control parameters {I_p = 4 A, T_{on} = 16 μs, and A_p = 20 g/L} for both material states is the favorite election.

17.4.4 Verification experiments of results

The experimental values of machining responses at the optimal control parameters are depicted in Table 17.10. The errors between the values of output attributes for the empirical and predicted models are within the tolerable assortment, with the largest error of 5.4% for the output attribute (TWR_{ht} obtained from NSGA II-EAMR) and the smallest error of 1% for the output response (TWR_{nht} procured from NSGA II-EAMR).

The results show that the NSGA II-EAMR gives the best results for R_{anht}, whereas the GRA produces the best results for R_{aht}. However, the surface features obtained from the optimum control parameters in the two material states must be explored for a more in-depth view. This assists technologists and researchers in making the right decision about which material states to choose. Hence the surface microdefects at the optimum technological modes achieved from two material states were explored and compared, as depicted in Fig. 17.10. Fig. 17.10A represents the surface structure at the optimum process variables attained by the heat-treated material state, whereas Fig. 17.10B expresses the surface structure at the optimum process variables by the nonheat-treated material state. It is clear that the defects of surfaces like the microcrack number (the track marked *green color*) and the void number (the circle marked *blue color*) of Fig. 17.10A are more than those of Fig. 17.10B, whereas the droplet number (the circle marked *red color*) and the debris globule number (the circle marked *yellow color*) of Fig. 17.10B are more than those of Fig. 17.10A. The fact that the surface defects of the two material states are different, which is related to the thermal and electrical conductivity, chemical composition, and formation of phase organization of the workpiece material. In addition, the optimum parameter value of A_p obtained from the NSGA II-EAMR and GRA is various. From there the discharge channel, heat release, debris release, and vaporization are formed differently (Straumal et al., 2012; Wang et al., 2023).

Upon observing the surface chemistry compositions, which are determined by the documents (Le, 2021a, 2022a, 2022b, 2023; Le & Banh, 2021; Le et al., 2024, 2025b, 2025d, 2025e, 2025a), a notable contrast in compositional chemistry was obtained from two material states at the optimal control parameters of NSGA II-EAMR and GRA algorithms, as illustrated in Fig. 17.11. In here, it is observed that the content of elements such as V, Si, Cr, Fe, Cu, Co, and W on surfaces achieved from the optimum machining parameters of the nonheat-treated state is higher compared to that of the heat-treated state.

TABLE 17.9 Comparison of optimum results.

Algorithm	Optimum control parameters			Machining attributes					
				The heat-treated state			The nonheat-treated state		
	T_{on} (μs)	I_p (A)	A_p (g/L)	MRR_{ht} (g/min)	TWR_{ht} (g/min)	R_{aht} (μm)	MRR_{nht} (g/min)	TWR_{nht} (g/min)	R_{anht} (μm)
GRA	16	4	10	0.0244	0.0021	2.14	0.0211	0.0025	1.94
NSGA II-EAMR	16	4	20	0.03416	0.001984	2.1528	0.03114	0.002719	1.8626
NSGA II-EAMR versus GRA (%)				40 (↑)	−5.52 (↓)	1.28 (↑)	47.58 (↑)	8.76 (↑)	−3.99 (↓)

GRA, *gray relational analysis;* NSGA, *nondominated sorted-genetic algorithm.*

TABLE 17.10 Verification empirics of the results.

Algorithm	Machining attributes	Optimum control parameters	PV	EV	Error (%)
GRA	MRR_{ht} (g/min)	$I_p = 4$ A, $T_{on} = 16$ μs, $A_p = 10$ g/L	0.0244	0.0254	4.1
	TWR_{ht} (g/min)		0.0021	0.00205	2.4
	R_{aht} (μm)		2.14	2.19	2.3
	MRR_{nht} (g/min)		0.0211	0.0219	3.8
	TWR_{nht} (g/min)		0.0025	0.00261	4.4
	R_{anht} (μm)		1.94	1.91	1.5
NSGA II-EAMR	MRR_{ht} (g/min)	$I_p = 4$ A, $T_{on} = 16$ μs, $A_p = 20$ g/L	0.03416	0.03282	3.9
	TWR_{ht} (g/min)		0.001984	0.002091	5.4
	R_{aht} (μm)		2.1528	2.1218	1.4
	MRR_{nht} (g/min)		0.03114	0.03256	4.6
	TWR_{nht} (g/min)		0.002719	0.002692	1
	R_{anht} (μm)		1.8626	1.82	2.3

Error = Abs (PV − EV)/PV × 100%

GRA, gray relational analysis; MRR, material removal rate; NSGA, nondominated sorted-genetic algorithm; TRL, tool wear rate.

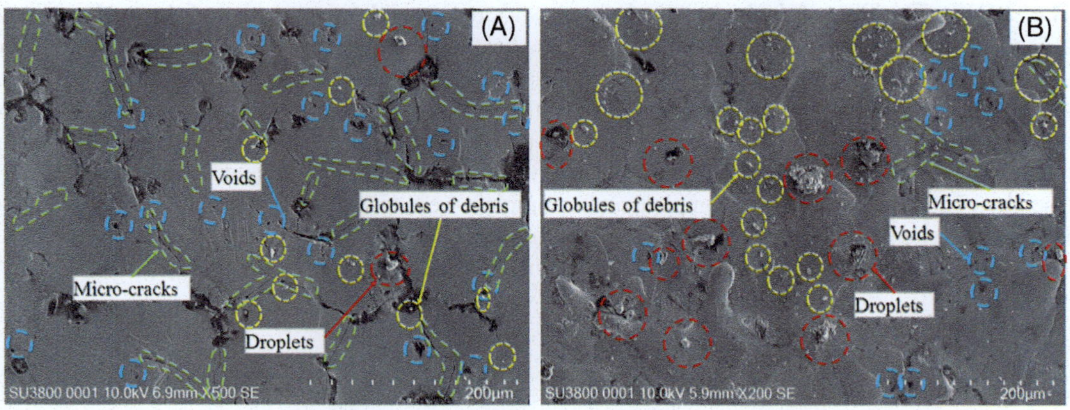

FIGURE 17.10 Surface defects at optimum control parameters: (A) GRA for heat-treated state and (B) NSGA II-EAMR for nonheat-treated state. *GRA*, Gray relational analysis; *NSGA*, nondominated sorted-genetic algorithm.

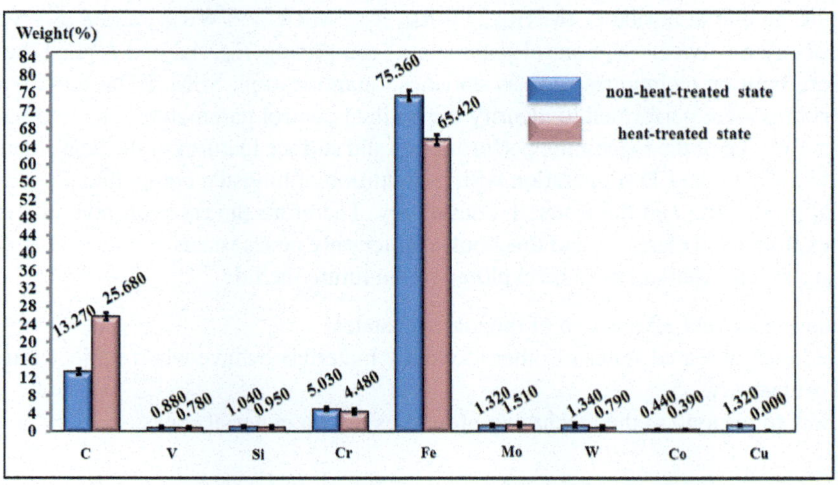

FIGURE 17.11 Compositional chemistry and element content on surfaces at optimum control parameters: NSGA II-EAMR for nonheat-treated state; GRA for heat-treated state. *GRA*, Gray relational analysis; *NSGA*, nondominated sorted-genetic algorithm.

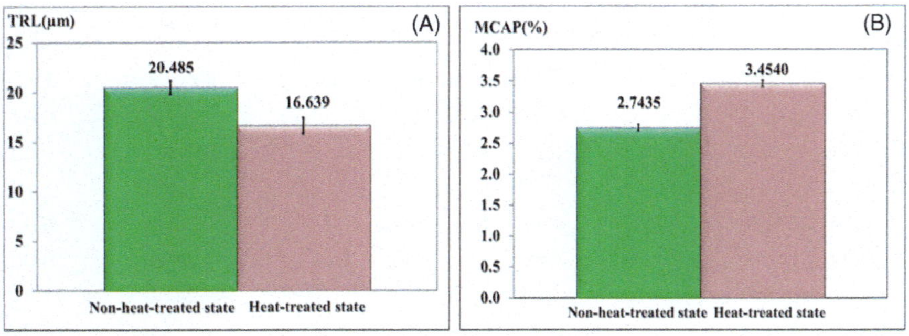

FIGURE 17.12 TRL and MCAP at optimal control parameters with NSGA II-EAMR for nonheat-treated state; GRA for heat-treated state: (A) TRL and (B) MCAP. *GRA*, Gray relational analysis; *MCAP*, microcrack acreage percentage; *NSGA*, nondominated sorted-genetic algorithm; *TRL*, tool wear rate.

Conversely the content of elements such as C and Mo on the machined surface obtained from the optimal process parameters of the heat-treated state is also greater than that of the nonheat-treated state. The reason for the contrast between the two material states relates to a difference in phase states (Bleys et al., 2006). In addition, another factor is the diversity of the formation discharge channel process from the optimum parameters of NSGA II-EAMR and GRA, which generates thermal fields that facilitate distinct chemical and physical processes (Gul et al., 2023; Le et al., 2026; Son et al., 2026).

In regard to the recast layer the method of determining the thickness of recast layer (TRL) is carried out according to the document (Le, 2021b). The TRL of the heat-treated state has a smaller thickness than that of the nonheat-treated state, as indicated in Fig. 17.12A. This is explained by the fact that the heat-treated material state has a higher thermal conductivity (Wu et al., 2023), so when the spark discharge process ends, the material melts and evaporates faster. This means that the molten material deposited on the surface is less than that of the nonheat-treated material state. For the microcrack acreage percentage (MCAP), this indicator is determined according to the document (Le, 2022a); the MCAP obtained at the optimum control parameters of the nonheat-treated state is lower than that of the heat-treated state, as indicated in Fig. 17.12B. This result is caused by grain boundaries in the complete and incomplete wetting states of the second solid phase; this generates between the ductile ferrite grains a network of brittle cementite layers (Straumal et al., 2012), as in Fig. 17.10A, that is the phase appearance of the dark gray and partially or fully delimits with the grains of the light gray phase.

17.5 Conclusion

Different results are obtained when analyzing operations for the treatment of heat-treated and nonheat-treated X40CrMoV51 alloy in the EDM process with tungsten carbide powder. Through comparative evaluation of criteria such as surface roughness (R_a), MRR, and electrode wear rate, it was found that the influence trend is different in these two material states. However, because of the heat generation in PMEDM, the thermal adsorption of the materials is different. Moreover, the prediction models of the output attributes (MRR_{ht}, TWR_{ht}, R_{aht}, MRR_{nht}, TWR_{nht}, and R_{anht}) were instituted to evolve the nonlinear-high association between the control parameters (comprising T_{on}, A_p, and I_p) and the output attributes. The evaluation methods were built on techniques such as statistics, quantification, SEM, EDS, and the experiment. GRA and NSGA II-EAMR algorithms were performed to identify the optimal control parameters. The findings of this study provide more comprehensive knowledge of the machining performances and surface features of the heat-treated and nonheat-treated states for X40CrMoV51 alloy in the EDM operation with the addition of tungsten compound powder. These could help the well-being of industrial production and the research community. Therefore the research potential of this effect trend still has many issues to be explored and clarified, and this book chapter only raises issues that need attention for researchers and technologists. Opportunities and challenges to be explored in the future include

- The change in insulating solvent affects two of the material states.
- How the heat-treated and untreated states of other tool alloy materials behave when subjected to the PMEDM process need to be studied extensively.
- Various powders: how to respond to the machining properties of two material states.

Acknowledgments

This research is funded by Vietnam National Foundation for Science and Technology Development (NAFOSTED) under grant number 107.99–2021.29.

References

Alam, S. T., Amin, A. N., Hossain, M. I., Huq, M., & Tamim, S. H. (2021). Performance evaluation of graphite and titanium oxide powder mixed dielectric for electric discharge machining of Ti–6Al–4V. *SN Applied Sciences, 3*(4). springer.com/snas. doi:10.1007/s42452-021-04450-6.

Amorim, F. L., Dalcin, V. A., Soares, P., & Mendes, L. A. (2017). Surface modification of tool steel by electrical discharge machining with molybdenum powder mixed in dielectric fluid. *International Journal of Advanced Manufacturing Technology, 91*(1–4), 341–350. http://www.springerlink.com/content/0268-3768. doi:10.1007/s00170-016-9678-x.

Bhattacharya, A., Batish, A., & Kumar, N. (2013). Surface characterization and material migration during surface modification of die steels with silicon, graphite and tungsten powder in EDM process. *Journal of Mechanical Science and Technology, 27*(1), 133–140. doi:10.1007/s12206-012-0883-8.

Bhattacharya, A., Batish, A., Singh, G., & Singla, V. K. (2012). Optimal parameter settings for rough and finish machining of die steels in powder-mixed EDM. *The International Journal of Advanced Manufacturing Technology, 61*(5–8), 537–548. doi:10.1007/s00170-011-3716-5.

Bleys, P., Kruth, J.-P., Lauwers, B., Schacht, B., Balasubramanian, V., Froyen, L., & Van Humbeeck, J. (2006). Surface and sub-surface quality of steel after EDM. *Advanced Engineering Materials, 8*(1–2), 15–25. doi:10.1002/adem.200500211.

Chaudhari, R., Ayesta, I., Doshi, M., Khanna, S., Patel, V. K., Vora, J., & De Lacalle, L. N. L. (2022). Effect of multi-walled carbon nanotubes on the performance evaluation of nickel-based super-alloy–Udimet 720 machined using WEDM process. *International Journal of Advanced Manufacturing Technology, 123*(5–6), 2087–2105. https://www.springer.com/journal/170. doi:10.1007/s00170-022-10209-z.

Chaudhari, R., Rehman, I. U., Khanna, S., Patel, V. K., Vora, J., Prakash, C., Campilho, R. D. S. G., Al-Sharif, M. S., Ali, E., & Ghoneim, S. S. M. (2024). A parametric study with experimental investigations of expanded graphite on performance measure of EDM process of Ni55.8Ti SMA. *Alexandria Engineering Journal, 87*, 164–174. https://www.sciencedirect.com/science/journal/11100168. doi:10.1016/j.aej.2023.12.013.

Chawla, G., & Mittal, V. K. (2023). Experimental investigation and optimization of process parameters of hybrid Al/SiC/B4C–MMCs finished by MAFM process using RSM modeling with supervised machine learning algorithm. *Sādhanā, 48*(2). doi:10.1007/s12046-023-02106-2.

Chen, W., Zhang, D., Yao, D., Zhang, S., & Wu, W. (2017). Investigations on microstructure and mechanical properties of containing-Si coatings. *Surface Engineering, 33*(7), 536–541. doi:10.1080/02670844.2017.1287837.

Davim, J. P. (2012). *Statistical and computational techniques in manufacturing* (pp. 1–285). Springer Berlin. doi:10.1007/978-3-642-25859.

Devgan, S., Mahajan, A., & Sidhu, S. S. (2021). Multi-walled carbon nanotubes in powder mixed electrical discharge machining: An experimental study, state of the art and feasibility prospect. *Applied Physics A: Materials Science and Processing, 127*(11). http://www.springer.com/materials/journal/339. doi:10.1007/s00339-021-04934-7.

Do, M. T., Le, V. T., & Nguyen, T. G. (2025). Machinability of 9CrSi steel as processed by powder-added electric discharge machining: Investigation and optimization for boosted machining characteristics. *Advances in Science and Technology Research Journal, 19*(8), 449–466. https://www.astrj.com/pdf-205764-126564?filename=Machinability%20of%209CrSi.pdf. doi:10.12913/22998624/205764.

Ekmekci, B., Ulusöz, F., Ekmekci, N., & Yaşar, H. (2015). Suspended SiC particle deposition on plastic mold steel surfaces in powder-mixed electrical discharge machining. *Proceedings of the Institution of Mechanical Engineers, B: Journal of Engineering Manufacture, 229*(3), 475–486. doi:10.1177/0954405414530902.

Ekmekci, B., Yaşar, H., & Ekmekci, N. (2016). A discharge separation model for powder mixed electrical discharge machining. *Journal of Manufacturing Science and Engineering, 138*(8). doi:10.1115/1.4033042.

Eyercioglu, O., & Gov, K. (2021). The effect of magnesium content on drilling of Al-Mg-Ti alloy by hole electrical discharge machining process. *Proceedings of the Institution of Mechanical Engineers, B: Journal of Engineering Manufacture, 235*(1–2), 125–133. doi:10.1177/0954405420949211.

Gao, R. X., Krüger, J., Merklein, M., Möhring, H. C., & Váncza, J. (2024). Artificial intelligence in manufacturing: State of the art, perspectives, and future directions. *CIRP Annals, 73*(2), 723–749. https://www.sciencedirect.com/science/journal/00078506. doi:10.1016/j.cirp.2024.04.101.

Gapontseva, T. M., Degtyarev, M. V., Voronova, L. M., Chashchukhina, T. I., & Pilyugin, V. P. (2021). Structural evolution and hardness of molybdenum deformed by high pressure torsion at 573 K. *International Journal of Refractory Metals and Hard Materials, 94*, 105389. doi:10.1016/j.ijrmhm.2020.105389.

Gerçekcioğlu, E., & Albaşkara, M. (2023). Multi-response optimization of electrical discharge machining of 17-4 PH SS using Taguchi-based grey relational analysis. *Archives of Metallurgy and Materials, 68*(3), 861–868. https://journals.pan.pl/dlibra/publication/145448/edition/128323/content. doi:10.24425/amm.2023.145448.

Ghorabaee, M. K., Zavadskas, E. K., Amiri, M., & Antucheviciene, J. (2016). Evaluation by an area-based method of ranking interval type-2 fuzzy sets (EAMRIT-2F) for multi-criteria group decision-making. *Transformations in Business and Economics, 15*(3), 76–95. http://www.transformations.khf.vu.lt/.

Gill, A. S., & Kumar, S. (2018). Micro-hardness evaluation for surface alloying of H11 die steel with Cu–Cr–Ni powder metallurgy tool in electrical discharge machining. *Proceedings of the Institution of Mechanical Engineers, B: Journal of Engineering Manufacture, 232*(3), 438–450. http://pib.sagepub.com/content/by/year. doi:10.1177/0954405416646027.

Gul, I. A., Abdul-Rani, A. M., Al-Amin, M., & Garba, E. (2023). Elucidating powder-mixed electric discharge machining process, applicability, trends and futuristic perspectives. *Machines, 11*(3), 381. doi:10.3390/machines11030381.

Hoseini Rahdar, M., Heidari, M., Ataei, A., & Choi, J. K. (2016). Modeling and optimization of R-717 and R-134a ice thermal energy storage air conditioning systems using NSGA-II and MOPSO algorithms. *Applied Thermal Engineering, 96*, 217–227. http://www.journals.elsevier.com/applied-thermal-engineering/. doi:10.1016/j.applthermaleng.2015.11.068.

Hosni, N. A. J., & Lajis, M. A. (2020). Experimental investigation and economic analysis of surfactant (Span-20) in powder mixed electrical discharge machining (PMEDM) of AISI D2 hardened steel. *Machining Science and Technology, 24*(3), 398–424. http://www.tandfonline.com/loi/lmst20. doi:10.1080/10910344.2019.1698609.

Ishfaq, K., Maqsood, M. A., Anwar, S., Harris, M., Alfaify, A., & Zia, A. W. (2022). EDM of Ti6Al4V under nano-graphene mixed dielectric: A detailed roughness analysis. *International Journal of Advanced Manufacturing Technology, 120*(11–12), 7375–7388. https://www.springer.com/journal/170. doi:10.1007/s00170-022-09207-y.

Jadam, T., Sahu, S. K., Datta, S., & Masanta, M. (2019). EDM performance of Inconel 718 superalloy: Application of multi-walled carbon nanotube (MWCNT) added dielectric media. *Journal of the Brazilian Society of Mechanical Sciences and Engineering, 41*(8). http://rd.springer.com/journal/40430. doi:10.1007/s40430-019-1813-9.

Jain, S., & Parashar, V. (2021). Critical review on the impact of EDM process on biomedical materials. *Materials and Manufacturing Processes, 36*(15), 1701–1724. doi:10.1080/10426914.2021.1942907.

Kamenskikh, A. A., Muratov, K. R., Shlykov, E. S., Sidhu, S. S., Mahajan, A., Kuznetsova, Y. S., & Ablyaz, T. R. (2023). Recent trends and developments in the electrical discharge machining industry: A review. *Journal of Manufacturing and Materials Processing, 7*(6). www.mdpi.com/journal/jmmp. doi:10.3390/jmmp7060204.

Kansal, H. K., Singh, S., & Kumar, P. (2007). Effect of silicon powder mixed EDM on machining rate of AISI D2 die steel. *Journal of Manufacturing Processes, 9*(1), 13–22. doi:10.1016/s1526-6125(07)70104-4.

Khan, O., Parvez, M., Kumari, P., Yadav, A. K., Akram, W., Ahmad, S., Parvez, S., & Idrisi, M. J. (2023). Modelling of compression ignition engine by soft computing techniques (ANFIS-NSGA-II and RSM) to enhance the performance characteristics for leachate blends with nano-additives. *Scientific Reports, 13*(1). https://www.nature.com/srep/. doi:10.1038/s41598-023-42353-1.

Korgal, A., Shettigar, A. K., Navin Karanth, P., & Prabhakar, D. A. P. (2024). Advances in micro electro discharge machining of biomaterials: A review on processes, industrial applications, and current challenges. *Machining Science and Technology, 28*(2), 215–265. http://www.tandfonline.com/loi/lmst20. doi:10.1080/10910344.2024.2311376.

Kumar, S., Ghoshal, S. K., Arora, P. K., Kumar, H., & Nagdeve, L. (2024). Unlocking AISI420 martensitic stainless steel's potential: Precision enhancement via S-EDM with copper electrodes and multivariate optimization. *Arabian Journal for Science and Engineering, 49*(8), 11457–11478. https://www.springer.com/journal/13369. doi:10.1007/s13369-024-08711-5.

Kumar, S., Goud, M., & Suri, N. M. (2022). Grey relational analysis-based multi-response optimization of magnetic-field-assisted powder-mixed electric discharge machining of Inconel 706. *Arabian Journal for Science and Engineering, 47*(7), 8315–8339. https://link.springer.com/journal/13369. doi:10.1007/s13369-021-06204-3.

Kuo, Y., Yang, T., & Huang, G. W. (2008). The use of grey relational analysis in solving multiple attribute decision-making problems. *Computers and Industrial Engineering, 55*(1), 80–93. doi:10.1016/j.cie.2007.12.002.

Le, V., Banh, T., Tran, X., & Nguyen, T. H. M. (2019b). Improving surface roughness by electrical discharge machining with tungsten powder. *ASEAN Engineering Journal, 9*(1), 44–53. doi:10.11113/aej.v9.15507.

Le, V. T. (2021a). Influence of processing parameters on surface properties of SKD61 steel processed by powder mixed electrical discharge machining. *Journal of Materials Engineering and Performance, 30*(4), 3003–3023. http://www.springerlink.com/content/1059-9495/. doi:10.1007/s11665-021-05584-9.

Le, V. T. (2021b). The machined performance and recast layer properties of AISI H13 steel processed by powder mixed EDM process: An investigation and comparison in fine-finishing and semi-finishing processes. *Journal of the Brazilian Society of Mechanical Sciences and Engineering, 43*(11). http://rd.springer.com/journal/40430. doi:10.1007/s40430-021-03243-7.

Le, V. T. (2021c). The role of electrical parameters in adding powder influences the surface properties of SKD61 steel in EDM process. *Journal of the Brazilian Society of Mechanical Sciences and Engineering, 43*(3). http://rd.springer.com/journal/40430. doi:10.1007/s40430-021-02844-6.

Le, V. T. (2022a). An investigation on machined performance and recast layer properties of AISI H13 steel by powder mixed-EDM in fine-finishing process. *Materials Chemistry and Physics, 276*, 125362. doi:10.1016/j.matchemphys.2021.125362.

Le, V.-T. (2023). New insights into the surface features of SKD61 steel at heat-treated and non-heat-treated states as processed by powder-mixed EDM. *Materials Letters, 352*, 135199. doi:10.1016/j.matlet.2023.135199.

Le, V. T. (2025a). Novel perspectives on machining and surface metallurgical characteristics of X40CrMoV51 in two different heat-treated states with finishing-PMEDM operation. *Materials and Manufacturing Processes, 40*(12), 1599–1620. www.tandf.co.uk/journals/titles/10426914.asp. doi:10.1080/10426914.2025.2535305.

Le, V.-T. (2025b). Comparison of electrode wear rate between graphite and copper materials during powder-mixed EDM process. In *Proceedings of the 4th Annual International Conference on Material, Machines, and Methods for Sustainable Development (MMMS2024)* (pp. 75–81). doi:10.1007/978-3-031-96122-9_10.

Le, V. T., & Banh, T. L. (2021). Evaluation of the influence of electrical parameters and powder concentration to the content of tungsten on the surface of SKD61 steel in EDM process. *Lecture notes in mechanical engineering*. Springer Science and Business Media Deutschland GmbH www.springer.com/series/11236. doi:10.1007/978-3-030-69610-8_24.

Le, V. T., Banh, T. L., Do, T. L., & Nguyen, T. H. (2022). Tool rotary assisted EDM processing of SKD61 steel: Influence and optimizing of main process parameters and surface roughness. In *Proceedings of the International Conference on Advanced Mechanical Engineering, Automation, and Sustainable Development 2021 (AMAS2021), vol. 2022* (pp. 316–322). doi:10.1007/978-3-030-99666-6_48.

Le, V. T., Banh, T. L., Nguyen, T. H. M., Hoang, T. D., Dang, V. T., & Phan, H. C. (2023). Modeling, impact evaluation, and optimization of machining performances of heat-treated SKD61 steel in a tungsten powder alloy mixed EDM process via the RSM-GRA methodology. *Science and Technology Development Journal*. doi:10.32508/stdj.v26i4.4222.

Le, V. T., Banh, T. L., Nguyen, T. H. M., Le, N. K., Hoang, T. D., Dang, V. T., & Han, H. C. (2025c). Graphite electrode wear rate in electro-discharge machining operation with mixed powder for machining X40CrMoV51 alloy steel: Investigation and optimizing technological parameters. In *Lecture notes in mechanical engineering* (pp. 67–73). Springer Science and Business Media Deutschland GmbH. https://www.springer.com/series/11236. doi:10.1007/978-3-031-96122-9_9.

Le, V. T., Banh, T. L., Tran, X. T., & Minh, N. T. H. (2019a). Surface modification process by electrical discharge machining with tungsten carbide powder mixing in kerosene fluid. *Applied Mechanics and Materials, 889*, 115–122. doi:10.4028/www.scientific.net/amm.889.115.

Le, V. T., Banh, T. L., Tran, X. T., Nguyen, T. H. M., & Le, V. T. (2020). Studying the microhardness on the surface of SKD61 in PMEDM using tungsten carbide powder. *International Journal of Modern Physics B, 34*(22–24). http://www.worldscinet.com/ijmpb/mkt/archive.shtml. doi:10.1142/S0217979220401645.

Le, V.-T., Dang, V. T., & Phan, H. C. (2025a). Investigation and optimization of machining peculiarities of heat treatment and non-heat treatment X40CrMoV51 alloy by powder-suspended electro-discharge machining in final-finishing operation. *Proceedings of the Institution of Mechanical Engineers, C: Journal of Mechanical Engineering Science, 239*(19), 7869–7887. doi:10.1177/09544062251350471.

Le, V. T., Do, M. T., Le, N. K., & Nguyen, T. G. (2025b). Evaluation of electrode wear rate as processed 9CrSi steel in heat-treated and non-heat-treated states during powder-mixed EDM process. In *Lecture notes in networks and systems* (pp. 331–337). Springer Science and Business Media Deutschland GmbH. https://www.springer.com/series/15179/1611. doi:10.1007/978-3-032-03859-3_35.

Le, V. T., Hoang, L., Ghazali, M. F., Le, V. T., Do, M. T., Nguyen, T. T., & Vu, T. S. (2024). Optimization and comparison of machining characteristics of SKD61 steel in powder-mixed EDM process by TOPSIS and desirability approach. *International Journal of Advanced Manufacturing Technology, 130*(1–2), 403–424. https://www.springer.com/journal/170. doi:10.1007/s00170-023-12680-8.

Le, V. T., & Le, V. T. (2021). Study and evaluation the influence of rotary tool on surface roughness of SKD61 steel in EDM process. In B. T. Long, Y. H. Kim, K. Ishizaki, N. D. Toan, I. A. Parinov, & N. P. Vu (Eds.), *Proceedings of the 2nd Annual International Conference on Material, Machines and Methods for Sustainable Development (MMMS2020). MMMS 2020. Lecture Notes in Mechanical Engineering* (pp. 558–563). Springer. doi:10.1007/978-3-030-69610-8_76.

Le, V. T., & Nguyen, H. Q. (2022). Comparison of influence of main process parameters on surface roughness of heated treatment and non-heated treatment SKD61 steels processed by PMEDM process. In B. T. Long, H. S. Kim, K. Ishizaki, N. D. Toan, I. A. Parinov, & Y.-H. Kim (Eds.), *Proceedings of the International Conference on Advanced Mechanical Engineering, Automation, and Sustainable Development 2021 (AMAS2021), vol. 2022* (pp. 16–22). Springer International Publishing. doi:10.1007/978-3-030-99666-6_3.

Le, V. T., Nguyen, T. H. M., & Hoang, T. D. (2025e). Machining characteristics of different tool electrodes during PMEDM for machining X40CrMoV51 alloy: An investigative and comparative study. *Ceramics International, 51*(29), 61550–61566. https://www.sciencedirect.com/science/journal/02728842. doi:10.1016/j.ceramint.2025.10.347.

Le, V-T, Nguyen, THM, & Hoang, TD (2026). Investigation and optimization for enhanced machining features of heat-treated SKD61 steel as processed by powder-mixed electro-discharge machining in finishing regime. *Vietnam Journal of Science and Technology, 64*(3), 576–593. doi:10.15625/2525-2518/21413.

Le, V. T., Nguyen, T. H. M., Hoang, T. D., & Nguyen, V. S. (2025d). Novel insights for machining and metallurgical surface features of the heat treatment and non-heat treatment X40CrMoV51 tool steel by powder-added EDM process. *International Journal of Advanced Manufacturing Technology, 137*(7), 4063–4088. https://www.springer.com/journal/170. doi:10.1007/s00170-025-15396-z.

Li, X., Wang, K., Liu, L., Xin, J., Yang, H., & Gao, C. (2011). Application of the entropy weight and TOPSIS method in safety evaluation of coal mines. *Procedia Engineering, 26*, 2085–2091. doi:10.1016/j.proeng.2011.11.2410.

Matys, A., Dadan, M., Witrowa-Rajchert, D., Parniakov, O., & Wiktor, A. (2022). Response surface methodology as a tool for optimization of pulsed electric field pretreatment and microwave-convective drying of apple. *Applied Sciences, 12*(7), 3392. doi:10.3390/app12073392.

Oghazi, B., Baseri, H., & Rabiee, M. (2023). Evaluation of modified Ti-6Al-4V implant surface through β-tricalcium phosphate by using additive mixed electrical discharge machining for biomedical application. *Journal of Materials Engineering and Performance, 33*(17). doi:10.1007/s11665-023-08570-5.

Ozay, C., Altug, M., & Ballikaya, H. (2021). A new optimization technique in examining the machinability of Sverker 21 steel: Gray relational analysis-based genetic algorithm. *Arabian Journal for Science and Engineering, 46*(12), 11783–11795. doi:10.1007/s13369-021-05721-5.

Pasandideh, S. H. R., Niaki, S. T. A., & Sharafzadeh, S. (2013). Optimizing a bi-objective multi-product EPQ model with defective items, rework and limited orders: NSGA-II and MOPSO algorithms. *Journal of Manufacturing Systems, 32*(4), 764–770. doi:10.1016/j.jmsy.2013.08.001.

Pedroso, A. F. V., Sebbe, N. P. V., Silva, F. J. G., Campilho, R. D. S. G., Sales-Contini, R. C. M., Costa, R. D. F. S., Barbosa, M. L. S., & Nogueira, F. R. (2024). A concise review on materials for injection moulds and their conventional and non-conventional machining processes. *Machines, 12*(4). www.mdpi.com/journal/machines. doi:10.3390/machines12040255.

Pervaiz, S., Kannan, S., Deiab, I., & Kishawy, H. (2020). Role of energy consumption, cutting tool and workpiece materials towards environmentally conscious machining: A comprehensive review. *Proceedings of the Institution of Mechanical Engineers, B: Journal of Engineering Manufacture, 234*(3), 335–354. doi:10.1177/0954405419875344.

Philip, J. T., Mathew, J., & Kuriachen, B. (2021). Transition from EDM to PMEDM—Impact of suspended particulates in the dielectric on Ti6Al4V and other distinct material surfaces: A review. *Journal of Manufacturing Processes, 64*, 1105–1142. http://www.elsevier.com/wps/find/journaldescription.cws_home/620379/description#description. doi:10.1016/j.jmapro.2021.01.056.

Rezaei Ashtiani, H. R., & Hojati, F. (2022). The influences of spark energy density on the electrical discharge machining (EDM). *Advances in Materials and Processing Technologies, 8*(3), 3165–3181. tandfonline.com/toc/tmpt20/current. doi:10.1080/2374068X.2021.1945304.

Ribeiro da Silva Junior, R., Totten, G. E., Carmo Vendramim, J., & de Campos Franceschini Canale, L. (2017). Thermal fatigue in hot work tool steels—A review. In *Heat Treat 2017: Proceedings from the 29th Heat Treating Society Conference and Exposition, vol. 2017* (pp. 457–468). doi:10.31399/asm.cp.ht2017p0457.

Sahu, D. R., & Mandal, A. (2020). Critical analysis of surface integrity parameters and dimensional accuracy in powder-mixed EDM. *Materials and Manufacturing Processes, 35*(4), 430–441. doi:10.1080/10426914.2020.1718695.

Sahu, S. K., Jadam, T., Datta, S., & Nandi, G. (2018). Effect of using SiC powder-added dielectric media during electro-discharge machining of Inconel 718 superalloys. *Journal of the Brazilian Society of Mechanical Sciences and Engineering, 40*(7). http://rd.springer.com/journal/40430. doi:10.1007/s40430-018-1257-7.

Sana, M., Asad, M., Umar Farooq, M., Anwar, S., & Talha, M. (2024). Sustainable electric discharge machining using alumina-mixed deionized water as dielectric: Process modelling by artificial neural networks underpinning net-zero from industry. *Journal of Cleaner Production, 441*, 140926. doi:10.1016/j.jclepro.2024.140926.

Schubert, A., Bui, V. D., Schaarschmidt, I., Berger, T., & Martin, A. (2022). Developments in powder mixed EDM and its perspective application for targeted surface modification. *Procedia CIRP, 113*, 100–119. http://www.sciencedirect.com/science/journal/22128271/113. doi:10.1016/j.procir.2022.09.134.

Shabarinathan, K. T., Senthilkumar, K., Kathiravan, N., & Udhayachandran, R. M. (2024). Investigation of enhancing EDM machining performance of INCONEL alloy using composite electrodes. *Materials Research Express, 11*(1), 016513. doi:10.1088/2053-1591/ad1b09.

Shard, A., Shikha, D., Gupta, V., & Garg, M. P. (2018). Effect of B4C abrasive mixed into dielectric fluid on electrical discharge machining. *Journal of the Brazilian Society of Mechanical Sciences and Engineering, 40*(12). doi:10.1007/s40430-018-1474-0.

Son, V.T., Tao, L.V., & Thanh, N.T. (2026). A short overview of powder-mixed electrical discharge machining in biomedical materials. International Journal of Applied Electromagnetics and Mechanics. Advance online publication. https://doi.org/10.1177/13835416261454490.

Srivastava, S., Vishnoi, M., Gangadhar, M. T., & Kukshal, V. (2023). An insight on powder mixed electric discharge machining: A state of the art review. *Proceedings of the Institution of Mechanical Engineers, B: Journal of Engineering Manufacture, 237*(5), 657–690. https://journals.sagepub.com/home/PIB. doi:10.1177/09544054221111896.

Straumal, B. B., Kucheev, Y. O., Efron, L. I., Petelin, A. L., Majumdar, J. D., & Manna, I. (2012). Complete and incomplete wetting of ferrite grain boundaries by austenite in the low-alloyed ferritic steel. *Journal of Materials Engineering and Performance, 21*(5), 667–670. doi:10.1007/s11665-012-0130-6.

Le, V. T. (2021d). The influence of additive powder on machinability and surface integrity of SKD61 steel by EDM process. *Materials and Manufacturing Processes, 36*(9), 1084–1098. doi:10.1080/10426914.2021.1885710.

Le, V. T. (2022b). The evaluation of machining performances and recast layer properties of AISI H13 steel processed by tungsten carbide powder mixed EDM process in the semi-finishing process. *Machining Science and Technology, 26*(3), 428–459. doi:10.1080/10910344.2022.2129983.

Tebassi, H., Yallese, M. A., & Belhadi, S. (2023). Optimization and machinability assessment at the optimal solutions across Taguchi OA, GRA, and BBD: An overall view. *Arabian Journal for Science and Engineering, 48*(9), 12455–12483. https://www.springer.com/journal/13369. doi:10.1007/s13369-023-07825-6.

Thakur, S. S., Patel, B., Upadhyay, R. K., Bagal, D. K., & Barua, A. (2023). Machining characteristics of metal matrix composite in powder-mixed electrical discharge machining—A review. *Australian Journal of Mechanical Engineering, 21*(5), 1755–1777. http://www.tandfonline.com/loi/tmec20#.VhTePLfos5g. doi:10.1080/14484846.2022.2030089.

Tufail, M. S., Giri, J., Makki, E., Sathish, T., Chadge, R., & Sunheriya, N. (2024). Machinability of different cutting tool materials for electric discharge machining: A review and future prospects. *AIP Advances, 14*(4). http://scitation.aip.org/content/aip/journal/adva. doi:10.1063/5.0201614.

Van Tao, L., Tien Long, B., Minh, N. T. H., Dung, H. T., Thuc, D. V., & Cuong, P. H. (2024). Prediction of machining performances in powder mixed electro-discharge machining to process SKD61 steel by response surface methodology. *Transport and Communications Science Journal, 75*(4), 1617–1630. doi:10.47869/tcsj.75.4.12.

Vijayakumar, R., Srirangarajalu, N., Santhanakumar, M., & Adalarasan, R. (2023). Investigation in μ-WEDM of Inconel 625 superalloy using RSM-CCD technique. *Materials and Manufacturing Processes, 38*(4), 449–460. doi:10.1080/10426914.2022.2116035.

Vinet, L., & Zhedanov, A. (2011). A 'missing' family of classical orthogonal polynomials. *Journal of Physics A: Mathematical and Theoretical, 44*(8), 085201. doi:10.1088/1751-8113/44/8/085201.

Wang, X. (2019). Application of grey relation analysis theory to choose high reliability of the network node. *Journal of Physics: Conference Series, 1237*(3), 032056. doi:10.1088/1742-6596/1237/3/032056.

Wang, X., Li, C., Guo, H., & Ding, S. (2023). Electrical discharge machining of polycrystalline diamond: A review. *Proceedings of the Institution of Mechanical Engineers, B: Journal of Engineering Manufacture, 237*(11), 1611–1627. doi:10.1177/09544054221136511.

Wong, C. W. Y., Lai, K. H., Shang, K. C., Lu, C. S., & Leung, T. K. P. (2012). Green operations and the moderating role of environmental management capability of suppliers on manufacturing firm performance. *International Journal of Production Economics, 140*(1), 283–294. doi:10.1016/j.ijpe.2011.08.031.

Wu, D., Morimoto, T., Sugio, K., & Sasaki, G. (2023). Preparation and thermal conductivity of copper plated carbon fiber dispersed steel matrix composites. *Materials Transactions, 64*(6), 1205–1209. doi:10.2320/matertrans.mt-d2022010.

Xie, B., Hou, J., Dong, B., Xu, C., Yao, R., & Zhang, Y. (2024). Thermal model of crater formation process in electrical discharge machining. *Case Studies in Thermal Engineering, 56*, 104303. doi:10.1016/j.csite.2024.104303.

Yan, B. H., Lin, Y. C., Huang, F. Y., & Wang, C. H. (2001). Surface modification of SKD 61 during EDM with metal powder in the dielectric. *Materials Transactions, 42*(12), 2597–2604. doi:10.2320/matertrans.42.2597.

Yildiz, Y., Sundaram, M. M., Rajurkar, K. P., & Altintas, A. (2017). Correlation of surface roughness and recast layer thickness in electrical discharge machining. *Proceedings of the Institution of Mechanical Engineers, E: Journal of Process Mechanical Engineering, 231*(3), 414–424. https://journals.sagepub.com/home/PIE. doi:10.1177/0954408915600949.

Zhong, Z. W. (2021). Processes for environmentally friendly and/or cost-effective manufacturing. *Materials and Manufacturing Processes, 36*(9), 987–1009. www.tandf.co.uk/journals/titles/10426914.asp. doi:10.1080/10426914.2021.1885709.

Chapter 18

Mechatronic system design and optimization

Devendrasinh Vashi, Sunil Gautam, Vrunda Mulivala and Aryan Parmar
Computer Science and Engineering, Nirma University, Ahmedabad, Gujarat, India

18.1 Introduction

The heart of the multidisciplinary discipline is mechatronics, which brings together all the traditional branches of engineering, including mechanical engineering, electrical engineering, computer science, control engineering, and system engineering. This creates intelligent systems or products that interact and see in the physical world, process information, and make decisions through mechanical components as well as electrical and computational systems using control algorithms. The outcome is, therefore, a system working with or without human intervention. Design and systems are two essentials in engineering that work hand-in-hand toward the creation of efficient, reliable, as well as functional products or processes. Let us dive deeper into it in the context of mechatronics. A system is defined to be a collection of interacting components or subsystems, which operate to accomplish a definite objective or to perform a specific function. It may be as simple or as complex and intricate as a robotic arm or a manufacturing process. The main aspects of the system consist of interconnection, functionality, feedback loops, and open versus closed systems. The open systems interact with external environments through sharing information and what is material and energy file closed systems are self-contained and only share energy or information. In such mechatronics systems as installed in robots, home automation devices, and manufacturing automation systems, mechanical, electrical, control, and computational components are integrated cohesively (Nakamura & Sato, 2009). An example of the mechanical, electrical, control, and computational layers is presented by the characterization of the robotic welding system. Design is an important engineering process; it "involves a drafted plan or drawing of a system or product whose purpose is to achieve certain goals or requirements." Engineers are inventive, iterative, and problem-solving in their work, describing how each part has to work to achieve a result. Of importance in design are the steps of conceptual design, detailed design, prototyping and testing, optimization, and iteration. Conceptual design creates the underpinning nature of the system and the functionality of the system (Zhang & Han, 2016). Detailed design describes the components, materials, and specifications of each subsystem. Testing ensures the design meets real-world conditions and is acceptable. Optimization is a change in parameters such as cost, performance, energy, and reliability as well as user experience. Designing has an iterative nature, which results in improvement through testing feedback or because of new insight. In summary, design leads to a system command that incorporates parts of the system for the purpose of achieving a specific goal. An actual working system forms what the system itself is, whereas the design gives a blueprint or a plan for its creation. The design considers parts, functionalities, and interdependencies. Design is motivated by identifying system requirements and constraints. Moreover, design involves planning, selection of parts, optimization of interactions, and testing to ensure performance. Good design leads to good systems. Mechatronic system design and optimization integrates mechanical engineering, electrical engineering, computer science, and control engineering to create systems that can perform complex tasks. These systems often include components such as sensors and embedded controllers.

18.1.1 History of mechatronics

Mechatronics goes as far back as the early 20th century with mechanical devices in automata, gear systems, and further early automation. Another significant stride in automation development took place during the Industrial Revolution itself in the early 20th century, when improvements occurred such as the assembly line pioneered by Henry Ford. The first industrially manufactured robots were those of the 1940s–1960s, capable of integrating mechanical, electrical, and control systems,

Computational Intelligence in Mechatronics: Solving Real-World Problems in Electronic Systems Design. DOI: https://doi.org/10.1016/B978-0-443-27580-7.00010-X

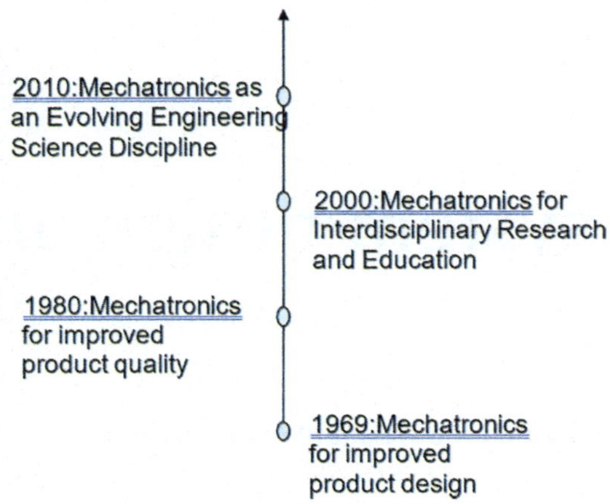

FIGURE 18.1 History and evolution of mechatronics (Book & Crandall, 2009).

which materialized in Unimate (Nagrath & Gopal, 2015). The term "mechatronics" was invented by Japanese engineer Tetsuro Mori in 1969 during its development in a multidisciplinary field that stemmed from mechanical engineering, electronics, and computer science (Lippiello et al., 2010; Mori, 1969). Further into the 1970s and 1980s the advancement of industrial automation took another whole dimension through the PLCs, microcontrollers, and all kinds of high-tech sensors. The product trends in these lines include the PUMA (Programmable Universal Machine for Assembly), which made the automation of even general tasks feasible in the manufacturing environment (Gupta & Ray, 2006). The two decades, the 1990s to 2000s, mark a century leap in the advancements of robotics and mechatronics. This is with artificial intelligence and machine learning that have made tremendous advancement at highly accelerated rates. This provided a ground for self-driving cars, robotic arms in manufacturing, and beginning smart factories (Shankar et al., 2011). AI added to mechatronics systems opened the scope of possibilities from advanced industrial automation to more intelligent consumer electronics and robotics. The Internet of Things, or IoT, had also emerged by the 2010s as one central component of the mechatronics landscape, which enabled interconnected systems to communicate, collect data, and optimize in real time. Such connectivity led to systems that are smarter and more interactive and, therefore, adaptable and capable of learning changes based on user inputs and environmental variations (Bastani et al., 2013). From a simple automation to intelligent and smart systems in general, mechatronics has revolutionized all industries and day-to-day life (Fig. 18.1).

18.2 Research challenges in optimization-based design methods

The complexity of design problems: In general design problems, there are so many variables whose numbers are usually interrelated, and most of the relationships are nonlinear. It is, thus, difficult to predict how the change in one variable will affect another (Saha & Zheng, 2014). The worst case is that this leads to several types of local minima, which have always turned out to be nightmares for algorithms in the quest for the global optimum (Saha & Zheng, 2014). This interplay of variables complicates the algorithms used in modeling such intricately complex interrelationships. The use of conventional methods of optimization collapses with the exponential increase of dimensions each time new variables get added to the research space. New strategies include hybrid techniques and genetic algorithms, and the focus of research is on improving exploration and convergence (Bastani et al., 2013) (Fig. 18.2).

Uncertainty and variability: The design parameters often become uncertain because of variation in material properties, manufacturing processes, and environmental conditions (Gupta & Ray, 2006). Considerably, performance and reliability are affected because of variability. Modeling uncertainties, thus, call for sophisticated statistical techniques. As optimization for performances achieved with a guarantee of robustness to variability, accounting frameworks are required in the cases of uncertainties (Fasse & Hossain, 2007). It can be used by designers to calculate the effect on performance caused by variations of an input, which can lead to decisions.

18.2.1 Optimization-based design process

System requirement analysis: Aspects that define system specifications and objectives, in terms of performance criteria, constraints, and user needs (Shankar et al., 2011; Sharma & Gupta, 2014).

Conceptual design: Generation of alternatives within some alternative concepts and evaluating their feasibility. This stage typically includes brainstorming and preliminary modeling (Bastani et al., 2013; Hanafiah & Ibrahim, 2016).

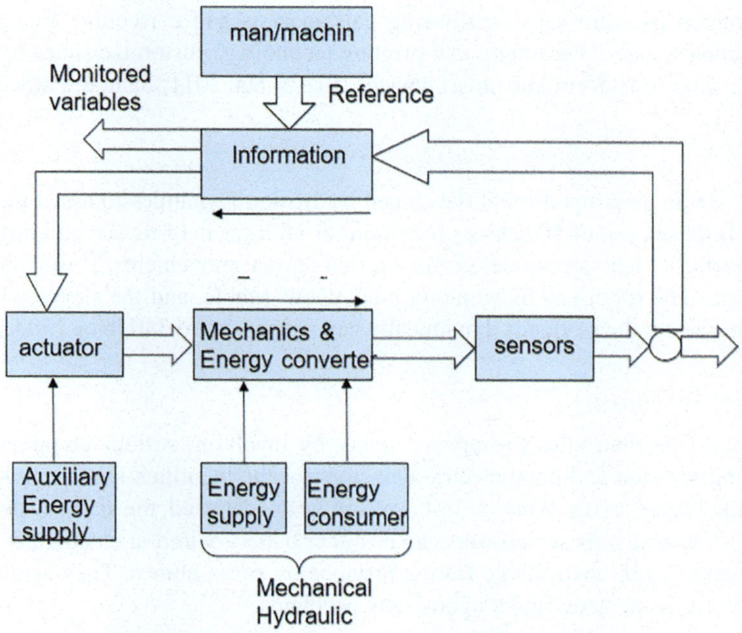

FIGURE 18.2 Mechatronic system architecture (Kuo & Wei, 2011; Patterson & Holtz, 2010).

Detailed design: All the constituent components' detailed designs are developed; this means a detailed mechanical structure, electronic circuits, and software algorithms. This can include CAD models for mechanical parts and circuit diagrams for electronics (Michalec & Wang, 2008; Vijayan & Saha, 2012).

Prototyping: Physical proofs of concept for testing and validation of design ideas. Rapid prototyping techniques, such as 3D printing, are widely applied (Saha & Zheng, 2014).

Testing and validations: It refers to evaluating the system designed to meet the requirements. It includes physical testing and simulation (Hanafiah and Ibrahim, 2016; McLean & Ng, 2017).

Tools and software:

- CAD software: Tools such as SolidWorks and AutoCAD for mechanical design.
- Simulation software: MATLAB/Simulink, ANSYS, or similar tools for dynamic modeling and system simulation.
- Control system's design: Tools for designing control algorithms, such as LabVIEW or specific control design toolboxes in MATLAB.

18.2.2 Modeling and simulation as prerequisites for the design process

It entails dynamic models and simulations in the analysis and synthesis of the systems. Methods can be refined based on the level of detail required in the analysis relevant to the problem (Ma & Miao, 2013; Yan & Sun, 2015). Simulations are applied where the real systems are either unavailable, expensive, or time-consuming to manage (Hanafiah & Ibrahim, 2016; Zhao & Wang, 2012). Such applications allow adjustment of control variables, state variables, and system parameters beyond the possible range of the actual system (Naoe & Fukuda, 2011). Simulations also damped some disturbances, which naturally would come about in the physical system and thus allow more profound insight into the behavior of a system (Mathews & Bellingham, 2011). On the other hand, simulations have some drawbacks: they are general and very easy to apply. They model the world under very limited experimental conditions, and they hardly ever illuminate. Analytical techniques reveal how the model behaves under arbitrary conditions; but one simulation run tells us only how the model behaves for one particular set of conditions applied during the simulation run (Johnson & Raja, 2009). Hence, whenever a valid alternative to simulation is available, it should be used.

Let us get a detailed view of some of the applications with their properties and usefulness (Sharma & Gupta, 2014).

18.2.2.1 Running shoes

Running shoes, which incorporate mechatronics systems, assure performance, comfort, and immunity against injuries. Moreover, the matrices under study, such as steps, stride, and impact, are monitored in real-time by sensors within the shoe to improve running techniques. The adaptive cushioning mechanisms adjust dynamically with activity type and terrain, whereas advanced smart insoles contain pressure sensors allowing gait analysis and correction. Energy harvesting mechanisms minimize charging requirements, and 3D scanning and printing technology customize shoes for individual biomechanics; thus the traditional running shoes transform into smart devices (Lee & Ha, 2011; Sahu & Patnaik, 2016).

18.2.2.2 BigDog

The BigDog is a dynamic, stable quadruped robot developed by Boston Dynamics to assist the soldier in carrying heavy loads over rugged terrain. It makes use of 50 sensors that monitor all aspects of its surroundings, including inertial, joint, hydraulic, and inertial sensors. Its legs are modeled from actual animal movements; it runs on a go-kart engine used to support the hydraulic system. The robot can move along mud, snow, rubble, and the slopes whose angles reach up to 35 degrees. In the mechatronic system the payloads dynamically can support up to 340 lb or 150 kg.

18.2.2.3 Smoke detector system

A smoke detector implements mechatronics to improve safety by involving various components for fire detection and alarming. Sensors used are ionization and photoelectric detectors for the identification of smoke and a microcontroller to process signals in monitoring smoke levels. If higher levels of smoke are detected, the alarm systems within the detectors are activated (Kalawsky, 2014). Generally, these detectors can be either battery-wired or electrically wired, and they now come with connectivity, which makes them controllable from a distance by smart homes. They also self-diagnose; this enables users to conveniently monitor system status and diagnose any problems.

18.2.2.4 Prosthetics

The prosthetic limbs are incorporated with mechanical, electronic, and control systems for improvement (Craig & Soo, 2012; Crandall & Hayward, 2014). They resemble human limbs in appearance when a user is living; their movements are controlled by myoelectric sensors and motors. Information from multiple sensors is gathered for the control of passive movement. Power management is through batteries; using them is supplemented by haptic feedback to make the user feel comfortable (Crandall & Hayward, 2014). It has capabilities to improve prosthesis use through AI and machine learning and developments in 3D to personalize neural interface design. Soft robotics is applied for smoother movements and improved comfort. These innovations enhance both function and comfort, increasing autonomy in users (Crandall & Hayward, 2014).

18.2.2.5 Automobiles

An example of mechatronics in modern vehicles is the mechanical, electronic, and control systems used, which have improved the performance and safety factors of modern cars (Craig & Soo, 2012). This includes ADAS, which integrates sensors and actuators for lane-keeping and adaptive cruise control; efficient battery management and motor control for electric and hybrid vehicles; information systems that incorporate touch screens and voice recognition with connectivity for navigation and entertainment; and even vehicle dynamics control that maintains the integrity of safety and handling (Zhou & Lin, 2015). It made the modern car intelligent, safe, and efficient. With AI, cameras, and sensors, navigation comes into the self-driving car of today. Electronic control units have ensured fuel efficiency, climate control, smart parking, and active safety systems.

18.2.2.6 High-speed trains

The very heart of high-speed trains is mechatronics, which brings together mechanical, electronic, control, and computer systems to provide performance, safety, and efficiency. Its features encompass control and automation in train propulsion systems, passenger comfort, energy efficiency, safety and collision avoidance, condition monitoring systems, passenger information, and track and route monitoring (Zhou & Lin, 2015). Automatic train control and a train signaling system also utilize sensors, communication technologies, and algorithms for control so that there is a safe run. High-performance inverter-based traction motors with electrification of the track offer precise speed regulation with comfort in acceleration and braking. There is regenerative braking that captures kinetic energy at braking. Energy-saving technology prevails because of an active suspension, whereby the real-time adjustment of train suspension reduces vibrations. HVAC systems are there with sensor-controlled climate and comfort. Today the aerodynamic design minimizes resistance to air, whereas the energy-efficient system reduces consumption of power. The passenger information system offers real-time schedules and delays. Advanced signaling systems monitor train locations safely over complex rail networks.

TABLE 18.1 Summary of related work.

S. no.	Author	Research paper/book	Overview
1	Nagrath and Gopal (2015)	Control Systems Engineering (Book)	Nagrath and Gopal's control system works from the basic framework of mechatronics, dealing with stability analysis, control feedback, and system dynamics, forming the basis for designing systems in terms of robotics, automation, and system performance
2	Gupta and Ray (2006)	Mechatronics: Principles and Applications (Book)	Gupta and Ray's book successfully integrates mechanical engineering, electronics, and computing into the theoretical principles and practical approaches to robotics, automated manufacturing and, more widely, embedded systems. Their work was one that helped draw the interdisciplinary boundaries defining mechatronics and prompts innovation for newer applications in automotive and aerospace
3	Shankar et al. (2011)	Design and Development of Mechatronic Systems for Industrial Automation (IEEE)	This paper concerned mechatronic system design and integration in industrial automation. It targeted integration of mechanical design with the embedded control systems and sensors to potentially improve the efficiency of the system. This, therefore, facilitated the pushing of the wider adoption of mechatronics more intensely into manufacturing
4	Asada and Hollerbach (1985)	Robot Analysis and Control (Book)	Asada and Hollerbach only discuss the analysis and control of robotic systems, including only kinematics and dynamics and control of robotic arms in their work. Their work formed the basic knowledge source for modern design in all industrial, medical, and consumer applications
5	Vijayan and Saha (2012)	Design and Analysis of Mechatronic Systems (International Journal of Mechatronics and Manufacturing Systems)	This paper describes design methodologies applied to mechatronic systems, as applied in the manufacturing and assembly process, with special points of optimization in the mechanical, electrical, and computational elements that characterize performance and reliability in real applications
6	Book and Crandall (2009)	Mechatronics: An Introduction (Book)	This textbook introduces mechatronic ideas, constituents, and applications that revolve around the merger of sensors, actuators, controllers, and mechanical systems. It set the first foundation to understand mechatronic system design as an integral but not an individual constituent

(continued on next page)

TABLE 18.1 Summary of related work—cont'd

S. no.	Author	Research paper/book	Overview
7	McLean and Ng (2017)	Real-Time Control of Mechatronic Systems Using Embedded Systems (Journal of Mechatronics)	The paper states the use of embedded systems for real-time control in mechatronics systems, roles of a microcontroller and real-time operating systems in control of complex systems in robotics and automated manufacturing
8	Bastani et al. (2013)	Optimization of Mechatronic Systems Design Using Artificial Intelligence (IEEE Transactions on Industrial Electronics)	This paper uses AI techniques, specifically genetic algorithms and neural networks, for the optimization of a mechatronic system. Performance improvement, efficiency, and self-optimization were to be achieved in dynamic environments toward intelligent mechatronic systems
9	D. M. Dawson and L. P. G. Morris (Johnson and Raja, 2009)	The Role of Sensors and Actuators in Mechatronic Systems (Sensors and Actuators Journal)	This paper considers the importance of sensors and actuators in mechatronic systems as it probes into the analysis of how technologies such as vision systems, pressure sensors, and accelerometers combined with actuators such as motors and pneumatics make for accurate control, feedback, and automation
10	B. L. Ooi, D. H. Lee, and S. W. Tan	Smart Mechatronics: A Review of Recent Trends in Smart Materials and Intelligent Systems (Journal of Intelligent Material Systems and Structures)	This paper considers the introduction of applications of active materials such as shape memory alloys and piezoelectric materials into enhancing the functions of mechatronic systems. Indeed, this kind of integration brings out more responsive, adaptive, and energy-efficient systems

18.3 Related work

18.4 Future scenarios

The future revolution for the industry is AI-driven autonomous systems, collaborative robots, smart materials, quantum computing for design optimization, 5G and edge computing, advanced human machine interfaces (HMIs), sustainability and green technologies, autonomous vehicles, and smart transportation. AI-driven autonomous systems can optimize manufacturing performance by learning about their environment and analyzing data to predict failures (Rook & Ayer, 2007). Collaborative robots interface with humans in manufacturing and healthcare settings (McLean & Ng, 2017); they bring about a reduction in risks as well as productivity improvements. Smart materials may allow soft robotics to recover from injuries, thus leading to delicate workloads in agriculture and surgery (Suh & Jung, 2010). The work of design optimization using quantum computing will help accelerate the development of energy-efficient systems and unique designs (Wang & Wang, 2013). With 5G, edge computing is integral to accelerate connections with minimal latency. In essence, this means processes at the edge or on location will be computed in mechatronic systems so that decision-making at the cloud computing level is no more imperative (Miura & Iwasaki, 2013). Advanced HMIs will provide intuitive control and real-time interaction with the mechatronic systems. Ethical and human-centered design: Mechatronics can bring improvement in human welfare and accessibility as well as safety to focus the capabilities not on building the well-being of social inequity (Table 18.1).

References

Asada, H., & Hollerbach, J. H. (1985). *Robot analysis and control*. Wiley.

Bastani, F. B., Azizi, M. S., & Tavakkol, M. R. (2013). Optimization of mechatronic systems design using artificial intelligence. *IEEE Transactions on Industrial Electronics, 60*(5), 2165–2173.

Book, W. J., & Crandall, B. C. (2009). *Mechatronics: An introduction*. Pearson Education.

Craig, J. J., & Soo, H. K. (2012). *Introduction to robotics: Mechanics and control*. Pearson Education.

Crandall, B. C., & Hayward, S. (2014). Robotic arm control using mechatronic systems. *Journal of Robotic Systems, 21*(1), 33–45.

Fasse, M., & Hossain, I. (2007). Mechatronics: A review of key systems, applications, and future trends. *Journal of Automation and Control Engineering, 5*(7), 1304–1310.

Gupta, R. K., & Ray, A. C. (2006). *Mechatronics: Principles and applications*. McGraw-Hill.

Hanafiah, M., & Ibrahim, M. (2016). Optimization techniques for mechatronic design. *Advances in Mechatronics Engineering, 5*(3), 277–284.

Johnson, M., & Raja, R. (2009). The role of sensors and actuators in mechatronic systems. *Sensors and Actuators A: Physical, 149*(2), 160–167.

Kalawsky, R. (2014). *Mechatronics: Principles and applications*. Butterworth-Heinemann.

Kuo, W., & Wei, J. (2011). Intelligent control algorithms for mechatronic systems. *Journal of Intelligent Systems, 25*(4), 223–237.

Lee, C., & Ha, Y. (2011). Real-time embedded systems for mechatronics. *IEEE Transactions on Industrial Electronics, 58*(7), 2795–2805.

Lippiello, V., Siciliano, B., & Siciliano, A. (2010). Design of mechatronic systems: Theory and practice. *Springer series in advanced manufacturing*. Springer.

Ma, J., & Miao, J. (2013). Modeling and control of mechatronic systems with embedded microcontrollers. In *Proceedings of the International Symposium on Mechatronics* (pp. 160–168). New York: IEEE.

Mathews, G., & Bellingham, J. (2011). Control design in mechatronics: Application of simulations. *International Journal of Control, 84*(12), 1967–1979.

McLean, J. M., & Ng, S. H. (2017). Real-time control of mechatronic systems using embedded systems. *Journal of Mechatronics, 49*, 76–84.

Michalec, M., & Wang, J. (2008). Optimization and performance design for mechatronic systems. In *Proceedings of the 3rd IEEE International Conference on Mechatronics and Automation* (pp. 217–223).

Miura, K., & Iwasaki, H. (2013). Robotic control and mechatronic systems: Recent advances and future trends. *International Journal of Robotics Research, 32*(5), 517–530.

Mori, T. (1969). Mechatronics: A new approach to mechanical-electrical engineering. *IEEE Transactions on Industrial Electronics, 16*(4), 261–270.

Nagrath, I. J., & Gopal, M. (2015). *Control systems engineering* (5th edition). Prentice Hall.

Nakamura, Y., & Sato, R. (2009). *Mechatronics and automation*. Springer.

Naoe, T., & Fukuda, T. (2011). Advanced mechatronic systems and control: Design for flexibility and robustness. *IEEE Transactions on Mechatronics, 16*(2), 100–108.

Patterson, M. R., & Holtz, R. S. (2010). A framework for mechatronics systems design in industrial automation. *Journal of Engineering Design, 21*(6), 789–802.

Rook, G., & Ayer, M. (2007). Designing mechatronic systems for high performance and reliability. *IEEE Transactions on Robotics, 23*(2), 397–405.

Saha, S. K., & Zheng, L. (2014). Design, analysis, and performance optimization of mechatronic systems. *Journal of Mechanical Engineering Science, 228*(2), 1985–1997.

Sahu, M. K., & Patnaik, S. (2016). Design and integration of mechatronic systems for industrial applications. *International Journal of Mechatronics, 8*(5), 417–426.

Shankar, S. S., Manjunath, M. S., & Shankar, R. S. (2011). Design and development of mechatronic systems for industrial automation. *IEEE Transactions on Industrial Electronics, 58*(9), 4020–4029.

Sharma, A., & Gupta, A. (2014). Optimization algorithms for mechatronic systems: Applications in industrial automation. *International Journal of Advanced Manufacturing Technology, 70*(1), 255–268.

Suh, I., & Jung, Y. (2010). Modeling and control of mechatronic systems with smart materials. *Materials Science and Engineering B, 171*(3), 202–212.

Vijayan, G. K., & Saha, S. K. (2012). Design and analysis of mechatronic systems. *International Journal of Mechatronics and Manufacturing Systems, 5*(3), 259–277.

Wang, X., & Wang, Q. (2013). Control and automation of mechatronic systems: Algorithms and approaches. *Journal of Automation, 47*(6), 345–357.

Yan, X., & Sun, H. (2015). Intelligent control and application of mechatronic systems. *IEEE Access, 3*, 1991–2000.

Zhang, X., & Han, Y. (2016). Control systems for mechatronic systems: Adaptive and robust techniques. *Control Engineering Practice, 59*, 113–123.

Zhao, Z., & Wang, H. (2012). Development of real-time systems for control and automation in mechatronics. *Journal of Mechatronics and Automation, 2*(3), 94–102.

Zhou, J., & Lin, J. (2015). Optimized design of mechatronic systems for robotics applications. *Robotica, 33*(8), 1939–1950.

Chapter 19

Leveraging convolutional neural networks for sentiment analysis on social media content

Devendrasinh Vashi, Sunil Gautam, Aarchi Shah and Kriyanshi Soni
Computer Science and Engineering, Nirma University, Ahmedabad, Gujarat, India

19.1 Introduction

In the digital era, almost every person has access to the internet and mobile technology, and so many products and services are easily available. Therefore there is an increasing interest in the understanding of how people communicate, in particular by text, and their willingness to express opinions, sentiments, and trends. Sentiment analysis—this term means that it is the study of the feelings of the individuals based on certain expressions of them, in other words, the abstract—positive, negative, or neutral. It has applications ranging from simple embracing or hailing of such events to gauging perception and attitudes extended toward brands, products, and services. The traditional approaches of sentiment analysis have focused on rather primitive levels, such as tone detection, which has been nowadays phased out, owing to the advancement of technology in the visual area of analysis, particularly, deep learning techniques.

Social media networks—in particular, Twitter, Instagram, and Facebook—have taken over contemporary interaction and provide useful unprocessed data on people's feelings. These systems allow the audience to express their thoughts and react to events happening in the world almost instantaneously. Owing to the acceptance of social media worldwide with over billions of users, content, which at any moment changes with society's views and even moods, is shared widely. It is yet easy to see that determining sentiment on social networks presents quite a number of difficulties. Responses are often short and relatable, involving much shorthand, emojis, pictures, hashtags, and even the use of irony, which makes understanding these texts rather difficult. Adding to that, social media is a conversation that is animated, thus turning easily, and may therefore need processing in real time or time close to real.

To extract more profound meanings from the intricate and rich but difficult emotions surrounding this environment, more sophisticated techniques are used. One such way is the utilization of techniques associated with convolutional neural networks (CNNs), a type of modeling technique that is mainly utilized for processing images. CNNs are pioneers in the recognition of intricate manifestations and patterns, which is a cardinal aspect in relationship management in social media collection of texts. CNN has been used by some researchers in recent years to do sentiment analysis, which is a task related to image processing and, of course, text containing pictures. Such techniques move above the basic methods, which only consider the positivity or negativity of specific terms along with their synonyms, leading to a more valued and delicate quantification of the impressions made on various sources of social media. The failure in traditional methods of sentiment analysis is learnt through the understanding of expression of sentiments in the content of the global social media conversations.

The development of neural networks may take-off with their simple-to-complex system. For instance the introduction of a neuron concept helped develop the McCulloch–Pitts neuron model in 1943, which is regarded as the beginning of neural network design. The idea of a perceptron was born in the 1960s when a simple structure with less computational complexity was introduced, which was a single-layer neural network that has the capability of performing pattern recognition. In the 1980s the neocognitron appeared, containing a multilayer network with features of overlapping convolutional layers in resource allocation, which forbade the present day CNN for pattern classification. In 1989 a turning point happened with the introduction of LeNet-1 by Yann LeCun as the first known CNN architecture. After a phase of slow development, the year

Computational Intelligence in Mechatronics: Solving Real-World Problems in Electronic Systems Design. DOI: https://doi.org/10.1016/B978-0-443-27580-7.00033-0

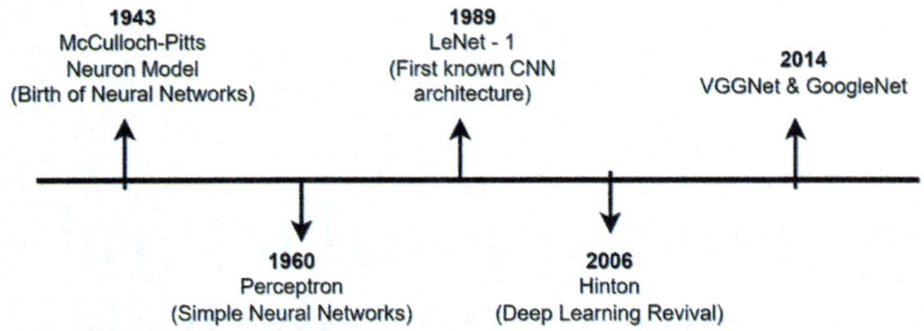

FIGURE 19.1 History of CNNs. *CNN*, Convolutional neural network.

2006 celebrated the return of the research on deep learning with Geoffrey Hinton's approach of deep neural networks creating attention for the complex models. In 2012 there was a shift from traditional machine learning models to the use of CNN with very dramatic results in the model named AlexNet, introduced by Alex Krizhevsky, that won the ImageNet challenge. The architectures of 2014—VGGNet or GoogleNet, which proposed 16–19 layers, further enhanced the capabilities of the models with a ceiling for even more complex structures. It is these recent improvements that have contributed greatly to the rapid advancement of applications that depend on neural networks, image processing, and all forms of artificial intelligence (AI) (Fig. 19.1).

19.2 Understanding convolutional neural network

19.2.1 Basic CNN architecture

Deep learning is a specialized skill set that has become extremely popular in the IT business in recent years. Deep learning is a branch of machine learning that includes algorithms modeled after neural networks or the way the human brain functions. Deep learning is a specialized skill set that has become extremely popular in the IT business in recent years. Deep learning is a branch of machine learning that includes algorithms modeled after neural networks or the way the human brain functions (Schmidhuber, 2015). These structures are known as neural networks. It educated the computer to do what people do instinctively. Artificial neural network, autoencoders, recurrent neural network, and reinforcement learning are some of the models used in deep learning. CNNs are also referred to as ConvNets.

The mathematical term "convolution" in CNN refers to a unique type of linear operation in which two functions are multiplied to create a third function that expresses how one function's shape is altered by another function. CNNs are deep learning models that use convolutional layers to extract features by fully connecting and pooling layers for tasks (LeCun et al., 2010).

A CNN architecture consists of two primary components: Feature extraction is the process of using a convolution tool to separate and identify the different features of an image for study. The feature extraction network is made up of numerous convolutional or pooling layer pairs. A fully connected layer that makes use of the convolutional process's output and forecasts the image's class using the features that were gathered in earlier phases.

Reducing the quantity of features in a dataset is the goal of this CNN feature extraction model. It generates new features by synthesizing the preexisting features found in an initial feature set (Heaton, 2016) (Fig. 19.2).

- **Convolutional layer:** Point birth from the input image is the responsibility of the convolutional subcaste, which is the first subcaste in a CNN. It uses the compilation process, which involves a fleck product between the image areas and a sludge (size $M \times M$) that slides across the image. A point chart that emphasizes significant features like corners and edges is the end result. After that, this feature map is sent to the next layer for more processing. As the image is processed through the network, the convolutional layer maintains the structure of the image by preserving the relations between pixels (Lecun et al., 1998).
- **Pooling layer:** This layer comes after the convolutional layer, contracts features map's size and lowers computing expenses. It summarizes the features that the convolutional layer has extracted and works independently on each featured map. A typical pooling technique consists of
  - Max pooling: Extracts feature map section's highest value.
  - Average pooling: Determines a section's average value.
  - Sum pooling: A section's values are added together using sum pooling.
  - This layer serves as a link between convolutional layer and fully connected layer.

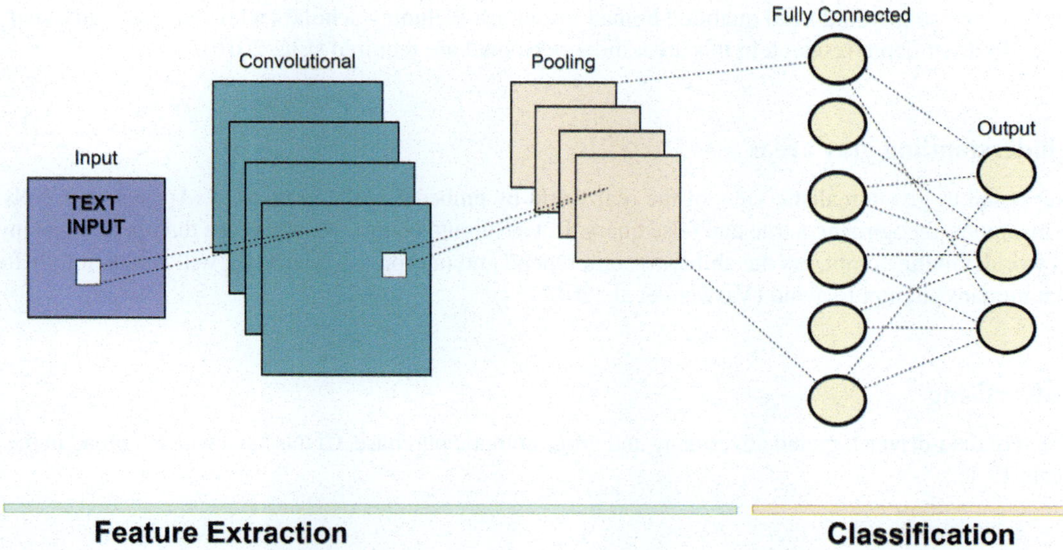

FIGURE 19.2 Architecture of CNN used for text classification. *CNN*, Convolutional neural network.

- **Fully connected layer:** By using weights and biases the fully connected layer joins neurons from various layers. In CNN, it generally comes before the output layer. This layer processes the output from earlier layers through a number of fully connected layers; many fully connected layers also need human network oversight (Krizhevsky et al., 2012).
- **Dropout:** Overtraining, which happens when a model performs well on training data but badly on fresh data, is avoided in CNNs by using a dropout layer. Dropout layer reduces the size of neurons during training (e.g., 30% with a dropout rate of 0.3). The model performs better on unseen data, which lessens the model's dependence on particular neurons (Srivastava et al., 2014).
- **Activation functions:** A CNN activation function is essential as it adds nonlinearity to decide which data should be sent across the network. The model learns about complicated correlations between variables. ReLU, Softmax, Tanh, and Sigmoid are examples of activation functions. Softmax is used for multiclass classification, and Sigmoid and Softmax are utilized for binary classification, whereas activation functions use the relevance of the input for prediction to determine whether a neuron should be engaged or not.

19.3 Usages of CNN

19.3.1 Understanding face recognition

A CNN divides facial recognition into the following parts: First it recognizes each face in the image, then pays attention to each face in spite of outside influences such as light, angle, and posture. After all this, matching face with a name by comparing all the data gathered with data that is already present in the database. The labeling of scenes is done in a similar manner (Triantafyllidou & Tefas, 2016).

19.3.2 Analyzing documents

Document analyzer is another application for CNNs. This has a remarkable impact on the recognizer in addition to being helpful for handwritten analysis. A machine needs to execute millions of commands per minute to read a person's writing and compare it with an extensive database. Through testing it has been observed that the error rate has been decreased to at least 0.4% at the character level through the use of CNN and algorithms (Marinai et al., 2005).

19.3.3 Understanding climate

CNN also plays a vital role in climate change, particularly in figuring out why the changes are so extreme and trying to figure out how to reduce their effects. More social and scientific insights are possible from the data in these natural history archives, but by doing so it would need qualified human resources, including scholars who can physically visit these kinds of repositories. To do in-depth research in this area, more personnel are required (Liu, 2016).

19.3.4 Understanding gray areas

CNNs are supposed to give a realistic view of the real world by embedding the gray areas. At present, CNNs essentially operate like machines, recognizing a true and false query. But as human beings, we are aware that there are many gray areas in the actual world. Giving computers the ability to comprehend and interpret fuzzier logic will enable it to better grasp the gray area that humans attempt to avoid (Vaswani et al., 2017).

19.3.5 Advertising

With the advent of data-driven targeted advertising and programmatic purchase, CNNs have widely spread in the advertising industry (Table 19.1).

19.4 Benefits and drawbacks of CNNs

TABLE 19.1 Advantages and disadvantages of CNNs.

Advantages	Disadvantages
Hierarchical feature learning: Focuses on the learning of low-level features such as edges to high-level features such as objects for improved understanding	**High computational cost:** Requires advanced and costly GPU power and time, including high inference cost, especially for applications that are real-time
Automatic feature extraction: Refers to technological processes that automatically identify, describe, and retrieve useful content from raw data, without any manual intervention in feature engineering	**Data hunger:** Dependent on large labeled data sets whose acquisition may be expensive and time-consuming
Scalability: The capacity to manage large amounts of data and complex structures, for instance, by applying transfer learning techniques	**Overfitting:** There is a tendency to overfit when training data available is less because it results in poorer generalization to new and unseen data
Flexibility: Applicable in several areas such as image processing, needs processing, and speech processing	**Lack of interpretability:** Issues surrounding the lack of clear guidance on how the models make the decisions they do creates lower assurance and higher difficulty in the debugging process
End-to-end learning: Makes it possible to learn without reference to any intermediate processes from the raw input to the target output	**Sensitivity to hyperparameters:** CNNs require attentive tuning of various hyperparameters to achieve optimal execution, which can increase the training complexity, requiring additional time and expertise for tuning
Advanced applications: In use, for instance, in applications such as self-driving cars, medical image processing, and even facial recognition, these techniques are paramount	

19.5 Challenges in analyzing social media sentiment

Assessing public attitudes through social media practices is a very alluring but also very strenuous one because of the high degree of informality, dynamism, and diversity in the nature of content from social media. Offered by the social networking sites is the undiluted and genuine expression of the public and feeling of the heart addressed to certain occurrences world over, entities, and ideologies. Nevertheless, in helping such emotions, understanding them becomes very tricky. The language employed in social media is informal, very contextual, and socioculturally thick as opposed to the written discourse where everything is formal. In the section below are some of the important reasons why social media sentiment analysis is fraught with difficulties, each of which calls for advanced methods.

19.5.1 Complex language and tone

Despite its frequent use in almost every text, social media poses a significant problem for sentiment analysis because of the informal language and constant change. In many cases, social media updates go beyond written content and the use of language. They include slang, shortcuts, markings, animation for emotional expressions, and many others. This capability often results in failures to analyze texts efficiently by machines. A study presented Valence Aware Dictionary and sEntimentReasoner (VADER), a model that incorporates emojis and adaptable elements of text, which also manifest in this growing problem (Hutto & Gilbert, 2014). In the last few years, models have had to improve consistently to cope with this sort of language. Exemplifying the challenges in sentiment detection that have arisen with increased usage of sarcasm and humor along with implied sentiments that require in-depth knowledge of the context in which they are used. This is especially genuine in the case of Twitter, where users have to express their feelings easily and briefly because of a character limitation (Medhat, 2023).

19.5.2 Massive amount of data

The growth of social media data presents a problem of scalability for sentiment analysis models because of the fact that they need to analyze vast amounts of data that are turned out daily. It is because every single day, there are millions of users who are uploading new content on the platforms. It requires efficient data storage and processing systems because of the large and fast generation of data. As noted in a research, because of the volume of data in real-time, traditional machine learning techniques usually do not perform well, even state-of-the-art deep learning models require large amounts of computational power (Kanungo, 2023). These scholars point out that effective management of such a data flow will also hinge on the utilization of distributed systems and parallel processing. Further noting that just increasing the resources available in terms of hardware might not be enough in the absence of some modifications to the algorithms for quicker and more efficient analysis (Sarker, 2021).

19.5.3 Data cleaning and preprocessing

Data cleaning and data preprocessing techniques, though necessary for social media sentiment analysis, consume a lot of time integrating social media, especially in sentiment analysis (Symeonidis, 2018). Activities in social media, however, are usually noisy, bringing in a lot of application symbols, web links, and even different languages, thus the need to clean and normalize content before analysis is carried out. According to a study, these techniques include tokenization, stop-word removal, and text normalization among many others that aid in processing of raw data to give out better results in detecting sentiments (Sarker, 2021). They go on to explain that data, which does not undergo sufficient processing will yield a wastage of resources because of high without errors rates and therefore preprocessing must be incorporated in any social media analytics system for checking sentiment on the contents. Equally, highlighting the challenge that multilingual data presents and also feeling that preparation goes beyond cleaning oneself. Scholars argue for sophisticated levels of processing applications considering the combined features that language of social media carries; for instance, the use of two or more languages and the use of phonetic spellings and slang terms (Devlin et al., 2018) (Fig. 19.3).

19.6 Why does CNN work well with sentiment analysis?

CNNs have demonstrated remarkable efficiency in natural language processing (NLP) tasks such as sentiment analysis, despite their traditional association with image processing. Their capacity to identify regional trends and characteristics in the text data accounts for this is flexible.

19.6.1 Capturing local feature

Detect *n*-grams: This makes them especially well-suited for tasks involving sentiment analysis and text classification. CNNs can achieve this in one of the ways that the network learns to recognize *n*-grams, which describes sequences of consecutive words in a text. Here, *n*-grams can be as small as individual words like unigrams; pairs of words like bigrams; or even longer sequences like trigrams and 4-grams. This crucial insight is that some *n*-grams carry sentiment information that can be used to predict the general sentiment of a text (Kim, 2014).

For example, in sentiment analysis, "absolutely loved" or "terrible service" can be indicative of high-pitched sentiment where CNNs can be trained to recognize these as indicators of the overall sentence or document. This is valuable because

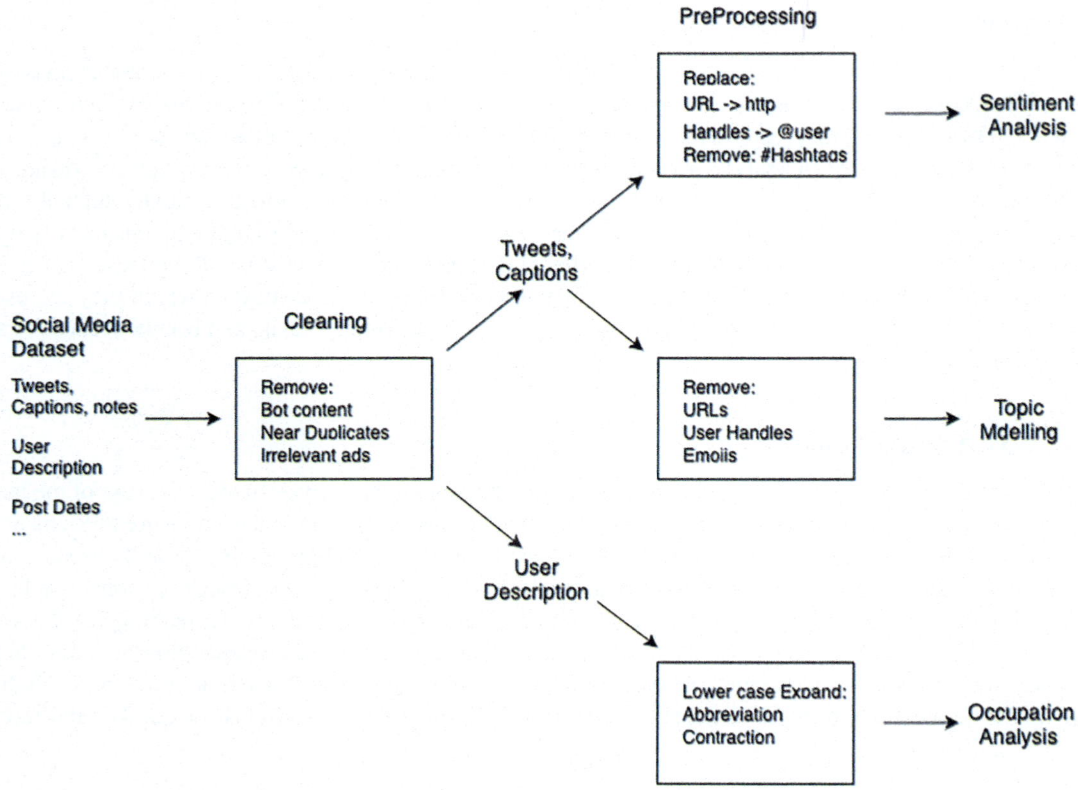

FIGURE 19.3 Data preprocessing in sentiment analysis on social media content.

the local patterns that CNNs are capable of capturing represent the sentiment of a majority of terms being encoded in combinations rather than being isolated terms.

Feature extraction: This allows CNNs to be an effective strategy for learning features in different granularities. The convolution allows the CNNs to learn meaningful features at different hierarchical levels within the input text. For example, at the lowest granularity, patterns of specific words, such as adjectives or verbs, that have a tendency to express certain sentiments can be captured by filters. In higher levels the filters could learn longer sequences of words or phrases contributing to richer meanings, such as sentiment-modifying words or phrases, such as "not bad" or "quite good."

The CNNs apply convolutional filters across the entire text to generate feature maps representing important local dependencies in the input sequence. These feature maps may then be fed into pooling layers that can identify the most salient features, which are used further up in the model to do the prediction. This multilevel approach then is what allows CNNs to learn and represent the relationship between individual words and longer sequences better and thus more accurately handle the intricate structures of natural language (Rakhlin, 2016) (Fig. 19.4).

19.6.2 Effective feature learning

One of the major advantages of using CNN in NLP is that it automatically extracts features directly. Unlike other approaches in machine learning, which usually require a series of manually crafted feature engineering steps, CNN learns how to find and extract such features, which may be word *n*-grams, syntactic structures, or specific domain-related features, instead applying convolutional filters over the data and automatically determining what important patterns exist within it. This automation of feature selection not only simplifies the model development process but also makes the workflows much more efficient (Kalchbrenner et al., 2014). This allows models to learn the most important features for a large dataset, avoiding potential biases or shortcomings in human-curated features. The result is CNN tends to outperform traditional methods. The ability to learn directly off of data also helps in enhancing the performance of this model, especially when working with huge amounts of unstructured data, where the sheer variety of possible features can overwhelm manual approaches (Kalchbrenner et al., 2014).

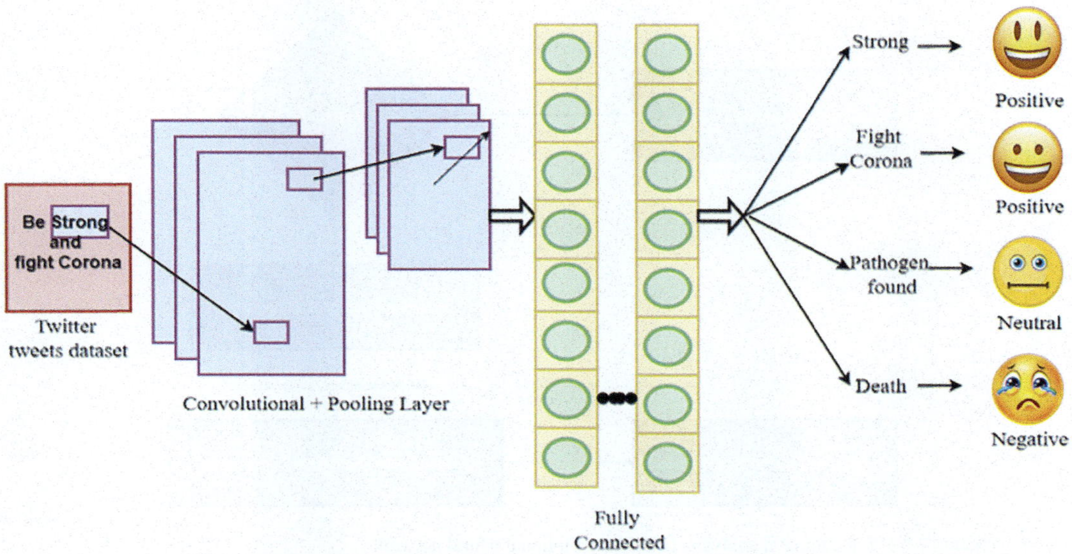

FIGURE 19.4 CNN for sentiment analysis of COVID-19-related tweets. *CNN*, Convolutional neural network.

19.6.3 Hierarchical feature learning

In addition to automatic feature extraction, CNNs also support hierarchical feature learning. Where a network learns features at different levels of abstraction. The shallower layers of CNNs would capture local features such as word-level-*n*-grams or even syllables or very short phrases. These are usually the basis of text meaning. And as the cascade flows down the layers, it starts to abstract local features and generalize more complex semantics and syntactic relationships among words, thereby allowing it "to pick up" such subtle nuances of language as negations, sentence structure, or context. This hierarchical approach allows the CNNs to grasp the fine details of words and the broad meaning of sentences or even entire documents, thus improving the ability of the model to process and interpret natural language more effectively (Conneau et al., 2017).

19.6.4 Processing in parallel

Effective computation: CNNs are very powerful with applications requiring high computational efficiency. The three most important strengths of CNNs occur in parallel processing, making the approach highly suitable for modern hardware accelerators such as graphics processing units (GPUs). This parallelism greatly reduces the training time as well as applications time and holds scope for fast execution along with efficient processing at large sizes.

CNNs are also good at effectively handling sequences of variable lengths of data. They are able to learn hierarchical representations, thus enabling them to handle text data: the ability to identify local patterns and the extraction of features within a sequence. An important feature in these capabilities makes them very useful in sentiment analysis, where the nuances are very important to understand, for instance, the sentiment-bearing words and their relationships within a sentence (Conneau et al., 2017).

Using strengths of CNNs related to parallel processing and feature extraction, we get highly accurate and efficient models of sentiment analysis. It is going to find a pretty wide range of practical applications in tasks like analysis of customer feedback all the way through social media monitoring trends to uncover some valuable insights into public sentiment and opinion; hence CNNs do prove robust solutions to this set of challenges in the domain of NLP and especially sentiment analysis (LeCun et al., 2015).

19.7 Related work in CNN for sentiment analysis

Within the existing literature on models of sentiment analysis, notable success has been achieved using CNNs when performing structured pattern recognition, which makes deploying them for a task such as text/classification/sentiment analysis very feasible. CNNs have, for instance, been considered a powerful option for social media sentiment analysis, and this is why they tend to capture local dependencies and hierarchical features of text data. For example, CNNs have been used to determine positive or negative sentiment in movie reviews captured in the Internet Movie Database (IMDB) dataset

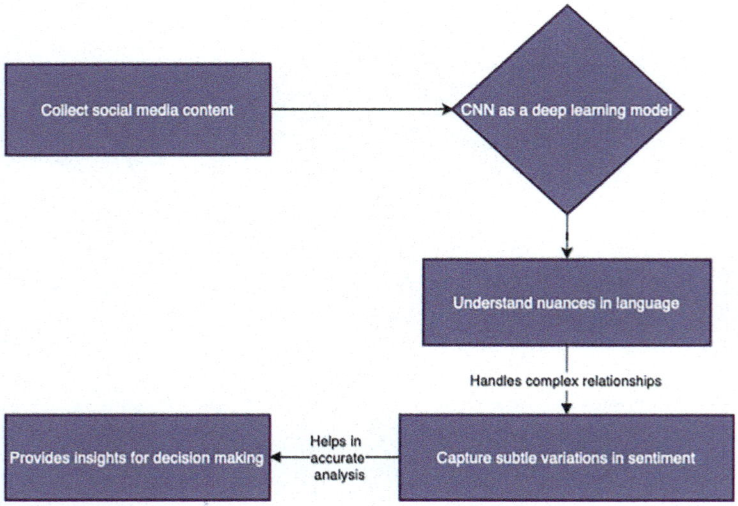

FIGURE 19.5 Flowchart of CNN for sentiment analysis. *CNN*, Convolutional neural network.

and achieved the best performance levels in comparison with other conventional machine learning models such as naive Bayes and support vector machines (Goodfellow, 2016).

The increasing penetration of social media has propelled the scientific community to carry out research on sentiment analysis. Efforts were made to improve methods of analyzing the users' sentiments with an emphasis on the intensity. However, this research illustrates the tendency toward considering other forms of assessment, not focusing on the positivity or negativity of responses, but rather analyzing them in aspects such as the strength and the specific feeling presented. This is especially important considering social media where such devotion is not only diverse but is also affected by interaction, whether active or passive, the trends at that time, or even the time itself. The paper explains why, to understand such phenomena, incorporation of fighting networks and Large Language Model (LLMs) becomes necessary. Social networking sites stand out as an important mode of communication today—this complements their comprehension by not only the message carried by the post but also different contextual connotations lending more perspectives to users' behavior and interaction on Twitter, Facebook, Instagram (Zhou et al., 2023).

NLP models, such as Bidirectional Encoder Representations from Transformers (BERT) and Robustly Optimized BERT Pretraining Approach (RoBERT), are more recent where it is possible to do very effective sentiment analysis, as they can comprehend and process the context of a longer piece of text because of the bidirectional attention mechanism they employ. Although transformer-infused deep learning models are known to offer high accuracy rates for sentiments expressed within even the most complex texts, they are, however, very power-consuming and computationally complex, which restricts their application in real-time applications (Long Tan et al., 2022) (Fig. 19.5).

Another study scrutinizes detection of hate speech on Twitter through sentiment analysis and machine learning techniques. By careful preprocessing—such as label verification and pattern elimination—and some exploratory forecasting, e.g., hypertag word cloud, they improved data quality for further model-training purposes. The researchers assessed the performance of four models: Random forest, support vector classifier (SVC), and logistic regression, and decision tree, indicating that the highest accuracy of 95% was reached by random forest and SVC models, followed closely by logistic regression and decision tree models. This work illustrates the applicability of sentiment analysis and machine learning for addressing issues of hate speech on the internet by assisting moderation of content in social networks (Shiri et al., 2023).

Study indicates that CNNs, which were initially developed for visual data, can easily transpose to a multimodal analysis. For instance an Instagram post can be a picture in an image and, in some of cases, Twitter containing a different message. The models that support such concepts can treat these domains independently—by employing conventional 2D CNNs for images and text-based CNNs or word vectors for captions and merging the results to predict the sentiment. This method allows for a much more complex appreciation of the sentiment being expressed in the multimedia posts (Nip & Berthelier, 2024).

Authoritatively reaching the advanced social media sentiments analysis by not only analyzing the text but also the images posted socially. The authors consider brand- or product-centric content and use a dataset to perform this classification through the use of deep CNNs (DCNNs), which allows for the detection and interpretation of the text embedded within images. The Stacked Deep Neural Network (SDNN) methodology in sentiment analysis proves to be very precise in classifying sentiment which brings out the power of D-CNN in multimodal analysis of social media posts (Paolanti et al., 2017).

TABLE 19.2 Comparative analysis of CNNs with other models.

Model	Examples	Description	Strengths	Limitations
Traditional models	Naive Bayes, SVM	Sentiment analysis, which is the detection of subjective opinions in natural language text, was first tackled using traditional feature-based machine learning models	Easier to implement, it works well given a small-sized and less complicated dataset	Fail to grasp the context fully; cannot avoid manual feature engineering, thus hampering scalability
RNNs	Vanilla RNNs	RNNs are specifically designed for sequence-based data, where the model remembers information over a sequence of text, making it valuable for predicting context in sequential text	Good for both simple sentiment transfer and sentence sequences	Takes longer to train since it is done sequentially; there is trouble dealing with the vanishing gradient's long-range dependencies
LSTM	LSTM, bi-LSTM	An RNN variant with memory cells that better capture long-term dependencies and therefore reduces the vanishing gradient problem	More effective in capturing context and dependencies compared to classic RNNs	Measurements using RNNs are, however, slower than training using CNNs and transformers, particularly on large datasets
Transformer models	BERT, GPT	The use of self-attention in transformer models allows for complex relationship capture where sequence is not necessary	Efficient to capture local and global dependencies very well; fast, as it can be trained in a parallel manner	Takes a lot of computing resources; DNN can be very complex, making it expensive and difficult to train the model

CNNs, convolutional neural networks; *LSTM*, long short-term memory; *RNN*, recurrent neural network; SVM, Support Vector Machine; DNN, Deep Neural Network.

19.8 Comparative analysis

Recent studies have shown that CNNs are commonly preferred for sentiment classification of text because of their speed and effectiveness, particularly when dealing with brief and focused content such as social media updates. Nevertheless, for tasks that call for comprehending dependencies across long distances, models such as BERT or Generative Pre-trained Transformer (GPT), which are based on transformers, gain popularity instead because of their self-attention mechanism at the core of the design, which enables them to tackle complex relations (Xu et al., 2022). Recurrent neural networks (RNNs) and long short-term memories (LSTMs), even though they are still inductively biased architectures, are becoming more and more theory-driven and slower because of their inherent sequential working (Table 19.2).

It has been proven in research that the transformer's capabilities are far superior to neural networks when used in a more subtle analysis of sentiment. CNN's, on the other hand, is efficient and works well when text is shorter than the resources available, in cases such as microblogs where posts are more than 160 characters such as Twitter or Instagram correlatively. Many also suggest different alternatives such as using simpler evolved forms of sentiment analysis or transformer models, which are not only well-shaped but also tuned to perform their functions effectively. References, such as comprehensive comparisons of deep learning models and their use in sentiment analysis, are also recommended for learning about these issues more thoroughly (Ram et al., 2024).

19.9 Future enhancements and advanced techniques

Outcomes like the outstanding steps that CNNs have made in sentiment analysis, as well as processing social media content, through their ability to capture spatial hierarchies, have borne fruits that are very great. While CNNs have put up quite good performances across many domains, areas in which improvements could be made and more sophisticated techniques adopted for more accurate classification of sentiments also exist. To-be-talked-about upgrades and advanced methods in the context of CNNs regarding sentiment analysis on social media content.

19.9.1 Combining CNNs with other models (hybrid approaches)

An exciting development is integrating CNNs with other architectural models, such as an RNN and transformers. CNNs can capture images effectively, whereas RNNs can recollect the order of images. However, transformers are good at remembering information that was present a long time ago. This could enhance the functionality of such a model in terms of sentiment analysis, as it would be able to look at the entire text comprehensively instead of just the content and structure, giving an understanding of optimized social media text (Wang & Singh, 2023).

CNN + RNN/LSTM: The differences are such that CNNs extract local features like *n*-grams and keywords, whereas RNNs/LSTMs capture sequential dependencies. Hence they excel at tasks where sentiment changes over time, like in the case of tweets or reviews. For example, a CNN–LSTM employs CNN layers and LSTM layers to extract local features and capture the sentiment flow within a sequence.

CNN + transformer: CNNs treat local text patterns, whereas transformers such as BERT provide a rich contextual understanding. This combination actually improves the sentiment analysis of complex and context-dependent social media content. For example: A CNN–BERT hybrid uses CNNs to extract more localized features like keywords and BERT for capturing broader sentence-level context to improve sentiment prediction.

19.9.2 Multimodal analysis

Sentiment analysis is improving in the way that it will not only be limited to text but also multimedia such as images and videos as well as audio that are becoming more common on services such as Instagram and TikTok. CNNs can be intended for multimodal analysis whereby aspects of sentiment are derived from visual images together with text. This is in coherency with the changing ways and the dynamics of content engagement and user participation in social networks (Al-Tameemi & Khalaf, 2024).

CNN for text + image analysis: Actually, CNNs were designed to be valid in the context of image classification but can easily be extended to posts with images and texts on social media. This manner captures the sentiment delivered within the visual and textual content. On Instagram a combination of an image and a caption can express a different sentiment than either could alone. Hybrid approaches can break CNN for text and image separately, generating one output that can be used to make final predictions about sentiment (Abdu et al., 2021).

CNNs for text + video: Other social video types have been deployed with audio cues along with visual cues. CNNs, especially 3D-CNNs for video, could leverage the spatial and temporal features of video frames toward indicating sentiment in speech, facial expression, and other visual cues. For example, an advertisement video may use words and body language to show feelings. A multimodal CNN model would consider the text, audio speaking tone, and visual body language or facial expression in determining a judgment of sentiment (Kim, 2022).

19.9.3 Ethical concerns with sentiment analysis

The expansion of sentiment analysis, particularly on social media sites, throws up important challenges to ethical concerns, including privacy, fairness in the use of data, and responsibility of algorithms. It is important to ensure ethics of artificial intelligence to deal with biases and to secure user privacy. The study also highlights these issues, recommending that clear and privacy-preserving models should be adopted considering the changes in data use policies and public concerns (Binns, 2018).

Bias and fairness: Anything that occurs by means of a sentiment model inherits the biases found in their training data to make wrong or unfair outcomes, especially concerning underrepresented groups. This is very important for fairness. For example, a CNN model trained on data from English-speaking regions might fail or be biased in non-Western cultures or other languages. To avoid such bias, multilanguage datasets must be used.

Responsibility in AI development: The development of AI includes ethical dilemmas, such as controlling sentiment models to avoid destructive usage toward public opinion formation or otherwise exploiting vulnerable entities. For example, in political context, sentiment analysis is used, but the risks of manipulation and misinformation are serious. Ethical AI development then has to adhere to principles such as transparency, accountability, and consent for the users.

19.10 Conclusion

Contrary to LSTM networks, which are based on recurrent networks and sequence relations for text classification tasks, CNNs have shown promising results for text sentiment classification, especially in social networks based on their benefits in automatic hierarchical feature extraction and complex text pattern learning. Processing rich unstructured data with little form of feature engineering is also another reason why such models are preferred to classifying general sentiments across diverse data sets. Nevertheless, it can be challenging because of factors such as noise and ambiguity of language. The increased performance on sentiment analysis, in as much as it involves social media content analysis calling for greater caution in the application of AI practices, will rely on improvements in hybrid models, multimodal approaches, and considerations of ethical issues.

References

Abdu, S. A., Yousef, A. H., & Salem, A. (2021). Multimodal video sentiment analysis using deep learning approaches, a survey. *Information Fusion, 76*, 204–226. doi:10.1016/j.inffus.2021.06.003.

Al-Tameemi, I., & Khalaf, S. (2024). A comprehensive review of visual–textual sentiment analysis from social media networks. *Journal of Computational Social Science, 3*, 1–72.

Binns, R. (2018). On the importance of transparency in machine learning and AI systems. In *Proceedings of the 2018 CHI Conference on Human Factors in Computing Systems* (pp. 1–12). Cambridge, MA, USA.

Conneau, A., Kiela, D., Schwenk, H., Barrault, L., & Bordes, A. (2017). Supervised learning of universal sentence representations from natural language inference data. In *EMNLP 2017—Conference on Empirical Methods in Natural Language Processing, Proceedings*. Association for Computational Linguistics (ACL) http://dblp.uni-trier.de/db/conf/emnlp/emnlp2017.html. doi:10.18653/v1/d17-1070 9781945626838 670 680.

Devlin, J., Chang, M. W., Lee, K., & Toutanova, K. (2018). BERT: Pre-training of deep bidirectional transformers for language understanding. *arXiv*. https://arxiv.org. doi:10.48550/arxiv.1810.04805.

Goodfellow, I. (2016). *Deep learning*.

Heaton, J. (2016). Ian Goodfellow, Yoshua Bengio, and Aaron Courville: Deep learning. In *Genetic programming and evolvable machines: 19* (pp. 305–307). The MIT Press. 800.

Hutto, C., & Gilbert, E. (2014). VADER: A parsimonious rule-based model for sentiment analysis of social media text. *Proceedings of the International AAAI Conference on Web and Social Media, 8*(1), 216–225. doi:10.1609/icwsm.v8i1.14550.

Kalchbrenner, N., Grefenstette, E., & Blunsom, P. (2014). A convolutional neural network for modelling sentences 1. In *52nd Annual Meeting of the Association for Computational Linguistics, ACL 2014—Proceedings of the Conference* (pp. 655–665). Association for Computational Linguistics (ACL). doi:10.3115/v1/p14-1062 9781937284725.

Kanungo (2023). A Feature Extraction Based Improved Sentiment Analysis on Apache Spark for Real-Time Twitter Data. Scalable Computing: Practice and Experience, 24.4, 847-855.

Kim, S.-M. (2022). Ethical AI for social media analysis: Addressing privacy and fairness concerns. *AI & Society, 37*(3), 459–470.

Kim, Y. (2014). Convolutional neural networks for sentence classification. In *EMNLP 2014—2014 Conference on Empirical Methods in Natural Language Processing, Proceedings of the Conference* (pp. 1746–1751). Association for Computational Linguistics (ACL). 9781937284961. doi:10.3115/v1/d14-1181.

Krizhevsky, A., Sutskever, I., & Hinton, G. E. (2012). ImageNet classification with deep convolutional neural networks. In *Advances in Neural Information Processing Systems, vol. 2* (pp. 1097–1105). Neural Information Processing Systems (NeurIPS) Foundation. 10495258.

Lecun, Y., Bottou, L., Bengio, Y., & Haffner, P. (1998). Gradient-based learning applied to document recognition. *Proceedings of the IEEE, 86*(11), 2278–2324. doi:10.1109/5.726791.

LeCun, Y., Kavukcuoglu, K., & Farabet, C. (2010). Convolutional networks and applications in vision. In *ISCAS 2010—2010 IEEE International Symposium on Circuits and Systems: Nano-Bio Circuit Fabrics and Systems* (pp. 253–256). doi:10.1109/ISCAS.2010.5537907.

LeCun, Y., Bengio, Y., & Hinton, G. (2015). Deep learning. *Nature, 521*(7553), 436–444. doi:10.1038/nature14539.

Liu, Y., Racah, E., Prabhat, Correa, J., Khosrowshahi, A., Lavers, D., Kunkel, K., Wehner, M., & Collins, W. (2016). Application of deep convolutional neural networks for detecting extreme weather in climate datasets. *arXiv*. doi:10.48550/arXiv.1605.01156.

Long Tan, K., Poo Lee, C., Sonai Muthu Anbananthen, K., Kian, M., & Lim, R. B.-L. (2022). A hybrid model for sentiment analysis with transformer and recurrent neural network. *IEEE Access, 10*, 21517–21525. doi:10.1109/access.2022.3152828.

Marinai, S., Gori, M., & Soda, G. (2005). Artificial neural networks for document analysis and recognition. *IEEE Transactions on Pattern Analysis and Machine Intelligence, 27*(1), 23–35. doi:10.1109/tpami.2005.4.

Medhat, W. (2023). Understanding sarcasm and ambiguity in social media text: A sentiment analysis challenge. *Journal of AI Research, 58*, 210–225.

Nip, J. Y. M., & Berthelier, B. (2024). Social media sentiment analysis. *Encyclopedia, 4*(4), 1590–1598. doi:10.3390/encyclopedia4040104.

Paolanti, M., Kaiser, C., Schallner, R., Frontoni, E., & Zingaretti, P. (2017). Visual and textual sentiment analysis of brand-related social media pictures using deep convolutional neural networks. In *Lecture notes in computer science (including subseries lecture notes in artificial intelligence and lecture notes in bioinformatics)* (pp. 402–413). Springer Verlag. http://springerlink.com/content/0302-9743/copyright/2005/10484. doi:10.1007/978-3-319-68560-1_36.

Rakhlin, A. (2016). Convolutional neural networks for sentence classification. *GitHub, 6*, 1746–1751.

Ram, N. R., Gautum, S., Jadeja, A., Joisar, H., & Rathore, N. (2024). Social media sentiment analysis using twitter dataset. *1st International Conference on Cognitive, Green and Ubiquitous Computing, IC-CGU 2024*. Institute of Electrical and Electronics Engineers Inc. http://ieeexplore.ieee.org/xpl/mostRecentIssue.jsp?punumber=10530616. doi:10.1109/IC-CGU58078.2024.10530694.

Sarker, A. (2021). Challenges in handling large-scale social media data for sentiment analysis. *Journal of Big Data, 8*(1).

Schmidhuber, J. (2015). Deep learning in neural networks: An overview. *Neural Networks, 61*, 85–117. doi:10.1016/j.neunet.2014.09.003.

Shiri, F. M., Perumal, T., Mustapha, N., & Mohamed, R. (2023). A comprehensive overview and comparative analysis on deep learning models: CNN, RNN, LSTM, GRU. *arXiv*. Malaysia arXiv https://arxiv.org . doi:10.48550/arXiv.2305.17473.

Srivastava, N., Hinton, G., Krizhevsky, A., Sutskever, I., & Salakhutdinov, R. (2014). Dropout: A simple way to prevent neural networks from overfitting. *Journal of Machine Learning Research, 15*, 1929–1958. http://jmlr.org/papers/volume15/srivastava14a/srivastava14a.pdf.

Symeonidis (2018). A comparative evaluation of pre-processing techniques and their interactions for twitter sentiment analysis. *Expert Systems with Applications, 110, 298-310.*

Triantafyllidou, D., & Tefas, A. (2016). Face detection based on deep convolutional neural networks exploiting incremental facial part learning. In *Proceedings—International Conference on Pattern Recognition* (pp. 3560–3565). Institute of Electrical and Electronics Engineers Inc. http://ieeexplore.ieee.org/xpl/conhome.jsp?punumber=10005450. doi:10.1109/ICPR.2016.7900186.

Vaswani, A., Shazeer, N., Parmar, N., Uszkoreit, J., Jones, L., Gomez, A. N., Kaiser, Ł., & Polosukhin, I. (2017). Attention is all you need. In *Advances in Neural Information Processing Systems* (pp. 5999–6009). Neural Information Processing Systems Foundation.

Wang, W., & Singh, A. (2023). Multimodal sentiment analysis for social media platforms: Leveraging visual and textual features. *Journal of Advanced Multimedia Processing, 28*(1), 45–57.

Xu, Q. A., Chang, V., & Jayne, C. (2022). A systematic review of social media-based sentiment analysis: Emerging trends and challenges. *Decision Analytics Journal, 3*, 100073. doi:10.1016/j.dajour.2022.100073.

Zhou, C., Li, Q., Li, C., Yu, J., Liu, Y., Wang, G., Zhang, K., Ji, C., Yan, Q., He, L., Peng, H., Li, J., Wu, J., Liu, Z., Xie, P., Xiong, C., Pei, J., Yu, P. S., & Sun, L. (2023). A comprehensive survey on pretrained foundation models: A history from BERT to ChatGPT. *arXiv* United States arXiv https://arxiv.org . doi:10.48550/arXiv.2302.09419.

Chapter 20

A compact square shaped dielectric resonator antenna used for early diagnosis of breast tumor

Gagandeep Kaur[a,b] and Komalpreet Kaur[b]

[a] *Department of Emerging Technologies, Chandigarh, Engineering College, Landran, Mohali, Punjab, India,* [b] *Department of Computer Science and Engineering, Chandigarh, Engineering College, Landran, Mohali, Punjab, India*

20.1 Introduction

20.1.1 Microstrip-fed dielectric resonator antenna and method of forming 2D image of breast's cancerous cell

20.1.1.1 Technical field

This invention relates in general to detection of breast cancer using antennas and more specifically to dielectric resonator (DR) antennas.

20.1.1.2 Background

Nowadays, breast cancer is one of the most fatal diseases among females. That mostly occurs when some unwanted malignant tissues start growing around the healthy body tissues. To detect the presence of the breast tumor, several clinical methods are available, such as X-ray mammography, ultrasound, and magnetic resonance imaging (MRI) (Adnan, 2010; Al-Zuhairi & Gahl, 2017; Matin & Mohd Ali, 2008). But all these existing methods have some shortcomings, like ionized exposure of the radiation along with painful, costly, and time-consuming treatment, etc (Kaur et al., 2024; Khosla & Malhi, 2022; Sharma & Dishant, 2018). Therefore these methods are not used widely for the detection of breast cancer. An alternate method, i.e., monostatic microwave imaging (MWI), is gaining the attention of numerous researchers because of its simple, nonionized exposure; cost-effectiveness; and high image resolution of the scanned tissues_(Ansari et al., 2008; Azim et al., 2016). It is essentially working on the principle of large deviation in dielectric properties of malignant and healthy tissue. This method is preferred basically because of its simplicity; cost-effectiveness; confronts (provides least amount anxiety to the patient); and availability to detect small-sized cancerous cells_(Azim et al., 2011; Bhaskaran & Krishnan, 2018; Bohra & Shaikh, 2016; Choudhury, 2014).

The proposed DRA is designed and modeled on a mechanically robust FR4 lossy substrate that has total dimensions of $22 \times 21 \times 1.57$ mm^3 (Fig. 20.1). It consists of four different layers; defected ground structure (DGS), followed by FR4 substrate ($\varepsilon_r = 4.4$); feedline; and rectangular-shaped DR of alumina ($\varepsilon_r = 9.8$). The initial dimensions for the proposed DR have been selected using dielectric waveguide model (DWM), as given in the following dominant mode (Eq. 20.1) (Moore, 2001) :

$$f_{mnl} = \frac{v}{2\pi\sqrt{\varepsilon_r}}\sqrt{k_x^2 + k_y^2 + k_z^2} \quad (20.1)$$

where$v = 3 \times 10^8$ m/s, $\varepsilon_r = 9.8$, and $k_x = m \times \pi/d1$, $k_y = n \times \pi/d2$, and $k_z = l \times \pi/2 \times d$ represent the wave number along x, y, and z axis, respectively, of the proposed DRA and get resonant frequency of 12 GHz.

Computational Intelligence in Mechatronics: Solving Real-World Problems in Electronic Systems Design. DOI: https://doi.org/10.1016/B978-0-443-27580-7.00025-1

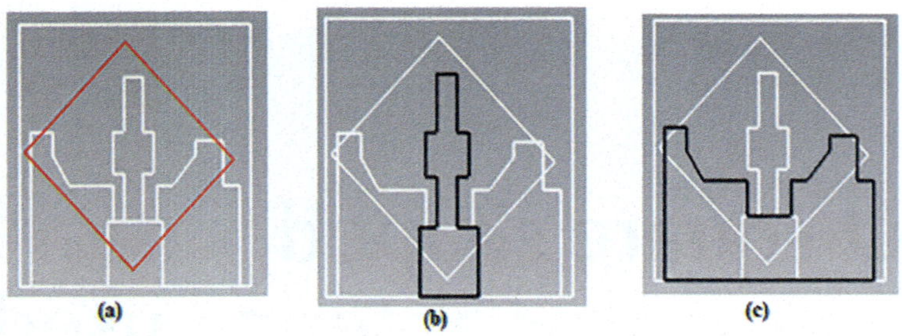

FIGURE 20.1 Different layers of proposed DRA: (A) dielectric resonator, (B) feedline, and (C) ground plane.

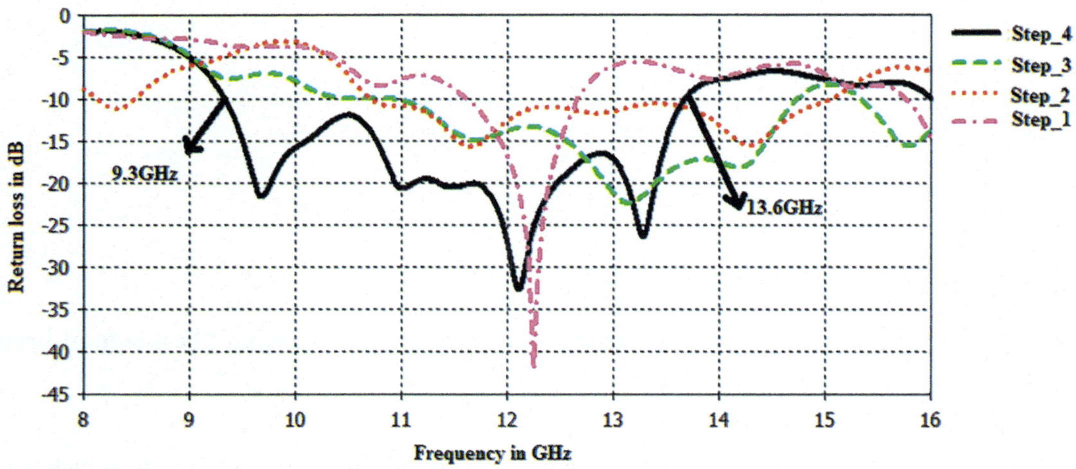

FIGURE 20.2 S_{11} parameter results of the four intermediate DRA geometries.

20.2 Design procedure followed to attain the desired UWB characteristics

Lots of intermediate steps are followed to achieve the preferred ultrawide band (UWB) characteristics (9.3–13.6 GHz) from the conventional DRA, as illustrated in Fig. 20.2 and their corresponding responses in terms of *S*-parameter, as illustrated in Fig. 20.3. The conventional geometry (Step_1), i.e., rectangular DR, simple microstrip feedline, and full ground plane (A) allow the antenna to excite only one resonances frequency at 12.2 GHz with 11.6–12.6 GHz of impedance bandwidth. For bandwidth improvement, two rectangular stubs of dimensions of 3.7 × 3.5 and 5 × 6.5 mm^2 are incorporated at the bottom endpoint of the feedline (Step_2). It allows the DRA to excite wider frequency band from 10.6 to 14.9 GHz but with poor impedance matching. To get better results in terms of bandwidth and return loss, DGS technique is used at the surface of ground plane (Step_3) as depicted in (C), which allows the antenna to excite 10.7–14.7 GHz frequency band. Finally the proposed rectangular DRA is rotated at angle of 45 degrees and placed over the modified feedline (Step_4). It allows the antenna to stimulate desired UWB characteristics from 9.3 to 13.6 GHz frequency band (*solid black color* plot).

20.3 Fabrication and testing of the proposed DRA

The proposed prototype of DRA is fabricated on a 22 × 28 mm^2 of FR4 substrate using wet etching photolithography technique. On bottom and top of the FR4 substrate a layer of copper material (0.035 mm) is deposited to be used as ground structure and feed point, respectively. On the top of the feedline a rotated square-shaped DRA of alumina (4 mm) is fixed as depicted in (A and B). The DRA is cut into the desired shape by water-jet cutting technique_(Mahmud, 2018; Mahmud et al., 2016). The validity of the simulated results in terms of *S*-parameter, voltage standing wave ratio (VSWR), gain, and radiation pattern is tested on Agilent's VNA (vector network analyzer) model no. E-5063A, having frequency range of 0.1 MHz to 18 GHz, by connecting a SMA female connector of 50 Ω impedance.

Fig. 20.4 shows the *S*-parameter comparison plot between measured *S*-parameter plots for two cases (1) with placement of tumor, i.e., 8.72–9.23 and 10.61–12.23 GHz (2) with placement of tumor, i.e., 8.64–9.45 and 10.5–12.3 GHz inside the

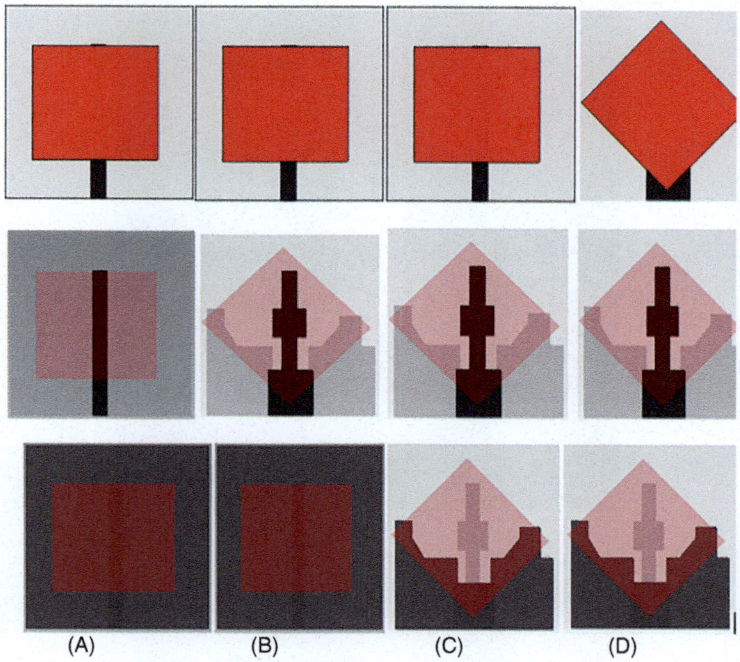

FIGURE 20.3 Intermediate design steps for the construction of proposed DRA: (a) basic design (b) modified feedline (c) DGS in ground plane (d) Proposed final prototype.

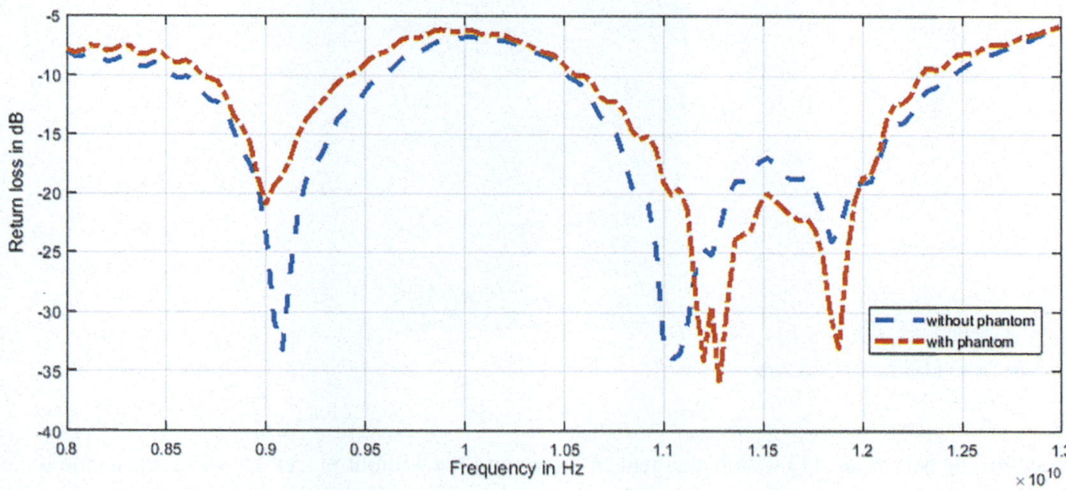

FIGURE 20.4 *S*-Parameter comparison plot between measured *S*-parameter plots for two cases: (1) with placement of tumor and (2) with placement of tumor.

artificial breast phantom. Although testing the slight discrepancies are observed between the simulated and measured result, this is because of some reflection of the object placed in close vicinity of the VNA or because of manufacturing errors.

20.3.1 *S*-Parameters responses

20.3.1.1 Monostatic MWI setup and reconstruction of the 2D image for the detection of breast tumor

For the simulation process in CST software the designed DRA is placed parallel to the breast phantom at a distance of 8 mm, as depicted in Figs. 20.5 and 20.6. The breast phantom is made up of agar–agar used as skin (4 mm thickness), followed by petroleum jelly as fat (34 mm thickness) and wheat as tumor (8 mm thickness). The proposed DRA is rotating around

FIGURE 20.5 Fabricated DRA sensor.

FIGURE 20.6 Side view of fabricated DRA sensor.

the breast phantom for two cases: (1) with placement of tumor and (2) without placement of tumor inside breast phantom from 0 to 2π in the azimuthal plane at approximate focus points represented by *black dots* such as $p1$, $p2$, $p3$... and so on, depicted in Fig. 20.7A, and 0–π in the elevation plane at focus points such as $r1$, $r2$, $r3$...., depicted in Fig. 20.7B, with fixed distance of 8.9 mm.

For the removal of unwanted clutter signals, i.e., skin and fat, the subtraction method is used as specified in Eq. (20.2) (Dowla, 2014). In this method a tumor of 8 mm is inserted in the phantom at (0, 0) coordinates and reflections are recorded at each position named $S_{11}{}^{\text{presence}}$. Then reflections that are recorded without the placement of any tumor in breast phantom named $S_{11}{}^{\text{absence}}$ (Karli & Ammor, 2014).

$$S_{11}(x, y) = S_{11}^{\text{presence}}(x, y) - S_{11}^{\text{absence}}(x, y) \quad (20.2)$$

These results are utilized in different beam-forming algorithms to get a clue about the presence of the breast tumor by amplifying only tumorous signals. But the subtraction method is not practically reasonable to distinguish the position of the breast tumor_(Lazaro et al., 2011; Pozar, 1985; Tiang et al., 2013). So, other interference removal beam-forming algorithms have been used to find out the position of the breast tumor that is explained briefly in subsection of this section (Fig. 20.8).

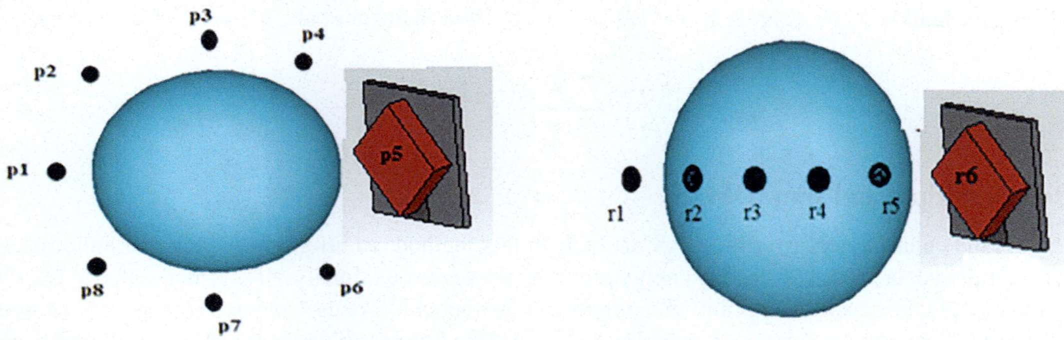

FIGURE 20.7 Microwave imaging setup in CST software: (A) azimuthal view and (B) elevation view.

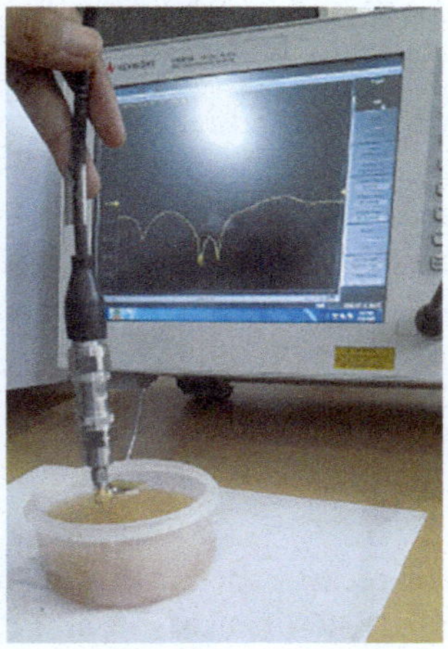

FIGURE 20.8 Testing of DRA with artifical sensor.

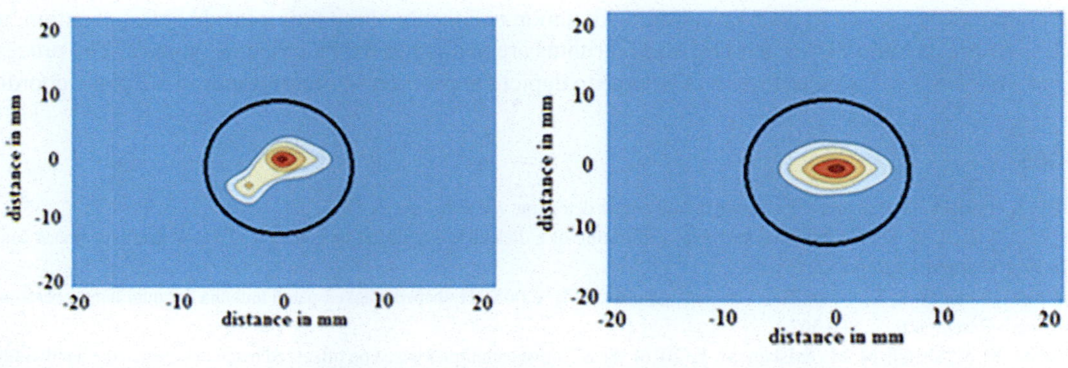

FIGURE 20.9 2D image reconstructed using (A) DAS and (B) DMAS algorithms. *DAS*, Delay and sum; *DMAS*, delay multiply and sum.

20.3.2 Delay and sum (DAS)

In this algorithm, initially the received *S* parameter responses are recorded at individual antenna positions (e.g., at $p1$, $p2$, $p3$, … and $r1$, $r2$, $r3$, …) by rotating it around the breast phantom and time delays for all received signals are calculated based on different position of transmitter and receiver DRA named S_{11} (x, y). Afterward, these time-delayed signals are summed together as $S(t)$ according to Eq. (20.3) (Ghassemi et al., 2008). And get a noticeable difference among the backscattered values of presence and absence of the tumor. That amplifies the energy level of the tumorous signals only in terms of return losses by adding signals them coherently, and *S*-parameter responses from other cells, i.e., healthy cells, are added incoherently_(Garg et al., 2011). Same procedure is done for all the focal positions within the breast phantom. Finally a 2D image of breast tumor is reconstructed in MATLAB as shown in Fig. 20.9A and observed position of the tumor is at

Fig. 20.9B (Bohra & Shaikh, 2016; Kaur & Kaur, 2017a, 2017b; Uma Shankar et al., 2014) in *XY* coordinates.

$$S(t) = \sum_{i=1}^{N}\sum_{j=1}^{N} s_{ij}(t + \text{Delay}) \tag{20.3}$$

20.3.3 Delay multiply and sum (DMAS)

DMAS method is the enhancement over the DAS method. In this method, addition of the pairwise multiplications of the recorded back-scattered data is used to amplify energy profile of the tumorous signals only, as mentioned in Eq. (20.4) (Kaur, 2015; Kaur et al., 2014). For each focal point, time alignment synthetically focuses received signals at that point. The 2D microwave image of breast tumor is reconstructed in the MATLAB software to get a clue about its position, as illustrated in Fig. 20.9B at (1.5, 0) coordinates in *XY* plane (Lazaro et al., 2011).

$$Y(t) = \sum_{i=1}^{(N\times N)-1}\sum_{j=2}^{N\times N}\{X_i(t)X_j(t)\} \tag{20.4}$$

where w is window size, N is number of DRA rotation, and i and j represent the different position of antenna around the breast model.

In Fig. 20.9A and B, the *red color* (existence of tumor) and *blue color* (absence of tumor) potion characterizes the utmost and least amount of intensity of the processed data, respectively_(Kaur et al., 2017; Malik et al., 2013; Patil, 2012; Paul et al., 2015). Moreover, it is observed that DMAS provides a 2D breast image with better quality and approximately same positions as that of actual placement of tumor, i.e., (0, 0) by removing the back clutter, i.e., skin or fat reflections, as compared to DAS algorithm.

20.4 Conclusion

The proposed UWB DRA for the detection of breast cancer using mono-static microwave imaging is proposed and investigated in this research article. The proposed DRA consists of an FR4 substrate, microstrip feedline, and a ground plane with DGS technique, designed to achieve 70.96% fractional bandwidth and 5 dB peak gain for resonant frequency of 10.8 GHz along with VSWR < 2. To detect the breast cancer, proposed DRA is rotated around the phantom with $\pi/6$ interval in elevation (0–π) and azimuthal planes (0–2π) to record the S_{11} signals. And a significant deviation in S_{11} responses is compared among two cases: first, breast phantom with tumor, and then without tumor inside breast phantom, which helps in the detection of breast tumor. DAS and DMAS algorithms are applied to the recorded S_{11} signals. The processed data is then plotted in MATLAB software to form a 2D image to depict the position of breast tumor in *XY* plane coordinates.

References

Adnan, S. (2010). A compact UWB antenna design for breast cancer detection. *PIER Letters, 6*, 129–132.

Al-Zuhairi, D. T., & Gahl, J. M. (2017). Simulation design and testing of a dielectric embedded tapered slot UWB antenna for breast cancer detection. *Progress in Electromagnetics Research, 79*, 1–15.

Ansari, J. A., Singh, P., Dubey, S. K., Khan, R. U., & Vishvakarma, B. R. (2008). H-shaped stacked patch antenna for dual band operation. *Progress in Electromagnetics Research B, 5*, 291–302.

Azim, R., Aldhaheri, R. W., Sheikh, M. M., & Islam, M. T. (2016). An effective techniques based on offset fed patch to enhance the bandwidth of microstrip planar antenna. *Microwave and Optical Technology Letters, 58*, 1221–1226.

Azim, R., Islam, M. T., & Misran, N. (2011). Design of a planar UWB antenna with new band enhancement technique. *Applied Computational Electromagnetics Society Journal, 26*, 856–862.

Bhaskaran, D., & Krishnan, R. (2018). Breast tissue tumor analysis using wideband antenna and microwave scattering. *IETE Journal of Research, 1*, 1–11.

Bohra, T., & Shaikh, T. (2016). UWB microstrip patch antenna for breast cancer detection. *International Journal of Advanced Research in Electronics and Communication Engineering, 1*, 88–91.

Choudhury, S. (2014). Effect of dielectric permittivity and height on a microstrip-fed rectangular patch antenna. *IJECT, 5*, 1–12.

Dowla, F. (2014). Ultra-wideband communication. In F Dowla (Ed.), *Handbook of RF and wireless technologies* (pp. 325–328). Newnes.

Garg, R., Bharti, P., & Bahl, I. (2011). *Microstrip antenna design handbook*. Artech House.

Ghassemi, N., Rashed-Mohassel, J., Neshati, M. H., Tavakoli, S., & Ghassemi, M. (2008). A high gain dual stacked aperture coupled microstrip antenna for wideband application. *Progress In Electromagnetics Research B, 9*, 127–135.

Karli, R., & Ammor, H. (2014). Miniaturized UWB microstrip antenna with T-Slot for detecting malignant tumors by microwave imaging. *International Journal of Microwave and Optical Technology, 9*(3), 214–220. http://ijmot.com/ijmot/uploaded/1452837518.pdf.

Kaur, A. (2015). Semi spiral G-shaped dual wideband microstrip antenna with aperture feeding for WLAN/WiMAX/U-NII band applications. *International Journal of Microwave and Wireless Technologies, 8*, 931–941.

Kaur, A., Khanna, R., & Kartikeyena, M. V. (2014). *A stacked rectangular MSA with defected ground structure for IEEE 802.11b/g bands and Wi-Max applications*. ICMARS (IEEE).

Kaur, A., Khanna, R., & Kartikeyena, M. V. (2017). A multilayer dual wideband circularly polarized microstrip antenna with DGS for WLAN/Bluetooth/ZigBee/Wi-Max/IMT band applications. *International Journal of Microwave and Wireless Technologies, 9*, 317–325.

Kaur, G., & Kaur, A. (2017a). Design of a slotted micro-strip patch antenna with DGS for an UWB applications. *International Conference on Advancements in Engineering and Technology, 5*, 39–41.

Kaur, G., & Kaur, A. (2017b). Design of a stacked aperture coupled microstrip antenna for UWB applications. *Journal of Telecommunication, Switching Systems and Networks, 4*, 12–16.

Kaur, G., Kaur, A., Kaur, S., et al. (2024). Dual port UWB MIMO Dielectric Resonator Antenna (DRA) with Airplane Shaped Defective Ground Plane (DGS) with Improved Isolation Level. *Wireless Pers Commun, 139*, 1225–1237. https://doi.org/10.1007/s11277-024-11664-5.

Khosla, D., & Malhi, K. S. (2022). An ultra-wideband dielectric resonator antenna for WSN based IoT applications in agriculture. *International Journal of Sensors Wireless Communications and Control, 12*(4), 281–291.

Lazaro, A., Villarino, R., & Girbau, D. (2011). Design of tapered slot Vivaldi antenna for UWB breast cancer detection. *Microwave and Optical Technology Letters, 53*(3), 639–643. doi:10.1002/mop.25770.

Mahmud, M. Z., Islam, M. T., & Samsuzzaman, M. (2016). A high performance UWB antenna design for microwave imaging system. *Microwave and Optical Technology Letters, 58*, 1824–1831.

Mahmud, M. Z. (2018). Microwave imaging for breast tumor detection using unipolar AMC based CPW fed Microstrip patch. *IEEE Access, 6*, 44763–44775.

Malik, J., Kalaria, P. C., & Kartikeyan, M. V. (2013). Complementary Sierpinski gasket fractal antenna for dual band WiMAX/WLAN (3.5/5.8 GHz) applications. *International Journal Microwave Wireless Technology, 5*, 499–505.

Matin, M. A., & Mohd Ali, M. A. (2008). Design of broadband stacked E-shaped patch antenna. *IEEE ICMMT Proceedings, 3*, 1–2.

Moore, S. K. (2001). Better breast cancer detection. *IEEE Spectrum, 38*(5), 50–54. doi:10.1109/6.920031.

Patil, P. V. (2012). Enhancement of bandwidth of rectangular patch Antenna using two square slots techniques. *International Journal of Engineering Sciences & Emerging Technologies, 3*, 1–12.

Paul, L. C., Hosain, M. S., Sarker, S., Prio, M. H., Morshed, M., & Sarkar, A. K. (2015). The effect of changing substrate material and thickness on the performance of inset feed microstrip patch antenna. *American Journal of Networks and Communications, 4*, 54–58.

Pozar, D. M. (1985). A micro strip antenna aperture coupled to a microstrip line. *Electronics Letters, 21*, 49–50.

Sharma, N., & Khosla, D. (2018). *A compact two-element U-shaped MIMO planar inverted-F antenna (PIFA) for 4G LTE mobile devices. 2018 Fifth International Conference on Parallel, Distributed and Grid Computing (PDGC)*, IEEE.

Tiang, S. S., Sadoon, M., Tareq, F. Z., Aim, F. M., & Abdullah, Z. M. (2013). Radar sensing featuring biconical antenna and enhanced delay and sum algorithm for early stage breast cancer detection. *Progress in Electromagnetics Research B, 46*, 299–316.

Uma Shankar, M., Gajanand, J., & Anubhav, K. (2014). E-Shaped multilayer aperture coupled patch antenna with notching characteristics for UWB applications. *International Journal of Computer Applications, 2*, 7–11.

Chapter 21

Aerosol-induced visibility reduction: Mechanisms, health impacts, and enhanced imaging techniques

Jhinuk De[a] and Tannistha Pal[b]

[a] *Department of Chemistry, National Institute of Technology Agartala (NITA), Agartala, Tripura, India,* [b] *Computer Science & Engineering Department, National Institute of Technology Agartala, Agartala, Tripura, India*

21.1 Introduction

Haze formation produces aerosol particles, which in turn affects fog. For instance, these particles impact a wide range of atmospheric constituents and processes, significantly changing the atmosphere (Mohan & Payra, 2014). They serve as water vapor condensation nuclei, promoting fog droplet formation and thickening fog. Aerosols improve the overall clarity of the air while scattering and absorbing incoming radiation, which reduces visibility in overcast conditions. It includes how they affect visibility in foggy conditions, which varies depending on size, composition, and concentration (Jung et al., 2015). Aerosols that absorb water may enlarge the fog particles as they take in moisture, lowering visibility. Correlations between aerosols and their fog slowly change the regional meteorological conditions, significantly interfering with air quality in relevant areas and human health and transportation security.

Given the climate effect, choking air quality, and fog visibility deterioration, it aimed to understand the intricacies of aerosol–fog interaction as a precursory step for formulating air-quality control strategies and location-specific urban weather forecasting models. The effect of aerosol particles on visibility is essential in many areas, such as environmental studies, and plays a critical role in the aviation and transportation sectors. Those particles are the subject of a study on how they either make for clearer air or blur our vision during fog. The impacts of this include the action of aerosol particles advected into a water droplet or in a fog, which can cause its dispersal and, therefore, reduced visibility. Therefore, in this study, we investigate aerosols' physical and chemical properties that can lead to efficient scattering and absorption over foggy conditions (Lyu et al., 2021). The work will also investigate how varying concentrations and compositions of aerosol particles impact the reduction in visibility resulting from fog more deeply. On a more granular level, it examines the processes by which these structures come together with chemistries and some of what that means. The roles of CO_2, O_2, and N_2 in these mechanisms are then considered. In their research the scientists analyzed how aerosol particles and water droplets in fog mix and how much this mixing scatters or absorbs light. It also studies the chemical interactions of aerosols with gases in the atmosphere and explores how such mechanisms could influence fog formation and life. The study also looks at possible consequences for human health and the environment in areas experiencing a decrease in visibility as a by-product of the interaction of aerosol and fog (Strauss et al., 2010).

21.1.1 Aerosol particles: Definition and sources

Aerosols are small, solid, or liquid droplets suspended in air. These particles are also from natural outcomes such as dust storms, sea spray, volcanic eruptions, and human activities like emissions because of industrial processes, etc., and biomass combustion. Aerosols differ largely in size, content, and concentration, resulting in different types of interactions and environmental consequences (Mahowald et al., 2017). Depending on their size and environment, they can stay in the air

Computational Intelligence in Mechatronics: Solving Real-World Problems in Electronic Systems Design. DOI: https://doi.org/10.1016/B978-0-443-27580-7.00014-7

long after leaving a body, for hours and even weeks. Aerosols (or atmospheric particles) are essential for other atmospheric processes such as cloud formation, precipitation distribution, and Earth's energy balance. These diverse sources and physical characteristics make aerosols an inherently complex but crucial component of the study of our atmosphere to evaluate climate change and air quality. These particles affect human health and the purity of the air, as fine particulates can penetrate deeper into human lungs, leading to respiratory problems. The aerosols would reduce the clarity of the atmosphere, leading to some haziness and blurriness in the sky. Dust, soot, and pollen particles are important, as they allow moisture to condense out onto clouds. However, other particles—like the fine particulate matter defining aerosols—absorb sunlight or reflect it into space, creating heat in the atmosphere. Aerosols greatly impact atmospheric chemistry through their involvement in intricate reactions that have important sink effects on trace atmospheric gases. These reactions can, among other things, create additional pollutants that are not emitted directly during the initial operation of a vehicle, and they can drive changes in the lifetimes of various airborne chemical compounds. Aerosols can potentially affect ecosystems through long-range transport and deposition in terrestrial and aquatic environments, where they can alter biodiversity and nutrient cycles (Palacios-Peña et al., 2020). As cloud condensation nuclei (CCN), aerosols can alter the processes of cloud-drop formation and lead to changes in the characteristics or number of clouds and rainfall patterns. Aerosols affect clouds by absorbing solar radiation; they lead to the heating of the surrounding air, but owing to their small size and short lifetime, diffuse light causes them to change and scatter colors. In addition, aerosols are an increasing health problem, as fine particulate matter can penetrate the respiratory system and cause breathing and heart problems.

21.2 Mechanisms of visibility degradation

21.2.1 Light scatters and absorption

Visibility is controlled by the scattering and absorption of the incident light, which has been scattered inside the aerosol particles. These particles scatter light in all directions as it hits them, both from sunlight and artificial lighting. As a result the amount of direct light that passes freely through the atmosphere is reduced, and this scattering decreases clarity, creating hazy shadows. Aerosols, likewise, act as CCNs that lead to expanded cloud cover and altered rainfall patterns and effectively change the Earth's climate system on a regional and global scale. Composition and size distribution control the optical properties of aerosols and, therefore, their impacts on visibility/atmospheric radiation balance (Lyu et al., 2022). However, we must understand that these processes should be included in climate models and utilized by policymakers to mitigate air pollution and its effects on human health, ecosystem services, etc. The complex interplay between aerosols and the atmosphere is another aspect of Earth's energy budget. Aerosols directly and indirectly impact radiative forcing by reflecting incoming solar radiation into space (which results in cooling) or absorbing light (which leads to warming). The oversimplified conclusion is that aerosols cool, but the net effect of all aerosols can be cooling or heating depending on what they are made out of, how large they are, and where they reside in the atmosphere.

Rayleigh scattering: Rayleigh scattering happens when particles in the air are much smaller than the wavelength of light, leading to shorter wavelengths being scattered much more. This explains why the sky appears blue, as air scatters blue light more effectively than red light. The same process produces red hues in sunsets, in which light from a setting (or rising) sun must travel through more of the atmosphere to reach you as its angle is very low, allowing the longer wavelengths to make it through. It is important to check Rayleigh scattering to build accurate atmospheric optics models and examine remote sensing data in environmental and climate studies. This is critical information for analyzing atmospheric processes and how they impact the environment.

Mie scattering: Spherical aerosol particles of similar diameter to the wavelength of light scatter light uniformly over all wavelengths. Because of how scattered light from Mie scattering is evenly distributed, we experience it as a white haze covering the sky. Mie scattering is essential for optical haze, fog, and atmospheric aerosols. In contrast to the randomly scatterable light, Mie is much more forward-biased in terms of scattering direction because it allows most of the light to be scattered forward. A unit volume of suspended illuminated particles is represented in Fig. 21.1.

21.3 Condensation nuclei in fog formation

Fog is mostly made up of aerosol particles. Fog is formed when water vapor cools and condenses upon particles in the air, creating tiny droplets. Aerosol particles, which are responsible for much of the chemical pollution in the atmosphere, have a major influence on droplet size and concentration (which, in their harmless state, predetermines the visibility range); the presence of an increased number of small droplets increases the light scattering and thus impacts further on fog particle opacity. Fog, or its formation and duration, is influenced by aerosols (Ziemann, 2011). In turn, these particles can act as nuclei

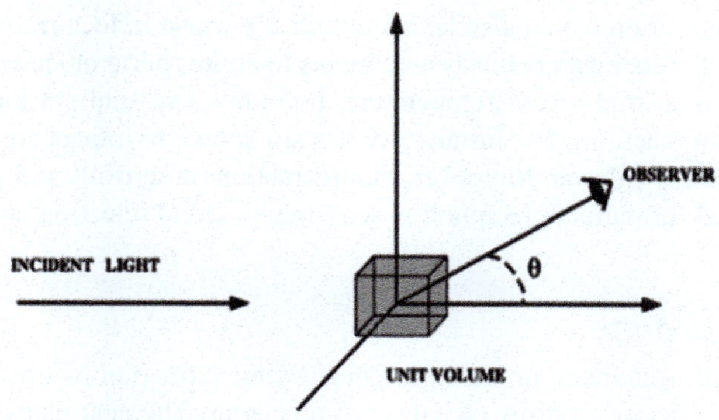

FIGURE 21.1 A unit volume of suspended particles illuminated.

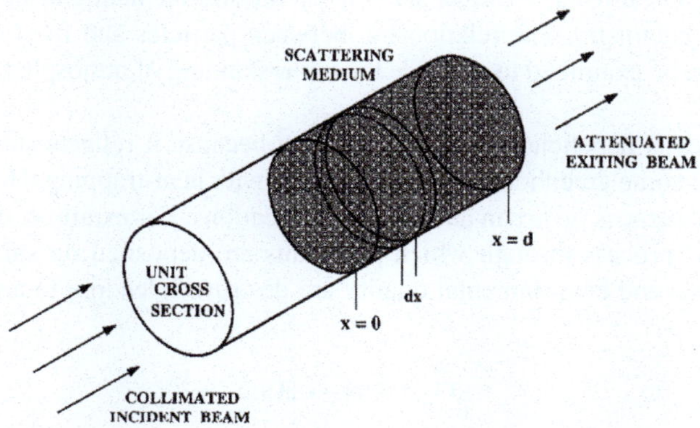

FIGURE 21.2 Attenuation of suspended particles.

for condensation and cloud formation, which likely influences the complete lifecycle of fog events. In addition, the chemical composition of aerosols can change the hygroscopic properties of fog droplets, changing the rates at which they grow and even their stability. The feedback effects because of the interaction of aerosols with fog droplets can, thus, lead to complex initiation and suppression of fog events, potentially lengthening the duration or further altering the spatial patterns of fog. In addition, fog layers' vertical configuration alongside the vertical distribution of aerosols in the atmosphere can modify their thickness and optical properties. The complex system of connections between aerosols and fog also demonstrates that it is necessary to consider both components and modeling of atmospheric systems and meteorological prediction schemes in such studies.

Hygroscopic: Many particle species in the atmosphere can take up (absorb) water from the gas phase and become increasingly heavier or express changes in phase, with apparent size changing as a function of humidity. Meanwhile, in foggy conditions with high relative humidity, these hygroscopic particles can swell dramatically through water uptake, boosting the ability to scatter light. It also makes the light scatter even more, reducing visibility. Because aerosols are hygroscopic, they can also modify the evaporation rate of fog droplets, thus lengthening the duration of a fog event. In addition, it can set off a feed-forward loop by absorbing more water, increasing its cloud condensation nucleus functionality. The complex synergy between aerosol composition and hygroscopicity with fog dynamics necessitates a deeper understanding of these drivers to predict perspectives accurately for effective management strategies. In addition to fog development the hygroscopic properties affect clouds' bulk properties and help control precipitation levels. Both significantly influence the water-vapor mixing ratio in the lower free troposphere, the hygroscopicity of coarse mode aerosols, and the subsequent combined enhancement effect from both fine and coarse particles. Secondary organic aerosol (SOA), on the other hand, is most likely produced by reactions between hygroscopic aerosols and air pollutants. This is problematic in parameterizing how air pollutants are associated with fog formation. The attenuation of suspended particles is depicted in Fig. 21.2.

21.3.1 Aerosols and gas-phase chemistry

Chemical reactions on aerosol particles in the atmosphere can alter their chemical properties, reducing visibility. Prominent among these are CO_2, O_2, and N_2. Some of these particles and gases combine through complex chemical processes to form SOAs, which will perturb fog formation and persistence in higher concentrations. If these aerosols have gas-phase chemistry, the atmosphere's composition would also be altered for the worse in localized or regional regions, as far as air quality is concerned. These detailed chemical pathways are inputs to atmospheric models that can help predict and control fog-related issues in urban and industrial areas. Temperature, humidity, and sunlight intensity are a few environmental factors influencing these chemical reactions. For instance, SOAs are known to impact atmospheric optical properties and, consequently, visibility and radiative balance. Moreover, the interaction of aerosols and gaseous-phase (I) chemistry has important consequences for cloud formation, precipitation scavenging, cloud selection, and climate dynamics on a global scale (Fan et al., 2022).

21.3.1.1 Sulfate aerosol build-up

Sulfate aerosols are predominantly generated in the atmosphere by the oxidation of sulfur dioxide released during fossil fuel combustion, volcanic eruptions, and an array of industrial processes. The next phase involves the conversion of SO_2 to sulfate aerosols through oxidation with oxygen; these particles, in turn, help in cloud condensation under appropriate conditions, hence causing smog. The effects of sulfate aerosols on the soluble nature of other pollutants in conversion with fog water shift further isotropic chemistry. The relationship between particles and fog formation and between gases and fog formation is complex. It must be examined to enhance the understanding of atmospheric quality in fog-prone cities and regions.

The formation of the particles is beneficial in regional cooling because it reflects sunlight into space, leading to less heat and hence reduced effects of some greenhouse gases associated with heat-trapping. However, the cooling effects of the sulfate aerosols have some consequences on urban and industrial locations. For example, the relationship between aerosols and fog formation speeds up the process through which pollutants are deposited on various solid surfaces. As a result, plant life, human-made structures, and environmental quality are destroyed, leading to artificial and natural effects on air pollution.

$$SO_2 + OH\cdot \rightarrow HSO_3$$

$$HSO_3\cdot + O_2 \rightarrow SO_3 + HO_2$$

$$SO_3 + H_2\,O \rightarrow H_2SO_4$$

21.3.1.2 Nitrate aerosol formation

Nitrogen oxides are produced by vehicles, and industrial emissions are being emitted into the environment. Then, they interact with atmospheric moisture to create nitric acid. This acid is followed by transformation into nitrate aerosols, which increase the acidity of fog water, enhancing the environmental effect of air pollution. The presented information demonstrates a close interaction between human-made emissions, atmospheric chemistry, and fog development. Therefore a comprehensive approach is required for inclusion in air quality management plans. These data are essential for designing effective preparatory measures and enhancing the quality of air quality forecasting in urban and industrial areas where the probability of fog occurrence is high. The creation of nitrates affects the comprehensive composition of environmentally hazardous particulate matter in the air, impacting air cleanliness and leading to large-scale climate change. Nitrate aerosols can serve as nuclei for condensation and cloud formation. In this case the properties of clouds may be altered, and precipitation changes can affect large territories. It is also noteworthy that the visibility in urban air is often reduced because of the generation of nitrates in the atmosphere, known as urban haze. The following equation describes the chemical reaction of nitrogen oxide with a hydroxyl radical:

$$NO_2 + OH\bullet \rightarrow HNO_3$$

Moreover, the reaction of nitrogen oxide with 1/2 molecule of molecular oxygen helps transform the first component into nitrogen dioxide—$NO + 1/2O_2 \rightarrow NO_2$. In this way, nitrogen oxides participate in synthesizing nitric acid and the oncoming creation of nitrate aerosols. These data indicate the complexity of the atmospheric chemical composition and its impact on air quality.

21.3.1.3 Organic aerosol formation

Natural and anthropogenic sources deliver volatile organic molecules into the atmosphere, which, because of diverse chemical transformations, may oxidize and establish SOAs. The referent process involves numerous interactions with hydroxyl radicals, nitrate radicals, and other pollutants, including ozone. The mechanism resulting in SOAs is vital in

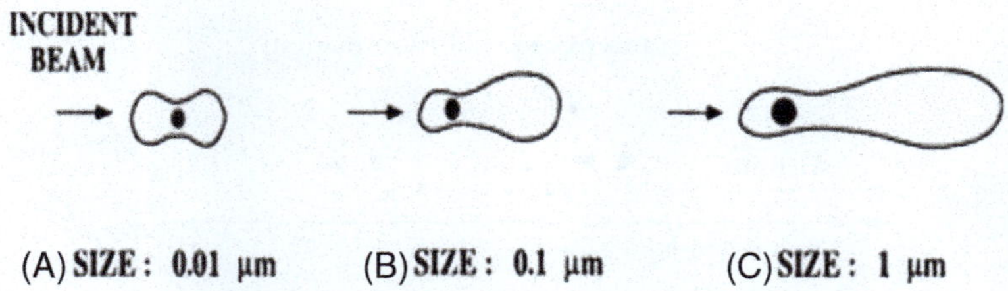

FIGURE 21.3 A particle in the way of an incident light wave where (A) denotes a particle size of 0.01 micrometer with a symmetrical outline, (B) denotes a particle size of 0.1 micrometer with an elongated outline and (C) denotes a particle size of 1 micrometer with a much larger elongated outline.

cloud formation and may seriously impact the climate. Volatile organic molecules' oxidation is a variable process affected by temperature, humidity, and other factors in the atmosphere. Furthermore the formation of SOAs may conditionally be covered in terms of fine particulate matter, the characteristic of which is central to human health and the urban environment. The complexity of organic aerosol formation is related to the likelihood of a multiphase process and the necessity to deal with simultaneous gas- and liquid-phase reactions. The variables yield highly complicated reactions and establish an array of compounds that may be oxidized to differing extents. The autoxidation pathways also condition their volatility. These processes are extremely important because they may lead to the rapid establishment of low-volatility products and contribute to the formation of SOAs (Szmigielski, 2012).

$$VOC + OH^{\cdot} \rightarrow RO_2^{\cdot} + H_2O$$

$$RO_2^{\cdot} + NO^{\cdot} \rightarrow RO^{\cdot} + NO_2^{\cdot}$$

$$RO^{\cdot} + O_2 \rightarrow R^{\cdot}\ (\text{Carbonyls}) + O_3$$

21.3.1.4 Impact on visibility

Atmospheric aerosol particles affect visibility in the case of fog principally by their optical extinction properties, which are determined by their light absorption and scattering properties. These depend on the chemical composition, humidity, and size distribution for organic and other aerosols. However, it is even more likely to form SOAs from carbonyl oxidation processes, which might dramatically affect their visibility. Moreover, these secondary aerosols can also take up ozone, meaning that they alter their physical and chemical properties and their light-scattering ability. Because of the intricate feedback between atmospheric processes and meteorological settings, their action probably becomes more impactful in reducing visibility, particularly in city areas with strong atmospheric pollution. The reaction between carbonyls and ozone forms secondary aerosols, which further impair visibility. During the production of these new particles, they act as condensation nuclei and contribute to fog and haze formation through favorable meteorological conditions. Ozone can also speed up the oxidation of aerosols already in the atmosphere, altering their light-scattering capabilities and making them more effective. Particles resulting from combustion smaller than 2.5 μm in diameter are also known as efficient light scatterers through their mass. Aerosol particles are sensitive to sulfate, nitrate, and organic compound radiative properties (Diachenko-Bohun & Timoshenko, 2023). A particle in the way of an incident light wave is shown in Fig. 21.3.

These particles are also hygroscopic; they form cloud droplets. They can act as CCN, affecting cloud formation and precipitation processes. The interactions between these substances and clouds change the albedo and persistence of clouds, which may affect the Earth's radiation balance. The ties among ozone, aerosols, and atmospheric chemistry will likely produce a feedback loop in which such relationships are enhanced or reduced. The flow of scattering of light by atmospheric suspended particles is represented in Fig. 21.4.

21.4 Literature review

21.4.1 Impact on transportation safety

Aerosol particles in the nonrespirable range of airflow drastically reduce atmospheric visibility, which is a vital component of vehicle safety. In air transportation, low visibility makes the pilot reroute the vehicle and take necessary measures for landing or take-off. The risks of aviation accidents increase, which are also associated with low visibility, which makes the pilot land incorrectly. Another example includes driving vehicle safety, where the risks of collision are increased, and low visibility leads to the establishment of adjusted safety measures such as slowed speed and fog lights. The application of fog

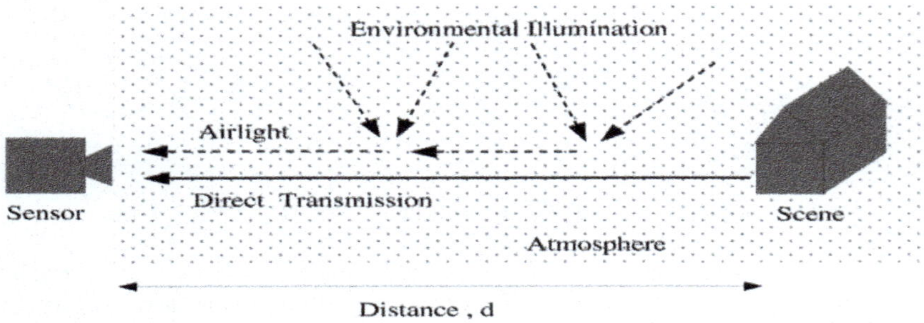

FIGURE 21.4 Flow of scattering of light by atmospheric suspended particles.

lights visually strengthens vehicles and helps other drivers to assess their distance. The lack of visibility in coastal areas and harbors increases the risks of collision or ground. Reduced visibility and associated traffic are economically detrimental because of fuel consumption rises and delays across all industrial domains.

21.4.2 Impact on environment and health

The impact of reduced visibility on the atmosphere is high because of its associated concentration. Thus it can be concluded that the particles are responsible for air acidification and impact regional climate formation. Notably, these chemicals affect all the explored areas of visibility, health, and the overall ecosystem, which stresses the necessity of atmospheric pollution management by addressing all types of pollutants. They are likely to have numerous conditions that negatively affect human health and cause respiratory and cardiovascular diseases. In the long run the pressure can lead to the rarefaction and death of some vegetation and the eruption of forest fires. The particles are known to accumulate within vegetative materials, which may lead to health conditions, including cancer and diabetes. The particles can be subducted and enter the atmosphere, initiating a chain reaction leading to precipitation rate changes and the lack of water for ice formation. Long-term exposure can also result in the particles' accumulation in food chains and bioaccumulation, causing animal and potential human deaths.

21.4.3 Meteorological and climate studies

Correcting for the loss because of aerosols is crucial in many meteorological and climatological studies because of its impact on visibility. The second critical is fog modeling, which depends on an accurate simulation of aerosols and their atmospheric dynamics to accurately model their evolution, dissipation, and changes in visibility (Pal & Bhattacharjee, 2022). Aerosols also alter visibility and climate, as well as have a significant impact on the formation and precipitation of clouds. These small particles act as seeds for water droplets and ice crystals to condense around, thus changing the properties and lifetime of clouds, ultimately affecting regional and global weather patterns. Indeed, aerosol research is necessary for better forecasting of air quality and global climate to employ refined models that enable scientists to predict the impact of atmospheric pollution on human health and ecosystems. It helps in understanding the long-range transport of pollutants and their export to remote ecosystems headed by analysis of aerosols. By analyzing aerosols' composition and global distribution, researchers can determine their origins and evaluate the large-scale impacts on remote areas. Furthermore, studying how aerosols interact with radiation is fundamental for use in predicting changes in the Earth's energy budget and climate forcing. As it interacts with carbon dioxide, oxygen, and nitrogen, the following characteristics are attributes of sebum:

21.4.3.1 CO_2 and urban fog

Urban areas are grappling with increasing CO_2 concentrations, a condition primarily facilitated by the upsurge in vehicular and industrial performances. However, although CO_2 is not an aerosol, its concentrations in most industrial zones usually increase humidity levels, which is a prerequisite to the frequent formation of fog. Such formulation processes are directly associated with these, including an increasing variety of aerosols. In contrast the efforts to reduce CO emissions will likely minimize aerosol pollution by applying appropriate filters in industrial zones. The relationship between CO_2 and the concentrations of aerosols produced is quite complex. These interactions have to be subjected to comprehensive atmospheric analyses. In urban zones, concentrations of CO_2 are trapping fog and other contaminants, causing these to occur frequently. These compounds can be precipitated to the earth to impact air quality throughout urban areas with high populations. On the one hand, although actions to reduce CO_2 can be implemented, there is a probable effect on the levels of aerosols. This factor necessitates the development of pollution management initiatives that are all-inclusive.

21.4.3.2 O_2 and the associated processes

In the production of the majority of secondary aerosols, oxygen is an essential component. For example, it oxidizes sulfur dioxide to sulfuric acid and nitrogen oxides to nitric acid. Carbon dioxide may also undergo transformation processes that yield organic acids and involve oxygen at certain stages. Increasing the production of water droplets, which manifest as range or fog, these aerosols mitigate visibility. Oxidation is similarly vital concerning the occurrence of secondary aerosols and plays a central role in atmospheric chemistry and climate. It depends on complex interactions between oxygen, other gases, and aerosols like feedback. These interactions may also determine such workings and impact radiative quality, cloudage, and balance. Thus the knowledge of it warrants consideration for developing proper methods for air pollution and refinement of the models used to analyze the climate.

21.4.3.3 Nitrogen compounds and emissions from agriculture

NH_3 is one of the most common elevated gases in agricultural operations. Ammonia (NH_3) reacts with this acid to form ammonium sulfate ($(NH_4)_2SO_4$), an unwanted fine aerosol smog pollutant, and ammonium nitrate (NH_4NO_3) aerosol, especially in southern Europe. Aerosols also strongly impact air quality and human health because of their contribution to fine particulate matter ($PM_{2.5}$). In turn, agricultural emissions may also affect the formation of clouds in a way that modifies regional precipitation patterns and crop yields.

More research is needed to understand the intricate interactions between agricultural practices, atmospheric chemistry, and climate change to develop better mitigation options. These emissions can have local effects on air quality and longer range impacts through the influence of cloud formation and precipitation on climate patterns. When these emissions are combined with other air pollutants, they can trigger particular chemical processes that aggravate rural and urban air quality problems.

These mutualistic relationships are key to creating agroecological strategies for sustainable agriculture that will meet our food production demands while minimizing environmental concerns and public health risks. Given the long-lasting impacts of agricultural emissions on the human and natural world, further work is required. Climate change can alter precipitation patterns and air quality, impacting plant species or animal populations. Aspects of agricultural policy and planning will have to be reassessed if changes in crop yields arise from emission-driven climate changes because of probable economic consequences.

These challenges are met by interdisciplinary integration of the many facets, including atmospheric scientists, agronomists, ecologists, and policymakers, to propose new ideas that benefit humanity on the one hand and protect our planet on the other hand, tackling these problems.

$$NH_3 + H_2SO_4 \rightarrow (NH_4)_2SO_4$$

$$NH_3 + HNO_3 \rightarrow NH_4NO_3$$

Hence future work should try and measure the relative effects of farm-related emissions on ecosystem components and create predictive models capable of assessing long-term consequences to environmental habitats (Diachenko-Bohun & Timoshenko, 2023). Novel technological interventions such as precision farming approaches or emission reduction strategies could be vital in reducing the detrimental effects of agricultural activities on air quality and climate. Investing in new farming practices and crop types suited to changing environmental conditions could also improve food security and lower air pollution generated by the agriculture sector. Economic analyses paired with climate models might provide additional insights into the trade-offs and cobenefits of mitigation strategies (Pal & Bhattacharjee, 2022). Understanding this would allow policymakers to help make it more economically viable for the farmers to adopt new methods by offering tailored incentives and restrictions. In addition, partnerships between the public and private sectors can accelerate innovation and deployment of new technologies that reduce emissions from agriculture without harming or increasing yields.

21.5 Proposed method

Visibility issues are thought to be a major contributor to traffic accidents. According to US Department of Transportation research, low visibility is a major contributing factor in 26% of accidents that happen each year. Therefore improving visibility quality is crucial for driver assistance systems to reduce the number of incidents. This research presents a dynamic real-time framework that can identify moving objects in the surrounding area and, at the same time, somewhat increase sight when driving in low-visibility circumstances. First, we provide a unique technique to dynamically categorize films into fog and nonfoggy categories. The video is then defogged using a defogging technique, which is based on the idea that fog obscures the dynamic real-time video sequences. Finally, moving objects are identified in the driver's surroundings to reduce the likelihood of collisions. Not only does the suggested framework for improving low-visibility situations meet human perception requirements, but it also streamlines the current approaches in the computer vision fields. Based on dynamic real-time films, experimental results show more promise than current state-of-the-art approaches, suggesting that this might be a viable route for driver assistance systems in fog, lowering safety risks. Because of variations in speed and scene size, object recognition in video is one of the most important areas of study. It has to do with many applications in real time, including vehicle perception, characteristics, and superficially trained architectures. It may easily be staggered by a complicated setting that combines various low-level imaging functions with a high-level object identifier and the context of scene categorization. The authors in Pal et al. (2023a) build a Region-Based Convolutional Neural Network (RCNN) detection framework by integrating conceptual frameworks with Convolutional Neural Networks (CNNs), which are noticeably better than conventional methods for object identification. The authors in Pal et al. (2023b) propose deep learning–based pooling layers and multitop loss functions to flatten the process and improve the extraction feature efficiency. An architecture for the Region Proposal Network is developed by Zhu et al. (2017), where fusion of images is sent to the prior channel to further increase the enhancement of the degraded images. This paper uses Fourier transform and Wiener filter to enhance the scenes affected by fog. These methods are not intended for a thorough investigation of contextual and temporal data; rather, they are aimed at discreet photographic detection. Underwater image enhancement methods are commonly classified into five categories based on how they handle color constancy, fusion, frequency domain, spatial domain, and deep learning. However, there are three main kinds of underwater picture restoration techniques: optics, polarization, and previous knowledge (Pal et al., 2015). One needs a basic grasp of the principles regulating light behavior in water to interpret underwater photographs.

Because of the physical properties of the material, underwater images are more prone to deterioration effects than photos shot from the air. These photos often have poor visibility thanks to light's exponential attenuation when it flows through water, resulting in blurry and poorly contrasted photographs. The authors in Pal (2019) improved picture quality by combining prior dark channel methods with contrast-limited adaptive histogram equalization (CLAHE). The new method is perfect for improving the quality of underwater photos when compared to the existing technologies in use (Pal, 2021). The contrast stretching technique was created by Pal (2021) to enhance the quality of underwater photographs. The authors in Shi et al. (2021) investigated the use of altered software to improve the look. Improved input for the image is provided by automated image processing techniques by the authors in Hertzberg and Schreuder (2016).

21.6 Result and discussion

Visual quality and quantitative measures were used to assess the dehazing of photos using the dark channel prior (DCP) approach. Cloudy photos become much more visible while using the DCP technique. Images displayed better contrast, natural color, and more details after dehazing. The method of eliminating haze sharpened and distinguished distant things that had before been muddled by fog. Pictures taken outside were more natural and realistic because the sky and horizon had no white cast from haze (Pal et al., 2015). Regaining scene brightness and restoring object colors and textures to their original state was accomplished by the algorithm with good results. The restored images showed higher color fidelity when comparing the recovered photographs to the initial blurry ones. Accurate estimation of the transmission map: Determining the precise quantity of haze at each pixel on the transmission map is a crucial component of the DCP approach. The transmission maps that were produced correctly indicated regions covered in strong fog and gave a decent approximation of the thickness of the haze across the image. This increased the dehazing procedure's overall efficacy. The experimental results of the input image and the enhanced output image are shown in Fig. 21.5.

21.7 Conclusion

The visibility hindrance because of fog mainly originates from aerosols, necessitating a meteorological and aerosol dynamic analysis. Aerosols can scatter light, serve as condensation nuclei, or undergo atmospheric growth in a humid environment to become cloud droplets. Aerosols are created or changed by several chemical reactions in a particular mixture, including nitrogen molecules, CO_2, O_2, and other components. Hence these factors must be carefully examined to develop effective methods for reducing various gases and liquids and ensuring transportation safety. The production of secondary aerosols, ammonium sulfate, and nitrate derivatives overcomplicates the atmospheric chemistry impacts on climate. These particles

FIGURE 21.5 (A) Input hazy image taken from benchmark images, (B) dark channel, (C) estimation map, (D) image radiant, and (E) output image.

reduce visibility and greatly affect the formation of clouds and radiation balance, hence having an important influence on local, regional, and even global trends in climate change. Understanding these complex pathways is necessary to create personalized treatments that ameliorate the negative consequences of agricultural emissions on air quality and climate. This study also discusses the approach, which has gained popularity as a dehazing technique. Its foundation is the observation that most natural images have at least one-color channel with low-intensity pixels in areas that are not foggy. However, the typical DCP approach is restricted and can produce artifacts or insufficient improvement in complex settings or weak conditions. To overcome these limitations the improved DCP method presented in this study combines advances such as enhanced estimations of atmospheric light, adaptive filtering techniques, and updated dark channel estimation. Contrast stretching and color correction are examples of postprocessing procedures that improve the final visual quality of photos and movies. Applications such as surveillance, outdoor photography, and autonomous driving are suitable for the suggested method.

References

Diachenko-Bohun, M., & Timoshenko, O. (2023). Modern aspects of the implementation of the course 'meteorology and climatology' in the educational process of students in the direction of preparation 101 ecology. *The Sources of Pedagogical Skills*, (31), 70–76. doi:10.33989/2075-146x.2023.31.283242.

Fan, W., Chen, T., Zhu, Z., Zhang, H., Qiu, Y., & Yin, D. (2022). A review of secondary organic aerosols formation focusing on organosulfates and organic nitrates. *Journal of Hazardous Materials, 430*, 128406. doi:10.1016/j.jhazmat.2022.128406.

Hertzberg, M., & Schreuder, H. (2016). Role of atmospheric carbon dioxide in climate change. *Energy & Environment, 27*(6–7), 785–797. doi:10.1177/0958305x16674637.

Jung, C. H., Lee, J. Y., & Kim, Y. P. (2015). Estimation of aerosol optical properties considering hygroscopicity and light absorption. *Atmospheric Environment, 105*, 191–201. doi:10.1016/j.atmosenv.2015.01.058.

Lyu, R., Wang, Y., Peng, Y., Gao, W., Yang, H., Tan, X., He, Q., Cheng, T., & Zhang, R. (2021). Long-term variation characteristics and influencing factors of low-visibility events on the coast of China. *Atmospheric Research, 257*, 105583. doi:10.1016/j.atmosres.2021.105583.

Lyu, R., Gao, W., Peng, Y., Qian, Y., He, Q., Cheng, T., Yu, X., & Zhao, G. (2022). Fog–Haze transition and drivers in the coastal region of the Yangtze River Delta. *International Journal of Environmental Research and Public Health, 19*(15), 9608. doi:10.3390/ijerph19159608.

Mahowald, N. M., Scanza, R, Brahney, J, Goodale, C. L., Hess, P. G., Moore, J. K., & Neff, J. (2017). Aerosol deposition impacts on land and ocean carbon cycles. *Current Climate Change Reports, 3*(1), 16–31. doi:10.1007/s40641-017-0056-z.

Mohan, M., & Payra, S. (2014). Aerosol number concentrations and visibility during dense fog over a subtropical urban site. *Journal of Nanomaterials, 2014*(1), 1–6. doi:10.1155/2014/495457.

Pal, T. (2019). A fast method for defogging of outdoor visual images. *Recent Advances in Computer Science and Communications, 14*(2), 416–428. doi:10.2174/2213275912666190819105422.

Pal, T. (2021). A robust method for dehazing of single image with sky region detection and segmentation. *International Journal of Image and Graphics, 21*(04). doi:10.1142/s0219467821500455.

Pal, T., & Bhattacharjee, D. (2022). Visibility enhancement of fog degraded images using adaptive defogging function. *Multimedia Tools and Applications, 81*(24), 35317–35347. doi:10.1007/s11042-022-12182-x.

Pal, T., Bhowmik, M. K., & Ghosh, A. K. (2015). Defogging of visual images using SAMEER-TU database. *Procedia Computer Science, 46*, 1676–1683. doi:10.1016/j.procs.2015.02.108.

Pal, T., Halder, M., & Barua, S. (2023a). A deep learning model to detect foggy images for vision enhancement. *The Imaging Science Journal, 71*(6), 484–498. doi:10.1080/13682199.2023.2185429.

Pal, T., Halder, M., & Barua, S. (2023b). A transmission model based deep neural network for image dehazing. *Multimedia Tools and Applications, 83*(14), 39255–39281. doi:10.1007/s11042-023-17010-4.

Palacios-Peña, L., Jiménez-Guerrero, P., Fast, J. D., & Pravia-Sarabia, E. (2020). *Sensitivity of aerosol optical properties to the aerosol size distribution over central Europe and the Mediterranean basin*. Copernicus GmbH. doi:10.5194/gmd-2020-68.

Shi, S., Zhang, Y., Zhou, X., & Cheng, J. (2021). A novel thin cloud removal method based on multiscale dark channel prior (MDCP). *IEEE Geoscience and Remote Sensing Letters, 19*, 1–5. doi:10.1109/lgrs.2021.3062853.

Strauss, N., Fricke-Begemann, C., & Noll, R. (2010). Size-resolved analysis of fine and ultrafine particulate matter by laser-induced breakdown spectroscopy. *Journal of Analytical Atomic Spectrometry, 25*(6), 867. doi:10.1039/b927493f.

Szmigielski, R. (2012). *Chemistry of organic sulfates and nitrates in the urban atmosphere* (pp. 211–226). *Springer Netherlands*. doi:10.1007/978-94-007-5034-0_17.

Zhu, M., He, B., & Wu, Q. (2017). Single image dehazing based on dark channel prior and energy minimization. *IEEE Signal Processing Letters, 25*(2), 174–178. doi:10.1109/lsp.2017.2780886.

Ziemann, P. J. (2011). Effects of molecular structure on the chemistry of aerosol formation from the OH-radical-initiated oxidation of alkanes and alkenes. *International Reviews in Physical Chemistry, 30*(2), 161–195. doi:10.1080/0144235x.2010.550728.

Chapter 22

Artificial neural networks in mechatronics

Van-Hai Nguyen [a,b] and Tien-Thinh Le [a,b]

[a] *Faculty of Mechanical Engineering and Mechatronics, Phenikaa University, Ha Dong, Hanoi, Vietnam,* [b] *A&A Green Phoenix Group JSC, Phenikaa Research and Technology Institute (PRATI), Cau Giay, Hanoi, Vietnam*

22.1 Introduction

The field of mechatronics has rapidly evolved with advancements in computational intelligence, which integrates artificial intelligence (AI) methods to solve complex engineering challenges across disciplines such as mechanical and electronic engineering, control systems, and computer science. Among computational intelligence techniques, artificial neural networks (ANNs) have proven particularly powerful for mechatronics applications, addressing complex tasks such as adaptive control, fault diagnosis, and optimization that traditional control methods struggle to handle effectively. ANNs can learn from data, adapt to new conditions, and generalize solutions for nonlinear problems, offering a flexible approach well-suited to dynamic and uncertain environments that characterize real-world mechatronic systems (Jin et al., 2018).

ANNs are biologically inspired models capable of processing vast amounts of data and identifying patterns in complex systems, making them invaluable for real-time decision-making in mechatronic applications (Liu et al., 2018). Unlike conventional algorithms, which require explicit programming for each potential scenario, ANNs allow for learning-based control, where systems can improve over time through exposure to data. This adaptability is crucial for managing the nonlinearities and uncertainties in complex systems such as robotics and autonomous vehicles. For instance, in robotic manipulators, ANNs enable the system to respond in real-time to environmental changes, ensuring higher precision and resilience in complex tasks (Ganie & Jagannathan, 2022).

The role of ANNs extends significantly into fault detection, where they facilitate predictive maintenance and early fault diagnosis in mechatronic systems. By analyzing real-time operational data, ANNs can identify subtle shifts in system behavior that may indicate potential failures, enabling proactive responses that reduce maintenance costs and prevent costly breakdowns. Studies indicate that ANN-based predictive maintenance frameworks can significantly reduce downtime in industrial machinery, often resulting in meaningful improvements to system reliability and component lifespan (Fordal et al., 2023; Serradilla et al., 2022).

Optimization is another critical area where ANNs provide significant benefits in mechatronics, as they can adjust system parameters dynamically to enhance performance and efficiency. By continuously analyzing operational data, ANNs optimize energy consumption, response speed, and accuracy, adapting to environmental changes without manual recalibration. In unmanned aerial vehicles (UAVs), for example, ANN algorithms have been used to optimize flight paths, balancing stability and power consumption to extend operational time (Su et al., 2024). This type of optimization is essential for UAVs and other autonomous systems that operate in resource-constrained environments (Zhang et al., 2024).

This chapter provides an overview of ANNs in mechatronic systems, beginning with foundational concepts such as ANN architecture, learning algorithms, and common network types relevant to mechatronics, including convolutional and recurrent neural networks (RNNs). It then explores the key applications of ANNs in mechatronics, focusing on their role in control, fault detection, and optimization through case studies and recent research findings. The chapter concludes by addressing challenges such as computational complexity and real-time processing constraints. It discusses future research directions, including hybrid approaches that combine ANNs with other computational intelligence methods such as fuzzy logic and genetic algorithms to create more robust and adaptable mechatronic systems. For example, combining fuzzy logic with ANNs enhances interpretability and adaptability in uncertain environments, whereas integrating genetic algorithms aids

Computational Intelligence in Mechatronics: Solving Real-World Problems in Electronic Systems Design. DOI: https://doi.org/10.1016/B978-0-443-27580-7.00003-2

in optimizing network parameters and architectures, improving performance in complex, real-time applications (Fordal et al., 2023; Serradilla et al., 2022).

Through this exploration the chapter aims to illustrate how ANNs have become indispensable in advancing the capabilities of modern mechatronic systems, providing intelligent solutions that meet the demands of increasingly complex and autonomous applications.

22.2 Fundamentals of artificial neural networks

22.2.1 Introduction of artificial neural networks

One of the most used methods in the field of data mining and machine learning is ANN, as introduced in 1943s by McCulloch and Pitts (1943). This concept, inspired by the biological neural networks of human brains, is an interconnected group of artificial nodes that could perform computational tasks (Aggarwal, 2018). Similar to a biological neuron that is composed of dendrites, cell body, axon, and synapse, an artificial neuron has inputs, variable weights, and outputs (Yaghoubi et al., 2024). Such an interconnected system of nodes is able to understand and solve complex problems that exhibit nonlinear relationships between causal factors and responses. Indeed the ANN model has relevant profits that are not found in conventional computational models. Hypotheses or preconstraints during modeling are not necessarily required when training the ANN model. The technique can analyze and discover complex relationships and processes in the data. From a computational point of view, ANN is very efficient for problems that have high dimensions because of its strong capacity for parallel processing (Kalogirou, 2001).

22.2.2 Architecture of artificial neural networks

ANNs are constructed from interconnected units called "neurons." Each artificial neuron receives input signals, processes them through a weighted sum, adds a bias, and applies an activation function to produce an output. These neurons are organized into layers: the input layer, which receives raw data; hidden layers, where computation and feature extraction occur; and the output layer, which provides the final prediction. The layers are connected through weights, representing the strength of connections, which are adjusted during training to minimize prediction error. Bias terms allow the model to shift activation functions, enhancing its learning ability. Together, these components form the foundation for the network's capability to process data and learn patterns.

For a problem with one output response the following nonlinear function is generalized by the ANN model (Inkworks, 2024):

$$f : X \in \mathbb{R}^N \longmapsto Y \in \mathbb{R}^M,$$

where X is the input vector (dimension N), and Y is the output vector (dimension M). The nonlinear function f could be fully detailed as follows:

$$X \longmapsto f(X) = f_o\{M[f_h(b + WX)] + b_o\},$$

where f_h, W, and b are the transfer function, weight matrix, and bias for neurons in the hidden layer. f_o, M, and b_o are the transfer function, weight matrix, and bias for neurons in the output layer (as shown in Fig. 22.1).

There are different types of ANN models, as in the following (Abhishek & Abdelaziz, 2023). Neural network architectures vary depending on the problem, each designed to handle specific data types and tasks. Feedforward neural networks (FNNs) are the simplest type, where information flows in one direction—from the input layer, through hidden layers, to the output layer. They are commonly used for tasks such as regression and classification. Next, RNNs introduce loops that allow the network to retain information from previous steps, making them suitable for sequential data such as time series or natural language. Variants such as long short-term memory (LSTM) networks and gated recurrent units (GRUs) address the vanishing gradient problem in standard RNNs. Convolutional neural networks (CNNs) are designed for spatial data, such as images, using convolutional layers to detect features such as edges, textures, and shapes. They excel in tasks such as image classification and object detection. Other advanced architectures include generative adversarial networks (GANs) for generating new data and autoencoders for dimensionality reduction and unsupervised learning. The choice of architecture depends on the nature of the data and the specific goals of the application.

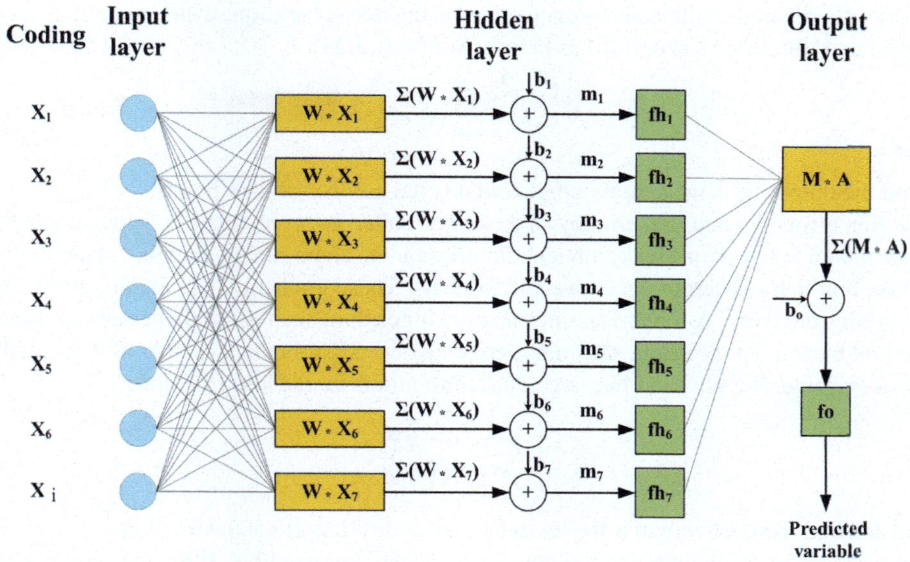

FIGURE 22.1 Illustration of the ANN model involving input, hidden, and output layers. *ANN*, Artificial neural networks.

TABLE 22.1 Different types of activation functions.

No.	Transfer function	Equation
1	Hard-limit	$n \longmapsto \text{hardlim}(n) = \begin{cases} 1, \text{if } n \geq 0 \\ 0, \text{otherwise} \end{cases}$
2	Linear	$n \longmapsto \text{purelin}(n) = n$
3	Log-sigmoid	$n \longmapsto \text{logsig}(n) = \frac{1}{1+\exp(-n)}$
4	Radial basis	$n \longmapsto \text{radbasn}(n) = \exp(-n^2)$
5	Normalized radial basis	$n \longmapsto \text{radbasn}(n) = \frac{\exp(-n^2)}{\sum \exp(-n^2)}$
6	Symmetric hard-limit	$n \longmapsto \text{hardlims}(n) = \begin{cases} 1, \text{if } n \geq 0 \\ -1, \text{otherwise} \end{cases}$
7	Positive-linear	$n \longmapsto \text{poslin}(n) = \begin{cases} n, \text{if } n \geq 0 \\ 0, \text{otherwise} \end{cases}$
8	Tan-sigmoid	$n \longmapsto \text{tansig}(n) = \frac{2}{1+\exp(-2n)} - 1$
9	Triangular basis	$n \longmapsto \text{tribas}(n) = \begin{cases} 1 - \lvert n \rvert, \text{if } -1 \leq n \leq 1 \\ 0, \text{otherwise} \end{cases}$
10	Satlin	$n \longmapsto \text{satlin}(n) = \begin{cases} 0, \text{if } n \leq 0 \\ n, \text{if } 0 \leq n \leq 1 \\ 1, \text{otherwise} \end{cases}$
11	Satlins	$n \longmapsto \text{satlins}(n) = \begin{cases} -1, \text{if } n \leq -1 \\ n, \text{if } -1 \leq n \leq 1 \\ 1, \text{otherwise} \end{cases}$
12	Compet	$n \longmapsto \text{compet}(n) = a_i = \begin{cases} 1, \text{if } i = i^* \\ 0, \text{if } i \neq i^* \end{cases}$, where $n_{i^*} \geq n_i, \forall i$ and $i^* \leq i, \forall n_i = n_{i^*}$
13	Softmax	$n \longmapsto \text{softmax}(n) = \frac{\exp(n)}{\sum_{j=0}^{m} \exp(n_j)}$

22.2.3 Training of artificial neural networks

ANNs learn from data through training, which involves iteratively adjusting their internal parameters—known as weights and biases—to minimize prediction errors (Deisenroth, 2020). This learning process begins with forward propagation, where input data passes through the network's layers. Each neuron computes a weighted sum of its inputs, applies a bias, and transforms the result using an activation function (see Table 22.1 for details of different activation functions). The final output of the network is compared to the actual target values using a loss function, which quantifies the error or difference between the predicted and true values, is defined as below (Murphy, 2023):

$$L = \frac{1}{n}\sum_{i=1}^{n}(y_i - \hat{y}_i)^2$$

where y_i is predicted the output, $\hat{y}_i$ is the actual output, and n is the number of data points.

To reduce prediction errors the network employs a technique called backpropagation, which calculates how much each weight and bias contributed to the error. Backpropagation relies on the chain rule of calculus to compute gradients—partial derivatives of the loss function concerning each weight and bias. These gradients indicate the direction and magnitude of changes needed to minimize errors. A numerical optimization algorithm, such as gradient descent, uses these gradients to update the weights and biases. The learning rate, a hyperparameter, controls the step size of these updates, balancing the speed and stability of learning. Weights and bias are adjusted to minimize the loss function:

$$\begin{cases} \omega = \omega - \eta\frac{\partial L}{\partial \omega} \\ b = b - \eta\frac{\partial L}{\partial b} \end{cases}$$

ω and b are weights and bias, respectively; η is the learning rate, controlling the step size of updates. $\frac{\partial L}{\partial \omega}$ and $\frac{\partial L}{\partial b}$ are gradients of the loss function concerning the weight and the bias, respectively. For example, if the learning rate is $\alpha = 0.1$, it means on each training iteration, the weight parameters are updated only 10% of the gradient term. On the one hand the model can learn more quickly with an excessively high learning rate, but because the weights are updated rapidly, it may be going beyond the minimum point. On the other hand the model can learn cautiously and gradually with a modest learning rate. It takes too long to converge because it makes little adjustment to the weight for every update (Yaghoubi et al., 2024).

In the training of neural networks, batch size and number of epochs are critical hyperparameters that influence how the model learns and converges (Aggarwal, 2023). The batch size refers to the number of training samples the network processes before its parameters are updated. Smaller batch sizes allow for more frequent updates, potentially leading to faster learning but with noisier gradients. In comparison, larger batch sizes provide more stable gradient estimates but require greater computational resources. Common choices for batch size include powers of two, such as 32, 64, or 128, because of memory optimization on modern hardware.

The number of epochs defines how often the entire training dataset is passed through the network. For example, if a dataset contains 1000 samples and the batch size is 100, one epoch consists of 10 iterations, each processing one batch of 100 samples. Training for multiple epochs allows the model to refine its parameters iteratively, improving performance over time. However, too many epochs can lead to overfitting, where the model learns patterns specific to the training data and performs poorly on unseen data. The optimal number of epochs is typically determined through validation, where the model's performance on a separate validation dataset is monitored to identify the point at which it begins to overfit. Batch size and number of epochs significantly balance training efficiency, convergence speed, and model generalization.

22.3 Applications in mechatronic systems

Integrating ANNs in mechatronic systems has transformed traditional control, maintenance, optimization, and real-time decision-making approaches. ANNs, with their ability to model complex, nonlinear systems and learn from data, have proven highly beneficial for applications that require adaptability, accuracy, and speed. This section explores how ANNs are used for adaptive control, predictive maintenance, optimization, and edge computing within mechatronic systems, highlighting their significant impact across various applications.

22.3.1 Using ANNs for adaptive control and real-time decision-making

Adaptive control is one of the most impactful applications of ANNs in mechatronics, where systems must respond dynamically to changing environments and uncertainties. Adaptive control is essential for complex systems, including robotics and autonomous vehicles, which operate in unpredictable settings. Unlike traditional control methods, which struggle to adapt to unexpected conditions, ANNs learn from data and adjust responses as needed, offering critical flexibility for real-time decision-making (Goriparthi, 2024).

In robotic applications, ANNs enhance the adaptability of manipulators, allowing them to respond precisely to environmental variations. For example, in robotic arms, ANN-based controllers can adjust movements in real time to accommodate variables such as load changes and interactions with unpredictable objects, ensuring high precision and robustness in demanding tasks (Gao et al., 2023). This adaptability is crucial for performing complex tasks where conventional, preprogrammed instructions may be insufficient. In addition, ANNs facilitate self-calibration, which reduces the need for human intervention, thereby improving efficiency in applications that require significant autonomy.

For autonomous vehicles, including UAVs, real-time decision-making powered by ANNs is vital. These systems require rapid sensory data processing to make immediate decisions, such as obstacle avoidance and path optimization. ANNs effectively process data from multiple sources, including cameras, lidar, and radar, enabling autonomous systems to respond quickly to external stimuli and operate safely in complex environments. This capability is especially important in applications where real-time adjustments are necessary for safety, as demonstrated in collision avoidance systems that utilize ANN-based models for predictive hazard assessment (Niture & Abdellatif, 2024).

22.3.2 Leveraging ANNs for early detection and predictive maintenance

Predictive maintenance has gained substantial importance in industrial applications because of the high financial costs associated with unplanned downtime from machinery failures. Based on scheduled inspections and routine repairs, traditional maintenance strategies often prove inefficient; they may delay addressing critical issues or result in premature repairs that increase operational costs without necessarily improving. ANNs offer an advanced solution for predictive maintenance, leveraging real-time data analysis to detect early signs of wear or potential failures, allowing industries to transition from reactive to proactive maintenance approaches (Ahmed Murtaza et al., 2024).

In predictive maintenance, ANNs are trained on historical data to learn normal and abnormal system behavior patterns, enabling them to recognize subtle anomalies that precede mechanical failures. For example, vibration and acoustic signals from industrial equipment can be processed by ANN models to detect deviations that suggest bearing wear, misalignment, or motor imbalance, which may otherwise go unnoticed in traditional inspections (Serradilla et al., 2022). This early detection allows maintenance teams to take preemptive actions, reducing unexpected failures, minimizing equipment downtime, and ultimately extending machinery life, all of which contribute to increased productivity and cost savings (Salawu et al., 2023).

Moreover, ANNs enhance fault detection accuracy by distinguishing between actual faults and false alarms, reducing unnecessary interventions and maintenance costs. This capability is critical in high-stakes environments, such as aerospace and power plants, where undetected minor faults can escalate into significant issues if left unaddressed. ANN-based predictive maintenance systems continuously monitor and analyze real-time data, providing maintenance teams with timely alerts about potential problems that may not be immediately visible through standard diagnostic methods (Mahfoud et al., 2024). This continuous monitoring approach prevents critical failures and helps optimize maintenance schedules, aligning maintenance activities with actual equipment conditions rather than arbitrary timelines.

Implementing ANN-based predictive maintenance systems has improved efficiency and operational safety in industrial environments by minimizing unplanned shutdowns and enhancing system reliability. Studies show that industries employing ANNs for predictive maintenance have reported reduced downtime and improved decision-making capabilities for maintenance teams, supporting long-term operational resilience (Lee et al., 2020).

22.3.3 ANN-based optimization techniques for improving system performance

Optimization is another critical area where ANNs enhance the performance of mechatronic systems. In applications ranging from UAVs to industrial automation, ANN-based optimization techniques maximize efficiency, minimize resource consumption, and improve overall system performance. Optimization is particularly important in mechatronic systems, where various parameters—such as energy consumption, response speed, and accuracy—must be carefully balanced to achieve optimal performance.

ANNs support optimization by continuously analyzing system parameters and identifying the best configuration for achieving desired outcomes. For example, in UAV operations, ANN-based algorithms can optimize flight paths to conserve battery life while ensuring stability, enabling longer missions, and reducing frequent recharging (Cheng et al., 2023). Similarly, in automated manufacturing processes, ANNs can optimize production lines by dynamically adjusting process parameters, such as speed and pressure, based on real-time data to maintain product quality and reduce waste.

Moreover, ANN-based optimization is valuable for energy management in resource-constrained environments. For instance, in electric vehicles (EVs), ANNs can optimize energy usage by adjusting throttle response and regenerative braking based on real-time driving conditions, thus extending battery life and increasing travel range. This type of optimization is essential for applications where power resources are limited, and maximizing efficiency is critical for the system's success.

Another notable application of ANN-based optimization is robotic process automation (RPA), where ANNs can optimize task sequences and resource allocation to improve efficiency and reduce operational costs. By continuously monitoring performance metrics and adapting task assignments, ANN-powered RPA systems can maximize output in assembly lines and other repetitive tasks, which is essential for meeting high production demands in industries such as automotive and electronics manufacturing (Nalluri et al., 2023).

22.3.4 Integrating neural networks into embedded systems for edge computing

Edge computing is an emerging trend in mechatronics, where data processing is performed locally on devices rather than relying on centralized servers. This is particularly important for applications that require real-time processing and low-latency responses, such as autonomous vehicles, robotics, and IoT devices in industrial automation. Integrating ANNs into embedded systems allows mechatronic devices to perform complex data analysis and decision-making tasks on the edge, improving responsiveness and reducing the need for continuous connectivity to cloud-based resources.

Embedding ANNs in edge devices allows for efficient and rapid data processing, enabling applications such as autonomous navigation, machine vision, and sensor fusion to be performed in real-time. In autonomous drones, for example, onboard ANN models process sensor data locally to make immediate adjustments in flight, allowing the UAV to operate effectively even in areas with limited network connectivity (Wilson et al., 2022). Similarly, in industrial IoT applications, embedded ANN models can analyze sensor data from machinery to detect anomalies, trigger alarms, or initiate corrective actions without relying on remote servers.

Integrating ANNs in embedded systems also enhances security and privacy by minimizing the amount of data sent to the cloud. This is especially relevant in applications where sensitive information is processed, such as medical devices or surveillance systems. By processing data locally, embedded ANNs reduce the risk of data breaches and ensure that critical information remains secure.

Challenges associated with implementing ANNs in embedded systems include computational limitations and power constraints, as traditional neural networks require significant processing power that may exceed the capabilities of small devices. However, advances in lightweight ANN architectures and specialized hardware, such as neural processing units (NPUs), have made deploying ANNs on edge devices feasible without compromising performance. Techniques such as model compression and quantization also help reduce the computational footprint of ANNs, making them more compatible with embedded platforms while maintaining high accuracy.

The application of ANNs in mechatronic systems has significantly improved adaptive control, predictive maintenance, optimization, and edge computing. By leveraging ANNs, mechatronic devices can operate more autonomously, adapt to changes in real-time, and optimize performance across various parameters. Integrating ANN models into embedded systems further enhances these capabilities, enabling real-time data processing and decision-making at the edge. As ANN technology continues to evolve, its impact on mechatronics will likely grow, with future advancements focusing on improving computational efficiency and exploring new hybrid approaches that combine ANNs with other AI techniques to address the demands of increasingly complex and dynamic applications in mechatronics.

22.4 Case study

22.4.1 Overview

Object detection is a cornerstone of autonomous driving systems, enabling vehicles to interpret their environment and react appropriately. Traffic lights are critical in ensuring safe and rule-abiding navigation among the objects requiring detection. This case study demonstrates the application of YOLOv8 (You Only Look Once version 8), a cutting-edge deep learning model, for traffic light detection. YOLOv8 is known for its efficiency in real-time object detection tasks, making it particularly suitable for autonomous driving systems.

22.4.2 Dataset preparation

The foundation of any deep learning task is a robust dataset. For this project a dataset containing images with annotated traffic lights in urban settings was curated. The photos were sourced from publicly available datasets such as the Bosch Small Traffic Lights Dataset (Behrendt, Novak, & Botros, 2017) and supplemented with custom data captured in diverse lighting and weather conditions. Each traffic light in the images was annotated with bounding boxes and labeled according to its state (for example, red, yellow, and green).

Preprocessing steps were applied to ensure consistent input to the YOLOv8 model. Images were resized to a standard resolution of 640 × 640 pixels and normalized pixel values, and the dataset was split into training, validation, and test sets, typically in a 70/20/10 ratio. These steps ensured that the model could learn effectively while preventing overfitting.

22.4.3 Model selection and training

YOLOv8 was selected because of its high accuracy and real-time performance, which are critical for autonomous driving applications. The model architecture balances lightweight design with powerful detection capabilities, making it ideal for deployment in resource-constrained environments like embedded systems in autonomous vehicles.

The model was fine-tuned on the traffic light dataset using transfer learning. A pretrained YOLOv8 model on a large general dataset like Common Objects in Context (COCO) provided a strong starting point. Fine-tuning was performed using a learning rate scheduler and data augmentation techniques, such as flipping, rotation, and brightness adjustments, to improve the models.

22.4.4 Transfer learning for fine-tuning

In this case study the training process for traffic light detection utilizes transfer learning with the YOLOv8 model pretrained on the COCO dataset, a comprehensive collection of images featuring 80 object categories. This pretrained model provides a robust foundation by leveraging generalized feature extraction capabilities, such as recognizing edges, shapes, and textures. These features are transferrable to detecting traffic lights, enabling the model to focus on task-specific details with reduced training data and computational requirements. Fine-tuning begins by adjusting the model's output layer to classify traffic lights into their respective states (red, yellow, and green) while retaining the bounding box regression layers for localization. The initial layers of the model, which contain general knowledge, are frozen, allowing the training process to focus on higher-level layers responsible for recognizing traffic light-specific patterns.

A reduced initial learning rate is applied to optimize the fine-tuning process, and a scheduler adjusts the rate dynamically based on validation performance. Data augmentation techniques, including flipping, brightness adjustment, and cropping, are incorporated to improve the model's robustness against variations in real-world scenarios. Early stopping is implemented to halt training when improvements plateau, preventing overfitting. The fine-tuned model significantly outperforms the baseline pretrained model, accurately detecting traffic lights and classifying their states while maintaining precise localization. This approach demonstrates the efficiency of transfer learning in adapting a general-purpose model like YOLOv8 to specialized tasks such as traffic light detection, a critical component for autonomous driving systems.

22.4.5 Evaluation and results

After training the model was tested on the held-out test set to evaluate its performance. YOLOv8 accurately detected traffic lights across various scenes, achieving an Mean Average Precision (mAP) of over 90%. The model could reliably differentiate between red, yellow, and green traffic lights and was robust against challenging scenarios such as occlusion and varying illumination (see Figs. 22.2 and 22.3 for examples of the image dataset).

Inference speed was another critical metric, as real-time detection is necessary for autonomous driving. YOLOv8 achieved inference times of less than 20 ms per frame on a mid-range GPU, demonstrating its suitability for integration into real-time systems.

Fig. 22.4 shows some recognition results on the test dataset. Note that the *yellow box* is predicted, and the *blue box* is the ground truth.

22.4.6 Conclusion

This case study highlights the effectiveness of YOLOv8 for traffic light detection, a critical task in autonomous driving. With its high accuracy and real-time performance, YOLOv8 is a reliable tool for enabling safe and efficient navigation in urban environments. The methodology and insights presented here can be a foundation for developing advanced object detection systems in mechatronics and autonomous vehicles.

FIGURE 22.2 Images of a dataset for traffic light detection. (*From Behrendt, K., Novak, L., & Botros, R. (2017). A deep learning approach to traffic lights: Detection, tracking, and classification. In* Proceedings—IEEE International Conference on Robotics and Automation *(pp. 1370–1377). Institute of Electrical and Electronics Engineers Inc. 10.1109/ICRA.2017.7989163*)

FIGURE 22.3 Images of a dataset for traffic light detection. (*From Behrendt, K., Novak, L., & Botros, R. (2017). A deep learning approach to traffic lights: Detection, tracking, and classification. In* Proceedings—IEEE International Conference on Robotics and Automation *(pp. 1370–1377). Institute of Electrical and Electronics Engineers Inc. 10.1109/ICRA.2017.7989163*)

22.5 Challenges and future directions

22.5.1 Challenges in implementing ANNs in real-world mechatronic systems

Despite their transformative potential, implementing ANNs in real-world mechatronic systems faces significant challenges. Addressing these challenges requires overcoming issues related to computational constraints, data requirements, reliability, and integration with existing infrastructure. Moreover, emerging application trends such as autonomous systems, industrial IoT, and smart manufacturing highlight the need for more efficient, scalable, and explainable ANN models.

FIGURE 22.4 Recognition results on the test dataset. *(From Behrendt, K., Novak, L., & Botros, R. (2017). A deep learning approach to traffic lights: Detection, tracking, and classification. In* Proceedings—IEEE International Conference on Robotics and Automation *(pp. 1370–1377). Institute of Electrical and Electronics Engineers Inc. 10.1109/ICRA.2017.7989163)*

22.5.1.1 Computational constraints

One of the main barriers to deploying ANNs in mechatronic systems is the intensive computational resources required for training and inference. Real-time applications, such as autonomous drones, robotics, and industrial automation, demand fast decision-making, but traditional ANNs often exceed the computational capacity of embedded systems.

Recent trends in edge computing and the development of specialized hardware such as tensor processing units (TPUs) and NPUs have shown promise in addressing these computational constraints. Lightweight models such as MobileNetV3 and EfficientNet, along with pruning and quantization techniques, are increasingly used to reduce model size and complexity without compromising accuracy (Chen et al., 2024; Dantas et al., 2024).

22.5.1.2 Data requirements and quality

ANNs require vast amounts of labeled data to achieve high accuracy, which presents a challenge in many mechatronic applications where acquiring comprehensive datasets is difficult. In predictive maintenance, for example, gathering failure data is challenging because of the infrequent occurrence of specific faults.

Trends in synthetic data generation, self-supervised learning, and federated learning are emerging to address these issues. Self-supervised learning allows models to learn useful features from unlabeled data, whereas federated learning enables decentralized data usage, ensuring privacy and security (Dai et al., 2024; Saeed et al., 2021).

22.5.1.3 Reliability and safety concerns

Reliability and safety are critical in mechatronic systems, particularly in aerospace, automotive, and healthcare sectors, where failures can have catastrophic consequences. ANNs, being probabilistic models, pose challenges in guaranteeing consistent and predictable behavior.

Explainable AI (XAI) techniques are gaining traction to enhance the interpretability of ANN models. Methods, such as SHAP (SHapley Additive exPlanations) and Grad-CAM (gradient-weighted class activation mapping), are increasingly used in industrial applications to provide insights into model decisions (Lundberg & Lee, 2017; Selvaraju et al., 2020). Moreover, hybrid approaches combining ANNs with rule-based systems can improve reliability in safety-critical applications (Aksjonov & Kyrki, 2022).

22.5.1.4 Integration with legacy systems

Many industries still rely on legacy systems not designed to interface with modern AI technologies. Retrofitting existing mechatronic systems with ANN-based solutions requires overcoming compatibility and infrastructure limitations.

Trends in modular AI frameworks and middleware solutions are emerging to bridge this gap. For instance, Open Neural Network Exchange (ONNX) and Robot Operating System (ROS) provide standardized interfaces that facilitate integration across different platforms and environments (Abawi et al., 2024; Grigorescu & Zaha, 2024).

22.5.2 Future research directions

As mechatronic systems evolve, future research will likely focus on improving ANN-based solutions' efficiency, reliability, and adaptability. Trends in hybrid intelligence, transfer learning, and ethical AI are expected to shape the next generation of mechatronic systems.

22.5.2.1 *Hybrid approaches*

Hybrid approaches integrating ANNs with other AI methods, such as fuzzy logic, genetic algorithms, and reinforcement learning, are gaining popularity. These approaches leverage the strengths of different techniques to improve adaptability and robustness.

For example, combining ANNs with fuzzy logic improves interpretability in uncertain environments, whereas genetic algorithms optimize network architectures. Reinforcement learning enables adaptive control in dynamic settings, such as autonomous navigation and robotic manipulation (Shan et al., 2020).

22.5.2.2 *Real-time processing and edge computing*

Edge computing is becoming increasingly important in applications that require real-time processing, such as autonomous vehicles and smart sensors. Edge computing reduces latency and enhances responsiveness by performing computations locally on devices.

Future research will focus on developing more efficient ANN models for edge devices, including techniques such as model distillation, pruning, and neural architecture search (NAS). These methods optimize performance while minimizing resource consumption (Klein et al., 2024; Salmani Pour Avval et al., 2025).

22.5.2.3 *Explainable artificial intelligence*

XAI will be critical in adopting ANN-based systems in safety-critical applications. Future research will explore advanced XAI methods to improve the transparency of complex models.

Recent studies have proposed interactive visualization tools and case-based reasoning systems to enhance human understanding of ANN decisions. These developments aim to build trust and ensure compliance with regulatory standards (Pradeep et al., 2024).

22.5.2.4 *Transfer learning and few-shot learning*

Transfer learning and few-shot learning are promising techniques for reducing the data dependency of ANN models. By leveraging pretrained models, these methods enable ANNs to adapt to new tasks with minimal additional training.

Recent research has applied transfer learning in predictive maintenance and autonomous systems, reducing the need for extensive labeled datasets. Few-shot learning techniques are also being explored to adapt to novel conditions in industrial environments rapidly (Song et al., 2023).

22.5.2.5 *Ethical and regulatory considerations*

The growing use of ANNs in mechatronic systems raises ethical and regulatory concerns. Future research will focus on developing frameworks to ensure fairness, accountability, and transparency in AI systems.

For example, bias detection tools and algorithmic audits are being developed to address potential biases in ANN models. Regulatory bodies are also establishing guidelines for the safe deployment of AI in industries such as automotive and healthcare.

22.5.2.6 *Lifelong learning and continuous adaptation*

Lifelong learning enables ANN models to adapt to changing conditions without completely retraining continuously. This capability is essential for mechatronic systems operating in dynamic environments.

Online learning and metalearning are being explored to facilitate lifelong adaptation. These methods allow ANN models to retain knowledge while integrating new information, improving their performance over time (Son et al., 2024).

22.6 Conclusion

22.6.1 Summary of key points

ANNs have emerged as a powerful tool in modern mechatronic systems, offering solutions to complex, nonlinear problems that traditional control methods cannot effectively address. Integrating ANNs in mechatronics provides significant advantages in adaptive control, predictive maintenance, optimization, and real-time decision-making. However, considerable challenges remain to overcome, including computational constraints, data requirements, reliability concerns, and integration with legacy systems.

This chapter also highlighted emerging trends and future research directions to address these challenges, such as hybrid approaches, real-time processing through edge computing, XAI, and transfer learning. Exploring lifelong learning and ethical considerations is also crucial to ensuring the safe and responsible deployment of ANN-based systems in industrial and autonomous applications.

22.6.2 Importance of artificial neural networks in the future of mechatronic systems

ANNs will be essential for the future of mechatronic systems, providing adaptive, real-time solutions in areas such as manufacturing, robotics, and autonomous vehicles. With their ability to optimize processes, reduce downtime, and enhance decision-making, ANNs will enable smarter, more efficient mechatronic devices. As AI technologies evolve, combining ANNs with other intelligence methods will create robust systems capable of handling uncertainties. Future research must focus on explainability and compliance with ethical standards to ensure their safe deployment, promoting trust and wider adoption across industries.

References

Abawi, F., Fu, D., Allgeuer, P., & Wermter, S. (2024). Wrapyfi: A Python wrapper for integrating robots, sensors, and applications across multiple middleware. In *ACM/IEEE International Conference on Human-Robot Interaction* (pp. 860–864). IEEE Computer Society. http://ieeexplore.ieee.org/xpl/conferences.jsp. doi:10.1145/3610977.3637471 21672148.

Abhishek, K., & Abdelaziz, D. M. (2023). *Machine learning for imbalanced data: Tackle imbalanced datasets using machine learning and deep learning techniques*. Packt Publishing.

Aggarwal, C. C. (2018). *Machine learning with shallow neural networks* (pp. 53–104). Springer Science and Business Media LLC. doi:10.1007/978-3-319-94463-0_2.

Aggarwal, C. C. (2023). *Neural networks and deep learning: A textbook* (pp. 1–529). Springer International Publishing. https://link.springer.com/book/10.1007/978-3-031-29642-0. doi:10.1007/978-3-031-29642-0.

Ahmed Murtaza, A., Saher, A., Zafar, M. H., Moosavi, S. K. R., Aftab, M. F., & Sanfilippo, F. (2024). Paradigm shift for predictive maintenance and condition monitoring from Industry 4.0 to Industry 5.0: A systematic review, challenges and case study. *Results in Engineering, 2*, 102935. doi:10.1016/j.rineng.2024.102935.

Aksjonov, A., & Kyrki, V. (2022). A safety-critical decision making and control framework combining machine learning and rule-based algorithms. *arXiv*. Finland arXiv https://arxiv.org . doi:10.48550/arxiv.2201.12819.

Behrendt, K., Novak, L., & Botros, R. (2017). *A deep learning approach to traffic lights: Detection, tracking, and classification* (pp. 1370–1377). Institute of Electrical and Electronics Engineers Inc. doi:10.1109/ICRA.2017.7989163.

Chen, F., Li, S., Han, J., Ren, F., & Yang, Z. (2024). Review of lightweight deep convolutional neural networks. *Archives of Computational Methods in Engineering, 31*(4), 1915–1937. https://www.springer.com/journal/11831. doi:10.1007/s11831-023-10032-z.

Cheng, N., Wu, S., Wang, X., Yin, Z., Li, C., Chen, W., & Chen, F. (2023). AI for UAV-assisted IoT applications: A comprehensive review. *IEEE Internet of Things Journal, 10*(16), 14438–14461. http://ieeexplore.ieee.org/servlet/opac?punumber=6488907. doi:10.1109/JIOT.2023.3268316.

Dai, C., Wei, S., Dai, S., Garg, S., Kaddoum, G., & Hossain, M. S. (2024). Federated self-supervised learning based on Prototypes clustering contrastive learning for internet-of-vehicles applications. *IEEE Internet of Things Journal*, 1. doi:10.1109/jiot.2024.3453336.

Dantas, P. V., Sabino da Silva, W., Cordeiro, L. C., & Carvalho, C. B. (2024). A comprehensive review of model compression techniques in machine learning. *Applied Intelligence, 54*(22), 11804–11844. https://www.springer.com/journal/10489. doi:10.1007/s10489-024-05747-w.

Deisenroth, M. P. (2020). *Mathematics for machine learning* (1st ed.). Cambridge University Press.

Fordal, J. M., Schjølberg, P., Helgetun, H., Skjermo, T. Ø., Wang, Y., & Wang, C. (2023). Application of sensor data based predictive maintenance and artificial neural networks to enable Industry 4.0. *Advances in Manufacturing, 11*(2), 248–263. https://www.springer.com/journal/40436. doi:10.1007/s40436-022-00433-x.

Ganie, I., & Jagannathan, S. (2022). Adaptive control of robotic manipulators using deep neural networks. *IFAC-PapersOnLine, 55*(15), 148–153. http://www.journals.elsevier.com/ifac-papersonline/55. doi:10.1016/j.ifacol.2022.07.623.

Gao, Y., Wang, S., Dragicevic, T., Wheeler, P., & Zanchetta, P. (2023). Artificial intelligence techniques for enhancing the performance of controllers in power converter-based systems—An overview. *IEEE Open Journal of Industry Applications*, 366–375. https://ieeexplore.ieee.org/xpl/RecentIssue.jsp?punumber=8782707. doi:10.1109/OJIA.2023.3338534.

Goriparthi, R. G. (2024). Adaptive neural networks for dynamic data stream analysis in real-time systems. *International Journal of Machine Learning Research in Cybersecurity and Artificial Intelligence, 1*, 689–709.

Grigorescu, S., & Zaha, M. (2024). CyberCortex.AI: An AI-based operating system for autonomous robotics and complex automation. *Journal of Field Robotics*. http://onlinelibrary.wiley.com/journal/10.1002/(ISSN)1556-4967. doi:10.1002/rob.22426.

Inkworks, C. (2024). Machine learning for beginners: Master fundamentals in NLP, ML algorithms, deep learning, and more with this Simple Introductory Guide. Learn ML techniques in less than 14 days.Self-published.

Jin, L., Li, S., Yu, J., & He, J. (2018). Robot manipulator control using neural networks: A survey. *Neurocomputing, 28*, 23–34. doi:10.1016/j.neucom.2018.01.002.

Kalogirou, S. A. (2001). Artificial neural networks in renewable energy systems applications: A review. *Renewable and Sustainable Energy Reviews, 5*(4), 373–401. doi:10.1016/s1364-0321(01)00006-5.

Klein, A., Golebiowski, J., Ma, X., Perrone, V., & Archambeau, C. (2024). Structural pruning of pre-trained language models via neural architecture search. *arXiv*. United States arXiv https://arxiv.org . doi:10.48550/arXiv.2405.02267.

Lee, J., Ni, J., Singh, J., Jiang, B., Azamfar, M., & Feng, J. (2020). Intelligent maintenance systems and predictive manufacturing. *Journal of Manufacturing Science and Engineering, 142*(11). doi:10.1115/1.4047856.

Liu, R., Yang, B., Zio, E., & Chen, X. (2018). Artificial intelligence for fault diagnosis of rotating machinery: A review. *Mechanical Systems and Signal Processing, 10*, 33–47. doi:10.1016/j.ymssp.2018.02.016.

Lundberg, S. M., & Lee, S. I. (2017). Neural information processing systems foundation United States A unified approach to interpreting model predictions. *Advances in Neural Information Processing Systems, 30*, 4766–4775.

Mahfoud, H., Moutaoukil, O., Toum Benchekroun, M., & Latif, A. (2024). Real-time predictive maintenance-based process parameters: Towards an industrial sustainability improvement. In *Lecture notes in networks and systems: 931* (pp. 18–34). Springer Science and Business Media Deutschland GmbH. https://www.springer.com/series/15179. doi:10.1007/978-3-031-54288-6_3.

McCulloch, W. S., & Pitts, W. (1943). A logical calculus of the ideas immanent in nervous activity. *The Bulletin of Mathematical Biophysics, 5*(4), 115–133. doi:10.1007/BF02478259.

Murphy, K. P. (2023). *Probabilistic machine learning: Advanced topics*. The MIT Press.

Nalluri, M., Reddy, S. R. B., Rongali, A. S., & Polireddi, N. S. A. (2023). Investigate the use of robotic process automation (RPA) to streamline administrative tasks in healthcare, such as billing, appointment scheduling, and claims processing. *Tuijin Jishu/Journal of Propulsion Technology, 44*, 2458–2468.

Niture, N., & Abdellatif, I. (2024). *A systematic review of factors, data sources, and prediction techniques for earlier prediction of traffic collision using AI and machine learning*. Multimedia Tools and Applications https://www.springer.com/journal/11042. doi:10.1007/s11042-024-19599-6.

Pradeep, P., Caro-Martínez, M., & Wijekoon, A. (2024). A practical exploration of the convergence of case-based reasoning and explainable artificial intelligence. *Expert Systems with Applications, 255*, 124733. doi:10.1016/j.eswa.2024.124733.

Saeed, A., Salim, F. D., Ozcelebi, T., & Lukkien, J. (2021). Federated self-supervised learning of multisensor representations for embedded intelligence. *IEEE Internet of Things Journal, 8*(2), 1030–1040. doi:10.1109/jiot.2020.3009358.

Salawu, E. Y., Awoyemi, O. O., Akerekan, O. E., Afolalu, S. A., Kayode, J. F., Ongbali, S. O., Airewa, I., & Edun, B. M. (2023). Impact of maintenance on machine reliability: A review. *E3S Web of Conferences*. EDP Sciences www.e3s-conferences.org/430. doi:10.1051/e3sconf/202343001226 22671242.

Salmani Pour Avval, S., Eskue, N. D., Groves, R. M., & Yaghoubi, V. (2025). Systematic review on neural architecture search. *Artificial Intelligence Review, 58*(3). https://www.springer.com/journal/10462. doi:10.1007/s10462-024-11058-w.

Selvaraju, R. R., Cogswell, M., Das, A., Vedantam, R., Parikh, D., & Batra, D. (2020). Grad-CAM: Visual explanations from deep networks via gradient-based localization. *International Journal of Computer Vision, 128*(2), 336–359. http://www.kluweronline.com/issn/0920-5691/. doi:10.1007/s11263-019-01228-7.

Serradilla, O., Zugasti, E., Rodriguez, J., & Zurutuza, U. (2022). Deep learning models for predictive maintenance: A survey, comparison, challenges and prospects. *Applied Intelligence, 52*(10), 10934–10964. Www.springer.com/journal/10489. doi:10.1007/s10489-021-03004-y.

Shan, Y., Zheng, B., Chen, L., Chen, L., & Chen, D. (2020). A reinforcement learning-based adaptive path tracking approach for autonomous driving. *IEEE Transactions on Vehicular Technology, 69*(10), 10581–10595. http://ieeexplore.ieee.org/xpl/tocresult.jsp?isnumber=8039128&;punumber=25. doi:10.1109/TVT.2020.3014628.

Son, J., Lee, S., & Kim, G. (2024). When meta-learning meets online and continual learning: A survey. *IEEE Transactions on Pattern Analysis and Machine Intelligence*. https://ieeexplore.ieee.org/servlet/opac?punumber=34. doi:10.1109/TPAMI.2024.3463709.

Song, Y., Wang, T., Cai, P., Mondal, S. K., & Sahoo, J. P. (2023). A comprehensive survey of few-shot learning: Evolution, applications, challenges, and opportunities. *ACM Computing Surveys, 55*(13s). http://dl.acm.org/citation.cfm?id=J204. doi:10.1145/3582688.

Su, G., Cheng, X., Feng, S., Liu, K., Song, J., Chen, J., Zhu, C., & Lin, H. (2024). Flight path optimization with optimal control method. *arXiv*. United States arXiv https://arxiv.org . doi:10.48550/arXiv.2405.08306.

Wilson, A. N., Kumar, A., Jha, A., & Cenkeramaddi, L. R. (2022). Embedded sensors, communication technologies, computing platforms and machine learning for UAVs: A review. *IEEE Sensors Journal, 22*(3), 1807–1826. http://ieeexplore.ieee.org/xpl/RecentIssue.jsp?punumber=7361. doi:10.1109/JSEN.2021.3139124.

Yaghoubi, E., Yaghoubi, E., Khamees, A., & Vakili, A. H. (2024). A systematic review and meta-analysis of artificial neural network, machine learning, deep learning, and ensemble learning approaches in field of geotechnical engineering. *Neural Computing and Applications, 36*(21), 12655–12699. https://www.springer.com/journal/521. doi:10.1007/s00521-024-09893-7.

Zhang, X., Wang, A., Sun, G., Liu, L., & Zhang, J. (2024). UAV deployment optimization in UAV-assisted wireless communications. *arXiv*. China arXiv https://arxiv.org . doi:10.48550/arXiv.2403.15411.

Chapter 23

Monostatic radar–based microwave imaging of breast cancer detection

Komalpreet Kaur[a] **and Gagandeep Kaur**[b]

[a] *Computer Science and Engineering, Chandigarh Engineering College CGC, Mohali, Punjab, India,* [b] *Emerging Technologies, Chandigarh Engineering College CGC, Mohali, Punjab, India*

23.1 Introduction

Cancer is a major global health issue, affecting millions of people and posing significant social, economic, and emotional challenges for both individuals and families (Bharmjeet & Das, 2023). In 2023 approximately 1.95 million new cancer cases were reported in the United States, with 1.01 million cases in men and 0.948 million in women (Siegel et al., 2023). Among the various types of cancer, breast cancer is the most prevalent, accounting for 0.3 million of all newly diagnosed cases. Breast cancer, like other cancers, is characterized by abnormal growth and uncontrolled division of cells (Wang & Wu, 2023). It can occur in any part of the breast tissue, with different forms emerging based on the site and characteristics of the abnormal growth.

In addition, to identify malignancies in breast tissues, dermatologists and practitioners typically rely on traditional methods such as biopsy and screening procedures (such as MRI, X-rays, CT scans, and ultrasound) (Kaur & Kaur, 2026; Hussain et al., 2022; Morrow et al., 2011; Sarno et al., 2023; Sun et al., 2023). However, the patient may experience difficulties because of these operations' discomfort and unpleasant side effects, such as nausea (Sun et al., 2023). Regular exposure to the ionizing radiation used in conventional imaging methods such as X-rays and CT scans can be harmful, especially for more vulnerable groups such as elderly patients or pregnant women. Researchers have worked on antennas and developed a nonionizing and noninvasive cancer screening tool called microwave imaging (MI) because of the limitations of ionizing radiation, subjectivity (biopsy), and time-consuming procedures (Kaur & Kaur, 2025). In addition to being safe for usage on people, it is essential to note that this method uses nonionizing electromagnetic (EM) impulses to penetrate through human tissues. Radar-based monostatic microwave imaging (RMMWI) relies on the analysis of backscattered signals to distinguish the different dielectric characteristics of the targeted tissues and, primarily, to detect tumor cells within the tissues. As the intermediaries for transmitting microwave impulses to human organs, antennas are crucial in MI (Kaur & Kaur, 2021, 2022a, 2022b). The antenna used in MI must adhere to specific specifications to effectively diagnose tumor cells, which have a higher dielectric constant than healthy skin tissues: compact size, higher gain with large bandwidth (BW), high efficiency, and the ability to operate in both lower and higher frequencies (better image resolution) (Latif et al., 2015; Lazaro et al., 2011; Qashlan et al., 2022).

In fact, researchers have suggested numerous kinds of antenna and methods to enhance the precision and efficacy of MI for the detection of breast cancer. Slot microstrip antennas, horn antennas, Vivaldi antennas, dielectric resonator antennas (DRA), bi-conical, log periodic, bow tie, Artificial Magnetic Conductor (AMC) integrated, etc. are some of the numerous types of antennas used for MI (Kaur & Kaur, 2025, 2022a; Kaur et al., 2024; Latif et al., 2015; Lazaro et al., 2011; Mersani et al., 2019; Syed et al., 2022; Tiang et al., 2013). The aforementioned antennas are difficult to integrate into microwave imaging (MI) systems due to their large profile, low gain, poor directivity, low efficiency, and broad operating bandwidth. Thus this article proposed a UWB cubical-shaped DRA (CSDRA) for MI of the breast cancer diagnosis to get over the shortcomings of the aforementioned antennas. The proposed CSDRA is designed on a Rogers RT5880 substrate of dimensions $16 \times 20 \times 0.8$ mm^3. It offers an impedance BW of 5.1 GHz from 4.2 to 9.39 GHz with high efficiency.

Computational Intelligence in Mechatronics: Solving Real-World Problems in Electronic Systems Design. DOI: https://doi.org/10.1016/B978-0-443-27580-7.00034-2

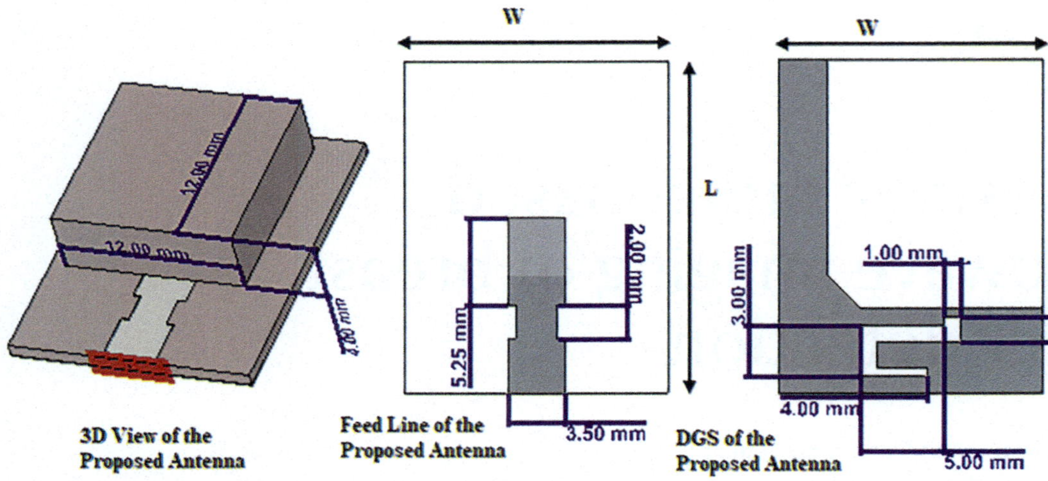

FIGURE 23.1 Perspective view, feed line and ground plane of the proposed CSDRA. *CSDRA*, Cubical-shaped dielectric resonator antenna; *DGS*, defective ground structure.

The scanning of the in silico biophantom has been done to obtain the essential data for imaging. Typically, this involves sending EM signals into biophantoms and employing CSDRA to capture the scattered or reflected signals (S_{11}). To produce the synthetic focus point, several measured signals are collected from various locations (along the x and y axes) of the biophantom. Following preprocessing, beam-forming algorithms employ these collected data to create a 2D image of the scanned area. The suggested MI technique makes use of beam-forming techniques to increase the resolution of the reconstructed picture of the scan area and make it possible to locate and identify breast cancers.

23.2 Geometry of the proposed antenna

This section of the chapter describes the design layout of a UWB CSDRA using a 3D EM program called Computer Simulation Technology (CST) Microwave Studio. The simulation was performed using a transient solver. Fig. 23.1 displays a 3D view of the proposed CSDRA, depicting its cubical DRA with defective ground structure (DGS) and a network of dumbbell-shaped microstrip feed lines. The CSDRA design utilizes a single layer of Rogers RT5880 substrate, which has a moderate permittivity (εr) of 2.2 and a loss tangent of 0.0009. The dimensions of the substrate are $16 \times 20 \times 0.8$ mm^3. The upper side of the substrate consists of dumbbell-shaped feed network, and lower side of the substrate consists of reduced ground with DGS, as depicted in Fig. 23.1, respectively. A cube-shaped alumina with dimensions $12 \times 12 \times 4$ mm^3 is placed on the top of the substrate. CSDRA is fed utilizing a dumb microstrip feed line to achieve good impedance matching and efficient performance throughout the UWB frequency range. The antenna's dimensions of the top (patch and feed line) and backside (DGS ground) are shown in Fig. 23.1, respectively.

23.3 Analysis of the proposed CSDRA for radiation characteristics

The proposed CSDRA is designed and analyzed for its reflection parameters in CST MWS. Simulated results demonstrate that the final CSDRA offers UWB operation with double resonance, as shown in a *solid red line* graph (Fig. 23.2). This graph indicates that the proposed CSDRA features an impedance BW of 11.325 GHz, ranging from 4.2 to 9.3 GHz. It depicts the efficiency of the proposed antenna, showing that the peak efficiency is 105% and the average efficiency is about 94% as depicted in Fig. 23.3, making it suitable for the suggested applications.

23.4 Breast phantom design and confocal MI algorithm

This MI technique analyses changes in the backscattering signal, particularly in the presence of high-dielectric objects such as tumors. The setup for the MI procedure is depicted in Fig. 23.4, where the antenna revolves around a breast cancer phantom to collect data for cases with and without a tumor. Data collection across the breast phantom is achieved by rotating the antenna in polar coordinates from 0 to 2π at 36 equally spaced locations, 10 degrees apart. This process is repeated for both azimuthal and elevation angles.

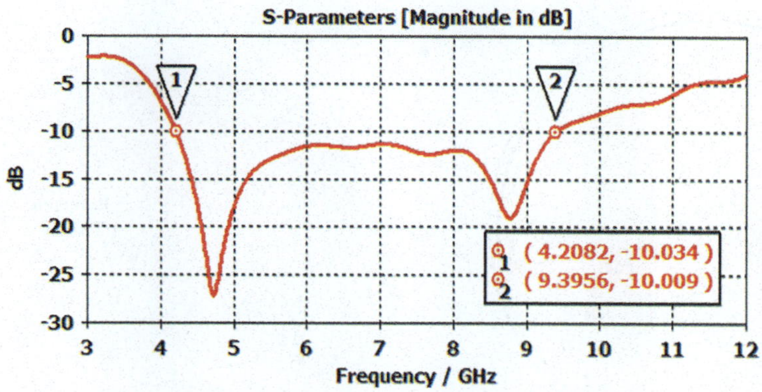

FIGURE 23.2 Reflection parameters of the proposed CSDRA. *CSDRA*, Cubical-shaped dielectric resonator antenna.

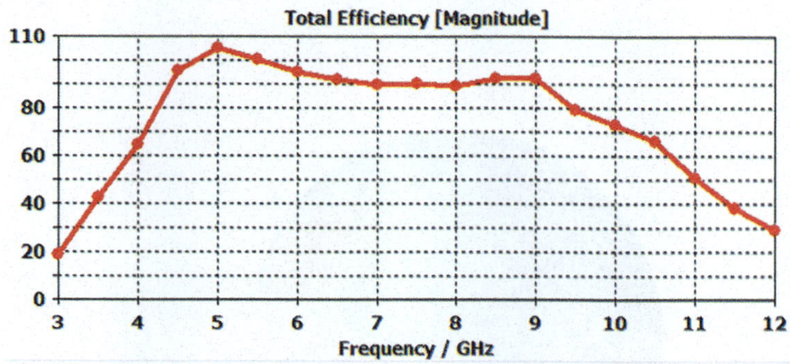

FIGURE 23.3 Efficiency of the proposed CSDRA. *CSDRA*, Cubical-shaped dielectric resonator antenna.

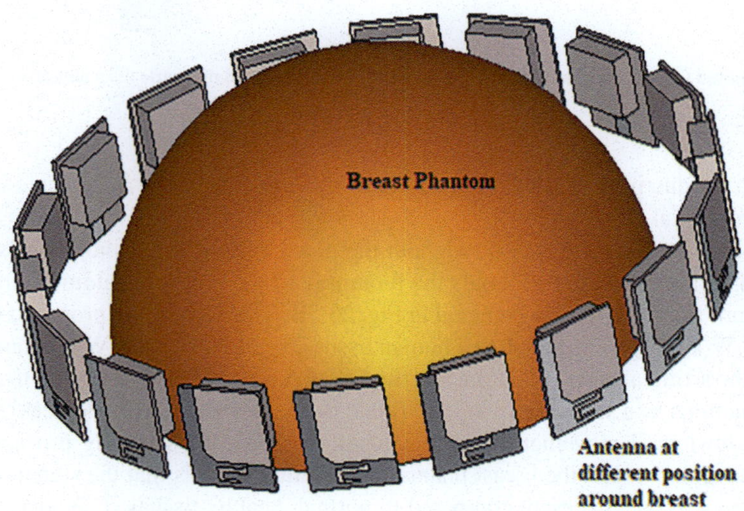

FIGURE 23.4 Setup of the microwave imaging using monostatic radar–based technique.

To assess the system's efficacy a homogeneous breast phantom containing a single tumor with a higher dielectric constant is utilized. This setup enables evaluating the system's performance in tumor localization and MI. The phantom dimension measures approximately 12×6 cm^2, containing a dielectric constant similar to that of a human breast and an 8-mm diameter tumor embedded within. The breast phantom with inserted tumor includes the skin layer and the fat layer as depicted in Fig. 23.5A. The skin layer possesses a dielectric constant of 38 and a conductivity of 1.49 S/m, with a thickness of 2 mm. The fat layer, with a maximum diameter of 7.6 cm, contains a dielectric constant of 5.14 and a conductivity of 0.141 S/m.

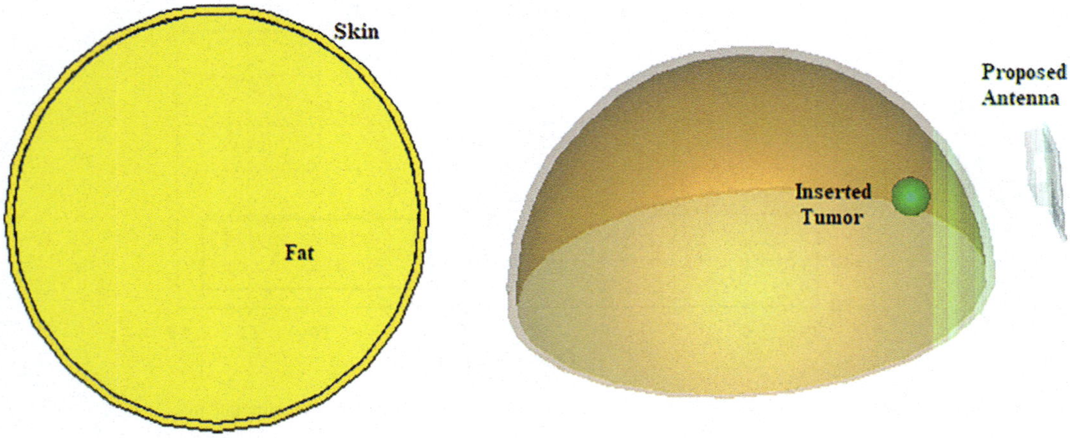

FIGURE 23.5 (A) Different layers of the breast phantom and (B) proposed antenna with designed breast phantom.

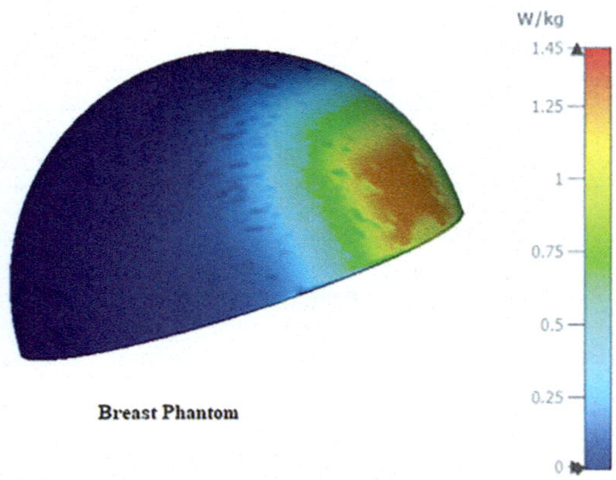

FIGURE 23.6 SAR of the proposed CSDRA with breast phantom. *CSDRA*, Cubical-shaped dielectric resonator antenna; *SAR*, specific absorption rate.

The tumor layer exhibits a conductivity of 3.1 S/m and a dielectric constant of 67, as depicted in Fig. 23.5B (Bhargava & Rattanadecho, 2022; Matrone et al., 2015).

The phantom rotation aids in gathering data throughout the phantom to identify the tumor's location. To maintain the specific absorption rate (SAR) within safe thresholds the biomimic of the breast phantom is strategically positioned at an optimized distance of 10 mm from CSDRA as depicted in Fig. 23.5B. The SAR of the proposed CSDRA in the presence of the breast phantom is 1.45 W/kg (Fig. 23.6) for 1 g of human tissues, i.e., less than 1.6 W/kg, thus safe for human exposure.

The distance between the artificial breast phantom and the CSDRA is maintained at 10 mm to ensure near-field radiation. The proposed CSDRA functions as a sensor, moving in 10-degree increments (0–360 degrees) within the elevation plane, followed by upward and downward movements of the antenna to repeat the scanning process. Fig. 23.7 illustrates the reflection parameters of the antenna with the breast phantom. The graph reveals that the signal reflected from the phantom is considerably higher when a tumor is present compared to normal, healthy tissues. This increase results from the higher water content and distinct electrical properties of tumorous tissues compared to healthy tissues.

The RMMWI technology is a noninvasive imaging technique that uses UWB EM waves to generate a comprehensive dielectric profile of malignant tissues beneath the skin layer. The EM signals that are reflected and backscattered are gathered by the same CSDRA, allowing for the detection of changes in the tissues' EM properties caused by malignant growths within the phantom. The backscattered signals are subjected to a conversion from the frequency domain to the time domain. In addition, the method involves using Hermitian signal preprocessing techniques on the acquired data to reduce the presence of signal artifacts.

Finally, eliminating artifacts from time-domain backscattered signals requires knowledge of a tumor-free forearm biomimic, as mentioned in Eq. (23.1). As described in references the malignant or tumor response is calculated by

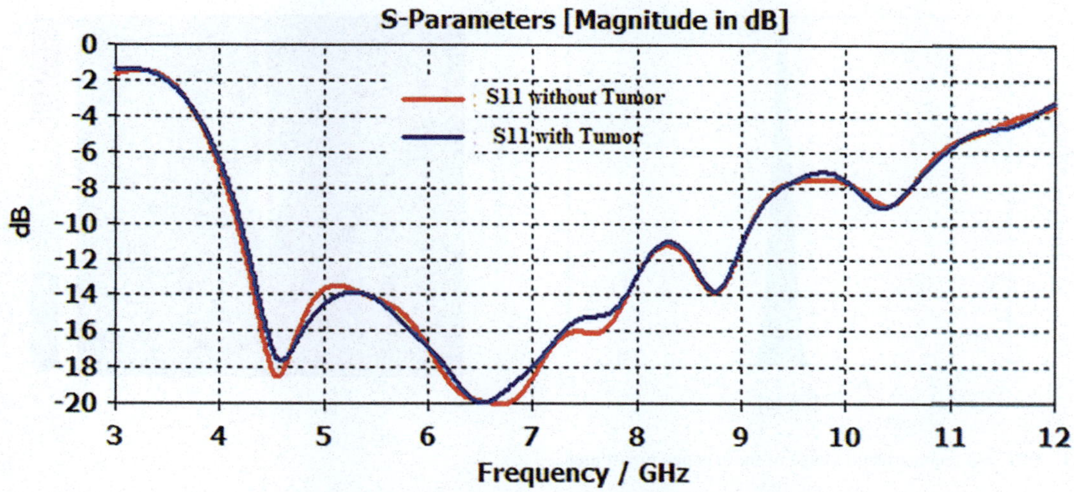

FIGURE 23.7 Reflection parameters of the CSDRA with the presence and absence of tumor inside the breast phantom. *CSDRA*, Cubical-shaped dielectric resonator antenna.

Kaur and Kaur (2022a), subtracting the backscattered signals from the tumor-bearing forearm phantom from the tumor-free one:

$$S_{ii}[t] = S_{ii}[t]_{tumor-bearing} - S_{ii}[t]_{tumor-free}$$

The preprocessed backscattered signal is subsequently directed to the beam-forming algorithm, which serves to diminish early-time content, mitigate clutter in late-time content, and amplify high-intensity focus (tumor) signals, ensuring dependable tumor detection within the reconstructed image of the scanned area. Beam-forming constitutes a spatial filtering technique involving the temporal integration of transmitted or received RF signals gathered at various time intervals, aiming to optimize both maximum and minimum energy levels.

23.4.1 Confocal MI algorithm

The imaging technique combines backscattered signals obtained by an antenna from different points to provide an image intensity of the scanned area, which helps identify the tumor's location. The tumor response is affected by clutter signals because of the dispersive nature of forearm tissues. The reconstruction algorithm plays a crucial role in accurately identifying tumors by reducing the influence of unwanted signals. The confocal imaging algorithm, initially utilized for breast cancer diagnosis, is currently undergoing clinical assessment. This technique utilizes the quantitative distribution of dielectric constant and permittivity to detect locations with significant reflection. In the confocal imaging algorithm coherent-factor delay multiply and sum (CF-DMAS) the process involves pairwise multiplication of time-delayed signals followed by summation (Kibria et al., 2019). Each antenna position calculates the round-trip time delay to a specific point in the test domain, and the corresponding signal time delays are aggregated. The iterative focusing procedure scans the focal point to a new location, ensuring the utilization of signals from a specific pixel. The resultant signals enhance high-intensity signals while reducing side-lobe and clutter effects. The collected scattering data are concentrated artificially at a specific location on the forearm phantom to generate an image. Subsequently, time delays are computed based on the distances between each antenna position and the focal point.

In this algorithm a grid matrix of maximum and minimum intensities based on the collected backscattered signals has been generated using CF-DMAS and displayed using MATLAB, providing a PDF and 2D picture of the scanned area region for different locations around the phantom while highlighting tumorous position and size as depicted in Fig. 23.8.

$$S_{DMAS} = \sum_{j=1}^{M-1} \sum_{k=j+1}^{M} \sqrt{S_j(t + \Delta t_j)}\sqrt{S_k(t + \Delta t_k)}$$

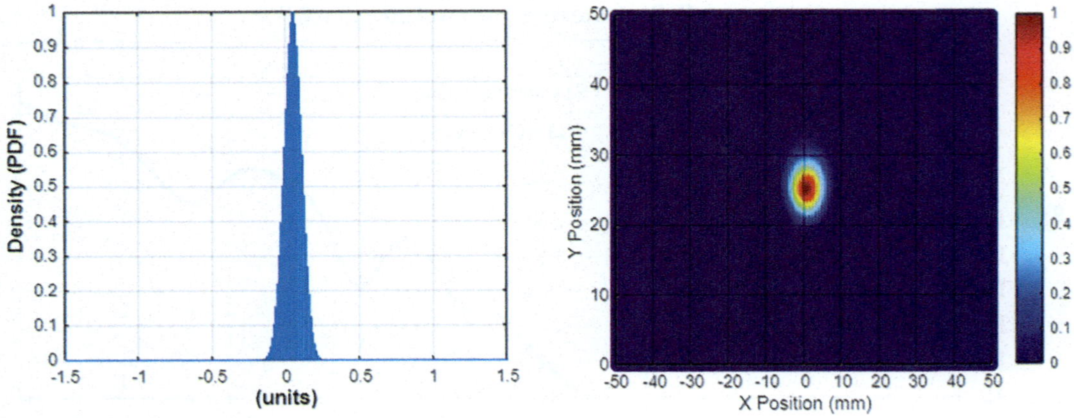

FIGURE 23.8 PDF and reconstructed image of the scanned breast area.

$$\mathrm{CF_{DMAS}} = \frac{\left(\sum_{j=1}^{M-1}\sum_{k=j+1}^{M}\sqrt{S_j(t+\Delta t_j)}\sqrt{S_k(t+\Delta t_k)}\right)^2}{\frac{M(M-1)}{2}\sum_{i=1}^{M-1}\sum_{j=i+1}^{M}\left|S_j(t+\Delta t_j)\right|\left|S_k(t+\Delta t_k)\right|}$$

Thus final DMAS-CF representation is given as

$$S_{DMAS-CF} = S_{DMAS}(t) \times CF_{DMAS}(t)$$

$$\mathrm{S_{DMAS-CF}} = \frac{\left[\left\{\left(\sum_{j=1}^{M}\sqrt{S_j(t+\Delta t_j)}\right)^2 - \sum_{j=1}^{M}\left|S_j(t+\Delta t_k)\right|\right\}/2\right]^3}{2M(M-1)\left[\sum_{j=1}^{M}\left|S_j(t+\Delta t_j)\right|^2 - \sum_{j=1}^{M}\left(\left|S_j(t+\Delta t_j)\right|\right)^2\right]}$$

where M represents number of iteration, $S_j(t+\Delta t_j)$ is the signal received by jth antenna in terms of S_{11} with resultant time delay of Δt_j.

23.5 Conclusion

The proposed chapter proposed a CSDRA that works as a sensor for microwave imaging of breast cancer. This technique is noninvasive and nonionizing in nature. It utilizes a single antenna for transmission and reception, simplifying the imaging process while maintaining accuracy. The proposed CSDRA is miniaturized with dimension of 16 × 20 mm^2 and designed on a Rogers RT5880 substrate. The proposed antenna demonstrates excellent performance for imaging applications with an impedance BW of 5.1 GHz from 4.2 to 9.39 GHz. This chapter also showcases an in silico approach for RMMWI using the CSDRA and a breast phantom, employing the CF-DMAS beam-forming algorithm to reconstruct dielectric profiles. This method ensures safety for human exposure, meeting SAR standards with values below 1.6 W/kg for 1 g of tissue. Overall the proposed system holds significant potential for advancing safe, effective, and accessible cancer detection technologies.

References

Bhargava, D., & Rattanadecho, P. (2022). Microwave imaging of breast cancer: Simulation analysis of SAR and temperature in tumors for different age and type. *Case Studies in Thermal Engineering, 31*, 101843. doi:10.1016/j.csite.2022.101843.

Bharmjeet, & Das, A. (2023). Racial disparities in cancer care, an eyeopener for developing better global cancer management strategies. *Cancer Reports, 6*(S1). doi:10.1002/cnr2.1807.

Hussain, S., Mubeen, I., Ullah, N., Ud Din Shah, S. S., Khan, B. A., Zahoor, M., Ullah, R., Khan, F. A., Sultan, M. A., & Li, C. (2022). Modern diagnostic imaging technique applications and risk factors in the medical field: A review. *BioMed Research International, 2022*, 1–19. doi:10.1155/2022/5164970.

Kaur, G., Kaur, R., Kamboj, P., & Kaur, H. (2024). A 2 × 2 compact and portable UWB MIMO dielectric resonator antenna (DRA) for future mobile handset with Improved Diversity and Isolation Level. In *Lecture Notes in Electrical Engineering* (pp. 625–635). Springer Science and Business Media GmbH. https://www.springer.com/series/7818/1191. doi:10.1007/978-981-97-2508-3_46.

Kaur, G., & Kaur, A. (2022a). Monostatic radar-based microwave imaging of breast tumor using an ultra-wideband dielectric resonator antenna (DRA) with a Sierpinski fractal defected ground structure. *MAPAN, 37*(4), 917–928. doi:10.1007/s12647-022-00536-7.

Kaur, K., & Kaur, A. (2022b). In vitro detection of skin cancer using an UWB stacked micro strip patch antenna with microwave imaging. *International Journal of RF and Microwave Computer-Aided Engineering, 32*(12). doi:10.1002/mmce.23407.

Kaur, K., & Kaur, A. (2025). A hybrid fractal microstrip patch antenna rooted from pythagoras-koch snowflake geometries for microwave imaging of skin and breast cancer. *Physica Scripta, 100*(8), 085547.

Kaur, K., & Kaur, A. (2026). Fractal dielectric resonator antenna for microwave imaging of skin tumors: A multimodal in-silico, in-vitro, and ex-vivo based investigation. *Measurement, 285*, 122302. https://doi.org/10.1016/j.measurement.2026.122302.

Kibria, S., Samsuzzaman, M., Islam, M. T., Mahmud, M. Z., Misran, N., & Islam, M. T. (2019). Breast phantom imaging using iteratively corrected coherence factor delay and sum. *IEEE Access, 7*, 40822–40832. http://ieeexplore.ieee.org/xpl/RecentIssue.jsp?punumber=6287639. doi:10.1109/ACCESS.2019.2906566.

Latif, S. I., Tapia, D. F., Herrera, D. R., Nepote, M. S., Pistorius, S., & Shafai, L. (2015). A directional antenna in a matching liquid for microwave radar imaging. *International Journal of Antennas and Propagation, 2015*, 1–8. doi:10.1155/2015/751739.

Lazaro, A., Villarino, R., & Girbau, D. (2011). Design of tapered slot Vivaldi antenna for UWB breast cancer detection. *Microwave and Optical Technology Letters, 53*(3), 639–643. doi:10.1002/mop.25770.

Matrone, G., Savoia, A. S., Caliano, G., & Magenes, G. (2015). The delay multiply and sum beamforming algorithm in ultrasound B-mode medical imaging. *IEEE Transactions on Medical Imaging, 34*(4), 940–949. http://ieeexplore.ieee.org/xpl/RecentIssue.jsp?punumber=42. doi:10.1109/TMI.2014.2371235.

Mersani, A., Osman, L., & Ribero, J. M. (2019). Flexible UWB AMC antenna for early stage skin cancer identification. *Progress in Electromagnetics Research M, 80*, 71–81. http://www.jpier.org/PIERM/pierm80/07.18121404.pdf. doi:10.2528/PIERM18121404.

Morrow, M., Waters, J., & Morris, E. (2011). MRI for breast cancer screening, diagnosis, and treatment. *The Lancet, 378*(9805), 1804–1811. doi:10.1016/s0140-6736(11)61350-0.

Qashlan, A. M., Aldhaheri, R. W., & Alharbi, K. H. (2022). A modified compact flexible Vivaldi antenna array design for microwave breast cancer detection. *Applied Sciences, 12*(10), 4908. doi:10.3390/app12104908.

Sarno, A., Valero, C., Tucciariello, R. M., Dukov, N., Costa, P. R., & Tomal, A. (2023). Physical and digital phantoms for 2D and 3D x-ray breast imaging: Review on the state-of-the-art and future prospects. *Radiation Physics and Chemistry, 204*, 110715. doi:10.1016/j.radphyschem.2022.110715.

Siegel, RL, Miller, KD, Wagle, NS, & Jemal, A (2023). Cancer statistics, 2023. *CA Cancer J Clin, 73*(1), 17–48. doi:10.3322/caac.21763.

Sun, M., Moquet, J., Ellender, M., Bouffler, S., Badie, C., Baldwin-Cleland, R., Monahan, K., Latchford, A., Lloyd, D., Clark, S., Anyamene, N. A., Ainsbury, E., & Burling, D. (2023). Potential risks associated with the use of ionizing radiation for imaging and treatment of colorectal cancer in Lynch syndrome patients. *Familial Cancer, 22*(1), 61–70. https://www.springer.com/journal/10689. doi:10.1007/s10689-022-00299-9.

Syed, A., Sobahi, N., Sheikh, M., Mittra, R., & Rmili, H. (2022). Modified 16-Quasi log periodic antenna array for microwave imaging of breast cancer detection. *Applied Sciences, 12*(1), 147. doi:10.3390/app12010147.

Tiang, S. S., Sadoon, M., Zanoon, T. F., Ain, M. F., & Abdullah, M. Z. (2013). Radar sensing featuring biconical antenna and enhanced delay and sum algorithm for early stage breast cancer detection. *Progress in Electromagnetics Research B*, (46), 299–316. http://www.jpier.org/PIERB/pierb46/16.12102201.pdf. doi:10.2528/PIERB12102201Malaysia.

Wang, J., & Wu, S. G. (2023). Breast cancer: An overview of current therapeutic strategies, challenge, and perspectives. *Breast Cancer: Targets and Therapy, 15*, 721–730. https://www.dovepress.com/getfile.php?fileID=93625. doi:10.2147/BCTT.S432526.

Chapter 24

Artificial intelligence methods for mechatronics: Enhancing functionality and efficiency

Lamia May[a] and Mohamed Arezki Mellal[b]
[a] *Laboratory of Mechanics, Materials and Energy (L2ME), Faculty of Technology, University of Bejaia, Bejaia, Algeria,*
[b] *LMSS, Faculty of Technology, M'Hamed Bougara University, Boumerdes, Algeria*

24.1 Introduction

Artificial intelligence (AI) has become a cornerstone of modern computational science, providing a range of sophisticated methodologies designed to replicate, enhance, and in some cases surpass human cognitive functions. These functions encompass data-driven learning, pattern recognition, decision-making, autonomous reasoning, predictive analytics, and adaptive control. AI techniques span a broad spectrum, from rule-based symbolic reasoning systems to advanced machine learning paradigms such as deep learning (DL) and reinforcement learning (RL). The continuous evolution of these techniques has had a profound impact on various engineering disciplines, particularly in the domain of mechatronics, where AI-driven innovations are redefining system autonomy and intelligence.

Mechatronics combines mechanical engineering, electronics, control theory, and computer science to develop intelligent systems. It has been pivotal in advancing automation, robotics, and cyber-physical systems, contributing to breakthroughs in a range of industries, from manufacturing to autonomous transportation. Nevertheless, conventional mechatronic systems, which are frequently designed with fixed control architectures and preprogrammed responses, are increasingly inadequate for addressing the challenges posed by dynamic, uncertain, and highly complex operating environments. AI methodologies provide a transformative alternative by enabling systems to learn from data, self-optimize in real-time, and exhibit adaptive behavior that extends beyond conventional deterministic programming.

The integration of AI into mechatronics signifies a transformation in the established paradigms of design and control. Conventional approaches rely heavily on predefined mathematical models, which often struggle to accommodate the stochastic and nonlinear characteristics of real-world environments. By contrast, AI-augmented mechatronic systems leverage data-driven learning mechanisms, enabling the inference of optimal control strategies, the detection of anomalies, and the execution of predictive adjustments without the need for human intervention. The efficiency of such techniques, exemplified by artificial neural networks (ANNs), deep RL, and evolutionary algorithms, has been demonstrated in the domains of robotics, precision manufacturing, and intelligent automation, resulting in enhanced efficiency, resilience, and autonomy. AI-driven predictive maintenance strategies further demonstrate this transformation by their ability to preempt failures and optimize operational uptime, thereby reducing costs and increasing reliability in industrial applications.

The historical evolution of AI in mechatronics has been characterized by a series of pivotal advancements that have progressively expanded the field's capabilities. Early AI implementation, such as expert systems and fuzzy logic controllers, established the foundation for intelligent decision-making and uncertainty management in control systems. The subsequent rise of computational power and the proliferation of sensor-driven data have paved the way for more advanced AI techniques, such as genetic algorithms (GAs), DL architectures, and neurofuzzy hybrid systems. These methodologies have facilitated unprecedented improvements in system autonomy, fault tolerance, and optimization, as well as innovations in autonomous robotics, self-learning adaptive controllers, and intelligent cyber-physical systems that operate with minimal human supervision.

Computational Intelligence in Mechatronics: Solving Real-World Problems in Electronic Systems Design. DOI: https://doi.org/10.1016/B978-0-443-27580-7.00008-1

Emerging AI paradigms are now addressing pressing challenges related to expansibility, security, and real-time performance in mechatronics. Explainable AI (XAI) has become an essential consideration for safety-critical applications such as medical robotics and autonomous vehicles, where transparent and interpretable decision-making is crucial. Concurrently the field is undergoing a paradigm shift through the introduction of edge AI, which is enabling decentralized, low-latency computation to be performed directly on embedded hardware. This shift is reducing reliance on cloud-based processing and enhancing system responsiveness. These advancements are poised to further enhance the intelligence and operational robustness of next-generation mechatronic systems.

This chapter presents a through examination of the progressive integration of AI into mechatronics, structured along a historical and technological continuum. It commences with an examination of foundational methodologies, including expert systems and fuzzy logic, which have historically influenced intelligent control. Subsequently, it explores contemporary advancements, such as DL, RL, and hybrid AI frameworks, which are driving significant enhancements in system autonomy and performance. The discussion extends to state-of-the-art innovations in explainable AI and edge computing, highlighting their potential to address contemporary challenges and unlock new frontiers in intelligent mechatronic systems.

By delineating the evolution and application of AI in mechatronics, this chapter seeks to provide a nuanced understanding of AI's transformative impact on the field. The synthesis of AI and mechatronics not only fosters technological breakthroughs but also redefines the interaction between autonomous systems and their environments. As AI continues to evolve, its integration into mechatronics will further revolutionize automation, intelligence augmentation, and human–machine collaboration, shaping the trajectory of future research and industrial innovation.

24.2 The role of AI in mechatronics: Challenges and emerging trends

As an interdisciplinary field, mechatronics integrates the principles of mechanical engineering, electronics, control systems, and computer science, resulting in the development of intelligent systems capable of performing complex tasks with precision and efficiency. The contemporary conception of mechatronic systems necessitates a high degree of adaptability and intelligence to function effectively in dynamic and unpredictable environment. Such systems may include autonomous robots, advanced manufacturing equipment, and cyber-physical systems (Habib & Davim, 2013).

In conventional mechatronic systems, mathematical models and rule-based control strategies are utilized. While the efficacy of these methods in controlled environments is well documented, they encounter difficulties when confronted with real-world uncertainties, system complexities, and large-scale automation. The advent of AI has precipitated a paradigm shift in mechatronics by virtue of the introduction of learning-based, data-driven, and adaptive solutions that enhance system intelligence, reliability, and efficiency (Brunton & Kutz, 2019).

24.2.1 Challenges in traditional mechatronic systems

Several fundamental challenges limit the effectiveness of conventional mechatronic control and optimization techniques:

- Nonlinear dynamics and modeling complexity
  It is evident that a considerable number of mechatronic systems exhibit highly nonlinear behavior, which renders them difficult to model accurately by means of conventional analytical approaches. To illustrate this point, one may consider the case of robotic manipulators in industrial automation, which are required to take into account variable loads, joint friction, and external disturbances. These factors, among others, serve to complicate model-based control (Slotine & Li, 1991). Consequently, there has been an exploration of alternative modeling techniques, such as AI-driven methods, including neural networks and RL, as a potential solution to the challenge posed by traditional physics-based models (Levine et al., 2016).
- Uncertainty and variability in system behavior
  A number of factors have been identified as introducing significant uncertainty into mechatronic systems. These include environmental disturbances, sensor noise, mechanical wear, and unexpected operational conditions. To provide an example, in the context of autonomous driving, sensor degradation because of adverse weather conditions can have a severe impact on perception and decision-making (Bansal et al., 2018). However, recent studies have demonstrated that AI techniques such as Bayesian inference and DL have the potential to enhance robustness of mechatronic systems by incorporating probabilistic reasoning and adaptive learning mechanisms (Papamarkou et al., 2024).
- Scalability and computational complexity
  As mechatronic systems grow in complexity, traditional rule-based and model-based methods become computationally expensive and difficult to scale. Intelligent manufacturing systems, for instance, require real-time optimization of

numerous interconnected processes. Consequently, there has been a notable shift toward the utilization of AI-driven optimization methods, such as deep RL and evolutionary algorithms, to enhance the efficiency of industrial production lines and robotic systems (Nelson et al., 2023; Tortorelli et al., 2022).

- Lack of adaptability in dynamic environments
  Conventional mechatronics control systems are unable to adapt to evolving conditions because of their reliance on static models and preprogrammed rules. In contrast, AI-driven controllers have the capacity to learn from experience and continuously improve system performance. The utilization of RL and DL facilitates real-time adaptation in robotics, autonomous vehicles, and intelligent sensor networks (Hamidaoui et al., 2025; Zaitceva & Andrievsky, 2022).

24.2.2 The convergence of AI and mechatronics

AI improves mechatronic systems by addressing these limitations through advanced learning, control, and optimization techniques:

- Data-driven learning and predictive analytics
  The integration of AI within mechatronic systems facilitates the analysis of substantial sensor data, the detection of patterns, and the subsequent formation of intelligent decisions. A prime illustration of this is the implementation of predictive maintenance, which utilizes AI algorithms to analyze vibration, temperature, and pressure data, thereby facilitating the prediction of failures before their occurrence. This, in turn, results in a reduction in downtime and maintenance cost (Ohalete et al., 2023; Yuan et al., 2020).
- Adaptive and intelligent control
  In dynamic environments where external conditions are subject to rapid and frequent changes, traditional PID (proportional-integral-derivative) controller systems frequently demonstrate an inability to respond adequately. Conversely, AI-based adaptive control techniques, including RL and fuzzy logic, empower robots and autonomous systems to perpetually enhance their control strategies through real-time feedback (Brunton & Kutz, 2019).
- Optimization of performance through AI-based methods
  The utilization of AI-driven optimization algorithms has been demonstrated to enhance the overall system efficiency through the dynamic fine-tuning of parameters. Within the context of smart grids, for instance, AI algorithms are employed to optimize energy distribution in accordance with demand and supply fluctuations, thereby minimizing power losses and operational expenses (Biswas et al., 2024; Yuan et al., 2020; Zulu et al., 2023).
- DL for perception and decision-making
  The advent of deep neural networks (DNNs) has precipitated a paradigm shift in the domain of perception systems in robotics and automation. The AI-driven computer vision models have enabled real-time object recognition, path planning, and autonomous decision-making in a range of applications, including robotic surgery, warehouse automation, and self-driving cars (Kendoul, 2012).

The integration of AI into mechatronics represents more than mere incremental improvement; rather, it signifies a transformative shift from rigid, rule-based systems to intelligent, self-learning, and adaptive solutions. This transformation has already begun to reshape a number of industries, including autonomous robotics, industrial automation, and cyber-physical systems. Nevertheless, challenges such as ensuring interpretability of AI systems, the capacity to process data in real-time, and assurance of safety remain active areas of research.

24.2.3 The evolution of AI in mechatronics

The adoption of AI in mechatronics can be traced back to wider evolution of AI itself. The early approaches, including symbolic AI and fuzzy logic, laid the foundation for intelligent control and decision-making. A notable application of fuzzy logic, for instance, has been in household appliances such as washing machines and air conditioners, enhancing both energy efficiency and user comfort (Yurkovich & Passino, 1999).

As computational power and data availability increased, more sophisticated techniques such as GAs, neural networks, and RL emerged, thus enabling significant advances in autonomy and adaptability. For instance, GAs have been utilized to optimize the design of mechatronic systems, including the suspension system of vehicles, thereby achieving enhanced performance and comfort (Goldberg, 1989).

Emerging trends in the field of AI, such as Explainable AI (XAI) and edge AI, are poised to address critical challenges related to transparency, safety, and real-time processing. This has the potential to further expand the scope and applications of AI in mechatronics. For instance, XAI techniques are being integrated into autonomous vehicles to facilitate interpretable

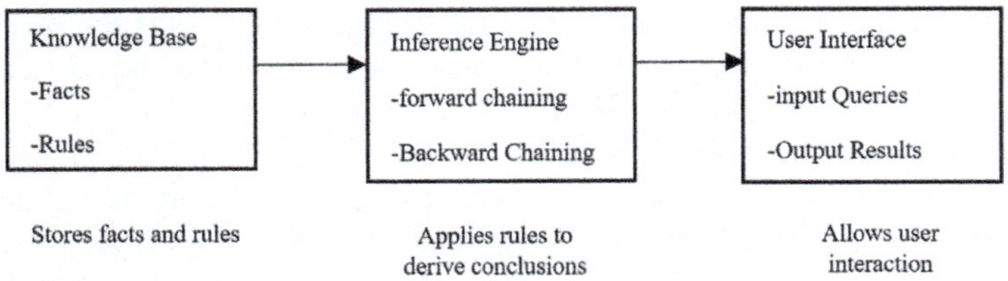

FIGURE 24.1 Structure of a symbolic AI system. Schematic representation of symbolic AI architecture, showcasing the knowledge base, inference engine, and interface for role-based reasoning in mechatronic applications. *AI*, Artificial intelligence.

decision-making processes, thereby enhancing safety and user trust (Barredo Arrieta et al., 2020). A similar approach is seen in edge AI, which facilitates real-time processing in embedded systems, such as industrial robots, by deploying lightweight AI models directly on the hardware.

This section has outlined the challenges faced by traditional mechatronic systems and the transformative potential of AI methods. In the following sections the chronological development of AI methods will be explored, alongside their applications in mechatronics and the opportunities they present for future innovation.

24.2.4 Chronological overview of AI methods

The evolution of AI methods has been marked by significant milestones, each contributing to the advancement of mechatronics. This section provides a chronological overview of key AI methods, highlighting their development, underlying principles, and applications in mechatronic systems.

24.2.4.1 Symbolic AI

Symbolic AI, also known as rule-based AI, was one of the earliest approaches to AI and emerged in the 1950s as a formal approach for replicating human reasoning using logic-based structures. It relied on predefined rules and formal logic to encode knowledge, thereby enabling expert systems and automated reasoning in early AI applications (Johnson et al., 1973). Despite its dominance until the late 1980s, its limitation such as poor scalability and the inability to learn, led to the rise of data-driven methods (Russell & Norvig, 2016).

24.2.4.1.1 Theoretical foundations

Symbolic AI is rooted in first-order logic (FOL), where knowledge is represented using constants (e.g., Robots *A*, Sensors *X*), variables (e.g., *x*, *y*), predicates (e.g., fault(*x*), meaning "*x* has a fault"), and inference rules. For example, an industrial diagnostic rule can be written as

$$\forall x Fault(x) \rightarrow Alert(x) \quad (24.1)$$

Rule-based systems apply inference through forward chaining (data-driven) or backward chaining (goal-driven), forming the basis of early expert systems like MYCIN (Shortliffe, 1976). These systems employed knowledge bases and inference engines to reason about structured rules, rendering them indispensable for fault diagnosis and industrial automation. The architecture of a typical symbolic AI framework is depicted in Fig. 24.1.

24.2.4.1.2 Applications in mechatronics

Symbolic AI had a profound impact on early mechatronic systems, particularly in fault diagnosis, control, automation, and planning. The utilization of expert systems for purpose of predictive maintenance in industrial machinery has been a prevalent practice, with the adaptation of rule-based reasoning (Shortliffe, 1976) being a prominent approach. This method has been employed to detect faults in Computer Numerical Control (CNC) machines and robotic manipulators. In the domain of manufacturing automation, programmable logic controllers (PLCs) have been instrumental in implementing rule-based decision-making methodologies to orchestrate assembly lines, sorting processes, and welding operations. This approach ensures optimal efficiency and precision. Early robotic systems utilized symbolic AI for path planning and motion control, thereby enhancing task scheduling and automated operations in structured environments (Russell & Norvig, 2016).

It also contributed to process control, where rule-based controllers adjusted operational parameters in chemical plants, HVAC systems, and power grids. In addition, it influenced early human–machine interaction, enabling vice-command recognition in industrial settings.

Nevertheless, the inherent rigidity of symbolic AI in dealing with uncertainty and inability to adapt in real time resulted in the adoption of more flexible approaches, such as fuzzy logic and probabilistic reasoning, in mechatronics.

24.2.4.1.3 Limitations and evolution

Despite its success, symbolic AI encountered issues with scalability, lack of adaptability, and management uncertainty. Rigid logic systems encountered challenges in dynamic environments, which led to the emergence of fuzzy logic (Zadeh, 1965) and probabilistic models to enhance flexibility. Hybrid AI approaches, combining symbolic reasoning with machine learning, have gained traction, particularly in modern neurosymbolic AI, which enhances interpretability and robustness in complex decision-making (Sarker et al., 2021).

24.2.4.2 Fuzzy logic (1960s–1970s)

The concept of fuzzy logic, introduced by Lotfi Zadeh in 1965, is an extension of classical Boolean logic to include the notion of reasoning under uncertainty and imprecision. In contradistinction to binary logic, where a variable is either strictly true (1) or false (0), fuzzy logic enables gradual truth values, rendering it ideal for modeling complex, ambiguous, or human-like reasoning (Zadeh, 1965). This paradigm has seen extensive adoption in fields such as control systems, robotics, and intelligent automation, where crisp classifications are impractical.

24.2.4.2.1 Theoretical foundations

Fuzzy logic is based on fuzzy set theory, where each element x in a universal set X is assigned a degree of membership $\mu_A(x)$ in a fuzzy set A, defend as

$$\mu_A(x) : X \rightarrow [0, 1] \tag{24.2}$$

This allows flexible classifications; for instance, in temperature control, "warm" might have partial membership in both "cool" and "hot" categories.

A fuzzy inference system (FIS) follows four main steps:

- Fuzzification: This process involves the conversion of fuzzy values through the utilization of membership functions, such as the Gaussian or triangular function.
- Rule base: It employs linguistic rules, e.g., IF temperature is HIGH THEN fan speed is FAST.
- Inference engine: Applies fuzzy operators (min–max, fuzzy AND/OR) to derive conclusions.
- Defuzzification: This process involves the conversion of fuzzy output into a crisp numerical value through the utilization of methodologies such as the centroid approach:

$$y^* = \frac{\sum \mu_B(y).y}{\sum \mu_B(y)} \tag{24.3}$$

Fuzzy control has been shown to be particularly powerful in the context of nonlinear systems, where the derivation of explicit mathematical model can be challenging (Ross, 2005). The general structure of fuzzy logic control system is illustrated in Fig. 24.2, highlighting its core components.

24.2.4.2.2 Applications in mechatronics

Fuzzy logic finds extensive application in robotics, automotive systems, and industrial automation.

Industrial control: Fuzzy controllers optimize temperature regulation, motor speed, and process stability (Yurkovich & Passino, 1999).

Automotive engineering: Used in antilock braking system (ABS), adaptive cruise control, and suspension damping, improving vehicle safety and comfort (Yager & Filev, 1994).

Robotics: Facilitates autonomous motion planning and obstacle avoidance, notably in uncertain environments, through the use of sensor data (Stoian & Ivanescu, 2008).

Predictive maintenance: Fuzzy models assess machine health by analyzing vibration and temperature data, reducing unexpected failures and downtime (Baban et al., 2016).

Washing machines: Fuzzy logic controllers optimize water usage, detergent levels, and wash cycles based on load size and fabric type. These systems improve energy efficiency and user convenience (Masood, 2017).

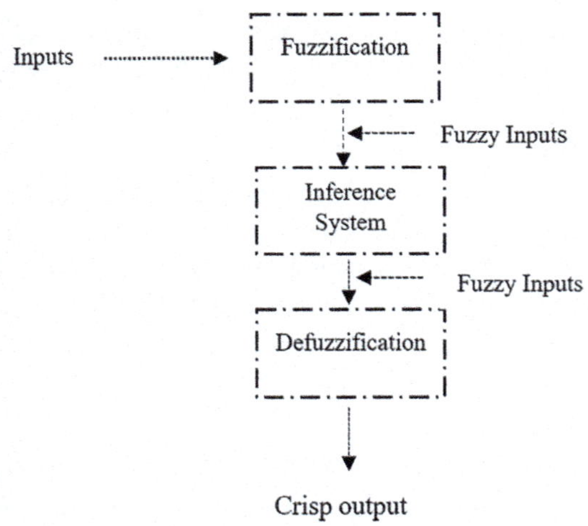

FIGURE 24.2 The structure of the fuzzy logic control. Structure of a fuzzy logic controller, illustrating fuzzification, rule base, inference mechanism, and defuzzification in control systems.

24.2.4.2.3 Limitation and evolution

Despite its advantages of fuzzy logic, challenges such as rule explosion (exponential growth of rule sets) and difficulty in parameter tuning are to be noted. To address these challenges, hybrid models such as fuzzy-neural networks (FFNs) and fuzzy-GAs (FGAs) have been proposed, integrating learning and optimization techniques (Cho, 2002). The integration of fuzzy logic with DL represents a nascent field, with the potential to enhance interpretability in AI-driven automation (Storn & Price, 1997).

24.2.4.3 Genetic algorithms (GAs) (1970s–1980s)

Introduced by Holland (1975), GAs represent a class of stochastic optimization techniques that draw inspiration from the process of natural selection. They are designed to solve complex, multiobjective optimization problems by evolving a population of candidate solutions over successive generations. GAs are a class of stochastic optimization techniques that draw inspiration from the process of natural selection. They are designed to solve complex, multiobjective optimization problems by evolving a population of candidate solutions over successive generations (Deb et al., 2002; Goldberg, 1989). In contradistinction to deterministic methodologies, GAs leverage randomness and selection pressure to efficiently explore vast, nonlinear spaces, rendering them especially well-suited for mechatronic system optimization.

24.2.4.3.1 Theoretical foundations

GAs are a computational model of biological evolution, incorporating operators such as selection, crossover, and mutation. The algorithm starts by encoding potential solutions as chromosomes, which are typically binary or real-valued strings, thereby forming an initial population. The fitness function, $f(x)$, evaluates the quality of each solution.

At each iteration (generation) the following steps occur:

- Selection: select parent solutions based on their fitness.
- Crossover (recombination): combine two parents to create offspring by exchanging genetic material:

$$\mathit{Offspring} = \mathit{Crossover}(\text{Parent}_1, \text{Parent}_2) \tag{24.4}$$

- Mutation: apply random alterations to offspring genes to maintain diversity:

$$\mathit{Mutated\ Offspring} = \mathit{Mutated}\left(\text{Offspring}\right) \tag{24.5}$$

- Replacement: the new population replaces the old one based on survival criteria.

Convergence occurs when the population reaches satisfactory fitness after a fixed number of generations. Mathematically, GAs aim to solve the following:

$$\max_{x \in S} f(x), S \subseteq \mathbb{R}^n \tag{24.6}$$

where S is the search space of feasible solutions.

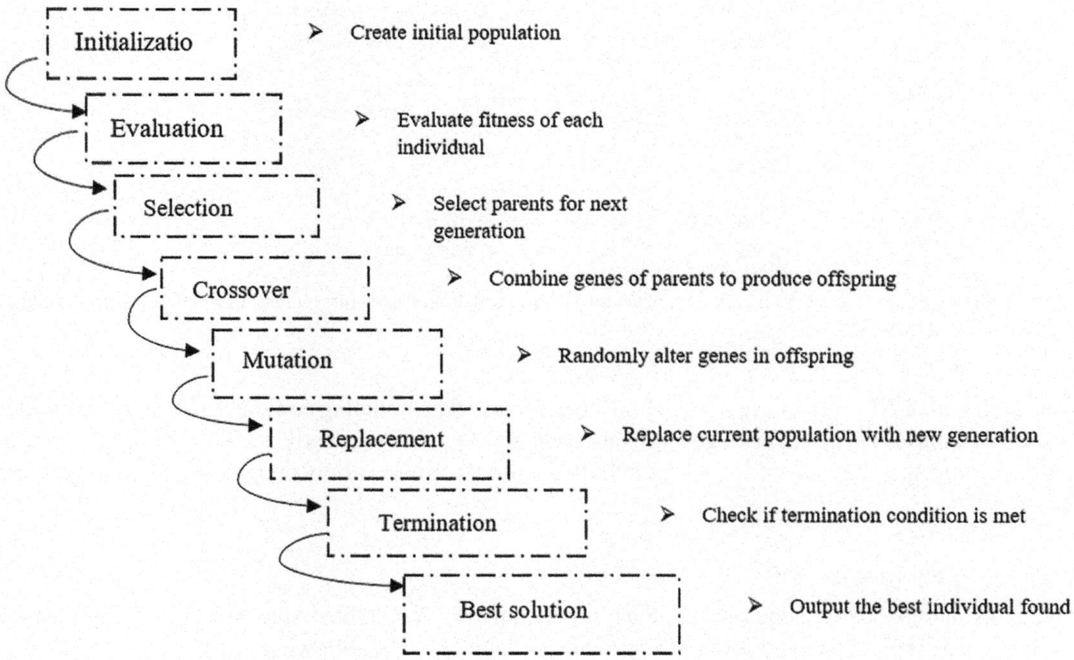

FIGURE 24.3 Structure of a genetic algorithm. Flowchart of a genetic algorithm process, showing initialization, selection, crossover, mutation, and fitness evaluation of optimization tasks.

Advanced variants include multiobjective algorithms (MOGAs) like Nondominated Sorting Genetic Algorithm (NSGA-II) (Deb et al., 2002) for Pareto-optimal solutions. Fig. 24.3 represents the structural workflow of a GA.

24.2.4.3.2 Applications in mechatronics

Mechanical design optimization: GAs are instrumental in the optimization of parameters such as mass, stiffness, and damping within the domain of automotive suspension systems. This process entails a balanced consideration of trade-offs between ride comfort and handling (Bouazara et al., 2007).

Robotics path planning: GAs optimize robot path planning and task scheduling in industrial automation. They compute collision-free, energy-efficient trajectories for industrial robots and autonomous mobile platforms in complex environments (Harun & Ibrahim, 2019; Santa et al., 2022).

Control parameter tuning: GAs fine-tune fuzzy logic controllers and PID gains for servo motors and robotic arms; this process serves to enhance precision and reduce oscillations (Alcalá et al., 2006; Jaen-Cuellar et al., 2013).

Structural health monitoring (SHM): In mechatronic structures such as bridges or aerospace components, they optimize sensor placement and damage-detection algorithms (Worden & Dulieu-Barton, 2004).

These applications highlight the ability of GAs to tackle the multimodal and highly constrained optimization problems that are inherent in mechatronic systems.

24.2.4.3.3 Limitation and evolution

Despite their apparent versatility, GAs have been observed to exhibit premature convergence and computational efficiency in the context of high-dimensional problems. Hybrid methods, incorporating a combination of GA with local search (memetic algorithms) or gradient-based reinforcements, have witnessed a marked rise in popularity (Neri & Cotta, 2012). More recent developments include biologically inspired algorithms such as differential evolution (Storn & Price, 1997) and evolutionary strategies tailored for real-time optimization in adaptive mechatronic systems.

24.2.4.4 Neural networks (1980s–1990s)

ANNs are a computational model inspired by the human brain. This model consists of interconnected layers of nodes (neurons) that learn patterns from data. The advent of backpropagation in the 1980s (Rumelhart et al., 1986) marked a pivotal development in the field, paving the way for the training of multilayer networks and catalyzing a substantial

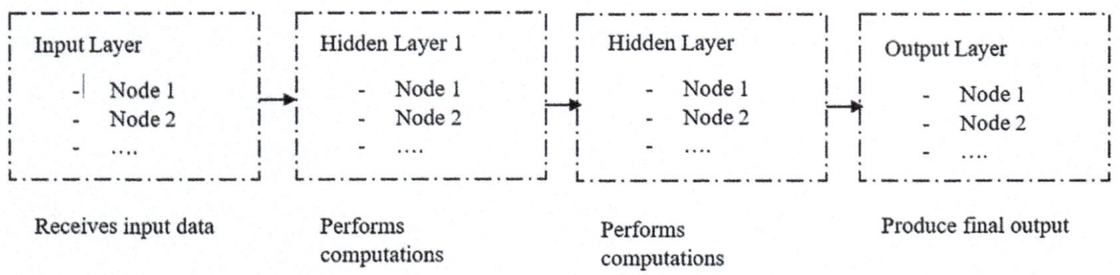

FIGURE 24.4 Neural network structure. Basic architecture of neural network, with layered structure representing input, hidden, and output neurons used in signal processing and classification.

progression in the realm of AI. ANNs have been demonstrated to excel at modeling complex, nonlinear relationships in high-dimensional datasets, thus offering superior generalization capabilities. Consequently, their application to mechatronics has enabled intelligent control, system identification, and fault diagnosis across a variety of dynamic and uncertain environments (Haykin, 2009).

24.2.4.4.1 Theoretical foundations

ANNs of layers of interconnected nodes, where each node computes a weighted sum of inputs followed by a nonlinear activation function. The typical feedforward ANN structure is defined by the following:

$$y = f\left(\sum_{i=1}^{n} w_i x_i + b\right) \tag{24.7}$$

where x_i are inputs features, w_i are synaptic weights, b is the bias, $f(.)$ is an activation function (e.g., sigmoid, ReLU, tanh), and y is the neuron output.

The training process is facilitated through the utilization of supervised learning methodologies, employing the back-propagation algorithm to minimize a loss function, which is most often the mean squared error (MSE) or cross-entropy, via the gradient descent approach.

$$w_i \leftarrow w_i - \eta \frac{\partial L}{\partial w_i} \tag{24.8}$$

where η is the learning rate and L is the loss.

Variants include

- Multilayer perceptrons (MLPs)
- Recurrent neural networks (RNNs)
- Radial basis function (RBF) networks

These architectures can approximate any continuous function (universal approximation theorem) (Hornik et al., 1989). Fig. 24.4 represents the structural architecture of a feedforward neural network.

24.2.4.4.2 Application in mechatronics

Nonlinear system identification: ANNs have been demonstrated to excel in the modeling of complex dynamics, such as those observed in actuators and robotic arms, where conventional models are unable to adequately capture the required complexity (Narendra & Parthasarathy, 1990).

Intelligent control: The integration of neural networks into adaptive and hybrid controllers (e.g., neuro-PID controllers) has been demonstrated to enhance performance in nonlinear and uncertain environments (Lewis et al., 1998).

Fault detection and diagnosis: ANNs identify anomalies in mechanical and electronic subsystems by learning normal and abnormal operating patterns from sensor data (Widodo & Yang, 2007).

Human–robot interaction (HRI): Used for gesture and voice recognition to enhance collaborative robotics in smart manufacturing environments (Zhang et al., 2022).

Energy-efficient motion planning: optimize robot trajectories and energy consumption in mobile and industrial robots (Airlangga et al., 2025).

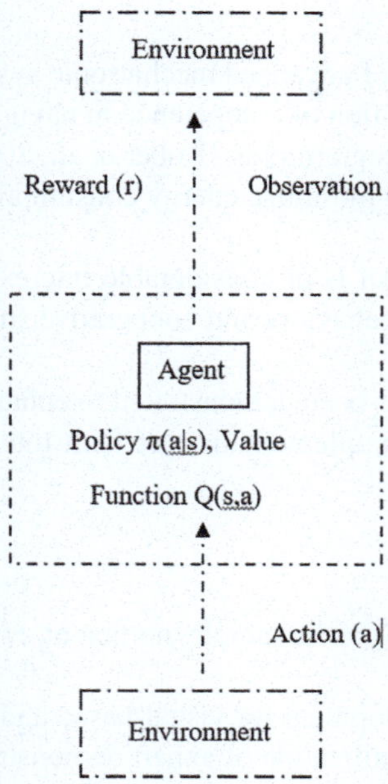

FIGURE 24.5 Reinforcement learning (RL) framework. RL framework, displaying agent–environment interaction, policy learning, and reward-based feedback for adaptive decision-making.

24.2.4.4.3 Limitations and evolution

Despite their success, ANNs are afflicted by overfitting, slow convergence, and a paucity of interpretability (the black-box problem). This has prompted research into regularization methods (e.g., dropout), deep architectures (DL), and hybrid systems combining ANNs with fuzzy logic or evolutionary algorithms (Huang, 2009).

The subsequent emergence of DNNs and convolutional architectures has enabled the recognition of increasingly complex patterns, particularly in the domain of vision-based robotics and autonomous vehicles (Alam et al., 2020).

24.2.4.5 Reinforcement learning (RL) (1980s–1990s)

RL is a branch of machine learning in which agents learn to make optimal decisions by interacting with their environment and receiving feedback in the form of rewards or penalties (Sutton & Barto, 1998). RL gained prominence in the fields of AI and control engineering during the 1990s, facilitating autonomous decision-making in complex, stochastic environments (Kaelbling et al., 1996). In the field of mechatronics, RL has been shown to offer robust solutions for adaptive control, path planning, and robot learning in dynamic and uncertain contexts (Kober et al., 2013).

24.2.4.5.1 Theoretical foundations

At its core, RL seeks to learn a policy, $\pi(a|s)$, that maximizes the expected cumulative reward.

$$\pi^* = \arg\max_{\pi} \mathbb{E}\left[\sum_{t=0}^{\infty} \gamma^t r_t\right] \tag{24.9}$$

where s is the state, a is the action, r_t is the reward at time t, $\gamma \in [0, 1)$ is the discount factor.

Common frameworks include

- Model-free methods, such are Q-learning (Watkins & Dayan, 1992) learn by iteraively updating the value function Q(s,a).

$$Q(s, a) \leftarrow Q(s, a) + \alpha\left[r + \max_{a'} Q(s', a') - Q(s, a)\right] \tag{24.10}$$

- Policy gradient methods: directly optimize $\pi_\theta(als)$ through gradient ascent on the expected reward.
- Actor-critic architectures: combine value function estimation with policy optimization.

The combination of RL with neural networks has been proven to result in the creation of a new form of machine learning known as deep RL (Mnih et al., 2015). Fig. 24.5 illustrates the structure of a typical ANN.

24.2.4.5.2 Applications in mechatronics

RL has made a significant contribution to the field of advanced mechatronic systems:

Autonomous robotics: RL facilitates the acquisition of competence in navigation, object manipulation, and cooperative tasks by robots without the necessity for explicit programming (Kober et al., 2013).

Adaptive control: RL has been demonstrated to minimize energy consumption in HVAC systems and robotic actuators while maintaining performance (Wei et al., 2017).

Human–robot collaboration (HRC): This subject is of considerable interest; research in this field has shown that RL agents can be trained to learn socially acceptable behaviors and cooperative strategies for collaborative robots (cobots) in industrial settings (Mnih et al., 2015).

RL in smart factories: The application of RL has been demonstrated to enhance various aspects of industrial operations, including scheduling, material handling, and the implementation of fault-tolerant control systems within the context of Industry 4.0 environments (Bonada et al., 2020).

24.2.4.5.3 Limitation and evolution

RL methods necessitate extensive training time and can be sample inefficient, especially in high-dimensional environments. Potential solutions are outlined below:

Deep RL: integration with DNNs is a key development for visual-based and high-dimensional state spaces.

The process of imitation learning involves the utilization of expert demonstrations for the pretraining of agents.

Safe RL: the incorporation of safety constraints designed to avert disastrous behaviors in critical systems (García & Fernández, 2015).

Contemporary trends in the field emphasize the utilization of multiagent and hierarchical RL methodologies for the resolution of large-scale mechatronic problems.

24.2.4.6 Deep learning (2000s–present)

The advent of DL can be traced back to the evolution of neural networks, which was propelled by the advancement in computational power, the availability of voluminous datasets, and the refinement of optimization algorithms (Lecun et al., 2015). DL employs multilayered architectures, frequently termed DNNs, to automatically extract hierarchical representations of data. In contradistinction to conventional machine learning models, DL systems demonstrate a high degree of proficiency in the learning of high-dimensional, nonlinear relationships, rendering them highly efficacious in tasks involving perception, control, and decision-making within mechatronics (Schmidhuber, 2015).

24.2.4.6.1 Theoretical foundations

DL is predicated on the premise of deep ANNs, wherein each layer is responsible for implementing a nonlinear transformation of its input. The standard feedforward DNN model can be described mathematically as follows:

$$h^{(l)}\left(w^{(l)}h^{(l-1)} + b^{(l)}\right), l = 1, \ldots, L \tag{24.11}$$

where $h^{(0)} = x$ is the input vector, $w^{(l)}$ and $b^{(l)}$ are the weights and biases of layer l, $f(.)$ is a nonlinear activation function such as ReLU or sigmoid, and L is the number of hidden layers.

Backpropagation is used to minimize a loss function (e.g., cross-entropy or MSE 0 via gradient descent.

$$\theta \leftarrow \theta - \eta \nabla_{\theta} L(y, \hat{y}) \tag{24.12}$$

where θ includes all $w^{(l)}$ and $b^{(l)}$, η is the learning rate, and $\mathcal{L}$ is the loss function comparing true labels y and predictions $\hat{y}$.

Key architectures relevant to mechatronics include

- Convolutional neural networks (CNNs) for image-based tasks such as object recognition and fault detection,
- RNNs and long short-term memory (LSTM) networks for time-series data like sensor readings,
- Autoencoders for dimensionality reduction and anomaly detection.

Deep RL combines deep networks with RL for autonomous control. Fig. 24.6 depicts the hierarchical structure of a DL model.

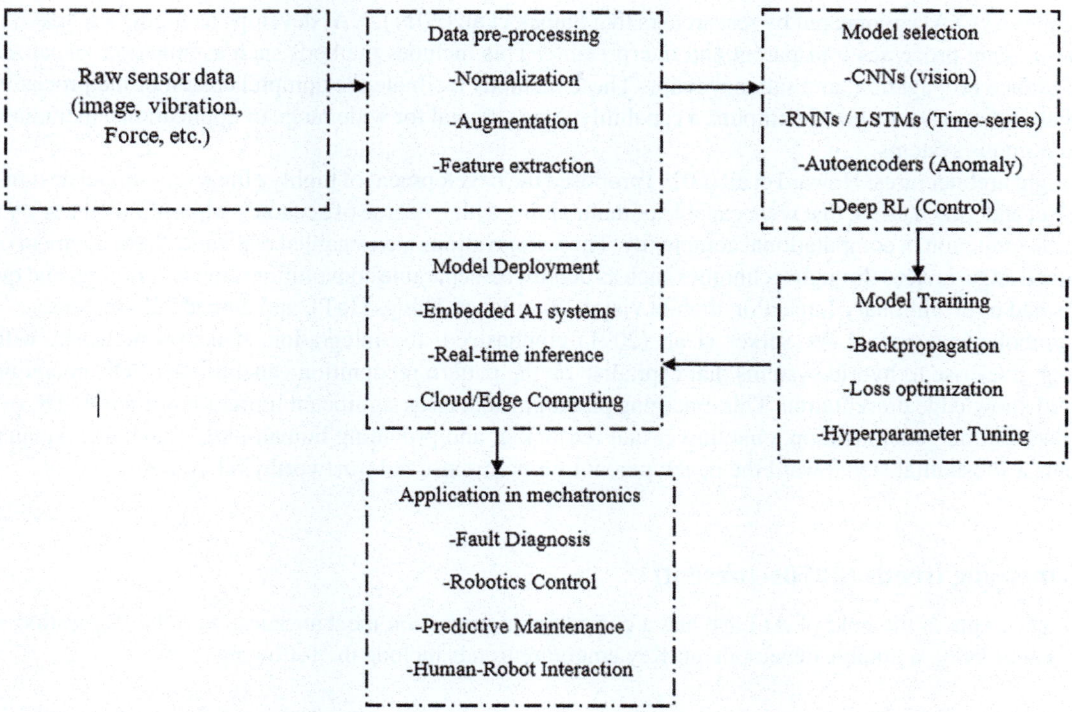

FIGURE 24.6 Deep learning pipeline for mechatronics. Deep learning pipeline in mechatronics, outlining stages from data preprocessing to model deployment for fault detection and autonomous control. *AI*, Artificial intelligence; *CNNs*, convolutional neural networks; *LSTMs*, long short-term memory; *RL*, reinforcement learning; *RNNs*, recurrent neural networks.

24.2.4.6.2 Applications in mechatronics

Robotics: CNNs are key component of vision systems, facilitating real-time object detection, pose estimation, and scene understanding (Krizhevsky et al., 2017). Conversely, deep RL has emerged as a significant paradigm for enabling adaptive control and decision-making in robotic arms and autonomous vehicles (Levine et al., 2016).

Fault diagnosis and prognostics: DL models, including stacked autoencoders and LSTMs, have been shown to learn complex degradation patterns from vibration and acoustic signals, facilitating predictive maintenance of motors, gearboxes, and bearings (Zhang et al., 2019).

Industrial automation: The utilization of CNN-based quality inspection systems has been demonstrated to enhance the efficiency of defect detection on production lines when contrasted with traditional image processing methodologies (Ren et al., 2015).

HRI: RNNs and LSTMs algorithms are employed to analyze temporal patterns of human motion, with the objective of enhancing collaborative robot behavior within manufacturing contexts (Franceschi et al., 2023).

Furthermore the integration of DL algorithms has been shown to facilitate sensor fusion, thereby combining visual, tactile, and force feedback to ensure robust control of unstructured environments (Lee et al., 2013).

24.2.4.6.3 Limitations and evolution

Despite the apparent success of the DL model, there are several challenges that require consideration:

Data dependency: the necessity of substantial, clearly labeled datasets for the successful generalization of DL models in niche mechatronic domains can be a costly or impractical undertaking.

Interpretability: the black-box nature of deep networks has significant implications for safety-critical applications, such as autonomous robotics.

Computational cost: the training and deployment of deep models necessitates substantial computing resources, particularly in the context of real-time embedded systems.

Recent research trends address these limitations through the following means:

- Explainable AI (XAI): pioneered by researchers like Samek et al. (2017), XAI develops techniques to make DL models' decision-making processes transparent and interpretable. This includes methods such as attention visualization, layer-wise relevance propagation, and saliency maps. These methods facilitate the comprehension of the processes by which neural networks arrive at specific outputs, a capability that is crucial for wide range of applications, including healthcare and autonomous systems.
- Lightweight architectures: Howard et al. (2017) proposed the development of highly efficacious model designs, including MobileNet and Tiny YOLO, that were capable of maintaining a high degree of accuracy while simultaneously achieving a substantial reduction in computational complexity. These innovations have enabled real-time AI deployment on resource-constrained edge devices through techniques such as depthwise separable convolutions, model pruning, and quantization. This has had a revolutionary impact on mobile vision, Internet of Things (IoT), and embedded systems.
- Neurosymbolic integration: As Sarker et al. (2021) emphasized, the integration of neural networks with symbolic reasoning gives rise to hybrid systems that capitalize on the pattern recognition capabilities of DL in conjunction with structured knowledge presentation. This emerging paradigm addresses significant limitations of purely DL by enhancing out-of-distribution generalization, enabling casual reasoning, and providing human-understandable explanations. This represents a substantial step toward the development of more robust and trustworthy AI systems.

24.2.5 Emerging trends (2010s–present)

Recent advancements in the field of AI have had a considerable impact on mechatronics, with the integration of learning-based approaches being a notable development. Key emerging trends include the following:

24.2.5.1 Neuromorphic computing

Neuromorphic computing replicates neural architectures through the use of specialized hardware, thus enhancing real-time processing and energy efficiency in embedded systems. Exemplifying this approach are the architectures developed by Intel (Loihi) and International Business Machines (IBM) (TrueNorth), which demonstrate the potential for application in domains such as autonomous robotics and sensor fusion (Davies et al., 2018).

24.2.5.2 Explainable AI (XAI)

Explainable AI (XAI) has been developed for the purpose of enhancing the interpretability of DL models, a factor that is of paramount importance in safety-critical applications in mechatronics, including autonomous vehicles and industrial automation. Approaches such as SHAP (shapley additive explanations) and LIME (local interpretable model-agnostic explanations) promote transparency in decision-making, thereby ensuring regulatory compliance and user trust (Samek et al., 2017).

24.2.5.3 Physics-informed machine learning (PIML)

The incorporation of physics-informed machine learning (PIML) has emerged as a pivotal integration of AI with physical models, thereby circumventing the constraints imposed by purely data-driven methodologies. The integration of governing equations into neural networks within the framework of PIML has been demonstrated to enhance system modeling, fault prediction, and control strategies for complex electromechanical systems (Raissi et al., 2017).

24.2.5.4 Self-learning robotic systems

To achieve the capacity for dynamic adaptation to new environment with a minimum of human intervention, robotic systems are equipped with the capability of self-learning through the implementation of two distinct methodologies, namely, deep RL and metalearning. Advancements in model-free and model-based RL, including PPO (proximal policy optimization) and SAC (soft actor-critic), have enhanced real-time learning capabilities in industrial and service robotics (Haarnoja et al., 2018).

These trends collectively shape the future of mechatronics, fostering more intelligent, adaptive, and efficient automation systems.

24.3 Applications and case studies

The integration of AI methods into mechatronics has precipitated a wave of innovation across a range of industries, engendering a transformation in the design, control, and optimization of systems. From manufacturing floors to autonomous vehicles, AI-powered mechatronic systems are rapidly pushing the boundaries of what is technologically feasible. This section will examine real-world applications and case studies that demonstrate how AI enhances functionality and efficiency, rendering systems smarter, more adaptable, and more reliable.

24.3.1 Robotics

Robots have advanced significantly from their initial capacity to execute rudimentary, repetitive tasks. Presently the advent of AI has empowered robots to execute intricate operations with exactitude and adaptability. This encompasses a wide range of applications, including the assembly of products in industrial settings, the welding of automotive components, and the collaborative operation alongside humans in designated environments (cobots).

24.3.2 Case study: FANUC's AI-driven robots

FANUC Corporation, a global leader in industrial robotics, has embraced AI to enhance the intelligence and efficiency of its robotic systems. The company has utilized the RL method, enabling its robots to acquire knowledge from their own actions and to enhance their movement capabilities over time. In manufacturing context, for instance, these robots can adapt to variations in part placement or assembly processes, thereby reducing errors and enhancing productivity. This capacity for learning and adaptation renders these robots highly advantageous in dynamic environments characterized by constant change (Kober et al., 2013; Soori et al., 2023).

24.3.3 Autonomous vehicles

Autonomous vehicles represent a particularly noteworthy application of AI in the field of mechatronics. These vehicles are reliant on AI to facilitate perception of their surroundings, enable the rapid formation of decisions, and ensure safe navigation through complex environments.

24.3.4 Case study: Waymo's self-driving cars

Waymo, a pioneering entity in the domain of autonomous driving, employs DL and computer vision to facilitate the operation of its self-driving vehicles. The vehicles are equipped with an array of sensors, cameras, and LiDAR systems that feed data into neural networks. These networks process the data in real-time to detect objects, predict movements, and plan safe routes. The extensive training of Waymo's AI algorithms on a substantial volume of driving data has enabled the vehicles to navigate urban environments with high degree of proficiency (Redmon et al., 2016).

- *Smart manufacturing*: In the domain of manufacturing, AI is a game that has emerged as a transformative force. It facilitates predictive maintenance, real-time quality control, and process optimization, thereby ensuring the smooth and efficient operation of manufacturing facilities.
- Case study: Siemens's AI-powered predictive maintenance strategy has capitalized on the potential of AI to anticipate equipment failures before they occur. By analyzing data from sensors embedded in gas turbines, the company's AI system can detect subtle changes in vibration, temperature, and other parameters. These early warnings enable technicians to address issues proactively, thereby preventing costly downtime and extending the lifespan of the equipment. In a notable instance the system's early identification of a potential turbine failure, a week in advance, resulted in significant cost savings, amounting to millions, because of the avoidance of repair expenses (Meriem et al., 2023).
- *Healthcare mechatronics:* AI is driving a significant transformation within the field of healthcare mechatronics, particularly evident in the domains of medical robotics and diagnostic systems. These technologies encompass a wide range of applications, from surgical robots to AI-powered imaging, and are collectively improving patient outcomes and transforming healthcare delivery.

24.3.5 Case study: da Vinci Surgical System

The da Vinci Surgical System is a prime example of AI's impact on the field of healthcare. This robotic-assisted surgical platform employs AI and machine learning to augment the precision and dexterity of surgeons. During surgical procedures the system's AI analyses real-time data from the surgical site, providing feedback to the surgeon and enabling minimally invasive operations with unparalleled accuracy. Notably the system has been employed in intricate surgical procedures, such as prostatectomies, where precision is paramount (Levine et al., 2016).

- *Energy and smart grids*

The role of AI in optimizing energy distribution and consumption is of crucial importance, with the potential to enhance the efficiency and sustainability of smart grids.

24.3.6 Case study: Google's DeepMind for energy optimization

Google's DeepMind has utilized AI to optimize energy usage in its data centers. Utilizing deep RL, the AI system is capable of predicting cooling requirements and adjusting energy consumption in real-time. This approach has been shown to engender a reduction in energy usage, with financial savings amounting to millions of dollars and a substantial decrease in the company's carbon footprint (Luo et al., 2022).

- *Consumer electronics*

AI is an integral component within numerous consumer electronics devices, ranging from smart household appliances to wearable technology, thereby optimizing both the functionality and the user experience.

- Case study: Samsung's AI-powered washing machines utilize fuzzy logic and machine learning algorithms to optimize wash cycles. The system employs a variety of analytical processes, including the consideration of factors such as load size, fabric type, and dirt level, to determine the most appropriate settings. The system is designed to learn over time, adapting to user preferences and providing personalized recommendations, with the aim of enhancing convenience and energy efficiency (Yurkovich & Passino, 1999).
- *Aerospace and defense*

In the domains of aerospace and defense, AI finds application in a number of areas. These include autonomous navigation, fault detection, and mission planning. The result is the exploration of remote and hazardous environments.

24.3.7 Case study: NASA's autonomous rovers

It is an irrefutable fact that the navigation and scientific analysis systems employed on board NASA's Mars rovers, including the curiosity and perseverance models, are reliant on AI technology. The rover's systems process data from onboard sensors to plan routes, avoid obstacles, and identify areas of scientific interest. For instance the navigation of Perseverance, driven by AI, enables the rover to traverse the rugged Martian terrain autonomously, thereby facilitating the collection of valuable data for researchers on Earth (Zhong et al., 2025).

24.4 Challenges and limitations in AI models

The integration of AI into mechatronics has yielded remarkable possibilities, yet it is not without its challenges. A multitude of factors, ranging from technical hurdles to ethical concerns, have the potential to hinder the widespread adoption and effectiveness of AI-driven mechatronic systems. This section undertakes a detailed examination of these challenges, offering a balance on the limitations that researchers and engineers must address to fully realize the potential of AI in mechatronics.

A significant challenge in the realm of AI is the computational complexity of numerous AI methodologies. DL and RL, for instance, necessitate considerable computational resources to train and operate effectively. In applications such as autonomous vehicles, where real-time decision-making is paramount, the requirement for high-performance hardware can escalate costs and energy consumption. The processing of the voluminous sensor data generated by self-driving cars, e.g., demands potent onboard computers, which may not always be viable in environments with limited resources (Jambotkar et al., 2021). This reliance on sophisticated hardware constrains the scalability and affordability of AI-powered systems, particularly in smaller scale or cost-sensitive applications.

Another significant challenge pertains to the reliance on substantial quantities of high-quality data. AI algorithms, particularly those founded on machine learning, demonstrate a substantial dependence on data for training and validation. However, acquiring such data can be arduous, especially in specialized or niche domains. A case in point is predictive maintenance in manufacturing: AI models necessitate extensive historical data on equipment failures to accurately predict future issues. In industries where equipment failures are infrequent the accumulation of sufficient and reliable data can pose a substantial challenge. The absence of sufficient and reliable data can compromise the efficacy of AI models, resulting in inaccurate or unreliable outputs (Rojek et al., 2023).

The issue of interpretability and transparency in AI methods also poses a challenge. DL models, for instance, frequently function as "black boxes," which hinders the comprehension of their decision-making processes, a problem that is especially salient in safety-critical domains such as healthcare mechatronics (Madumal et al., 2018). In the context of robotics-assisted surgeries, e.g., surgeons must place their trust in the AI system's decisions, and if the rationale behind these decisions is opaque, it becomes challenging to achieve user acceptance. This dearth of transparency has the potential to hinder trust and adoption, especially in fields where accountability and safety are paramount.

Real-time performance is another critical concern. A significant number of mechatronic systems are operated within environments in which delays can have severe, significant ramifications. The capacity of AI algorithms to process data and make decisions within rigorous time constrains is a matter of great complexity. Within industrial robotics, e.g., even a minor delay in the processing of sensor data can result in errors in assembly or indeed accidents. This underscores the importance of real-time performance when it comes to the implementation of AI in applications where there is a requirement for time sensitivity.

The integration of AI into mechatronics gives rise to a number of ethical and legal concerns, including questions of accountability, privacy, and bias. To illustrate this point, consider the example of autonomous vehicles: determining liability in the event of an accident involving an AI-driven car is a complex legal issue (Lovell, 2024). Similarly, AI systems that collect and process personal data raise concerns about privacy. These ethical and legal challenges have the potential to impede the adoption of AI technologies, thus necessitating the development of a novel regulations and frameworks to effectively address these concerns.

The integration of AI into existing legacy systems represents a substantial challenge. Many contemporary mechatronic systems were not conceptualized with the intention of incorporating AI, and the process of retrofitting them with AI capabilities can be technically demanding and financially burdensome. In the context of older manufacturing facilities, for instance, the enhancement of machinery with AI-related functionalities necessitates substantial alterations to both hardware and software components. This expense and complexity can act as a deterrent to the adoption of AI within well-established industrial sectors.

Energy consumption is also a significant concern, particularly in the context of battery-powered or energy-sensitive applications. The utilization of AI algorithms, particularly those involving DL, can be a substantial energy consumer. In wearable devices, for instance, the substantial energy consumption of AI algorithms can result in reduction of battery life, thereby limiting the device's practical application (Ippolito et al., 2020). Consequently, energy efficiency emerges as a pivotal consideration in the design and implementation of AI-powered mechatronic systems.

Finally, ensuring the robustness and reliability of AI is a major challenge, especially in the context of safety-critical applications. It is imperative that AI models demonstrate consistent performance under all conditions; however, attaining this level of reliability remains a formidable challenge. To illustrate this point, consider the necessity of autonomous drones to respond effectively to unexpected environmental changes, such as sudden wind gusts or the presence of obstacles. A deficiency in robustness can result in system failures, thereby compromising safety and reliability.

Despite the significance of these challenges, ongoing research and technological advancements are helping to address them. For instance, edge AI (deploying lightweight AI models on edge devices) can reduce computational and energy demands. Similarly, Expandable AI (XAI) is rendering AI decisions more interpretable, improving trust and transparency. In cases where data is scarce, synthetic data generation can help overcome limitations. In addition, establishing clear regulatory frameworks can address ethical and legal concerns, paving the way for responsible AI adoption.

24.5 Future directions

As AI continues to evolve, its integration with mechatronics is poised to unlock even greater possibilities. Advancements in AI algorithms, alongside the emergence of novel hardware and frameworks, are poised to steer the future of AI in mechatronics toward a period of substantial potential. This section explores emerging trends and future directions that could shape the next generation of intelligent mechatronic systems, driving innovation across industries and overcoming current limitations.

Explainable AI (XAI) represents a significant development in the field of AI, with particular relevance to the growing complexity of AI systems. In safety-critical applications such as autonomous vehicles and healthcare robotics, it is imperative to understand the decision-making processes of these systems become more sophisticated. XAI aims to address this need by enhancing the transparency and interpretability of AI models, thereby enabling users to trust and verify the output of these systems. In the context of robotic surgery, for instance, XAI could provide surgeons with clear explanations of the AI's recommendations, leading to safer and more reliable procedures. This enhanced transparency is poised to not only foster user confidence but also streamline regulatory approval processes, thereby expediting the integration of AI in high-stakes domains.

Another promising trend is the rise of edge AI, which involves the deployment of AI algorithms directly on edge devices as opposed to reliance on cloud-based processing (Shi et al., 2020). This approach has the advantage of reducing latency, enhancing privacy, and lowering energy consumption, making it ideal for real-time applications. In mechatronics, edge AI could enable autonomous drones to process sensor data on-board, allowing them to make split-second decisions without reliance on external servers. A similar potential is seen in the field of smart manufacturing, where the implementation of edge AI could empower robots to optimize their operations in real-time, even in environments characterized by limited connectivity. The integration of AI with the IoT is also set to revolutionize mechatronics. By combining AI's analytical capabilities with IoT's connectivity, mechatronic systems can become more intelligent and interconnected. For instance, in a smart factory, AI-powered IOT devices could monitor equipment health, predict failures, and coordinate repairs autonomously. This seamless integration could lead to fully autonomous production lines, where machines communicate and collaborate without human intervention.

Advancements in the field of RL are precipitating new possibilities for autonomous systems. RL algorithms, which learn through trial and error, are becoming more efficient and scalable, thus enabling robots and other mechatronic systems to master complex tasks. For instance, RL could be used to train robots for intricate assembly tasks or to develop adaptive control systems for autonomous vehicles. As these algorithms become more robust, they could enable mechatronic systems to operate in highly dynamic and unpredictable environments, such as disaster zones or deep-sea exploration.

The development of neuromorphic computing represents a significant frontier with considerable potential. Inspired by human brain, neuromorphic chips are designed to process information in a manner that emulates neural networks, thereby offering substantial enhancement in energy efficiency and processing speed. In the domain of mechatronics, neuromorphic computing has the capacity *p* facilitate real-time AI processing on low-power devices, such as wearable sensors or autonomous drones. This could result in a new generation of energy-efficient, intelligent systems that operate seamlessly in real-world environments.

The advent of AI-driven generative design is poised to effect a transformation in the realm of mechatronic system development. Through the utilization of AI to explore a myriad of design possibilities, engineers are empowered to formulate optimized solutions that align with specific performance criteria. A salient illustration of this is the potential for generative design to facilitate the development of lightweight, high-strength components for aerospace applications or the creation of energy-efficient actuators for robotics. This approach not only expedites the design process but also engenders innovative solutions that may not be attainable through conventional methodologies.

The future of AI in mechatronics is also contingent on the resolution of ethical and societal challenges, as pervasiveness of AI systems necessitates the establishment of frameworks for accountability, the rectification of biases in AI algorithm, and the protection of user privacy. To illustrate this point, consider the case of autonomous vehicles, which will require ethical guidelines to govern decision-making in critical situations, such as accident avoidance. By proactively addressing these challenges the mechatronics community can build trust and ensure the sustainable adoption of AI technologies.

Recent developments in the field of AI have begun to intersect with other emerging technologies, including 5G connectivity, quantum computing, and advanced materials. This convergence has the potential to unlock entirely new capabilities. For instance the ultralow latency and high bandwidth of 5G could facilitate real-time collaboration between AI-powered mechatronic systems. In addition, quantum computing could address complex optimization problems that are currently intractable. Similarly, advancements in materials science could result in smarter, more adaptive mechatronic components that respond dynamically to their environment (Nnodim et al., 2021).

The future of AI in mechatronics appears to be a promising one, with a multitude of opportunities for innovation and growth. Key emerging trends, such as XAI, edge computing, neuromorphic chips, and generative design, are poised to address existing limitations and unlock new possibilities. By embracing these advancements and addressing the associated ethical and societal challenges the mechatronics community can be at the forefront of a new era of intelligent, adaptive, and efficient systems that have the potential to transform industries and improve lives.

24.6 Conclusion

The integration of AI into mechatronics signifies a paradigm shift in the design, control, and optimization of intelligent systems. From robotics and autonomous vehicles to smart manufacturing and healthcare, AI-enabled mechatronic systems are driving innovation across multiple sectors, improving functionality, efficiency, and adaptability. This chapter has examined the evolution of AI methodologies, their applications, the challenges they face, and the exciting future directions that lie ahead.

The study commenced with examination of the chronological progression of AI methodologies, commencing from early symbolic AI and fuzzy logic, progressing to contemporary DL and RL. Each significant milestone in the evolution of AI has facilitated novel capabilities within mechatronics, thereby empowering systems to acquire knowledge from data, adapt to fluctuating environments, and enhance their performance in real-time. These advancements have metamorphosed conventional mechatronic systems into intelligent, autonomous entities that are proficient in the adept execution of intricate tasks with precision and dependability.

Through the analysis of real-world applications and case studies, it has been demonstrated how AI is affecting a revolution within various industries. Within the field of robotics, AI has been instrumental in enabling machines to perform intricate tasks and collaborate safely with humans. In the domain of autonomous vehicles, AI facilitates real-time decision-making, thereby enhancing the safety and efficiency of transportation. In the manufacturing sector, AI-driven predictive maintenance has been shown to reduce downtime and costs. Similarly, in the field of healthcare, AI has been found to enhance the precision and effectiveness of medical robotics. These examples underscore the transformative potential of AI in mechatronics, demonstrating its ability to solve real-world problems and drive innovation.

Nonetheless, the endeavor is not without its challenges. Such challenges include, but are not limited to, computational complexity, data dependency, lack of interpretability, and ethical concerns. To fully realize the potential of AI in mechatronics, these issues must be addressed. Owing to ongoing research and technological advancements, such as edge AI, explainable AI, and neuromorphic computing, solutions to these challenges are becoming increasingly apparent. This ensures that AI-powered systems are robust, reliable, and trustworthy.

In the forthcoming period the potential of AI in the field of mechatronics appears to be flourishing. Emerging trends, such as generative design, AI-driven IoT, and the convergence of AI with quantum computing and advanced materials, hold the promise of unveiling new capabilities and applications. As these technologies reach maturity, they will facilitate the development of increasingly sophisticated, adaptable, and efficient mechatronic systems, with the potential to transform various industries and enhance human lives.

In conclusion the synergy between AI and mechatronics is reshaping the technological landscape, creating new opportunities for innovation and growth. By embracing these advancements and addressing the associated challenges, researchers, engineers, and industry leaders can realize the full potential of AI in mechatronics, thereby entering a new era of intelligent systems that redefine what is possible. The journey has only just begun, and the future of AI-powered mechatronics looks bright.

References

Airlangga, G., Bata, J., Nugroho, O. I. A., & Sugianto, L. F. (2025). Neural network architectures for UAV path planning: A comparative study with *A** algorithm as benchmark. *International Journal of Robotics and Control Systems, 5*(1), 625–639. https://pubs2.ascee.org/index.php/IJRCS/article/download/1757/pdf. doi:10.31763/ijrcs.v5i1.1757.

Alam, M., Samad, M. D., Vidyaratne, L., Glandon, A., & Iftekharuddin, K. M. (2020). Survey on deep neural networks in speech and vision systems. *Neurocomputing, 417*, 302–321. doi:10.1016/j.neucom.2020.07.053.

Alcalá, R., Alcalá-Fdez, J., Gacto, M. J., & Herrera, F. (2006). Fuzzy rule reduction and tuning of fuzzy logic controllers for a HVAC system. In *Studies in fuzziness and soft computing: 201* (pp. 89–117). http://www.springer.com/sgw/cda/frontpage/0,11855,1-40356-69-1191779-0,00.html?changeHeader=true. doi:10.1007/3-540-33517-x_3.

Baban, C. F., Baban, M., & Suteu, M. D. (2016). Using a fuzzy logic approach for the predictive maintenance of textile machines. *Journal of Intelligent & Fuzzy Systems, 30*(2), 999–1006. doi:10.3233/ifs-151822.

Bansal, M., Krizhevsky, A., & Ogale, A. (2018). ChauffeurNet: Learning to drive by imitating the best and synthesizing the worst. *arXiv* United States arXiv https://arxiv.org . doi:10.48550/arxiv.1812.03079.

Barredo Arrieta, A., Díaz-Rodríguez, N., Del Ser, J., Bennetot, A., Tabik, S., Barbado, A., Garcia, S., Gil-Lopez, S., Molina, D., Benjamins, R., Chatila, R., & Herrera, F. (2020). Explainable artificial intelligence (XAI): Concepts, taxonomies, opportunities and challenges toward responsible AI. *Information Fusion, 58*, 82–115. http://www.elsevier.com/inca/publications/store/6/2/0/8/6/2/index.htt. doi:10.1016/j.inffus.2019.12.012.

Biswas, P., Rashid, A., Biswas, A., Nasim, M. A. A., Chakraborty, S., Gupta, K. D., & George, R. (2024). AI-driven approaches for optimizing power consumption: A comprehensive survey. *Discover Artificial Intelligence, 4*(1). doi:10.1007/s44163-024-00211-7.

Bonada, F., Echeverria, L., Domingo, X., & Anzaldi, G. (2020). *AI for improving the overall equipment efficiency in manufacturing industry*. IntechOpen. doi:10.5772/intechopen.89967.

Bouazara, M., Gosselin-Brisson, S., & Richard, M. J. (2007). Design of an active suspension control for a vehicle model using a genetic algorithm. *Transactions of the Canadian Society for Mechanical Engineering, 31*(3), 317–333. doi:10.1139/tcsme-2007-0021.

Brunton, S. L., & Kutz, J. N. (2019). *Data-driven science and engineering: Machine learning, dynamical systems, and control* (pp. 1–472). Cambridge University Press. http://dx.doi.org/10.1017/9781108380690. doi:10.1017/9781108380690.

Cho, S. B. (2002). Fusion of neural networks with fuzzy logic and genetic algorithm. *Integrated Computer-Aided Engineering, 9*(4), 363–372. https://content.iospress.com/journals/integrated-computer-aided-engineering/Pre-press/Pre-press. doi:10.3233/ica-2002-9405.

Davies, M., Srinivasa, N., Lin, T. H., Chinya, G., Cao, Y., Choday, S. H., Dimou, G., Joshi, P., Imam, N., Jain, S., Liao, Y., Lin, C. K., Lines, A., Liu, R., Mathaikutty, D., McCoy, S., Paul, A., Tse, J., Venkataramanan, G., ... Wang, H. (2018). Loihi: A neuromorphic manycore processor with on-chip learning. *IEEE Micro, 38*(1), 82–99. doi:10.1109/MM.2018.112130359.

Deb, K., Pratap, A., Agarwal, S., & Meyarivan, T. (2002). A fast and elitist multiobjective genetic algorithm: NSGA-II. *IEEE Transactions on Evolutionary Computation, 6*(2), 182–197. doi:10.1109/4235.996017.

Franceschi, P., Bertini, F., Braghin, F., Roveda, L., Pedrocchi, N., & Beschi, M. (2023). Learning human motion intention for PHRI assistive control. In *IEEE International Conference on Intelligent Robots and Systems* (pp. 7870–7877). Institute of Electrical and Electronics Engineers Inc. doi:10.1109/IROS55552.2023.10342014.

García, J., & Fernández, F. (2015). A comprehensive survey on safe reinforcement learning. *Journal of Machine Learning Research, 16*, 1437–1480. http://jmlr.csail.mit.edu/papers/volume16/garcia15a/garcia15a.pdf.

Goldberg, D. E. (1989). *Genetic algorithms in search, optimization, and machine learning*. Addison-Wesley.

Haarnoja, T., Zhou, A., Abbeel, P., & Levine, S. (2018). Soft actor-critic: Off-policy maximum entropy deep reinforcement learning with a stochastic actor. In *35th International Conference on Machine Learning, ICML, vol. 5* (pp. 2976–2989). International Machine Learning Society (IMLS).

Habib, M. K., & Davim, J. P. (2013). Preface. In *Interdisciplinary mechatronics: Engineering science and research development* (p. xvii). http://onlinelibrary.wiley.com/book/10.1002/9781118577516. doi:10.1002/9781118577516.

Hamidaoui, M., Talhaoui, M. Z., Li, M., Midoun, M. A., Haouassi, S., Mekkaoui, D. E., Smaili, A., Cherraf, A., & Benyoub, F. Z. (2025). Survey of autonomous vehicles' collision avoidance algorithms. *Sensors, 25*(2). http://www.mdpi.com/journal/sensors. doi:10.3390/s25020395.

Harun, S., & Ibrahim, M. F. (2019). A genetic algorithm based task scheduling system for logistics service robots. *Bulletin of Electrical Engineering and Informatics, 8*(1), 206–213. doi:10.11591/eei.v8i1.1437.

Haykin, S. (2009). *Neural networks and learning machines* (Third edition). Prentice Hall.

Holland, J. H. (1975). *Adaptation in natural and artificial systems*. University of Michigan Press.

Hornik, K., Stinchcombe, M., & White, H. (1989). Multilayer feedforward networks are universal approximators. *Neural Networks, 2*(5), 359–366. doi:10.1016/0893-6080(89)90020-8.

Howard, A. G., Zhu, M., Chen, B., Kalenichenko, D., Wang, W., Weyand, T., Andreetto, M., & Adam, H. (2017). MobileNets: Efficient convolutional neural networks for mobile vision applications. *arXiv*. undefined arXiv https://arxiv.org . doi:10.48550/arxiv.1704.04861.

Huang, Y. (2009). Advances in artificial neural networks—Methodological development and application. *Algorithms, 2*(3), 973–1007. doi:10.3390/algor2030973.

Ippolito, D., Duckworth, D., Callison-Burch, C., & Eck, D. (2020). Automatic detection of generated text is easiest when humans are fooled. In *Proceedings of the Annual Meeting of the Association for Computational Linguistics* (pp. 1808–1822). Association for Computational Linguistics (ACL). undefined. https://aclweb.org/ .

Jaen-Cuellar, A. Y., Romero-Troncoso, R. J., Morales-Velazquez, L., & Osornio-Rios, R. A. (2013). PID-controller tuning optimization with genetic algorithms in servo systems. *International Journal of Advanced Robotic Systems, 10*(9). doi:10.5772/56697.

Jambotkar, S., Guo, L., & Jia, Y. (2021). *Adaptive optimization of autonomous vehicle computational resources for performance and energy improvement*. IEEE. doi:10.1109/IROS51168.2021.9635828.

Johnson, D. M., Newell, A., & Simon, H. A. (1973). Human problem solving. *The American Journal of Psychology, 86*(2), 449. doi:10.2307/1421456.

Kaelbling, L. P., Littman, M. L., & Moore, A. W. (1996). Reinforcement learning: A survey. *Journal of Artificial Intelligence Research, 4*, 237–285. doi:10.1613/jair.301.

Kendoul, F. (2012). Survey of advances in guidance, navigation, and control of unmanned rotorcraft systems. *Journal of Field Robotics, 29*(2), 315–378. doi:10.1002/rob.20414.

Kober, J., Bagnell, J. A., & Peters, J. (2013). Reinforcement learning in robotics: A survey. *International Journal of Robotics Research, 32*(11), 1238–1274. doi:10.1177/0278364913495721.

Krizhevsky, A., Sutskever, I., & Hinton, G. E. (2017). ImageNet classification with deep convolutional neural networks. *Communications of the ACM, 60*(6), 84–90. doi:10.1145/3065386.

Lecun, Y., Bengio, Y., & Hinton, G. (2015). Deep learning. *Nature, 521*(7553), 436–444. http://www.nature.com/nature/index.html. doi:10.1038/nature14539.

Lee, J., Lapira, E., Bagheri, B., & Kao, H.-a. (2013). Recent advances and trends in predictive manufacturing systems in big data environment. *Manufacturing Letters, 1*(1), 38–41. http://www.journals.elsevier.com/manufacturing-letters/. doi:10.1016/j.mfglet.2013.09.005.

Levine, S., Finn, C., Darrell, T., & Abbeel, P. (2016). End-to-end training of deep visuomotor policies. *Journal of Machine Learning Research, 17*. http://jmlr.csail.mit.edu/papers/volume17/15-522/15-522.pdf.

Lewis, F. L., Jagannathan, S., & Yesildirak, A. (1998). *Neural network control of robot manipulators and non-linear systems*. CRC Press.

Lovell, J. J. (2024). Legal aspects of artificial intelligence personhood: Exploring the possibility of granting legal personhood to advanced AI systems and the implications for liability, rights and responsibilities. *International Journal of Artificial Intelligence and Machine Learning, 4*(2), 23–40. doi:10.51483/ijaiml.4.2.2024.23-40.

Luo, J., Paduraru, C., Voicu, O., Chervonyi, Y., Munns, S., Li, J., Qian, C., Dutta, P., Davis, J. Q., Wu, N., Yang, X., Chang, C. M., Li, T., Rose, R., Fan, M., Nakhost, H., Liu, T., Kirkman, B., Altamura, F., … Mankowitz, D. J. (2022). Controlling commercial cooling systems using reinforcement learning. *arXiv* United Kingdom arXiv https://arxiv.org . doi:10.48550/arXiv.2211.07357.

Madumal, P., Newn, J., Singh, R., & Vetere, F. (2018). Association for computing machinery australia interaction design for explainable AI. In *ACM International Conference Proceeding Series* (pp. 607–608). http://portal.acm.org/. doi:10.1145/3292147.3293450.

Masood, R. F. (2017). Application of fuzzy logic in design of smart washing machine. *arXiv*. doi:10.48550/arXiv.1701.01654.

Meriem, H., Nora, H., & Samir, O. (2023). Predictive maintenance for smart industrial systems: A roadmap. *Procedia Computer Science, 220*, 645–650. doi:10.1016/j.procs.2023.03.082.

Mnih, V., Kavukcuoglu, K., Silver, D., Rusu, A. A., Veness, J., Bellemare, M. G., Graves, A., Riedmiller, M., Fidjeland, A. K., Ostrovski, G., Petersen, S., Beattie, C., Sadik, A., Antonoglou, I., King, H., Kumaran, D., Wierstra, D., Legg, S., & Hassabis, D. (2015). Human-level control through deep reinforcement learning. *Nature, 518*(7540), 529–533. http://www.nature.com/nature/index.html. doi:10.1038/nature14236.

Narendra, K. S., & Parthasarathy, K. (1990). Identification and control of dynamical systems using neural networks. *IEEE Transactions on Neural Networks, 1*(1), 4–27. doi:10.1109/72.80202.

Nelson, J. P., Biddle, J. B., & Shapira, P. (2023). Applications and societal implications of artificial intelligence in manufacturing: A systematic review. *arXiv*. United States arXiv https://arxiv.org . doi:10.48550/arXiv.2308.02025.

Neri, F., & Cotta, C. (2012). Memetic algorithms and memetic computing optimization: A literature review. *Swarm and Evolutionary Computation, 2*, 1–14. doi:10.1016/j.swevo.2011.11.003.

Nnodim, C. T., Arowolo, M. O., Agboola, B. D., Ogundokun, R. O., & Abiodun, M. K. (2021). Future trends in mechatronics. *IAES International Journal of Robotics and Automation (IJRA), 10*(1), 24. doi:10.11591/ijra.v10i1.pp24-31.

Ohalete, N. C., Aderibigbe, A. O., Ani, E. C., Ohenhen, P. E., & Akinoso, A. E. (2023). Advancements in predictive maintenance in the oil and gas industry: A review of AI and data science applications. *World Journal of Advanced Research and Reviews, 20*, 167–181.

Papamarkou, T., Skoularidou, M., Palla, K., Aitchison, L., Arbel, J., Dunson, D., Filippone, M., Fortuin, V., Hennig, P., Hernández-Lobato, J. M., Hubin, A., Immer, A., Karaletsos, T., Khan, M. E., Kristiadi, A., Li, Y., Mandt, S., Nemeth, C., Osborne, M. A., … Zhang, R. (2024). Position: Bayesian deep learning is needed in the age of large-scale AI. *arXiv*. United Kingdom arXiv https://arxiv.org . doi:10.48550/arXiv.2402.00809.

Raissi, M., Perdikaris, P., & Karniadakis, G. E. (2017). Physics informed deep learning (Part I): Data-driven solutions of nonlinear partial differential equations. *arXiv*. United States arXiv https://arxiv.org . doi:10.48550/arxiv.1711.10561.

Redmon, J., Divvala, S., Girshick, R., & Farhadi, A. (2016). You only look once: Unified, real-time object detection. In *Proceedings of the IEEE Computer Society Conference on Computer Vision and Pattern Recognition* (pp. 779–788). IEEE Computer Society. doi:10.1109/CVPR.2016.91.

Ren, S., He, K., Girshick, R., & Sun, J. (2015). Faster R-CNN: Towards real-time object detection with region proposal networks. *Advances in Neural Information Processing Systems, 28*, 91–99.

Rojek, I., Jasiulewicz-Kaczmarek, M., Piechowski, M., & Mikołajewski, D. (2023). An artificial intelligence approach for improving maintenance to supervise machine failures and support their repair. *Applied Sciences, 13*(8), 4971. doi:10.3390/app13084971.

Ross, T. J. (2005). *Fuzzy logic with engineering applications*. John Wiley & Sons.

Rumelhart, D. E., Hinton, G. E., & Williams, R. J. (1986). Learning representations by back-propagating errors. *Nature, 323*(6088), 533–536. doi:10.1038/323533a0.

Russell, S. J., & Norvig, P. (2016). *Artificial intelligence: A modern approach*. Prentice Hall.

Samek, W., Wiegand, T., & Müller, K. R. (2017). Explainable artificial intelligence: Understanding, visualizing and interpreting deep learning models. *arXiv*. Germany arXiv https://arxiv.org . doi:10.48550/arxiv.1708.08296.

Santa, F. M., Sarmiento, F. H. M., & Ariza, H. M. (2022). Path optimization for mobile robots using genetic algorithms. *International Journal of Advanced Computer Science and Applications, 13*(2), 657–662. http://thesai.org/Publications/Archives?code=IJACSA. doi:10.14569/IJACSA.2022.0130277.

Sarker, M. K., Zhou, L., Eberhart, A., & Hitzler, P. (2021). Neuro-symbolic artificial intelligence. *AI Communications, 34*(3), 197–209. http://www.iospress.nl/. doi:10.3233/AIC-210084.

Schmidhuber, J. (2015). Deep learning in neural networks: An overview. *Neural Networks, 61*, 85–117. doi:10.1016/j.neunet.2014.09.003.

Shi, Y., Yang, K., Jiang, T., Zhang, J., & Letaief, K. B. (2020). Communication-efficient edge AI: Algorithms and systems. *IEEE Communications Surveys and Tutorials, 22*(4), 2167–2191. http://ieeexplore.ieee.org/xpl/RecentIssue.jsp?punumber=9739. doi:10.1109/COMST.2020.3007787.

Shortliffe, E. H. (1976). *MYCIN: Computer-based medical consultations*. Elsevier.

Slotine, J. J. E., & Li, W. (1991). *Applied nonlinear control*. Prentice Hall.

Soori, M., Arezoo, B., & Dastres, R. (2023). Artificial intelligence, machine learning and deep learning in advanced robotics, a review. *Cognitive Robotics, 3*, 54–70. doi:10.1016/j.cogr.2023.04.001.

Stoian, V., & Ivanescu, M. (2008). *Robot control by fuzzy logic*. InTech. doi:10.5772/6329.

Storn, R., & Price, K. (1997). Differential evolution—A simple and efficient heuristic for global optimization over continuous spaces. *Journal of Global Optimization, 11*(4), 341–359. www.kluweronline.com/issn/0925-5001/. doi:10.1023/A:1008202821328.

Sutton, R. S., & Barto, A. G. (1998). Reinforcement learning: An introduction. *IEEE Transactions on Neural Networks, 9*(5), 1054 1054. doi:10.1109/TNN.1998.712192.

Tortorelli, A., Imran, M., Priscoli, F. D., & Liberati, F. (2022). A parallel deep reinforcement learning framework for controlling industrial assembly lines. *Electronics, 11*(4), 539. doi:10.3390/electronics11040539.

Watkins, C. J. C. H., & &Dayan, P. (1992). *Q-learning. Machine Learning, 8*(3), 279–292. doi:10.1007/BF00992698.

Wei, T., Wang, Y., & Zhu, Q. (2017). Deep reinforcement learning for building HVAC control. In *Proceedings—Design Automation Conference*. Institute of Electrical and Electronics Engineers Inc. doi:10.1145/3061639.3062224.

Widodo, A., & Yang, B.-S. (2007). Support vector machine in machine condition monitoring and fault diagnosis. *Mechanical Systems and Signal Processing, 21*(6), 2560–2574. doi:10.1016/j.ymssp.2006.12.007.

Worden, K., & Dulieu-Barton, J. M. (2004). An overview of intelligent fault detection in systems and structures. *Structural Health Monitoring, 3*(1), 85–98. doi:10.1177/1475921704041866.

Yager, R. R., & Filev, D. P. (1994). *Essentials of fuzzy modeling and control*. John Wiley & Sons.

Yuan, Y., Ma, G., Cheng, C., Zhou, B., Zhao, H., Zhang, H.-T., & Ding, H. (2020). A general end-to-end diagnosis framework for manufacturing systems. *National Science Review, 7*(2), 418–429. doi:10.1093/nsr/nwz190.

Yurkovich, S., & Passino, K. M. (1999). A laboratory course on fuzzy control. *IEEE Transactions on Education, 42*(1), 15–21. doi:10.1109/13.746327.

Zadeh, L. A. (1965). Fuzzy sets. *Information and Control, 8*(3), 338–353. doi:10.1016/s0019-9958(65)90241-x.

Zaitceva, I., & Andrievsky, B. (2022). Methods of intelligent control in mechatronics and robotic engineering: A survey. *Electronics, 11*(15), 2443. doi:10.3390/electronics11152443.

Zhang, S., Li, Y., Zhang, S., Shahabi, F., Xia, S., Deng, Y. u, & Alshurafa, N. (2022). Deep learning in human activity recognition with wearable sensors: A review on advances. *Sensors, 22*(4), 1476. doi:10.3390/s22041476.

Zhang, W., Yang, D., & Wang, H. (2019). Data-driven methods for predictive maintenance of industrial equipment: A survey. *IEEE Systems Journal, 13*(3), 2213–2227. doi:10.1109/jsyst.2019.2905565.

Zhong, P., Wang, L., Zhang, G., Li, X., Xu, J., Sun, Q., Wang, S., Zhang, S., Wang, C., Chen, L., Yang, X. u, Xu, K., Ding, X., & Zhang, T. (2025). Thermal-vacuum regolith environment simulator for drilling tests in lunar polar regions. *Acta Astronautica, 229*, 13–26. doi:10.1016/j.actaastro.2025.01.002.

Zulu, M. L. T., Carpanen, R. P., & Tiako, R. (2023). A comprehensive review: Study of artificial intelligence optimization technique applications in a hybrid microgrid at times of fault outbreaks. *Energies, 16*(4). http://www.mdpi.com/journal/energies/. doi:10.3390/en16041786.

Chapter 25

A review of soft pneumatic actuator applications in real-world engineering challenges

Mingtai Chen
Wichita State University, Wichita, KS, United States

25.1 Introduction

In robotics, actuators play a pivotal role in providing movement and control, with each type offering unique advantages for different applications. Electric actuators, including direct current (DC), stepper, and servo motors, are widely used in robotic systems because of their precision, reliability, and ease of control. These actuators are ideal for tasks that require continuous or discrete motion, such as robotic arms, mobile robots, and grippers (Almurib et al., 2012; Li et al., 2012). Pneumatic actuators, which utilize compressed air to generate movement, are commonly employed in applications requiring rapid motion and lower forces, such as lightweight grippers or soft robotics. Pneumatic actuators are fast and relatively easy to design, though they may not provide the high force output of hydraulic actuators. Hydraulic actuators, relying on pressurized fluid for motion, are well-suited for applications that require high force, such as industrial robots and heavy-duty tasks in construction, manufacturing, and material handling (Jang et al., 2012). Conversely, shape memory alloy (SMA) actuators, which change shape in response to temperature changes, are compact, lightweight, and ideal for small or soft robotic systems. SMAs are widely used in miniature robots and applications that require lightweight and compact actuators (Kheirikhah et al., 2011). Piezoelectric actuators, known for their high precision and rapid response, are often utilized in microrobotic applications, particularly for tasks requiring fine manipulation or precise positioning (Kanchan et al., 2023). Magnetic actuators, which use magnetic fields to generate movement, are highly effective for contactless motion and high-precision tasks. These actuators are often used in specialized applications such as magnetic levitation systems or environments requiring cleanliness (Luan et al., 2022). Finally, electrostatic actuators are used in systems requiring micromotions or high-frequency actuation, such as microrobots or sensors (Zhou et al., 2023).

Soft actuators in soft robotics have gained attention for their flexibility and compliance, with pneumatic, hydraulic, and electroactive polymer (EAP) actuators being the most prominent. These actuators mimic biological systems, allowing robots to conform to their environment and perform delicate tasks, such as handling fragile objects. Soft pneumatic actuators (SPAs), in particular, are commonly used in soft robotics for their ability to create adaptive and flexible movements (Calisti et al., 2011; Laschi et al., 2016). SPAs can be actuated using positive or negative pressures. Negative pressure actuation provides a fail-safe feature, along with improved lifetime and durability. Vacuum actuators are suitable for applications with constrained volumes, as they shrink when actuated (Tawk & Alici, 2021; Zaidi et al., 2021). In addition, the performance of SPAs can be enhanced by utilizing a combination of both positive and negative pressure (Kalisky et al., 2017; Xavier et al., 2021). SPAs can exhibit various types of motions, such as bending, extension, contraction, and twisting (Gorissen et al., 2017; Hines et al., 2017; Polygerinos et al., 2017). SPAs can be fabricated using a molding process (Marchese et al., 2015; Schmitt et al., 2018) or directly 3D-printed using flexible filaments or elastomeric resins (Gul et al., 2018; Wallin et al., 2018). Despite recent breakthroughs, SPAs and robots experience challenges and limitations related to autonomy, portability, scalability, noise, repeatability, reproducibility, durability, accessibility, impact, complex modeling, integrated sensing, and intelligent control.

Given the large number of recent publications on SPAs, both newcomers and established researchers often find it challenging to assess the current state of the art. This chapter aims to provide a preliminary overview of SPAs, with

Computational Intelligence in Mechatronics: Solving Real-World Problems in Electronic Systems Design. DOI: https://doi.org/10.1016/B978-0-443-27580-7.00001-9

particular emphasis on recent advancements in real-world engineering applications. The chapter is structured as follows: Section 25.2 introduces the fundamental principles of SPAs, whereas Section 25.3 discusses intelligent control strategies for SPAs. Section 25.4 explores their first applications in robotics and automation, followed by Section 25.5, which examines their use in aerospace applications. Finally, Section 25.6 highlights biomedical and assistive technology applications before discussing challenges and future directions in Section 25.7 and concluding the chapter in Section 25.8.

25.2 Fundamental principles of SPAs

SPAs are a type of flexible actuator that uses compressed air to produce movement in deformable materials, often utilized in soft robotics, prosthetics, and various adaptive systems. The fundamental principles of SPAs involve the careful selection of materials, mechanisms of deformation, and control strategies, which enable their unique capabilities in handling complex and dynamic tasks.

25.2.1 Materials and fabrication techniques

The materials used in SPAs significantly affect their performance, including flexibility, stretchability, and durability. Commonly employed materials include elastomers, composites, and 3D-printed structures:

- **Elastomers**: Elastomers, such as silicone rubber and polyurethane, are popular choices because of their flexibility and ability to stretch without permanent deformation. These materials allow SPAs to endure large deformations and maintain their structural integrity, which is crucial for applications where softness and adaptability are required (Sun et al., 2013).
- **Composites**: To improve the mechanical properties of SPAs, elastomers are often combined with reinforcing materials, such as carbon fibers or Kevlar. These composites help in controlling the direction and magnitude of deformation, making actuators stronger without sacrificing flexibility (Polygerinos et al., 2017).
- **3D-printed structures**: Additive manufacturing enables the creation of complex geometries and multimaterial structures, offering high customization and precise control over actuator design. Techniques, such as fused deposition modeling (FDM) or stereolithography (SLA), allow for the fabrication of intricate lattices and internal channels that can optimize the actuator's deformation behavior (Xavier et al., 2021).

25.2.2 Mechanisms of deformation

SPAs are designed to deform in specific ways when inflated, allowing them to generate motion. The three primary modes of deformation include

- **Bending**: This is the most common mode for soft actuators, where inflation causes the actuator to bend because of pressure differences across its cross-section. These actuators are used for tasks requiring flexible manipulation, such as in soft robotic arms or grippers (Liu et al., 2022).
- **Twisting**: Some SPAs are designed to twist when inflated, allowing for rotational movement. These actuators are useful for applications that require rotation, such as in specific robotic end-effectors (Chen et al., 2021a).
- **Expansion**: Inflation can also cause actuators to expand or contract in volume. Actuators with this capability are used in applications such as artificial muscles or soft grippers, where volume changes are essential for handling delicate objects or performing precise movements (Elsamanty et al., 2023).

25.2.3 Control strategies for SPAs

The control of SPAs can be divided into three main categories:

- **Open-loop control**: Open-loop control systems involve preprogrammed commands where actuator pressure is set based on anticipated behavior. Although easy to implement, this method does not allow for real-time adjustments in response to varying conditions or feedback (García-Samartín et al., 2024).
- **Closed-loop control**: Closed-loop control utilizes sensors (e.g., pressure sensors or strain gauges) to monitor the actuator's state and adjust inputs accordingly. This type of feedback allows for more precise control and compensation for unexpected disturbances, enabling more stable and reliable operation in dynamic environments (Yu et al., 2020).
- **AI-driven adaptive control**: In more advanced systems, artificial intelligence (AI) and machine learning (ML), particularly reinforcement learning (RL), are used to optimize the actuator's behavior. RL allows SPAs to adapt their

movement strategies through trial and error based on feedback from the environment. This is particularly useful for complex robotic systems or when operating in uncertain or dynamic environments (Kozhubaev et al., 2023).

25.3 Computational intelligence for SPAs

The control of SPAs is complex because of their nonlinear, highly deformable nature. Computational intelligence techniques, such as machine learning, RL, fuzzy logic, and bioinspired control approaches, are increasingly being integrated to enhance the performance and adaptability of SPAs. These techniques enable more sophisticated control and real-time adaptability, improving their functionality in dynamic environments.

25.3.1 Machine learning-based modeling and control

ML is a powerful tool for modeling the behavior of SPAs because of their highly nonlinear and dynamic responses to inputs, such as pressure. Traditional modeling approaches often fail to capture the complexity of SPA behavior, but ML algorithms, such as supervised learning, neural networks, and support vector machines, can learn from experimental data to predict actuator behavior and optimize control strategies (Youssef et al., 2022). For instance, deep learning methods can be used to predict the displacement or force generated by an actuator for a given pressure, enabling real-time control adjustments. This data-driven approach allows for more accurate modeling of the actuators' nonlinearities compared to classical physical models.

25.3.2 Reinforcement learning for real-time adaptability

RL, a subset of machine learning, has shown significant potential in providing real-time adaptability for SPAs. RL allows the actuator to "learn" optimal control strategies through trial and error, improving its performance over time based on feedback from the environment. This is particularly useful in applications requiring complex, dynamic responses, where the SPA must adjust to environmental changes or unexpected obstacles. By leveraging RL algorithms such as deep *Q*-learning or proximal policy optimization (PPO), soft actuators can autonomously learn to optimize parameters, such as pressure and deformation, to achieve specific tasks such as grasping or moving objects (Viquerat et al., 2021). RL-based systems enable SPAs to continuously adapt to varying conditions, making them highly suitable for unpredictable environments.

25.3.3 Fuzzy logic and bioinspired control approaches

Fuzzy logic and bioinspired control strategies offer an alternative approach to control, focusing on imitating the adaptability and robustness found in natural organisms. Fuzzy logic provides a way to handle the uncertainty and imprecision inherent in controlling soft actuators by using linguistic variables and rule-based systems to approximate human reasoning (Ramos-Arreguin et al., 2010). This approach is effective when modeling systems that exhibit uncertain behavior, which is often the case with SPAs. In addition, bioinspired control techniques, such as neuromorphic engineering and genetic algorithms, have been applied to SPAs. These methods mimic biological neural networks or evolutionary processes to optimize control strategies. By simulating natural processes, bioinspired control algorithms can improve the SPA's ability to adapt and optimize its behavior based on continuous feedback, leading to more robust and efficient operation in real-time scenarios.

25.3.4 Integration with sensors and feedback systems

For SPAs to operate efficiently and with high precision, they must be integrated with sensors and feedback systems that provide real-time data on their state (e.g., pressure, displacement, and strain). Feedback control loops, often using closed-loop control systems, enable continuous monitoring and adjustment of actuator performance based on sensor inputs. Sensors, such as pressure sensors, strain gauges, and optical sensors, are commonly used in SPA systems to provide feedback on the actuator's performance. This feedback can be used to fine-tune control inputs, adjust inflation rates, and ensure that the actuator reaches the desired position or force (Xavier et al., 2022). In combination with machine learning or fuzzy logic controllers, sensors enable the actuator to operate more precisely and respond to environmental changes in real-time.

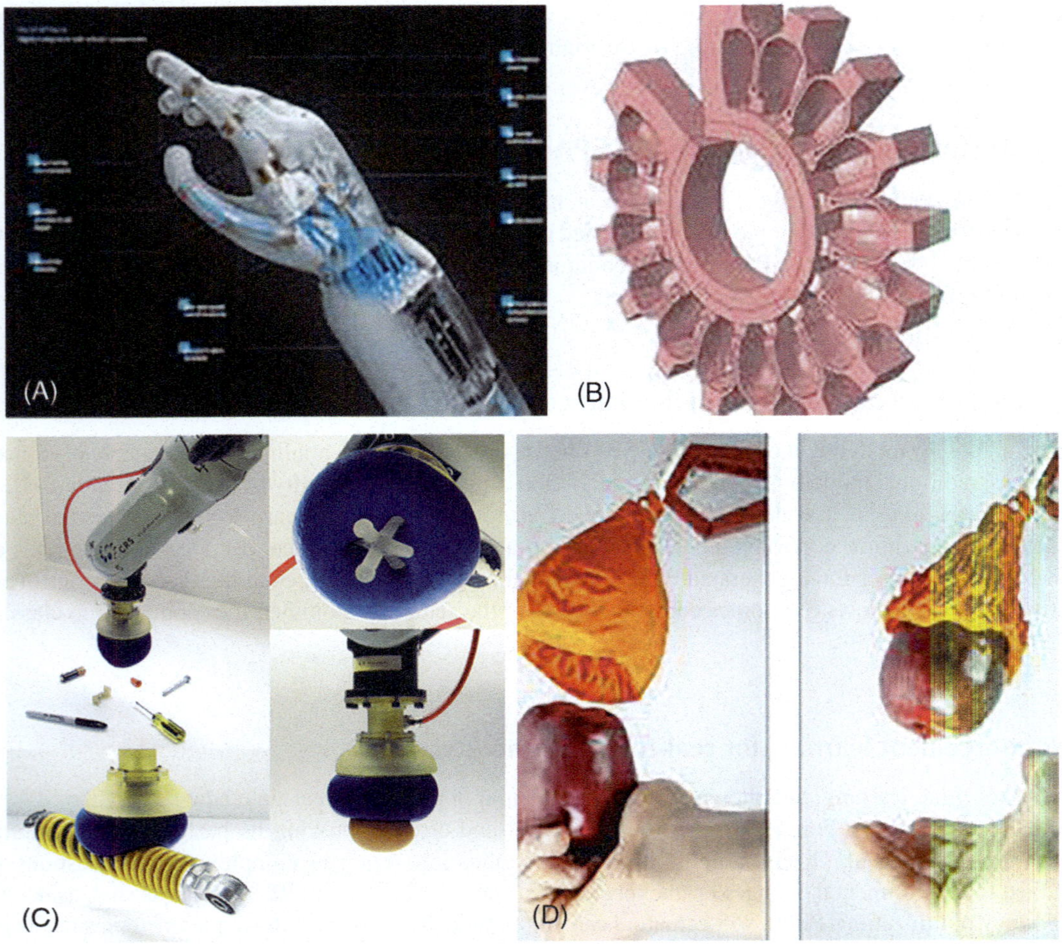

FIGURE 25.1 Soft robotic grippers driven by SPAs: (A) Festo's BionicSoftHand, (B) Harvard's PneuNet gripper, (C) universal jamming gripper, and (D) MIT's soft origami gripper. *SPA*, Soft pneumatic actuator. *((A) https://www.festo.com/; (B) https://softroboticstoolkit.com/; (C) https://www.creativemachineslab.com/; (D) https://news.mit.edu/2019/new-robot-hand-gripper-soft-and-strong-0315.)*

25.4 Applications in robotics and automation

25.4.1 Soft robotic grippers for industrial handling

Soft robotic grippers using pneumatic actuation are widely employed in industrial settings because of their ability to handle delicate, deformable, and irregularly shaped objects. Unlike rigid robotic end-effectors, these grippers leverage air-pressure-driven deformation to achieve adaptive and conformal grasping, making them ideal for handling fragile materials, such as food, textiles, and electronics. The working principle of pneumatic soft grippers varies depending on the design. Fiber-reinforced pneumatic actuators (FRPAs) use internal chambers that expand when pressurized, whereas fiber constraints force a predefined bending motion. A well-known example is Festo's BionicSoftHand, shown in Fig. 25.1A, which mimics human-like gripping and is commonly used in assembly lines for packaging and object sorting. Another design, known as PneuNet (pneumatic networks) grippers, as seen in Fig. 25.1B, relies on multiple interconnected chambers that inflate sequentially to create controlled bending. Harvard's PneuNet Gripper is an excellent demonstration of this concept, widely applied in the food industry, where gentle contact is essential.

Granular jamming-based pneumatic grippers operate using a flexible membrane filled with granular material, such as coffee grounds, that conforms to an object. When a vacuum is applied, the gripper stiffens, enabling it to securely hold objects of various sizes and textures. A good example is the "universal jamming gripper," presented in Fig. 25.1C (Brown et al., 2010), which is highly effective for pick-and-place tasks in warehouses and e-commerce fulfillment centers. Another innovative approach is the origami-inspired pneumatic gripper, which uses folding structures that inflate to grasp objects through shape-morphing principles. MIT's soft origami gripper, depicted in Fig. 25.1D, for instance, can pick up delicate

FIGURE 25.2 Autonomous robotic systems with SPAs: (A) Harvard's octobot, (B) Ex-Robots's humanoid robot, and (C) biomimetic fish. *SPA*, Soft pneumatic actuator. *((A) https://wyss.harvard.edu/news/the-first-autonomous-entirely-soft-robot/; (B) https://www.reuters.com/technology/china-robots-conference-spotlights-changing-face-humanoids-2024-08-26/; (C) From Chen et al., 2021b.)*

fruits like grapes without causing damage, making it ideal for applications in agriculture and food processing. Pneumatic soft grippers offer several advantages. They are highly adaptive, capable of conforming to objects of different shapes and sizes. Their soft nature ensures safe human interaction, making them suitable for collaborative robotics environments. In addition, they are cost-effective, as they rely on simple air pumps rather than complex motors or gears, and they are lightweight and energy-efficient, reducing overall system weight compared to rigid robotic arms.

Industrially, these grippers have been integrated into various sectors. In food handling, they are used to grip fragile items, such as fruits, vegetables, and baked goods without bruising. In automated packaging, they provide flexible grasping solutions for sorting and packing in logistics. Manufacturing industries use them to handle deformable or irregularly shaped parts in assembly lines, whereas the medical field has explored their potential in soft exoskeletons and assistive devices for rehabilitation. Despite these advantages, there are still challenges to address. The durability of soft elastomers can degrade over time because of repeated pressurization, requiring improvements in material selection. Control precision remains an issue, as fine-tuned force modulation necessitates advanced sensing and feedback control. Looking ahead, integrating AI and machine learning will enable adaptive grasping and real-time adjustments, enhancing the efficiency and versatility of these systems.

25.4.2 Autonomous robotic systems with compliant actuation

Autonomous robotic systems equipped with SPAs have gained prominence because of their adaptability, safety, and efficiency in interacting with diverse environments. These actuators, typically composed of flexible materials, utilize pressurized air to produce movement, enabling robots to perform tasks that require gentle handling or navigation through irregular terrains. The ability of these actuators to deform and recover their shape makes them particularly suitable for applications where traditional rigid robots would struggle, such as in handling fragile objects, traversing unpredictable landscapes, or operating in confined spaces. One significant application of SPAs is in peristaltic robots, which are inspired by the locomotion of earthworms, like Harvard's octobot in Fig. 25.2A. These robots generate wave-like movements through sequential inflation and deflation of pneumatic chambers, allowing them to navigate confined or complex environments. Their flexibility and compliance enable effective movement across diverse terrains, making them ideal for pipeline inspections, medical endoscopy, and search-and-rescue operations in debris-filled environments. The adaptability of peristaltic robots enhances their functionality in situations where conventional robotic systems may be hindered by rigid structures.

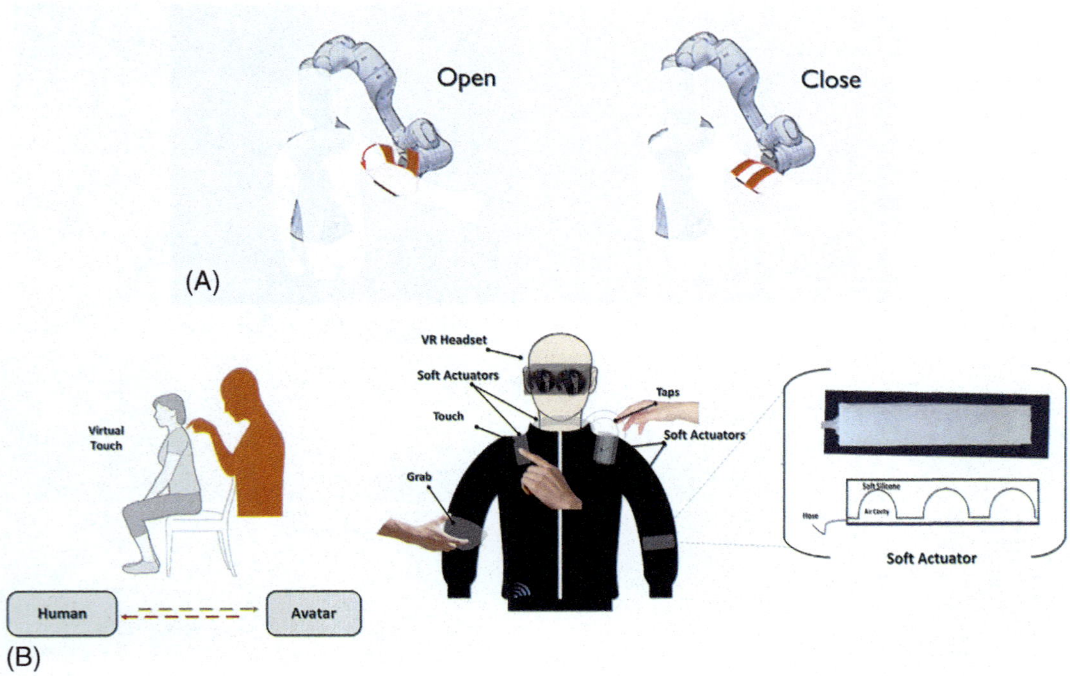

FIGURE 25.3 Human–robot interaction and wearable robotics: (A) SPA strap and (B) soft pneumatic haptic suit. *SPA*, Soft pneumatic actuator; *VR*, virtual reality. *((A) From van Vlerken et al. (2023); (B) From Talhan et al. (2024).)*

Another emerging area is the development of flexible robotic manipulators, which leverage SPAs to provide a high degree of dexterity and adaptability. At the World Robot Conference in Beijing, researchers showcased a humanoid robot in Fig. 25.2B, powered by pneumatic artificial muscles (PAMs) and 3D-printed materials. These designs offer cost-effective and versatile solutions for robotic manipulation, reducing reliance on traditional motors and transmission devices.

Recent research has explored the broader capabilities of SPAs in autonomous systems. The paper *Pneumatic Soft Robots: Challenges and Benefits* (Martinez-Sanchez et al., 2023) discusses the advantages and challenges associated with pneumatic soft robots, emphasizing their potential applications and technical limitations. Another study, *Water Hydraulic Soft Actuators for Underwater Autonomous Robotic Systems* (Chen et al., 2021b), investigates how water-powered soft actuators can enhance the capabilities of underwater autonomous robots, as shown in Fig. 25.2C. These studies highlight the need for advancements in control systems, materials, and actuation methods to optimize performance and energy efficiency. The integration of SPAs into autonomous robotic systems significantly enhances their adaptability and safety, making them suitable for a wide range of applications, from medical rehabilitation to industrial automation. Soft robotics research continues to evolve, addressing challenges, such as durability, response time, and precise control, ultimately pushing the boundaries of what autonomous robots can achieve in dynamic and unstructured environments.

25.4.3 Human–robot interaction and wearable robotics

SPAs have gained significant attention in physical human–robot interaction because of their compliance, adaptability, and ability to safely interact with humans. Two key developments in this field are SPA straps and soft pneumatic haptic wearables, both of which contribute to enhanced interaction experiences by leveraging air pressure modulation and deformable materials to mimic human touch. SPA straps are designed for applications in wearable robotics and collaborative robotics, offering dynamic support and force feedback. Fig. 25.3A illustrates an application of SPA straps (van Vlerken et al., 2023). The potential advantages of SPA bending in strapping procedures include user comfort and safety. Future research could lead to the development of more effective and safe strapping interfaces in wearable robotics, thereby improving the overall performance of these devices and increasing user acceptance.

Soft pneumatic haptic wearables, such as gloves and suits, create a realistic sense of human touch by generating localized pressure changes. This capability is essential for applications in virtual reality (VR), telepresence, and immersive training, where users need realistic force feedback, presented in Fig. 25.3B (Talhan et al., 2024). By carefully controlling air pressure distribution, these wearables can simulate sensations such as grasping, tapping, or stroking, allowing users to

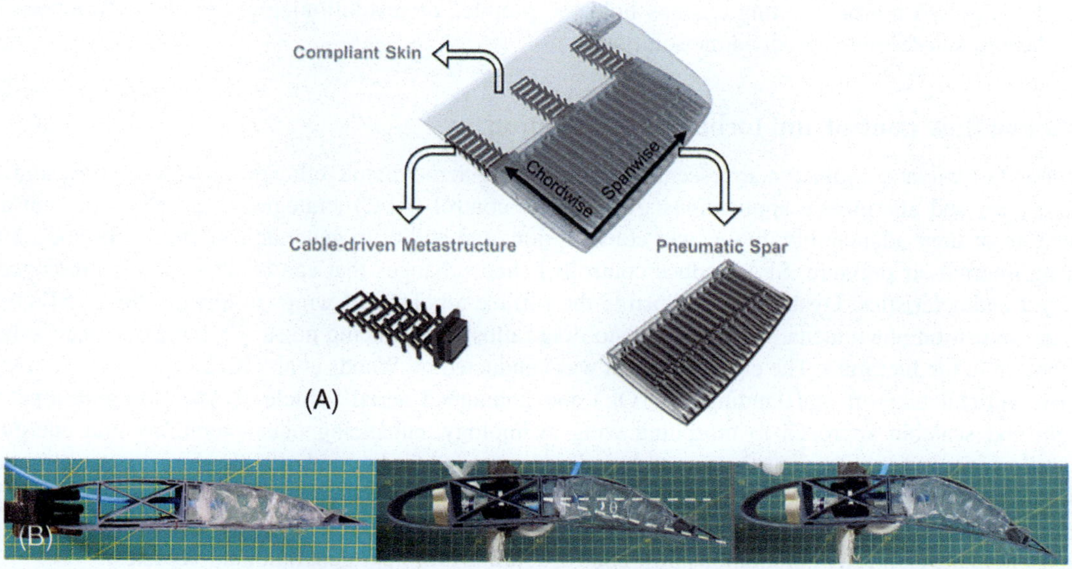

FIGURE 25.4 Morphing wing driven by SPAs: (A) morphing aerofoil and (B) morphing wing with tunable stiffness. *SPA*, Soft pneumatic actuator. *((A) From Xiao et al. (2022); (B) From De Silva et al. (2023).)*

experience physical feedback from remote robotic interactions or virtual environments. Beyond VR, these haptic wearables also play a crucial role in enhancing remote robotic operation by enabling intuitive control mechanisms. Operators can receive real-time force feedback when manipulating robotic arms or drones, improving dexterity and situational awareness in space exploration, hazardous environments, and precision assembly tasks. Although these technologies offer exciting possibilities, challenges remain in improving control accuracy, response time, and durability. Integrating advanced materials, AI-driven control strategies, and sensor feedback mechanisms will be the key to advancing their capabilities in human–robot collaboration and wearable robotics.

25.5 Aerospace applications

25.5.1 Morphing wings for aerodynamic efficiency

Morphing wings that adapt their shape during flight can significantly enhance aerodynamic efficiency. SPAs have emerged as a promising technology for achieving such adaptability because of their flexibility and lightweight properties. These actuators offer a smooth and continuous way to modify the shape of wings in real-time, optimizing performance across various flight conditions without the need for traditional mechanical components like hinges. SPAs are constructed from compliant materials that deform when pressurized. This deformation allows the wing surface to adjust its camber, twist, or other aerodynamic parameters, making it ideal for applications requiring dynamic shape adaptation. One application of SPAs in morphing wings is the development of adaptive compliant wings. These wings utilize SPAs to achieve variable camber configurations, effectively acting like traditional flaps but without discrete segments and gaps. The continuous surface adjustments lead to reduced aerodynamic drag and improved fuel efficiency. Fig. 25.4 presents two prototypes of morphing driven by SPAs (De Silva et al., 2023; Xiao et al., 2022).

The integration of SPAs into wing structures offers several advantages over conventional actuation mechanisms. By eliminating complex mechanical linkages, SPAs reduce weight and minimize potential points of failure. In addition, they provide smooth, gap-free surfaces during morphing, which minimizes flow disturbances and aerodynamic drag. Optimizing wing shape for various flight regimes also leads to significant fuel savings, further improving the overall efficiency of an aircraft. Despite their potential, several challenges must be addressed to fully realize the benefits of SPAs in morphing wings. Material durability is a key concern, as these actuators must withstand cyclic loading conditions over extended periods. In addition, developing responsive control systems that can adjust wing shape in real-time based on flight dynamics remains a challenge. Another issue is scalability, as adapting SPA technology for larger aircraft while maintaining effectiveness and efficiency requires further research and testing. Ongoing research aims to overcome these challenges by focusing on advanced materials, robust control algorithms, and comprehensive testing protocols. The continued development and integration of SPAs into morphing wing designs hold the promise of revolutionizing aircraft performance by offering adaptable, efficient, and resilient aerodynamic solutions.

25.5.2 Active flow control and turbulence mitigation

Flow control and turbulence mitigation are essential for enhancing aerodynamic efficiency, reducing drag, and suppressing noise in aerospace and automotive applications. Active flow control (AFC) strategies using SPAs present a promising solution because of their adaptability, low power consumption, and ability to generate smooth, continuous deformations. By leveraging internal air pressure, SPAs induce controlled shape changes that can be strategically employed to modify boundary layer characteristics. Dynamically adjusting the surface topology of wings or airfoils allows SPAs to influence separation and transition phenomena, improving lift-to-drag ratios and reducing noise. Currently, no successful prototype has been reported in the literature. The closest attempt was conducted by Woods et al. (2012).

In electric vertical take-off and landing (eVTOL) and unmanned aerial vehicle (UAV) configurations, SPAs offer a lightweight and scalable approach to morphing wing technology, enhancing maneuverability and energy efficiency. Their integration into rotor-wing interfaces could mitigate rotor-induced turbulence, leading to quieter operation and reduced vibrational loads. Advancements in AI-driven control algorithms, such as deep RL, are expected to improve SPA responsiveness by optimizing actuation patterns in real time. Future research aims to integrate multichannel SPAs with feedback sensors to develop fully autonomous flow control solutions for next-generation air vehicles.

25.6 Biomedical and assistive technology applications

25.6.1 Soft exoskeletons and rehabilitation devices

Soft exoskeletons and rehabilitation devices utilizing SPAs have emerged as promising solutions for assisting individuals with mobility impairments. Unlike traditional rigid exoskeletons, soft exoskeletons employ compliant materials and pneumatic actuators to provide adaptive and comfortable assistance, reducing the risk of injury and enhancing user comfort. SPAs function by inflating and deflating air chambers to produce controlled movements. These actuators can generate bending, twisting, or extending motions, mimicking natural muscle behavior. By regulating air pressure, they can provide variable assistance tailored to the user's specific needs, making them ideal for rehabilitation applications where gradual strength recovery is required.

Soft exoskeletons integrated with SPAs are widely used in rehabilitation for stroke survivors, individuals with spinal cord injuries, and patients recovering from orthopedic surgeries. These devices assist in gait training, arm mobility enhancement, and hand function restoration by providing adjustable levels of assistance. Several real-world examples highlight the effectiveness of these devices. Harvard's soft exosuit (Wehner et al., 2013), developed by the Harvard Biodesign Lab in Fig. 25.5A, uses textile-based pneumatic actuators to assist in walking for stroke patients, reducing muscle fatigue and improving gait mechanics. The soft robotic glove, designed by MIT (Polygerinos et al., 2015) and the Wyss Institute (Nuckols et al., 2020) and presented in Fig. 25.5B and C, employs SPAs to support finger flexion and extension, aiding in hand function recovery for stroke survivors. In addition, the PAM exoskeleton (Lo Piccolo et al., 2022), as shown in Fig. 25.5D, enhances movement for patients with neuromuscular disorders through soft actuators.

Soft exoskeletons offer several advantages over rigid alternatives. They are lightweight and comfortable, as their soft materials conform to the user's body, reducing discomfort and allowing for longer wear durations. Their compliance enhances safety and adaptability, minimizing the risk of injury, making them particularly suitable for elderly users and patients with fragile musculoskeletal structures. Furthermore, they are cost-effective, requiring fewer mechanical components, which reduces production costs and makes them more accessible for widespread rehabilitation use. Despite their advantages, soft pneumatic exoskeletons face challenges, such as limited actuation speed, air leakage, and the need for external air supply systems. Future research aims to develop more efficient, portable, and self-contained pneumatic systems, integrating advanced materials and AI to enhance control and adaptability. The continued advancement of soft pneumatic exoskeletons holds great promise for revolutionizing rehabilitation by providing effective, user-friendly, and accessible assistive devices.

25.6.2 Prosthetics with enhanced dexterity and comfort

Advancements in prosthetic technology have led to the development of SPAs that enhance both the dexterity and comfort of prosthetic limbs. Unlike traditional rigid prosthetics, SPAs utilize flexible materials and pneumatic systems to closely mimic the natural movements of human limbs, offering users a more intuitive and comfortable experience. SPAs are constructed

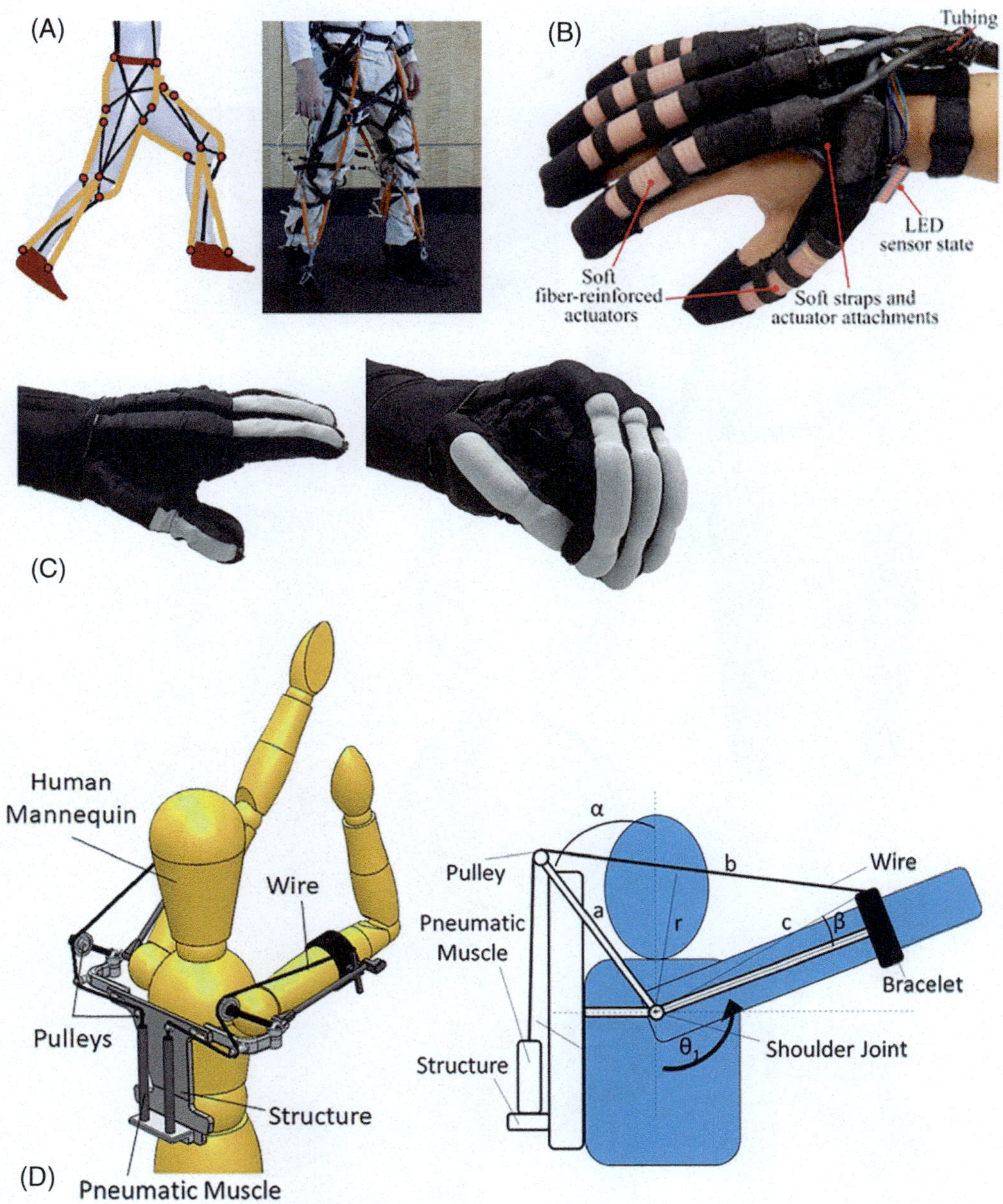

FIGURE 25.5 Soft exoskeletons and rehabilitation devices: (A) Harvard's soft exosuit, (B) MIT's soft robotic glove, (C) the Wyss Institute's soft robotic glove, and (D) PAM exoskeleton. (*(A) From Wehner et al. (2013); (B) From Polygerinos et al. (2015); (C) From Nuckols et al. (2020); (D) From Lo Piccolo et al. (2022).*)

from soft, compliant materials that can undergo significant deformation when pressurized air or fluids are introduced. This design allows for movements that closely resemble the bending and flexing of natural joints. A preliminary demonstration of this technology is in Fig. 25.6A. In a more advanced prototype the soft prosthetic in Fig. 25.6B can perform dexterous operations adaptively by controlling the internal air pressure. In addition, it achieves various tactile sensing functions through simple artificial nerve innervation. Thanks to its sensing and actuating capabilities, the soft hand has been successfully applied in grasping and human–machine interface (HMI) scenarios (Zhao et al., 2016).

The flexibility of SPAs contributes significantly to user comfort. By conforming to the residual limb's shape and accommodating slight movements, these actuators reduce pressure points and the risk of skin ulcerations, a common issue with rigid prosthetics. A study utilizing microfluidic-based soft robotics in Fig. 25.6C demonstrated a reduction in skin ulcerations and pain for below-knee amputees, highlighting the comfort benefits of soft prosthetic designs. The adaptability of SPAs allows for a greater range of motion and more precise control, enhancing the functionality of prosthetic limbs.

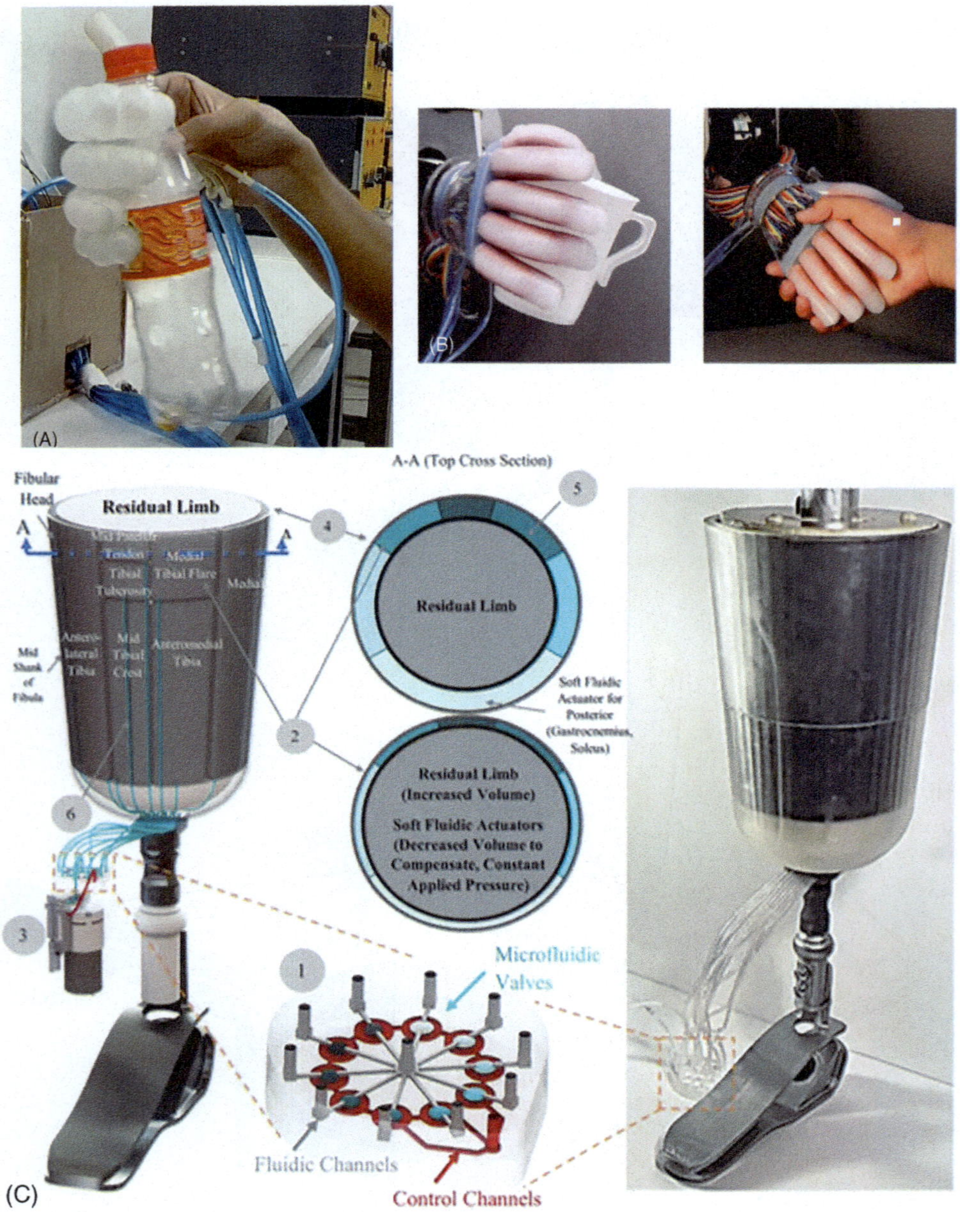

FIGURE 25.6 Prosthetics with enhanced dexterity and comfort: (A) soft prosthetic robotic hand, (B) innervated prosthetic hand, and (C) microfluidic-based soft robotic prosthetics. (*(A) https://www.youtube.com/watch?v=m7bChemOzog; (B) From Zhao et al. (2016); (C) https://www.medicaldesignbriefs.com/component/content/article/46694-microfluidic-based-soft-robotic-prosthetics.*)

The integration of SPAs into prosthetic devices represents a significant leap forward in assistive technology. Ongoing research focuses on refining actuator designs to further enhance durability, control, and user feedback mechanisms, aiming to create prosthetics that are indistinguishable from natural limbs in both function and comfort. In summary the adoption of SPAs in prosthetic limbs offers promising improvements in user comfort and limb functionality, paving the way for more natural and effective assistive devices.

25.6.3 Wearable assistive technologies for mobility improvement

Wearable assistive technologies employing SPAs have emerged as a promising avenue for enhancing mobility in individuals with movement impairments. These devices, characterized by their flexibility and adaptability, offer significant advantages over traditional rigid exoskeletons, providing more natural and comfortable assistance. SPAs are constructed from compliant

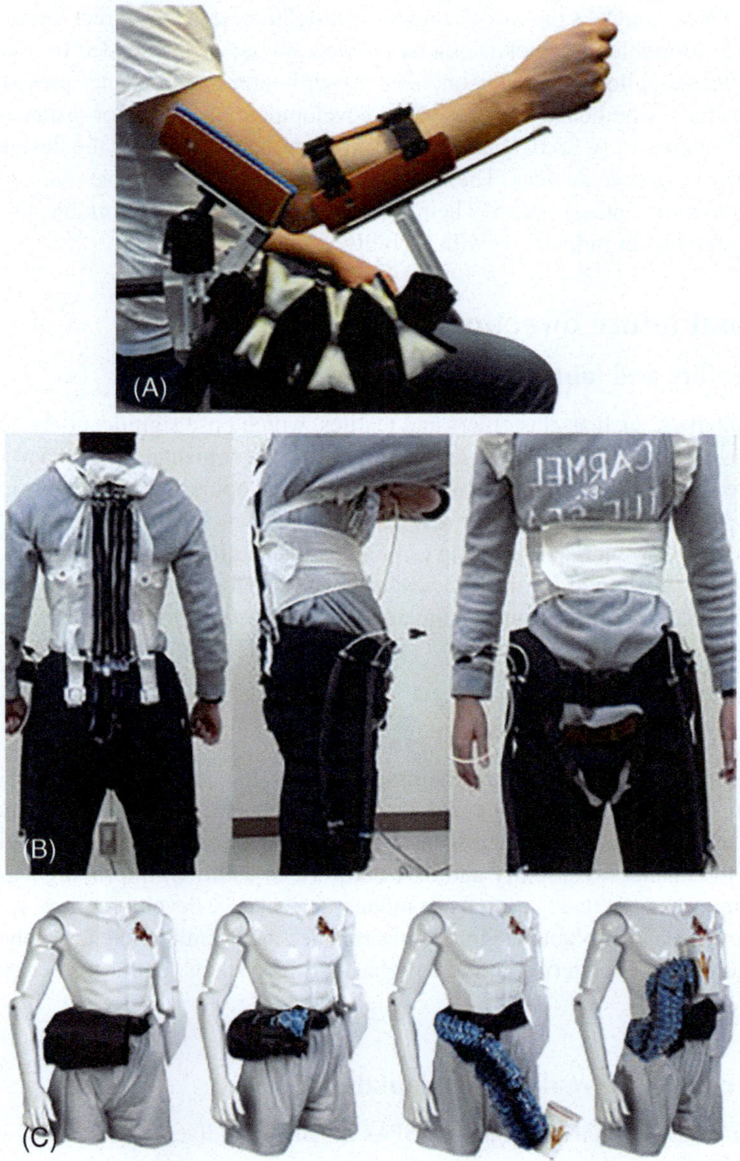

FIGURE 25.7 Wearable assistive technologies for mobility improvement: (A) soft elbow trainer, (B) soft power back assistance device, and (C) soft robotic limb. (*(A) From Wilkening et al. (2015); (B) From Cho et al. (2017); (C) From Nguyen et al. (2019).*)

materials that can mimic the natural movements of human muscles. When pressurized air is introduced, these actuators expand or contract, producing motion that can assist or augment human movement. Their inherent flexibility allows for safe interaction with users, reducing the risk of injury and enhancing comfort.

A soft elbow trainer using SPAs as bending joints was designed (Wilkening et al., 2015), as shown in Fig. 25.7A. The device consists of a bending joint, a passive rotation joint, and a passive translation joint, allowing motion in both extension and flexion while aligning naturally with the polycentric movement of a human joint. In addition, lower back pain (LBP), a common issue associated with repetitive and heavy lifting, can be mitigated using soft wearable power-assist devices made from SPAs. A powered assistive device was designed to prevent LBP by providing support during lifting and holding tasks (Cho et al., 2017), as shown in Fig. 25.7B. This device integrates two types of SPAs—an elongation-type pneumatic actuator and a layer-type pneumatic actuator—to deliver assistive force to the lower back. Lastly, innovations, such as fabric-based soft robotic limbs (Nguyen et al., 2019), have been developed to aid in posture correction and movement assistance, as shown in Fig. 25.7C. These devices utilize SPAs to provide the necessary support for maintaining or improving posture, benefiting individuals with musculoskeletal conditions.

The compliant nature of SPAs reduces the risk of injury, making them safer for direct interaction with the human body. Their flexibility ensures a comfortable fit, which is crucial for wearable devices intended for extended use. These actuators can be tailored to individual needs, allowing for personalized rehabilitation and assistance programs. Advancements in rapid design and fabrication tools have significantly improved the development of SPAs. For instance, MIT's Computer Science and Artificial Intelligence Laboratory (CSAIL) has created a system that streamlines the design process, enabling quicker iteration and customization of assistive devices. The integration of SPAs into wearable assistive technologies represents a significant advancement in mobility enhancement. Their ability to provide safe, comfortable, and effective assistance opens new possibilities for daily support for individuals with mobility impairments.

25.7 Challenges and future directions

25.7.1 Material durability and fabrication challenges

SPAs rely on compliant materials, such as elastomers and textiles, which pose significant challenges in terms of durability and long-term performance. Frequent actuation cycles, exposure to environmental factors, and material fatigue can lead to mechanical degradation, reducing the lifespan of these actuators. Developing more resilient materials with improved tear resistance, fatigue resistance, and self-healing properties is crucial for enhancing the longevity of SPAs. Fabrication methods also remain a critical challenge. Current manufacturing techniques, such as molding, 3D printing, and soft lithography, often lack precision and repeatability, leading to inconsistencies in actuator performance. Scalable and high-precision fabrication processes, such as multimaterial 3D printing and automated casting, need to be further explored to ensure reliability and reproducibility in mass production.

25.7.2 Power efficiency and miniaturization

Wearable assistive devices require compact, lightweight, and energy-efficient power sources to operate SPAs effectively. However, current pneumatic systems often rely on bulky air compressors or external power sources, limiting portability and usability. Miniaturization of pneumatic components, including compact compressors, lightweight tubing, and integrated air reservoirs, is essential to enhance wearability and user comfort. In addition, improving the energy efficiency of SPAs is a key research focus. Optimizing actuator designs to reduce air leakage, developing low-power control strategies, and integrating energy-recycling mechanisms could significantly enhance the overall efficiency of these devices. The integration of novel power sources, such as soft batteries and energy-harvesting systems, could further enable standalone and long-lasting wearable solutions.

25.7.3 AI-enhanced control for real-world adaptability

AI-driven control algorithms are crucial for enabling SPAs to adapt to diverse user needs and real-world conditions. Traditional control methods often struggle with nonlinearities in soft actuator behavior, making precise and responsive control challenging. Machine learning techniques, particularly deep RL and adaptive control algorithms, offer promising solutions by allowing SPAs to learn optimal actuation patterns based on user movement and environmental interactions. Moreover, real-time sensing and feedback integration play a vital role in enhancing control accuracy. Embedding soft sensors, vision-based tracking, and electromyographic (EMG) signals into wearable systems could enable AI-driven adjustments that enhance performance and user experience. Future research should focus on improving the robustness of AI-based controllers, ensuring real-time adaptability, and reducing computational demands to facilitate onboard processing in wearable applications.

25.7.4 Hybrid soft-rigid actuation for enhanced performance

Combining soft and rigid actuation mechanisms presents an opportunity to address the limitations of purely soft actuators while retaining their inherent advantages. Hybrid soft-rigid actuators can offer improved load-bearing capacity, faster response times, and greater stability in wearable assistive devices. This approach is particularly relevant for applications requiring both flexibility and strength, such as exoskeletons for mobility assistance and rehabilitation. Innovations in materials and structural design, such as variable-stiffness composites and shape-memory alloys, could enable seamless transitions between soft and rigid states. In addition, integrating hybrid actuation with AI-driven control strategies could enhance adaptability, allowing devices to switch between different stiffness levels based on user activity and external forces.

Future research should explore scalable fabrication techniques, optimized actuator architectures, and user-centric testing to validate the effectiveness of hybrid systems in real-world applications.

25.8 Conclusion

SPAs have demonstrated remarkable potential in a wide range of applications, from industrial automation and aerospace engineering to biomedical and assistive technologies. Their compliance, lightweight nature, and adaptability enable safer and more efficient interactions in environments where traditional rigid actuators struggle. Advances in computational intelligence, including machine learning and RL, have further enhanced their control and functionality, allowing for real-time adaptability in complex and dynamic settings. In addition, innovations in materials and fabrication techniques, such as 3D printing and composite reinforcements, have expanded the design possibilities for SPAs, improving their durability and performance across different use cases.

Despite these advancements, several challenges remain, including material degradation over time, energy efficiency constraints, and the complexity of achieving precise control. Future research must focus on improving material longevity, miniaturizing pneumatic systems for portability, and developing hybrid actuation approaches that combine the strengths of both soft and rigid components. The integration of AI-driven control strategies and embedded sensing will further enhance SPA performance, making them more intelligent and responsive. As these challenges are addressed, SPAs are expected to play an increasingly critical role in next-generation robotics, aircraft morphing structures, and advanced prosthetics, driving innovation in automation and human–machine interaction.

References

Almurib, H. A. F., Al-Qrimli, H. F., & Kumar, N. (2012). A review of application industrial robotic design. In *Ninth International Conference on ICT and Knowledge Engineering* (pp. 105–112). doi:10.1109/ICTKE.2012.6152387.

Brown, E., Rodenberg, N., Amend, J., Mozeika, A., Steltz, E., Zakin, M. R., Lipson, H., & Jaeger, H. M. (2010). Universal robotic gripper based on the jamming of granular material. *Proceedings of the National Academy of Sciences, 107*(44), 18809–18814. doi:10.1073/pnas.1003250107.

Calisti, M., Giorelli, M., Levy, G., Mazzolai, B., Hochner, B., Laschi, C., & Dario, P. (2011). An octopus-bioinspired solution to movement and manipulation for soft robots. *Bioinspiration & Biomimetics, 6*(3), 036002. doi:10.1088/1748-3182/6/3/036002.

Chen, F., Miao, Y., Gu, G., & Zhu, X. (2021a). Soft twisting pneumatic actuators enabled by freeform surface design. *IEEE Robotics and Automation Letters, 6*(3), 5253–5260. doi:10.1109/LRA.2021.3072813.

Chen, G., Yang, X., Zhang, X., & Hu, H. (2021b). Water hydraulic soft actuators for underwater autonomous robotic systems. *Applied Ocean Research, 109*, 102551. https://www.sciencedirect.com/science/article/pii/S0141118721000286. doi:10.1016/j.apor.2021.102551.

Cho, F., Sugimoto, R., Noritsugu, T., & Li, X. (2017). Improvement of wearable power assist wear for low back support using pneumatic actuator. *IOP Conference Series: Materials Science and Engineering, 249*(1), 012004. doi:10.1088/1757-899X/249/1/012004.

De Silva, K., Maduranga, S., Chamika, V., De Silva, L., Kulasekera, A. L., Samaraweera, N., & Jayaweera, N. D. (2023). Design and experimental characterization of a soft bending actuator for morphing aerofoils. In *Moratuwa Engineering Research Conference (MERCon)* (pp. 684–689). doi:10.1109/MERCon60487.2023.10355480.

Elsamanty, M., Hassaan, M. A., Orban, M., Guo, K., Yang, H., Abdrabbo, S., & Selmy, M. S. (2023). Soft pneumatic muscles: Revolutionizing human assistive devices with geometric design and intelligent control. *Micromachines, 14*(7), 1431. doi:10.3390/mi14071431.

García-Samartín, J. F., Rieker, A., & Barrientos, A. (2024). Design, manufacturing, and open-loop control of a soft pneumatic arm. *Actuators, 13*(1), 36. doi:10.3390/act13010036.

Gorissen, B., Reynaerts, D., Konishi, S., Yoshida, K., Kim, J.-W., & De Volder, M. (2017). Elastic inflatable actuators for soft robotic applications. *Advanced Materials, 29*(43), 1604977. doi:10.1002/adma.201604977.

Gul, J. Z., Sajid, M., Rehman, M. M., Siddiqui, G. U., Shah, I., Kim, K.-H., Lee, J.-W., & Choi, K. H. (2018). 3D printing for soft robotics – A review. *Science and Technology of Advanced Materials, 19*(1), 243–262. doi:10.1080/14686996.2018.1431862.

Hines, L., Petersen, K., Lum, G. Z., & Sitti, M. (2017). Soft actuators for small-scale robotics. *Advanced Materials, 29*(13), 1603483. doi:10.1002/adma.201603483.

Jang, S., Lee, G., Kim, H., Ahn, K., Park, J., & Ryew, S. (2012). Hydraulic actuators in application of robot manipulator. In *IEEE International Conference on Automation Science and Engineering (CASE)* (pp. 924–925). doi:10.1109/CoASE.2012.6386486.

Kalisky, T., Wang, Y., Shih, B., Drotman, D., Jadhav, S., Aronoff-Spencer, E., & Tolley, M. T. (2017). Differential pressure control of 3D printed soft fluidic actuators. In *IEEE/RSJ International Conference on Intelligent Robots and Systems (IROS)* (pp. 6207–6213). doi:10.1109/IROS.2017.8206523.

Kanchan, M., Santhya, M., & Bhat, R. N. N. (2023). Application of modeling and control approaches of piezoelectric actuators: A review. *Technologies, 11*(6), 2227–7080. doi:10.3390/technologies11060155.

Kheirikhah, M., Mahdi, R., Samaneh, E., & Mohammad, E. (2011). *A review of shape memory alloy actuators in robotics* (pp. 206–217). Springer Berlin Heidelberg. 978-3-642-20217-9.

Kozhubaev, Y., Ovchinnikova, E., Viacheslav, I., & Krotova, S. (2023). Incremental machine learning for soft pneumatic actuators with symmetrical chambers. *Symmetry, 15*(6), 1206. doi:10.3390/sym15061206.

Laschi, C., Mazzolai, B., & Cianchetti, M. (2016). Soft robotics: Technologies and systems pushing the boundaries of robot abilities. *Science Robotics, 1*(1), eaah3690. doi:10.1126/scirobotics.aah3690.

Li, S. A., Weng, C. W., Chen, Y. H., Lo, C. H., Yang, M. H., Lin, Y. C., Hsieh, M. H., & Wong, C. , C. (2012). Servo motor controller design for robotic manipulator. In *2012 International Symposium on Intelligent Signal Processing and Communications Systems* (pp. 254–257). doi:10.1109/ISPACS.2012.6473490.

Liu, C. H., Chen, L. J., Chi, J. C., & Wu, J. Y. (2022). Topology optimization design and experiment of a soft pneumatic bending actuator for grasping applications. *IEEE Robotics and Automation Letters, 7*(2), 2086–2093. doi:10.1109/LRA.2022.3142910.

Lo Piccolo, M. V., Muscolo, G. G., & Ferraresi, C. (2022). Use of pneumatic artificial muscles in a passive upper body exoskeleton. In *New trends in medical and service robotics* (pp. 78–85). Springer International Publishing. 978-3-030-76147-9.

Luan, B., Zhang, X., Xu, F., Yang, G., Jin, J., Xu, C., Sun, F., & Oka, K. (2022). High precision magnetic levitation actuator for micro-EDM. *Actuators, 11*(12), 361. doi:10.3390/act11120361.

Marchese, A. D., Katzschmann, R. K., & Rus, D. (2015). A recipe for soft fluidic elastomer robots. *Soft Robotics, 2*(1), 7–25. doi:10.1089/soro.2014.0022.

Martinez-Sanchez, D. E., Sandoval-Castro, X. Y., Cruz-Santos, N., Castillo-Castaneda, E., Ruiz-Torres, M. F., & Laribi Med, A. (2023). Soft robot for inspection tasks inspired on annelids to obtain peristaltic locomotion. *Machines, 11*(8), 779. doi:10.3390/machines11080779.

Nguyen, P. H., Imran Mohd, I. B., Sparks, C., Arellano, F. L., Zhang, W., & Polygerinos, P. (2019). Fabric soft poly-limbs for physical assistance of daily living tasks. In *International Conference on Robotics and Automation (ICRA)* (pp. 8429–8435). doi:10.1109/ICRA.2019.8794294.

Nuckols, K., Hohimer, C. J., Glover, C., de Lucena, D. S., Moyo, W., Wagner, D., Cloutier, A., Lin, D. J., & Walsh, C. J. (2020). Effects of a soft robotic glove using a high repetition protocol in chronic stroke: A pilot study. In *8th IEEE RAS/EMBS International Conference for Biomedical Robotics and Biomechatronics (BioRob)* (pp. 428–433). doi:10.1109/BioRob49111.2020.9224291.

Polygerinos, P., Wang, Z., Galloway, K. C., Wood, R. J., & Walsh, C. J. (2015). Soft robotic glove for combined assistance and at-home rehabilitation. *Wearable Robotics, 73*, 135–143. https://www.sciencedirect.com/science/article/pii/S0921889014001729. doi:10.1016/j.robot.2014.08.014.

Polygerinos, P., Correll, N., Morin, S. A., Mosadegh, B., Onal, C. D., Petersen, K., Cianchetti, M., Tolley, M. T., & Shepherd, R. F. (2017). Soft robotics: Review of fluid-driven intrinsically soft devices; manufacturing, sensing, control, and applications in human-robot interaction. *Advanced Engineering Materials, 19*(12), 1700016. doi:10.1002/adem.201700016.

Ramos-Arreguin, J.-M., Guillen-Garcia, E., Canchola-Magdaleno, S. L., Pedraza-Ortega, J.-C., Gorrostieta-Hurtado, E., Aceves-Fernández, M.-A., & Ramos-Arreguin, C.-A. (2010). *Fuzzy logic hardware implementation for pneumatic control of one DOF pneumatic robot* (pp. 500–511). Springer Berlin Heidelberg. 978-3-642-16773-7.

Schmitt, F., Piccin, O., Barbé, L., & Bayle, B. (2018). Soft robots manufacturing: A review. *Frontiers in Robotics and AI, 5*. https://www.frontiersin.org/journals/robotics-and-ai/articles/10.3389/frobt.2018.00084.

Sun, Y., Song, Y. S., & Paik, J. (2013). Characterization of silicone rubber based soft pneumatic actuators. In *2013 IEEE/RSJ International Conference on Intelligent Robots and Systems* (pp. 4446–4453). doi:10.1109/IROS.2013.6696995.

Talhan, A., Yoo, Y., & Cooperstock, J. R. (2024). Soft pneumatic haptic wearable to create the illusion of human touch. *IEEE Transactions on Haptics, 17*(2), 177–190. doi:10.1109/TOH.2023.3305495.

Tawk, C., & Alici, G. (2021). A review of 3D-printable soft pneumatic actuators and sensors: Research challenges and opportunities. *Advanced Intelligent Systems, 3*(6), 2000223. doi:10.1002/aisy.202000223.

Viquerat, J., Rabault, J., Kuhnle, A., Ghraieb, H., Larcher, A., & Hachem, E. (2021). Direct shape optimization through deep reinforcement learning. *Journal of Computational Physics, 428*, 110080. ttps://www.sciencedirect.com/science/article/pii/S0021999120308548. doi:10.1016/j.jcp.2020.110080.

van Vlerken, C., Ballen-Moreno, F., Roels, E., Ferrentino, P., Langlois, K., Vanderborght, B., & Verstraten, T. (2023). Development of an active physical interface for physical human-robot interaction: Investigation of soft pneumatic actuator straps for automatic enclosure system. *Actuators, 12*(6), 241. doi:10.3390/act12060241.

Wallin, T. J., Pikul, J., & Shepherd, R. F. (2018). 3D printing of soft robotic systems. *Nature Reviews Materials, 3*(6), 84–100. doi:10.1038/s41578-018-0002-2.

Wehner, M., Quinlivan, B., Aubin, P. M., Martinez-Villalpando, E., Baumann, M., Stirling, L., Holt, K., Wood, R., & Walsh, C. (2013). A lightweight soft exosuit for gait assistance. In *2013 IEEE International Conference on Robotics and Automation* (pp. 3362–3369). doi:10.1109/ICRA.2013.6631046.

Wilkening, A., Stöppler, H., & Ivlev, O. (2015). Adaptive assistive control of a soft elbow trainer with self-alignment using pneumatic bending joint. In *2015 IEEE International Conference on Rehabilitation Robotics (ICORR)* (pp. 729–734). doi:10.1109/ICORR.2015.7281288.

Woods, B. K. S., Gentry, M. F., Kothera, C. S., & Wereley, N. M. (2012). Fatigue life testing of swaged pneumatic artificial muscles as actuators for aerospace applications. *Journal of Intelligent Material Systems and Structures, 23*(3), 327–343. doi:10.1177/1045389X11433495.

Xavier, M. S., Tawk, C. D., Yong, Y. K., & Fleming, A. J. (2021). 3D-printed omnidirectional soft pneumatic actuators: Design, modeling and characterization. *Sensors and Actuators A: Physical, 332*, 113199. https://www.sciencedirect.com/science/article/pii/S0924424721006622. doi:10.1016/j.sna.2021.113199.

Xavier, M. S., Fleming, A. J., & Yong, Y. K. (2022). Model-based nonlinear feedback controllers for pressure control of soft pneumatic actuators using on/off valves. *Frontiers in Robotics and AI, 9*. https://www.frontiersin.org/journals/robotics-and-ai/articles/10.3389/frobt.2022.818187.

Xiao, L., Zhao, H., Xu, Z., Li, X., Shen, C., Wang, K., & Zhang, L. (2022). A new architecture of morphing wing based on hyperelastic materials and metastructures with tunable stiffness. *Frontiers in Mechanical Engineering, 7*. https://www.frontiersin.org/journals/mechanical-engineering/articles/10.3389/fmech.2021.814446.

Youssef, S. M., Soliman, M. A., Saleh, M. A., Mousa, M. A., Elsamanty, M., & Radwan, A. G. (2022). Modeling of soft pneumatic actuators with different orientation angles using echo state networks for irregular time series data. *Micromachines, 13*(2), 216. doi:10.3390/mi13020216 2072-666X.

Yu, Z., Gao, Z., Li, Y., & Pan, J. (2020). Closed-loop control of a pneumatic soft gripper for fast and accurate response. In *2020 8th International Conference on Power Electronics Systems and Applications (PESA)* (pp. 1–5). doi:10.1109/PESA50370.2020.9344000.

Zaidi, S., Maselli, M., Laschi, C., & Cianchetti, M. (2021). Actuation technologies for soft robot grippers and manipulators: A review. *Current Robotics Reports, 2*(3), 355–369. doi:10.1007/s43154-021-00054-5.

Zhao, H., O'Brien, K., Li, S., & Shepherd, R. F. (2016). Optoelectronically innervated soft prosthetic hand via stretchable optical waveguides. *Science Robotics, 1*(1), eaai7529. doi:10.1126/scirobotics.aai7529.

Zhou, Y., Ye, M., Hu, C., Qian, H., Nelson, B. J., & Wang, X. (2023). Stimuli-responsive functional micro-/nanorobots: A review. *ACS Nano, 17*(16), 15254–15276. doi:10.1021/acsnano.3c01942.

Chapter 26

Some aspects of sensitivity analysis of induction generator for wind power applications

Adhir Baran Chattopadhyay[a], Sunanda Hazra[b], Sandip Ojha[b], Suman Chakraborty[a], Mohamed Arezki Mellal[c] and Lamia May[d]

[a] *Department of Electrical Engineering, Budge Budge Institute of Technology, Budge Budge, West Bengal, India,*
[b] *Department of Electrical Engineering, Haldia Institute of Technology, Haldia, West Bengal, India,* [c] *LMSS, Faculty of Technology, M'Hamed Bougara University, Boumerdes, Algeria,* [d] *Laboratory of Mechanics, Materials and Energy (L2ME), Faculty of Technology, University of Bejaia, Bejaia, Algeria*

26.1 Introduction

The search for alternate methods for generating electrical energy has been started since last 20 or 30 years. This has led to more development in the research and industrialization of solar power. But the activities in the area of wind energy applications have started late as compared to solar energy field. This may be probably because of fixing the strategy for handling the uncertainty of wind speed. Later it was established that induction generators can play a crucial role for generation of electrical energy, whereas mechanical systems based on wind power (e.g., wind mill) act as its prime mover. It is well known that the wind turbines currently provide approximately 33% of the worldwide power supply. Wind energy is considered to be one of the fastest growing renewable energy technologies all over the world because wind as a source of energy is renewable, widely distributed, and plentiful. In countries such as Denmark, Spain, Portugal, and Germany, spread of wind-driven power applications roughly lay in the range of 19%, 9%, 9%, and 6% of the required electric power, respectively, during 2000–07. From 2000 to 2007 the global wind power generation increased to approximately five times its previously recorded capacity. Still a good number research papers are being published in the areas of modeling, design, fabrication, and testing of induction generators, including self-excited induction generators (Akbarali et al., 2020; Giribabu et al., 2016; Murali Krishna et al., 2023; Peter & Iderus, 2018; Sarathbabu Duvvuri et al., 2022). The mathematical model of induction generator developed by the group of researchers (Adkins & Harley, 1975; Bhadra et al., 2005; Kothari & Nagrath, 1985) is very exhaustive so far, the involvement of different variables and parameters, based on the equivalent circuit of a three phase self-excited induction generator, are concerned. Such model draws considerable number of attention from the researchers, especially those who are more inclined to the modeling and simulation aspects. They have nicely derived the expression for per-unit frequency generated as a function of the "ratio of the reactance of the excitation capacitor at the base frequency to the load resistance." Such expression (Pannell, 1997; Rajendran et al., 2022; Yeo, 1991) involves a lot of machine parameters, but nowhere have the sensitivity aspects been touched upon. In general, the research papers based on the modeling or experimental works related to sensitivity aspects of wind driven induction generators are rarely available in the literature (Khaligh & Onar, 2010; Velmurugan & Chattopadhayay, 2020a, 2020b). Renewable energy has been deployed in load dispatch problem for economic emission minimization problems (Hazra & Roy, 2019, 2020, 2023; Hazra et al., 2024; Paul et al., 2024). The authors feel that sensitivity analysis and computations may help any researcher to decide the strategy of experimentation and also to design a new machine. The present chapter tries to investigate the sensitivity of the threshold per-unit frequency generated by a capacitor-excited induction generator driven by a windmill, because of the tuning of some selected parameters. The details of problem formulations and subsequent results and discussions have been presented in the appropriate sections. Before concentrating on the actual problem formulation, it is felt that there is a need to explain the technical part of the self-excitation process of an induction generator. It is presented in the next section.

Computational Intelligence in Mechatronics: Solving Real-World Problems in Electronic Systems Design. DOI: https://doi.org/10.1016/B978-0-443-27580-7.00004-4

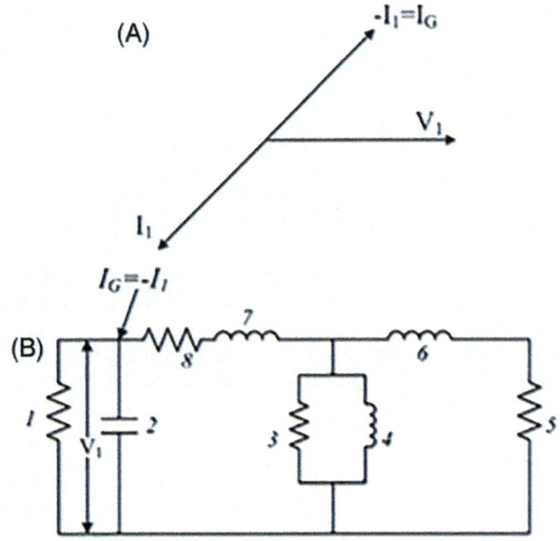

FIGURE 26.1 Self-excitation of induction generator: (A) phasor diagram and (B) equivalent circuit.

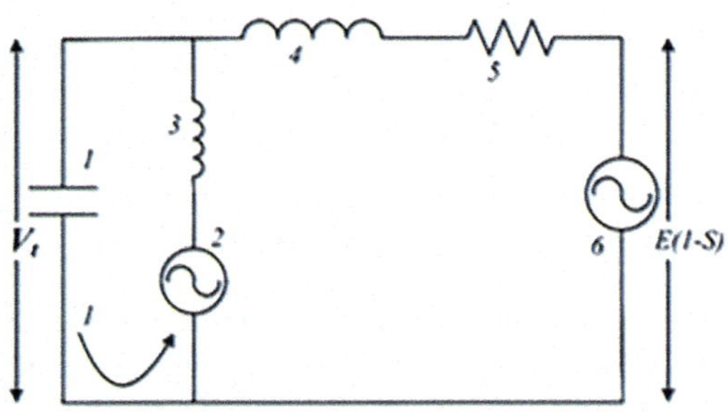

FIGURE 26.2 Self-excitation of induction generator: simplified equivalent circuit.

26.2 Explanation of the self-excitation process of an induction generator

The core of the generator retains some residual flux. A capacitor bank of capacity of supplying the required reactive power or the magnetizing current to the stator winding of the induction generator must be connected to the machine terminals. The existence of the residual flux is a must in such case because the nonlinear magnetization characteristic must start from a certain flux level, which is nothing but the flux supplied by the initial condition, or the residual property of the core. This phenomenon of induction generator is very close to the ever-known voltage build-up process of a dc shunt generator. To have a feel for the basic self-excitation process, Fig. 26.1A and B is drawn below.

In Fig. 26.1B, 1—R_L, 2—C, 3—R_c, 4—X_m, 5—R_r/s, 6–X_lr, 7—R_s, 8—X_ls, I_G—Generator current, V_1—Voltage across capacitor (C). With reference to the phasor diagram in Fig. 26.1A, it may be noted that when the slip of a three-phase induction motor becomes negative, the machine's operating status converts from motor mode to generator mode and as a convention the terminal current of a generator is assumed to be negative, as compared to motor mode. In other words, motor input current will lag behind the supply voltage by an angle greater than $\pi/2$. Such operating condition naturally implies the induction machine to be a source of electrical power. Therefore, based on this concept, it can be expected that this machine can feed power to a parallel combination of load resistance and capacitance, and conceptually it is presented as an equivalent circuit of induction generator shown in Fig. 26.1B. Neglecting the stator resistance and leakage reactance (as, in general, they are of small values) in the equivalent circuit in Fig. 26.1B, the simplified equivalent circuit (mostly suitable for no-load condition) is presented in Fig. 26.2. Such circuit will be very helpful to feel the voltage build-up phenomenon of three phase induction generator, because as per the established concepts of electrical machines, the magnetizing phenomenon becomes dominant only under no-load condition of any machine. The voltage build-up process based on the magnetization characteristic is separately taken up for explanation in the next subsection so that the readers may put crystallized attention on this positive feedback process.

In Fig. 26.2, 1—Capacitor (C), 2—Residual voltage or airgap voltage of synchronous generator mode (E_r), 3—L_m, 4—sX_lr, 5—R_r, 6—Induced emf source [$E(1-s)$], I—Current in the $L\_m - C$ loop

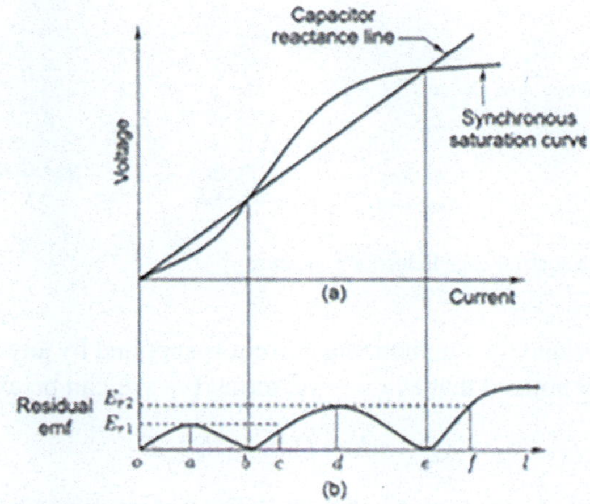

FIGURE 26.3 Building up of voltage process of a self-excited induction generator: (A) the capacitor load line and the magnetization characteristic (the synchronous saturation curve) and (B) the difference between them.

26.2.1 Explanation of the voltage build-up process of an induction generator

With reference to Fig. 26.2, it is realized that corresponding to any residual magnetic flux (φ_r) in the core, there is produced a residual voltage (E_r) which, in turn, sets up a current (I) in the $L\_m - C$ loop. The voltage across L_m in instantaneous form basically can be considered the phasor voltage induced across jX_m, which is the magnetizing reactance. On the other side of the $L\_m - C$ loop, capacitor voltage versus current acts as a load line. If we assume L_m as a driver and C as a load, then the voltage versus current profile of the "driver" must intersect that profile of the "load" at one or more points to ensure the successful operation of the voltage build-up process. Such intersection points are generally known as "operating points" or "quiescent points" or "equilibrium points" whenever any interaction process at steady-state is analyzed (similar to diode load line or transistor load lines and the corresponding Q-points). These profiles are shown in Fig. 26.3A and B.

One important point in such context is that as Fig. 26.2 corresponds to no-load operation. Generally, at no-load condition of any induction machine, the difference between the synchronous speed and the rotor speed is very small. In other words the no-load slip is approximately equal to zero. As a consequence, rotor induced current becomes zero leading to the operation of the machine as a synchronous generator. At the points, "b" and "e" (Fig. 26.3A and B), it is observed that $1/(\omega\_nC) = \omega\_n \times L\_m$, where ω_n is the natural frequency of oscillation, in electrical radians/second. In Fig. 26.2, if we consider the $L\_m - C$ loop, then we can write the current dynamics as

$$E\_r - V\_C = L\_m dI/dt \tag{26.1}$$

where

$$V\_C = V\_t \tag{26.2}$$

The above-said mathematical equations lead to one important physical concept that whenever $E_r < V_C$, magnetizing current cannot build-up, hence, also, the induction generator mode cannot exist. So, during the whole dynamics, the voltage build-up process can start and continue satisfactorily in the asynchronous mode (or induction generator mode) only when the synchronous saturation curve lies above the capacitor reactance line, which is exactly happening in the zone "b–e" in Fig. 26.3A and B. It is also interesting to note that at the point "e," as the driver voltage (E_r) equals the load voltage (V_C), the dynamic condition stops leading to the constant flow of magnetizing current. Basically, as per the tradition of any research work on any type of dynamics, the terminologies "driver" and "load" play dominant roles. This analogy is exactly brought into the present explanation. Another explanation is needed to clarify the fact what is happening exactly in the zone "b–e" so that the induction generator action becomes feasible and this explanation is presented as a separate subsection.

26.2.2 Feasibility study on the induction generator

Between the points "b" and "e," the synchronous reactance because of L_m is greater than the capacitive reactance. That means, virtually, the magnetizing inductance (L_m) increases, leading to the drop in the value of the natural frequency of oscillation (ω_n). Hence the natural frequency becomes lower than the rotational (i.e., synchronous frequency) leading to the development of a negative slip. This proves that during this zone (b–e), induction generator operation is feasible (because,

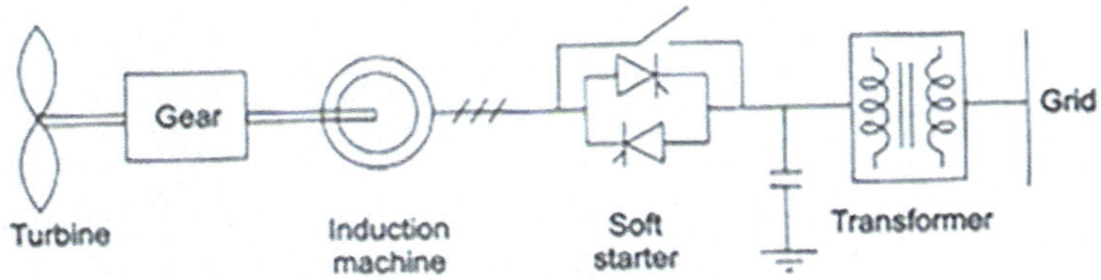

FIGURE 26.4 The fixed-speed system with a squirrel-cage induction generator.

conceptually, the slip value is negative and the magnetizing current is supplied by any external element). Furthermore, with reference to Fig. 26.2, it can be easily noticed that as $s = -\text{ve}$, term $(1 - s)E$ can be expressed as

$$(1 - s)E = (1 + |s|)E > E \tag{26.3}$$

where |s| indicates only the magnitude (or modulus) of "slip." The observation on Eq. (26.3) is that the rotational emf "$(1 - s)E$" drives a current into the stator circuit of the induction generator, building up the terminal voltage. The above-said explanation gives a picture of the dynamics happening in the magnetizing circuit of the induction generator. However, this explanation may get more light if some practical examples of wind-power-driven induction generator are cited, and one such application is explained in the next subsection.

26.2.3 An example of the application of induction generator driven by wind power

Fig. 26.4 illustrates a simple scheme of a wind-power-driven induction generator. The windmill speed is controlled through reduction gearbox to make it compatible with the machine. As the generator is basically a squirrel-cage induction machine, the stator winding generates voltage along with drawing magnetizing current through the same terminals. So, it is basically a closed-loop system. As the induction generator is coupled to the grid, the available speed variation is approximately 1% above the synchronous speed. As this negative slip variation is not much, such system is somewhat called a "fixed speed system." The generally used fixed-speed wind turbines are needed to limit the power at high wind speeds. That is why they mostly employ either blade pitch regulation or stall regulation. The thyristorized soft starter is provided to allow bi-directional power flow. Usually the wind turbine accelerates the induction generator to synchronous speed, and then this machine is coupled to the infinite bus or grid. Usually the generator draws lagging reactive power from the bus, and as a compensation technique, the capacitor is installed as shown in Fig. 26.4.

26.3 Problem formulations

The problem of formulating sensitivity of an objective function with respect to any tuning parameter can be looked upon as a very wide region of research and development problem covering engineering, social dynamics, economic scenarios, health sector, legal sectors, etc. (Pannell, 1997; Velmurugan & Chattopadhayay, 2020a, 2020b; Yeo, 1991). Which tuning parameter should be chosen for such formulation is again a question to be answered by the concerned researcher based on his or her own observations on the system dynamics. Sometimes mere practical experience also helps a lot to make a decision about selecting a particular tuning parameter. This particular problem of sensitivity aspects of induction generator has been formulated as a two-part problem, presented in the following sections.

26.3.1 Part-I: Problem formulation for sensitivity of the threshold per unit frequency generated, with respect to the ratio of the reactance of the excitation capacitor at the base frequency to the load resistance

The established equivalent circuit per phase of a capacitor-excited induction generator is shown in Fig. 26.5, corresponding to the threshold per unit frequency generated (F).

In Fig. 26.5, 1—Load reactance ((*jX*)_Lb), 2—Load resistance (*R_L*), 3—Adjusted excitation capacitive reactance at base frequency((−*jX*_cb)/*F*^2), 4—Magnetizing reactance (*jX_m*), 5—Rotor resistance referred to and adjusted (*R_r*/(*F* − *v*)), 6—Rotor leakage reactance (referred) (*jX*_lr), 7—Stator leakage reactance (*jX*_ls), 8—Adjusted stator resistance (*R_s*/*F*); Additional symbols used: "*v*"—the per unit speed of wind (speed of prime mover/windmill), *V_t*—Terminal voltage, *V_g*—Generated voltage, *I*_1—Current through stator branch. The threshold per unit frequency (F)

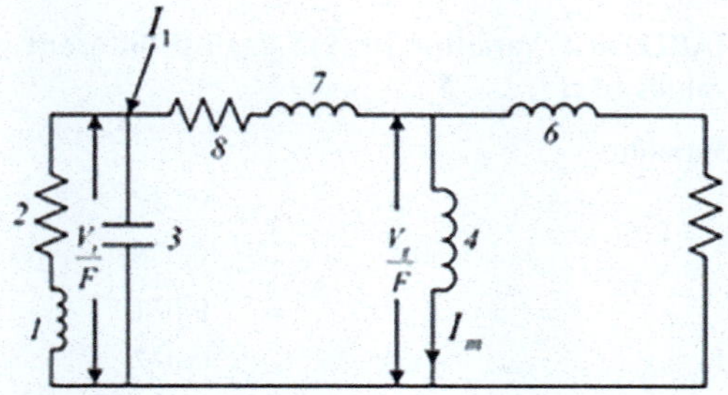

FIGURE 26.5 Circuit model for a self-excited induction generator at stator frequency.

generated by an SEIG can be expressed as (Bhadra et al., 2005)

$$F = v[2b(X_r)^3 + R_r(X_m)^2X_r]/2[b(X_r)^3 + R_r(X_m)^2X_r] \tag{26.4}$$

where

$$b = \tau X\_\sigma + R\_s \tag{26.5}$$

With reference to Eqs. (26.1) and (26.2) the symbols X_r, X_σ are the rotor winding self-reactance and transient reactance of induction machine, respectively. Furthermore the symbol "τ" expressed in Eq. (26.2) can be expressed as (Rajendran et al., 2022).

$$\tau = X\_cb/R\_L \tag{26.6}$$

where X_cb is the reactance of the excitation capacitor at the base frequency and R_L is the load resistance. τ is not a direct physical variable. The same value of τ may be obtained by different combinations of X_cb and R_L. Hence the maximum values of τ do not bring out the critical values of X_cb or, $R\_L.\tau = 0$ indicates either $X\_cb = 0$ or, $R\_L = \infty$. As $X\_cb = 0$ indicates a short-circuit condition, $\tau = 0$ implies a no-load condition. Eq. (26.1) may be expressed as

$$F = (v/2)(N/D) \tag{26.7}$$

where

$$N = [2b(X\_r)^3 + R\_r(X\_m)^3X\_r] \tag{26.8}$$

and

$$D = \left[b(X\_r)^3 + R\_r(X\_m)^2X\_r\right] \tag{26.9}$$

Sensitivity of "F" with respect to "τ" is expressed as

$$\left(\tfrac{F}{\tau}\right)S = (\partial F/\partial \tau)(\tau/F) \tag{26.10}$$

Based on Eq. (26.4), $\partial F/\partial \tau$ comes out to be

$$\partial F/\partial \tau = [(v/2)(1/D)(\partial N/\partial \tau)] - [(vN/(2D^2))(\partial D/\partial \tau)] \tag{26.11}$$

On the basis of Eqs. (26.2) and (26.5) it yields

$$\partial N/\partial \tau = 2(X\_r)^3(\partial b/\partial \tau) = 2(X\_r)^3X\_\sigma \tag{26.12}$$

$$\partial D/\partial \tau = (X\_r)^3(\partial b/\partial \tau) = 1/2(\partial N/\partial \tau) \tag{26.13}$$

From Eqs. (26.7) and (26.9), it comes out as

$$\partial F/\partial \tau = v(\partial N/\partial \tau)[(1/2D) - (N/(4D^2))] \tag{26.14}$$

TABLE 26.1 Variation of $\binom{F}{\tau}S$ against different values of τ.

Serial no.	τ	
1	0.2	0.037
2	0.4	0.058
3	0.6	0.071
4	0.8	0.077
5	0.9	0.080
6	1	0.081
7	1.1	0.084
8	1.2	0.083
9	1.3	0.083
10	1.4	0.083

26.3.1.1 Machine data

1. Data in actual unit—[parameters in ohms per phase]: KW 30, Voltage 440 volts: Stator winding resistance (R_s)—0.54, Rotor winding resistance, referred to stator (R_r)—0.81, Magnetizing reactance (X_m)—48.6, Rotor winding leakage reactance, referred to stator (X_lr)—3.24.
2. Conversion to per units: There is a serious need to convert the given data in actual units to those in per unit. The reason behind such requirement lies in the following fact: If the rotor winding self-reactance can be measured directly, then the problem of conversion may be avoided, but generally self-reactance data of any machine winding are not supplied by the manufacturer. As an alternative, generally leakage reactance data (properly referred to) are supplied. Now the question raises how to calculate self-reactance based on leakage reactance and magnetizing reactance. As a matter of fact, in such context, Adkins and Harley (1975) clearly believe that the self-inductance of a coil obtained by adding the mutual and leakage inductances would not be true if the quantities are expressed in actual units. Based on this concept and using the basic definitions, the method of conversion from actual units to per units is shown in detail in the next paragraph [point no. (3)] to compute the per unit values of all machine parameters concerned.
3. Calculations for per unit conversions: Base KW = 30. Power factor (assumed) = 0.8; Hence, base KVA = 30/0.8 = 37.5; Base MVA = 37.5/1000 = 0.0375; Base kV = 0.44; Base impedance = Z_base=(BasekV)^2/(Base MVA) = (((0.44)^2))/0.0375 = 5.16 Ω; R_s = 0.54/5.16 = 0.106 p.u.; R_r = 0.81/5.16 = 0.157 p.u.;X_m = 48.6/5.16 = 9.42 p.u.; X_lr = 3.24/5.16 = 0.628 p.u.;X_r = X_m + X_lr = 9.42 + 0.628 = 10.05 p.u.
4. Wind speed: Per unit speed (v)—0.8 p.u.; Range of values of transient reactance of induction machine (X_σ)—0.15 to 0.25 p.u.

26.3.1.2 Sample calculation of (_τ^F)S for different values of τ

Calculation for "τ = 0.2": X_σ = 0.15 b = τX_σ + R_s = (0.2)(0.15) + (0.105) = 0.135 N = [2b(X_r)^3 + R_r (X_m)^2 X_r] = 2(0.135) (10.05)^3 + 140.01 = 414.08 D = [b(X_r)^3 + R_r (X_m)^2 X_r] = N − b(X_r)^3 = 414.08 − {(0.135) (10.05)^3} = 277.04 $\partial N/\partial\tau$ = 2{(X_r)^3} X_σ = 2{(10.05)^3}(0.15) = 304.52 $\partial F/\partial\tau$ = {v($\partial N/\partial\tau$)}[(1/2D) − (N/(4D^2))] = (0.8)(304.52)[(1/(2 × 277.04)) − {414.08/4{(277.04)^2}}] = 0.1096 [(_τ^F)S]_(τ = 0.2) = [($\partial F/\partial\tau$)(τ/F)]_(τ = 0.2) = [0.2/{(0.8/2)(414.08/277.04)}][0.1096] = 0.037

26.3.1.3 Results and discussions

Variation of $\binom{F}{\tau}S$ against different values of τ: It reveals that this sensitivity function (sensitivity of per unit frequency generated, "F" w.r.t. the parameter "τ") is needed to be plotted so that such plot can be used further for detailed design of induction generator. However, for initial feeling about the formulation and the resulting computation work (Table 26.1). The table data has been converted into a graphical image shown in Fig. 26.6.

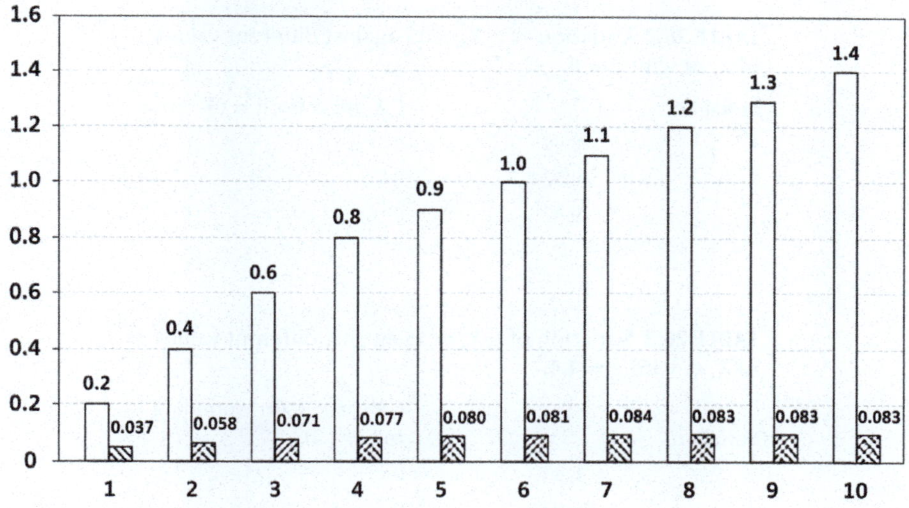

FIGURE 26.6 Figure of against different values of τ.

Results of Table 26.1 show that just by anticipation, any amount of loading in terms of "X_cb/R_L"-ratio cannot be put on the induction generator because it has a limiting value. However, the initial approximate monotonic rise of the plot/profile indicates that the excitation capacitance value influences the frequency of generation to a larger extent from the start to the stage before saturation. Conceptually also, this should be the real tendency because initially the reactive power provided by the capacitor supplies the magnetizing current smoothly to make interaction with the stator windings for materializing the voltage-buildup process.

26.3.2 Part-II: Problem formulation for sensitivity of the threshold per unit frequency (*F*) with respect to the magnetizing reactance (*X_m*)

Conceptually, it is well known that the threshold per unit frequency generated by the self-excited induction generator depends mostly on wind speed (or prime mover speed) and parameter "τ." So conceptually, "F" should remain insensitive to any change in the magnetizing reactance. However, such conclusion is not supposed to be drawn by any researcher without proper analytical or numerical investigations. Based on such motivation the following analysis is presented: Sensitivity of "F" with respect to "X_m" is expressed as

$$(\_X\_m^F)S = (\partial F/(\partial X\_m))(X\_m/F) \tag{26.15}$$

In this part of analysis, one critical aspect is observed, and it is centered around the functional nature of X_r when X_m is treated as an independent variable. Apparently, it seems that X_r is constant with respect to X_m but in per unit system, X_r is the arithmetic sum of X_m and X_lr (rotor winding leakage reactance, referred to stator) (Adkins & Harley, 1975). Hence, based on this fact, the mathematical equations are derived and are presented as follows:

$$F = (v/2)(N\_1/D\_1) \tag{26.16}$$

where

$$N\_1 = \left[2b((X\_m + X)\_\mathrm{lr})^3 + R\_r\,(X\_m)^2\,((X\_m + X)\_\mathrm{lr})\right] \tag{26.17}$$

$$D\_1 = \left[b((X\_m + X)\_\mathrm{lr})^3 + R\_r\,(X\_m)^2\,((X\_m + X)\_\mathrm{lr})\right] \tag{26.18}$$

$$X\_r\,((= X)\_m + X)\_\mathrm{lr} \tag{26.19}$$

The term "$\partial F/(\partial X\_m)$" should be derived as a first step, and it comes out to be

$$\partial F/(\partial X\_m) = [(v/2)((\partial N\_1)/(\partial X\_m))] - [((vN\_1)/2((D\_1)^2))((\partial D\_1)/(\partial X\_m))] \tag{26.20}$$

TABLE 26.2 Variation of $(\_X\_m^F)S$ against different values of X_m with $\tau = 0$.

Serial no.		$(\_X\_m^F)S$ (for $\tau = 0$)
1	8.42	0
2	8.92	0
3	9.42	0

TABLE 26.3 Variation of $(\_X\_m^F)S$ against different values of X_m with $\tau = 1.4$.

Serial no.		$(\_X\_m^F)S$ (corresponding to $\tau = 1.4$)
1	8.42	0
2	8.92	0
3	9.42	0

The other equations are derived as follows:

$$(\partial N\_1)/(\partial X\_m) = 6b(X\_r)^2 + 2R\_r\, X\_m + R\_r\, (X\_m)^2 \tag{26.21}$$

$$(\partial D\_1)/(\partial X\_m) = 3b(X\_r)^2 + 2R\_r\, X\_m + R\_r\, (X\_m)^2 \tag{26.22}$$

26.3.2.1 *Sample calculation of (_X_m^F) S for different values of* X_m *with* $\tau = 0$

Calculation for "X_m = 9.42 p.u.": The same machine data used in part-1 formulation has been used here also. b = 0.105; N_1 = 353.18; D_1 = 246.59; F = $(v/2)(N\_1/D\_1)$ = (0.8/2)(353.18/246.59) = 0.573 $(\partial N\_1)/(\partial X\_m)$ = 6(.105) (10.05)^2 + 2(0.157) (10.05) (9.42) + (.157) (9.42)^2 = 107.28 $(v/2)((\partial N\_1)/(\partial X\_m))$ = 0.174; $(\partial D\_1)/(\partial X\_m)$ = $(\partial N\_1)/(\partial X\_m)$ − $3b(X\_r)$^2 = 107.28 − [3(0.105) (10.05)^2] = 75.46 $((vN\_1)/2((D\_1)^2))((\partial D_1)/(\partial X_m))$ = [(0.8 × 353.18)/2{(246.59)^2}][75.46] = 0.175 Hence, $\partial F/(\partial X\_m)$ = $[(v/2)((\partial N\_1)/(\partial X\_m))]$ − $[((vN\_1)/2((D\_1)^2))((\partial D_1)/(\partial X_m))]$ = 0.174 − 0.175 ≈ 0 Therefore, $([(\_X\_m$^$F)S])\_(X\_m = 9.42) = 0$

26.3.2.2 *Results and discussions*

Variation of $(\_X\_m$^$F)S$ against different values of $X\_(m)$ with $\tau = 0$: It reveals that the concept of "sensitivity of the threshold per unit frequency generated, "F" with respect to the magnetizing reactance, ("X)_(m)" does not exist in reality. However, to verify such statement, Table 26.2 is presented below. As the parameter "τ" was taken into account as a main variable during the part-1 formulation, logically the same parameter should be treated in the current formulation (part-2 formulation) as a dummy variable (or fixed parameter). So logically this current formulation should be completed for different values of "τ." To start with, $\tau = 0$ is considered in Table 26.2 calculations.

Table 26.2 results clearly show that "F" is not sensitive to any change in "X_m." But this was shown for $\tau = 0$. For other values of τ also, it is established that "F" is not sensitive to any change in "X_m." These results are shown in Table 26.3.

26.4 Conclusion

1. After certain value of "τ," the sensitivity of "F" with respect to "τ" approximately saturates. Its physical significance is that there is a limit on the capacitance value to supply the required reactive power to the machine.
2. Just by using any empirical formula, any parameter cannot be picked up for carrying on the sensitivity analysis. It needs some physical conception or practical experience about the concerned system to identify the tuning parameter.
3. The threshold per unit frequency generated is not sensitive to any change in magnetizing reactance. It does not necessarily mean that functional dependence of the former variable on the latter parameter will not exist.

Nomenclature

C	Capacitance
E_r	Residual voltage or airgap voltage in synchronous generator mode
F	Per unit threshold frequency
I	Current in the $L_m - C$ loop
KVA	Kilo volt-ampere
L_m	Magnetizing inductance
MVA	Mega volt-ampere
p.u.	Per unit
Q-point	Quiescent point
R_c	Resistance per phase indicating core loss
R_L	Load resistance per phase
R_r	Rotor resistance per phase referred to stator
R_s	Stator winding resistance per phase
v	The per unit speed of wind (speed of prime mover/windmill)
V_g	Generated voltage
V_t	Terminal voltage or, voltage across capacitor
X_{lr}	Rotor leakage reactance referred to stator
X_{ls}	Stator winding leakage reactance per phase
X_m	Magnetizing reactance per phase
τ	Capacitive reactance to load resistance ratio

References

Adkins, B., & Harley, R. G. (1975). *The general theory of alternating current machines application to practical problems* (pp. 14–16). Chapman and Hall Limited. ISBN 0 412 15560 5.

Akbarali, M. S., Subramanium, S. K., & Natarajan, K. (2020). Modeling, analysis, and control of wind-driven induction generators supplying DC loads under various operating conditions. *Wind Journal, 45*, 197–211. doi:10.1177/0309524X2092538.

Bhadra, S. N., Kastha, D., & Banerjee, S. (2005). *Wind electrical systems* (pp. 180–221). Oxford University Press. ISBN-13:978-0-19-567093-6.

Giribabu, D., Das, M., & Kumar, A. (2016). Comparative study of control strategies for the induction generators in wind energy conversion system. *Wind and Structures, 22*, 635–662. doi:10.12989/was.2016.22.6.635.

Hazra, S., & Roy, P. K (2019). Quasi-oppositional chemical reaction optimization for combined economic emission dispatch in power system considering wind power uncertainties. *Renewable Energy Focus, 31*, 45–62. doi:10.1016/j.ref.2019.10.005.

Hazra, S., & Roy, P. K. (2020). Optimal dispatch using moth-flame optimization for hydro-thermal-wind scheduling problem. *International Transactions on Electrical Energy, 30*(8), e12460. doi:10.1002/2050-7038.12460.

Hazra, S., & Roy, P. K. (2023). Solar-wind-hydro-thermal scheduling using moth flame optimization. *Optimal Control Applications and Methods, 44*(2), 391–425. doi:10.1002/oca.2783.

Hazra, S., Roy, P. K., & Paul, C. (2024). State of the art for moth-flame optimization applied electric vehicles–solar–wind–hydro–thermal power system. *Electrical Engineering*. doi:10.1007/s00202-024-02573-8.

Khaligh, A., & Onar, O. C. (2010). *Energy harvesting solar wind and ocean energy conversion systems* (pp. 101–103). CRC Press (Taylor and Francis Group). ISBN 978-1-4398-1508-3. doi:10.1201/9781439815090.

Kothari, D. P., & Nagrath, I. J. (1985). *Electric machines* (pp. 625–627). Tata Mc-Graw Hill Publishing Limited. ISBN 0-07-058377-3.

Murali Krishna, V. B., Sandeep, V., Narendra, B. K., & Prasad, K. R. K. V. (2023). Experimental study on self-excited induction generator for small-scale isolated rural electricity applications. *Results in Engineering, 18*, 1–8. doi:10.1016/j.rineng.2023.101182.

Pannell, D. J. (1997). Sensitivity analysis of normative economic models: Theoretical framework and practical strategies. *Agricultural Economics, 16*, 139–152. doi:10.1016/S0169-5150(96)01217-0.

Paul, C., Sarkar, T., Dutta, S., Hazra, S., & Roy, P. K. (2024). Optimal power flow of multi-objective combined heat and power with wind-solar-electric Vehicle-Tidal using hybrid evolutionary approach. *Process Integration and Optimization for Sustainability*. doi:10.1007/s41660-024-00431-5.

Peter, G., & Iderus, S. B. (2018). Comparative performance study of a self-excited three phase induction generator with a three phase synchronous generator. *Journal of Advanced Research in Dynamical and Control Systems, 10*, 41–47 ISSN:1943-023X.

Rajendran, S., Diaz, M., Cárdenas, R., Espina, E., Contreras, E., & Rodriguez, J. (2022). A review of generators and power converters for multi-MW wind energy conversion systems. *Processes, 10*, 1–24. doi:10.3390/pr10112302.

Sarathbabu Duvvuri, S. S. S. R., Sandeep, V., Yadlapati, K., & Murali Krishna, V. B. (2022). Research on induction generators for isolated rural applications: State of art and experimental demonstration. *Measurement: Sensors, 24*, 100541. doi:10.1016/j.measen.2022.100541.

Velmurugan, P., & Chattopadhayay, A. B. (2020a). Rigorous mathematical steps for sensitivity analysis of high impedance ground fault detection in power distribution systems. *Cogent Engineering, 7*, 1–16. doi:10.1080/23311916.2020.1747373.

Velmurugan, P., & Chattopadhayay, A. B. (2020b). Sensitivity analysis of a single phase to ground fault system in connection with high impedance faults: A case study. *Cogent Engineering, 7*, 1–23. doi:10.1080/23311916.2020.1770916.

Yeo, K. T (1991). Project cost sensitivity and variability analysis. *International Journal of Project Management, 9*, 111–116. doi:10.1016/0263-7863(91)90070-C.

Chapter 27

Study of methods for improving the reliability of passively redundant aircraft subsystems taking into account tolerances

Vu Tuan and Duong Vu
School of Engineering Technology, Duy Tan University, Da Nang, Vietnam

27.1 Introduction

The reliability of passively redundant aircraft subsystems, taking into account tolerances in case of sudden failures, is not a widely discussed topic in reliability theory. However, the needs of practice require a deep and comprehensive study of this issue. The subject of this work is the pairs of subsystems interconnected by the output/input parameter W.

The feature of the first subsystem is the inadmissibility of even short-term breaks in work and the failure to meet the requirements for its reliability. This leads to the need to introduce redundancy into it using the methods of multiple and nonmultiple redundancy. A feature of the second subsystem is the possibility of deviating its input parameter W downwards within the tolerance determined by the design and technical documentation. The structure of the passive redundancy of the first subsystem (total number of elements n, number of main elements m, and number of reserve elements r) is chosen in such a way that in case of failure of all its reserve elements, the required level of its failure-free is ensured and, at the same time, the reduction of its output parameter W entering the input of the second subsystem does not exceed the tolerance limits.

Examples of pairs of related subsystems are fuel supply system—aircraft engine (Antseliovich, 1985), power booster—steering gear, steering gear—steering device of aircraft control circuits (Redko, 2002), Alternating current (AC) and Direct current (DC) power supply system on aircraft—electricity consumers (Dedkov & Tatuev, 2010), etc. The output parameter can be the total capacity of the fuel pumps of the aircraft engine fuel supply system, the total output power of synchronous generators of the aircraft side AC system, etc. A similar situation occurs in related subsystems: power amplifier of the control loop—steering gear, steering gear—aircraft control, etc.

The purpose of the work is to find ways to increase the reliability of passively redundant subsystems, taking into account tolerances, by studying the impact of significant factors on failure-free operation.

27.2 Problem motivation

Usually the research model consists of factors that most strongly affect the purpose of the work. These factors are called essential. The purpose of the work, as noted, is to search for ways to increase the reliability of subsystems of the class under consideration. An analysis of the set of visible factors affecting the reliability of these subsystems shows that three factors are essential: the failure of the elements of the redundant subsystem, the multiplicity of redundancy, and the value of tolerance for deviation of its output parameter.

The value of the tolerance for the deviation of the output parameter is a discrete value determined by the redundancy structure, i.e., the parameters n, m, and r:

$$n = m + r. \tag{27.1}$$

Computational Intelligence in Mechatronics: Solving Real-World Problems in Electronic Systems Design. DOI: https://doi.org/10.1016/B978-0-443-27580-7.00011-1

It is convenient to specify the structure of the reservation by the parameters n and m because they simultaneously determine not only the structure but also the multiplicity of the reservation (Polovko & Gurov, 2006).

$$K = \frac{n}{m}. \tag{27.2}$$

The tolerance that is defined by the redundancy structure is the realizable tolerance. It is convenient to set it in a relative representation, expressed as a percentage of the ratio of the number of reserve elements to the total number of elements:

$$dW_p = \frac{r}{n} \cdot 100\%. \tag{27.3}$$

This tolerance defines the amount of tolerance of the output parameter of the passively reserved subsystem in relative terms, expressed as a percentage:

$$dW_p = \frac{r \cdot W_e}{n \cdot W_e} \cdot 100\% \tag{27.4}$$

where w_e is the value of the output parameter of one element of the passively reserved subsystem. In the development of passively redundant aircraft subsystems, taking into account tolerances, the realized tolerance, dW_p, must be no greater than the tolerance of dW_n assigned to the adjacent subsystem in accordance with the regulatory and technical documentation, i.e.

$$dW_p \leq dW_n. \tag{27.5}$$

Let us introduce the following notations. Let us take as a criterion indicator $P_c(t_{\text{nom}}) = P_c$ is the probability of fail-safe operation of the passively redundant subsystem during the execution of the task, and as an indicator of the failure of an element of the passively redundant subsystem, $p_c(t_{\text{nom}}) = p$ is the probability of its failure-free operation during the execution of the task.

One of the important features of the research and development of passively redundant subsystems, taking into account the tolerances described in Epifanov (1975), is that the vast majority of realized tolerances are provided by multiple-redundancy methods (Bessonov & Moroz, 1984), and only a very small number of realizable tolerances are provided by multiple-redundancy methods. Systems with multiple redundancy are characterized by the presence of so-called critical probabilities, p_{kr}, similar to some types of active redundancy (Grishin & Ko, 2009).

The value p_{kr} defines the overcritical region ($p_{kr} - 1$), where the failure of the redundant subsystem is higher than the failure of the nonredundant subsystem, and the subcritical region ($0 - p_{kr}$), where this type of redundancy is not beneficial. Critical probability is another, not immediately visible, essential factor that affects the reliability of a passively redundant subsystem. Thus the dependence of the criterion on essential factors is as follows:

$$P_c = P_c\left(p, dW_p, dW_c, n, m, p_{kr}\right) \tag{27.6}$$

To achieve the goal set in the work, it is necessary to study the dependence of the criterion on the essential parameters, taking into account the restrictions imposed. In this regard, there is a need to solve the problem, which in a formalized form looks like this:

$$P_c = P_c\left(p, dW_p, dW_c, n, m, p_{kr}\right)$$

With restrictions:

$$dW_n \geq dW_p \tag{27.7}$$

$$p > p_{kr} \tag{27.8}$$

$$m,\ n > 0,\ \text{intergers}. \tag{27.9}$$

27.3 Methodology

Indicator (8) will be calculated according to the formula of the binomial distribution law (Yu et al., 2005), which is valid for both subsystems with multiple and nonmultiple redundancy.

$$P_c = \sum_{i=m}^{n} C_n^i p^i.(1-p)^{n-i} \tag{27.10}$$

TABLE 27.1 Failure-free values P_c of subsystems depending on the multiplicity of K_i, the failure-free of the elements p, and the realized first-level tolerances.

dW_p(%)	Individual multiplicity	Probabilities of elements p						
		0.470	0.500	0.550	0.600	0.650	0.700	0.750
50.0	$K_2 = 4/2$	0.641	0.687	0.758	0.821	0.873	0.916	0.949
	$K_4 = 8/4$	0.569	0.637	0.740	0.826	0.894	0.942	0.973
	$K_6 = 12/6$	0.530	0.613	0.739	0.842	0.915	0.961	0.986
	$K_8 = 16/8$	0.502	0.598	0.744	0.858	0.933	0.974	0.992
	$K_{10} = 20/10$	0.480	0.588	0.751	0.872	0.947	0.983	0.996
33.3	Individual multiplicity	Probabilities of elements p						
		0.691	0.725	0.750	0.775	0.800	0.825	0.850
	$K_2 = 6/4$	0.727	0.789	0.831	0.868	0.901	0.929	0.953
	$K_4 = 12/8$	0.700	0.787	0.842	0.890	0.927	0.956	0.976
	$K_6 = 18/12$	0.692	0.797	0.861	0.912	0.949	0.973	0.988
	$K_8 = 24/16$	0.691	0.810	0.879	0.930	0.964	0.984	0.994
	$K_{10} = 30/20$	0.693	0.823	0.894	0.944	0.974	0.990	0.997
25.0	Individual multiplicity	Probabilities of elements p						
		0.804	0.815	0.830	0.845	0.860	0.875	0.890
	$K_2 = 8/6$	0.825	0.842	0.863	0.884	0.903	0.921	0.938
	$K_4 = 16/12$	0.805	0.833	0.868	0.898	0.925	0.947	0.965
	$K_6 = 24/18$	0.817	0.850	0.890	0.923	0.949	0.969	0.982
	$K_8 = 32/24$	0.833	0.869	0.911	0.943	0.966	0.982	0.991
	$K_{10} = 40/30$	0.848	0.886	0.927	0.957	0.977	0.989	0.996
20.0	Individual multiplicity	Probabilities of elements p						
		0.871	0.880	0.890	0.900	0.910	0.920	0.930
	$K_2 = 10/8$	0.872	0.891	0.912	0.930	0.946	0.960	0.972
	$K_4 = 20/16$	0.894	0.917	0.939	0.957	0.971	0.982	0.989
	$K_6 = 30/24$	0.917	0.939	0.959	0.974	0.985	0.992	0.996
	$K_8 = 40/32$	0.935	0.956	0.973	0.985	0.992	0.996	0.998
	$K_{10} = 50/40$	0.949	0.968	0.982	0.991	0.996	0.998	0.999

(continued on next page)

TABLE 27.1 Failure-free values P_c of subsystems depending on the multiplicity of K_i, the failure-free of the elements p, and the realized first-level tolerances—cont'd

dW_p(%)	Individual multiplicity	Probabilities of elements p						
16.7	Individual multiplicity	Probabilities of elements p						
		0.917	0.925	0.930	0.935	0.940	0.945	0.950
	$K_2 = 12/10$	0.929	0.944	0.953	0.961	0.968	0.975	0.980
	$K_4 = 24/20$	0.956	0.970	0.977	0.983	0.987	0.991	0.994
	$K_6 = 36/30$	0.973	0.984	0.989	0.992	0.995	0.997	0.998
	$K_8 = 48/40$	0.984	0.991	0.994	0.997	0.998	0.999	0.999
	$K_{10} = 60/50$	0.990	0.995	0.997	0.998	0.999	1.000	1.000
14.3	Individual multiplicity	Probabilities of elements p						
		0.943	0.947	0.951	0.955	0.959	0.963	0.967
	$K_2 = 14/12$	0.958	0.965	0.971	0.977	0.982	0.986	0.990
	$K_4 = 28/24$	0.980	0.985	0.989	0.992	0.995	0.997	0.998
	$K_6 = 42/36$	0.991	0.994	0.996	0.997	0.999	0.999	1.000
	$K_8 = 56/48$	0.996	0.997	0.998	0.999	1.000	1.000	1.000
	$K_{10} = 70/60$	0.998	0.999	0.999	1.000	1.000	1.000	1.000
12.5	Individual multiplicity	Probabilities of elements p						
		0.958	0.960	0.962	0.964	0.966	0.968	0.970
	$K_2 = 16/14$	0.972	0.976	0.979	0.982	0.984	0.987	0.989
	$K_4 = 32/28$	0.990	0.992	0.993	0.995	0.996	0.997	0.998
	$K_6 = 48/42$	0.996	0.997	0.998	0.998	0.999	0.999	0.999
	$K_8 = 64/56$	0.999	0.999	0.999	1.000	1.000	1.000	1.000
	$K_{10} = 80/70$	0.999	1.000	1.000	1.000	1.000	1.000	1.000

(continued on next page)

11.1	Individual multiplicity	Probabilities of elements p						
		0.968	0.970	0.972	0.974	0.976	0.978	0.980
	$K_2 = 18/16$	0.981	0.984	0.987	0.989	0.991	0.993	0.995
	$K_4 = 36/32$	0.994	0.996	0.997	0.998	0.998	0.999	0.999
	$K_6 = 54/48$	0.998	0.999	0.999	1.000	1.000	1.000	1.000
	$K_8 = 72/64$	1.000	1.000	1.000	1.000	1.000	1.000	1.000
	$K_{10} = 90/80$	1.000	1.000	1.000	1.000	1.000	1.000	1.000
10	Individual multiplicity	Probabilities of elements p						
		0.975	0.978	0.981	0.984	0.987	0.990	0.993
	$K_2 = 20/18$	0.987	0.991	0.994	0.996	0.998	0.999	1.000
	$K_4 = 40/36$	0.997	0.998	0.999	1.000	1.000	1.000	1.000
	$K_6 = 60/54$	0.999	1.000	1.000	1.000	1.000	1.000	1.000
	$K_8 = 80/72$	1.000	1.000	1.000	1.000	1.000	1.000	1.000
	$K_{10} = 100/90$	1.000	1.000	1.000	1.000	1.000	1.000	1.000

TABLE 27.2 Subsystem failure-free values P_c as a function of K_i multiplicity, element failure-free p, and second-level tolerances realized.

dW_p(%)	Individual multiplicity	Probabilities of elements p						
		0.269	0.300	0.400	0.500	0.600	0.700	0.800
66.7	$K_3 = 6/2$	0.511	0.580	0.767	0.891	0.959	0.989	0.998
	$K_5 = 12/4$	0.411	0.507	0.775	0.927	0.985	0.998	1.000
	$K_7 = 18/6$	0.350	0.466	0.791	0.952	0.994	1.000	1.000
	$K_9 = 24/8$	0.306	0.435	0.808	0.968	0.998	1.000	1.000
	$K_{11} = 30/10$	0.271	0.411	0.824	0.979	0.999	1.000	1.000
50	Individual multiplicity	Probabilities of elements p						
		0.485	0.543	0.601	0.659	0.717	0.775	0.833
	$K_3 = 8/4$	0.603	0.726	0.828	0.904	0.954	0.983	0.995
	$K_5 = 16/8$	0.550	0.725	0.860	0.943	0.983	0.996	1.000
	$K_7 = 24/12$	0.522	0.736	0.888	0.966	0.993	0.999	1.000
	$K_9 = 32/16$	0.502	0.748	0.910	0.979	0.997	1.000	1.000
	$K_{11} = 40/20$	0.487	0.760	0.927	0.987	0.999	1.000	1.000
40	Individual multiplicity	Probabilities of elements p						
		0.610	0.650	0.690	0.730	0.770	0.810	0.850
	$K_3 = 10/6$	0.658	0.751	0.832	0.896	0.943	0.973	0.990
	$K_5 = 20/12$	0.631	0.762	0.866	0.936	0.975	0.993	0.999
	$K_7 = 30/18$	0.622	0.780	0.895	0.961	0.989	0.998	1.000
	$K_9 = 40/24$	0.619	0.798	0.917	0.975	0.995	0.999	1.000
	$K_{11} = 50/30$	0.618	0.814	0.934	0.984	0.998	1.000	1.000
33.3	Individual multiplicity	Probabilities of elements p						
		0.691	0.700	0.738	0.776	0.814	0.852	0.890
	$K_3 = 12/8$	0.700	0.724	0.817	0.891	0.945	0.977	0.993
	$K_5 = 24/16$	0.691	0.725	0.848	0.931	0.977	0.995	0.999
	$K_7 = 36/24$	0.696	0.737	0.876	0.957	0.990	0.999	1.000
	$K_9 = 48/32$	0.704	0.749	0.899	0.972	0.996	1.000	1.000
	$K_{11} = 60/40$	0.712	0.762	0.917	0.982	0.998	1.000	1.000

(continued on next page)

28.6	Individual multiplicity	Probabilities of elements p						
		0.755	0.780	0.810	0.840	0.870	0.900	0.930
	$K_3 = 14/10$	0.756	0.824	0.891	0.941	0.973	0.991	0.998
	$K_5 = 28/20$	0.770	0.857	0.931	0.974	0.993	0.999	1.000
	$K_7 = 42/30$	0.789	0.885	0.957	0.988	0.998	1.000	1.000
	$K_9 = 56/40$	0.808	0.908	0.972	0.995	0.999	1.000	1.000
	$K_{11} = 70/50$	0.825	0.926	0.982	0.998	1.000	1.000	1.000
25	Individual multiplicity	Probabilities of elements p						
		0.803	0.826	0.849	0.872	0.895	0.918	0.940
	$K_3 = 16/12$	0.807	0.869	0.919	0.956	0.979	0.993	0.998
	$K_5 = 32/24$	0.837	0.909	0.957	0.984	0.995	0.999	1.000
	$K_7 = 48/36$	0.864	0.937	0.977	0.994	0.999	1.000	1.000
	$K_9 = 64/48$	0.887	0.956	0.988	0.998	1.000	1.000	1.000
	$K_{11} = 80/60$	0.906	0.969	0.993	0.999	1.000	1.000	1.000
22.2	Individual multiplicity	Probabilities of elements p						
		0.843	0.860	0.880	0.900	0.920	0.940	0.950
	$K_3 = 18/14$	0.860	0.904	0.944	0.972	0.988	0.997	0.998
	$K_5 = 36/28$	0.900	0.944	0.976	0.992	0.998	1.000	1.000
	$K_7 = 54/42$	0.928	0.967	0.990	0.998	1.000	1.000	1.000
	$K_9 = 72/56$	0.948	0.980	0.996	0.999	1.000	1.000	1.000
	$K_{11} = 90/70$	0.963	0.988	0.998	1.000	1.000	1.000	1.000
20	Individual multiplicity	Probabilities of elements p						
		0.871	0.884	0.897	0.910	0.923	0.936	0.950
	$K_3 = 20/16$	0.894	0.926	0.952	0.971	0.984	0.993	0.997
	$K_5 = 40/32$	0.935	0.963	0.981	0.992	0.997	0.999	1.000
	$K_7 = 60/48$	0.960	0.981	0.993	0.998	0.999	1.000	1.000
	$K_9 = 80/64$	0.975	0.990	0.997	0.999	1.000	1.000	1.000
	$K_{11} = 100/80$	0.984	0.995	0.999	1.000	1.000	1.000	1.000
18.2	Individual multiplicity	Probabilities of elements p						
		0.892	0.900	0.910	0.920	0.930	0.940	0.950
	$K_3 = 22/18$	0.919	0.938	0.958	0.973	0.984	0.991	0.996
	$K_5 = 44/36$	0.957	0.972	0.985	0.993	0.997	0.999	1.000
	$K_7 = 66/54$	0.977	0.987	0.995	0.998	0.999	1.000	1.000
	$K_9 = 88/72$	0.988	0.994	0.998	0.999	1.000	1.000	1.000
	$K_{11} = 110/90$	0.993	0.997	0.999	1.000	1.000	1.000	1.000

Consideration of the constraint (7) is obvious, and to take into account the constraint (8), it is necessary to determine the right boundary (RB) and the left boundary (LB) of the interval of change in the probabilities of the elements of the passively reserved subsystem. The right-hand boundary is tolerance-dependent and is determined by the element probabilities at which the element uptime reaches 0.99.

The LB is determined by the critical probability, which depends on the tolerance value and the redundancy multiplicity, which are interrelated with each other. Let us examine this relationship and how the results of solving the problem depend on it.

Let us consider a discrete series of realizable tolerances formed at $r = 1$ and the variation m from 1 to 9. It looks like this: 50.0%, 33.3%, 25.0%, 20.0%, 16.7%, 14.3%, 12.5%, 11.1%, and 10.0%. Let us call this series of first-level tolerances. A similar series of discrete realizable tolerances at $r = 2$ is as follows: 66.7%, 50.0%, 40.0%, 33.3%, 28.6%, 25.0%, 22.2%, 20.0%, and 18.2%. Let us call this series the second-level tolerances. The series is considered to form the main grid of tolerances because they correspond to the smallest redundancy structures and because the total number of elements n increases sharply with an increase in r.

The considered tolerance grid corresponds to the minimum redundancy multiples. The fact is that each realized tolerance can be represented by an increasing number of individual multiples, starting with the minimum. For example, a tolerance of 50.0% of the first level is realized by individual multiplicities: $K_1 = 2/1 = 2$; $K_2 = 4/2$; $K_3 = 6/3$, etc. The tolerance of 33.3% of the first level is realized by individual redundancy multiples: $K_1 = 3/2$; $K_2 = 6/4$; $K_3 = 9/6$, etc. The tolerance of 66.7% of the second level is realized by individual multiples: $K_2 = 3$; $K_3 = 6/2$; $K_4 = 9/3$, etc. The tolerance of 40.0% of the second level is realized by individual multiples: $K_2 = 5/3$; $K_3 = 10/6$; $K_4 = 15/9$, etc.

This circumstance makes it possible to calculate indicator (6) for each tolerance at different individual multiplicities inherent in this tolerance and, consequently, at different passive reserve structures, revealing the degree of influence of the reserve structure on the accepted indicator. It should be noted that the value of $K_1 = 2/1 = 2$ with a tolerance of 50.0% of the first level and the value of $K_2 = 3/1 = 3$ with a tolerance of 66.7% of the second level correspond to cases of multiple redundancy. Therefore they are written as integer values. The remaining multiplicity values correspond to nonmultiple reservations, and therefore, they are represented as fractional values.

Because of the need to conduct studies in the probability range $p > p_{kr}$ and p_{kr}, as shown in Grishin and Tuan (2018), depending on the redundancy multiples and the values of the realized tolerances, it is first necessary to determine the maximum value of the critical probability for each tolerance.

$$p_{krm} = \max_i p_{kr}(K_i) \tag{27.11}$$

The value p_{krm} determines the initial values of variable probabilities in the range (p_{krm}–RB). As preliminary calculations showed, the right border of the range varies depending on the tolerance from 0.750 to 0.993. In this range, for each tolerance, we will specify 7 approximately evenly distributed probabilities of the elements of the passively reserved behavior analysis subsystem P_c.

27.4 Results of solving the problem

In accordance with the above methodology, we will conduct a study of the failure index of a passively reserved subsystem. The results of the preparation of the initial data and the calculations of the indicator for the first-level tolerances are presented in Table 27.1.

Tables 27.1 and 27.2 show that with a decrease (tightening) of the realized tolerances of one level, the passively redundant subsystem index decreases with the same failure-free elements and minimum redundancy multiplicities, K_i. For example, at 50%, $K_2 = 4/2$ and $p = 0.750 \rightarrow P_c = 0.949$, and at $dW_p = 33.3\%$, $K_2 = 6/4$ and $p = 0.750 \rightarrow P_c = 0.831$. Similarly, at 25.0%, $K_2 = 8/6$ and $p = 0.890 \rightarrow P_c = 0.938$, and at $dW_p = 20.0\%$, $K_2 = 10/8$ and $p = 0.890 \rightarrow P_c = 0.912$. The same is true for the second-level tolerances that are being implemented. This is because of the fact that when the realized tolerances of one level decrease (with constant r), the values of n and m increase, approaching each other and, thus, reducing the multiplicity of redundancy.

The calculations made it possible to establish that the minimum redundancy multiples provide the best failure rates only at four tolerances (out of the entire set of considered tolerances of two levels) in narrow ranges of probabilities of elements:

- $dW_p = 50\%$, $p_{kr} = 0.470 < p < 0.600$, including multiple redundancy
- $dW_p = 33.3\%$, $p_{kr} = 0.691 < p < 0.725$ (multiple-redundancy)
- $dW_p = 66.7\%$, $p_{kr} = 0.269 < p < 0.400$, including multiple redundancy
- $dW_p = 40\%$, $p_{kr} = 0.610 < p < 0.650$ (multiple-redundancy)

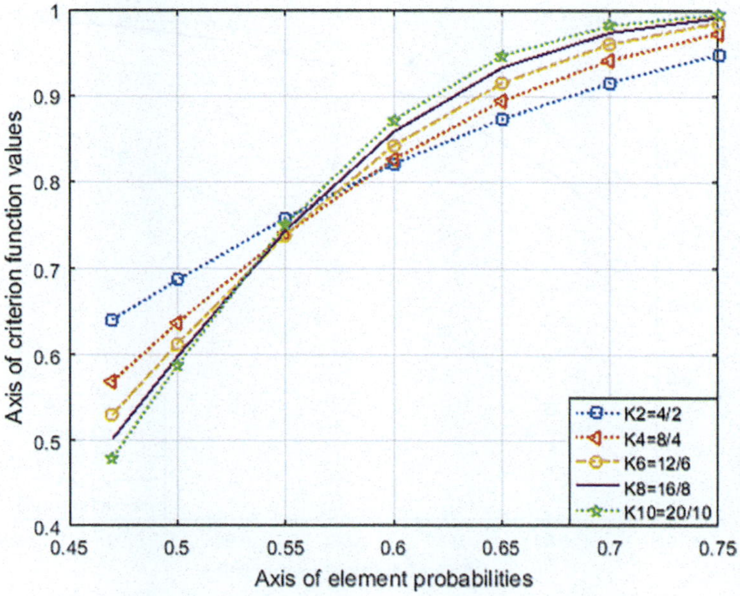

FIGURE 27.1 Graphs of dependencies of the indicator, P_c from p for = 50%.

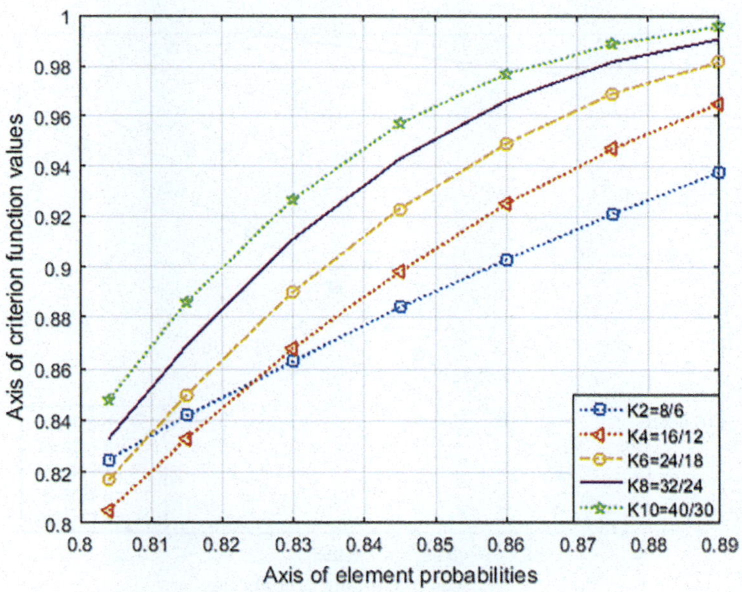

FIGURE 27.2 Graphs of dependencies of the indicator, P_c from p for = 25%.

Moreover, two ranges (the first and the third) include one dependency corresponding to the multiple reservation, and two ranges (the second and fourth) in which all dependencies correspond to the nonmultiple reservation. For the remaining probabilities in the supercritical region (where this type of redundancy is beneficial) of the above four tolerances, an increase in the redundancy ratio allows for higher uptime.

With other tolerances considered in the work an increase in the redundancy multiplicity increases the reliability in the entire probability range of supercritical regions. The first and third ranges show that multiple redundancy is more profitable than nonmultiple only at low fault-free values of the elements: at $p < 0.600$ for the first range and at $p < 0.400$ for the second range. Moreover, these values decrease with an increase in the level of tolerances (0.400 at $r = 2$, less than 0.600 at $r = 1$). Thus it is advantageous to use the minimum values of subsystem redundancy at large tolerances and when using unreliable elements in redundancy structures. In the overwhelming majority of cases, to increase the reliability of subsystems, it is

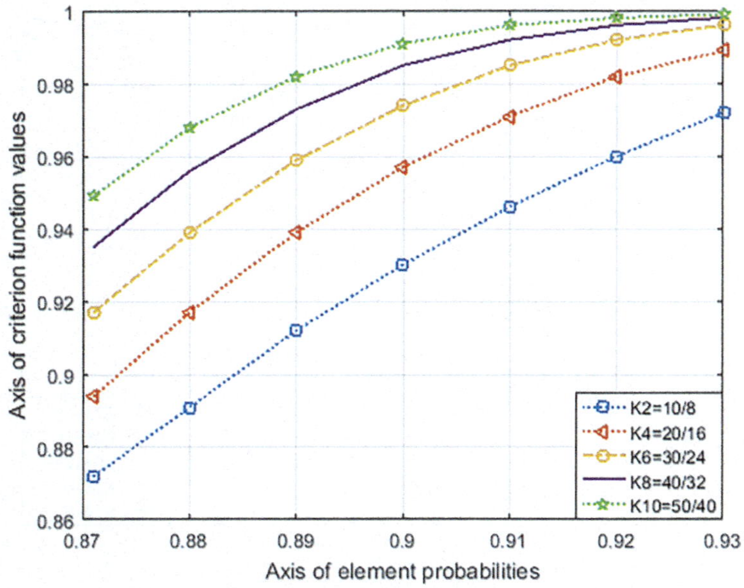

FIGURE 27.3 Graphs of dependencies of the indicator, P_c from p for = 20%.

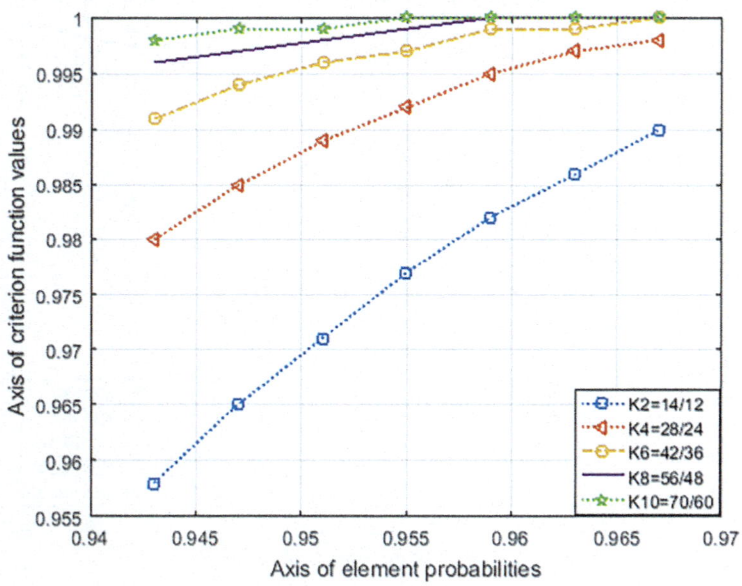

FIGURE 27.4 Graphs of dependencies of the indicator, P_c from p for = 14.3%.

advantageous to increase the redundancy multiplicity. Let us illustrate in the form of graphs some of the results presented in Tables 27.1 and 27.2.

Figs. 27.1–Figs. 27.4 show some characteristic dependencies of the $P - c$ indicator. Not for first-level tolerances, and in Figs. 27.5 and 27.6, similar dependencies for second-level tolerances. From the presented tables and graphs, it is clear how the nature of the dependencies of the indicator P_c changes on the reliability of elements and redundancy multiplicities at tolerances of the first and second levels exceeding 25% and less than this value. Indicator dependencies: P_c of the failure-free elements has a nonlinear, strictly concave character at any tolerances and redundancy ratios. This indicates the existence of a different extremum ($P_c - p$), which makes it possible to set the task of developing highly reliable passively redundant aircraft subsystems, taking into account tolerances from unreliable elements.

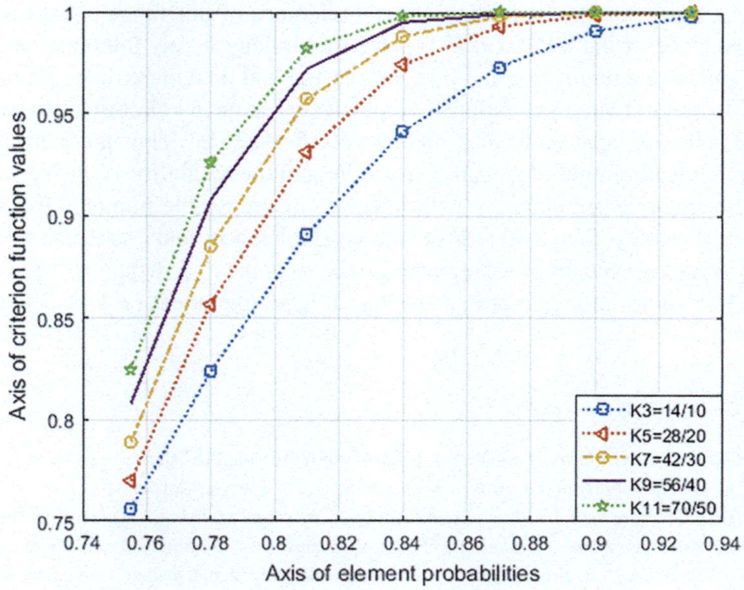

FIGURE 27.5 Graphs of dependencies of the indicator, *Pc* from *p* for = 28.6%.

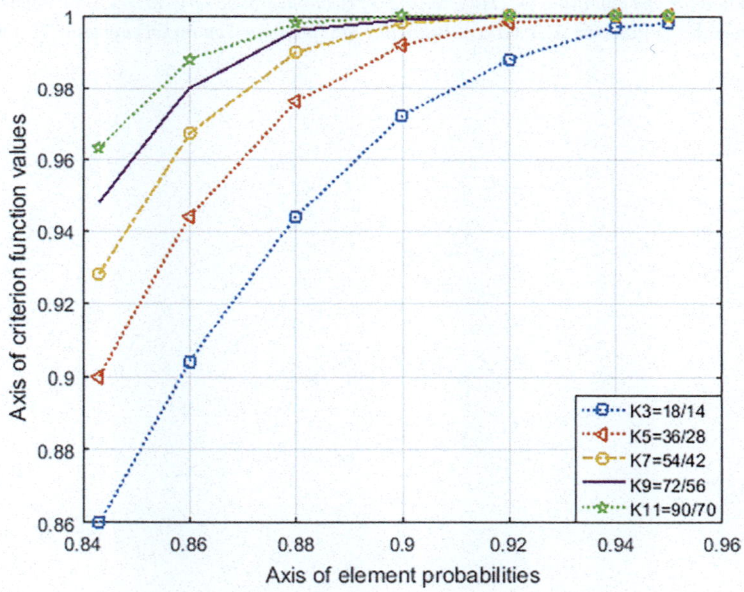

FIGURE 27.6 Graphs of dependencies of the indicator, P_c from p for = 22.2%.

27.5 Conclusion

Along with the failure of the elements the most important characteristic that makes it possible to increase the reliability of passively redundant aircraft subsystems at a given assigned tolerance is the redundancy multiplicity. On the nature of the dependence of the indicator, P_c, the value of the realized tolerances and the reliability of the elements have a significant impact on the redundancy multiplicity. The influence of the failure-free elements affects the dependence of the P_c indicator of the redundancy factor only with large tolerances of the first and second levels of at least 25%. With realized tolerances less than 25% the dependence of the indicator on P_c multiplicity is always of an increasing nature, so the redundancy multiplicity is the main means of increasing the reliability of passively redundant subsystems of the class under consideration at "hard" tolerances. Within a realized tolerance range of at least 25% in each tolerance, there are failure-free values of the elements that change the nature of the dependence of the indicator, P_c, on multiplicity: e.g., with the failure of elements $p \geq 0.650$, the dependence of the indicator, P_c, on the multiplicity at any tolerance of this range always has an increasing character, and at $p < 0.650$, the dependence of the indicator, P_c, on the multiplicity at any tolerance of a given range can both be monotonically decreasing and have a nonmonotonic (first decreasing and then increasing) character. The minimum values of redundancy ratios provide the best values of failure-free passively redundant aircraft subsystems of this class only with unreliable elements ($p < 0.600$) and large realizable tolerances (at least 25%). Thus there are no tolerances at which it is always advantageous to apply the minimum redundancy (regardless of the reliability of the elements). Multiple redundancy provides higher uptime than multiple redundancy (with unequal, incomparable numbers n and m) only at low fail-safe elements of redundant subsystems (less than 0.400 for second-level tolerances and less than 0.600 for first-level tolerances); increasing the reserve ratio is the main means of achieving high values of the P_c indicator at "hard" realizable tolerances (less than 25%) and any failure-free values of p elements, as well as at large tolerances (at least 25%) and failure-free elements greater than 0.650.

References

Antseliovich, L. L. (1985). *Reliability and survivability of the aircraft* p. 98. Mashinostroenie Publishers.

Bessonov, A. A., & Moroz, A. V. (1984). *Reliability of automatic control systems* p. 36. Energoatomizdat.

Dedkov, V. K., & Tatuev, A. I. (2010). *Ensuring the reliability of technical objects by stages of their life cycle* p. 76. Mashinostroenie–Polet.

Epifanov, A. D. (1975). *Nadezhnost' sistem upravleniya [Reliability of control systems]* p. 45. Mashinostroenie Publishers.

Grishin, V. M., & Ko, P. M. (2009). Optimization of aircraft control systems reliability with active loaded redundancy. *Vestnik, 16*(5), 52–59.

Grishin, V. M., & Tuan, V. T. (2018). Specific features of research and development of the passive redundant subsystems of the aircraft with due consideration of tolerances. *Journal of Mechanical Engineering Research and Developments, 41*(4), 82–87. doi:10.26480/jmerd.04.2018.82.87.

Polovko, A. M., & Gurov, S. I. (2006). *Fundamentals of the theory of reliability* p. 247. BHV-Petersburg.

Redko, P. G. (2002). *Improving reliability and improving the characteristics of electrohydraulic servo drives* p. 59. Publishing Center of MSTU "Stankin".

Yu, M. M., Yu, B. A., Kavitskaya, V. K., & Erenburg, E. I. (2005). *Reliability of solid propellant rocket engines* p. 73. OOO Eureka.

Chapter 28

Design and implementation of a smart pillbox system: Integrating IoT, chatbot AI, and real-time mobile monitoring for enhanced healthcare

Houda Daoud[a], Naourez Ben Hadj[b], Maryem Kotti[a] and Rihem Walha[a]
[a] *Department of Engineering of Electronic Communication Systems, National School of Electronics and Telecommunications of Sfax, University of Sfax, Sfax, Tunisia,* [b] *Department of Electrical Engineering, National Engineering School of Sfax, Information Technologies and electronics Laboratory, University of Sfax, Sfax, Tunisia*

28.1 Introduction

Over the past few decades the medical field has witnessed a profound technological transformation. The transition from basic manual tools to sophisticated digital systems has significantly enhanced diagnostic, monitoring, and therapeutic capabilities (Shaik et al., 2023). Early electronic diagnostic devices have evolved into advanced systems powered by artificial intelligence (AI), capable of processing large-scale, real-time data streams to support clinical decision-making.

The integration of Internet of Things (IoT) and AI technologies into healthcare systems has paved the way for innovative solutions that address longstanding challenges in patient care and resource management (Gupta et al., 2024; Rao et al., 2023). Connected medical devices now enable continuous, real-time monitoring of patient health parameters, ensuring timely interventions and improved medication adherence (Webster et al., 2022). These technologies also facilitate personalized treatment strategies, adapting interventions to the specific physiological and behavioral profiles of individual patients. Furthermore, IoT enhances the operational efficiency of healthcare delivery by enabling intelligent data management across both patient records and medical assets. Real-time tracking of equipment status and patient conditions supports informed clinical decision-making and predictive maintenance of medical infrastructure. The data generated through these systems contributes to a higher quality of care through optimization of clinical workflows, treatment protocols, and both in-person and remote healthcare services (Anjaneyulu et al., 2023). Medication nonadherence remains a significant challenge in modern healthcare systems, particularly among elderly patients and those undergoing complex polypharmacy regimens.

To address these issues, smart pill dispensers have emerged as a promising technological intervention (Karagiannis et al., 2022; Saran et al., 2025). These IoT-enabled devices are designed to automate medication administration by dispensing prescribed doses at scheduled intervals, while simultaneously issuing auditory, visual, or digital reminders to the patient. As part of the broader digital health ecosystem, smart pill dispensers play a critical role in enhancing remote patient monitoring, facilitating communication among patients, families, and care providers, and promoting a more proactive and data-driven approach to healthcare delivery. According to the literature, several pillbox systems have been studied and developed. In fact, Nedelcea and Albu (2020) propose a warning system to improve medication adherence among patients of varying ages and health conditions. It is designed for flexible use both in hospitals and at home to support timely medication intake. But the system does not verify actual medication intake; it only provides alerts without robust security. It lacks integration with healthcare providers for remote monitoring. User interface adaptability for elderly or cognitively impaired patients is not addressed. Moreover, the design of an IoT-enabled smart pillbox is presented in Selvabharathi et al. (2022) to

Computational Intelligence in Mechatronics: Solving Real-World Problems in Electronic Systems Design. DOI: https://doi.org/10.1016/B978-0-443-27580-7.00009-3

enhance medication adherence, particularly for patients with chronic diseases. The device provides scheduled auditory and visual reminders for medication intake. Unfortunately the system lacks high security and does not address usability for elderly or cognitively impaired patients. Besides, Nasir et al. (2023) present the design of an IoT-based smart medical box that automates pill dispensing and monitors patient health. It provides a user interface for managing patient data and scheduling. Nevertheless, the system is limited, especially in terms of consumption and cost. In addition, the work presented in Siddiqha et al. (2024) is focused on the design of a smart pill dispenser that combines embedded systems and IoT technologies to improve medication adherence and patient monitoring. It aims to ensure secure, timely medication delivery while providing real-time health data for caregivers or medical staff. However, the technological complexity may limit adoption among patients with low technical literacy. In this work, we propose the design of a secure and connected system architecture that ensures the integrity, precision, and accessibility of medical data while facilitating the efficient remote management of healthcare services for patients with chronic diseases. In fact, this chapter aims to address a few issues. On the one hand, its first objective is to design a low-cost, low-power, secure, and user-friendly smart pill dispenser, providing patients with an effective means to adhere to their treatment through automated reminders and connectivity to a mobile application. The second target of this study is to integrate the AI into the mobile application through a chatbot to enhance the user experience and make interaction more intuitive. This feature allows users to communicate easily with the application and receive instant responses to their inquiries.

This chapter is organized as follows: In Section 28.2 the system overview is explained. The smart intelligent pillbox design is presented in Section 28.3. After, we illustrate the intelligent pillbox system implementation, results, and evaluation in Section 28.4. Finally, some concluding remarks are given in Section 28.5.

28.2 System overview

The proposed system is an innovative IoT–AI-based solution designed to seamlessly connect three main users in the healthcare ecosystem: the patient, the family member, and the doctor.

The system provides personalized patient support through automated alerts and reminders for medication intake, significantly reducing the risk of nonadherence and improving treatment compliance. An integrated AI-powered chatbot enables direct interaction with the patient, allowing them to ask basic questions related to their therapy, receive preliminary guidance, and report symptoms or changes in their health status in real time. Simultaneously the family member is kept informed via synchronized notifications, enabling continuous remote monitoring of the patient's condition and timely intervention when necessary. This layer of oversight enhances safety and promotes proactive involvement in the patient's care. For the doctor a dedicated mobile application facilitates remote monitoring of the patient's health metrics, supports control over specific system functionalities, and allows for real-time adjustment of treatment plans or medical instructions based on dynamic health data. Overall, this IoT-driven intelligent architecture fosters seamless communication, real-time data exchange, and enhanced coordination among all users. It contributes to improved healthcare delivery, timely medical responses, and strengthened patient engagement, particularly in managing chronic conditions and at-home care scenarios (Fig. 28.1).

28.3 Smart intelligent pillbox design

The smart pillbox is designed to improve patient autonomy, particularly for elderly individuals living alone or those with chronic illnesses, by facilitating home-based care and enabling continuous monitoring, thereby keeping both healthcare providers and family members informed of the patient's health status.

28.3.1 Pillbox architecture operation principle

Fig. 28.2 demonstrates an illustration of the proposed system design, which includes an ESP32 microcontroller connected to a power supply, temperature and humidity sensor, radio frequency identification (RFID) module, buzzer, LCD display, LEDs, and servo motor. The microcontroller will acquire, process, and send sensor data to a database. A mobile application including an AI-powered chatbot has been designed and used to monitor medication adherence. The fundamental operating principle of this system is detailed as follows. At the scheduled time the buzzer sounds, the red LED turns on, and a notification is sent as a reminder. If the user is awake, it must use the RFID card to open the pillbox. If the card is correct, the servo motor unlocks the box and opens it, whereas the green LED lights up to indicate that the pillbox is open. If the card is incorrect, a notification is sent to alert that someone is trying to access the box. If the user does not take the medication,

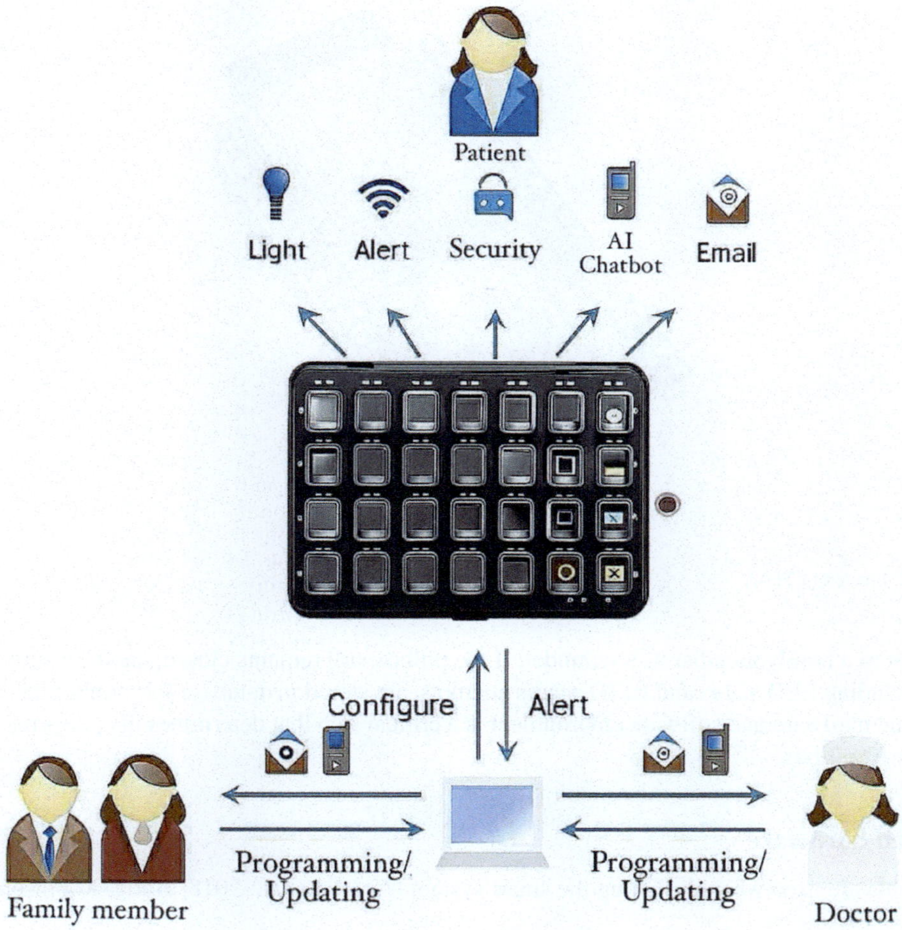

FIGURE 28.1 System operation mode: image of system operation mode.

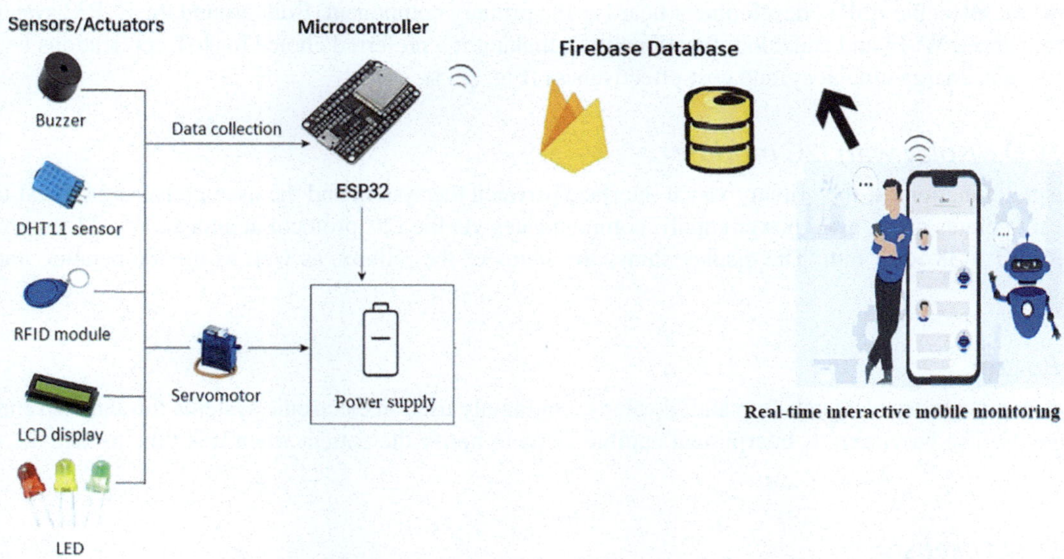

FIGURE 28.2 Architecture of the connected smart intelligent pillbox: image of architecture of the connected smart intelligent pillbox. *RFID*, Radio frequency identification.

FIGURE 28.3 ESP32: image of ESP32.

a notification is sent to a family member as a reminder. If the pillbox still remains closed, another notification is sent to the doctor. All data, including LED status and RFID access attempts, are stored in database for monitoring and record-keeping.

The choice of the hardware and software environment is a critical step that determines the performance, reliability, and efficiency of the smart pillbox.

28.3.2 Required hardware

Different components are used when designing the smart system (Antoun et al., 2018; Kuria, Robinson, & Gabriel, 2020; Oner, 2023).

28.3.2.1 ESP32 microcontroller

For this work, we select the ESP32 development board as the primary component. Built around the ESP32 system-on-chip, it integrates advanced Wi-Fi and Bluetooth functionalities, making it a preferred choice for IoT applications because of its streamlined design, energy efficiency, and cost-effectiveness (Fig. 28.3).

28.3.2.2 LCD display with I2C module

The LCD display functions as the primary visual interface between the system and the user, delivering critical information in real time. It operates under a 5V power supply, communicates via the I2C protocol at address 0×27, and has physical dimensions of $80 \times 38 \times 18$ mm. The display shows the status of the pillbox, as well as the temperature and humidity (Fig. 28.4).

28.3.2.3 Buzzer

The buzzer converts electrical signals into audible alerts, commonly used in electronic systems for alarms, reminders, and user notifications. The buzzer emits intermittent audible alerts to notify the patient when it is time to take the medication (Fig. 28.5).

28.3.2.4 DHT11 sensor

To maintain optimal medication storage conditions the system integrates a DHT11 temperature and humidity sensor. This cost-effective digital sensor delivers precise measurements of ambient temperature and humidity inside the pillbox, and it is selected for its reliable calibration and minimal power consumption, ensuring both accuracy and energy efficiency (Fig. 28.6).

FIGURE 28.4 LCD: image of LCD.

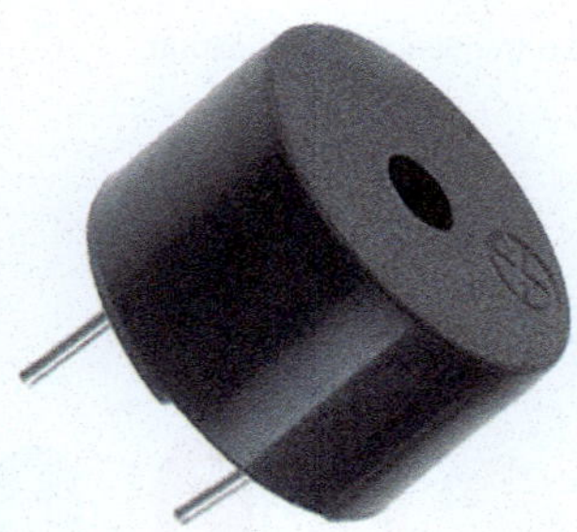

FIGURE 28.5 Buzzer: image of buzzer.

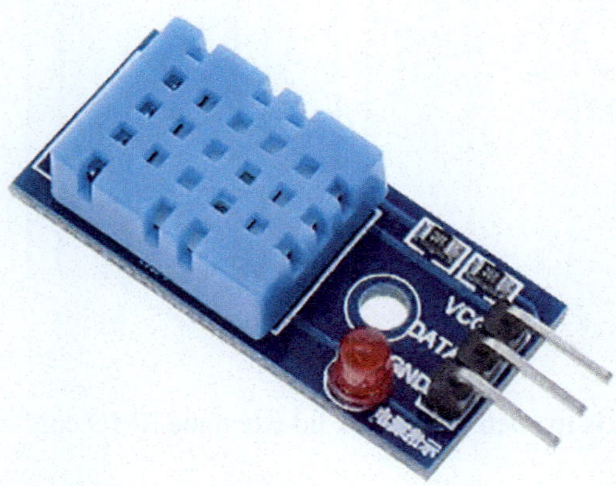

FIGURE 28.6 DHT11: image of DHT11.

28.3.2.5 RFID tag

The RFID module comprises an integrated electronic chip for data storage and an antenna for wireless communication with the reader. The RFID system ensures the security of our pillbox by controlling its opening (Fig. 28.7).

28.3.2.6 LEDs

An LED is a semiconductor-based optoelectronic device that emits visible light when forward-biased by an electric current. In this system, visual medication alerts are provided through alarm LEDs integrated into the pillbox, ensuring clear and immediate user notification. The LED indicates whether the medication has been taken or not (Fig. 28.8).

28.3.2.7 Servo motor

A servo motor is an electromechanical actuator designed to maintain precise positional control under static load conditions. It operates using continuous feedback, allowing real-time position monitoring and dynamic correction based on sensor

FIGURE 28.7 RFID tag: image of RFID tag. *RFID*, Radio frequency identification.

FIGURE 28.8 LED: image of LED.

measurements. The servo motor's role is to open the pillbox lid when the RFID card is valid and it is the correct time to take the medication (Fig. 28.9).

28.3.3 Required software

The design and implementation of the smart pillbox require the use of various development tools to meet the system's functional and performance objectives.

28.3.3.1 Arduino IDE

This development environment enables programming of microcontrollers like the ESP32 using C++, providing extensive compatibility with a wide range of libraries and sensor modules to facilitate efficient and versatile application development (Abiodun & Okpe, 2024).

28.3.3.2 Android Studio

Primarily utilizing Java and Kotlin, this integrated development environment offers several tools for designing, coding, testing, and deploying applications. Android Studio also supports integration with microcontrollers such as the ESP32, enabling seamless communication between mobile applications and external sensors (Runsewe et al., 2024).

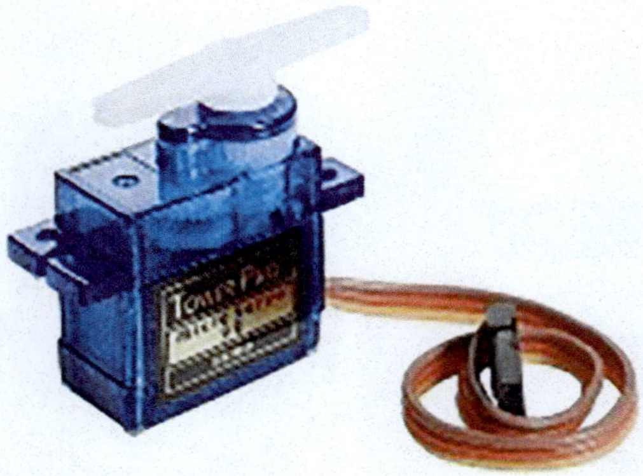

FIGURE 28.9 Servo motor: image of servo motor.

28.3.3.3 Flutter

Developed by Google, this open-source framework enables the creation of cross-platform mobile, web, and desktop applications from a unified codebase. Leveraging the Dart programming language, it is fully supported within Android Studio, providing a robust and efficient integrated development environment (Sultan et al., 2020).

28.3.3.4 Firebase

Firebase is a comprehensive development platform that provides an integrated suite of backend services to accelerate mobile application development. Its capabilities include secure data storage, robust authentication management, scalable content hosting, real-time push notifications, and advanced analytics for user behavior monitoring (Milojković et al., 2024).

28.3.3.5 Dialogflow

For the development of the chatbot, Dialogflow is selected because of its robust capabilities in creating intelligent conversational agents. Utilizing advanced natural language processing, it accurately interprets user inputs and delivers context-aware responses. Its intuitive interface and seamless integration with mobile applications, web platforms, and Google Assistant make it an ideal choice for enhancing user interaction (Kokane et al., 2024).

28.4 Intelligent pillbox implementation, test, and evaluation

Following the design phase, all selected components were integrated into the system. Fig. 28.10 illustrates the smart pillbox prototype.

28.4.1 Pillbox prototype test

This subsection presents the validation and performance testing of the smart pillbox prototype. In fact, when the pillbox is closed, the RFID system is disabled, the servo motor remains in its initial position, and both LEDs are turned off (Fig. 28.11). When it is time to take the medication, the red LED turns on, the buzzer sounds, and the LCD displays real-time temperature and humidity values (Fig. 28.12). At this moment the patient opens the pillbox using the RFID system; the servo motor rotates 90 degrees, and the green LED lights up (Fig. 28.13).

The backend of the Flutter application is powered by Firebase, which manages user authentication, medical data storage, and notification delivery. The ESP32 board is programmed via the Arduino IDE to collect sensor data and transmit it in real time to the mobile application via Wi-Fi (Fig. 28.14). When each user (patient, family member, or doctor) enters its email, it is recorded in Firebase Authentication (Fig. 28.15). Once the user completes the health profile form, all the information is stored in the Firestore database (Fig. 28.16). Using the analytics dashboard, it is possible to track whether the application has been employed by a user (Fig. 28.17).

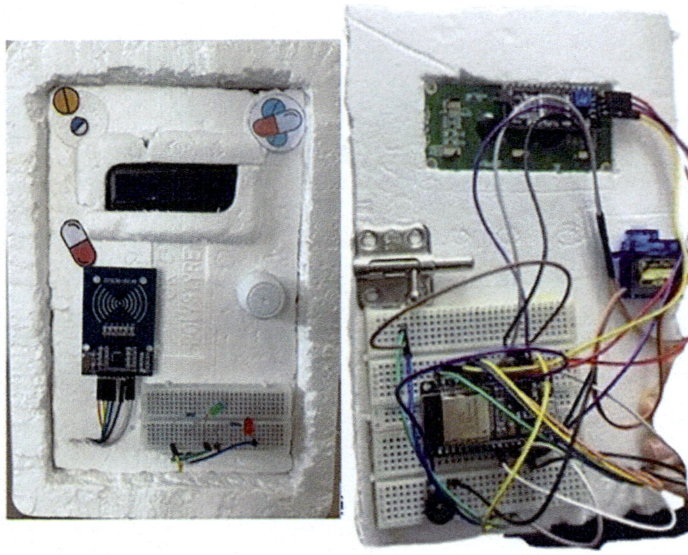

FIGURE 28.10 Smart pillbox prototype: image of smart pillbox prototype.

FIGURE 28.11 Closed pillbox: image of closed pillbox.

28.4.2 Integration of an AI-powered chatbot into the mobile application

During the modeling phase of the mobile application, we opt for the Unified Modeling Language (UML) employed to visualize, specify, design, and document software systems (Oliveira, 2023). By representing multiple dimensions of a system through a set of structured diagrams, UML enhances clarity and fosters efficient communication among developers, analysts, and stakeholders. It enables a comprehensive understanding of the system's architecture and behavior prior to

FIGURE 28.12 Medication reminder: image of medication reminder.

implementation, thereby supporting more informed and reliable software engineering decisions. The developed mobile application is designed to provide comprehensive control and real-time monitoring of the pillbox system for three users: the patient, the family member, and the doctor. Each user has a personalized interface tailored to their role and specific needs. In fact the application launches on the home page, providing access to the other sections of the interface (Fig. 28.18). Next, each user creates an account on the authentication page by completing the required fields for registration (Fig. 28.19). To create a profile the user fills out a form that includes age, height, weight, and the designated role (Fig. 28.20). After that the patient's home interface provides real-time monitoring of temperature and humidity, as well as a list of medications to be taken (Fig. 28.21). Then, medication reminders and alert notifications are triggered upon the opening of the pillbox (Fig. 28.22). In addition, the doctor and the family member can access their interfaces and receive notifications (Figs. 28.23–28.27).

Besides the mobile application incorporates various intuitive interfaces to enhance user experience and ensure seamless user interaction and system accessibility. In fact, AI is integrated into the mobile application through a chatbot. We develop a platform that enables the design of conversational agents or chatbots capable of understanding and responding in natural language. By default, Dialogflow provides certain preconfigured agents, such as "welcome" and "sorry," which are used to greet the user and handle situations where the chatbot does not understand the request, respectively. After exploring these default agents, we add new custom agents to handle specific question-and-answer scenarios. The chatbot is then trained with these interactions to respond accurately to user queries. To integrate the chatbot into our application, we used the URL provided by Dialogflow, allowing the application to communicate directly with the agent and retrieve responses in real time, thus offering an interactive user experience. From this figure, we notice the training process of Dialogflow, which allows the chatbot to effectively respond to various question formulations (Fig. 28.28). Fig. 28.29 illustrates the patient's inquiries conveyed via the chatbot interface.

FIGURE 28.13 Pillbox opening: image of pillbox opening.

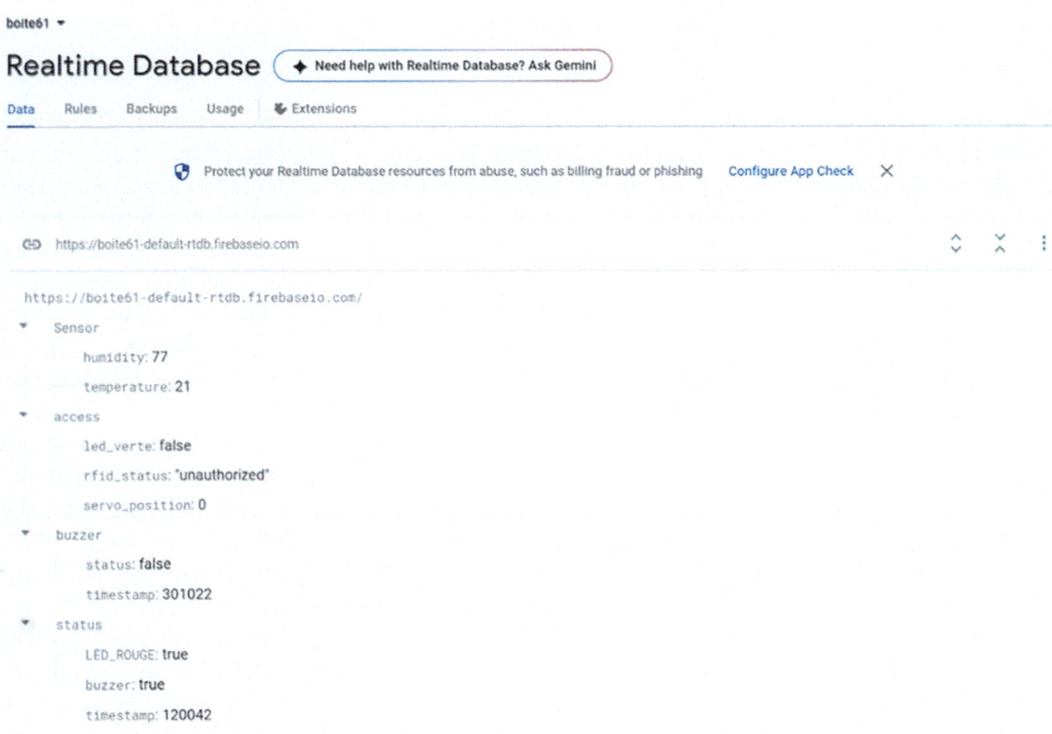

FIGURE 28.14 Real-time database results: image of real-time database results.

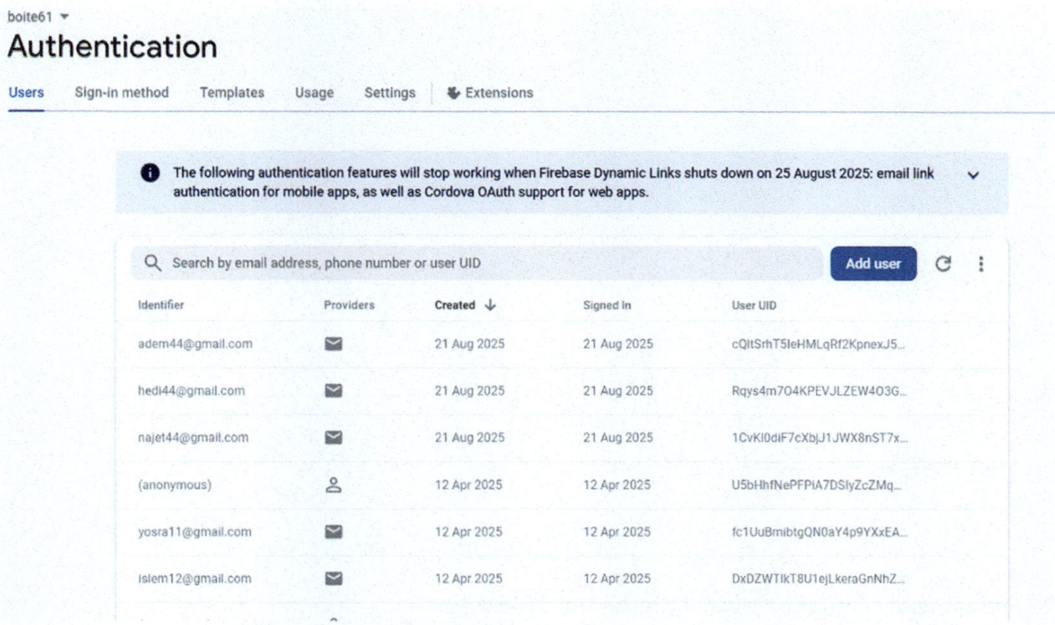

FIGURE 28.15 Firebase authentication: image of Firebase authentication.

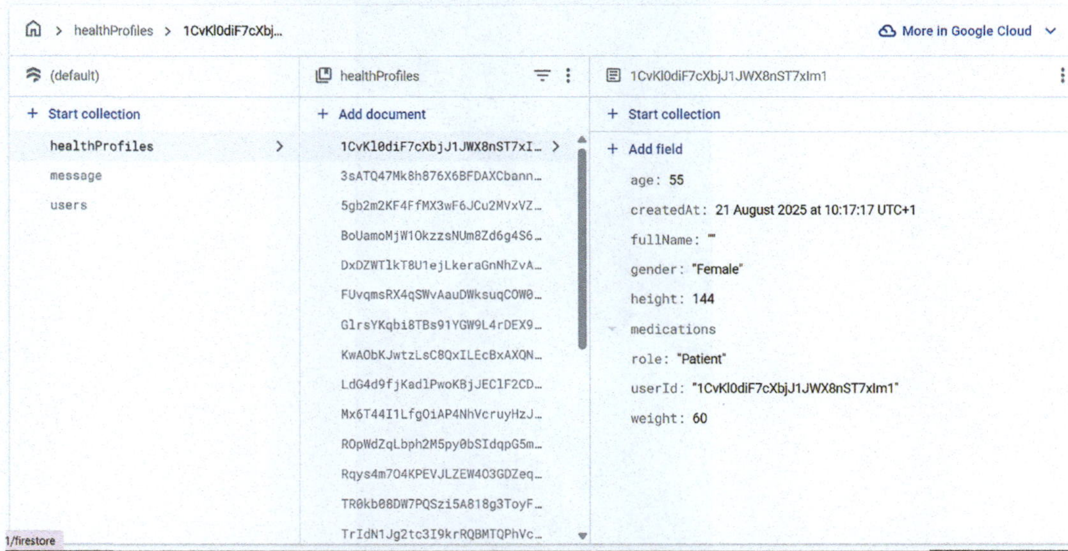

FIGURE 28.16 Firebase Firestore database for the patient: image of Firebase Firestore database for the patient.

28.4.3 System evaluation

To evaluate our prototype performances a comparison between different pillboxes is established (Daoud et al., 2023; Hamim, Ehsan, & Ali, 2023; Nasir et al., 2023; Nedelcea & Albu, 2020; Selvabharathi et al., 2022; Siddiqha et al., 2024). From Table 28.1, we can notice, on the one hand, that some pillboxes are implemented with low and medium security that can affect the patient's life. On the other hand, all pillbox prototypes are implemented without AI chatbot. Our smart pillbox seems the best prototype, particularly in terms of simplicity, security, cost, and interactive mobile application. In fact, our system supports the generation of different alerts and facilitates secure data sharing with both caregivers and healthcare professionals. The proposed pillbox system distinguishes itself from the presented solutions through the integration of an interactive chatbot, a feature that remains largely unexplored in current pill dispenser technologies. This intelligent conversational interface provides real-time guidance, allowing users to manage their medication schedules more

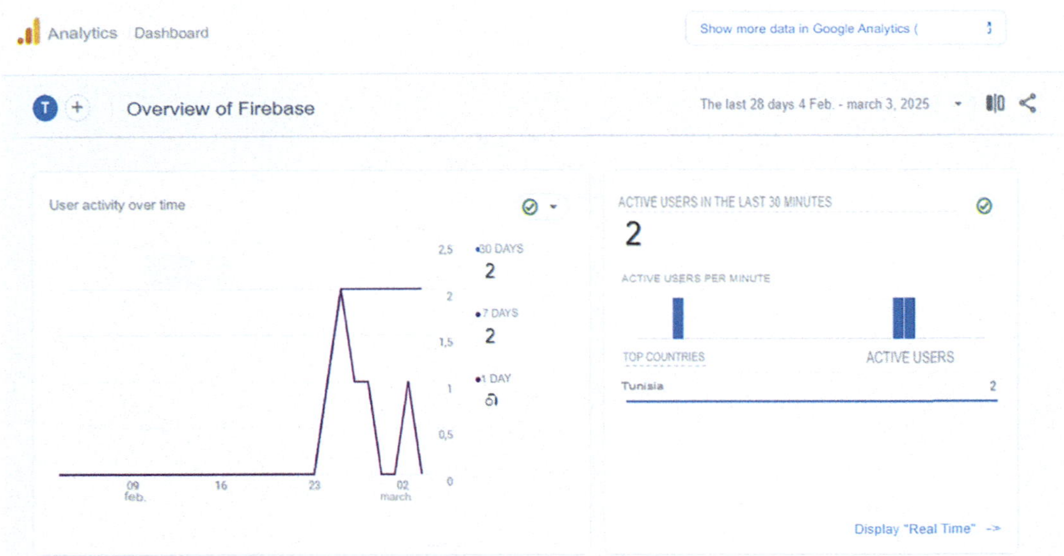

FIGURE 28.17 Firebase Analitycs dashboard: image of Firebase Analitycs dashboard.

FIGURE 28.18 Home page: image of home page.

effectively and intuitively. In addition, the system addresses common risks associated with improper usage or dosing confusion by incorporating a robust security mechanism: The pillbox remains securely locked outside of scheduled times and automatically unlocks only when it is time to take the prescribed medication. This combination of enhanced safety, intelligent user interaction, and operational simplicity offers a low-cost significant advancement over traditional pillbox designs.

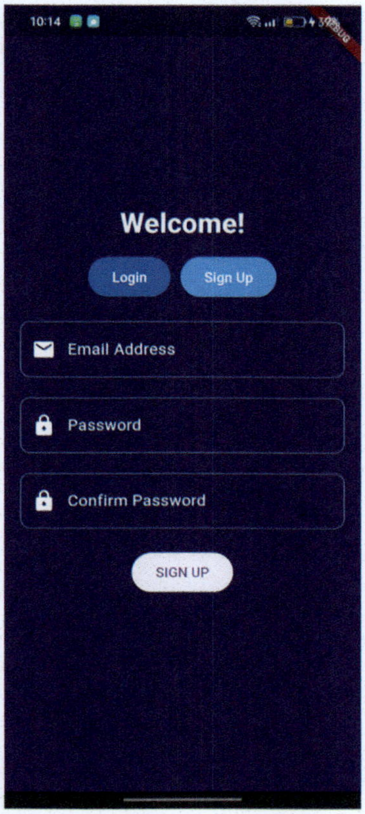

FIGURE 28.19 Authentication page: image of authentication page.

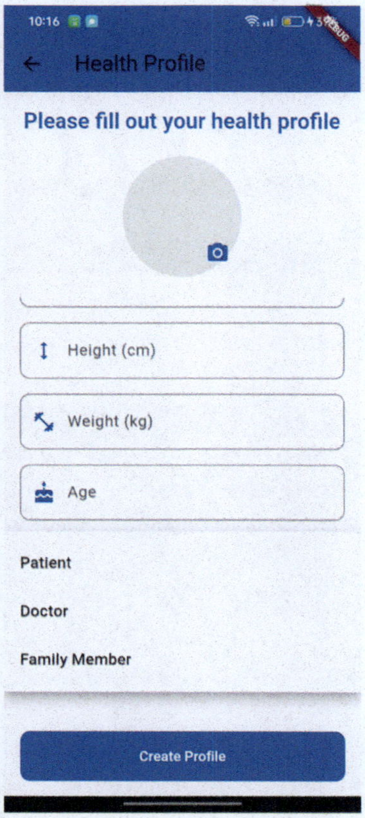

FIGURE 28.20 Profile form: image of profile form.

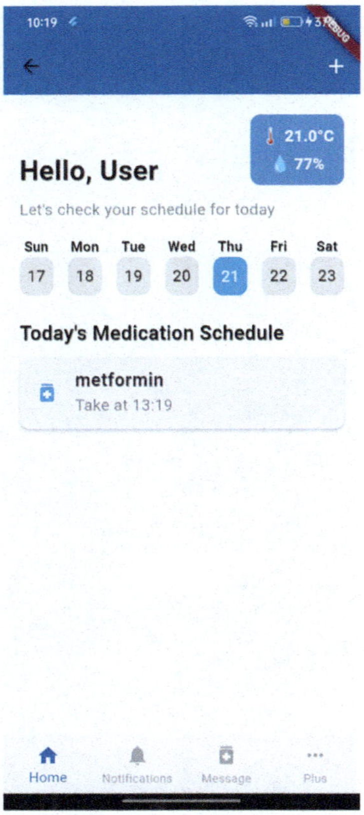

FIGURE 28.21 Patient's home interface: image of patient's home interface.

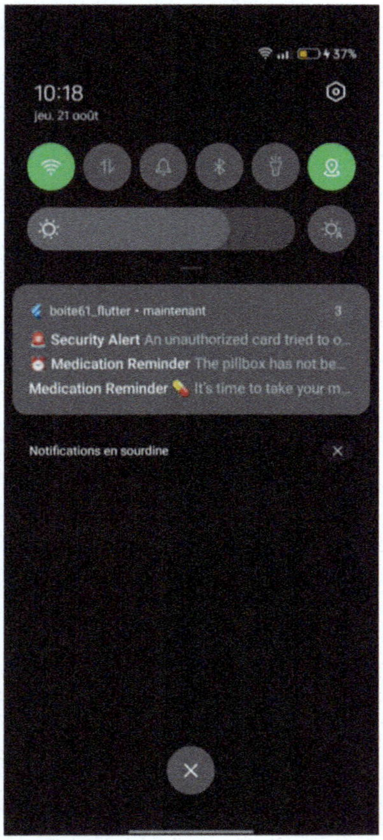

FIGURE 28.22 Medication reminder: image of medication reminder.

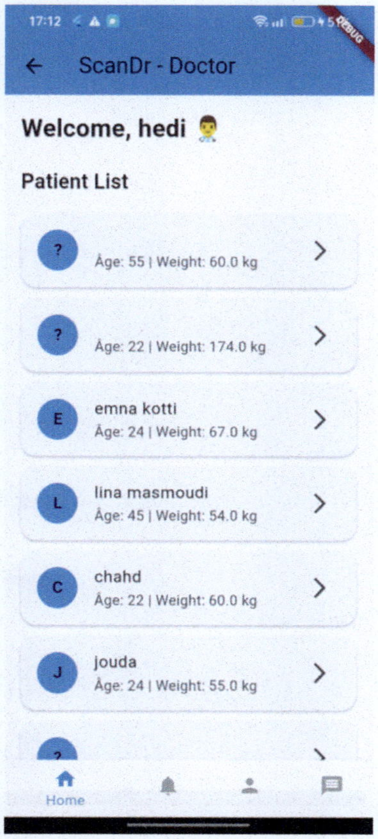

FIGURE 28.23 Doctor's home interface: image of doctor's home interface.

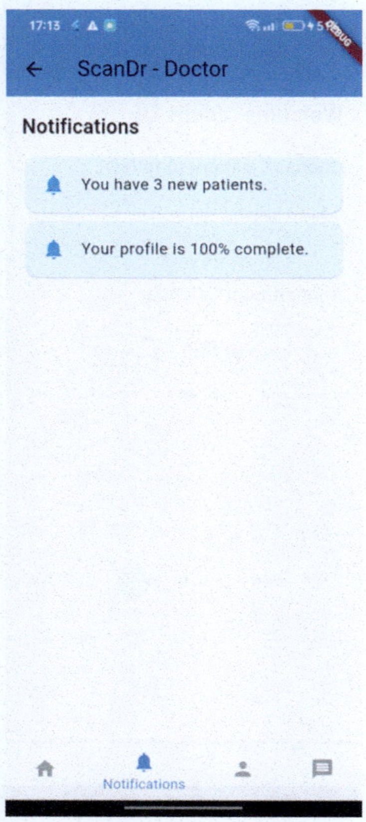

FIGURE 28.24 Doctor's notification interface: image of doctor's notification interface.

FIGURE 28.25 Doctor's messages interface: image of doctor's messages interface.

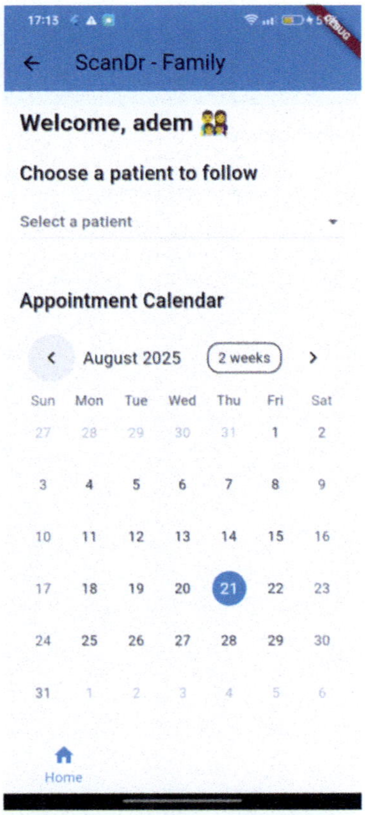

FIGURE 28.26 Family member's home interface: image of family member's home interface.

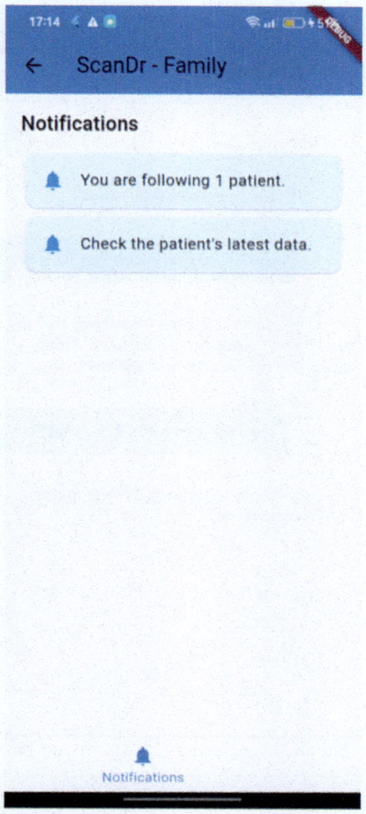

FIGURE 28.27 Family member's notification interface: image family member's notification interface.

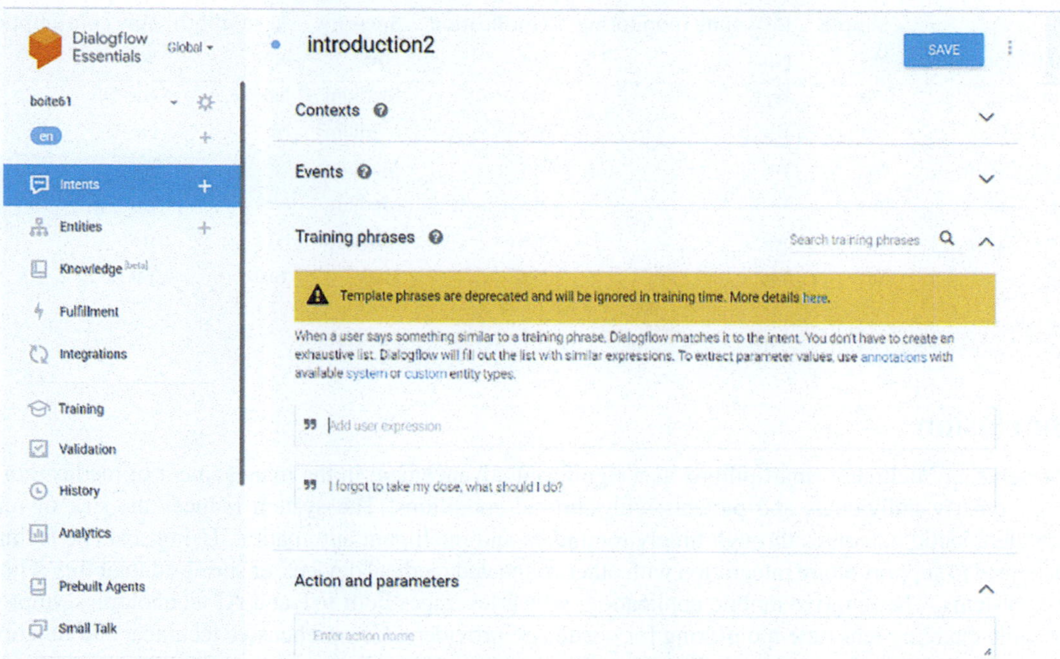

FIGURE 28.28 Dialogflow's training process: image of Dialogflow's training process.

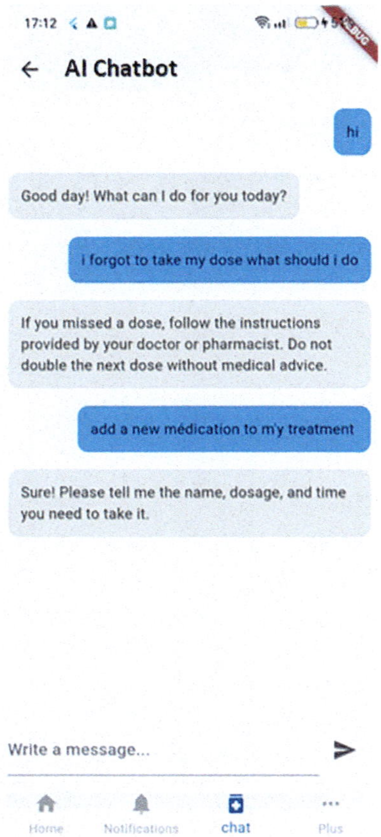

FIGURE 28.29 Patient chatbot's interface: image of patient chatbot's interface.

TABLE 28.1 Comparison with recent pillbox systems.

	Simple	Real-time monitoring	Notification	Security	AI chatbot	Low consumption	Cost
Nedelcea and Albu (2020)	Yes	No	Yes	Low	No	Yes	Low
Selvabharathi et al. (2022)	Yes	Yes	Yes	Medium	No	Yes	Low
Nasir et al. (2023)	No	Yes	Yes	Medium	No	No	High
Siddiqha et al. (2024)	No	Yes	Yes	Medium	No	Yes	Low
Daoud et al. (2023)	Yes	Yes	Yes	Medium	No	Yes	Low
Hamim et al., (2023)	Yes	Yes	Yes	Medium	No	Yes	Low
This work	**Yes**	**Yes**	**Yes**	**High**	**Yes**	**Yes**	**Low**

AI, artificial intelligence.

28.5 Conclusion

This work presents an intelligent smart pillbox as a significant advancement in the management of medication adherence, particularly for elderly individuals and patients with chronic conditions. The system reduces the risk of missed doses and enhances therapeutic outcomes through timely reminders and intelligent automation. Using scalable technologies the solution is designed to support future integration with other connected medical devices, ensuring adaptability within evolving healthcare ecosystems. The intuitive mobile application, which leverages both IoT and AI technologies, improves patient engagement while enabling real-time monitoring for healthcare providers. Comprehensive technical and functional evaluations have demonstrated the robustness and reliability of the system, positioning it as a good candidate for broader adoption in smart healthcare environments. This prototype can be further enhanced by including AI algorithms to dynamically adapt medication schedules based on user behavior, daily routines, and medical history. Besides the integration of blockchain technology into smart pillbox systems presents a transformative approach to ensure the integrity, transparency, and security of adherence data.

References

Abiodun, O. J., & Okpe, O. A. (2024). Smart home security using Arduino-based Internet of Things (IoTs) intrusion detection system. *World Journal of Advanced Research and Reviews, 22*(3), 857–864. doi:10.30574/wjarr.2024.22.3.2000.

Anjaneyulu, S. S. K. S. R., Anjaneyulu, S. S. R., & Anjaneyulu, S. S. R. (2023). Internet of Things (IoT)-based smart healthcare system for efficient diagnostics of health parameters of patients in emergency care. *IEEE Access, 1*, 12345–12356.

Antoun, W., Abdo, A., Al-Yaman, S., Kassem, A., Hamad, M., & El-Moucary, C. (2018). Smart medicine dispenser (SMD). In *Middle East Conference on Biomedical Engineering, MECBME* (pp. 20–23). IEEE Computer Society. http://ieeexplore.ieee.org/xpl/conhome.jsp?punumber=1800336. doi:10.1109/MECBME.2018.8402399.

Daoud, H., Daoud, M., Jardak, N., & Albouzidi, C. (2023). Design and implementation of smart pill box for healthcare applications. In *20th International Multi-Conference on Systems, Signals and Devices, SSD 2023* (pp. 1046–1050). Institute of Electrical and Electronics Engineers Inc. http://ieeexplore.ieee.org/xpl/mostRecentIssue.jsp?punumber=10411147. doi:10.1109/SSD58187.2023.10411150.

Gupta, S. K., Karras, D. A., & Natarajan, R. (2024). Revolutionizing healthcare: AI integration with IoT for enhanced patient outcomes. *Information Systems Engineering and Management*. doi:10.1007/978-3-031-65022-2.

Hamim, M. R. K., Ehsan, M. T., & Ali, M. H. (2023). In *Design and implementation of an IoT-based smart pillbox for improving medical adherence: 2023* (pp. 26–30). Kazakhstan: Kaskelen. https://doi.org/10.1109/ICECCO58239.2023.10147144.

Karagiannis, D., Mitsis, K., & Nikita, K. S. (2022). Development of a low-power IoMT portable pillbox for medication adherence improvement and remote treatment adjustment. *Sensors, 22*(15). http://www.mdpi.com/journal/sensors. doi:10.3390/s22155818.

Kokane, K., Varale, R. K., & Hegde, S. S. (2024). NLP Chatbot for order assistance using Dialogflow. *Journal of Electrical Systems, 2*, 1–9.

Milojković, K., Zivković, M., & Bačanin (2024). Agile multi-user Android application development with firebase: Authentication, authorization, and profile management. In *Proceedings of the 2024 International Conference on Mobile Computing and Networking* (pp. 123–130).

Nasir, Z., Asif, A., Nawaz, M., & Ali, M. (2023). Design of a smart medical box for automatic pill dispensing and health monitoring. The 2nd International Conference on Emerging Trends in Electronic and Telecommunication Engineering, Karachi, Pakistan, 15–16 March 2023. *Engineering Proceedings, 32*, 7.

Nedelcea, M. L., & Albu, A. (2020). Warning solution for medication administration. In *SACI 2020—IEEE 14th International Symposium on Applied Computational Intelligence and Informatics, Proceedings* (pp. 95–98). Institute of Electrical and Electronics Engineers Inc. http://ieeexplore.ieee.org/xpl/mostRecentIssue.jsp?punumber=9115579. doi:10.1109/SACI49304.2020.9118842.

Oliveira, E. (2023). *UML-based software product line engineering with SMarty* (pp. 1–509). Springer International Publishing. https://link.springer.com/book/10.1007/978-3-031-18556-4. doi:10.1007/978-3-031-18556-4.

Kuria, P. K., Robinson, O. O., & Gabriel, M. M. (2020). Monitoring temperature and humidity using Arduino Nano and module-DHT11 sensor with real time DS3231 data logger and LCD display. *International Journal of Engineering Research & Technology (IJERT), 9*(12), 416–422.

Oner, V. O. (2023). Developing IoT projects with ESP32. *Unlock the full potential of ESP32 in IoT development to create production-grade smart devices* (2nd ed.). Birmingham, England: Packt Publishing.

Rao, S. S., Fernandes, S. L., Singh, C., Gatti, R. R., Harisha, A., & Gatty, R. R. (2023). AIoT and big data analytics for smart healthcare applications. *IoT and big data analytics*. doi:10.2174/97898151960541230501.

Runsewe, O., Osundare, O. S., Folorunsho, S. O., & Akwawa, L. A. (2024). Innovations in android mobile computing: A review of best practices and emerging technologies. *World Journal of Advanced Research and Reviews, 23*(2), 2687–2697. doi:10.30574/wjarr.2024.23.2.2634.

Saran, S. P., Nishanth, K. T., Sathiya, S. P., & Suthanthira, V. N. (2025). Integration of embedded and IoT for smart pill dispenser in healthcare. *i-manager's Journal on Embedded Systems, 13*(2), 37. doi:10.26634/jes.13.2.21589.

Selvabharathi, A., & Brundha, J. S. (2022). Design of IoT based smart pill box for healthcare applications. *International Journal of Electrical Engineering and Technology (IJEET), 1*, 61–66.

Shaik, T., Tao, X., Higgins, N., Li, L., Gururajan, R., Zhou, X., & Acharya, U. R. (2023). Remote patient monitoring using artificial intelligence: Current state, applications, and challenges. *WIREs Data Mining and Knowledge Discovery, 13*(2). doi:10.1002/widm.1485.

Siddiqha, S., Nadig, N., Kruthi, K. V. P., Jain, A. N. M., & Divya, D. R. (2024). IoT based pill reminder and monitoring system. *International Journal of Research in Applied Science & Engineering Technology (IJRASET), 12*.

Sultan, A., Mahamudul, H., Gouse Pasha, M., Mohammad, S., Tasnia, T., & Mustafa Wasif, A. (2020). IoT Based Pill Reminder and Monitoring System. *International Journal of Computer Science and Network Security (IJCSNS), 20*(7), 152–158.

Webster, C. S., Scheeren, T. W. L., & Wan, Y. I. (2022). Patient monitoring, wearable devices, and the healthcare information ecosystem. *British Journal of Anaesthesia, 128*(5), 756–758. https://www.journals.elsevier.com/british-journal-of-anaesthesia. doi:10.1016/j.bja.2022.02.034.

Chapter 29

Toward sustainable and reliable hybrid vehicles: Fault diagnosis of permanent magnet motors using FEM and advanced analysis

Naourez Benhadj[a], **Houda Daoud**[b], **Manel Krichen**[c] **and Rafik Neji**[a]

[a] *Department of Electrical Engineering, National Engineering School of Sfax, Information Technologies and electronics Laboratory, Sfax University, Sfax, Tunisia,* [b] *Department of Engineering of Electronic Communication Systems, National School of Electronics and Telecommunications of Sfax, University of Sfax, Sfax, Tunisia,* [c] *Electrical department, National Engineering School of Sfax, Information Technologies and electronics Laboratory, Sfax University, Sfax, Tunisia*

29.1 Introduction

Throughout the last decades the reliability of automotive applications and the assurance of operational safety have emerged as key concerns within the research community. These challenges are strongly motivated by the increasing energy dependence of modern societies and the enforcement of strict CO_2 emission regulations. Within this framework the central focus of the research was upon transportation problems, mainly hybrid electric vehicles (HEVs). This was achieved through addressing the challenges of electrical vehicles (EVs) control strategies and the traction chain model, as well as optimization and the new technologies for energy storage. Yet the complexity of the drivetrains in HEV makes them vulnerable at the level of availability. Therefore, to be more competitive, the reliability of the electrical drivetrain is considered by researchers as one of the most intrinsic hardship of the HEV. This study addresses the availability challenge of HEV drivetrains through permanent magnet synchronous motor (PMSM) fault diagnosis, in line with the stringent requirements of availability, reliability, and safety in industrial processes. PMSMs are extensively employed in HEVs, serving as traction motors or steering assistance units, because of their numerous advantages such as high torque and power density, compact rotor design, low losses, superior dynamic response, precise control, high reliability, and simplified maintenance (Grunditz & Thiringer, 2016; Ullah & Hur, 2018; Youssef et al., 2025). Resting upon the previously stated benefits, the production of PMSMs encounters some obstacles, such as asymmetry in the air gap between rotor and stator, possible permanent magnet (PM) demagnetization, operating temperature sensitivity, a higher price with respect to the use of costly PMs, and the requisite position sensors for their control.

Within this situation the PMSM needs to develop and evolve to respond to the socioeconomic constraints. In fact a deeper and more comprehensive study of PMSM behavior in extreme environment, or under magnetic, electrical, and mechanical fault (MEF) conditions is much more intrinsic than ever (temperatures, wide speed ranges, vibrations, demagnetization, voltage fluctuation, and eccentricity). Initially an overview of faults diagnosis methods in PMSM is reported. Resting upon the different diagnosis methods elaborated in state-of-the-art works, fast Fourier transform (FFT) has been chosen as a convenient technique in terms of stationary state operation. Section 29.3 introduces the suggested methodology of surface-mounted PMSM (SPMSM) analytical modeling, analyzed by applying finite element modeling (FEM). In addition, demagnetization and eccentricity faults have been checked. Analytical and simulation models have been widely proposed to characterize faults in SPMSMs. Section 29.4 displays the signal processing (SP) analysis for SPMSM demagnetization and eccentricity faults. This permits more significant and larger diagnosis knowledge about SPMSM faults. Section 29.5 provides outstanding concluding remarks and offers new perspectives for future work.

Computational Intelligence in Mechatronics: Solving Real-World Problems in Electronic Systems Design. DOI: https://doi.org/10.1016/B978-0-443-27580-7.00006-8

29.2 Related works

Currently, both automotive industrials and researchers have been motivated to select traction motors for HEVs to find the most convenient electric propulsion system. By scrutinizing electrical traction chain, it is noteworthy that works on cage induction motor (CIM) and PMSM are highly growing, whereas those on switched reluctance motors (SRMs) are drawing considerable interest. On the other side, those on DC motors are decreasing. The PMSM is beneficial with regard to efficiency and power density; however, the high cost of magnets stands as a deficiency. Besides, as the control of induction motor (IM) is more intricate and the torque ripple of SRM is more conspicuous, PMSM is regarded as a relevant solution for EV motors. Using PMSM displays a good reliability in HEVs. Yet, some shortcomings are overcome under continuous speed and load changes and under conditions of continuous operation. Faults in PMSMs branch out into magnetic faults (MAFs), electrical faults (ELFs), and MEF. IEEE-IGA and Electric Power Research Institute (EPRI) carried out a report on the main component faults of motors (Yao & Mahesh, 2010).

MEF involves essentially shaft bending, magnet damage, air gap eccentricity, bolt loosening, and bearing faults. The last fault yields about 40%–50% of motor faults (Hang et al., 2016; Kim, 2011). Faults of eccentricity involve static eccentricity (SE), mixed eccentricity (ME), and dynamic eccentricity (DE). Faults in eccentricity can trigger noise, supplementary vibration, and ripple torque. Although eccentricity is significant, it can harm the stator or rotor core, affecting the motor's normal operation (Ebrahimi et al., 2014; Karami et al., 2014). PMs can be demagnetized, referring to such factors as magnet harm, high temperature, and faults in stator.

Basically, there are three motor fault diagnosis techniques. Usually, signal-based fault diagnosis methods (SFDM) are invested to remove and process from the motor signals a wide range of characteristics. These characteristics are subsequently identified manually using either knowledge or experience-based approaches. Model-based fault diagnosis methods (MFDM) estimate signal behavior by comparing the predicted motor response from a mathematical model with the actual measured output. The data-driven intelligent diagnosis methods (DDIDM) can automatically assess the fault state grounded on expert knowledge. However, in view of work in machine learning and artificial intelligence, data-driven smart diagnostic systems, which do not rely on previous knowledge, DDIDM have in recent times displayed wide application perspectives (Chen et al., 2019; Chowdhury, 2016; Elbouchikhi et al., 2017; Jokic et al., 2018; Ullah & Hur, 2018). For demagnetization, short-time Fourier transform (STFT) features combined with lightweight learners, such as K-Nearest Neighbors (KNN) and multilayer perceptron (MLP), have demonstrated reliable performance on experimental setups (Yu et al., 2023). For eccentricity, data-driven approaches increasingly exploit richer signal sources: branch-current signatures have shown superior sensitivity compared with phase currents, enabling joint eccentricity–demagnetization diagnosis with accuracies above 98% (Lin et al., 2025), whereas stray-flux mapping with mobile sensors provides noninvasive eccentricity detection validated through experiments (Chen et al., 2024).

It is worth noting that time–frequency analysis, frequency-domain, and time-domain approaches all fall under the category of SP methods. Among these the FFT remains the classical technique for frequency-domain analysis. The signal stands for certain sine or cosine functions' superposition. This indicates clearly the frequency distribution of the signal, where the frequency and amplitude of harmonic components can be invested as the characteristics of faults (Faiz & Exiri, 2016; Lin et al., 2018; Mortezaei et al., 2024). MEF triggers distortions in the flux distribution within the motor, which yields, as a matter of fact, harmonics in the stator current (Ojaghi & Bahari, 2017; Rosero et al., 2009a, 2009b). Yet the distributed magnetomotive force (MMF) will not be sinusoidal in case of demagnetization fault (DF). Therefore the failed and normal portions will generate current with multiple frequencies. This implies that low frequency components will emerge near the fundamental wave in current signal (Alameh et al., 2015; Melecio et al., 2019). In this context an elaborated method grounded on FFT is employed, which is STFT (Ilioudis, 2020), fulfilling time–frequency analysis (Chen et al., 2019; Rosero et al., 2009a). Continuous wavelet transform (CWT) and discrete wavelet transform (DWT) are categories of wavelet transform (WT) (Chen et al., 2019; Rosero et al., 2009a) that have been used to tackle MEF and to assess the degree of demagnetization in PMSM by computing the energy in the DWT detail. To identify the features of the PMSM interturn short circuit fault, authors adopted the wavelet packet band energy analysis technique (Yan & Gao, 2006).

The Hilbert–Huang transform (HHT) associates both empirical mode decomposition (EMD) and Hilbert transform. This time–frequency analysis technique settled the deficiencies of WT and STFT. In addition, HHT was employed to specify the DF, and the findings demonstrated that there were still convenient results (Chen et al., 2019). The joint time and frequency resolution offered by the Wigner–Ville distribution (WVD) proves to be better than other time–frequency analysis techniques (Haddad et al., 2016). Certain scientists have associated WVD with EMD to analyze the current by handling the demodulated internal mode functions (IMFs) with WVD (Fitouri et al., 2016). Moreover, recent contributions emphasize motor current signature analysis using analytically derived sideband maps (Zhang et al., 2024) and stray-flux mapping with mobile sensors (Chen et al., 2024), both validated experimentally, enabling noninvasive detection of mixed and static eccentricities. Beyond electrical signals, alternative sensing strategies, including acoustic image classification using deep transformers, have been introduced as low-cost complementary approaches (Sharma et al., 2025).

Data acquisition is an intrinsic step as far as fault diagnosis is concerned. In Fitouri et al. (2016), temperature measurement is carried out so as to monitor bearing faults. In Torregrossa et al. (2012), researchers manipulated vibration and acoustic emissions in terms of eccentricity defect and partial demagnetization (PD) (Alameh et al., 2015; Lin et al., 2018; Yu et al., 2014; Zhu et al., 2017). As a matter of fact, torque monitoring was invested in to trace PMSM faults [demagnetization (Zhu et al., 2019), short circuit fault (Fitouri et al., 2016), and SE fault]. In Yao and Mahesh (2010) the measured flux was also used to diagnose SE in PMSM. Although these techniques are significant, their implementation in real-world conditions remains challenging, as the required sensors increase both the cost and the complexity of the monitoring system. Multiple methods for detecting PM demagnetization are applied in research works. Urresty et al. (2009) examined the stator current of a PM motor through the use of HHT, a method based on back-electromotive force (EMF), zero-sequence voltage component, and box-counting fractal dimension. Pietrzak et al. estimated the stator current after chipping off a part of the magnet (Pietrzak & Wolkiewicz, 2023). In Liu and Zhu (2014) the magnetization can be measured from the back EMF. In addition, to specify the magnet temperature generated by the demagnetization, one potential way is an accurate flux observer (Ishikawa et al., 2017). Furthermore, applying the invasive method, it is possible to estimate the effect of the flux variation on the motor's magnetic circuit. Moreover, Barmpatza and Kappatou (2019) studied the spectra of stator current and the EMF for the faulty cases of axial flux permanent magnet (AFPM) synchronous generator prototype. In Zhu et al. (2018), tracing uniform permanent-magnet demagnetization rests upon acoustic noises so as to build up a reliable PMSM controller. The developed PM demagnetization detection approach was experimentally assessed.

Investigating conditional maintenance techniques requires a noninvasive, less expensive, and more efficient approach to condition monitoring. Among the most promising techniques is the one resting on current or EMF monitoring for control and protection objectives. From this perspective, certain research activities have centered around current-based fault detection. In PMSM (Gherabi et al., 2019; Quintal-Palomo, 2019; Toumi et al., 2019), measurements of voltage were also explored (Zhang et al., 2021). Song et al. (2020b) addressed the local DF of permanent magnet synchronous linear motor (PMSLM). Experimental findings corroborate that this approach can detect accurately the PMSLM faults (Song et al., 2020a) (with 100% recognition rate). In Song et al. (2020b), an approach based on time–time transform (TT) and extreme learning machine (ELM) for DF recognition and classification in double-side PMSLM (DPMSLM) prototype was introduced. In addition, several studies have shown the effectiveness of time–frequency features combined with lightweight classifiers, such as STFT-based KNN/MLP and wavelet–Principal Component Analysis (PCA)–Support Vector Machine (SVM) pipelines, achieving accuracies above 90% on experimental testbenches (Mohanty et al., 2024; Pate, 2024; Yu et al., 2023). Optimization-enhanced models, such as PCA–improved sparrow search algorithm (ISSA)–probabilistic neural network (PNN), further improved classification performance to nearly 96% (Liu et al., 2024), whereas branch-current analysis offered even higher diagnostic sensitivity, with reported accuracies exceeding 98% (Lin et al., 2025). Physics-guided harmonic selection has also been proposed to enhance the robustness of current-based indicators across varying loads (Martinez et al., 2024). A method in Haddad et al. (2016), grounded upon applying direction of the shift in the commanded d and q axis voltages, is foregrounded to detect and separate PMSM faults: SE, turn-to-turn short circuit, and demagnetization. Experimental tests were conducted and compared to finites elements approach (FEA) for the same machine under two faults (turn-to-turn short circuit as well as demagnetization).

29.3 SPMSM modeling and analysis

29.3.1 Methodological framework

The proposed methodology, illustrated in the subsequent Fig. 29.1, is designed to analyze the behavior of the SPMSM under both healthy and faulty operating conditions. It relies on a three-step framework that integrates FEM with FEMM and advanced processing and simulation capabilities in MATLAB, thereby enabling a comprehensive investigation of machine performance and fault characteristics (Fig. 29.1).

Initially the first step involves sizing the geometric parameters of the SPMSM after modeling analytically the torque needs as well as electromagnetics operations (this stands for an inverse backward approach: starting with the required vehicle speed, the calculation is carried out from the wheels to the powertrain). The set forward 2D finite element model is elaborated to determine eccentricity or demagnetization phenomenon. Simulation results provide key data for a deeper and fuller understanding of the effects of these faults on the SPMSM's basic parameters. In fact, FEM is conducted to investigate and estimate the electromagnetic quantities of the motor so as to validate an analytical model. Step 2 is governed by three phases: Preprocessing, by MATLAB script, which involves granting the needed information for the SPMSM geometry

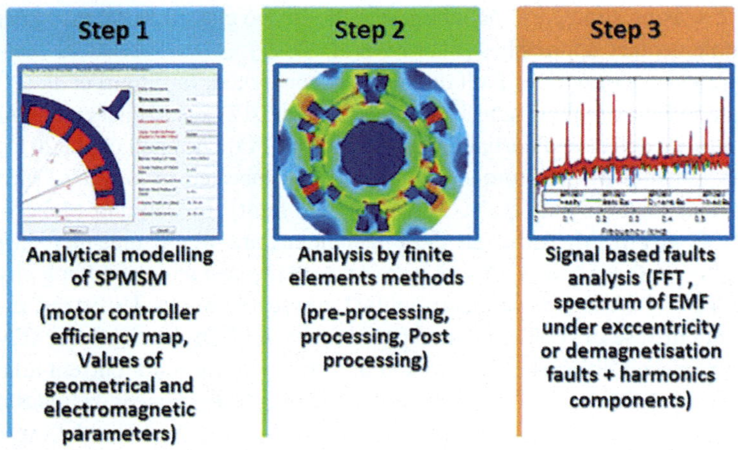

FIGURE 29.1 SPMSM modeling and analysis schematic presentation. *EMF*, Electromotive force; *FFT*, fast Fourier transform; *SPMSM*, surface-mounted permanent magnet synchronous motor.

TABLE 29.1 Features of the SPMSM.

Parameters and values		Parameters and values	
Magnet height (mm)	7.76	Stator exterior diameter (mm)	174
Core stack length (mm)	250	Poles number	8
rotor exterior diameter (mm)	82	Terminal current (A)	59.5
Rotational speed (rpm)	2824	Slots number	12
Stator interior diameter (mm)	102	Supply frequency (Hz)	188.3

SPMSM, surface-mounted permanent magnet synchronous motor.

representation. Processing using FEMM software consists in settling the magnetics problem. Postprocessing permits the calculation of multiple quantities, including torque, inductance, flux linkage, and currents. Notably, EMF spectrum can identify eccentricity or magnets' DFs.

The SPMSM is simulated and scrutinized under healthy and fault conditions to obtain an appropriate data set to establish an efficient fault detection method. In Step 3, the obtained flux linkage, EMFs and stator currents are investigated with demagnetization or eccentricity faults. The power spectral density (PSD) of the EMFs is calculated using FFT. Afterward the EMFs spectrum is assessed and compared to accomplish a reliable procedure that would allow the detection of the different fault cases. A novel frequency pattern in eccentricity or DF cases is incorporated, and the amplitude of sideband components (ASBCs) is invested to identify different fault cases.

The analyzed SPMSM in this research exhibits a power of 62 kW, a concentrated winding, six main teeth, and four pairs of poles. To increase the wave shape and to reduce the leakage risk, an interposed tooth is placed between the main teeth. In this research the height of the PM is 7.76 mm, and the air gap is 2 mm. The characteristics of the SPMSM are illustrated in Table 29.1. The characteristics of used PMs correspond to the relative permeability $\mu_r = 1.044$, coercivity $H_c = -895$ kA/m, and the remanence $B_r = 1.16\ T$. The field lines distribution is outlined in Fig. 29.2. The suggested SPMSM was studied at a first stage under eccentricity fault and then at a later stage under DF (Table 29.1) (Fig. 29.2).

29.3.2 SPMSM analytical computation

For the SPMSM the torque T_m, the angular speed W_m, and the power P_m are indicated by the three following equations (Abdelmoula et al., 2017; Ben Hadj et al., 2018), where the vehicle propulsion force F is expressed in terms of:

$$T_m = \frac{FR_{\mathrm{wh}}}{k_r} \tag{29.1}$$

$$W_m = \frac{V}{R_{\mathrm{wh}}}k_r \tag{29.2}$$

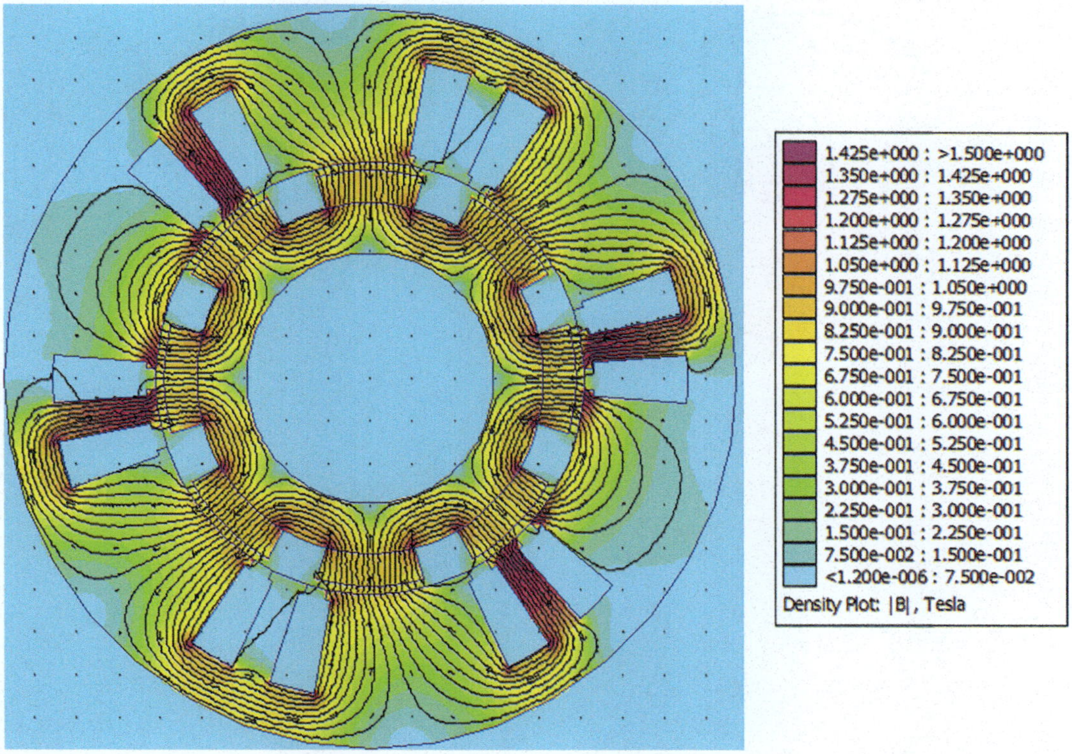

FIGURE 29.2 Surface PMSM field lines distribution for a rotor position. *PMSM*, Permanent magnet synchronous motor.

$$P_m = T_m W_m \tag{29.3}$$

$$F = mgC_{\mathrm{rr}} + \frac{1}{2}\rho C_D A_f V^2 + ma + mg\sin(\theta) \tag{29.4}$$

$$L_{\mathrm{enc}} = \frac{D_m + e + H_d}{2} A_{\mathrm{enc}} \tag{29.5}$$

$$S_{\mathrm{di}} = \frac{D_m + e}{2} A_{\mathrm{denti}} L_m \tag{29.6}$$

$$S_d = \frac{D_m + e}{2} A_{\mathrm{dent}} L_m \tag{29.7}$$

The slot average width L_{enc} is denoted by Eq. (29.5). The integrated tooth section S_{di} and the main tooth section S_d are identified by Eqs. (29.6) and (29.7), respectively.

The slot section S_e is expressed by

$$S_e = \frac{1}{2}\left[\frac{2\pi}{N_d} - A_{\mathrm{dent}} - A_{\mathrm{denti}}\right]\frac{D_m + e}{2} L_m \tag{29.8}$$

$$H_a = \frac{\mu_a B_e e}{M(T_a) - \frac{B_e}{K_{\mathrm{fu}}}} \tag{29.9}$$

By applying the Ampere theorem, the magnet height H_a is obtained through

$$M(T_a) = M[1 + \alpha_m(T_a - 20)] \tag{29.10}$$

$$H_{\mathrm{cr}} = \frac{B_e S_d}{2\mathrm{B}_{\mathrm{cs}} L_m K_{\mathrm{fu}}} \tag{29.11}$$

At T_a°C the magnet remanent induction $M(T_a)$ is denoted by Eq. (29.10). The rotor yoke thickness H_{cr} is indicated by Eq. (29.11).

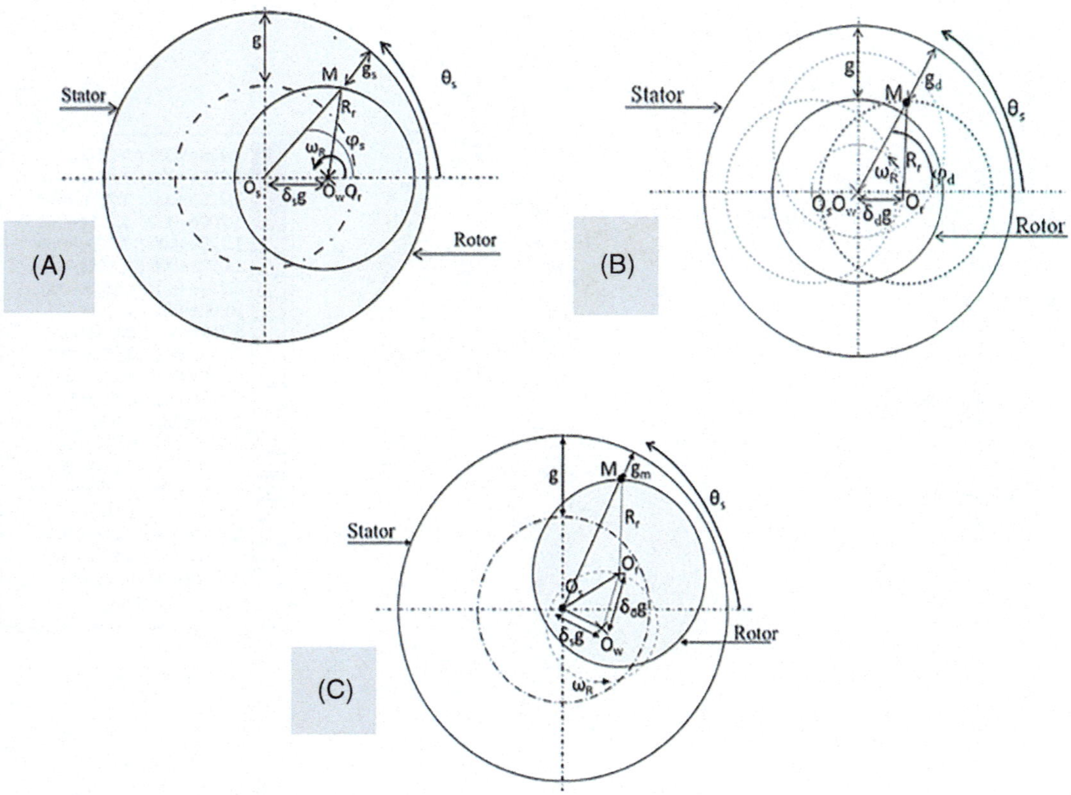

FIGURE 29.3 PMSM geometry (A), static eccentricity fault, (B) dynamic eccentricity fault, and (C) mixed eccentricity fault. *PMSM*, Permanent magnet synchronous motor.

29.3.3 FEA and investigation for SPMSM eccentricity faults

In a healthy motor the stator geometric center coincides with the rotor rotation center. In the case of SE ($O_r = O_w$) the SE ratio δs and the air gap width g_s are defined by Eqs. (29.12) and (29.13), respectively. Fig. 29.3A depicts the distance between an arbitrary point M on the rotor surface and the stator center, providing a geometric reference essential for analyzing rotor–stator interactions under both healthy and faulty conditions. In addition, it reveals that θ_s, which corresponds to the stator reference angle, is equal to φ_s. In dynamin excentricity (Krichen et al., 2019), ($O_s = O_w$), Eqs. (29.14) and (29.15) indicate, respectively, the DE ratio δ_d and the air gap width g_d. The PMSM geometry with DE is depicted in Fig. 29.3B. Besides, departing from the reference point of the stator frame, it is quite evident that the air gap width depends largely on pole pairs number p as well as electrical rotational speed ω_S (at a fixed stator angle) (Fig. 29.3).

$$\delta_s = \frac{|O_s O_w|}{g} \tag{29.12}$$

$$g_s(\theta) = g(1 - \delta_s \cos(\theta)) \tag{29.13}$$

$$\delta_d = \frac{|O_r O_w|}{g} \tag{29.14}$$

$$g_d = g\left(1 - \delta_d \cos\left(\frac{\omega_s}{p}t - \theta_s\right)\right) \tag{29.15}$$

The stator center is different from the rotor center and the rotation center in ME. Eqs. (29.16) and (29.17) indicate, respectively, the ME ratio δ_m and the air gap width g_m. Under ME, it is suggestive that the air gap broadness changes on the rotor's mechanical angle. Fig. 29.3C portrays the motor geometry with ME. The flux linkage displayed in Fig. 29.4 refers basically to four cases. They correspond to the case under 40% SE, the case under 40% DE, the case 40% ME (40% SE and 40% DE), and the health case. Nonuniform magnetic flux density arises around the rotor under various eccentricity faults, causing flux perturbations. In contrast, for a healthy SPMSM, SE has a minimal impact, with flux amplitude variations of approximately 0.69%. Moreover, for DE, the flux amplitude exhibits a significant variation of about 5.74% compared to healthy SPMSM, underscoring the pronounced impact of DE on the machine's magnetic behavior. The flux amplitude varies

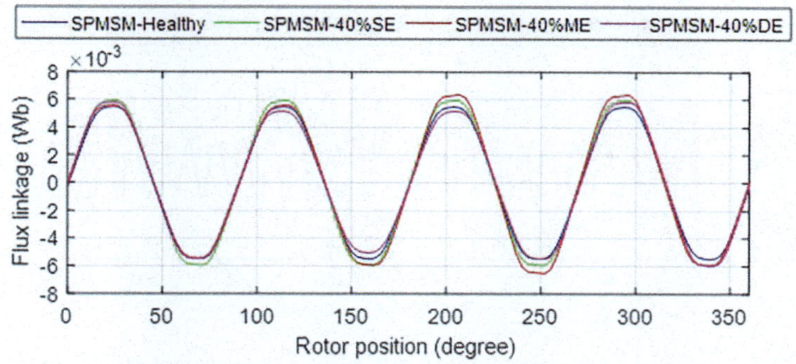

FIGURE 29.4 Flux linkage for healthy and faulty SPMSM. *DE*, Dynamic eccentricity; *ME*, mixed eccentricity; *SE*, static eccentricity; *SPMSM*, surface-mounted permanent magnet synchronous motor.

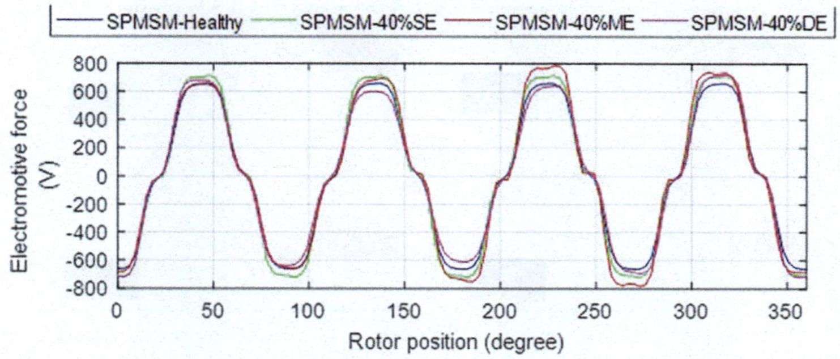

FIGURE 29.5 Electromotive force for healthy and faulty SPMSM. *DE*, Dynamic eccentricity; *ME*, mixed eccentricity; *SE*, static eccentricity; *SPMSM*, surface-mounted permanent magnet synchronous motor.

more significantly depending upon the one used in the case of static and DE. (Compared to the healthy case, amplitude flux variation is equal to 7.19%) (Figs. 29.4 and 29.5).

$$\delta_m = \delta_s + \delta_d = \sqrt{\delta_s^{\,2} + \delta_d^{\,2} + 2\delta_s\delta_d\cos(\theta)} \tag{29.16}$$

$$g_m(t) = R_s - \delta_m g\cos\left(\frac{\omega_s t}{p} - \varphi_m\right) - \sqrt{R_r^{\,2} - \delta_m^{\,2} g^2 \sin^2\left(\frac{\omega_s t}{p} - \varphi_m\right)} \tag{29.17}$$

$$E_{\text{ph},i} = -N\frac{d\varphi_{\text{phase},i}}{dt} \tag{29.18}$$

$$C_{\text{em}}(t) = \frac{E_{\text{ph1}}(t) \times I_{\text{ph1}}(t) + E_{\text{ph2}}(t) \times I_{\text{ph2}}(t) + E_{\text{ph3}}(t) \times I_{\text{ph3}}(t)}{\Omega(t)} \tag{29.19}$$

The per-phase back EMFs are computed with all coils connected in parallel. The back EMF $E_{\text{ph},i}$ in *i*th coil (Rosero et al., 2009a) is defined in Eq. (29.18) and illustrated in Fig. 29.5. The EMF magnitude varies with the air gap width, with more pronounced changes observed in the presence of ME. The SPMSM electromagnetic torque (EMT) is shown in Fig. 29.6. For the healthy motor the peak-to-peak torque ripple is 15.17%, increasing to 40.48% under ME, 37.27% under DE, and 23.08% under SE. In comparison with the healthy case an effect of eccentricities' faults can be detected in the PMSM torque. This refers basically to the sensitivity of the EMFs and the currents used for evaluating the EMT by Eq. (29.19) (Fig. 29.6).

29.3.4 FEA and investigation for SPMSM demagnetization faults

This section centers around the uniform demagnetization (UD) faults, which are embodied in three fault scenarios, namely, UD_{ArBpm}, UD_{Arpmaa}, and UD_{urBpm} with Bpm is the permanent magnet flux density. Moreover, as far as PD faults are concerned, our central focus is upon three fault scenarios, which are PD_{ArBpm}, PD_{urBpm}, and PD_{Arpmaa}. Fig. 29.7 displays illustrations and abbreviations of the PM DF.

In the four cases UD_{urBpm}, UD_{ArBpm}, PD_{ArBpm}, and PD_{urBpm} the selected magnet material was replaced with a material exhibiting the same relative permeability and conductivity as the healthy magnets but with a reduced permanent magnetic

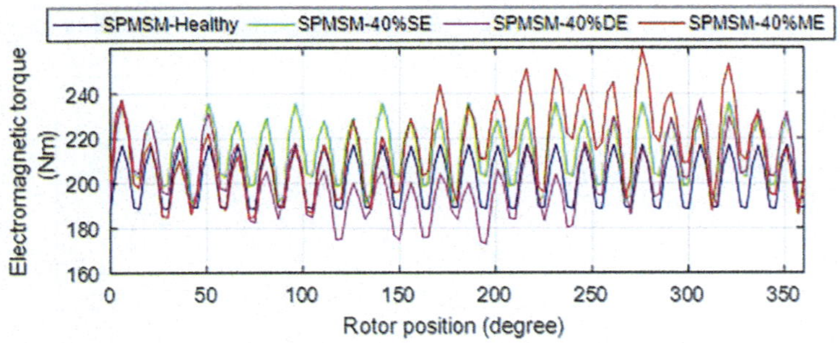

FIGURE 29.6 Electromagnetic torque for healthy and faulty SPMSM. *DE*, Dynamic eccentricity; *ME*, mixed eccentricity; *SE*, static eccentricity; *SPMSM*, surface-mounted permanent magnet synchronous motor.

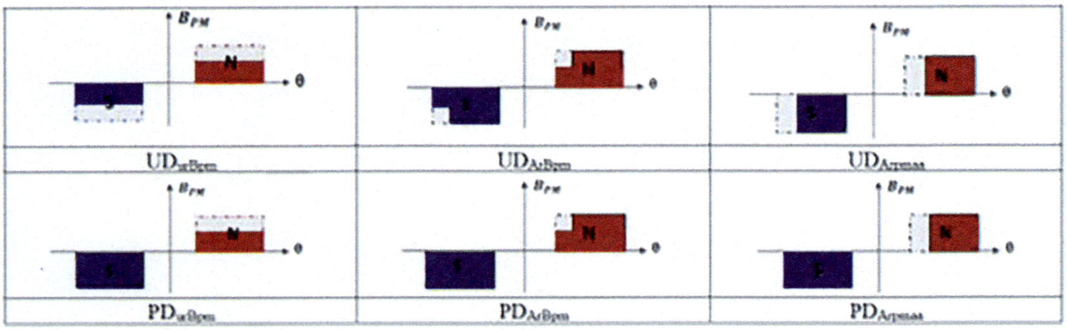

FIGURE 29.7 Illustrations and abbreviations of PM demagnetization fault cases. *PD*, Partial demagnetization; *PM*, permanent magnet; *UD*, uniform demagnetization.

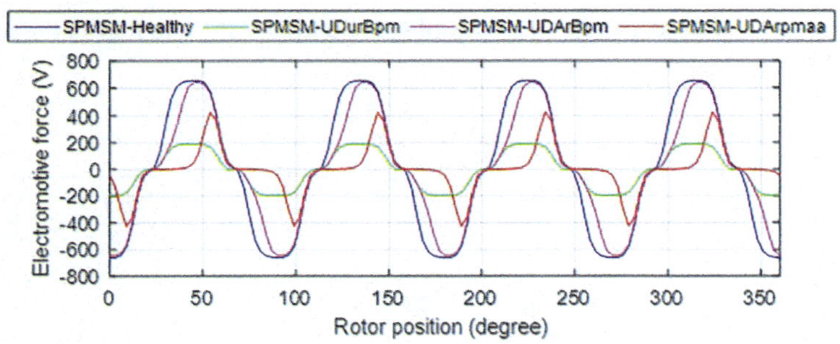

FIGURE 29.8 Electromotive force for healthy and faulty SPMSM. *SPMSM*, Surface-mounted permanent magnet synchronous motor; *UD*, uniform demagnetization.

flux density (B_{PM}). A 75% reduction of the healthy B_{PM} was considered to represent UD and PD, respectively ($M = 0.25$). For PD$_{Arpmaa}$ and UD$_{Arpmaa}$, magnet damage was represented by geometric changes in the PM segments. A 75% reduction of the healthy magnet arc angle was applied to simulate the decrease in arc length for PD and UD cases ($X = 0.25$). As prominent findings of the UD cases with 75% fault severity are compared to healthy cases, Figs. 29.8 and 29.9 highlight the EMFs and the flux versus the rotor position.

For UD$_{urBpm}$ case the demagnetization disrupts the air gap flux waveform. The defective curve displays the same shape but a lower magnitude value compared to the healthy state. In UD$_{ArBpm}$ case the asymmetric reduction of B_{PM} permits an asymmetric reduction of the sinusoids' widths since the fault is distributed in an unbalanced way throughout the magnet. The findings for the case of an asymmetric breakage (UD$_{Arpmaa}$) yield the same shape as the healthy state with lower amplitudes and an asymmetrical decrease of the sinusoids' widths. Furthermore the flux decreases in the UD cases compared to the healthy condition. The reductions are 69.3% for UD$_{UrBpm}$, 19.72% for UD$_{ArBpm}$, and 73.77% for UD$_{Arpmaa}$. In comparison with healthy case the simulation findings of PD$_{urBpm}$ case, presented in Fig. 29.10, display a similar shape but a smaller magnitude value, close to the damaged magnet. In PD$_{ArBpm}$ case the asymmetric reduction of B_{PM} in a single magnet results

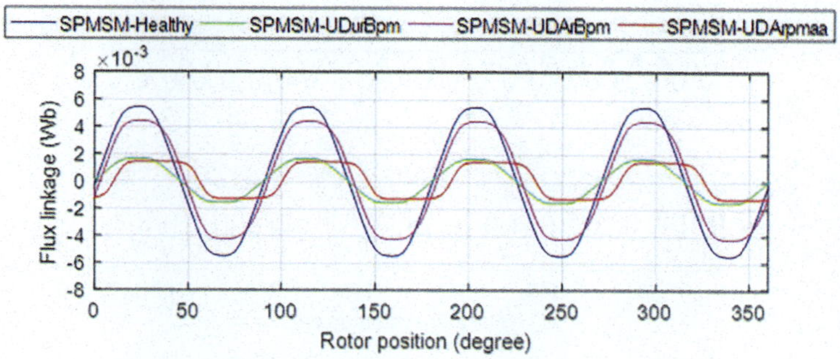

FIGURE 29.9 Flux linkage for healthy and faulty SPMSM. *SPMSM*, Surface-mounted permanent magnet synchronous motor; *UD*, uniform demagnetization.

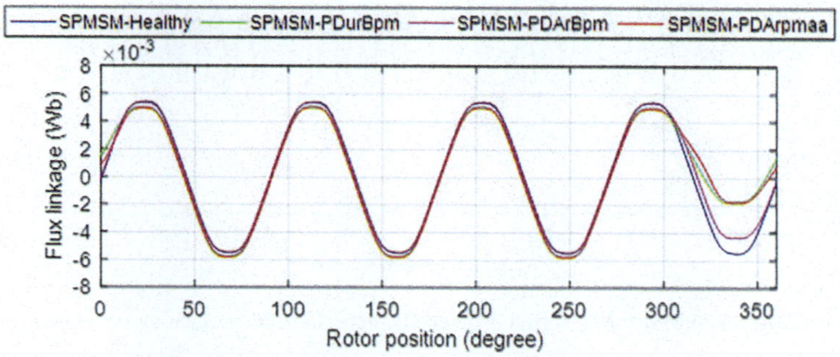

FIGURE 29.10 Flux linkage for healthy and faulty SPMSM. *PD*, Partial demagnetization; *SPMSM*, surface-mounted permanent magnet synchronous motor.

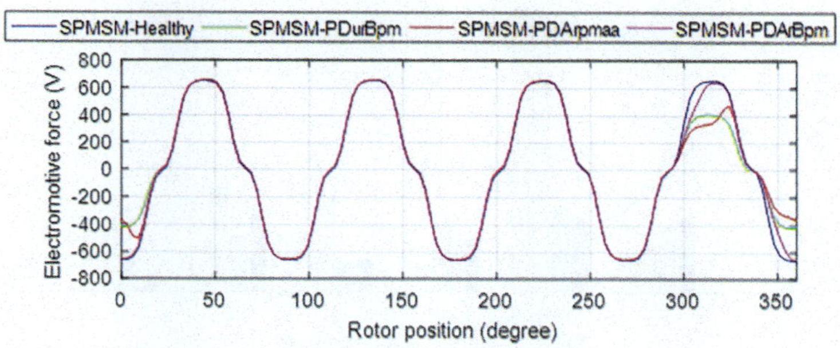

FIGURE 29.11 Electromotive force for healthy and faulty SPMSM. *PD*, Partial demagnetization; *SPMSM*, surface-mounted permanent magnet synchronous motor.

in an asymmetric decrease of the sinusoidal flux width. In the PD_{Arpmaa} case an asymmetric loss of ¾ of the magnet (75% severity) was analyzed. The obtained results are suggestive of the same shape with a lower magnitude value and widths of sinusoids, which asymmetrically decrease close to the position of the rotor where the fault is achieved (Fig. 29.10).

According to Fig. 29.10 the PMs partially demagnetized generate less flux and undermine their sinusoidal form. The EMFs are most weakened in the PD_{urBpm} and PD_{Arpmaa} cases. In these scenarios the magnet arc angle has a greater impact compared to PD_{ArBpm} (Fig. 29.11). In comparison with its value in the healthy case, it is worth nothing that the flux in three cases of PD drops (the variation in PD_{UrBpm} is 08.36%, in PD_{ArBpm} is 02.03%, and in PD_{ARpmaa} is 08.3%) (Fig. 29.11).

Fig. 29.12 shows the EMT as function of rotor position. It demonstrates that UD generates less torque and more undulations than healthy case, mostly in the cases of UD_{Arpmaa} and UD_{urBpm}. Yet, in the cases of PD, Fig. 29.13 reveals that the PD generates less torque than healthy case but more torque than the UD cases (UD_{urBpm} and UD_{Arpmaa}). We deduce that UD_{Arpmaa} and UD_{urBpm} correspond to more severe faults for UD cases, and PD_{Arpmaa} stands for the most severe one in PD cases. The peak-to-peak torque ripple equals 6.84% in the healthy case. It rises to 8.68% in UD_{ArBpm}, to 16% in UD_{urBpm}, and to 97.9% in UD_{Arpmaa} (Figs. 29.12 and 29.13).

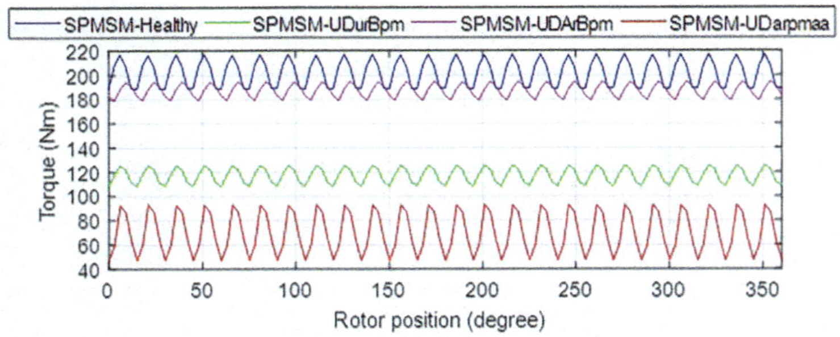

FIGURE 29.12 Electromagnetic torque for SPMSM. *SPMSM*, Surface-mounted permanent magnet synchronous motor; *UD*, uniform demagnetization.

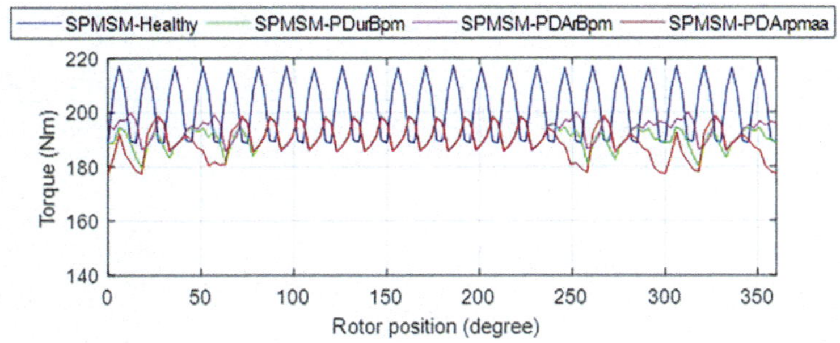

FIGURE 29.13 Electromagnetic torque for SPMSM. *PD*, Partial demagnetization; *SPMSM*, surface-mounted permanent magnet synchronous motor.

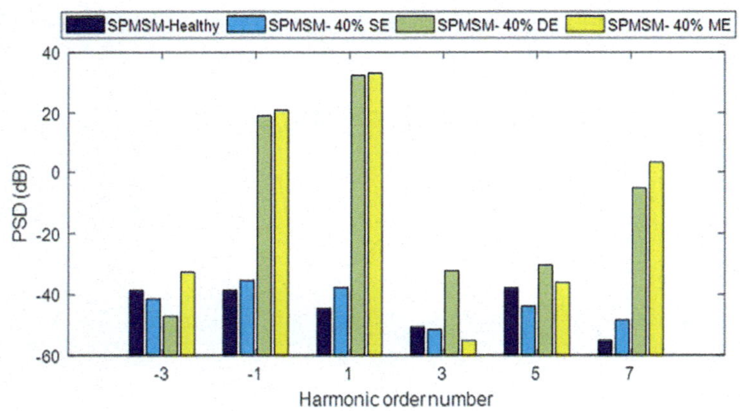

FIGURE 29.14 EMF harmonic components. *DE*, Dynamic eccentricity; *EMF*, electromotive force; *ME*, mixed eccentricity; *SE*, static eccentricity.

29.4 Results of signal processing analysis applied for SPMSM

29.4.1 Eccentricity faults case study

At this stage of analysis, to process the EMFs, the FFT is invested. Therefore a new frequency pattern is considered for SE, DE, and ME. To diagnose the wide range of fault eccentricities the ASBCs were used at frequencies obtained from the frequency pattern. Relying on research (Chen et al., 2019; Ebrahimi et al., 2014) the characteristic frequencies associated with eccentricity faults in PMSMs can be expressed according to Eq. (29.20).

$$f_{ecc} = f_s\left(1 \pm \frac{2k-1}{p}\right) \tag{29.20}$$

In ME fault the ASBC was higher in comparison with that recorded in SE and DE faults. To sum up, we would assert that the most risky fault is ME in comparison with SE and DE. Moreover, referring to Fig. 29.14, the SPMSM under ME is more influenced than that in SE and DE. The frequencies $[(1 + 5/P)f_s]$ and $[(1 + 3/P)f_s]$ are more related to SE and

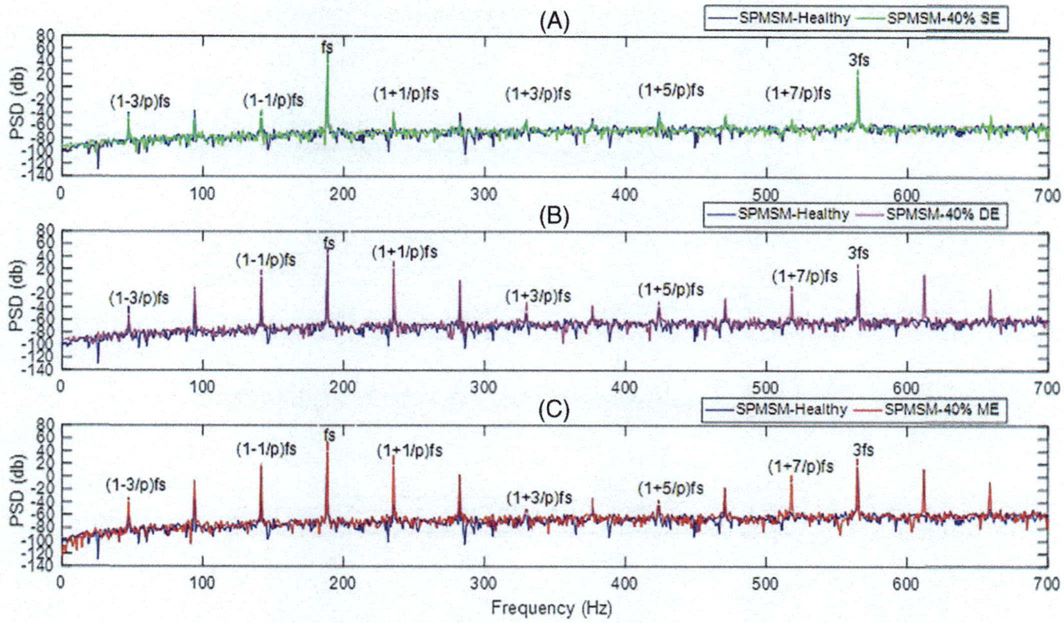

FIGURE 29.15 Electromotive forces spectrum for healthy and faulty SPMSM: (A) 40% SE (B) 40% DE (C) 40% ME. *DE*, Dynamic eccentricity; *ME*, mixed eccentricity; *SE*, static eccentricity; *PSD*, power spectral density; *SPMSM*, surface-mounted permanent magnet synchronous motor.

TABLE 29.2 Comparative analysis of EMF harmonic components resulting from eccentricity faults (in db).

Frequency	SPMSM Healthy	SPMSM40% SE	SPMSM40% DE	SPMSM40% ME
47.075	−38.64	−41.44	−47.16	−32.88
141.225	−38.70	−35.41	18.91	20.82
235.375	−44.58	−37.76	32.22	33.00
329.5	−50.62	−51.50	−32.34	−55.09
423.6	−37.91	−43.92	−30.55	−36.20
517.8	−55.03	−48.28	−5.27	3.16

EMF, electromotive force; *ME*, mixed eccentricity; *SPMSM*, surface-mounted permanent magnet synchronous motor.

DE, respectively. The $[(1-3/P)f_s]$ is more associated with ME. Fig. 29.15 portrays the spectrums of the EMF under the healthy case, 40% DE, 40% SE, and 40% ME, with considerable amplitudes at different frequencies. The ASBC of the aforementioned pattern frequency can be exploited to identify the different types of eccentricity faults. The ASBCs owing to SE, DE, and ME are illustrated in Table 29.2 (Figs. 29.14 and 29.15).

29.4.2 Demagnetization faults case study

The FFT was applied to the EMF signals to extract their spectral content. To identify and distinguish among various DFs a frequency pattern was defined, and the ASBC technique was employed. In the case of a healthy rotor, Melecio et al. (2019) reported the magnetic flux density distribution $B(\theta_r)$, expressed in Eq. (29.21). The air gap flux distribution is characterized as a periodic odd function with half-cycle symmetry, consisting exclusively of odd harmonic components. In the stator current signal (Krichen et al., 2020) the fundamental frequency is expressed by Eq. (29.22) (Hadj et al., 2020).

The harmonics of the PM field disturbance induced by the UD fault do not exhibit half-cycle symmetry or an odd periodic spatial distribution. Consequently, this fault only alters the amplitudes of the harmonic orders defined in Eq. (29.21). For

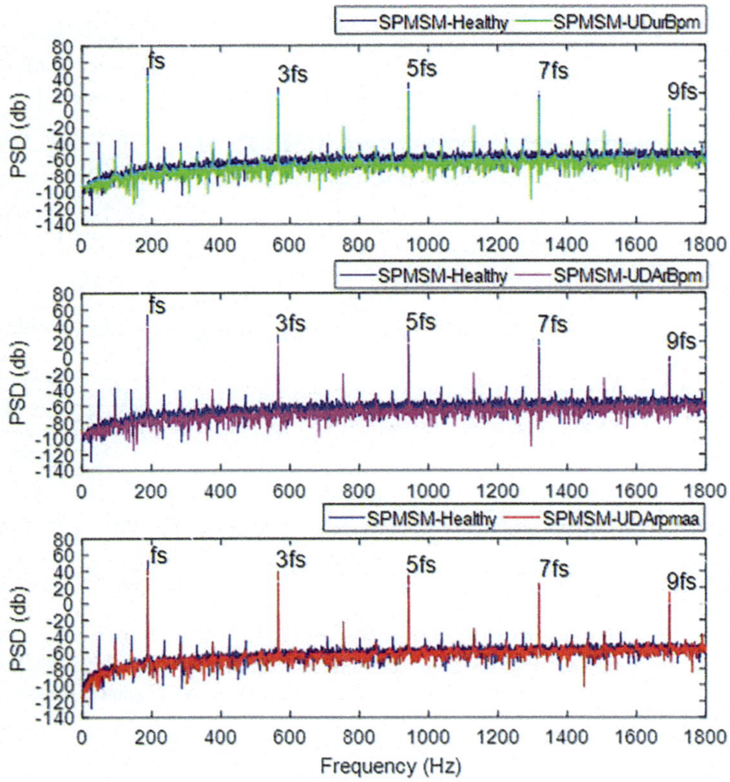

FIGURE 29.16 Spectrum of electromotive forces for healthy and faulty SPMSM: (A) $\mathrm{UD_{urBpm}}$ (B) $\mathrm{UD_{ArBpm}}$ (C) $\mathrm{UD_{Arpmaa}}$. *PSD*, Power spectral density; *SPMSM*, surface-mounted permanent magnet synchronous motor; *UD*, uniform demagnetization.

instance, in the case of $\mathrm{UD_{urBpm}}$, it can be revealed that the harmonic series in Eq. (29.21) will be rectified into Eq. (29.24). For a UD fault the possible EMF spectral frequencies, f_{UD}, will be the same as those existing in the healthy machine (Hadj et al., 2020).

$$B(\theta_r) = \sum_{n=1,3,5,\ldots}^{\infty} (B_{\mathrm{PM}})_n \sin(np\theta_r) \tag{29.21}$$

$$(B_{\mathrm{pm}})_n = \frac{4B_{\mathrm{PM}}}{\pi n}\left[\sin\left(\frac{n\pi}{2}\right)\sin\left(\frac{np\alpha_M}{2}\right)\right]$$

$$f_{\mathrm{Healthy}} = (2k-1)f_s \tag{29.22}$$

In PD faults (Melecio et al., 2019), f_{PD} represents the EMF spectral frequencies.

$$f_{\mathrm{UD}} = (2k-1)f_s \tag{29.23}$$

$$B_{\mathrm{UD1}}(\theta_r) = \sum_{n=1,3,5,\ldots}^{\infty} \frac{4MB_{\mathrm{PM}}}{\pi n}\left[\sin\left(\frac{n\pi}{2}\right)\sin\left(\frac{np\alpha_M}{2}\right)\right]\sin(np\theta_r) \tag{29.24}$$

$$f_{\mathrm{PD}} = \left(1 \pm \frac{2k-1}{p}\right)f_s \tag{29.25}$$

For a bandwidth of 0–1800 Hz, Fig. 29.16 exhibits the simulation findings for EMF spectrum for UD cases considered in Table 29.3 and for healthy SPMSM. The harmonic frequencies invested as typical frequencies for faults in demagnetization are indicated by Eq. (29.23). Table 29.3 foregrounds at frequencies defined in Eq. (29.23) the ASBCs for the faulty case with 75% of demagnetization severity as well as healthy SPMSM. In comparison with the healthy case of SPMSM, Fig. 29.16 unveils that for $\mathrm{UD_{urBpm}}$ case, the ASBCs rise is of about 20.83% for $7f_s$ frequency. For $\mathrm{UD_{ArBpm}}$ case, the ASBCs rise is about 39.52% for $3f_s$ frequency, whereas, in the case of $\mathrm{UD_{Arpmaa}}$, the ASBCs go up for both frequencies, $3f_s$ and $9f_s$. The fault signature patterns associated with UD fault types are characterized by recurring features in the EMF signal. In comparison with the healthy case the permanent magnet UD produces no additional fault-specific harmonics. Moreover,

TABLE 29.3 Comparison of EMF harmonic components comparison owing to UD and PD faults (db).

EMF harmonic components comparison owing to UD faults (db)						
Frequency	188.3	564.9	941.5		1318.1	1694.7
Healthy SPMSM	53.83	28.36	34.36		−53.24	1.80
UD_{urBpm}	43.29	20.47	25.79		−64.33	0.58
UD_{ArBpm}	52.22	39.57	27.52		−50.13	−4.52
UD_{Arpmaa}	42.96	39.72	34.08		−45.83	14.28
EMF harmonic components comparison owing to PD faults (db)						
Frequency	47.0	141.2	235.3	329.5	423.6	517.8
Healthy SPMSM	−38.64	−38.70	−44.58	−50.62	−37.91	−55.03
PD_{urBpm}	26.04	33.27	31.61	22.09	13.66	13.80
PD_{ArBpm}	15.10	22.92	22.60	15.54	10.82	−18.49
PD_{Arpmaa}	26.08	33.63	32.67	24.35	18.49	24.14

EMF, electromotive force; *PD*, partial demagnetization; *SPMSM*, surface-mounted permanent magnet synchronous motor.

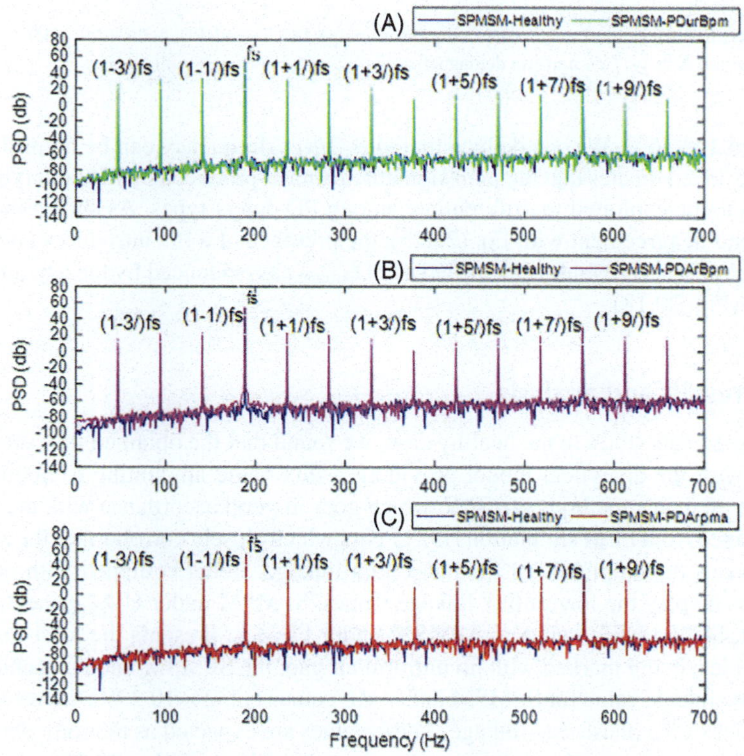

FIGURE 29.17 Spectrum of electromotive forces for healthy and faulty SPMSM: (A) PD_{urBpm} (B) PD_{ArBpm} (C) PD_{Arpmaa}. *PD*, Partial demagnetization; *PSD*, power spectral density; *SPMSM*, surface-mounted permanent magnet synchronous motor.

the magnitudes of the existing harmonics display specific alterations in the EMF signal, suggesting that EMF SP is beneficial in terms of noticing any modification in the present odd harmonics. In a nutshell the most dangerous fault corresponds to the UD fault through asymmetric reduction of the PM arc angle in comparison with other UD cases (Figs. 29.16 and 29.17).

Table 29.3 plots the ASBCs at the frequencies already defined in the case of healthy SPMSM owing to the three cases of PD. Fig. 29.17 illustrates the simulated EMF spectra within the 0–700 Hz range, comparing healthy and faulty SPMSM cases. Fig. 29.17 indicates that the PD of PM rises to the ASBC at all frequencies. In comparison with the healthy case, it is obvious that partial DF cases can yield a certain rise in ASBCs values at all frequencies. At $(1 + 7/p)\,f_s$ frequency the increase was about 125.07% for PD_{urBpm} case and about 135.19% for PD_{ArBpm} case, whereas, in the case of PD_{Arpmaa},

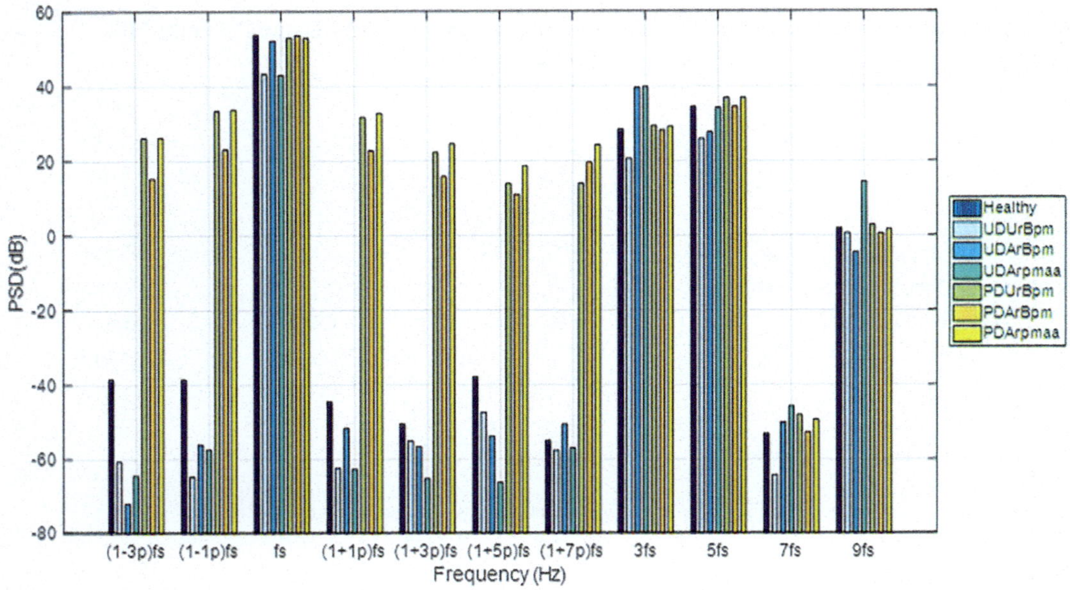

FIGURE 29.18 Comparison of EMF harmonic components comparison owing to UD and PD faults (db). *EMF*, Electromotive force; *PD*, partial demagnetization; *PSD*, power spectral density; *UD*, uniform demagnetization.

the ASBCs rise was around 143.86%. We can deduce that $(1 + 7/p)\, f_s$ frequency can be helpful and workable in terms of detecting the PD types. The data indicates that the fault signature patterns produced by PD fault types exhibit distinct features in the EMF signal, which can be exploited to differentiate among PD defect types. As shown in Table 29.3 and illustrated in Figs. 29.17 and 29.18, and in agreement with Eq. (29.25), the presence of a PD introduces additional harmonics into the EMF spectrum. Overall the most severe fault is identified as PD_{Arpmaa}, as evidenced by the asymmetric reduction of a single PM arc angle (Table 29.3) (Fig. 29.18).

29.4.3 Comparative results and analysis

After implementing finite elements study in the healthy case, we found that the obtained flux is equal to 0.0055 Wb, which goes in good conformity with the analytical model providing a flux value amounting to about 0.00589 Wb. Moreover, the EMF obtained by finite elements methods is 652 V, which goes in good accordance with the analytical model (697 V). Furthermore the EMT obtained from FEM simulations is 217 Nm, which closely corroborates the analytical model prediction of 210 Nm. Table 29.4 reveals the amplitude of the phase flux linkage. Being equipped with demagnetized magnets, the six considered cases of DFs display the lowest flux linkage values. SPMSM under UD_{Arpmaa} exhibits the least flux linkage value, which amounts to 1.44310–3 Wb, whereas SPMSM under PD_{urBpm} presents the least flux linkage value, which is about 5.042 10–3 Wb. Besides, being marked with no uniform air gap, the SPMSM under eccentricity faults shows highest flux linkage values. The flux linkage value for SPMSM under ME equals 6.232×10^{-3} Wb. However, the SPMSM case under SE displays the lowest values of the phase flux linkage. These values are expected as the rotor center and the rotation center coincide, being moved to a fixed distance from the stator's center. In addition, Table 29.4 depicts the maximum amplitude of the phase EMFs of the considered cases of SPMSM faults. The three cases of SPMSM under eccentricity faults exhibit a highest maximum EMF. The highest value is around 784,533 V for SPMSM under ME. The six cases of SPMSM under DFs show lower values of EMFs' maximum amplitude. SPMSM under UD_{urBpm} presents the least maximum amplitude of EMFs value, which equals 196,138 V, and SPMSM under PD_{Arpmaa} case yields the least maximum amplitude of EMFs value, reaching 653,366 V.

Table 29.4 outlines the peak-to-peak values and the mean values of EMT for the different considered SPMSM. As for the SPMSM affected by eccentricities faults, SPMSM under ME displays the highest peak-to-peak torque, amounting to 74.82 Nm, and the highest mean torque, reaching 214.82 Nm. Concerning the SPMSM affected by DFs, the case UD_{Arpmaa} generates more peak to peak torque equaling 46.18 Nm and less mean torque, amounting to 70.26 Nm. At this stage of analysis, it is noteworthy that uniformly demagnetized magnets generate less flux than healthy magnets and weak EMFs. The influence of the UD fault on the healthy SPMSM is more severe in comparison with the partial DF. One or more demagnetized

TABLE 29.4 Comparative results in SPMSM cases.

Cases of SPMSM	Amplitude flux max (Wb)	Amplitude FEM max (V)	Mean torque (Nm)	Torque peak to peak (Nm)
Healthy	5.489E − 03	664,259	202,262	28,646
UD_{urBpm}	1.689E − 03	196,138	117,373	17,365
UD_{ArBpm}	4.417E − 03	642,141	187,497	15,572
UD_{Arpmaa}	1.443E − 03	424,336	70,267	46,184
PD_{urBpm}	5.042E − 03	654,790	190,757	18,834
PD_{ArBpm}	5.391E − 03	654,379	192,725	14,811
PD_{Arpmaa}	5.045E − 03	653,366	189,299	21,452
SE	5.527E − 03	714,644	214,217	44,264
DE	5.805E − 03	723,381	202,754	64,425
ME	6.232E − 03	784,533	214,829	74,823

DE, dynamic eccentricity; *FEM*, finite element modeling; *ME*, mixed eccentricity; *SE*, static eccentricity; *PD*, partial demagnetization; *SPMSM*, surface-mounted permanent magnet synchronous motor; *UD*, uniform demagnetization.

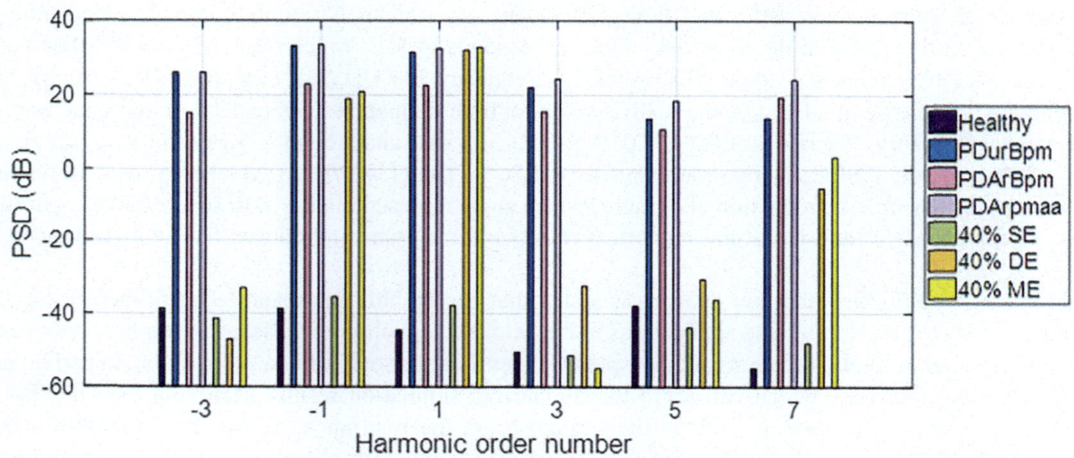

FIGURE 29.19 EMF harmonic components under healthy and faulty SPMSM (eccentricity and PD faults). *DE*, Dynamic eccentricity; *EMF*, electromotive force; *ME*, mixed eccentricity; *SE*, static eccentricity; *PD*, partial demagnetization; *SPMSM*, surface-mounted permanent magnet synchronous motor.

magnets generate less flux than healthy magnets, weak, EMFs, and rising stator current, which decreases EMT and further affects the performance of the SPMSM. Yet, eccentricity fault in SPMSM generates more flux and much increase in the EMFs. Eccentricity faults exert a pronounced impact even on healthy motors, leading to an increase in peak-to-peak electric machine (EM) and adversely affecting the overall performance of the SPMSM (Barmpatza & Kappatou, 2020; Gao et al., 2019; Hadj et al., 2020). Departing from Fig. 29.19, we infer that PD can be easily detected for diagnosis through the use of harmonic components. Moreover, UD does not produce asymmetry and, as a matter of fact, the harmonic component will not exist in the SPMSM (there is no unbalance) (Table 29.4).

Thus the EMF SP method is irrelevant as far as detecting and diagnosing a UD in the SPMSM is concerned. This refers basically to the electromotive harmonic composition analysis, which does not entail obvious results. Fig. 29.19 also shows that PD faults generate more harmonic components than eccentricity faults. Furthermore the PD_{Arpmaa} fault exhibits a greater severity compared to other PD types. Both PD and eccentricity faults can be effectively identified for diagnostic purposes using the harmonic components. Such defects generate the harmonic component and asymmetry (there is an imbalance). Hence the method of EMF SP proves to be appropriate in terms of detecting and diagnosing a PD as well as eccentricity in the SPMSM owing to the fact that electromotive harmonic composition analysis offers obvious findings. The findings of our research were assessed in comparison to those of recent papers. Among the main conclusions drawn from Gao et al. (2019), we report that more alterations in the harmonic components of EMF occur with rising demagnetization degree in SPMSM. Moreover, Barmpatza and Kappatou (2020) investigated the combined PD and SE fault in AFPM synchronous generator.

FFT analysis of the EMF spectrum was performed to investigate the harmonics associated with faults. Referring to the above, this study reveals that we can achieve a database for intelligent and more advanced diagnostic methods (Fig. 29.19).

29.5 Conclusion and future perspectives

Fault diagnosis in SPMSMs has attracted significant attention from both researchers and industry, particularly in the context of automotive systems. Indeed the detection and analysis of faults contribute to safer, more reliable, and sustainable operation. Section 29.1 was dedicated to PMSM faults. A significant portion of this section is dedicated to reviewing the most prominent fault diagnosis approaches, including various SP techniques, model-based methods, and data-driven diagnostic algorithms. This discussion highlights that fault diagnosis and analysis remain challenging, complex, and far from being fully comprehensive or exhaustive. Relying upon analytical computation, demagnetization and eccentricity faults of the SPMSM were investigated by FEM. The obtained findings by simulation under different conditions were recorded and discussed to validate the analytical values of EMF, flux, and EMT.

As initial results the EMF variation is more pronounced in the case of ME. Furthermore the peak-to-peak torque ripple is 15.17% for the healthy motor, increasing to 40.48% under ME conditions. As secondary results the flux decreases in the presence of UD and PD faults compared to the healthy case, with reductions of 73.77% for UD_{Arpmaa} and 8.3% for PD_{Arpmaa}. Demagnetized PMs exhibit reduced torque and increased torque ripple relative to healthy PMs, especially for UD_{Arpmaa} and UD_{UrBpm} faults. Afterward the SP techniques were invested as a diagnosis technique. Therefore this work includes essentially signal spectrum characteristics and fault performance as well as harmonic component and time–frequency analysis techniques. In this research the most risky fault corresponds to ME, which decreases the SPMSM performance. For UD_{urBpm} case, ASBCs increase is about 20.83% for $7\,f_s$ frequency. For UD_{ArBpm} case the ASBCs rise 39.52% for $3\,f_s$ frequency, whereas, in the case of UD_{Arpmaa}, the ASBCs go up for both frequencies, $3f_s$ and $9f_s$. In the same way, in the case of PD, at frequency $(1 + 7/p)f_s$, the rise is about 125.07% for PD_{urBpm} case and around 135.19% for PD_{ArBpm} case, whereas, in the case of PD_{Arpmaa}, the ASBCs increase amounts to 143.86%. The $(1 + 7/p)f_s$ frequency proves useful for PD type identification, with the asymmetric reduction of a single PM arc angle representing the most severe fault, impairing machine performance. Comprehensive fault cases and numerical results offer a promising dataset for future advanced diagnostic studies.

At this level, we would affirm that our analysis is a step that can be built upon and taken further. In fact, it lays the ground for future study to enact promising applications in the field, explore obscure areas, and decipher certain ambiguities. Among the most prominent future directions and new perspectives that this study offers, we mention, in particular, studying eccentricity fault experimentally, which can serve to validate the simulation results. Handling the DDIDM remains a relatively new research area, as they are still in their infancy and the road ahead is still long to fulfill a constructive contribution to the field. Estimating the fault severity and ultimately fault-tolerant control is another outstanding extension of this work.

Acknowledgments

This research was supported by the Ministry of Higher Education and Scientific Research of Tunisia.

List of symbols and abbreviations

Symbols	Abbreviations
PMSM	permanent magnet synchronous motor
PMSLM	permanent magnet synchronous linear motors
SPMSM	surface-mounted permanent magnet synchronous motor
EVs	electrical vehicles
HEV	hybrid electric vehicle
AFPM	axial flux permanent magnet
PM	permanents magnets
EM	electric machines
FFT	fast Fourier transform
SP	Signal Processing
ASBCs	amplitude of sideband components
CIM	cage induction motor
IM	induction motor
SRM	switched reluctance motor

ELF	electrical faults
MAF	magnetic faults
MEF	mechanical faults
EPRI	Electric Power Research Institute
DE	dynamic eccentricity
DF	demagnetization fault
B_{PM}	permanent magnet flux density
SE	static eccentricity
X	magnet arc angle fault severity indices
PD	partial demagnetization
UD	uniform demagnetization
MMF	magnetomotive force
ME	mixed eccentricity
DDIDM	data-driven intelligent diagnosis methods
STFT	short-time Fourier transform
SFDM	signal-based fault diagnosis methods
DWT	discrete wavelet transform
MFDM	model-based fault diagnosis methods
CWT	continuous wavelet transform
WT	wavelet transform
WPT	wavelet packet transform
EMD	empirical mode decomposition
WVD	Wigner–Ville Distribution
HHT	Hilbert–Huang Transform
DPMSLM	double-side permanent magnet synchronous linear motors
IMF	internal mode functions
SPWVD	smoothed pseudo Wigner–Ville distribution
TT	time-transform
ZAM	Zao-Atlas-Marks distribution
ELM	extreme learning machine
FEM	Finites Elements Methods
FEA	finites elements approach
EMF	electromotive force
EMT	Electromagnetic Torque
PSD	power spectral density
FEMM	element method magnetics software
V	rated vehicle speed
k_r	reduction ratio
R_{wh}	wheel radius
e	air gap thickness
g	vehicle gravity
C_{rr}	rolling resistance coefficient
A_f	vehicle frontal area
ρ	vehicle air density
C_D	coefficient of aerodynamic drag
m	vehicle mass
Θ	grade
a	vehicle acceleration
D_m	average diameter of the SPMSM
A_{enc}	slot angular width
H_d	tooth height
B_e	magnetic induction in air gap
L_m	motor length
A_{dent}	principal tooth angular width
N_d	number of principal tooth
A_{denti}	inserted tooth angular width
μ_a	relative permeability of magnets
k_{fu}	leakage flow coefficient

B_{cs}	magnetic induction in stator yoke
α_m	temperature coefficient
T_a	Magnet temperature
O_s	stator center
k	a positive integer ($k = 1, 2, 3\ldots$),
g	uniform air gap
O_w	rotation center
θ	angular position around the air gap
θ_s	fixed stator angle in DE
φ_m	ME transfer angle
R_r	rotor radius
O_r	rotor center
R_s	stator radius
$\varphi_{\text{phase},i}$	flux linkage of the ith coil
N	turns number per phase
f_s	power supply frequency equals 188.3 Hz
p	number of pole pairs
n	harmonic order number ($n = 1, 3, 5\ldots$),
M	flux magnitude fault severity indices, which can take values from 0 in UD to 1 in healthy PM
θ_r	rotor angular position
$(B_{\text{PM}})_n$	nth order PM flux density distribution Fourier coefficient,
α_M	magnet arc angle
UD_{urBpm}	uniform demagnetization fault by uniform reduction of the B_{pm} magnet
UD_{ArBpm}	uniform demagnetization fault by asymmetric reduction of the B_{pm} magnet
$\text{UD}_{\text{Arpma}a}$	uniform demagnetization fault by asymmetric reduction of the PM arc angle
PD_{urBpm}	partial demagnetization fault by uniform reduction of a single B_{pm} magnet
PD_{ArBpm}	partial demagnetization fault by asymmetric reduction of a single B_{pm} magnet
$\text{PD}_{\text{Arpmaa}}$	partial demagnetization fault by asymmetric reduction of a single PM arc angle

References

Abdelmoula, R., Hadj, N. B., Chaieb, M., & Neji, R. (2017). Reducing torque ripples in permanent magnet synchronous motor. *Journal of Electrical Systems, 13*(3), 528–542. http://journal.esrgroups.org/jes/papers/13_3_9.pdf.

Alameh, K., Cité, N., Hoblos, G., & Barakat, G. (2015). Vibration-based fault diagnosis approach for permanent magnet synchronous motors. *IFAC-PapersOnLine, 48*(21), 1444–1450. doi:10.1016/j.ifacol.2015.09.728.

Barmpatza, A. C., & Kappatou, J. C. (2019). A study of static angular and axis eccentricity in a double-sided rotor AFPM generator using 3D-FEM. In *Proceedings of the 2019 IEEE 12th International Symposium on Diagnostics for Electrical Machines, Power Electronics and Drives, SDEMPED 2019* (pp. 94–100). Institute of Electrical and Electronics Engineers Inc. http://ieeexplore.ieee.org/xpl/mostRecentIssue.jsp?punumber=8856151. doi:10.1109/DEMPED.2019.8864807.

Barmpatza, A. C., & Kappatou, J. C. (2020). Study of a combined demagnetization and eccentricity fault in an AFPM synchronous generator. *Energies, 13*(21). https://www.mdpi.com/1996-1073/13/21/5609. doi:10.3390/en13215609.

Ben Hadj, N., Abdelmoula, R., Chaieb, M., & Neji, R. (2018). Permanent magnet motor efficiency map calculation and small electric vehicle consumption optimization. *Journal of Electrical Systems, 14*(2), 127–147. http://journal.esrgroups.org/jes/papers/14_2_11.pdf.

Chen, L., Ma, J., & Zhou, P. (2024). Stray flux mapping with a rotating sensor for eccentricity detection in PMSMs. *Journal of Magnetism and Magnetic Materials, 586*, 171248.

Chen, Y., Liang, S., Li, W., Liang, H., & Wang, C. (2019). Faults and diagnosis methods of permanent magnet synchronous motors: A review. *Applied Sciences, 9*(10), 2116. doi:10.3390/app9102116.

Chowdhury, M. H. (2016). Modeling of faults in permanent magnet synchronous machines. In *IEEE Transportation Electrification Conference and Expo, Asia-Pacific, ITEC Asia-Pacific 2016* (pp. 246–250). Institute of Electrical and Electronics Engineers Inc. doi:10.1109/ITEC-AP.2016.7512956.

Ebrahimi, B. M., Javan Roshtkhari, M., Faiz, J., & Khatami, S. V. (2014). Advanced eccentricity fault recognition in permanent magnet synchronous motors using stator current signature analysis. *IEEE Transactions on Industrial Electronics, 61*(4), 2041–2052. http://ieeexplore.ieee.org/xpl/tocresult.jsp?isnumber=5410131. doi:10.1109/TIE.2013.2263777.

Elbouchikhi, E., Choqueuse, V., Auger, F., & Benbouzid, M. E. H. (2017). Motor current signal analysis based on a matched subspace detector. *IEEE Transactions on Instrumentation and Measurement, 66*(12), 3260–3270. doi:10.1109/TIM.2017.2749858.

Faiz, J., & Exiri, S. A. H. (2016). Short-circuit fault diagnosis in permanent magnet synchronous motors-an overview. In *Joint International Conference—ACEMP 2015: Aegean Conference on Electrical Machines and Power Electronics, OPTIM 2015: Optimization of Electrical and Electronic Equipment and ELECTROMOTION 2015: International Symposium on Advanced Electromechanical Motion Systems* (pp. 18–27). Institute of Electrical and Electronics Engineers Inc. doi:10.1109/OPTIM.2015.7427038.

Fitouri, M., Bensalem, Y., & Abdelkrim, M. N. (2016). Modeling and detection of the short-circuit fault in PMSM using finite element analysis. *IFAC-PapersOnLine, 49*(12), 1418–1423. http://www.journals.elsevier.com/ifac-papersonline/49. doi:10.1016/j.ifacol.2016.07.769.

Gao, C., Nie, Y., Si, J., Fu, Z., & Feng, H. (2019). Mode recognition and fault positioning of permanent magnet demagnetization for PMSM. *Energies, 12*(9), 1644. doi:10.3390/en12091644.

Gherabi, Z., Benouzza, N., Toumi, D., & Bendiabdellah, A. (2019). Eccentricity fault diagnosis in PMSM using motor current signature analysis. In *Proceedings 2019 International Aegean Conference on Electrical Machines and Power Electronics, ACEMP 2019 and 2019 International Conference on Optimization of Electrical and Electronic Equipment, OPTIM 2019* (pp. 205–210). Institute of Electrical and Electronics Engineers Inc. http://ieeexplore.ieee.org/xpl/mostRecentIssue.jsp?punumber=8988143. doi:10.1109/ACEMP-OPTIM44294.2019.9007164.

Grunditz, E. A., & Thiringer, T. (2016). Performance analysis of current BEVs based on a comprehensive review of specifications. *IEEE Transactions on Transportation Electrification, 2*(3), 270–289. https://www.ieee.org/membership-catalog/productdetail/showProductDetailPage.html?product=PER473-ELE&utm_source=Mainsite_CSE&utm_medium=CSE_Promotion&utm_campaign=Catalog_Promotion-PER473. doi:10.1109/TTE.2016.2571783.

Haddad, R. Z., Lopez, C. A., Foster, S. N., & Strangas, E. G. (2016). A voltage based approach for fault detection and separation in permanent magnet synchronous machines. In *ECCE 2016—IEEE Energy Conversion Congress and Exposition, Proceedings*. Institute of Electrical and Electronics Engineers Inc. doi:10.1109/ECCE.2016.7854806.

Hadj, N. B., Krichen, M., & Neji, R. (2020). Finites elements and signal processing analysis of eccentricity and demagnetization faults in PMSM drivetrains: Approach for diagnosis. *arXiv*. https://arxiv.org. doi:10.48550/arxiv.2011.04353.

Hang, J., Zhang, J., Cheng, M., & Ding, S. (2016). Detection and discrimination of open-phase fault in permanent magnet synchronous motor drive system. *IEEE Transactions on Power Electronics, 31*(7), 4697–4709. http://ieeexplore.ieee.org/xpl/tocresult.jsp?isnumber=4712525. doi:10.1109/TPEL.2015.2479399.

Ilioudis, V. C. (2020). Sensorless control of permanent magnet synchronous machine with magnetic saliency tracking based on voltage signal injection. *Machines, 8*(1), 14. doi:10.3390/machines8010014.

Ishikawa, T., Igarashi, N., & Kurita, N. (2017). Failure diagnosis for demagnetization in interior permanent magnet synchronous motors. *International Journal of Rotating Machinery, 2017*. http://www.hindawi.com/journals/ijrm/contents/. doi:10.1155/2017/2716814.

Jokic, S., Cincar, N., & Novakovic, B. (2018). The analysis of vibration measurement and current signature in motor drive faults detection. In *17th International Symposium on INFOTEH-JAHORINA, INFOTEH 2018—Proceedings* (pp. 1–6). Institute of Electrical and Electronics Engineers Inc. http://ieeexplore.ieee.org/xpl/mostRecentIssue.jsp?punumber=8337878-2018. doi:10.1109/INFOTEH.2018.8345531.

Karami, M., Mariun, N., Mehrjou, M. R., Ab Kadir, M. Z. A., Misron, N., & Mohd Radzi, M. A. (2014). Diagnosis of static eccentricity fault in line start permanent magnet synchronous motor. In *Conference Proceeding—2014 IEEE International Conference on Power and Energy, PECon 2014* (pp. 83–86). Institute of Electrical and Electronics Engineers Inc. doi:10.1109/PECON.2014.7062418.

Kim, K. H. (2011). Simple online fault detecting scheme for short-circuited turn in a PMSM through current harmonic monitoring. *IEEE Transactions on Industrial Electronics, 58*(6), 2565–2568. doi:10.1109/TIE.2010.2060463.

Krichen, M., Chaieb, M., Elbouchikhi, E., Hadj, N. B, Benbouzid, M., & Neji, R. (2019). Eccentricity faults diagnosis in permanent magnet synchronous motors: A finite element-based approach. *International Journal on Energy Conversion (IRECON), 7*(6), 207. doi:10.15866/irecon.v7i6.18602.

Krichen, M., Elbouchikhi, E., Benhadj, N., Chaieb, M., Benbouzid, M., & Neji, R. (2020). Motor current signature analysis-based permanent magnet synchronous motor demagnetization characterization and detection. *Machines, 8*(3), 35. doi:10.3390/machines8030035.

Lin, F., Zuo, S., & Deng, W. (2018). Impact of rotor eccentricity on electromagnetic vibration and noise of permanent magnet synchronous motor. *Journal of Vibroengineering, 20*(2), 923–935. https://www.extrica.com/journal/jve/browse. doi:10.21595/jve.2017.18436.

Lin, J., Yu, W., & Wang, S. (2025). Branch-current signatures for robust detection of demagnetization and eccentricity faults in PMSMs. *World Electric Vehicle Journal, 16*, 45.

Liu, H., Zhang, Y., & Li, K. (2024). PCA–ISSA–PNN-based diagnosis of demagnetization faults in interior PMSMs from stator current signals. *Scientific Reports, 14*(2), 1–12.

Liu, K., & Zhu, Z. Q. (2014). Online estimation of the rotor flux linkage and voltage-source inverter nonlinearity in permanent magnet synchronous machine drives. *IEEE Transactions on Power Electronics, 29*(1), 418–427. doi:10.1109/TPEL.2013.2252024.

Martinez, A., Morales, R., & Romero, E. (2024). Enhanced harmonic-based indicators for demagnetization fault detection in PMSMs under variable load. *Mechanical Systems and Signal Processing, 220*, 111980.

Melecio, J. I., Djurović, S., & Schofield, N. (2019). FEA model study of spectral signature patterns of PM demagnetisation faults in synchronous PM machines. *The Journal of Engineering, 2019*(17), 4127–4132. doi:10.1049/joe.2018.8048.

Mohanty, F., Sahoo, A., & Subudhi, B. (2024). Physics-guided current signature analysis for reliable demagnetization fault detection in permanent magnet synchronous motors. *Mechanical Systems and Signal Processing, 213*, 111212.

Mortezaei, S. R., Aliabadi, M. H., & Javadi, S. (2024). Eccentricity fault detection in surface-mounted permanent magnet synchronous motors by analytical prediction, FEM evaluation, and experimental magnetic sensing of the stray flux density. *Journal of Magnetism and Magnetic Materials, 589*, 171441. doi:10.1016/j.jmmm.2023.171441.

Ojaghi, M., & Bahari, V. (2017). Rotor damping effects in dynamic modeling of three-phase synchronous machines under the stator interturn faults-winding function approach. *IEEE Transactions on Industry Applications, 53*(3), 3020–3028. doi:10.1109/TIA.2016.2647704.

Pate, M. (2024). Demagnetization detection in PMSMs using wavelet packet transform and PCA–SVM classifier: Experimental validation. *International Journal of Electrical Power & Energy Systems, 158*, 109930.

Pietrzak, P., & Wolkiewicz, M. (2023). Demagnetization fault diagnosis of permanent magnet synchronous motors based on stator current signal processing and machine learning algorithms. *Sensors, 23*(4), 1757. doi:10.3390/s23041757.

Quintal-Palomo, R. E. (2019). Analysis of current signals in a partially demagnetized vector controlled interior permanent magnet generator. *Power Electronics and Drives, 4*(1), 179–190. doi:10.2478/pead-2019-0012.

Rosero, J., Romeral, L., Rosero, E., & Urresty, J. (2009a). Fault detection in dynamic conditions by means of discrete wavelet decomposition for PMSM running under bearing damage. In *Conference Proceedings—IEEE Applied Power Electronics Conference and Exposition—APEC*. doi:10.1109/APEC.2009.4802777 951 956.

Rosero, J. A., Romeral, L., Ortega, J. A., & Rosero, E. (2009b). Short-circuit detection by means of empirical mode decomposition and Wigner–Ville distribution for PMSM running under dynamic condition. *IEEE Transactions on Industrial Electronics, 56*(11), 4534–4547. doi:10.1109/tie.2008.2011580.

Sharma, S.K., Singh, P., Kumar, V. (2025). Acoustic signal-based fault diagnosis of PMSMs using deep transformer networks. Preprint arXiv : 0501.04562

Song, J., Zhao, J., Dong, F., Zhao, J., Xu, L., & Yao, Z. (2020a). A new demagnetization fault recognition and classification method for DPMSLM. *IEEE Transactions on Industrial Informatics, 16*(3), 1559–1570. http://ieeexplore.ieee.org/xpl/RecentIssue.jsp?punumber=9424. doi:10.1109/TII.2019.2928008.

Song, X., Zhao, J., Song, J., Dong, F., Xu, L., & Zhao, J. (2020b). Local demagnetization fault recognition of permanent magnet synchronous linear motor based on S-transform and PSO-LSSVM. *IEEE Transactions on Power Electronics, 35*(8), 7816–7825. https://ieeexplore.ieee.org/xpl/mostRecentIssue.jsp?punumber=63. doi:10.1109/TPEL.2020.2967053.

Torregrossa, D., Khoobroo, A., & Fahimi, B. (2012). Prediction of acoustic noise and torque pulsation in PM synchronous machines with static eccentricity and partial demagnetization using field reconstruction method. *IEEE Transactions on Industrial Electronics, 59*(2), 934–944. doi:10.1109/TIE.2011.2151810.

Toumi, S., Elbouchikhi, E., Amirat, Y., Benbouzid, M., & Feld, G. (2019). Magnet failure-resilient control of a direct-drive tidal turbine. *Ocean Engineering, 187*, 106207. doi:10.1016/j.oceaneng.2019.106207.

Ullah, Z., & Hur, J. (2018). A comprehensive review of winding short circuit fault and irreversible demagnetization fault detection in PM type machines. *Energies, 11*(12), 3309. doi:10.3390/en11123309.

Urresty, J., Riba, J., Romeral, L., Rosero, J., & Serna, J. (2009). Stator short circuits detection in PMSM by means of Hilbert-Huang transform and energy calculation. *IEEE International Symposium on Diagnostics for Electric Machines, Power Electronics and Drives, SDEMPED 2009*. doi:10.1109/DEMPED.2009.5292789.

Yan, R., & Gao, R. X. (2006). Hilbert–Huang transform-based vibration signal analysis for machine health monitoring. *IEEE Transactions on Instrumentation and Measurement, 55*(6), 2320–2329. doi:10.1109/tim.2006.887042.

Yao, D., & Mahesh, K. (2010). Novel fault diagnostic technique for permanent magnet synchronous machines using electromagnetic signature analysis 2010. In *IEEE Vehicle Power and Propulsion Conference* (pp. 1–6). doi:10.1109/VPPC.2010.5729130.

Youssef, M. Z., Elahi, A., & Ihsan, A. (2025). An optimal permanent magnet motor for pumps in electric/hybrid vehicles: Design and experimentation. *IEEE Access, 13*, 54028–54039. http://ieeexplore.ieee.org/xpl/RecentIssue.jsp?punumber=6287639. doi:10.1109/ACCESS.2025.3551606.

Yu, Y., Bi, C., Jiang, Q., Lin, S., Aung, N. L. H., & Mamun, A. A. (2014). Vibration study and classification of rotor faults in PM synchronous motor. *Microsystem Technologies, 20*, 1653–1659. http://www.springerlink.com/content/0946-7076-20. doi:10.1007/s00542-014-2206-8.

Yu, Z., Xu, L., & Yu, H. (2023). Demagnetization fault diagnosis of PMSMs using short-time Fourier transform and machine learning classifiers. *Sensors, 23*(15), 6425–6437.

Zhang, J., Tounzi, A., Benabou, A., & Menach, Y. L. (2021). Detection of magnetization loss in a PMSM with Hilbert Huang transform applied to non-invasive search coil voltage. *Mathematics and Computers in Simulation, 184*, 184–195. https://www.journals.elsevier.com/mathematics-and-computers-in-simulation. doi:10.1016/j.matcom.2020.02.009.

Zhang, Y., Yang, Z., & Gerada, M. (2024). Non-invasive detection of mixed eccentricity faults in LSPMSMs using motor current signature analysis. *Measurement, 220*, 113487.

Zhu, M., Hu, W., & Kar, N. C. (2017). Torque-ripple-based interior permanent-magnet synchronous machine rotor demagnetization fault detection and current regulation. *IEEE Transactions on Industry Applications, 53*(3), 2795–2804. doi:10.1109/TIA.2016.2634518.

Zhu, M., Hu, W., & Kar, N. C. (2018). Acoustic noise-based uniform permanent-magnet demagnetization detection in SPMSM for high-performance PMSM drive. *IEEE Transactions on Transportation Electrification, 4*(1), 303–313. doi:10.1109/tte.2017.2755549.

Zhu, M., Yang, B., Hu, W., Feng, G., & Kar, N. C. (2019). Vold-Kalman filtering order tracking based rotor demagnetization detection in PMSM. *IEEE Transactions on Industry Applications, 55*(6), 5768–5778. https://ieeexplore.ieee.org/servlet/opac?punumber=2855. doi:10.1109/TIA.2019.2932692.

Chapter 30

AI-driven self-healing calibration through digital twin verification and reinforcement learning

R. Deepa[a], Sushil Kumar Mishra[b], Arnav Prateek[b], Divynash Kumar[b] and A. Bennet Prabhu[c]

[a]*Department of Computing Technologies, School of Computing, SRM Institute of Science and Technology, KTR Campus, Chennai, Tamil Nadu, India,* [b]*Department of Computer Science and Engineering, School of Computing, SRM Institute of Science and Technology, KTR Campus, Chennai, Tamil Nadu, India,* [c]*Research & Development, Congruent Solutions Private Ltd, Chennai, Tamil Nadu, India*

30.1 Introductions

The manufacturing industry is experiencing a radical transformation as it moves from traditional labor-intensive processes to much more automated (Ogunmolu et al., 2025) data-driven systems that define the times of Industry 4.0. This paradigm shift utilizes the Internet of Things (IoT) and artificial intelligence (AI) types of technologies to improve the process of operation and quality of products (Mani & Attaranasl, 2025). Nevertheless, these smart manufacturing environments have become more complex, revealing the weakness of the traditional maintenance strategies (Guo et al., 2025). Conventional reactive or schedule-based systems tend to lead to unscheduled downtimes, loss of productivity, and major losses, and these issues create a dire necessity to seek smarter and more autonomous solutions that may guarantee both accuracy and uptime (Mengesha, 2025).

In the past, manufacturers used to manage maintenance through prevention, whereby machines were recalibrated at intervals without regard to the state of these machines (El Ahmadi & El Abbadi, 2025). Although this strategy can eliminate disastrous failures, it is inefficient most of the time, which results in either unnecessary servicing or unreliable failures in between the checks (Najafzadeh & Yeganeh, 2025). As a reaction, condition-based maintenance (CBM) and predictive maintenance (PdM) have developed, based on sensor data and machine learning (ML) to predict possible faults (Alsubaih et al., 2025). These data-based approaches have enhanced maintenance scheduling yet are still characterized by a major shortcoming that they completely rely on human factors in the form of corrective actions (Saxena et al., 2008). The existing systems are able to communicate to the operators about the imminent collapse, but still the diagnosis (Jardine et al., 2006) and recalibration are manual operations that require the skill of a good technician, which creates delays (Selcuk, 2016), bottlenecks in the operation process, and the possibility of human error (Carvalho et al., 2019). This is the loophole that does not fulfill the vision of a truly independent factory (Liu, 2018).

This study is going beyond mere fault prediction to propose an all-encompassing self-healing mechanism, one that automatically identifies calibration drift as well as automatically fixes it (Wang et al., 2018). Project Sentinel aims to fill this gap by proposing a closed-loop, self-healing calibration system that can bolster system resilience with minimum human intervention (Lee et al., 2015). The two important components of the system architecture are described below.

Anomaly detection layer: The system is based on a high-accuracy random forest model to conduct fault detection in real time by studying the multivariate sensor data and detecting an anomaly in the nominal operation of machines (Monostori, 2014). This layer is the most important step toward detecting when a machine is not functioning as expected.

Cycle-based correction algorithm: This study uses a new algorithm, which is based on the modeling of the optimal wear of a component, rather than on static rules (Kagermann et al., 2013). It uses the result of separate usage cycles in comparison with baseline data to construct an ideal model of expected behavior and corrects autonomously for drift by

Computational Intelligence in Mechatronics: Solving Real-World Problems in Electronic Systems Design. DOI: https://doi.org/10.1016/B978-0-443-27580-7.00024-X

comparing performance with this ideal in real time (Peres et al., 2020). This is a proactive correction that will help to avoid small deviations that could lead to major failure (Zhou et al., 2021).

Such an integrated system allows establishing a complete self-regulating system of detection and correction to enhance machine availability and quality products (Zhang & Zhou, 2019). The efficacy of the framework is confirmed with the help of the AI4I 2020 Predictive Maintenance corpus that includes dense sensor readings that reflect the real-life conditions of an industrial setting (Rauba, 2024). This study is quantitatively able to reduce calibration error, which shows that this type of self-healing system can build on the future of the intelligent, autonomous factory and will become the basis of future improvements in adaptive correction schemes in more complex multiparameter drift conditions (Balasubramanian & Gurushankar, 2019).

30.2 Literature review

30.2.1 Theoretical foundation

The history of modern manufacturing can be characterized by the unceasing quest for increased efficiency, reliability, and autonomy. In the past, industrial systems were either based on a regular maintenance program or were responsive in response to a failure that had already taken place. Although these tactics worked well previously in the industrial environment, they are slowly becoming unsuitable in the modern high-speed, information-driven production lines (Lundberg & Lee, 2017). Uncertainties in machine calibration caused by circumstances such as the wear of the tools and the sudden failure of the components result in the waste of resources, deterioration of the quality of the products, and the loss of much time. Industry 4.0 has become a paradigm revolution that combines cyber-physical systems, ML, and the IoT to transform factories into responsive systems and active, smart, and adaptable ecosystems (Ribeiro et al., 2016).

The central change here is the idea of PdM that uses powerful data analysis and ML algorithms to predict possible failures, even when they do not occur physically (Breiman, 2001). Having shown a promising reduction in maintenance costs and downtime, large-scale predictive systems have been widely studied and applied to essential industries, including the aviation and energy industries. The idea of digital twins, that is, the virtual representation of real-life systems, which can be simulated and monitored in real time, complements PdM. Digital twins enable engineers to test the corrective measures safely, forecast failures, and optimize operations without interfering with live operations (Goodfellow et al., 2016).

The other important theoretical foundation of this research is the new discipline of self-healing systems, which finds its inspiration in the biological processes. Within such systems, machines are envisioned as systems that are able to diagnose themselves, determine what corrective measures should be taken, and apply them without human intervention.

The combination of the three notions, PdM, digital twins, and self-healing systems, constitute the theoretical basis of this study. They are the next generation of simply avoiding failures to actively anticipate and rectify them and make it possible to have machines that are able to adjust their performance in real time. Although predictive and digital twin-based systems have already been established in the literature, a combination of a fully closed-loop self-healing calibration mechanism in the manufacturing setting is a relatively unexploited area with great potential for innovation (Sutton & Barto, 2018).

30.2.2 Models used in maintenance and anomaly detection

There is a broad range of computational models and frameworks of machine health monitoring, PdM, and adaptive control, which are provided in the literature. Statistical methods and ML models can be broadly considered the most common methods to detect calibration drift and other anomalies.

First attempts at fault detection were based on the use of statistical process control (SPC). Shewhart control charts, cumulative sum (CUSUM), and exponentially weighted moving averages (EWMA) were adopted to identify any discrepancies by comparing the observed values with the set statistical limits. These methods can be computationally efficient but can be ineffective in dynamic manufacturing settings where the drift of calibration may be slow, multidimensional, and dependent on complex interacting factors (Ong et al., 2020).

With the emergence of ML the state-of-the-art in the field of anomaly detection has advanced considerably. When there is availability of labeled data, failure classification has been done using supervised models such as logistic regression and support vector machines (SVMs).

Isolation forests, one-class SVMs, and autoencoders are unsupervised methods that, in particular, are quite effective at spotting diverse types of anomalies and do not need prelabeled failure data. The models have the ability to capture high-dimensional deviations that cannot be easily captured by statistical methods (Hevner et al., 2004). An example is isolation forests, which determine anomalies by partitioning the data recursively, whereas autoencoders, based on deep learning models, are trained to rebuild normal input data with a small error. The difference between the reconstruction errors can then be identified as an anomaly.

Although these models are useful in detection, a mechanism of correction is not inherent in them. This has prompted the study of reinforcement learning (RL) toward the development of self-healing abilities. The RLagents are educated on the best corrective methods via trial and error by engaging with an environment (or a simulated digital twin) and maximizing a cumulative reward signal (Rudin 2019). Negative rewards can be taken to be deviations from an ideal machine state and positive rewards to be successful corrections in a calibration scenario. With time an RL agent would be able to create a complex policy of changing machine parameters in real time to keep the performance optimal.

30.2.3 Fault detection and diagnosis

The role of fault detection and diagnosis (FDD) has been the primary long-term interest of manufacturing research because machine reliability directly influences the efficiency of production and the quality of products (Tao et al., 2018). There are three broad approaches to FDD, which are outlined in the literature: model-based approaches, data-driven approaches, and hybrid approaches.

Model-based approaches are based on the development of a highly accurate mathematical model of the dynamics of a machine. Deviations between the output of the model and actual sensor measurements can be used to show a fault by constructing equations to model the expected behavior of a system. Although these are very interpretable, they may need a lot of domain knowledge and may be impractical when dealing with very complex machines with nonlinear dynamics.

Such data-driven approaches, however, use sensor data directly and do not require an explicit system model. As the IoT-enabled sensors are increasingly common, data-driven solutions have become very popular (Aldrini et al., 2024). The ML models are developed by the use of large historical datasets of labeled or unlabeled operational data to predict trends that indicate the presence of faults. Clustering, classification, and deep learning techniques have proved to be effective in extracting the minor trends that can lead to failure. These approaches have been benchmarked on publicly available datasets, including the AI4I 2020 Predictive Maintenance Dataset, which can offer useful ideas about how machine degradation should behave and how the degradation state can be predicted. The standard classification metrics that are used to measure performance are based on the confusion matrix (Gao & Liu, 2025).

Although FDD techniques are well-known, the majority of the current research ends at the interpretation and diagnosis stage, with the next corrective action being left in the hands of human operators. These points to a large gap in the literature, namely, the lack of automated and integrated self-healing calibration mechanisms, which is the main topic of our study.

30.3 Methodology

The proposed system would be based on an autonomous, closed-loop system that combines anomaly detection with a self-healing calibration system. There are four major layers in the methodology that make a cycle of data acquisition, preprocessing, fault detection, and intelligent correction.

30.3.1 Data acquisition layer

Fig. 30.1 illustrates that the principle upon which the framework is built is the continuous gathering of multivariate sensor data of the machine components. The publicly available AI4I 2020 Predictive Maintenance Dataset, which is a proxy of real-time industrial data, is used in this research. Operational parameters, such as air temperature, process temperature, rotational speed, torque, and tool wear, are critical parameters included in this dataset, and all of them reflect the state of the machine. The data represents real operational states and other types of failure conditions, which offers a realistic training and validation situation of the fault detection and correction models.

Self-healing calibration update rule D_{u+1} is expressed in Eq. (30.1).

$$D_{u+1} = D_u + \pi(S_u - P_u) \tag{30.1}$$

This equation expresses the adaptive correction rule for recalibrating the system in real time. It adjusts the calibration coefficient incrementally based on the observed discrepancy.

D_u is the current calibration coefficient, π represents the learning or adaptation rate, S_u denotes the reference measurement, and P_u corresponds to the observed measurement.

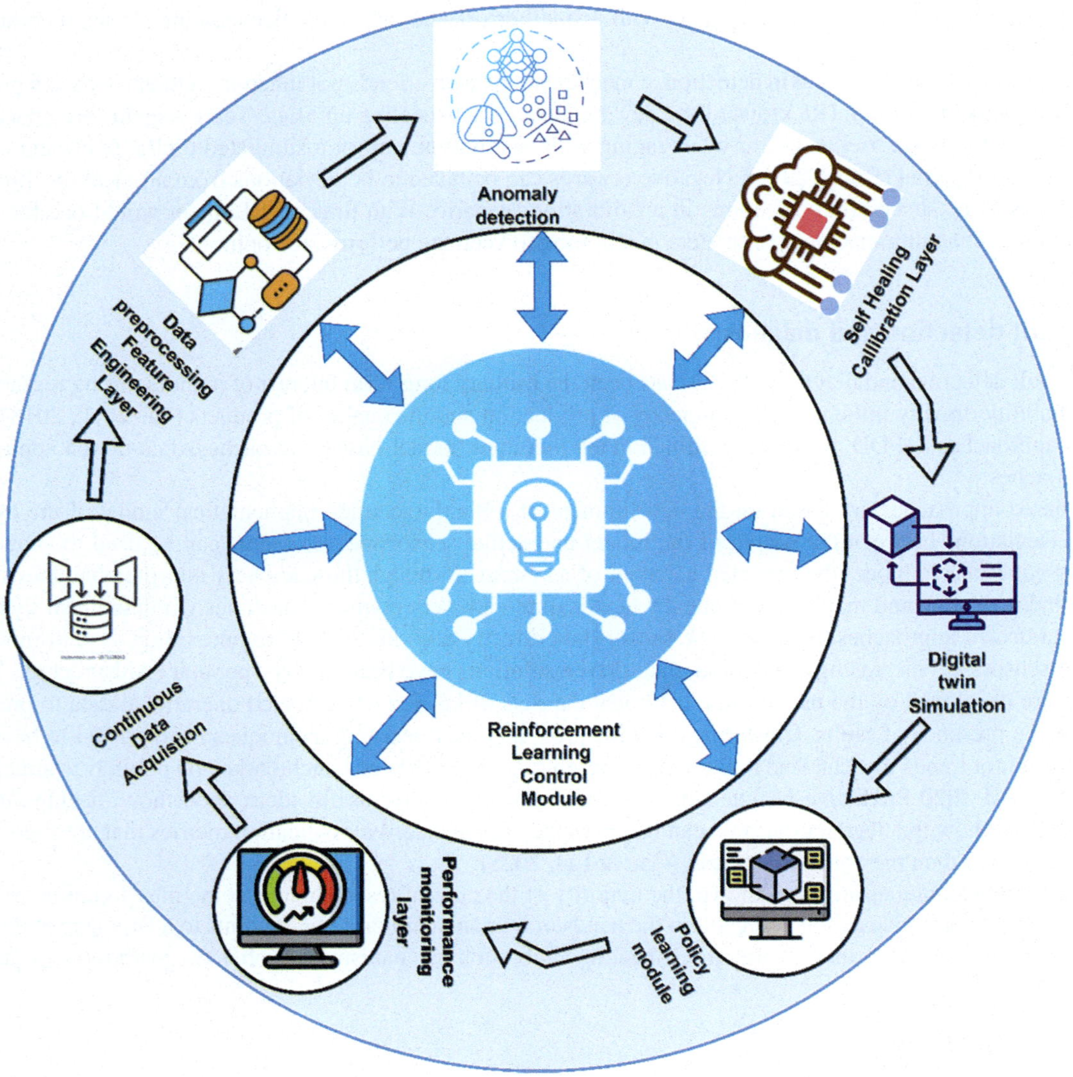

FIGURE 30.1 Project Sentinel self-healing calibration framework.

Drift compensation function C_u is expressed in Eq. (30.2).

$$C_u = \partial C_{u-1} + (1 - \partial)E_u \tag{30.2}$$

This recursive formulation smooths the temporal fluctuations in fault deviation. It ensures that short-term noise does not falsely trigger recalibration while preserving responsiveness to long-term drift.

C_{u-1} is the previous drift-compensated value, ∂ is the smoothing factor, and E_u denotes the fault deviation index at time u.

The normalized operational parameter transformation $M_{i,u}$ is expressed in Eq. (30.3).

$$M_{i,u} = \frac{(Q_{iu} - \pi_i) + \varepsilon_i(U_{iu} - U) + \delta_i(S_{iu} - S)}{\sqrt{\rho_i^2 + \sigma_i^2}} \tag{30.3}$$

This equation transforms each operational parameter into a normalized form that accounts for sensor variability and cross-influence of correlated parameters.

Q_{iu},t represents the raw sensor reading, π_i is the average value of the parameter, ε_i is the temperature sensitivity coefficient, U_{iu} represents the instantaneous temperature, U is the mean temperature, δ_i is the rotational sensitivity factor, S_{iu} represents the rotational speed, S is the mean rotational speed, ρ_i^2 is the standard deviation of the parameter, and σ_i^2 is the uncertainty variance component.

Predictive health score function I_t is expressed in Eq. (30.4).

$$I_t = u_1 1 - \frac{E_t}{E_{max}} + u_2 f^{-\gamma^{Ct}+} - u_3 \frac{D_t}{D_{max}} \quad (30.4)$$

This equation produces a comprehensive health score by combining normalized fault deviation, drift suppression, and calibration stability.

The predictive health score at the time t, u_1, u_2, and u_3 are weighting constants for each contributing component, E_t is the fault deviation index, E_{max} is the maximum permissible fault level, γ is the exponential decay constant,Ct is the drift-compensated value, D_t is the current calibration coefficient, and D_{max} is the maximum stable calibration value c.

This Algorithm 30.1 detects the calibration errors in real-time by employing multihead attention. It interprets the incoming sensor data, contrasts them with reference calibration patterns, and marks deviations. The result is a score on error and a feature attention map that can be used in further correction procedures to improve the sensitivity of the anomalies and lower the false detection values.

30.3.2 Preprocessing and feature engineering layer

Raw sensor data usually has noise, discrepancies, or useless features that might compromise the performance of the model. This layer includes preparing the data in terms of analysis by a set of standard preprocessing actions.

The steps will start with data cleaning and normalization so that all the features are on a similar scale. After this, meaningful features are created to increase the representation of the machine state and indicate possible drift. This involves the computation of moving averages and standard deviations to record the trend and volatility of sensor readings. To reduce the dimensions and to determine the most important parameters, one may also use such methods as principal component analysis (PCA) to remove the noise and alleviate redundancy in the data.

Self-healing correction model Y_{corr} is expressed in Eq. (30.5).

$$Y_{corr} = Y + \Delta Y - Z + h(Y, y; \emptyset) \quad (30.5)$$

The self-healing correction model predicts the required calibration adjustments and applies them dynamically.

Y is the raw sensor data matrix, ΔY represents the correction component, h is the correction function generated by the learning model, Y refers to the detected machine state, and $\emptyset$ represents the parameters of the correction model.

Self-healing adjustment model D_{u+1} is expressed in Eq. (30.6).

$$D_{u+1} = D_u - \delta_1 F_u + \delta_2 C_t - \delta_3 (U_u - U) \quad (30.6)$$

This equation defines how the calibration coefficient updates over time to self-correct detected drift.

In this equation, D_{u+1} is the updated calibration coefficient, D_u is the current coefficient, δ_1,δ_2,δ_3 are control gains, F_u is the calibration error, C_t is the drift component, U_u is the current temperature, and U is the average temperature.

Calibration efficiency index CE_J is expressed in Eq. (30.7).

$$CE_J = u_1 1 - \frac{F}{F_{max}} + u_{2f^{-\partial C_t}} \quad (30.7)$$

This efficiency index integrates both calibration accuracy and drift suppression into a unified score.

In this equation, CE_J denotes the calibration efficiency index, u_1 and u_2 are weighting factors, F is the mean calibration error, F_{max} is the maximum allowable error, ∂ is the drift decay constant, and C_t represents the drift compensation value.

Dynamic correction factor update rule ΔD_u is expressed in Eq. (30.8).

$$\Delta D_u = \pi_1 (S_u - P_u) + \pi_2 (U_u - U) + \in \frac{cF_u}{cu} \quad (30.8)$$

This rule models the dynamic update of the calibration correction factor based on real-time deviations.

In this equation, ΔD_u is the change in calibration correction, π_1 and π_2 are learning coefficients, S_u and P_u are the reference and observed readings, U_u is the instantaneous temperature, U is the mean temperature, ε is a correction sensitivity constant, and cF_u is the rate of change in calibration error.

These preprocessing steps are crucial for ensuring the reliability and accuracy of the subsequent detection and correction models.

This Algorithm 30.2 is a digital twin employed to simulate and test corrective measures prior to deployment. It calculates the similarity of the virtual and actual sensor values, which allows applying only the safe and effective updates to the calibration and reduces the risk and allows keeping the process running in industrial contexts.

30.3.3 Anomaly detection and fault diagnosis layer

The processed information is then input into a fault detection model to detect any abnormal operational modes. Although some classical ML algorithms can be used as a benchmark, this study employs a random forest classifier, which is an ensemble model characterized by high accuracy and the ability to deal with complex data of high dimension. The model is trained on a previous data set to gain knowledge about the patterns of various failure modes, such as calibration drift, sensor anomalies, and component wear.

After a deviation or an anomaly has been identified, the system identifies the type of fault. The main objective of this layer is to have high-accuracy, real-time detection of any machine state that is not at the optimal calibration, which, in turn, results in the activation of the self-healing mechanism. The results of this detection model are measured through cross-validation and standard classification measures, which are described in the results section.

30.3.4 Self-healing mechanism layer

When a drift in calibration is detected, the system will start its self-healing process, which is autonomous in nature. This study is based on a new, cycle-based correction algorithm, rather than traditional, nonadaptive adjustment approaches (e.g., setting an offset in case a threshold is violated).

The algorithm is intended to comprehend and characterize the natural operational cycles of a machine component, e.g., the predictable wear-and-reset behavior of a tool. The process is as follows:

Ideal model creation: This algorithm uses the nondrifted baseline data to find a model of the ideal wear pattern of the component during a typical usage cycle.

DE_R

$$\mathrm{DE}_R = 1 - \frac{E_{\mathrm{SHCS}}}{E_{\mathrm{PdM}}} \times 100 \tag{30.9}$$

This ratio shows how much calibration error the self-healing system removes compared to PdM. It serves as a direct performance metric for quantifying the percentage reduction in overall calibration bias.

In this equation, DE_R is the calibration error reduction ratio, E_{SHCS} represents the average error under the self-healing system, and E_{PdM} represents the average error under PdM.

Mean-squared error calculation M_{SE} is expressed in Eq. (30.10).

$$\mathrm{MSE} = \frac{1}{m}(x_i - \hat{x}_i)^2 \tag{30.10}$$

The mean squared error (MSE) quantifies the average of the squared differences between the observed and predicted values.

In this equation, MSE represents the mean squared error, m denotes the number of samples, x_i refers to the actual observed value, and $\hat{x}_i$ is the predicted or reconstructed value of the signal.

Dynamic correction gain H_d is expressed in Eq. (30.11).

$$H_d = \frac{\Delta x}{\Delta f}\frac{x_{u+1} - y_u}{y_u - x_u} \tag{30.11}$$

The dynamic correction gain expresses how effectively the self-healing calibration system adjusts predictions in response to deviations.

In this equation, x_{u+1} and y_u are the predicted outputs at successive time steps, and x_u represents the actual observed value at time u.

Drift calculation: This is performed in real time to compare the measured (drifted) signal with this ideal model to accurately compute the value and direction of the calibration drift.

Corrective action: The signal is then corrected automatically by the system through the calculation of the drift that is subtracted.

This RL Algorithm 30.3 is a dynamic calibration refinement algorithm that uses rewards to update calibration parameters. It is an optimization that uses the Q-value to maximize calibration and minimize system error through feedback. The model is self-healing and autonomously corrects the industrial systems, which requires no human assistance to maximize its benefits.

Such a method is highly adaptive and context-aware, adjusting to the root cause of the calibration error. This self-healing loop keeps the machine calibration in the desired position automatically once detected and corrected, eliminating the need for manual intervention to fix a machine and increasing the system's resilience and uptime.

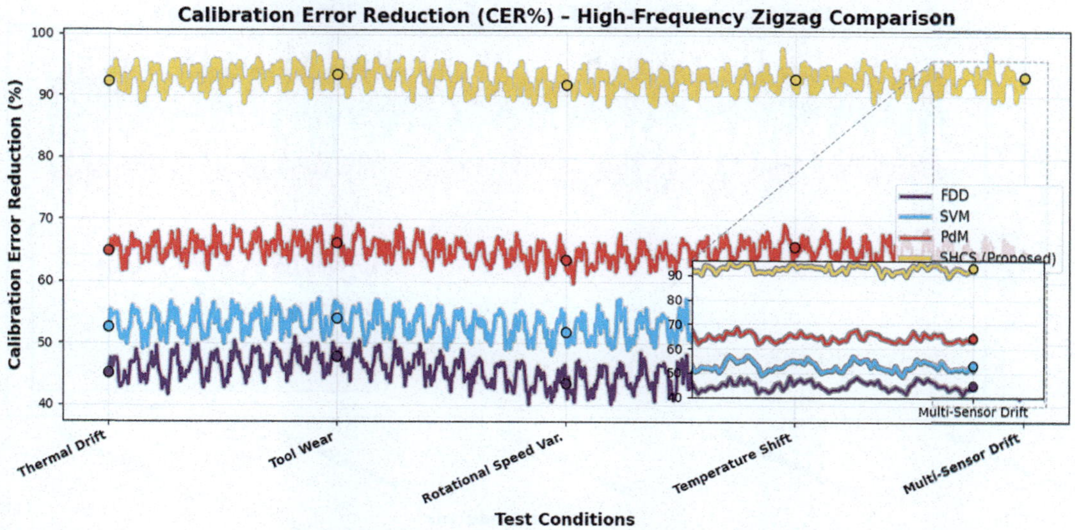

FIGURE 30.2 Analysis of calibration error reduction. *FDD*, Fault detection and diagnosis; *PdM*, predictive maintenance; *SVM*, support vector machine.

30.4 Results and discussion

Project Sentinel self-healing calibration system (SHCS) evaluation aims at using eight major performance parameters as compared to the current practices, including FDD, SVM, and PdM. These parameters gauge accuracy, efficiency, and reliability, which prove that SHCS is much better at attaining autonomous calibration, increased product quality, minimized downtime, and maximized energy consumption.

30.4.1 Dataset

The AI4I Predictive Maintenance Dataset (Artificial Intelligence for Industry 4.0) includes 10,000 samples that describe the actual state of the industrial machine together with the parameters air temperature, process temperature, rotational speed, torque, and tool wear. Every record is marked with the conditions of the machine failure, no failure or wear of the tool, heat loss, power loss, and overstrain. The data is suitable for supervised learning, anomaly detection, fault prediction, and maintenance modeling tasks and is thus best for testing self-healing and predictive calibration systems such as Sentinel (Amazon Web Services).

30.4.2 Calibration error reduction (CER%)

Fig. 30.2 illustrates that SHCS recorded a mean calibration error of only 93%, which was much higher than the 65% that PdM had. It means that sensor drift and process variations were effectively removed in real time by the self-healing process, restoring accurate machine calibration across various conditions and ensuring consistent product quality under variable industrial conditions.

Analysis of calibration error reduction efficiency μ_{err} is expressed in Eq. (30.12).

$$\mu_{err} = 1 - \frac{MS_{model}}{MS_{baseline}} \times 100 \tag{30.12}$$

Error reduction efficiency quantifies the percentage decrease in reconstruction error compared to a baseline approach.

In this equation, MS_{model} represents the MSE obtained from the proposed SHCS model, and $MS_{baseline}$ corresponds to the MSE of the benchmark system such as PdM or FDD.

30.4.3 Mean squared error (MSE)

Fig. 30.3 describes the planned SHCS had an average MSE of 40, as opposed to PdM of 870 and FDD of 1280. This 95% decrease is evidence of the capability of this model to reduce signal reconstruction errors. The cycle-based correction

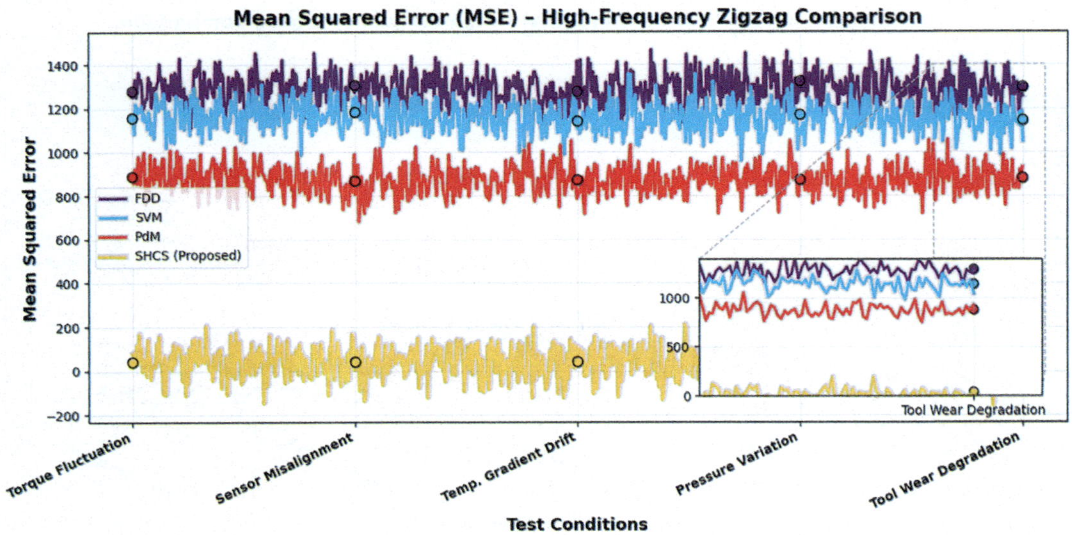

FIGURE 30.3 Analysis of mean squared error. *FDD*, Fault detection and diagnosis; *PdM*, predictive maintenance; *SVM*, support vector machine.

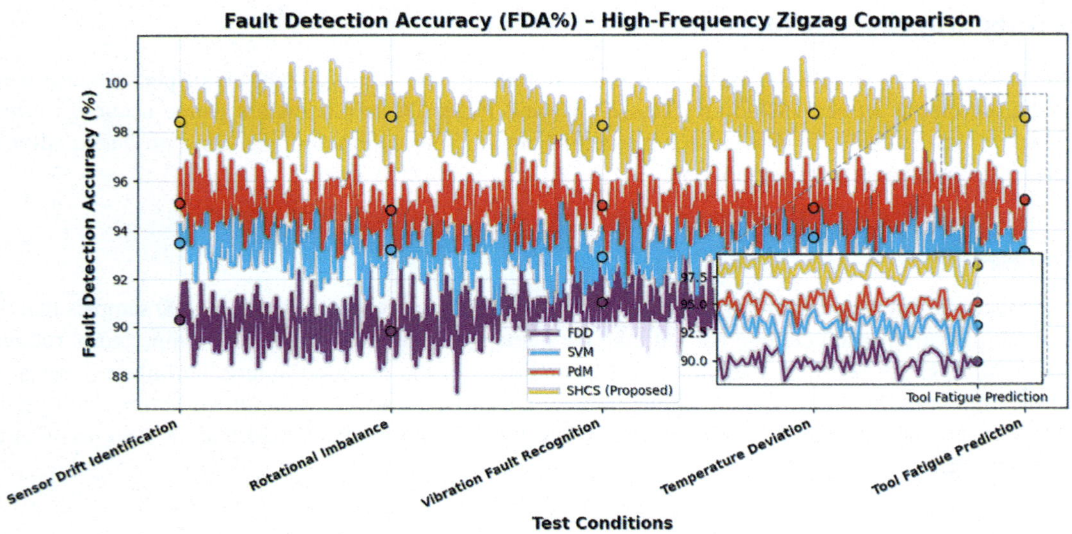

FIGURE 30.4 Analysis of fault detection. *FDD*, Fault detection and diagnosis; *PdM*, predictive maintenance; *SVM*, support vector machine.

algorithm would continuously control deviations dynamically to provide proper alignment of actual and ideal machine conditions with minimum noise effects.

Root MSE RM_{SE} is expressed in Eq. (30.13).

$$RM_{SE} = \sqrt{\frac{1}{m}(x_i - y_i)^2} \tag{30.13}$$

The root MSE expresses the standard deviation of prediction errors, offering a direct measure in the same units as the observed data.

In this equation, m is the total number of data samples, x_i represents the true measured value, and y_i is the estimated or predicted value.

30.4.4 Accuracy of fault detection (FDA%)

Fig. 30.4 shows that SHCS achieved an excellent 98.5% fault detection rate, even after PdM reached 95%. The great accuracy

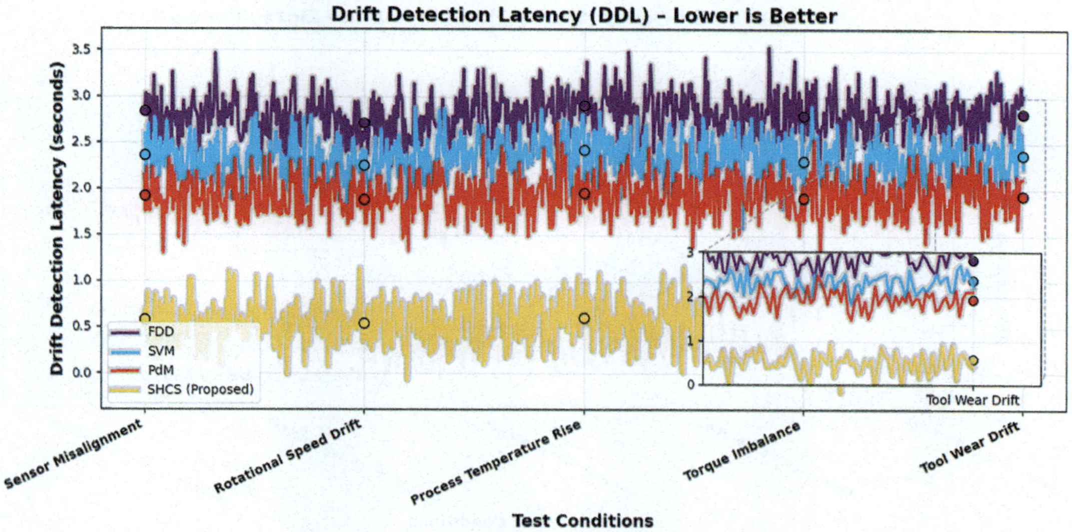

FIGURE 30.5 Analysis of drift detection. *FDD*, Fault detection and diagnosis; *PdM*, predictive maintenance; *SVM*, support vector machine.

was due to the embedded random forest anomaly detector and digital twin validation loop, which allowed detecting the calibration drifts and anomalies at an earlier stage with higher confidence, minimizing false alarms and false nondetections.

Accuracy of fault detection μ_{err} is expressed in Eq. (30.14).

$$\mu_{err} = 1 - \frac{MS_{model}}{MS_{baseline}} \times 100 \tag{30.14}$$

Accuracy of false deduction quantifies the percentage decrease in reconstruction error compared to a baseline approach. In this equation, MS_{model} represents the MSE obtained from the proposed SHCS model, and $MS_{baseline}$ corresponds to the MSE of the benchmark system such as PdM or FDD.

30.4.5 Drift detection latency (DDL)

Fig. 30.5 illustrates that SHCS had the lowest average drift detection latency (DDL) of only 0.56 seconds (PDM was 1.9 seconds). The quick reaction is due to the real-time sensor fusion and incorporation of RL that speeds up the detection and correction loop. This is because this fast latency will allow the uptime to be close to continuous and the response to arising calibration problems to be timely.

Analysis of drift detection FD_R is expressed in Eq. (30.15).

$$FD_R = \frac{U_Q}{U_Q + F_M} \times 100 \tag{30.15}$$

The fault detection rate determines the percentage of true faults correctly identified by the model. It measures the model's sensitivity and ability to recognize faults without overlooking actual anomalies.

In this equation, FD_R denotes the fault detection rate expressed as a percentage, U_Q represents the number of true positive fault detections, and F_M stands for the number of false negatives or missed faults.

30.4.6 Self-healing efficiency (SHE%)

Fig. 30.6 illustrates that SHCS showed an extraordinary efficiency of 95% self-healing as compared to PdM, which is only 69%. This enhancement validates the fact that the RL agent in Sentinel corrected calibration faults independently without the need for an operator. The model ensured the high stability of the performance of the machine by using intelligent recalibration, successfully reaching the conditions of close automation of the maintenance cycle.

Analysis of self-healing efficiency S_u is expressed in Eq. (30.16).

$$S_u = \alpha \times (1 - CMJ_u) - \beta \times DDM_u \tag{30.16}$$

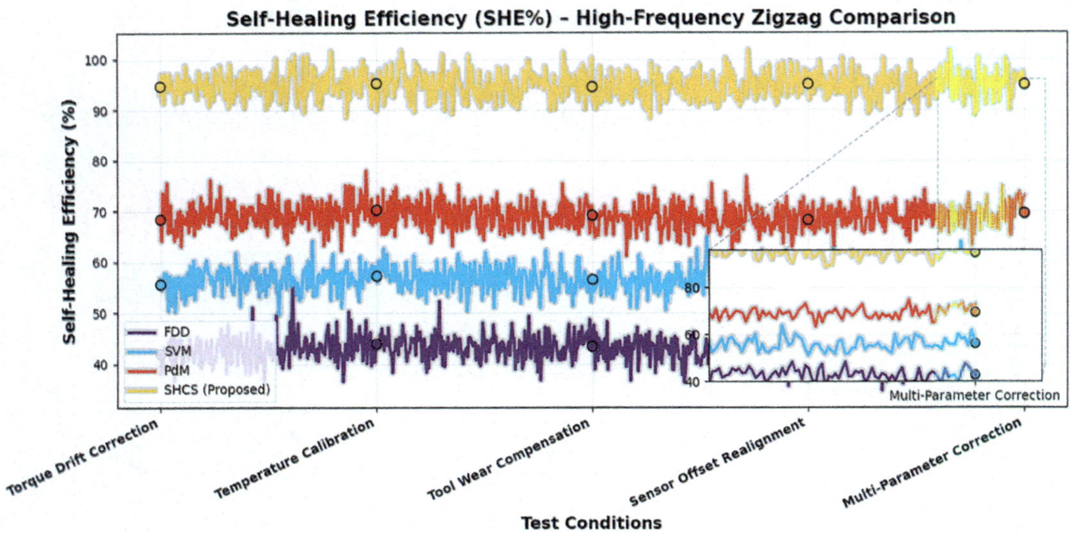

FIGURE 30.6 Analysis of self-healing efficiency. *FDD*, Fault detection and diagnosis; *PdM*, predictive maintenance; *SVM*, support vector machine.

TABLE 30.1 Analysis of process quality index.

Test condition	FDD	SVM	PdM	SHCS
Tool precision retention	0.76	0.81	0.85	**0.97**
Dimensional tolerance stability	0.75	0.82	0.86	**0.98**
Surface finish consistency	0.77	0.80	0.85	**0.97**
Temperature-driven variance	0.74	0.82	0.86	**0.98**
Speed-dependent variability	0.76	0.81	0.85	**0.97**

FDD, fault detection and diagnosis; *PdM*, predictive maintenance; *SVM*, support vector machine.

TABLE 30.2 Analysis of SDR.

Test condition	FDD	SVM	PdM	SHCS
Scheduled calibration	55.3	62.4	70.8	**94.5**
Unplanned maintenance	56.7	61.9	71.2	**95.0**
Thermal fault recovery	55.9	63.1	70.5	**94.7**
Rotational fault repair	57.1	62.8	71.0	**95.3**
Sensor failure recovery	56.2	62.6	70.9	**94.9**

FDD, fault detection and diagnosis; *PdM*, predictive maintenance; *SDR*, system downtime reduction; *SVM*, support vector machine.

The reinforcement-based correction reward evaluates the model's performance in minimizing drift and latency simultaneously.

In this equation, CMJ_u represents the drift magnitude index at time u, DDM_u is the drift detection latency at time u, α is the weight coefficient for accuracy, and β is the penalty coefficient for latency.

Reinforcement reward function S_u is expressed in Eq. (30.17).

$$S_u = \partial(\mathrm{SH}_u + \mathrm{PQ}_u) - \gamma \times F_u \tag{30.17}$$

TABLE 30.3 Analysis of energy efficiency gain.

Test condition	FDD	SVM	PdM	SHCS
Motor torque optimization	14.2	19.8	26.7	**43.5**
Cooling system regulation	13.9	20.1	25.9	**44.0**
Process load balancing	14.5	19.6	26.2	**43.2**
Temperature control loop	13.8	20.0	26.4	**44.3**
Speed synchronization	14.3	19.9	26.5	**43.9**

FDD, fault detection and diagnosis; *PdM*, predictive maintenance; *SVM*, support vector machine.

ALGORITHM 30.1 Discrepancy detection using multihead attention encoder.

Input:

$X \in \mathbb{R}^{T \times d_x} \# Real - timesensordatamatrix$

$P \in \mathbb{R}^{T \times d_x} \# Positionalencodings$

$H \# Numberofattentionheads$

$d_h \# Dimensionperhead$

$d_{model} = H \times d_h \# Totalmodeldimension$

Output:

$y \in \mathbb{R}\hat{\ } T \# Discrepancyprobabilitysequence$

Steps:

1: *Initializeparameters*:

$W^E \in \mathbb{R}^{d_x \times d_{model}}$

$For i = 1.H$:

$W_i^Q, W_i^K, W_i^V \in \mathbb{R}^{d_{model} \times d_h}$

$W^O \in \mathbb{R}^{d_{model} \times d_{model}}, W^c \in \mathbb{R}^{d_{model} \times 1}, b^c \in \mathbb{R}$

$2 : S \leftarrow X \cdot W^E + P \# Inputembedding + positionalencoding$

3: $For i = 1 to H do$

$4 : Q_i \leftarrow S \cdot W_i^Q$

$5 : K_i \leftarrow S \cdot W_i^K$

$6 : V_i \leftarrow S \cdot W_i^V$

7: $For t = 1 to T do$

8: $For u = 1 to T do$

$9 : score_{\{t,u\}}^{\{(i)\}} \leftarrow \frac{\left(Q_{i[t]} \cdot K_{i[u]}^T\right)}{\sqrt{d_h}}$

10: *endfor*

$11 : \alpha_{i[t,:]} \leftarrow softmax\left(score_{\{t,1.T\}}^{\{(i)\}}\right) 12 : head_{i[t]} \leftarrow \Sigma_{\{u=1\}_{i[t,u]}^{\{T\}\alpha}} \cdot V_{i[u]} 13 : endfor$

14: *endfor*

15: $H_{cat} \leftarrow [head_1 || head_2 || \ldots || head_H]$

16: $A \leftarrow H_{cat} \bullet W^O$

17: $S' \leftarrow LayerNorm(S + A)$

18: $S_{ff} \leftarrow FFN(S')$

19: $S_{out} \leftarrow LayerNorm(S' + S_{ff})$

20: $For t = 1 to T do$

21: $y[t] \leftarrow \sigma(S_{out[t]} \bullet W^c + b^c)$

22: *endfor*

23: *returny*

ALGORITHM 30.2 Digital-twin virtual calibration validation.

Input:

s_{real}#*Currentmachinestatevector*

$\mathcal{A} = \{a^1, a^2, \ldots, a^M\}$#*Setofcandidatecorrectiveactions*

τ_{safe}#*Safetythresholdforerror*

H#*Simulationhorizon*

$\lambda_e, \lambda_c, \lambda_s$#*Weightingcoefficientsforscoring*

Output:

$a*$ #*Safestvalidatedcorrectiveaction*

R#*Validationreportwithscoresandmetrics*

Steps:

$best_{score} \leftarrow -\infty$

$a^* \leftarrow NULL$

$R \leftarrow []$

ForeachainAdo

$s \leftarrow s_{real}$

$err_{sum} \leftarrow 0$

$cost_{sum} \leftarrow 0$

Forh = 1*toHdo*

$(s', metrics) \leftarrow SIM(s, a)$

$err_{sum} \leftarrow err_{sum} + metrics.error$

$cost_{sum} \leftarrow cost_{sum} + metrics.energy$

$s \leftarrow s'$

endfor

$score \leftarrow -(\lambda_e \bullet err_{sum} + \lambda_c \bullet cost_{sum}) + \lambda_s \bullet (1 - metrics.violation)$

Append$(a, err_{sum}, cost_{sum}, score)$*toR*

If$err_{sum} \leq \tau_{safe}$*andscore* > $best_{score}$*then*

$best_{score} \leftarrow score$

$a^* \leftarrow a$

endif

endfor

return(a^*, R)

The reinforcement reward function expresses the balance between the healing efficiency and restored performance, adjusted for the energy or computational cost incurred.

∂ is the weighting factor emphasizing performance restoration and healing efficiency, SH_u and PQ_u are the self-healing efficiency and performance restoration index at time u, respectively, γ is the penalty coefficient, and F_u represents the total energy or computational expenditure at time u.

30.4.7 Process quality index (PQI)

The proposed system had an outstanding process quality index (PQI) of 0.97, which was better than the PdM of 0.85. This value represents the very inert manufacturing quality since SHCS managed to retain the levels of precision, finish on the surface, and tolerance in changeable working environments. Improved efficiency in calibration led to easier production, as shown in Table 30.1.

PQI D_{ql} is expressed in Eq. (30.18).

$$D_{ql} = \min \frac{VS_M - \partial}{3\rho} \times \frac{\partial - MSM}{3\rho} \tag{30.18}$$

ALGORITHM 30.3 Reinforcement-learning-based adaptive calibration.

Input:
ENV#Realorsimulatedcalibrationenvironment
θ, φ*#Policyandvaluenetworkparameters*
$\mathcal{D}$*#Replaybuffer*
$\alpha_\theta, \alpha_\varphi$*#Learningrates*
γ*#Discountfactor*
EvalInterval#Evaluationfrequency
τ_{safe}*#Safetythresholdforvalidation*
Output:
$\pi_{(\theta*)}$*#Trainedpolicynetwork*
$Executed_{Actions}$*#Listofsafedeployedactions*
Steps:
$InitializePolicyNetwork_{Transformer(s;\ \theta)}$
$InitializeValueNetwork(s;\ \varphi)$
whiletrainingdo
$s \leftarrow ENV.reset()$
repeat
$a \sim \pi_{\theta(\cdot|s)}$
$(s', r, done) \leftarrow ENV.step(a)$
$Store(s, a, r, s', done) in \mathcal{D}$
$s \leftarrow s'$
untildone
$SamplebatchBfrom\mathcal{D}$
$Foreach\left(s_i, a_i, r_i, s_i'\right)inBdo$
$G_i \leftarrow r_i + \gamma \cdot V_{\varphi(s_i')}$
endfor
$L_{actor} \leftarrow -\Sigma_i\left[\log \pi_{\theta(a_i|s_i)} \cdot (G_i - V_{\varphi(s_i)})\right]$
$L_{critic} \leftarrow \Sigma_i(V_{\varphi(s_i)} - G_i)^2$
$\theta \leftarrow \theta - \alpha_\theta \bullet \nabla_\theta L_{actor}$
$\varphi \leftarrow \varphi - \alpha_\varphi \bullet \nabla_\varphi L_{critic}$
$IfiterationmodEvalInterval = 0then$
$(a_{valid}, R) \leftarrow DigitalTwin_{Validation(current_{state}, \{a_{pred}\}, \tau_{safe})}$
$Ifa_{valid} \neq NULLthen$
$Deploy(\pi_\theta)$
$RecordExecuted_{Actions} \leftarrow Executed_{Actions} \cup \{a_{valid}\}$
endif
endif
endwhile
$return(\pi_{\{\theta^*\}}, Executed_{Actions})$

The PQI measures how centered the process mean is within the specified limits. It accounts for both variability and mean shift.

In this equation, VS_M and MSM are the upper and lower specification limits, respectively. ∂ represents the mean of the process outputs, and ρ signifies the standard deviation of those outputs.

30.4.8 System downtime reduction (SDR%)

SHCS decreased the downtime of its system by 95%, whereas PdM achieved 71%. This is a major drop considering that automated, real-time fault correction and digital twin validation help to eliminate manual recalibration and result in fewer interruptions during the maintenance process, thus providing the ability to maintain the continuous operation of the machine with increased productivity, as shown in Table 30.2.

Weighted system downtime reduction (SDR) W_{SD} is expressed in Eq. (30.19).

$$W_{SD} = 101\log_{10}\frac{nu_i r_i^2}{nu_i(r_i - s_i)^2} \tag{30.19}$$

The weighted signal-to-distortion ratio accounts for varying sensor importance or reliability by applying adaptive weights.

In this equation, W_{SD} represents the weighted signal-to-distortion ratio in decibels,u_i denotes the confidence weight assigned to each signal sample, r_i is the true signal value, and s_i is the corresponding reconstructed signal.

30.4.9 Energy efficiency gain (EEG%)

The SHCS framework realized 44% energy efficiency, which is higher than 26% of PdM. The decrease in rework, undue recalibration, and optimization of the machine torque cycles resulted in significant energy savings. This shows that SHCS is making a contribution toward cost-effective and sustainable manufacturing in line with Industry 4.0 goals, as shown in Table 30.3.

Energy efficiency gain S_E is expressed in Eq. (30.20).

$$S_E = \beta \times \mathrm{EE}_t - \partial \times \mathrm{EC}_t \tag{30.20}$$

The reinforcement energy efficiency gain evaluates the system's adaptive energy optimization performance at a given time.

In this equation, S_E denotes the reinforcement reward associated with energy efficiency, EE_t is the energy efficiency gain at time t, EC_t is the energy consumption rate at time t, β represents the weight for efficiency emphasis, and ∂ is the penalty factor for excessive consumption.

SDR improvement factor D_{rimp} is expressed in Eq. (30.21).

$$D_{\text{rimp}} = \frac{D_{\text{model}} - D_{\text{baseline}}}{D_{\text{baseline}}} \times 100 \tag{30.21}$$

The SDR improvement factor measures the relative enhancement achieved by the proposed model compared to the baseline approach.

D_{model} represents the SDR achieved by the SHCS framework, and D_{baseline} corresponds to the SDR recorded under the traditional or PdM model.

The comparative study indicates that SHCS has a high level of performance in comparison with the traditional systems in all the parameters. The framework, with increased accuracy; speed in detecting drift; enhanced self-healing performance; and high energy saving, is a means to autonomously and in real-time control calibration. These findings confirm that Sentinel can transform PdM and make the industrial performance in-line with Industry 4.0 goals.

30.5 Conclusion and recommendations

The presented Sentinel framework shows a tremendous breakthrough in contemporary manufacturing by embedding self-healing calibration, digital twin simulation, and adaptive control based on RL. The AI4I Predictive Maintenance Dataset can be experimentally validated to demonstrate significant gains in the calibration accuracy, fault detection, and operational reliability and a significant reduction of DDL, system downtime, and energy consumption. Sentinel provides autonomous detection and correction of the discrepancies, which will result in continuous uptime, consistency in processes, and a minimum amount of human intervention, which fits the goals of the Industry 4.0 smart, connected, and resilient production systems perfectly.

Practically the study confirms the possibility of using AI-based maintenance insights together with virtual twin settings in the creation of sustainable and autonomous manufacturing ecosystems. But in the real-world application, there is a need to expand the framework to support cross-cell synchronization, multiasset interaction, and hybrid edge-cloud infrastructure.

Future applications are suggested to also consider real-time integration with edge computing, explainable AI in calibration choices, and self-supervised learning methods to improve flexibility and visibility. Furthermore the use of cybersecurity measures in the calibration loop and large-scale industrial tests is also possible to enhance the role of robustness in systems. In general, Sentinel forms a solid base of autonomous, energy-saving, and self-optimizing industrial processes for the future, the following generation of intelligent manufacturing systems.

References

Aldrini, J., Chihi, I., & Sidhom, L. (2024). Fault diagnosis and self-healing for smart manufacturing: A review. *Journal of Intelligent Manufacturing, 35*(6), 2441–2473. doi:10.1007/s10845-023-02165-6.

Alsubaih, A. A. S., Sepehrnoori, K., Delshad, M., & Alsaedi, A. (2025). A comprehensive review of well integrity challenges and digital twin applications across conventional, unconventional, and storage wells. *Energies, 18*(17). http://www.mdpi.com/journal/energies/. doi:10.3390/en18174757.

Amazon Web Services. (n.d.). Guidance for Self-Calibrating Level 4 Digital Twins on AWS. *GitHub*. https://github.com/aws-solutions-library-samples/guidance-for-self-calibrating-level-4-digital-twins-on-aws.

Balasubramanian, A., & Gurushankar, N. (2019). AI-powered hardware fault detection and self-healing mechanisms. *IJCEM, 6*, 23–30.

Breiman, L. (2001). Random forests. *Machine Learning, 45*(1), 5–32. doi:10.1023/a:1010933404324.

Carvalho, T. P., Soares, F. A. A. M. N., Vita, R., Francisco, R. P., Basto, J. P., & Alcalá, S. G. S. (2019). A systematic literature review of machine learning methods applied to predictive maintenance. *Computers & Industrial Engineering, 137*, 106024. doi:10.1016/j.cie.2019.106024.

El Ahmadi, S. E. A., & El Abbadi, L. (2025). A predictive self-healing model for optimizing production lines: Integrating AI and IoT for autonomous fault detection and correction. *Engineering Proceedings., 97*(1). Www.mdpi.com/journal/engproc. doi:10.3390/engproc2025097006.

Gao, A., & Liu, J. (2025). STEAD: Spatio-temporal efficient anomaly detection for time and compute sensitive applications. *arXiv*. United States arXiv https://arxiv.org . doi:10.48550/arXiv.2503.07942.

Goodfellow, I., Bengio, Y., & Courville, A. (2016). *Deep learning*. MIT Press.

Guo, L., Hu, H., & Chen, F. (2025). Research on intelligent fault diagnosis and self-healing control for smart grids based on digital twin. *7th International Conference on Energy Systems and Electrical Power, ICESEP 2025* (pp. 1096–1101). Institute of Electrical and Electronics Engineers Inc. doi:10.1109/ICESEP66633.2025.11155702.

Hevner, A. R., March, S. T., Park, J., & Ram, S. (2004). Design science in information systems research. *MIS Quarterly, 28*(1), 75–106. doi:10.2307/25148625.

Jardine, A. K. S., Lin, D., & Banjevic, D. (2006). A review on machinery diagnostics and prognostics implementing condition-based maintenance. *Mechanical Systems and Signal Processing, 20*(7), 1483–1510. doi:10.1016/j.ymssp.2005.09.012.

Lee, J., Bagheri, B., & Kao, H. A. (2015). A cyber-physical systems architecture for industry 4.0-based manufacturing systems. *Manufacturing Letters, 3*, 18–23. http://www.journals.elsevier.com/manufacturing-letters/. doi:10.1016/j.mfglet.2014.12.001.

Liu, R., et al. (2018). Data-driven predictive maintenance and anomaly detection in industrial informatics. *IEEE Transactions on Industrial Informatics,1*.

Lundberg, S. M., & Lee, S. I. (2017). A unified approach to interpreting model predictions. *Advances in Neural Information Processing Systems* (pp. 4766–4775). Neural Information Processing Systems Foundation.

Kagermann, H., Wahlster, W., & Helbig, J. (2013). Recommendations for implementing the strategic initiative INDUSTRIE 4.0: Securing the future of German manufacturing industry: Final report of the Industrie 4.0 Working Group. Forschungsunion

Mani, N., & Attaranasl, S. (2025). Self-healing digital twins: Hybrid generative and privacy-preserving AI for adaptive wellness platforms. *Proceedings—2025 IEEE/ACM International Conference on Connected Health: Applications, Systems and Engineering Technologies, CHASE 2025* (pp. 424–429). Institute of Electrical and Electronics Engineers Inc. doi:10.1145/3721201.3725427.

Mengesha, G. (2025). Beyond repair: A critical review of smart, sustainable, and AI-driven strengthening techniques for aging civil infrastructure. *International Journal of Emerging Science and Engineering (IJESE), 13*(6), 10–19.

Monostori, L. (2014). Cyber-physical production systems: Roots, expectations and R&D challenges. *Procedia CIRP, 17*, 9–13. doi:10.1016/j.procir.2014.03.115.

Najafzadeh, M., & Yeganeh, A. (2025). AI-driven digital twins in industrialized offsite construction: A systematic review. *Buildings, 15*(17), 2997. doi:10.3390/buildings15172997.

Ogunmolu, A. M., Olaniyi, O. O., Popoola, A. D., Olisa, A. O., & Bamigbade, O. (2025). Autonomous artificial intelligence agents for fault detection and self-healing in smart manufacturing systems. *Journal of Energy Research and Reviews, 17*(8), 20–37. doi:10.9734/jenrr/2025/v17i8445.

Ong, K. S. H., Niyato, D., & Yuen, C. (2020). Predictive maintenance for edge-based sensor networks: A deep reinforcement learning approach. *2020 IEEE 6th World Forum on Internet of Things (WF-IoT)* (pp. 1–6). IEEE.

Peres, R. S., Jia, X., Lee, J., Sun, K., Colombo, A. W., & Barata, J. (2020). Industrial artificial intelligence in industry 4.0—Systematic review, challenges and outlook. *IEEE Access, 8*, 220121–220139. http://ieeexplore.ieee.org/xpl/RecentIssue.jsp?punumber=6287639. doi:10.1109/ACCESS.2020.3042874.

Ribeiro, M. T., Singh, S., & Guestrin, C. (2016). "Why Should I Trust You?": Explaining the predictions of any classifier. *KDD Conference*.

Rudin, C. (2019). Stop explaining black box models and use interpretable models instead. *Nature Machine Intelligence, 1*(5), 206–215.

Saxena, A., Goebel, K., Simon, D., & Eklund, N. (2008). NASA turbofan engine degradation simulation data set (C-MAPSS). *NASA Ames prognostics data repository* https://www.nasa.gov/centers/ames/research/engineering/technology/prognostics.

Selcuk, S. (2016). Predictive maintenance, its implementation and latest trends. *Proceedings of the Institution of Mechanical Engineers, Part B: Journal of Engineering Manufacture, 231*(9), 1670–1679.

Sutton, R. S., & Barto, A. G. (2018). *Reinforcement learning: An introduction* (2nd ed.). MIT Press.

Tao, F., Qi, Q., Liu, A., & Kusiak, A. (2018). Data-driven smart manufacturing. *Journal of Manufacturing Systems, 48*, 157–169. doi:10.1016/j.jmsy.2018.01.006.

Wang, J., Ma, Y., Zhang, L., Gao, R. X., & Wu, D. (2018). Deep learning for smart manufacturing: Methods and applications. *Journal of Manufacturing Systems, 48*, 144–156. http://www.elsevier.com. doi:10.1016/j.jmsy.2018.01.003.

Zhang, Y., & Zhou, Z.-H. (2019). *Deep learning-based anomaly detection: A survey*. ACM Computing Surveys.

Zhou, Q., et al. (2021). *Robotics and computer-integrated manufacturing: Integration for intelligent manufacturing*. Robotics & Computer-Integrated Manufacturing.

Rauba, et al. (2024). Self-Healing Machine Learning: A Framework for Autonomous Adaptation in Real-World Environments. *Neural Information Processing Systems*. doi:10.52202/079017-1336.

Chapter 31

Intelligent remote actuation and control strategies for electrical tilt systems

Sujata N Bhavikatti
Electronics and Communication Engineering, KLS Gogte Institute of Technology, Belagavi, Karnataka, India

31.1 Introduction

The applications—such as smart traffic management, smart public transportation, smart management of energy, waste, water, etc., environmental monitoring, smart governance/e-governance, and smartness applied to security and surveillance, buildings and infrastructure, smart healthcare, smart parking, smart education, and disaster management systems—are developed and deployed as integral components of the smart city ecosystem. Among these, certain applications, such as smart buildings, hospitals, industrial systems, water management, and smart parking, typically do not require wide-area information exchange, as their data flow is confined to localized environments. The smart air monitoring system can be deployed city-wide without geographical limitations, and data flow occupies large geographical area.

Electrical tilt mechanism is essential in modern engineering applications where positioning of the system is to be adjusted in accordance with other parameters. In these applications tilt mechanisms are required to achieve precise positioning and orientation. This is critical to ensure optimal system performance (Issa et al., 2025; Paliyal et al., 2024). In solar systems the solar panels are focused toward the sun, which includes adjusting an antenna's beam angle to enhance signal coverage and aligning a robotic sensor. The electrical tilt system provides the fine control required to adapt to the dynamic environment. They function as an interface between the electrical signals and mechanical movement, translating the digital commands into accurate angular adjustments (Ding et al., 2026; Thungsuk et al., 2023).

In communication infrastructure, in particular cellular base station antennas, electrical tilt systems play a central role in controlling the antenna's vertical radiation angle. This process, referred to as remote electrical tilt (RET), allows network operators to modify the tilt of an antenna electronically to optimize coverage areas, manage interference, and adapt to shifting network demands (Mendo et al., 2023; Vannella et al., 2022). RET technology eliminates the need for manual adjustment by enabling electronic actuation through a feedback-controlled motor system. The integration of RET units has become standard in advanced 4G and 5G networks, supporting the growing need for intelligent and remotely manageable systems (Ericsson, 2024; Nikou et al., 2021). Similarly, in renewable energy applications such as solar tracking systems, electrical tilt mechanisms continuously adjust the orientation of photovoltaic modules to capture maximum sunlight, thereby enhancing energy yield and operational efficiency (Kumba et al., 2024; Paliyal et al., 2024).

The conventional control methods of existing tilt systems face limitations when they are operating in real-world conditions. Factors such as time delays, mechanical disturbances, and system nonlinearities can degrade positioning accuracy and dynamic response (Ding et al., 2026). Traditional proportional integral derivative (PID) controllers are effective in linear and predictable environments. They fail to maintain stability and precision under complex, rapidly changing conditions (Fan et al., 2025). These challenges necessitate the need for intelligent control strategies for adaptive compensation with uncertainties and disturbances. The remote accessibility reduces the need for manual intervention, simplifies maintenance operations, and allows rapid reconfiguration in response to positioning demands (Vannella et al., 2022). In telecommunications, operators can remotely adjust antenna tilts to optimize network coverage without deploying field technicians, significantly lowering operational costs and improving response times (Mendo et al., 2023; Nikou et al., 2021).

Computational Intelligence in Mechatronics: Solving Real-World Problems in Electronic Systems Design. DOI: https://doi.org/10.1016/B978-0-443-27580-7.00019-6

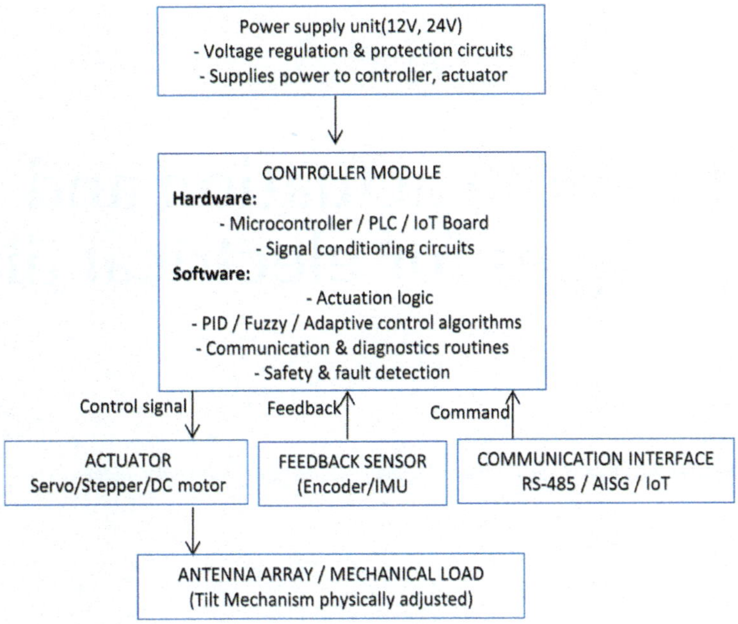

FIGURE 31.1 Block diagram of electrical tilt system. *IoT*, Internet of Things; *PID*, proportional integral derivative.

By combining intelligent control algorithms and remote actuation technologies the electrical tilt systems achieve the next level of autonomy and adaptability. These enhancements improve the system reliability and contribute to reduced maintenance costs and enhanced energy efficiency (Paliyal et al., 2024). Moreover, the integration of real-time feedback sensors and predictive control algorithms enables proactive fault detection, preventing potential failures and extending system lifespan (Ding et al., 2026). The intelligent computational strategies improve the key performance parameters such as accuracy, responsiveness, and reliability. The inclusion of remote control capabilities ensures operational flexibility and scalability. These together position the intelligent electrical tilt systems as foundational elements in the development of next-generation smart adaptive engineering infrastructures (Ding et al., 2026; Mendo et al., 2023; Vannella et al., 2022).

31.1.1 Functional block diagram of an electrical tilt system

The overall architecture and signal flow of an RET system is the interaction between hardware and software components. The electrical power, control signals, and feedback data circulate through the various blocks of the system in a closed-loop configuration. The functional blocks are the power supply unit, controller, feedback sensor, actuator, and communication interface. These elements work together to achieve precise antenna tilt adjustment. Block diagram of typical electrical tilt system to achieve adaptive and precise angular positioning is shown in Fig. 31.1.

The power supply unit provides required electrical energy for control electronics and motors. Depending on the application, ensure stable power delivery of 12 or 24 V to the actuation and control modules (Archiver Market Research, 2025). In intelligent RET systems the regulated power management is critical to maintain voltage stability during remote operations and in varying load conditions and (Vannella et al., 2022).

Controller module is the brain of the system, executing actuation logic and decision-making algorithms. It processes the inputs from feedback sensors and communication interfaces to generate control signals for the actuator. Controllers can be simple analog circuits or digital microcontrollers, Programmable Logic Controllers (PLCs), or Internet of Things (IoT)-enabled boards that are capable of advanced computation and remote interfacing (Ericsson, 2024; Nikou et al., 2021). The recent designs incorporate adaptive and fuzzy logic algorithms to enhance control precision under nonlinear conditions (Ding et al., 2026; Thungsuk et al., 2023).

The actuator is meant to convert electrical energy into mechanical movement. DC, servo, and stepper motors are employed based on the precision and torque requirements (Fan et al., 2025). Motors with integrated encoders are preferred, as it provides direct feedback on velocity angular position. In RET-based antenna systems, servo actuators along with position feedback loops are standard to ensure accurate beam alignment and stability in the presence of wind or vibration disturbances (Mendo et al., 2023; Vannella et al., 2022).

The feedback sensor is essential to ensure the accuracy and stability of the system by providing the real-time data on tilt position. Sensors, such as inertial measurement units (IMUs), potentiometers, optical encoders, and Hall effect sensors, are employed depending on the system requirements (Kumba et al., 2024). The feedback enables the controller to compare the actual position with desired setpoint and adjust the motor operation accordingly. Intelligent RET implementations combine multiple sensors to provide feedback for improved fault tolerance (UoKerbala, 2025).

The communication interface enables devices and operators to send the control commands and receive system feedback. Depending on the system requirements, communication is either wired (RS-485, Controller Area Network (CAN), and Antenna Interface Standards Group (AISG)) or wireless (Bluetooth, Wi-Fi, or LoRa) (Ericsson, 2024; Archiver Market Research, 2025). The choice of interface between tilt mechanism and controller determines the system's range, reliability, and compatibility with existing network infrastructure. In IoT-enabled RET systems, secure wireless communication allows cloud-based control, which significantly improves scalability (Paliyal et al., 2024).

These functional blocks interact in a closed-loop configuration. The actuator's output (tilt position) is continuously monitored by the sensors and communicated to the controller. The corrective actions are taken by the controller to minimize errors (Musa et al., 2023). This feedback mechanism is essential for achieving precise, stable tilt performance, in systems exposed to environmental variations and mechanical disturbances (Ding et al., 2026).

Further, the chapter flow includes Chapter 2, which introduces the electrical tilt systems, emphasizing its principal blocks, control mechanisms, the existing RET models in the market, and the limitations of conventional designs. Chapter 3 includes the integration of recent technologies, the IoT, fuzzy logic control, and cloud-based performance monitoring in the improved RET architecture. The hardware and software implementation aspects are highlighted. The chapter concludes with summary and remarks.

31.2 Improving the system performance

Performance enhancement of the electrical tilt system is realized through the integration of intelligent control strategies (Fan et al., 2025). The traditional control methods were effective in simple environments. The limitation is they struggle to maintain stability and precision in nonlinear conditions and variable loads (Thungsuk et al., 2023). Intelligent control methods introduce adaptability and learning capability to the system, which allows it to dynamically compensate for uncertainties (Ding et al., 2026).

The integration of recent technologies such as the IoT, fuzzy logic, and performance monitoring through cloud computing has significantly enhanced the efficiency and reliability of electrical tilt systems. These technologies enable real-time data collection, intelligent decision-making, and adaptive control to improve the performance parameters and optimized system operation.

The tilt system can correct positional errors dynamically by incorporating feedback sensors, microcontrollers, and machine learning–based prediction and optimize the motor performance in real time (UoKerbala, 2025). These enhancements lead to reduced power consumption, smoother movement, and faster response to the input commands. The temperature-compensated control ensures consistent operation across environmental fluctuations. The adaptive gain adjustment fine-tunes the response of the controller to prevent overshoot or oscillation (Mendo et al., 2023). Over long-term deployment the intelligent strategies improve the reliability and reduce the wear of mechanical components (Paliyal et al., 2024).

31.2.1 Market-available RET models and their limitations

Several leading manufacturers across the globe have developed commercial RET systems, which provide reliable electronic tilt adjustment for base station antennas. These systems integrate motorized actuators, feedback sensors, and standard communication interfaces such as AISG, enabling operators to manage antenna tilt remotely.

The key vendors of RET systems in the global market are listed in Table 31.1. They offer motorized actuators with feedback-controlled positioning. Communication Components Inc. provides RET actuators for the Base Station Antenna (BSA) RET200, using DC motors, encoders, and RS485/AISG interfaces. John Mezzalingua Associates (JMA) Wireless/Talley supplies AISG 2.0 compliant RET assemblies with calibration and fault reporting features. CommScope/Andrew RET supports antenna products for seamless remote tilt control. Amphenol Antennas offer RET components and guides to ensure multivendor interoperability. Moog/Thomson delivers improved precision brushless servo and linear actuators and magnetic encoders, leveraging aerospace-grade motion control technology that is adaptable for advanced RET applications.

Despite the widespread adoption of existing RET systems, they exhibit notable limitations such as using proprietary communication protocols, limiting interoperability across the brands and increasing deployment complexity (Mendo et al., 2023). The communication range in these systems is often restricted to short distances. System upgrades are hindered

TABLE 31.1 RET system vendors and specifications.

Model	Representative product	Interface
Communication Components Inc.	BSA-RET200	DC motor + encoder; RS-485/AISG interface
JMA Wireless/Talley	RET-200 and similar assemblies	AISG 2.0 compatible motor assemblies, calibration, and fault reporting
CommScope/Andrew	RET-enabled antennas and integrated actuator assemblies	Integrated RET actuator solutions for remote tilt control
Amphenol Antennas/other antenna Original Equipment Manufacturers (OEMs)	RET system guides and components	RET components for multivendor deployments
Moog/Thomson	High-performance rotary/linear servo actuator modules	BLDC/servo; magnetic absolute encoders; position-loop electronics—aerospace-grade tech applicable to RET

BLDC, brushless DC; *RET*, remote electrical tilt.

TABLE 31.2 Hardware and software details of improved architecture.

Technology	Hardware components	Software/algorithms
IoT integration	IoT-enabled sensors (tilt, temperature, and power), communication modules (4G/5G, LoRa, and Wi-Fi), microcontroller/embedded system	Message Queuing Telemetry Transport (MQTT)/Hypertext Transfer Protocol (HTTP) protocols, real-time data acquisition and transmission software
Fuzzy logic control	DSP/MCU	FIS, membership functions for signal strength, interference, and coverage metrics
Cloud monitoring	Edge gateway, server, or virtual cloud infrastructure	Data analytics software, predictive maintenance algorithms, KPI dashboards, and performance visualization tools

DSP, digital signal processor; *FIS*, fuzzy inference system; *IoT*, Internet of Things; *KPI*, key performance indicator; *MCU*, microcontroller unit.

by closed hardware architectures (Archiver Market Research, 2025). These commercial units typically lack AI-based optimization and adaptive control algorithms, which makes them less effective under nonlinear and disturbed operating conditions (Ding et al., 2026). Absence of environmental adaptability leads to no compensation for temperature changes or mechanical wear. This reduces long-term reliability (Fan et al., 2025). High cost of these commercial solutions restricts their use in experimental and small-scale projects (Paliyal et al., 2024).

31.3 Practical implementation of proposed feature enhancements

The versatility of electrical tilt (Tables 31.1 and 31.2) systems allows it to be applied in diversified engineering domains where precise angular control, remote actuation, and environmental adaptability are essential. Among these, two applications, antenna tilt mechanism in mobile communication base stations and solar tracking system in renewable energy installations, are considered in this chapter. Both applications use the same core architecture, consisting of power supply, actuator, controller, feedback sensor, and communication interface. These differ with respect to environmental conditions, control objectives, and required precision levels. Practical case studies of both systems are discussed, followed by discussion on intelligent feature enhancement and broader applications of electrical tilt technology (Tables 31.1 and 31.2).

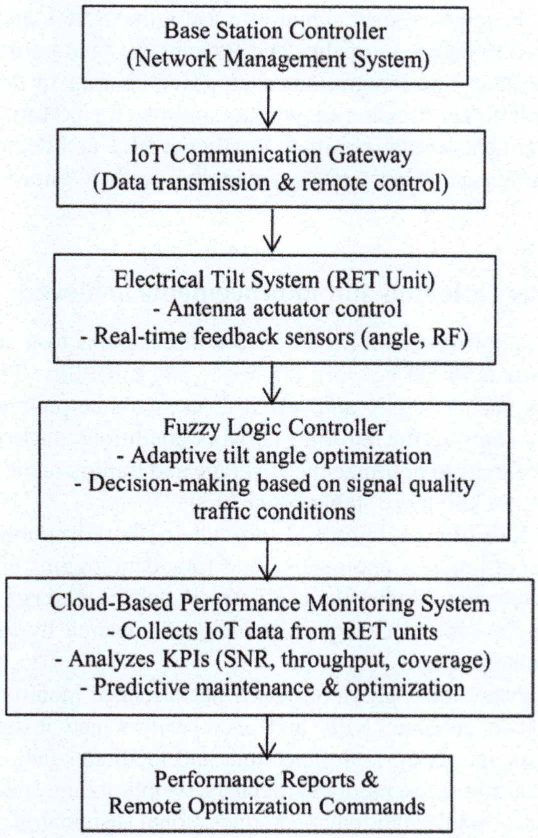

FIGURE 31.2 Improved architecture of RET system. *IoT*, Internet of Things; *KPIs*, key performance indicators; *RET*, remote electrical tilt; *SNR*, signal-to-noise ratio.

The improved RET system shown in Fig. 31.2 incorporates three functional blocks Table 31.1 that enhance its adaptability, overall intelligence, and efficiency beyond the existing actuator and controller setup. The IoT module enables remote connectivity, real-time data exchange, and centralized monitoring of RET units and network controllers and ensures seamless communication and operational visibility. The fuzzy logic control unit introduces intelligent and adaptive decision-making capabilities. This allows dynamic adjustment of antenna tilt angles depending on varying environmental and network conditions to optimize coverage and signal quality.

The cloud-based performance monitoring platform facilitates predictive maintenance based on data analysis and performance optimization by leveraging analytics and cloud computing, thereby improving the reliability, reducing downtime, and enhancing the system performance. These details are listed in Table 31.2.

The realization of the improved electrical tilt system involves the hardware and software components corresponding to each of the three technological enhancements, IoT, fuzzy logic, and cloud-based monitoring. The IoT module includes IoT-enabled sensors such as temperature sensors, tilt angle sensors, and power monitoring units to capture real-time operational parameters from the RET system. These sensors are interfaced with embedded systems equipped with communication modules such as 4G/5G, LoRa, or Wi-Fi for reliable data transmission. The software includes communication protocols such as MQTT or HTTP for efficient data exchange (Fig. 31.2; Table 31.1) between the RET units and network controllers to enable real-time monitoring and remote control functionality.

The fuzzy logic unit is realized using digital signal processor (DSP) or microcontroller unit (MCU) to execute the fuzzy inference algorithms. A fuzzy inference system (FIS) is the software component, which defines membership functions required for key performance indicators (KPIs) such as signal strength, interference, and traffic load. Depending on these inputs the fuzzy logic algorithm identifies the optimal antenna tilt angle in real time. The decision-making process ensures an adaptive adjustment in changing environment and network conditions, thus enhancing the overall signal quality and coverage efficiency.

The cloud platform is used to monitor the performance of multiple RET units using an edge gateway. It aggregates the data from multiple RET units and transmits it to a cloud server. On the software side, cloud-based analytics tools process and visualize performance power consumption, data signal-to-noise ratio (SNR), and system availability. The predictive maintenance algorithms are employed to detect anomalies and forecast the faults. Interactive dashboard provides network operators with real-time system insights. The integration of cloud computing in the RET system enhances scalability, supports data storage, and enables intelligent decision-making to optimize the performance.

The hardware and software components create a unified, intelligent RET architecture, which is capable of autonomous operation, adaptive control, and remote performance management that leads to improvements in efficiency, reliability, and operational intelligence.

31.3.1 Performance parameters affected and improvements achieved

The IoT, fuzzy logic, and cloud monitoring technologies have a direct impact on several key performance parameters, which leads to significant improvements in the network efficiency and reliability. The primary parameter signal quality, which is measured through metrics such as SNR and signal-to-interference-plus-noise ratio (SINR). The fuzzy logic controller regularly and continuously analyzes the real-time network conditions, such as interference levels and traffic load, to adaptively set the antenna tilt angle so as to optimize the coverage and minimize the interference. This intelligent control ensures clearer signals, fewer call drops, and more stable connections.

The **system availability and reliability** are improved through IoT-based monitoring, which allows the tracking of operational parameters and detection of faults or anomalies. Real-time data logging and remote diagnostics reduce system downtime and enable proactive maintenance, which enhances network uptime. **Energy efficiency** benefits from the adaptive rotation of the tilt actuator. The precise and necessary tilt movements ensured by fuzzy logic lead to minimized power consumption and extended component lifespan.

The **operational efficiency** is elevated through cloud-based performance monitoring. Large-scale data from multiple RET units is analyzed periodically to generate KPIs and also performance insights. The cloud integration reduces manual supervision of the tilt system, speeds up fault detection, and facilitates data-driven decision-making for network optimization. These improvements contribute to a more intelligent, self-optimizing, and resource-efficient tilt control system, which ensures improved network performance and long-term operational sustainability.

30.3.2 Calculations and mathematical modeling

The mathematical modeling of improved electrical tilt architecture focuses on quantifying the relationship between control inputs, the network parameters, and the resulting tilt angle optimization achieved. The **fuzzy logic control** unit forms the core of the adaptive mechanism. The input variables SNR, user traffic load, and interference level are processed by FIS. The predefined membership functions and rule sets are used by the system to determine optimal antenna tilt angle (θ), which maximizes the network performance. The defuzzification process, implemented using the centroid method, computes the control output as per Eq. (31.1).

$$\theta^* = \frac{\sum (\mu_i \cdot \theta_i)}{\sum \mu_i} \tag{31.1}$$

where μ_i represents the membership degree and θ_i denotes the tilt angle adjustment associated with each rule. The IoT communication efficiency in the RET system is represented by the Shannon Hartley theorem given by Eq. (31.2).

Here R is the achievable data rate, and B is the bandwidth. This relation evaluates the system's ability to transmit real-time monitoring data effectively. The cloud performance monitoring employs KPIs system availability, defined by

$$R = B\log_2(1 + SNR) \tag{31.2}$$

To ensure accurate calculations of these metrics, continuous IoT-based data acquisition is adapted to predictive performance assessment.

$$\text{Availability} = \frac{\text{uptime}}{\text{total time}} \times 100 \tag{31.3}$$

These mathematical relations provide a strong analytical foundation required for evaluation and optimization of intelligent RET systems. They facilitate the prediction of system behavior in varying network and environmental conditions. This enables data-driven fine-tuning of control parameters and ensures the optimal signal coverage, energy efficiency, and long-term reliability.

31.3.3 Applications and specific performance requirements

The scope of electrical tilt systems extends to Heating, Ventilation, and Air Conditioning (HVAC) louvers, robotic arms, surveillance camera gimbals, and intelligent lighting systems (Musa et al., 2023). Each application demands distinct performance requirements: control precision, actuation speed, robustness, and environmental resilience.

1. Cellular antennas require high precision (±0.1 degree) and stable operation under environmental conditions. Intelligent control ensures regular alignment and reduced interference.
2. Solar trackers demand slow, continuous, and energy-efficient actuation, with weather resistant hardware and long-term reliability.
3. HVAC louvers emphasize energy efficient control and rapid actuation for airflow regulation in dynamic environments.
4. Robotic arms require high-speed, high-precision control with low latency and strong disturbance rejection.
5. Lighting systems benefit from wireless adjustability and adaptive positioning to optimize illumination or reduce glare in dynamic spaces.

In all these domains, integrating computational intelligence techniques, adaptive control, fuzzy logic, and optimization-based algorithms enhance the system's resilience, ensure precise control under uncertainty, and allow predictive adaptation to disturbances. These methods enable real-time decision-making and self-tuning capabilities, which the traditional control systems lack.

31.4 Summary

The antenna tilt systems and solar trackers demonstrate the transformative potential of intelligent control in electromechanical actuation. The embedding of computational intelligence, real-time feedback processing, and remote communication into electrical tilt systems evolves the system into adaptive, efficient, and reliable components of modern engineering infrastructure. The proposed enhancements, such as adaptive control algorithms, wireless connectivity, and predictive fault management, represent critical advancements in the development of intelligent remote actuation systems. These innovations improve the precision and performance. They contribute to operational flexibility, energy efficiency, and reduced maintenance cost across diverse applications.

31.5 Conclusion

This chapter presents a comprehensive study of intelligent remote actuation and control strategies adapted in electrical tilt systems. The role of computational intelligence in enhancing precision, adaptability, and reliability is explored. Integration of IoT, fuzzy logic control, and cloud-based performance monitoring led to an improvement in RET architecture. It achieves real-time communication, adaptive decision-making, and predictive maintenance. These enhancements improve upon the limitations of traditional control methods and ensure better compensation for disturbances, nonlinearities, and environmental variations. The resulting system shows significant improvements in key performance parameters such as signal quality, energy efficiency, and operational reliability. The incorporation of remote accessibility and cloud computing reduces manual intervention and maintenance costs. The proposed framework is an intelligent electrical tilt system, as it is an essential enabler of smart, self-optimizing engineering infrastructure. It contributes to sustainable technological advancements across communication networks, renewable energy systems, and automated control applications.

References

ACM Digital Library. (2025). *Remote electrical tilt optimization via safe reinforcement learning WCNC* (pp. 2021–2022). ACM Digital Library.

Al-Ibrahime, Z., & Korkmaz, F. (2023). Using the fuzzy logic algorithm in the high-precision solar tracking system. *International Research Journal of Innovations in Engineering and Technology (IRJIET), 7*(12), 149–157. doi:10.47001/IRJIET/2023.712021.

Cognitive Market, Research (2025). Archive Market ResearchRemote electrical tilt control cable module market report 2025: Trends, growth analysis, and forecast. https://www.cognitivemarketresearch.com/. (Accessed 16 July 2026).

Ericsson. (2024). Beam Tilt. Stockholm, Sweden: Ericsson. https://www.ericsson.com/en/portfolio/networks/ericsson-radio-system/antenna-system/antenna-system-accessories/ret-components. (Accessed: Jul. 16, 2026)

Ding, Z., Yao, Y., Gao, S., Yang, X., Li, C., Ren, J., … Liu, X. (2026). Research on variable universe fuzzy adaptive PID control system for solar panel sun-tracking. *Sustainability, 18*(3), 1503. doi:10.3390/su18031503.

Ericsson. *Ericsson's Auptim RET rApp (AI/ML Remote Electrical Antenna Tilt)*. (2024).

Fan, S., Wang, H., Zuo, C., & Han, J. (2025). Fuzzy adaptive PID-based tracking control for autonomous underwater vehicles. *Actuators, 14*(10), 470. doi:10.3390/act14100470.

Feng, Y., Hu, Q., Qu, K., Yang, W., Zheng, Y., & Chen, K. (2023). Reconfigurable intelligent surfaces: Design, implementation, and practical demonstration. Electromagnetic Science, 1(2), 0020111. doi:10.23919/EMSCI.2022.0011.

Issa, H. A., Abdali, L. M., Alhusseini, H., & Velkin, V. I. (2025). Design, modeling, and control of a dual-axis solar tracker using fractional order PID controllers for enhanced energy efficiency. *Results in Engineering, 27*, 106073. doi:10.1016/j.rineng.2025.106073.

Jin, Y., Vannella, F., Bouton, M., Jeong, J., & Al Hakim, E. (2021). A graph attention learning approach to antenna tilt optimization. arXiv. doi:10.48550/arXiv.2112.14843.

Khalaf, H.M. (2025). Controlling the pneumatic solar tracker by using adaptive neuro-fuzzy inference system (Master's thesis, University of Kerbala). College of Engineering, Department of Electrical and Electronics Engineering, Karbala, Iraq.

Kumba, K., Upender, P., Buduma, P., Sarkar, M., Saravanan, R., & Sundar, R. S. (2024). Solar tracking systems: Advancements, challenges, and future directions: A review. *Energy Reports, 12*, 3566–3583. doi:10.1016/j.egyr.2024.09.038.

Li, F., Pan, T., Li, W., Peng, Z., Guo, D., Jia, X., … Lin, Y. (2025). Flexible intelligent microwave metasurface with shape-guided adaptive programming. *Nature Communications, 16*(1), 3161. https://doi.org/10.1038/s41467-025-58249-9.

Li, J. (2025). Low-cost wideband tilted beam antenna for millimeter-wave vehicle applications. IEEE Transactions on Antennas and Propagation. Advance online publication. doi: 10.48550/arXiv.2506.08239

Mendo, A., Outes-Carnero, J., Ng-Molina, Y., & Ramiro-Moreno, J. (2023). Multi-agent reinforcement learning with common policy for antenna tilt optimization. *arXiv*. https://arxiv.org. doi:10.48550/arXiv.2302.12899.

Musa, A., Alozie, E., Suleiman, S. A., Ojo, J. A., & Imoize, A. L. (2023). A review of time-based solar photovoltaic tracking systems. *Information, 14*(4), 211. doi:10.3390/info14040211.

Nikou, A., Mujumdar, A., Sundararajan, V., Orlic, M., & Feljan, A. V. (2021). Safe RAN control: A symbolic reinforcement learning approach. *arXiv*. https://arxiv.org. doi:10.48550/arxiv.2106.01977.

Paliyal, P. S., Mondal, S., Layek, S., Kuchhal, P., & Pandey, J. K. (2024). Automatic solar tracking system: A review pertaining to advancements and challenges in the current scenario. *Clean Energy, 8*(6), 237–262. doi:10.1093/ce/zkae085.

Paliyal, P. S., Mondal, S., Layek, S., Kuchhal, P., & Pandey, J. K. (2024). Automatic solar tracking system: A review pertaining to advancements and challenges in the current scenario. *Clean Energy, 8*(6), 237–262. https://academic.oup.com/ce/pages/About. doi:10.1093/ce/zkae085.

Prasad, D., Kumar, N., Sharma, R., Malik, H., García Márquez, & Pinar-Pérez, J. M. (2023). A novel ANROA based control approach for grid-tied multi-functional solar energy conversion system. *Energy Reports, 9*, 2044–2057. doi:10.1016/j.egyr.2023.01.039.

Reddy, V. S. S., Kruthika, U., & Paneerselvam, S. (2025). *Optimal PID tuning for solar tracking system using bio-inspired algorithms. In Trends in sustainable computing and machine intelligence (Algorithms for Intelligent Systems* (pp. 25–42). Singapore: Springer. doi:10.1007/978-981-96-1452-3_3.

Thungsuk, N., Tanaram, T., Chaithanakulwat, A., Savangboon, T., Songruk, A., Mungkung, N., Maneepen, T., Arunrungrusmi, S., Poonthong, W., Kasayapanand, N., Nilwhut, S., Kinoshita, H., & Yuji, T. (2023). Performance analysis of solar tracking systems by five-position angles with a single axis and dual axis. *Energies, 16*(16), 5869. doi:10.3390/en16165869.

Vannella, F., Jeong, J., & Proutiere, A. (2023). Off-policy learning in contextual bandits for remote electrical tilt optimization. *IEEE Transactions on Vehicular Technology, 72*(1), 546–556. doi:10.1109/TVT.2022.3202041.

Vannella, F., Proutiere, A., Jedra, Y., & Jeong, J. (2022). Learning optimal antenna tilt control policies: A contextual linear bandit approach. In *Proceedings—IEEE INFOCOM* (pp. 740–749). Institute of Electrical and Electronics Engineers Inc. http://ieeexplore.ieee.org/2022-. doi:10.1109/INFOCOM48880.2022.9796783.

Chapter 32

Novel approach on optimizing MPPT using IC and PID controller

S. Sivarajeswari, Valarmathi Govindarajan, A. Jenith Joseph, T.K. Niranjan and Sasikumar Potluri
Sri Sairam Institute of Technology, Chennai, Tamil Nadu, India

32.1 Introduction

Solar energy has experienced rapid adoption on a global scale. This represents a key milestone in the domain. By the completion of 2024, total installed capacity by solar globally surpassed 2 TW. It is estimated that it will grow by another terawatt nominally every year until 2030. In the first half of 2025, around 380 GW of new capacity was added globally. This represents a 64% growth relative to the same time in 2024. Finally, solar power generation exceeded 104 billion KWh in 2024, demonstrating a growing role in the global energy mix. However, solar irradiance is intermittent by nature. Advanced maximum power point tracking (MPPT) algorithms are vital to improving energy extraction efficiency (Abssane et al., 2024; Chellakhi et al., 2024). The current work reports on a comparison of two commonly used MPPT algorithms, perturb and observe (P&O) and incremental conductance (IC)-based techniques (Alsulami et al., 2024; Elgendy et al., 2012), evaluated on four performance metrics: controller effort index (CEI), energy harvesting time (EHT), computational time index (CTI), and steady-state ripple (SSR) (Koutroulis & Kalaitzakis, 2006; López Seguel et al., 2021). The evaluation was conducted using models run in MATLAB and Simulink, using a photovoltaic (PV) buck–boost system under a PID-controlled MPPT algorithm in a simulated environment. The work also presents useful information on performance/reliability and efficiency of both algorithms, as well as the capability to model and perform real-time algorithmic adjustments relative to real-time PV ecosystems.

32.2 Methodology

The suggested Fig. 32.1 was designed and modeled in MATLAB/Simulink to assess the performance of several MPPT algorithms under different exposure to solar conditions. The entire system includes a PV array, a DC–DC buck–boost converter, an MPPT controller, and connected to a load is the Proportional Integral Derivative (PID)-controlled Pulse Width Modulation (PWM) generator (Patel & Agarwal, 2008). The PV array is modeled under standard test conditions of irradiance = 1000 W/m^2 and temperature = 25°C and has the potential to output 25 kW of energy, which is an industrial-scale output. The buck–boost converter was chosen because of its capability of stepping up and down the voltage, enabling the output of the PV source to be regulated in a more adaptable manner. The inductor, capacitor, and switching parameters of the converter were tuned to minimize ripple output and response transient stability. The overall layout of the Simulink model of the proposed MPPT–PID-based PV system is shown in Fig. 32.1, depicting the interconnecting subsystems of the PV source, converter, controller, and load. Two separate MPPT control algorithms have been executed to test their efficacy within the same architecture; the algorithms were P&O and incremental conductance (IC). Each MPPT algorithm produces a reference voltage equating to the maximum power point (MPP), which is controlled by the PID controller via a PWM pulse to the switch of the converter. The PID gains were optimized for reduced settling time and oscillations around the MPP. Simulations were performed with different irradiances varying from 1000 → 700 → 500 W/m^2 and constant temperature of 25°C. The dynamic response of the respective algorithm was assessed in terms of controller effort index (CEI), EHT, CTI,

Computational Intelligence in Mechatronics: Solving Real-World Problems in Electronic Systems Design. DOI: https://doi.org/10.1016/B978-0-443-27580-7.00027-5

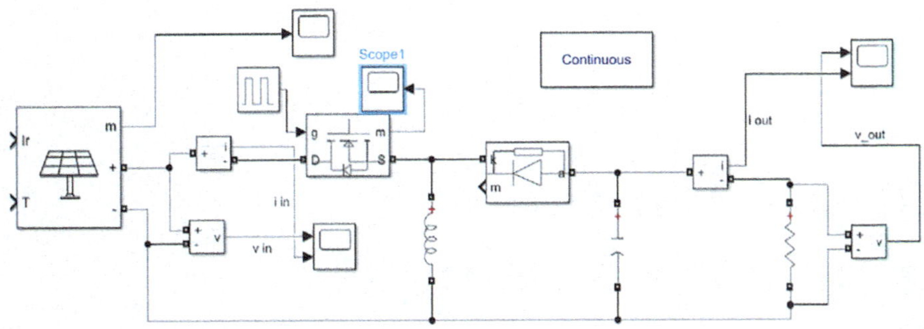

FIGURE 32.1 Circuit diagram of PV system. The circuit was simulated using the MATLAB app, with the PV panel with the MPPT and the output connected to a buck–boost converter. *MPPT*, Maximum power point tracking; *P&O*, perturb and observe; *PV*, photovoltaic.

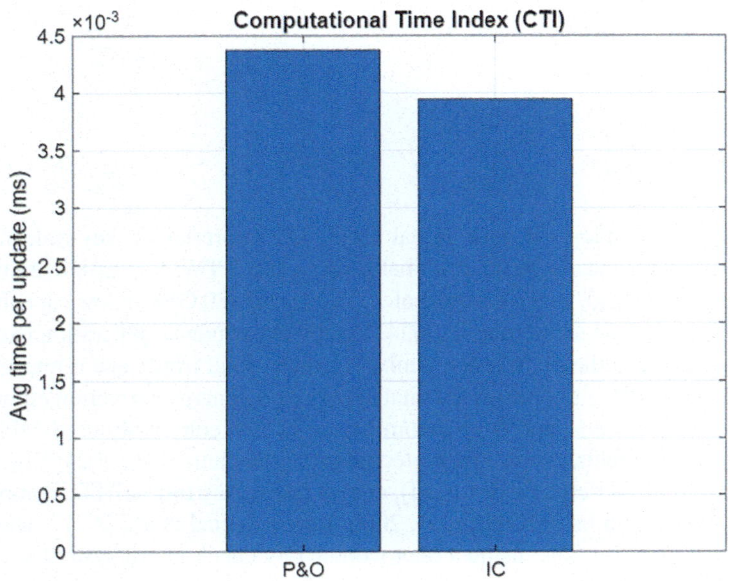

FIGURE 32.2 Computational time index. The time taken for the computation of both algorithms for input change, measured in "ms." *P&O*, Perturb and observe.

and the steady-state ripple factor (RSF) (Subudhi & Pradhan, 2013). These figures give a clear comparative picture of the efficiency, speed, and stability of each of the algorithms in a real-time PV energy application.

32.3 Analysis

32.3.1 Computational time index

The CTI refers to the duration the MPPT algorithm takes to carry out one iteration or cycle of computing. This includes the input request (e.g., voltage and current) to the output, sending the control command to the power converter and making calculations to establish the MPP of a PV system (Zulhelmi, 2021). It constitutes an important performance metric that determines the efficiency of the algorithm in terms of speed and responsiveness. Its importance has an effect in response time, which is the time taken to respond to input variations, and real-time operation for a rapid adjustment of system and energy efficiency to maximize the effective time of MPP level of the system. The simulation of the MPPT algorithms in this aspect produced the graph in Fig. 32.2, showcasing that P&O took a longer time for the calculation for the input variation in comparison to the IC algorithm. This is evident from the result that the IC is faster in this test, making it suitable for a faster response, which is critical, especially in the rapidly varying environmental conditions.

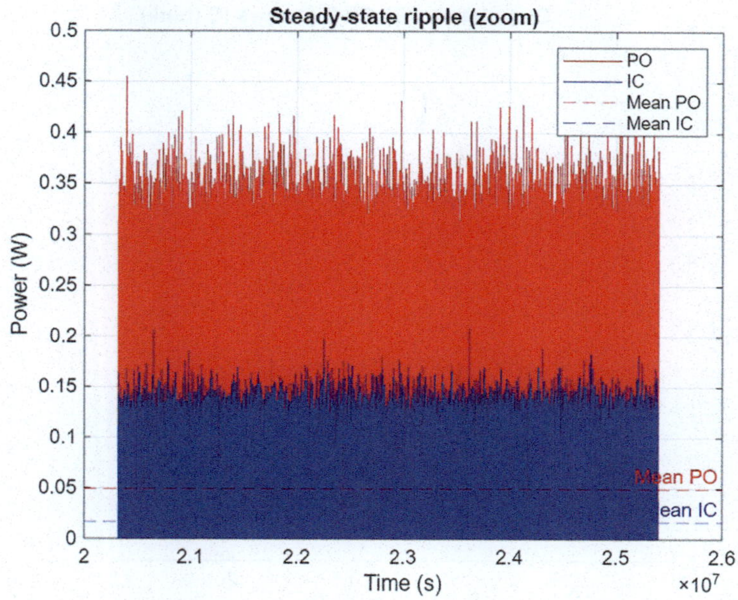

FIGURE 32.3 Steady-state ripple. The ripple present in the output of the PV system with the algorithm's usage. *PV*, Photovoltaic.

32.3.2 Steady-state ripple

The RSF quantifies the amount of oscillation around the MPP after the MPPT controller has reached steady state. Fig. 32.3 demonstrates a comparison of the RSF characteristics of the P&O and IC algorithms over the simulated 25 KW PV system. Noticeably, after the MPP is approached, the P&O algorithm (*red waveform*) continues to generate much larger oscillations in output power during the steady-state period, which indicates that the output is adapted to continuous perturbations from the operating point. This continued perturbation ultimately produces small but notable energy loss and makes the P&O algorithm less suitable for industrial systems that need high fidelity (Wang & Sun, 2024). On the other hand the IC algorithm (*blue waveform*) exhibits a small-amplitude ripple, ensuring a more constant operating point with limited changing values around the MPP. Overall the mean power variation in the IC method is comparatively limited relative to that of P&O. Thus the IC method exhibits the best performance in steady-state operation and convergence. The ripple of the output associated with IC-based control will result in demand for better energy quality, improved efficiency of the converter, and better translated life cycles of the components when operated at an industrial scale.

32.3.3 Control effect index

The CEI is defined as a quantitative measure of how intensely the MPPT controller is compelled to change the duty cycle to find and keep the system at the MPP. A higher CEI indicates a larger or more frequent duty cycle change, resulting in more effort and higher losses associated with the conditioning. Fig. 32.4 shows cumulative CEI information for the P&O and IC algorithms in the 25 KW PV system. The P&O algorithm (*red curve*) displayed a dramatic rise in CEI during the transient period of the simulation, as the algorithm continued to perturb the duty cycle when the system was nearing the MPP. The IC algorithm (*blue curve*) showed a near-linear and constant low CEI throughout the simulation. This indicates a smoother control action and less stress were placed on the converter. The low-effort action demonstrated the stability and efficiency of the IC controller under changing irradiance conditions. Therefore, for large-scale, or industrial, PV systems, the IC method is more efficient with energy, and the hardware is less stressed than the P&O method.

32.3.4 Energy harvesting time

The EHT measures the time that the MPPT algorithm takes to reach 98% of its MPP (P_{max}) after the beginning of a disturbance or startup and directly measures the efficiency of dynamic response. As shown in Fig. 32.5, the IC algorithm (*blue curve*) showed an earlier time in reaching 98% of P_{max} (approximately 0.23×10^7 s) than the P&O algorithm (shown as the *red curve* at nearly 0.43×10^7), which confirms an improvement in dynamic response time. The IC algorithm is able to calculate the slope of the power–voltage plot in the instantaneous moment, meaning it is able to converge to maximum

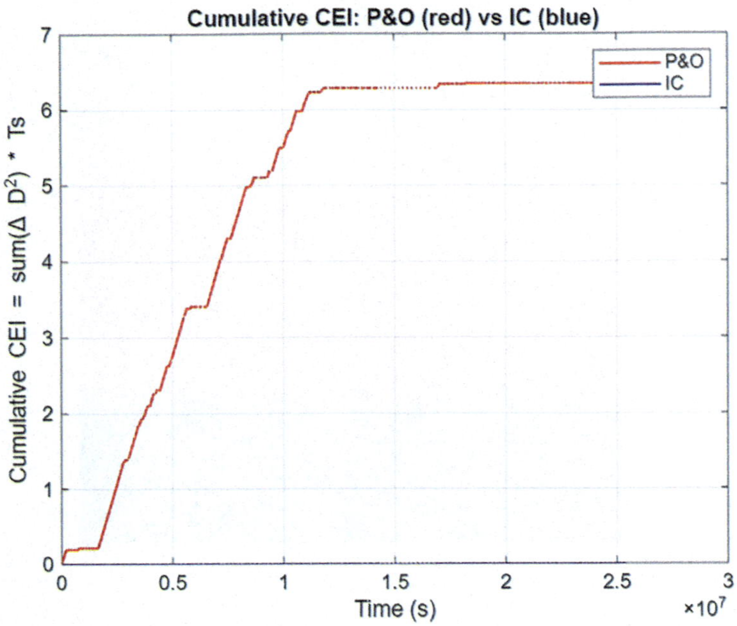

FIGURE 32.4 Control effect index. The effort taken by the algorithm to control the MPP value for the time duration. *MPP*, Maximum power point; *P&O*, perturb and observe.

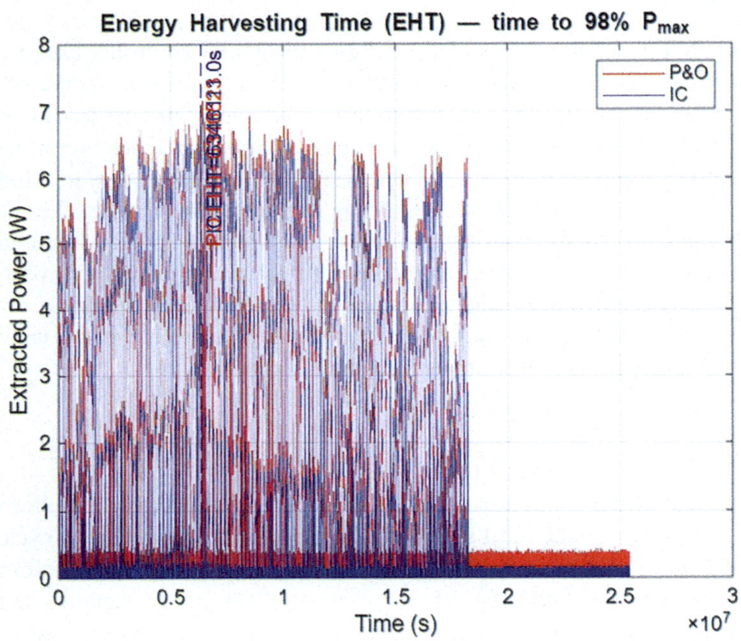

FIGURE 32.5 Energy harvesting time. The variation of the energy with time reference for both algorithms. *P&O*, Perturb and observe.

power without needing to perturb continuously; the P&O method, on the other hand, begins to periodically perturb the output voltage and is slow in convergence while causing slight power oscillation. Thus the IC algorithm demonstrates better tracking speed and adaptation under rapidly changing irradiance, making it much more effective and efficient when scaling for large-scale PV systems when rapid stabilization is needed to maximize power production.

32.4 Conclusion

The examination of the four critical metrics, CEI, EHT, CTI, and RSF, highlights the importance of understanding each one in the effort to improve the performance of MPPT-based PV systems for industrial processes. CEI provides the totality of conversion efficiency, which is imperative with energy systems of considerable size. EHT considers the system capability to achieve MPP quickly while operating under a time-varying irradiance condition, which is beneficial for industries with varying load behavior patterns. CTI describes how precise and stable the control operation is and provides assurance of reliable operation of automated power systems, and RSF refers to smoothness of power output, often reducing stress on the system's components in addition to improving inverter performance. Compared against each other the results indicate that IC outperformed P&O in CEI, EHT, and CTI because of the IC's better adaptation to irradiance conditions, whereas P&O did better in RSF but not obviously enough to negate IC's advantages, making it suitable for use in industrial-scale PV applications.

References

Abssane, S., Outzourhit, A., Amatoul, F.Z., 2024. Comparison between incremental conductance and perturb and observe algorithms in photovoltaic system under low temperature and irradiation levels. International Journal of Electrical and Computer Engineering 14 (5), 4897–4906. doi:10.11591/ijece.v14i5.pp4897-4906. https://ijece.iaescore.com/index.php/IJECE/article/view/35441.

Alsulami, A.A.G., Alhussainy, A.A., Allehyani, A., Alturki, Y.A., Alghamdi, S.M., Alruwaili, M., Alharthi, Y.Z., 2024. A comparison of several maximum power point tracking algorithms for a photovoltaic power system. Frontiers in Energy Research 12. doi:10.3389/fenrg.2024.1413252. http://www.frontiersin.org/Energy_Research/.

Chellakhi, A., Ouhrouche, M., Mehdaoui, A., Bossoufi, B., 2024. High-efficiency MPPT strategy for PV systems: Ripple-free tracking. Renewable Energy 210, 86–96. doi:10.1016/j.renene.2024.05.179.

Wang, Y., Sun, L., 2024. Photovoltaic maximum power point tracking technology based on improved perturbation observation method and backstepping algorithm. Electronics 13 (19), 3960. doi:10.3390/electronics13193960.

López Seguel, J., Seleme Jr, S.I., Moráis, L.M.F., 2021. Comparison of the performance of MPPT methods applied in converters buck and buck-boost for autonomous photovoltaic systems. Ingeniare Revista chilena de ingeniería 29 (2), 229–244. doi:10.4067/s0718-33052021000200229.

Zulhelmi, T.A., 2021. Comparisonal analysis of incremental conductance and perturb and observe methods as MPPT algorithm in photovoltaic system. Jurnal Mekintek 12 (1), 17–22. doi:10.32722/jm.v12i1.4182.

Subudhi, B., Pradhan, R., 2013. A comparative study on maximum power point tracking techniques for photovoltaic power systems. IEEE Transactions on Sustainable Energy 4 (1), 89–98. doi:10.1109/TSTE.2012.2202294.

Elgendy, M.A., Zahawi, B., Atkinson, D.J., 2012. Assessment of perturb and observe MPPT algorithm implementation techniques for PV pumping applications. IEEE Transactions on Sustainable Energy 3 (1), 21–33. doi:10.1109/TSTE.2011.2168245.

Patel, H., Agarwal, V., 2008. Maximum power point tracking scheme for PV systems operating under partially shaded conditions. IEEE Transactions on Industrial Electronics 55 (4), 1689–1698. doi:10.1109/TIE.2008.917118.

Koutroulis, E., Kalaitzakis, K., 2006. Design of a maximum power tracking system for wind-energy-conversion applications. IEEE Transactions on Industrial Electronics 53 (2), 486–494. doi:10.1109/TIE.2006.870658.

Chapter 33

Optimal UAV design configuration with multiple-choice components using PSO and GWO

Akram Fraih[a], **Mohamed Arezki Mellal**[a], **Edward J. Williams**[b,c], **Vipin Kumar Sharma**[d] and **Lamia May**[e]

[a] *LMSS, Faculty of Technology, M'Hamed Bougara University, Boumerdes, Algeria,* [b] *Industrial and Manufacturing Systems Engineering Department, College of Engineering and Computer Science, University of Michigan, Dearborn, MI, United States,* [c] *Decision Sciences, College of Business, University of Michigan, Dearborn, MI, United States,* [d] *Department of Mechanical Engineering, Meerut Institute of Engineering and Technology, Meerut, Uttar Pradesh, India,* [e] *L2ME, Faculty of Technology, University of Bejaia, Bejaia, Algeria*

33.1 Introduction

The increasing rate and intensity of natural disasters and wildfires present serious challenges to social safety, infrastructure, and environment. Unmanned aerial vehicles (UAVs), unmanned aerial systems (UASs), or drones, the three terms used, have rapidly ascended from novel gadgets to a crucial and cost-effective tool for fast emergency response. These terms denote an aircraft that flies with no human presence onboard, neither pilot nor passengers. Moreover, they may be either remotely controlled by human pilots sitting on ground-based stations or be autonomous flying vehicles using preprogramed flight paths or using AI-powered control software that treats data captured by drone sensors and makes the right decisions aiming to complete the given goals of that mission.

A typical UAS may include the following elements (Bayou et al., 2025):

An aircraft with no pilot on board.
A remote pilot station.
A command and control link.
Transmission devices for transmitting data.

A payload specific to the intended application/operation is aboard; this payload often includes specialized cameras or other sensors that collect data for near-term analysis. Drones increase safety for personnel, equipment, and infrastructure while offering numerous capabilities where human intervention is difficult or impossible on a timely basis. One of the primary features when using drones as an emergency tool, like firefighting drones, is their reliability, which ensures their capability for completing their crucial tasks in critical conditions. So the optimization of their reliability became essential for successful intervention. Drones are classified into different categories based on different characteristics: size, weight, range, morphology (shape), and power source.

The use of traditional optimization techniques for reliability–redundancy optimization has become inefficient for large and complex systems and requires a significant amount of time to complete, which makes it necessary to use modern, more effective, and more adequate methods like artificial intelligence (AI) tools. Nature-inspired optimization algorithms, which belong to AI methods, such as particle swarm optimization (PSO) and graywolf optimizer (GWO), play a key role in AI models and are good optimization techniques that offer accurate results and faster computational capabilities.

Computational Intelligence in Mechatronics: Solving Real-World Problems in Electronic Systems Design. DOI: https://doi.org/10.1016/B978-0-443-27580-7.00013-5

33.2 Fire detection and firefighter drones

Fires are one of the most serious disasters that claim human lives around the world, as well as cause severe damage to infrastructure and threaten the balance of the ecosystem on the planet. The frequency of fires has increased in recent years around the world; the causes of these fires are varied, between natural causes such as climate change, droughts, and extreme heat waves, which have increased in intensity and frequency and duration since the second half of the 20th century, and human causes such as the dumping of waste and flammable materials in dry areas or lack of attention to fire prevention measures in urban areas and failure to carry out frequent preventive maintenance of industrial facilities. These fires can cause significant damage to people and facilities, as well as severe environmental damage. The Times of India published a report that 27,027 people died because of fire accidents in India in 2017, equivalent to the death of 35 people daily. The National Crime Records Bureau (NCRB) recorded about 9 million fire incidents worldwide, causing the death of 120,000 people in the same year, and whereas Canada experienced catastrophic wildfire season in 2023 that led to the production of the highest carbon emissions recorded in Canada since 2003, wildfires destroyed about 384 Mha of land in that year and 367 Mha in 2024 (Rogers, 2024). In the Mediterranean basin, particularly in France, the year 2022 was the most catastrophic year since 1949 in terms of wildfire impacts, when 30,000 ha were destroyed in Gironde department and a total area of 508,260 ha in the European Union between June and September of the same year, according to the European Forest Fire Information System (EFFIS) (Bayou et al., 2025). In January 2025, fires spread through the Los Angeles metropolitan area and California overall.

The catastrophic consequences of the fire accidents led to serious research for highly efficient, low-cost, and safe methods for firefighting. Coinciding with the revolution of drone technology by the end of the 20th century and the beginning of the 21st century, it was clear that drones can play a vital role in firefighting operations around the world; their primary advantages, low cost, safety, and rapid deployment, make them an ideal choice for small and remote fires.

Fire-extinguishing drones are specially designed UAVs; they are mainly designed for firefighting purposes, especially in areas that are difficult to reach by traditional methods because of mountains, forests, high-rise buildings, and remote islands, and also in industrial infrastructures where short response time is critical for safety and environmental impact, such as oil and gas extraction sites, offshore platforms, cargo ships, high-voltage transformers, and power plants. Often in such places human intervention is extremely dangerous and needs advanced training and safety equipment. Enhanced by AI-powered models, they can even predict fire behavior by analyzing wind patterns, fuel loads, predictive temperature, historical data, etc. (Bayou et al., 2025).

These drones are designed taking into account the ability to fly in the harshest and most difficult conditions, such as extreme temperatures that can reach 700°C; high humidity in coastal areas; or rapid atmospheric pressure changes, in addition to their resistance to dust, fumes, as well as chemicals carried on board the drone itself. Because of the harsh conditions that fire-extinguishing drones have to withstand, it is necessary to design and manufacture them using composite materials that have high temperature resistance and high oxidation resistance while maintaining lightweight structure to increase their operational range and their payload capacity. These material and performance requirements, along with required experience and knowledge of mechanical design, computational fluid dynamics (CFD), finite element analysis (FEA), lead to higher research, development, and manufacturing costs with such specialized drones.

Fire-detecting and reconnaissance drones are mainly designed for early fire detection purposes and situation awareness and could be used for tracking individuals in fire scenes and data collection and real-time wildfire footage in dangerous and inaccessible areas. Because of the unpredictability of wildfires, these types of drones should have long endurance, with flight time often exceeding several hours. Moreover, they need better aerodynamic shape and advanced control system to maintain drone stability through powerful winds and air vortices. In addition, they must have very sensitive, high-quality sensors to cover vast areas, along with advanced and high-speed telemetry devices and telecommunication systems for real-time data transfer and analysis. These requirements make fixed-wing and hybrid VTOL designs the perfect structure morphologies for this type of drones. And because of the fact that these drones avoid direct flying through wildfires, this makes either gasoline or hydrogen better choices as power source for this type, but the low cost of gasoline-powered drones makes them economical and reliable for these missions.

33.3 Reliability optimization problems

Because of extremely harsh conditions that face the firefighting drone during its deployment, it is essential to design the drone taking into account its reliability as the primary feature. Optimizing the reliability of a firefighting drone is a complex task because of the complexity and variety of its subsystems, as well as the constraints that make this task particularly challenging. These constraints include economic factors (ensuring the drone remains more cost-effective and affordable than traditional firefighting methods), performance limitations such as weight—which affects payload capacity and reduces operational range, leading to the need for a large number of drones to cover vast areas—and logistical challenges, such as storage volume, which impacts the ability to store and transport these drones in and from fire departments.

The complexity of the drone's subsystems and their configurations makes it very difficult for engineers to optimize system reliability and forces them to implement AI tools and algorithms to increase computational efficiency and decrease optimization time. Nature-inspired optimization algorithms are considered good choices to solve reliability and redundancy allocation problems for their accuracy and speed. The naturally inspired algorithms, such as PSO and GWO, took their place as two of the best optimization algorithms for monoobjective nonlinear problems.

Reliability is a very important criteria when designing a firefighting drone; the crucial role played by these aerial systems in emergencies and dangerous situations makes it essential to ensure the proper work and the ability to perform their related function with no (or minimal) failures. In addition, maintaining high levels of reliability leads to short response times when timely intervention is critical for life-saving, equipment safety, environment protection, and cost reduction (Jin et al., 2024).

Reliability–redundancy allocation problems (designated as RRAP) are optimization problems that focus on maximizing overall system reliability by allocating reliabilities and redundancies of different subsystems. However, at the same time, it is essential to ensure that the considered constraints, such as cost, weight, and volume, are not violated. Some RRAPs may represent a multiobjective optimization problem, e.g., maximizing overall system reliability while minimizing total system cost, weight, or volume. Another common problem is maximizing overall system reliability while maximizing power efficiency of the system. This type of problem seems to be very complex to solve because it includes allocation of both the optimal number of redundant subsystems and their corresponding reliability for each subsystem, respectively (Gholinezhad, 2024). Also, because their objective function (for this problem: overall system reliability) is a nonlinear function, which means that the conventional linear optimization techniques are highly inefficient.

33.3.1 Reliability allocation problem

The objective of RAP is to determine the optimal reliabilities of individual subsystems that lead to maximizing the overall reliability of the system, while ensuring nonviolation of the given constraints. It can be described mathematically using the following model (Mellal & Zio, 2016):

$$\text{Maximize}\, R(r) = R_s(r_1, r_2, \ldots, r_m) \tag{33.1}$$

$$\begin{aligned} &\text{Subject to}\, g(r_1, r_2, \ldots r_m) \le b \\ &0 \le r_i \le 1, i = 1, 2, \ldots, m; r \in \mathbb{R} \end{aligned} \tag{33.2}$$

where $R(r)$ is the system reliability function of the reliabilities vector r, g is the design constraint, m is the number of subsystems in the system, and b is the resource limit.

33.3.2 Redundancy allocation problem

The objective of redundancy allocation problem (RdAP) is to determine the optimal redundancies of individual subsystems that lead to maximizing the overall reliability of the system while ensuring nonviolation of the given constraints. It can be described mathematically using the following model (Mellal & Zio, 2016):

$$\text{Maximize}\, R(n) = R_s(n_1, n_2, \ldots, n_m) \tag{33.3}$$

$$\begin{aligned} &\text{Subject to}\, g(n_1, n_2, \ldots, n_m) \le b \\ &1 \le n_i \le n_{\max}; i = 1, 2, \ldots, m; n \in \mathbb{Z} \end{aligned} \tag{33.4}$$

where $R(n)$ is the system reliability function of the reliabilities vector n, g is the design constraint, m is the number of subsystems in the system, b is the resource limit, and $n_{\max}$ is the maximum number of redundant components.

33.3.3 Reliability–redundancy allocation problem

RRAPs are a combination of both reliability allocation and redundancy allocation, aiming to define the optimal reliability and redundancy at each subsystem to maximize the system reliability without violating the given constraints. It is described

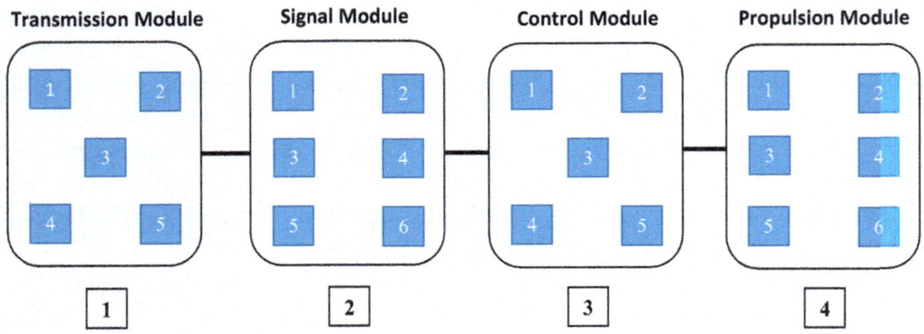

FIGURE 33.1 Block diagram of the electrical part of the studied firefighting drone.

by the following mathematical model (Mellal & Zio, 2016):

$$\text{Maximize } R(r, n) = R_s(r_1, r_2, \ldots, r_m, n_1, n_2, \ldots, n_m) \tag{33.5}$$

$$\begin{aligned} &\text{Subject to } g(r_1, r_2, \ldots, r_m, n_1, n_2, \ldots, n_m) \le b \\ &0 \le r_i \le 1; 1 \le n_i \le n_{max}; i = 1, 2, \ldots, m \\ &r \in \mathbb{R}, n \in \mathbb{Z} \end{aligned} \tag{33.6}$$

33.3.4 Multichoice reliability–redundancy allocation problem

Multichoice RRAPs that involve different design choices for each subsystem mean that each subsystem has more than one choice to consider, and each choice includes different values of constraint parameters.

To model multichoice RRAPs, it is necessary to add a new variable, z, that indicates the optimal choice chosen to get the optimal system reliability. The mathematical model of this type of RRAP can be written as follows:

$$\text{Maximize } R(r, n, z) = R_s(r_1, r_2, \ldots, r_m, n_1, n_2, \ldots, n_{m\ldots}) \tag{33.7}$$

$$\begin{aligned} &\text{Subject to } g(r, n, z) \le b \\ &0 \le r_i \le 1; 1 \le n_i \le n_{\max}; 1 \le z_i \le z_{\max} \\ &i = 1, 2, \ldots, m, r \in \mathbb{R}, n \in \mathbb{Z}, z \in \mathbb{Z} \end{aligned} \tag{33.8}$$

where z_i is the optimal choice for the ith subsystem and r_i and n_i are its corresponding reliability and redundancy, respectively. $Z_{\max}$ and $n_{\max}$ are the number of possible choices and maximum possible redundancy of at the ith subsystem, respectively. Table 33.1 summarizes some previous works on system reliability optimization.

33.4 Block diagram of the firefighting drone

Fig. 33.1 shows the block diagram of the study's firefighting drone.

33.4.1 Subsystem 1: Transmission module

The transmission module is the main device that allows remote piloting of the drone by coding the input signal from the operator and sending it via radio waves to the drone. It has several characteristics:

- **Control inputs:** refer to the inputs of the transmission module, such as buttons, switches, joysticks, and trackpad, allowing maneuverability and speed control.
- **Frequency modulation:** the transmission module modulates its data at specified frequencies to avoid signal interference with other signals and enhance communication reliability.
- **Transmission channels:** using multiple channels allows simultaneous control of different functions of the drone, such as flight navigation, speed control, object tracking, camera options, and fire-extinguishing systems. It reduces response time and enhances the drone's performances.

TABLE 33.1 Summary of previous works on system reliability optimization.

References	Method	Reliability allocation	Redundancy allocation	Reliability–redundancy allocation	Strategy
Bellman and Dreyfus (1958)	Dynamic programming		X		Active, homogeneous components
Tillman and Liittschwager (1967)	Integer programming		X		Active, homogeneous components
Tillman et al. (1968)	Computational procedure		X		Active, homogeneous components
Agarwal and Sharma (2010)	Ant colony optimization with an adaptive penalty function (ASACO)		X		Active, homogeneous components
Dohi et al. (2006)	Genetic algorithms with dynamic adaptive penalty strategy (GAs)		X		Active, homogeneous components
Zou et al. (2011)	Novel modified differential evolution (NMDE)			X	Active, homogeneous components
Garg (2015)	Cuckoo search with a penalty function (CS)		X	X	Active, homogeneous components
Mellal and Zio (2016)	Penalty guided stochastic fractal search (PSFS)	X	X	X	Active, homogeneous components
Mellal and Zio (2017)	Penalty-guided stochastic fractal search (PSFS)			X	Active, homogeneous components
Mellal and Williams (2018a)	Cuckoo optimization algorithm (COA)			X	Active, homogeneous components
Mellal and Zio (2019)	Adaptive particle swarm optimization (ADAP-PSO)	X		X	Active, homogeneous components
Taghiyeh et al. (2020)	α-Level cuts			X	Active, homogeneous components
Sharma (2021)	Bidirectional butterfly optimization algorithm (BBOA)		X	X	Active, homogeneous components
Coit (2001)	Integer programming		X		Cold-standby, homogeneous components

(continued on next page)

TABLE 33.1 Summary of previous works on system reliability optimization—cont'd

References	Method	Reliability allocation	Redundancy allocation	Reliability–redundancy allocation	Strategy
Mellal and Zio (2020)	Enhanced nest cuckoo optimization algorithm (ENCOA)			X	Cold-standby, homogeneous components
Abouei Ardakan and Zeinal Hamadani (2014)	Modified version of the genetic algorithm (GA)			X	Cold-standby, homogeneous components
Juybari et al. (2019)	Penalty-guided fractal search algorithm (SFS)			X	Cold-standby, homogeneous components
Dobani et al. (2019)	Hybrid stochastic fractal search with genetic algorithms (hSFS-GA)			X	Active, homogeneous components
Mellal et al. (2020)	Hosted cuckoo optimization algorithm (HO-COA)			X	Active, homogeneous components
Ouyang et al. (2019)	Particle swarm optimization with stochastic perturbation nature (SPPSO)			X	Mixed, homogeneous components
Mellal et al. (2021)	Shuffled frog-leaping algorithm (SFLA)			X	Active, homogeneous components
Aqel and Mellal (2023)	Adaptive particle swarm optimization (ADAP-PSO)			X	Active, homogeneous components
Saheb and Mellal (2024)	Particle swarm optimization (PSO)	X		X	Active, homogeneous components
Bayou and Mallik (2023)	Particle swarm optimization (PSO)			X	Active, homogeneous components
This dissertation	Result is mentioned in chapter IV			X	Active, homogeneous components

- **Effective range:** it designates the maximum distance that the drone is able to receive radio signals with the minimum allowed power for the signal module to work properly.

33.4.2 Subsystem 2: Signal module

It treats received signals from the transmission module via an antenna and decodes them, then transfers them to their corresponding device. The incoming signal follows the following sequence:

- **Signal reception and filtering:** the signal module receives radio signals emitted by transmission module via the antenna, filters it by eliminating the noises to make it readable, then decodes it to extract useful instructions, and transfers it to the flight computer.
- **Signal processing:** the flight computer gets the decoded data and executes preprogramed instructions to calculate flight parameters and execute demanded tasks. In addition, it treats and codes collected data from onboard sensors to send it to the ground control station.
- **Control data distribution:** the flight computer transfers the orders via data bus to the control module to reach the desired parameters.

33.4.3 Subsystem 3: Control module

Usually a proportional-integral-derivative (PID) controller, it enhances the stability and optimal control of the drone parameters (position, speed, and orientation) during flight time to reach the desired parameters sent by the operator.

- **Proportional response:** depends only on the difference between set point and process variable. It is proportionally the inverse of the response time of the system.
- **Integral response:** it sums the error terms over time to eliminate the steady-state error.
- **Derivative response:** it is proportional to the rate of change of the process parameters; it reduces the response time but makes the system more sensitive to the error changes.

33.4.4 Subsystem 4: Propulsion system

Generally, drones use brushless motors because of their power efficiency, low maintenance required, compact size, and relatively good reliability rate.

- **Thrust generation:** controlling the generated thrust from brushless motor is possible by controlling the speed of the motors.
- **Torque generation:** it is crucial for better maneuverability; it is possible by separate speed control of each motor.

33.5 Mathematical model of RRAP with multichoice of the firefighting drone

The mathematical model of this problem can be described as follows:

$$\begin{aligned}&\text{Maximize}: R_S(r,n,z)=\prod_{i=1}^{4}\left(1-(1-r_i)^{n_i}\right)\\&\text{with } r_i\in\mathbb{R}\,\&\,n_i\in\mathbb{Z}\,\&\,z_i\in\mathbb{Z}\end{aligned} \tag{33.9}$$

$$\text{Subject to } g_1=\sum_{i=1}^{4}v_{i,z}n_i^2\leq V \tag{33.10}$$

$$g_2=\sum_{i=1}^{4}\alpha_{i,z}\left(\frac{-T}{\ln(r_i)}\right)^{\beta_{i,z}}r_in_i\exp\left(\frac{n_i}{4}\right)\leq C \tag{33.11}$$

$$g_3=\sum_{i=1}^{4}w_{i,z}n_i\exp\left(\frac{n_i}{4}\right)\leq W \tag{33.12}$$

TABLE 33.2 Design parameters of the firefighting drone.

Design parameter	Value
V	250
C	400
W	700
T	1000

While considering that $0.5 \leq r_1 \leq 0.98$

$$0.5 \leq r_2 \leq 0.96$$

$$0.5 \leq r_3 \leq 0.99$$

$$0.5 \leq r_4 \leq 0.97$$

$$1 \leq n_1 \leq 3$$

$$1 \leq n_2 \leq 3$$

$$1 \leq n_3 \leq 4$$

$$1 \leq n_4 \leq 2$$

$$1 \leq z_1 \leq 5$$

$$1 \leq z_2 \leq 6$$

$$1 \leq z_3 \leq 5$$

$$1 \leq z_4 \leq 6$$

where α and β represent cost factors to calculate the overall cost of the drone, w_i is used to calculate the total weight, and v_i is used to calculate the total volume of the drone.

Table 33.2 reports the design specifications of the firefighting drone like budget of manufacturing C, maximum allowed volume V, maximum allowed weight W, and total mission time of the drone T.

Table 33.3 reports the possible design choices for every subsystem. α and β are real coefficients used to calculate the overall system cost, v is the volume coefficient, and w is the weight coefficient. The parameters are in arbitrary units.

33.6 Nature-inspired optimization algorithms for RRAP

To optimize the overall system reliability of the firefighting drone the naturally inspired algorithms, so-called metaheuristic algorithms are considered good techniques that offer reliable and accurate results in a reasonable time. There are many metaheuristic algorithms, such as GWO (Mirjalili et al., 2014), flower pollination algorithm (Mellal et al., 2023), PSO (Kennedy & Eberhart, 1995; Saheb & Mellal, 2024), genetic algorithms (Mellal et al., 2013), ant colony optimization (Kuo & Wan, 2007; Mellal & Williams, 2018b), artificial bee colony (Promkaew et al., 2024), cuckoo algorithms

TABLE 33.3 Multichoice design data of the firefighting drone.

Subsystem		1	2	3	4
Choice 1	v	1	2	3	2
	$10^5\alpha$	1	2.3	0.3	2.3
	β	1.5	1.5	1.5	1.5
	w	6	6	8	7
Choice 2	v	2	1	2	3
	$10^5\alpha$	1.5	1.8	0.7	2.6
	β	1.5	1.5	1.5	1.5
	w	5.5	6.2	7.8	7.5
Choice 3	v	1.5	2.5	3.2	2
	$10^5\alpha$	1	2.8	0.8	3
	β	1.5	1.5	1.5	1.5
	w	5	6	8	8
Choice 4	v	1.4	2.4	3.9	2.7
	$10^5\alpha$	2	1.7	1	3
	β	1.5	1.5	1.5	1.5
	w	5	6	7	7
Choice 5	v	0.8	2.7	3	2.5
	$10^5\alpha$	1.8	3	1.2	3.5
	β	1.5	1.5	1.5	1.5
	w	4.8	6.4	7.4	7
Choice 6	v	–	1	–	3
	$10^5\alpha$	–	2.8	–	3.2
	β	–	1.5	–	1.5
	w	–	5.7	–	7.3

(Mellal & Williams, 2017; Mellal & Zio, 2022; Templos-Santos et al., 2019), firefly algorithm (Zhang et al., 2022), and whale optimization algorithm (Nazari-Heris et al., 2019). In this chapter the efficiency and precision of two of the most known and practical algorithms, namely, the PSO and GWO, are compared.

33.6.1 Particle swarm optimization

PSO is a computational method and a population-based metaheuristic optimization algorithm created to solve complex optimization problems through emulating swarm behaviors observed in swarm animals.

33.6.1.1 Choosing criteria

The PSO has been chosen according to the following criteria:

- Few parameters to tune make it easy to use and apply for variety of optimization problems.
- Balance between exploration and exploitation allows them to find local and global solution and provide fast convergence in low-dimensional search spaces (Gad, 2022).
- Multiple interactions between neighbor agents allowing information transfers in the swarm to be used by individual agents.

33.6.1.2 Pseudo code of PSO

The pseudo code of the PSO algorithm is illustrated as follows:

```
1. Initialize the objective function
2. Set parameters:
- Number of decision variables
```

```
- Variable bounds (lower and upper)
- Number of iterations
- Population size
- PSO parameters: inertia weight, damping ratio personal and global learning coefficients
- Velocity limits (maximum and minimum)
3. Initialize particles:
- Each particle has position, velocity, best position, and best evaluation value.
- Set global best evaluation value to infinity
4. For each particle:
- Randomize initial position within bounds
- Set initial velocity to zero
- Evaluate position using the objective function
- Set best position and best evaluation value
- Update global best if the current particle's evaluation is better
5. Store the best evaluation values for each iteration
6. Main loop (for each iteration):
For each particle:
- Update velocity using PSO formula
- Apply velocity and position limits
- Evaluate new position
- Update best position and evaluation value if improved
- Update global best if improved
- Store and display the current iteration's global best evaluation value and solution
- Dampen the inertia weight
7. Set and display the best solution and elapsed time
```

33.6.2 Gray wolf optimizer

GWO is a single-objective metaheuristic optimization algorithm. The social hierarchy and the hunting method of gray wolves' inspiration are as follows:

- Tracking, chasing, and approaching the prey.
- Pursuing, encircling, and harassing the prey until it stops moving.
- Attack upon the prey.

33.6.2.1 Choosing criteria

- Few parameters make it easy to implement.
- Simple principles.
- Preserving information about the search space for every iteration and saving the best solution.

33.6.2.2 Pseudocode of GWO

The pseudocode of GWO is as follows:

```
Initialize the grey wolf population
Initialize parameters
Calculate the fitness of each grey wolf
Find the best search grey wolf
Find the second best search grey wolf
Find the third best search grey wolf
while (Maximum number of iterations not reached)
for grey wolf
Update the position of the current grey wolf
end for
```

```
Update the three parameters
Calculate the fitness of all grey wolves
Update the three parameters
Update the current iteration
end while
return the first parameter
Display the optimal results and elapsed time
```

33.7 Results and discussion

For the firefighting drone design, we considered two different motor-based configurations: single-motor configuration and dual-motor configuration. Each of the two configurations has advantages and disadvantages. Here we cite a comparison between single and dual-motor drone configurations:

- Single-motor configuration is less expensive than dual-motor configuration.
- Single-motor configuration provides lower maintenance cost, and it is easier to maintain than dual-motor configuration.
- Dual-motor configuration offers better reliability compared to their equivalent single-motor configuration.
- Dual-motor configuration increases the payload capacity and the effective range of the drone.
- Dual-motor configuration allows for better maneuverability and shorter response time.
- Dual-motor configuration is useless in case of failure of at least one motor, because the necessity for yaw control forces the flight computer to shut off some other motors to balance torques and thrust, which makes these stopped motors a deadweight for the drone and significantly reduces its performance like range and maneuverability.
- The solution approaches mentioned in this chapter were executed on MATLAB R2015a using personal computer equipped with the following configuration:
- Intel® Core™ i5-8350u, an x64 based architecture CPU at clock speed of 1.70 GHz up to 3.60 GHz using Intel turbo technology.
- 8 GB of RAM DDR4 single channel @2400 MHz.
- UHD 620 Integrated GPU @300 MHz up to 1100 MHz.
- Windows 10 Pro, a 64-bits operating system.

This configuration was chosen because of its improved reliability and robustness and performance in computational tasks and its capability of handling small-scale simulations.

33.7.1 Single-motor configuration

In the following results the number of redundant components of subsystem 4 (propulsion system) is fixed at one ($n_4 = 1$).

33.7.1.1 Obtained solutions using PSO

After running 10 executions of the PSO code, we obtained the following results shown in Table 33.4.

Table 33.3 shows the optimal reliability reached by PSO after 7 executions is 0.9697692 after an elapsed time of 0.591678 seconds; the number of function evolutions needed to find the optimal reliability reaches 7800.

We can observe from Table 33.3 that PSO is very stable in solving this problem, and we can see that in all executions, it results in the same optimal redundancy for all the subsystems in every execution and the same optimal reliability for subsystem 4 (motors) and the same choice (first choice) for subsystems 3 and 4.

In addition, the choice of subsystem 1 fluctuates between first choice and second choice, whereas PSO goes with the first choice for the best execution.

Ranging between 4300 and 7800, the number of function evaluations is relatively low. The maximum number of iterations is 78 iterations, which means the PSO method exceeds stopping criteria. We can see graphically the evolution of the fitness value for each iteration in Fig. 33.2.

33.7.1.2 Obtained solutions using GWO

Rebuilding the optimization program using the GWO algorithm and the same parameters, running 10 executions, we obtain the results in Table 33.5.

TABLE 33.4 Obtained results using PSO.

Run #	z_1	z_2	z_3	z_4	n_1	n_2	n_3	r_1	r_2	r_3	r_4	R_s	CPU time (s)	number of function evaluations (NFE)
1	3	4	1	1	3	3	4	0.955384	0.949005	0.937127	0.97	0.9697690	0.908102	6900
2	3	4	1	1	3	3	4	0.955211	0.949449	0.934795	0.97	0.9697690	0.403735	5100
3	1	4	1	1	3	3	4	0.955307	0.949076	0.937132	0.97	0.9697691	0.465449	6600
4	3	1	1	1	3	3	4	0.951899	0.941555	0.932660	0.97	0.9696774	0.405214	5500
5	3	2	1	1	3	3	4	0.954649	0.947892	0.935635	0.97	0.9697546	0.575671	7300
6	3	4	1	1	3	3	4	0.955008	0.949175	0.938109	0.97	0.9697689	0.544743	7200
7	**1**	**4**	**1**	**1**	**3**	**3**	**4**	**0.955080**	**0.949234**	**0.937298**	**0.97**	**0.9697692**	**0.591678**	**7800**
8	3	4	1	1	3	3	4	0.955345	0.949055	0.937049	0.97	0.9697691	0.513838	6100
9	1	4	1	1	3	3	4	0.954792	0.949419	0.937593	0.97	0.9697691	0.537991	7500
10	1	2	1	1	3	3	4	0.954287	0.948135	0.935937	0.97	0.9697546	0.282046	4300

PSO, particle swarm optimization.

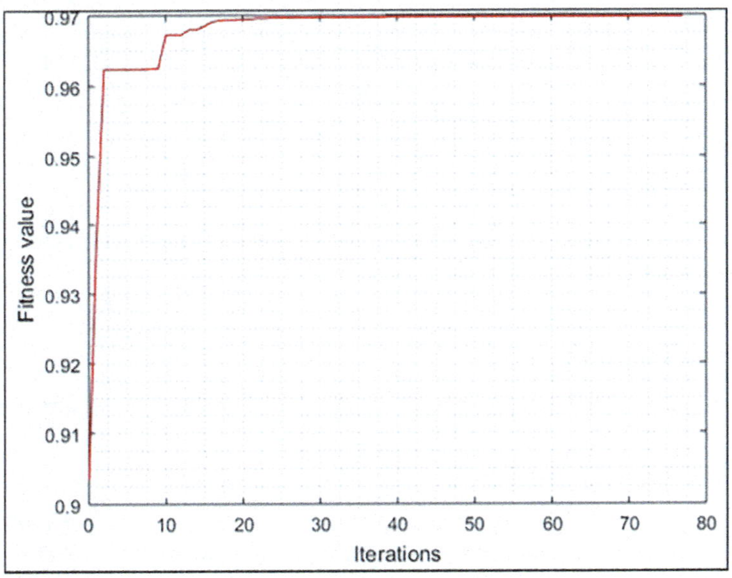

FIGURE 33.2 Convergence curve of PSO. *PSO*, Particle swarm optimization.

Table 33.5 shows the optimal reliability reached by GWO in the ninth executions is 0.969753 after an elapsed time of 0.395561 seconds; the number of function evolutions needed to find the optimal reliability reaches the maximum of 20,000.

We notice from Table 33.5 that GWO is relatively stable while solving this problem, and we notice that in all executions, it outcomes the same optimal redundancy for all the subsystems in every execution except the last execution that choose the second choice as optimal choice for subsystem 1, and it found the same optimal reliability for subsystem 4 (motors).

We can see absolute domination of first choice as the optimal choice for subsystems 1, 3, and 4 in majority of executions, including the best one, and, furthermore, domination of second choice as the optimal choice for subsystem 2, including the best execution.

The number of function evaluations reaches its maximum value (200 iterations) in all executions. We can see the convergence curve of GWO execution in Fig. 33.3.

TABLE 33.5 Obtained results using GWO.

Run #	z_1	z_2	z_3	z_4	n_1	n_2	n_3	r_1	r_2	r_3	r_4	R_s	CPU time (s)	NFE
1	2	4	1	1	3	3	4	0.948441	0.944486	0.925385	0.97	0.9696701	0.400611	20,000
2	3	2	2	1	3	3	4	0.952037	0.945459	0.921102	0.97	0.9696971	0.387279	20,000
3	2	2	1	1	3	3	4	0.944447	0.945333	0.934499	0.97	0.9696564	0.390130	20,000
4	1	2	1	1	3	3	4	0.954884	0.947570	0.934757	0.97	0.9697525	0.390478	20,000
5	1	2	1	1	3	3	4	0.955642	0.946835	0.935643	0.97	0.9697520	0.399930	20,000
6	1	2	2	1	3	3	4	0.951792	0.945113	0.924451	0.97	0.9696984	0.408126	20,000
7	1	1	3	4	3	3	4	0.944339	0.930363	0.902496	0.97	0.9694166	0.421317	20,000
8	1	4	1	1	3	3	3	0.952621	0.947258	0.962017	0.97	0.9697004	0.391070	20,000
9	1	2	1	1	3	3	4	0.953556	0.948299	0.937536	0.97	0.9697530	0.395561	20,000
10	1	2	1	1	2	3	4	0.973727	0.932799	0.918982	0.97	0.9689935	0.401129	20,000

GWO, grey wolf optimizer.

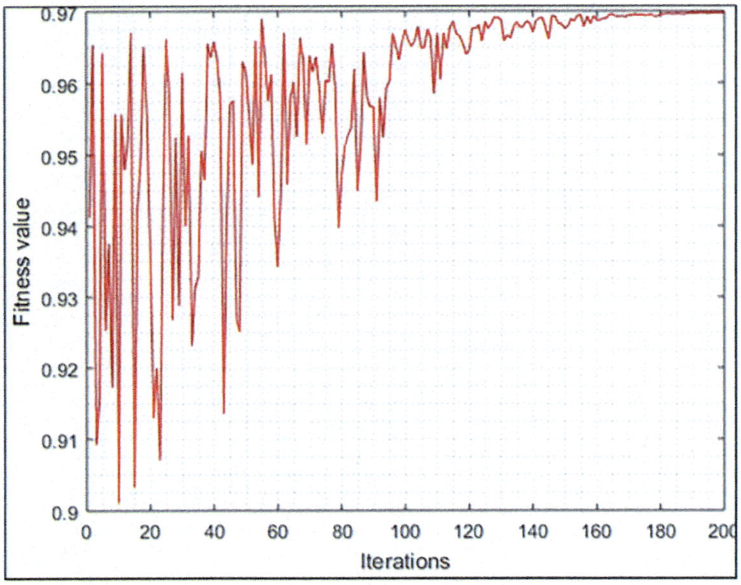

FIGURE 33.3 Convergence curve of GWO. *GWO*, Grey wolf optimizer.

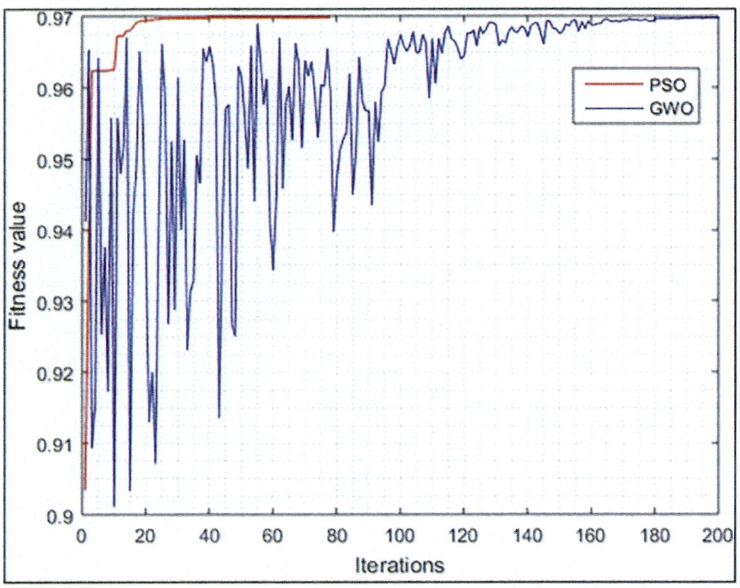

FIGURE 33.4 Convergence curve of PSO and GWO—comparison. *GWO*, Grey wolf optimizer; *PSO*, particle swarm optimization.

33.7.1.3 Best obtained solution of each method

In Table 33.6, we compare the results in best execution of each algorithm.

The evolution of the reliability in each execution can be represented graphically by combining the convergence curves of the two methods in Fig. 33.4.

From Table 33.6, we can compare both PSO and GWO solutions for this multichoice RRAP.

The optimal reliability achieved by PSO method is 0.969769, demonstrating an improvement over the GWO method. The improvements in reliability of PSO method over the GWO method are 0.00165%.

The PSO method reaches its optimal reliability of 0.969769 with an elapsed time of 0.5917 seconds, compared to 0.3956 seconds token by GWO method to reach its optimal reliability of 0.969753. Which makes GWO method 33.14% faster than PSO method. Therefore the PSO method provides slightly higher reliability over the GWO method while sacrificing longer elapsed time.

TABLE 33.6 Comparison between best run of PSO and GWO.

Best run	z_1	z_2	z_3	z_4	n_1	n_2	n_3	r_1	r_2	r_3	r_4	Best R_s	Mean R_s	CPU time (s)	NFE	SD
PSO	1	4	1	1	3	3	4	0.95508	0.949234	0.937298	0.97	0.969769	0.969757	0.5917	7800	2.9E − 05
GWO	1	2	1	1	3	3	4	0.95356	0.948299	0.937536	0.97	0.969753	0.969609	0.3956	20000	0.00024

GWO, grey wolf optimizer; *PSO*, particle swarm optimization; *SD*, standard deviation.

TABLE 33.7 Constraint values of best run of PSO and GWO.

Constraints	Volume	Cost	Weight
PSO	80.6	383.2273	172.1852
GWO	68	382.2922	173.4554

GWO, grey wolf optimizer; *PSO*, particle swarm optimization.

TABLE 33.8 Resource consumption of PSO and GWO.

Constraint	Volume (%)	Cost (%)	Weight (%)
PSO	32.24	95.81	24.60
GWO	27.20	95.57	24.78

GWO, grey wolf optimizer; *PSO*, particle swarm optimization.

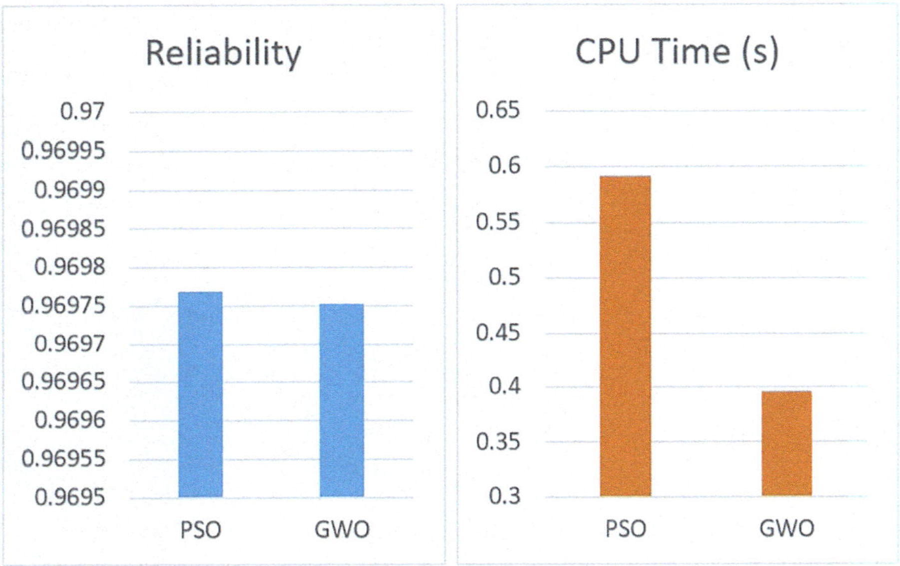

FIGURE 33.5 System reliability and elapsed time comparison.

From Table 33.6, we can observe the stability and convergence speed of both methods. With standard deviation (SD) of 0.00024, GWO method has an SD eight times larger than the SD of PSO method, which shows the more dispersion of fitness value of GWO method compared to PSO method. In addition, it is noticeable that PSO method has faster convergence than GWO method, needing just 78 iterations to achieve optimal reliability, whereas the GWO method exceeds its maximum iteration limits (200 iterations), and despite that, it achieved a slightly less optimal reliability value. Convergence rates are clearly appearing in the convergence curves in Fig. 33.4.

33.7.1.4 Checking the constraints

In Table 33.7, we verify nonviolation of the constraints by the two methods.

As we can observe in Table 33.7 the GWO method led to better resource management compared to PSO method in terms of volume (constraint g_1) and cost (constraint g_2). However, the PSO method performed better in terms of weight reduction (constraint g_3). The ratio of resource consumption is reported in Table 33.8.

33.7.1.5 Comparison histograms between PSO and GWO

Figs. 33.5 and 33.6 show a comparison histogram between PSO and GWO.

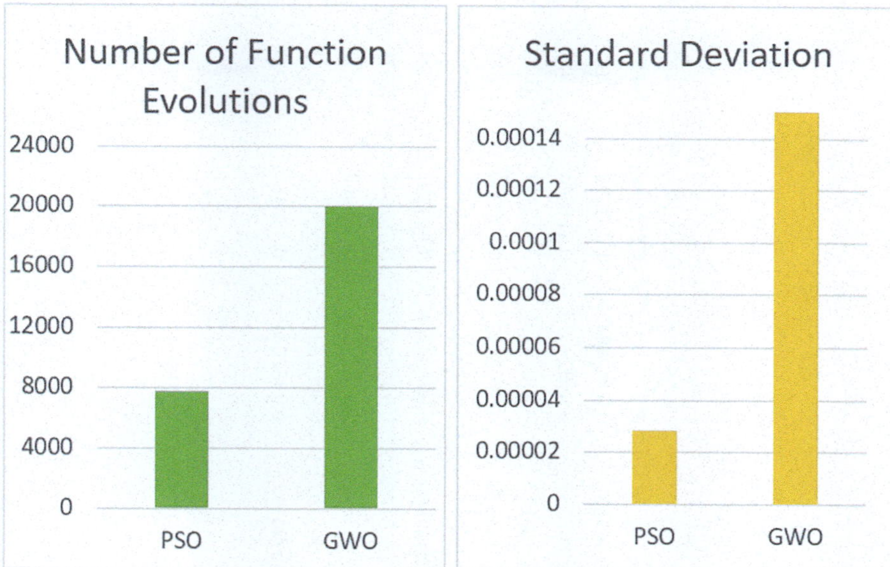

FIGURE 33.6 NFE and SD comparison.

33.7.2 Dual-motor configuration

In the following results the number of redundant components of subsystem 4 (motor) is fixed at two ($n_4 = 2$).

33.7.2.1 Obtained solutions using PSO

Following the same procedure as single-motor configuration but now n_4 is fixed at two, we obtained the results shown in Table 33.9.

Table 33.9 shows that the optimal system reliability reached by PSO on the fourth execution is 0.99828321 and lasting an elapsed time of 0.509080 seconds; the correspondent number of function evolutions needed to achieve the optimal reliability was 6300, which was the lowest between all executions.

We can observe from Table 33.9 that PSO is very stable in solving this problem, and we can see that in all executions, it results the same optimal redundancy for all the subsystems, and the same choice (first choice) for subsystems 3 and 4.

In addition, the choice of subsystem 1 fluctuates between first choice and third choice, whereas PSO goes with the third choice for the best execution.

For subsystem 2 the choice fluctuates between the second and fourth choice, and the fourth was chosen in the best execution. Varying between 6300 and 9300 the necessary number of function evaluation of PSO needed to complete its tasks stays relatively low. The convergence curve of the PSO method is shown it Fig. 33.7. It can be observed that PSO method has a very fast convergence rate.

33.7.2.2 Obtained solutions using GWO

After executing GWO script for the dual-motor configuration, we see noticeable differences as shown in Table 33.10.

Table 33.10 shows the optimal reliability reached by GWO after two executions is 0.998244992, achieved after 1.045658 seconds; the number of function evolutions needed to find the optimal reliability reaches its maximum limit of 20,000.

We can observe from Table 33.10 that GWO is relatively stable while running different executions, and it is noticeable that in all executions, it outcomes the same optimal redundancy for all the subsystems in every execution, including the best one, except last execution, which chooses three redundant components as optimal redundancy of subsystem 3.

Moreover, we can observe that the first choice is chosen as the optimal choice for subsystem 4 in all executions and 3 and 4 in majority of executions, including the best one and domination of second choice as the optimal choice for subsystem 2, including the best execution. The number of function evaluations reaches its maximum value (200 iterations) in all executions. Fig. 33.6 shows the convergence curve of the fitness function of the GWO method.

From Fig. 33.8, we observe high rate of dispersion during the first 100 iterations. After 100 iterations, the GWO method becomes more stable, and the dispersion decreases until it enters a convergence interval (>160 iterations).

TABLE 33.9 Results of PSO executions.

Run #	z_1	z_2	z_3	z_4	n_1	n_2	n_3	r_1	r_2	r_3	r_4	R_s	CPU time	NFE
1	1	2	1	1	3	3	4	0.939760	0.931058	0.920445	0.965867	0.99824842	1.268653	9300
2	1	2	1	1	3	3	4	0.939719	0.931834	0.919802	0.965737	0.99824874	0.511844	7300
3	3	4	1	1	3	3	4	0.939605	0.932215	0.920157	0.966170	0.99828296	0.615393	8100
4	3	4	1	1	3	3	4	0.940094	0.933140	0.919577	0.965928	0.99828321	0.509080	6300
5	3	4	1	1	3	3	4	0.940258	0.932703	0.920983	0.965947	0.99828320	0.759469	9600
6	1	2	1	1	3	3	4	0.939968	0.931006	0.919751	0.965869	0.99824866	0.503351	6700
7	1	4	1	1	3	3	4	0.941107	0.933069	0.919566	0.965777	0.99828262	0.562045	7300
8	3	2	1	1	3	3	4	0.939402	0.931790	0.919526	0.965805	0.99824870	0.756063	9200
9	1	4	1	1	3	3	4	0.939446	0.932336	0.919209	0.966197	0.99828275	0.591666	9300
10	1	4	1	1	3	3	4	0.939717	0.932477	0.919682	0.966117	0.99828315	0.588469	8200

PSO, particle swarm optimization.

TABLE 33.10 Results of GWO executions.

Run #	z_1	z_2	z_3	z_4	n_1	n_2	n_3	r_1	r_2	r_3	r_4	R_s	CPU time (s)	NFE
1	1	1	3	1	3	3	4	0.935348	0.923123	0.90377	0.964119	0.997902516	1.230589	20,000
2	3	2	1	1	3	3	4	0.938858	0.931372	0.920812	0.965887	0.998244992	1.045658	20,000
3	5	2	1	1	3	3	4	0.926984	0.927075	0.91983	0.964293	0.997906883	1.031113	20,000
4	3	2	1	1	3	3	4	0.938655	0.933106	0.926798	0.965328	0.998238736	1.173643	20,000
5	1	4	1	1	3	3	4	0.938077	0.934579	0.936963	0.964982	0.998240294	1.080513	20,000
6	1	4	4	1	3	3	4	0.940561	0.933198	0.897904	0.963528	0.998053011	1.061868	20,000
7	1	2	1	1	3	3	4	0.941557	0.931019	0.932907	0.964945	0.998222819	1.115638	20,000
8	1	4	2	1	3	3	4	0.939819	0.930524	0.892564	0.965616	0.998131186	1.097596	20,000
9	1	2	2	1	3	3	4	0.937522	0.931983	0.906636	0.964588	0.998111421	0.897666	20,000
10	1	6	1	1	3	3	3	0.941015	0.917089	0.950442	0.962852	0.99772368	1.138142	20,000

GWO, grey wolf optimizer.

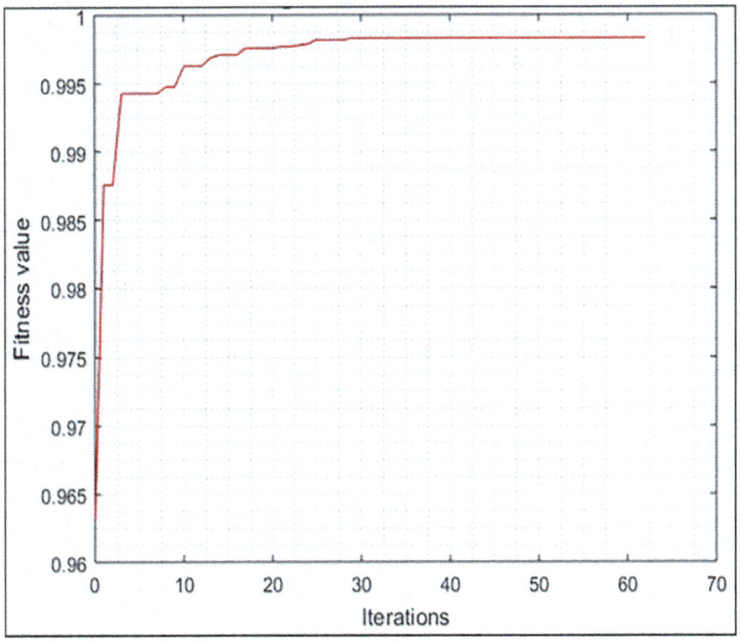

FIGURE 33.7 Convergence curve of PSO—dual motor. *PSO*, Particle swarm optimization.

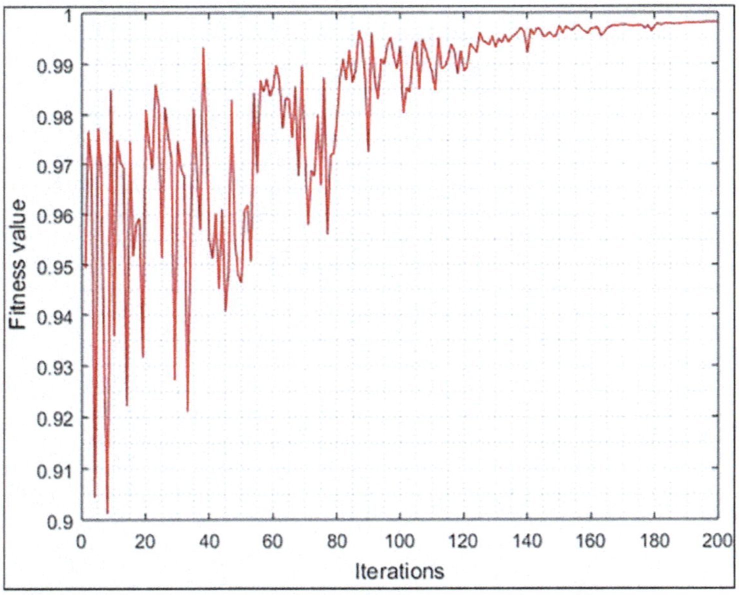

FIGURE 33.8 Convergence curve of GWO—dual motor. *GWO*, Grey wolf optimizer.

33.7.2.3 Best obtained solution of each method

Table 33.11 shows results of the best run of each method. The evolution of the reliability in the best executions of both PSO and GWO methods can be represented graphically by combining the convergence curves of the two methods (see Fig. 33.9).

From Table 33.11, we can compare both PSO and GWO solutions for this multichoice dual-motor RRAP. The optimal reliability achieved by PSO method is 0.9982832, which has clearly outperformed the reliability of 0.9982449 achieved by the GWO method. The optimal reliability by PSO method is higher than the GWO one by 0.00383%.

The PSO method reached its optimal reliability of 0.9982832 after an elapsed time of 0.50908 seconds, compared to 1.045658 seconds needed for GWO method to reach its optimal reliability of 0.9982449, which means that the PSO method

TABLE 33.11 Comparison between the results of the best run of PSO and GWO methods.

Method	z_1	z_2	z_3	z_4	n_1	n_2	n_3	r_1	r_2	r_3	r_4	Best R_s	Mean R_s	CPU time (s)	NFE	SD
PSO	3	4	1	1	3	3	4	0.940094	0.93314	0.919577	0.965928	0.9982832	0.9982692	0.509080	6300	1.77384E − 05
GWO	3	2	1	1	3	3	4	0.938858	0.931372	0.920812	0.965887	0.9982449	0.9980775	1.045658	20,000	0.000179883

GWO, grey wolf optimizer; *PSO*, particle swarm optimization.

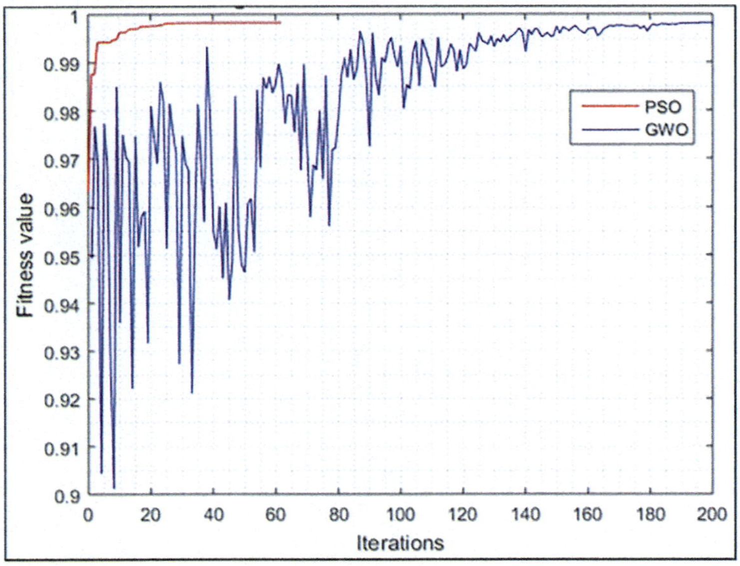

FIGURE 33.9 Convergence curves of PSO and GWO—dual motor. *GWO*, Grey wolf optimizer; *PSO*, particle swarm optimization.

TABLE 33.12 Constraint values of best run of PSO and GWO.

Constraints	Volume	Cost	Weight
PSO	91.1	381.427	179.9281
GWO	78.5	380.734	181.1983

GWO, grey wolf optimizer; *PSO*, particle swarm optimization.

is 51.31% faster than the GWO method. Therefore the PSO method provides slightly higher reliability than the GWO method while maintaining significantly shorter elapsed time.

From Table 33.11, we can observe the stability and convergence speed of both methods. With SD of 0.000179883 the GWO method has an SD that is 10 times larger than that of the PSO method, which shows the more dispersion of fitness value of the GWO method compared to the PSO method. In addition, it is seen that the PSO method has faster convergence than the GWO method, needing just 63 iterations to achieve optimal reliability, whereas the GWO method exceeds its maximum iteration limits (200 iterations), and despite that, it achieved slightly less optimal reliability value. Convergence rates are clearly appearing in the convergence curves in Fig. 33.9.

33.7.2.4 Checking the constraints

In Table 33.12, we verify nonviolation of the constraints by the two methods.

As we can observe in Table 33.12 the GWO method led to better resource management compared to PSO method in terms of volume (constraint g_1) and cost (constraint g_2). However, the PSO method performed better in terms of weight reduction (constraint g_3). The ratio of resource consumption is reported in Table 33.13.

33.7.2.5 Comparison histograms between PSO and GWO

Figs. 33.10 and 33.11 are histograms representing key differences between the best execution of both methods.

33.7.3 Best block diagram of the firefighting drone

Fig. 33.12 shows the best obtained block diagram for the single-motor configuration, whereas Fig. 33.13 shows the results

TABLE 33.13 Resource consumption in the best run of PSO and GWO.

Constraints	Volume (%)	Cost (%)	Weight (%)
PSO	36.44	95.36	25.70
GWO	31.40	95.18	25.89

GWO, grey wolf optimizer; *PSO*, particle swarm optimization.

FIGURE 33.10 NFE and SD comparison—dual motor.

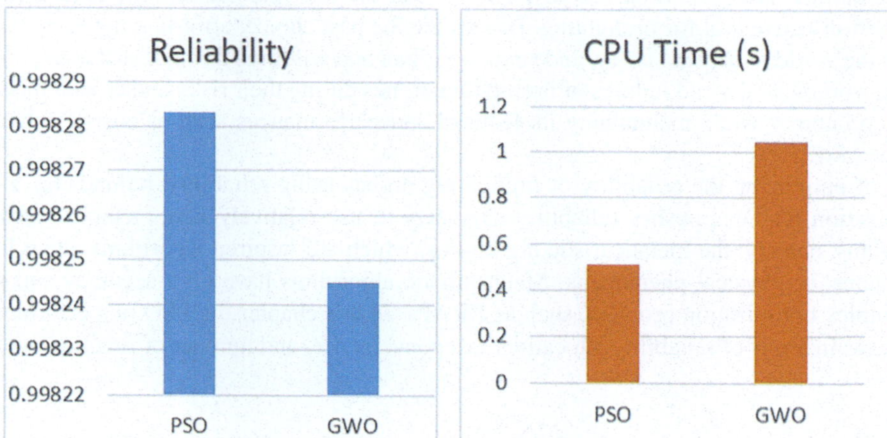

FIGURE 33.11 Reliability and CPU time comparison—dual motor.

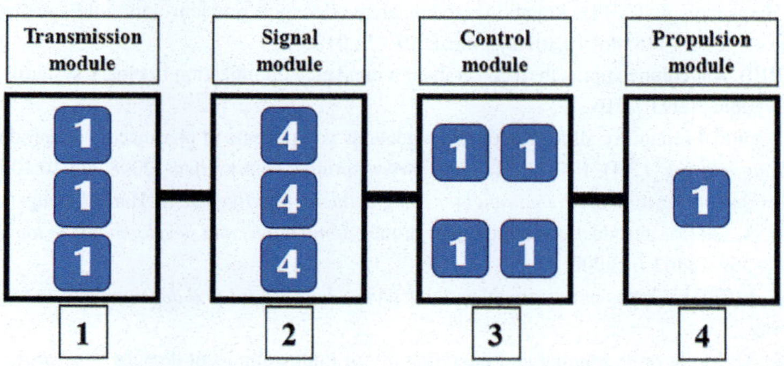

FIGURE 33.12 Best block diagram for the single-motor configuration.

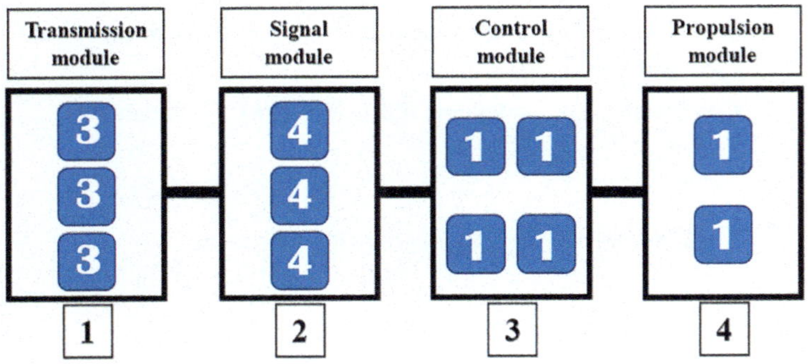

FIGURE 33.13 Best block diagram for the dual-motor configuration.

for the dual-motor configuration.

Fig. 33.13 shows the best dual-motor configuration (results obtained using PSO).

33.8 Conclusion

UAVs or drones are one of the greatest innovations in history. They use cutting-edge technology to perform different tasks with high accuracy and precision and get better results over traditional techniques. They play crucial role in several civilian fields such as photography, aerial light shows, agriculture, traffic control, mapping, industrial inspection, search and rescue, delivery, and scientific research and environmental monitoring.

Drones can be implemented in firefighting and other natural disasters; they have several advantages over traditional methods such as ability to perform tasks in extremely harsh conditions and for several hours at once, avoiding human risks and reducing fatalities. The increasing rate of wildfires over the last years defines the need for more efficient, safe, fast-response, and cost-effective tool for firefighting. Drones are the best choice, affording the three features, allowing fire departments around the world to detect, survey, and suppress urban and wildfires through vast areas.

To ensure proper working and avoid failures of firefighting drones during their risky and crucial missions, it is necessary to optimize drone's reliability while maintaining its features and performances such as cost-effectiveness, maneuvering capabilities, and ease of deploy.

The complexity of optimizing the reliability of firefighting drones using reliability–redundancy allocation model and conventional optimization methods pushes reliability engineers to use relatively new methods, such as nature-inspired optimization algorithms, namely, the metaheuristic algorithms, which are inspired algorithms either by evolution, swarm intelligence, ecosystems, or physical phenomena. Metaheuristic algorithms have high accuracy and precision, and offer fast solution for complex optimization problem, such as RRAPs. In this chapter the PSO outperformed the GWO for two different configurations in terms of reliability allocation, fast convergence, and number of function evolutions.

References

Abouei Ardakan, M., & Zeinal Hamadani, A. (2014). Reliability–redundancy allocation problem with cold-standby redundancy strategy. *Simulation Modelling Practice and Theory, 42*, 107–118. doi:10.1016/j.simpat.2013.12.013.

Agarwal, M., & Sharma, V. K. (2010). Ant colony approach to constrained redundancy optimization in binary systems. *Applied Mathematical Modelling, 34*(4), 992–1003. doi:10.1016/j.apm.2009.07.016.

Aqel, I., & Mellal, M. A. (2023). Optimal reliability allocation of heterogeneous components in pharmaceutical production plant. *International Journal on Interactive Design and Manufacturing, 17*(4), 1711–1720. https://www.springer.com/journal/12008. doi:10.1007/s12008-023-01256-1.

Bayou, S., & Mallik, I. (2023). *Design optimization of drone firefighter using artificial intelligence*. M'Hamed Bougara University.

Bayou, S., Mallik, I., & Mellal, M. A. (2025). Introduction to firefighter drones. *Innovations and developments in unmanned aerial vehicles* (pp. 19–28). IGI Global. doi:10.4018/979-8-3693-8462-6.ch002.

Bayou, S., Mallik, I., & Mellal, M. A. (2025). Types of unmanned aerial vehicles. *Innovations and developments in unmanned aerial vehicles* (pp. 1–18). IGI Global Scientific Publishing.

Bellman, R., & Dreyfus, S. (1958). Dynamic programming and the reliability of multicomponent devices. *Operations Research, 6*(2), 200–206. doi:10.1287/opre.6.2.200.

Coit, D. W. (2001). Cold-standby redundancy optimization for nonrepairable systems. *IIE Transactions, 33*(6), 471–478.

Dobani, E. R., Ardakan, M. A., Davari-Ardakani, H., & Juybari, M. N. (2019). RRAP-CM: A new reliability-redundancy allocation problem with heterogeneous components. *Reliability Engineering and System Safety, 191*. https://www.journals.elsevier.com/reliability-engineering-and-system-safety. doi:10.1016/j.ress.2019.106563.

Dohi, T., Kaio, N., Young Yun, W., Agarwal, M., & Gupta, R. (2006). Genetic search for redundancy optimization in complex systems. *Journal of Quality in Maintenance Engineering, 12*(4), 338–353. doi:10.1108/13552510610705919.

Gad, A. G. (2022). Particle swarm optimization algorithm and its applications: A systematic review. *Archives of Computational Methods in Engineering, 29*(5), 2531–2561. doi:10.1007/s11831-021-09694-4.

Garg, H. (2015). An approach for solving constrained reliability-redundancy allocation problems using cuckoo search algorithm. *Beni-Suef University Journal of Basic and Applied Sciences, 4*(1), 14–25.

Gholinezhad, H. (2024). A new model for reliability redundancy allocation problem with component mixing. *Reliability Engineering & System Safety, 242*, 109815. doi:10.1016/j.ress.2023.109815.

Jin, J., Kim, S., & Moon, J. (2024). Development of a firefighting drone for constructing fire-breaks to suppress nascent low-intensity fires. *Applied Sciences (Switzerland), 14*(4). Www.mdpi.com/journal/applsci/. doi:10.3390/app14041652.

Juybari, M. N., Abouei Ardakan, M., & Davari-Ardakani, H. (2019). A penalty-guided fractal search algorithm for reliability–redundancy allocation problems with cold-standby strategy. *Proceedings of the Institution of Mechanical Engineers, O: Journal of Risk and Reliability, 233*(5), 775–790. http://pio.sagepub.com/content/by/year. doi:10.1177/1748006 ž 19825707.

Kennedy, J., & Eberhart, R. (1995). Particle swarm optimization. *Proceedings of the IEEE International Conference on Neural Networks, 4*, 1942–1948.

Kuo, W., & Wan, R. (2007). Recent advances in optimal reliability allocation. *IEEE Transactions on Systems, Man, and Cybernetics A: Systems and Humans, 37*(2), 143–156. doi:10.1109/TSMCA.2006.889476.

Mellal, M. A., Adjerid, S., Benazzouz, D., Berrazouane, S., & Williams, E. J. (2013). Obsolescence optimization of electronic and mechatronic components by considering dependability and energy consumption. *Journal of Central South University, 20*(5), 1221–1225. doi:10.1007/s11771-013-1605-9.

Mellal, M. A., Al-Dahidi, S., & Williams, E. J. (2020). System reliability optimization with heterogeneous components using hosted cuckoo optimization algorithm. *Reliability Engineering and System Safety, 203*. https://www.journals.elsevier.com/reliability-engineering-and-system-safety. doi:10.1016/j.ress.2020.107110.

Mellal, M. A., Frik, A., & Boutiche, R. (2021). Reliability optimization of power plant safety system using grey wolf optimizer and shuffled frog-leaping algorithm. In M. A. Mellal, & M. G. Pecht (Eds.), *Nature-inspired computing paradigms in systems: Reliability, availability, maintainability, safety and cost (RAMS+C) and prognostics and health management (PHM)* (pp. 1–13). Academic Press.

Mellal, M. A., Khitous, M., & Zemmouri, M. (2023). Combined heat and power economic dispatch problem with binary method using flower pollination algorithm and differential evolution. *Electrical Engineering, 105*(4), 2161–2168. https://www.springer.com/journal/202. doi:10.1007/s00202-023-01801-x.

Mellal, M. A., & Williams, E. J. (2017). "Chapter 14 – The cuckoo optimization algorithm and its applications," P. Samui, S. Sekhar, & V. E. Balas Eds. Handbook of neural computation. Academic Press, pp. 269–277.

Mellal, M. A., & Williams, E. J. (2018a). Large-scale reliability-redundancy allocation optimization problem using three soft computing methods. *Modeling and simulation based analysis in reliability engineering*. CRC Press.

Mellal, M. A., & Zio, E. (2016). A penalty guided stochastic fractal search approach for system reliability optimization. *Reliability Engineering and System Safety, 152*, 213–227. doi:10.1016/j.ress.2016.03.019.

Mellal, M. A., & Zio, E. (2017). System reliability-redundancy allocation by evolutionary computation. In *2nd International Conference on System Reliability and Safety, ICSRS 2017* (pp. 15–19). Institute of Electrical and Electronics Engineers Inc. 9781538633229 15 19. doi:10.1109/ICSRS.2017.8272790.

Mellal, M. A., & Zio, E. (2019). An adaptive particle swarm optimization method for multi-objective system reliability optimization. *Proceedings of the Institution of Mechanical Engineers, O: Journal of Risk and Reliability, 233*, 990–1001.

Mellal, M. A., & Zio, E. (2020). System reliability-redundancy optimization with cold-standby strategy by an enhanced nest cuckoo optimization algorithm. *Reliability Engineering and System Safety, 201*. https://www.journals.elsevier.com/reliability-engineering-and-system-safety. doi:10.1016/j.ress.2020.106973.

Mellal, M. A., & Williams, E. J. (2018b). *A survey on ant colony optimization, particle swarm optimization, and cuckoo algorithms* (pp. 37–51). IGI Global. doi:10.4018/978-1-5225-2990-3.ch002.

Mellal, M. A., & Zio, E. (2022). Multi-objective availability and cost optimization by PSO and COA for series-parallel systems with subsystems failure dependencies. *Microprocessors and Microsystems, 89*, 104422. doi:10.1016/j.micpro.2021.104422.

Mirjalili, S., Mirjalili, S. M., & Lewis, A. (2014). Grey wolf optimizer. *Advances in Engineering Software, 69*, 46–61. http://www.journals.elsevier.com/advances-in-engineering-software/. doi:10.1016/j.advengsoft.2013.12.007.

Nazari-Heris, M., Mehdinejad, M., Mohammadi-Ivatloo, B., & Babamalek-Gharehpetian, G. (2019). Combined heat and power economic dispatch problem solution by implementation of whale optimization method. *Neural Computing and Applications, 31*(2), 421–436. http://link.springer.com/journal/521. doi:10.1007/s00521-017-3074-9.

Ouyang, Z., Liu, Y., Ruan, S. J., & Jiang, T. (2019). An improved particle swarm optimization algorithm for reliability-redundancy allocation problem with mixed redundancy strategy and heterogeneous components. *Reliability Engineering and System Safety, 181*, 62–74. doi:10.1016/j.ress.2018.09.005.

Promkaew, N., Thammawiset, S., Srisan, P., Sanitchon, P., Tummawai, T., & Sukpancharoen, S. (2024). Development of metaheuristic algorithms for efficient path planning of autonomous mobile robots in indoor environments. *Results in Engineering, 22*, 102280. doi:10.1016/j.rineng.2024.102280.

Rogers, J. P. (2024). *Handbook of drone warfare*. De Gruyter.

Saheb, T., & Mellal, M. A. (2024). Improvement of system reliability in a natural gas processing facility by PSO and DE. *International Journal on Interactive Design and Manufacturing, 18*, 167–176.

Sharma, T. K. (2021). Enhanced butterfly optimization algorithm for reliability optimization problems. *Journal of Ambient Intelligence and Humanized Computing, 12*(7), 7595–7619.

Taghiyeh, S., Mahmoudi, M., Fadaie, S., & Tohidi, H. (2020). Fuzzy reliability-redundancy allocation problem of the overspeed protection system. *Engineering Reports, 2*(8). https://onlinelibrary.wiley.com/journal/25778196. doi:10.1002/eng2.12221.

Templos-Santos, J. L., Aguilar-Mejia, O., Peralta-Sanchez, E., & Sosa-Cortez, R. (2019). Parameter tuning of PI control for speed regulation of a PMSM using bio-inspired algorithms. *Algorithms, 12*(3). https://res.mdpi.com/algorithms/algorithms-12-00054/article_deploy/algorithms-12-00054.pdf?filename=&attachment=1. doi:10.3390/A12030054.

Tillman, F. A., Hwang, C. L., Fan, L. T., & Balbale, S. A. (1968). Systems reliability subject to multiple nonlinear constraints. *IEEE Transactions on Reliability, R-17*(3), 153–157. doi:10.1109/TR.1968.5216929.

Tillman, F. A., & Liittschwager, J. M. (1967). Integer programming formulation of constrained reliability problems. *Management Science, 13*(11), 887–899. doi:10.1287/mnsc.13.11.887.

Zhang, T. W., Xu, G. H., Zhan, X. S., & Han, T. (2022). A new hybrid algorithm for path planning of mobile robot. *Journal of Supercomputing, 78*(3), 4158–4181. http://www.springerlink.com/content/0920-8542. doi:10.1007/s11227-021-04031-9.

Zou, D., Liu, H., Gao, L., & Li, S. (2011). A novel modified differential evolution algorithm for constrained optimization problems. *Computers and Mathematics With Applications, 61*(6), 1608–1623. doi:10.1016/j.camwa.2011.01.029.

Chapter 34

Future trends and research directions in automation of mechatronic control software engineering

Husain Mehdi[a], Pankaj Kumar Gupta[a], Vipin Kumar Sharma[a] and Mohamed Arezki Mellal[b]

[a] *Mechanical Engineering Department, Meerut Institute of Engineering and Technology, Meerut, Uttar Pradesh, India,*

[b] *LMSS, Faculty of Technology, M'Hamed Bougara University, Boumerdes, Algeria*

34.1 Introduction

Mechatronic products can only be designed through intensive and continuous collaboration between engineers in mechanical, control, electronic, and software engineering (Pahl et al., 2007). Among the inherent issues with these systems is that design decisions cannot be made independently in one field without considering their potential consequences in other fields. However, the industrial world continues to be dominated by traditional sequential model designs, with mechanical parts being finished first, and then control and software development commence. The reason behind the process is that manufacturing factors demand early hardware stabilization. Software development, on the other hand, is often left to accommodate the requirements that are introduced late to counter the flaws of previous design stages. Such a gradual approach will hardly yield the best system performance, as it will not be able to reflect the rapid interaction of mechanical dynamics, electronic behavior, and control algorithms. It also underestimates the importance of software that is gradually influencing the system's functionality, flexibility, and performance. To overcome such inadequacies, concurrent engineering processes must be well-integrated into development strategies (Wikander et al., 2001). In this case, design decisions would be evaluated based on multidisciplinary goals and constraints before committing to a design freeze. This holistic style is commonly referred to as the mechatronic design approach, which has been well-documented in the literature (Schöner, 2004). The fact that mechatronic systems should be handled as integrated systems has been well established for some time. Ziegler and Nichols (1943) emphasized the idea that the control systems and the physical processes should be considered one, with each factor making an equal contribution to the total system performance. It is based on this that the issues of adapting mechatronic design solutions (in this case, the development of control software) are analyzed in the current chapter. The current methods and tools are reviewed, and potential extensions and combinations are considered. The discussion highlights the significance of upperlevel forms of abstraction and the practical issues of implementation in a real-world environment. Furthermore, it is shown that the hardware and software should be streamlined and checked using the aid of corresponding modeling paradigms and tool support. Based on these observations a mechatronic system design support structure is incorporated. The framework employs a typical high-level system model, which also facilitates effective communication between designers and the exchange of information between tools, particularly during the initial design phases, without skewing the scheme toward a specific control paradigm, industry, or product type.

34.2 Challenges in mechatronic design

A substantial body of academic literature and industrial applications has highlighted that significant challenges remain in the design and development of mechatronic systems.

Computational Intelligence in Mechatronics: Solving Real-World Problems in Electronic Systems Design. DOI: https://doi.org/10.1016/B978-0-443-27580-7.00012-3

- Among the problems that are most commonly reported is the effective communication of design models and data between various engineering sectors (Schöner, 2004).
- A facet of this problem is the necessity of efficient collaboration and communication between engineers working in the fields of mechanical, electronic, control, and software engineering, which, in most cases, is impaired by the lack of compatibility between the tools used and the terminologies, as well as the development processes (Perrin, 2009).
- The second significant difficulty is related to multidisciplinary modeling, where heterogeneous components of the system must be modeled coherently and consistently, despite their varying physical behaviors, time scales, and levels of abstraction (Perrin, 2009).
- This is also made more challenging because design decisions in one or more disciplines must continually be assessed and coordinated with other fields, rather than being made individually, to ensure that system-level goals and constraints are met (Jackson, 2006b).
- Testing and verification of integrated system behavior in the early stages are also issues, as most verification activities are not conducted early in the development cycle, thereby increasing the likelihood of expensive redesigns (Perrin, 2009).
- Nevertheless, despite the identified advantages of integrated development, sequential design patterns still prevail in most industrial settings, depriving them of the opportunity for early feedback and cross-domain optimization.
- This is further compounded by the fact that few tools and methodologies specifically facilitate multidisciplinary design processes and the associated workflows (Jackson, 2006b).
- In addition, the systematic support for the design and implementation of control software is a significant concern, especially with the progressive rise in software complexity and functional accountability.

Further discussion of these concerns leads to the identification of three underlying problems that primarily influence the development of mechatronic systems, including the provision of a proper combination of design, appropriate checks of sound design, and the automatic production of control software. The key areas of concern are addressed in the following subsections.

34.2.1 Design integration

It is estimated that modern mechatronic systems will offer a vast scope of functions, operating within more challenging parameters in terms of energy consumption, space, mass, and time of development (Wikander et al., 2001). To meet these demands, expertise in any of the engineering disciplines should be effectively integrated into a single process of product development. This has led to the need for higher level design integration, which will not only concern the technical aspects of the tools and methodologies but also the proper organization of design teams and their interrelations. There are still significant discrepancies between the various design stages and tools that assist in many modern development practices. Design teams are typically organized based on specialized skills and can be located far apart, thus making coordination even more challenging. This results in the system-level integration being pushed to a later date when physical prototypes are available. Integration-related challenges also exist even when partial automation is introduced during the design process. Considering the example of using interconnected domain-specific component libraries, checks are necessary to ensure that these libraries are consistent and complete. Integration has become a universally recognized necessary task both in industry and academia. In the design of mechatronic systems, industrial research has identified integration as a prime direction of research and a key success factor (Perrin, 2009). Similar concerns, including collaborative design practice, efficient data sharing, knowledge management, project coordination, and parallel development across multiple domains, as well as the miniaturization of the control algorithm and the physical system, have also been highlighted in the literature. According to the recent research on the industrial practices the principal manufacturers are more concerned with the integration strategies to coordinate the specialized engineers as well as use tools that can assist in the management of information at the system level, as opposed to using monolithic tools that seek to facilitate all the detailed designing activities. Such tools are expected to manage high-level system information, track requirements, and efficiently handle design changes. However, achieving effective integration also requires appropriate design methodologies and active involvement of human designers, as clear communication across disciplines remains essential.

34.2.1.1 Design methods

Despite the existence of several studies that have strived to develop holistic theoretical frameworks of engineering design, a holistic solution to mechatronic systems has not been achieved so far. Being interconnected, as demonstrated in Fig. 34.1, activities in the design of such systems tend to be very interdependent, with decisions made in one area relying heavily on information produced in another location. For example, the development of controllers often requires accurate information

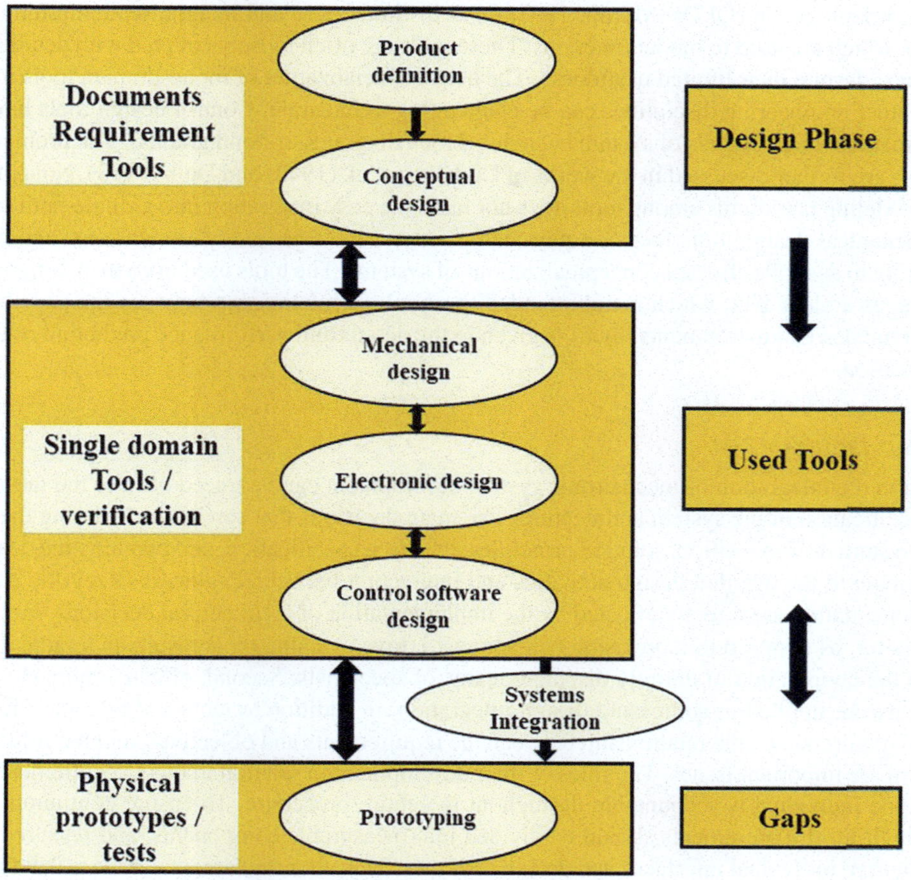

FIGURE 34.1 Common current design practice phases and tools.

about the physical characteristics of the mechanical system. There are two general engineering design methodologies: sequential and concurrent process structures. Nonetheless, sequential design has received considerable criticism because of its rigidity, resulting in higher development time and costs. This evaluation aligns with the opinion of working engineers in the industry. Rzevski (2003) opines that to overcome these shortcomings, the conventional end-to-end design approach should be replaced with a concurrent engineering approach tailored to the needs of mechatronic systems. Classic concurrent engineering is a method that considers all life cycle phases of a product at an early stage in the design process, thereby minimizing issues during manufacturing, operation, and disposal (Martin, 1997). According to Rzevski (2003), mechatronic systems are frequently built by incorporating single-domain subsystems via the designed interfaces. Although this strategy can provide functional integration, it rarely yields synergistic behavior.

34.2.1.2 Design tools

A recent exploration of practices in industrial design has revealed that the development of mechatronic products frequently relies on a diverse range of tools to manage design data. Another point highlighted in these studies is that the lack of tools that facilitate the easy integration of data and allow for sharing is one of the most significant challenges in this sphere (Jackson, 2006b). As each designer in various fields of engineering tends to rely on highly specialized software applications, as shown in Fig. 34.1, the across-the-board designers tend to rely on software applications that are highly specialized and specific to the areas of expertise. In-line with the nomenclature, these tools can be categorized as domain-specific, as they primarily facilitate activities in a single field of study. Common ones include SolidWorks, used in mechanical design, and Synopsys and OrCad, used in electronic design, as well as Matlab/Simulink, used in the development of control systems. Although these tools have been handy in their respective fields, few of them can accommodate the initial stages of conceptualizing design besides running smoothly into subsequent levels of development. This restriction, in most instances, is likely because of poor use rather than inherent flaws in the tools. Such tools used to target early-stage design include requirement management tools, such as Rational DOORS, as well as requirement capture and analysis tools, like quality function deployment (QFD) (Martin, 1997; QFD Institute, n.d.) and integration definition (Knowledge Based Systems Inc., n.d.), which are used to model functions. These tools are often loosely coupled with detailed, domain-specific design environments, despite their limited usefulness. The heavy specialization of mono-domain tools in itself ensures that information from other engineering disciplines can be challenging to integrate. Control design tools have more flexibility; they are based on mathematical models of system behavior (block diagrams or bond graphs). The problems associated with integrating the tools are further discussed in the works of Cutkosky et al. (1993) and Dolk (1993). Nonetheless, the essential heterogeneity of modeling paradigms among tools does not facilitate easy integration into a single familiar environment. As an illustration, mechanical design tools focus on physical aspects such as geometry, dimensions, and material properties, making it challenging to describe abstract conceptualizations of systems. The tools used in control design typically represent the physical system as a black box, which is not intended to reveal the relationship between the physical structure and its behavior. In electronic design environments, focus is given to the details and performance prediction of the implementation level through simulation.

34.2.1.3 Human factors

The main problem in the integration of mechatronic system development can be traced back to the initial stages of design, specifically the specification of the system architecture. The main decisions that can be made during the conceptual design phase include the system decomposition, solution principles, interface specification, and planning the design process, which significantly contribute to the detailed design activities and interaction between engineers (Rzevski, 2003). The hardware and software implementation are directly affected by the implementation of architectural decisions. Indicatively the choice of using a linear motor, as opposed to a precision ball screw, to drive machine tools results in a radically different system design. It requires the cooperation of entirely disparate teams of specialists. Second, effective human communication and collaboration can also be applied in addressing design integration, in addition to technical factors. The articulation of the system's goals and requirements, the relationship between the requirements and objectives, and the solutions adopted in the architectural system are important issues. The roles of their development in relation to these requirements should be clearly explained so that one individual is accountable throughout the entire procedure. To ensure continuous monitoring of the requirements the high-level requirements should be divided into measurable components, and resource budgets should be allocated to the extent of individual interfaces and design activities. The other fundamental issue is that the designers are left uninformed about how their local design decisions will affect other components in the system. The system performance may be jeopardized by the optimizations that engineers make in their respective fields inadvertently. Consequently the evolved design must be continually tested and verified to ensure consistency and suitability throughout the development life cycle. These problems highlight the need for effective tools and methods to support systems engineering and architectural design procedures, as well as to share and distribute relevant data to multidisciplinary teams.

34.2.2 Lack of interdisciplinary verification

Traditionally, four well-established methods of system verification exist, i.e., demonstration, testing, inspection, and analysis (Martin, 1997). Among these, demonstration, testing, and inspection typically require the development of physical prototypes, as well as analysis, which involves the application of mathematical models of the system. Another critical issue in mechatronic design is developing relevant models and respective verification platforms that can test various aspects of a system, such as control software. Theoretically, multiple models are typically created at different stages to support specific verification procedures. Because of the popularity of domain-specific modeling (DSM) tools, these models often represent rather solitary views, such as mechanical or electrical behavior or continuous dynamics versus discrete-event logic (Jackson, 2006a). Nonetheless, these partial representations are insufficient to describe the coupled and synergistic behavior that characterizes mechatronic systems. The restriction is further exacerbated in the analysis of systems with many operating modes, which are controlled by state-machine logic, as such systems require flexible, multidomain concepts that are seldom provided. Cosimulation and model exchange methods aim to integrate data from multiple analysis tools into control design settings, including the popular MATLAB/Simulink platform (Ramos-Hernandez et al., 2005). However, these unified models are often simplified to input–output models, such as transfer functions, and thus lose connections to design parameters. As a result, control software verification remains a heavily relied-upon method, often involving the use of breadboards or hardware prototypes, which results in high costs and time investments. Although full-system prototypes enable multidisciplinary verification, their construction requires a high level of design maturity. Partial prototype methods, such as the Iron Bird concept in the aerospace industry, eliminate the complexity of the system level but still require a significant amount of resources and attention to the detailed implementation of subsystems.

34.2.3 Lack of automation in control software design

In industrial applications, approximately 20%–40% of the total software development for mechatronic products is typically dedicated to developing control systems (Heck et al., 2003; Object Management Group, 2009). The modern computer-aided control system design (CACSD) toolsets, such as MATLAB/Simulink and dSPACE, along with software engineering tools like Rational Rose, provide means of transforming control logic, modeled as block diagrams or state-transition models, into executable code. Automation of code generation significantly reduces human errors in programming, increases reliability, and reduces development time. These advantages, however, are more relevant in the later stages of development, as most efforts in control system design are concentrated on constructing valid formal models and finding the correct parameter values to be used with the desired control components. The existing tools are primarily oriented toward translating current formal descriptions into source code and do not facilitate the generation of the descriptions. The use of model-based software development methods, in which code is generated from high-level models such as block diagrams or Unified Modeling Language (UML) models (Object Management Group, 2009), is thus limited. Although examples of leading companies using such approaches are available, they still fall short of the standards for developing mechatronic systems. Moreover, the modeling primitives that are used tend to be biased toward particular specialist perspectives and are therefore less compatible with interdisciplinary cooperation. To enhance integration, control models should be built on concepts relevant to a more diverse group of stakeholders engaged in system design, beyond control engineers. The generation of a model for code generation requires that the control structure and control strategy be clearly defined. The control structure describes how control inputs and outputs are organized into hardware or software units, and the control strategy determines the type of algorithm and control sequences required to meet the system's behavior and performance needs. After defining these elements, one can use optimization techniques to optimize the controller parameters, provided that the requirements are formulated in the correct form.

34.3 Review of available approaches

The intended purpose of these approaches is to enhance integration, verification, and automation of the mechatronic design process. Instead of creating a comprehensive survey the following section presents a focused review of some of the selected solutions, reflecting current research and practice. To make the reviewed methods and tools understandable and consistently used for organizing them, allowing for a direct comparison between the identified challenges and the approaches underlying their development.

34.3.1 Design integration

34.3.1.1 Design methods

Some of the design approaches focus on modeling functions, requirements, and other types of information that are typically outlined during the conceptual stage of product development. This information must be explicitly documented so that designers can have a system-level picture and can trace the development of design choices to the development process. Some authors have suggested their own model-based functional methods to assist and enhance decision-making in early design phases, such as the function-behavior-state (FBS) framework (Yoshioka et al., 2004), functional representation, Schemebuilder, and MACE. The purpose of these techniques is to capture the knowledge that concerns what a system is supposed to do, how it realizes its functions, and what physical or software components are used to accomplish them. Functional models frequently contain complementary information on how system functions are achieved. This information in certain methods is qualitative in nature, e.g., the Qualitative SIMulator (Barr & Cohen, 1989) and qualitative process theory, whereas others are based on a quantitative description, e.g., in the form of differential equations or bond-graph models. Erden et al. (2008) have talked about practical applications of such functional modeling techniques. An example of this is the FBS methodology, which has been applied to the Knowledge Intensive Engineering Framework (KIEF). It is designed to facilitate the exchange of information among engineering fields, including the integration of tools across various engineering areas. Other methods define the functionality of a system in terms of functional flow or block diagrams, where functions are described as changes in matter, energy, or information (European Cooperation for Space Standardization, 1999). Under these techniques, functional descriptions are held apart from other system models. A formalized representation of functional flows is available through the Integration definition for function modeling (IDEF)0 method (National Institute of Standards and Technology, 1993). However, other IDEF languages (Knowledge Based Systems Inc., n.d.) are more concentrated on describing specifics of a system that can be more closely related to domain-specific representations. Nevertheless, explicit links between various IDEF models are often unclear and implicit. The Functions and

Key Drivers (FunKey) model (Bonnema, 2008) proposes the idea of budgeting resource allocations to system functions, in which budgets of architectural decisions are documented, and alternatives between product architectures can be compared. This is largely because it is challenging to formalize abstract functional representations in a way that allows for automatic processing. This issue is further complicated by the fact that there are no commonly accepted definitions and formalisms in the design research community. The second weakness of most functional modeling techniques is that the requirements information is not adequately addressed. Another notable exception is FunKey, which directly deals with system-level budgeting. QFD involves capturing customer requirements and their relationship to quantifiable characteristics of the system, which reflects the fulfillment of these requirements. Muller (2004) suggested a customer, application, functional, conceptual, and realization (CAFCR) system to break down product architectures into five different perspectives. As a result of this decomposition, individual representation of customer needs, system functions, and conceptual and realization-level designs are possible. The primary objective of CAFCR is to support the management of stakeholder issues, including safety, usability, and performance, throughout the design lifecycle, ensuring that system goals and implementation decisions align. Although CAFCR offers valuable advice on the type of information that should be considered in architectural design, the mechanisms proposed for implementing this information are not well interrelated. The flexibility provided by the wide variety of methods available in CAFCR could be beneficial. Yet, it also imposes a significant burden on system architects who must choose the most suitable techniques to apply in a particular situation. Structuring development activities has also been suggested with the help of process-oriented design models. The V-model (Stevens et al., 1998) presents a generic model in which every phase of product definition is linked to its respective verification and validation exercise, with progressive integration of subsystems. Although the V-model provides a clear logic for development, it is abstract and offers limited practical guidance on how to implement it. The entire model lacks complete tool support, and companies must build their own toolchains. In line with the V-model the spiral model (Boehm, 1986) focuses on the development of prototypes to test the designs successively. Lastly, there is axiomatic design, coined by Suh (1990), which supposedly results in optimal designs as functional independence is achieved among the system elements. The approach employs assessment principles based on independence and information axioms and has been effectively applied in several situations (Suh, 1990). One of the most important contributions of axiomatic design is its focus on the connection between high-level functional requirements and those of the implementation-specific design parameters. Nevertheless, the trend of increasing functional density and integration of modern mechatronic systems renders functional independence increasingly difficult to attain. In general, although several approaches highlight the stages of integration and information management, significant gaps remain between the preliminary conceptual design and the specific stages outlined in the design. To effectively support interdisciplinary design activities, capturing, structuring, and managing engineering information are necessary to enable knowledge reuse (McMahon et al., 2009).

34.3.1.2 Tool integration

Citherlet et al. (2001) classify the multidisciplinary tool integration into four main approaches: stand-alone, interoperable, coupled (or linked), and fully integrated programs. Stand-alone tools are self-contained and do not facilitate communication; hence they are the least preferable. Interoperable tools, in their turn, allow transferring data and models across applications. Some tools of control design already adopt either the interoperable or coupled approach. Complete integration programs aim to facilitate multiple design areas within a single software suite. This has been adopted by mechanical CAD vendors, including the new CATIA release, which introduces the concept of modeling electronics, systems, and controls combined with embedded control code generation. In coupled and integrated tools, predicting system behavior at the detailed design phase is possible; however, such tools do not have a direct relationship with higher level design information, including the goals and functional descriptions from the early conceptual phase. Some of the interoperable methods include the KIEF and the virtual reality ship (VRS) systems project (VRS ROPAX, n.d.). KIEF utilizes a pluggable metamodel system, which connects objects from different modeling tools based on similar physical phenomena in a process-based approach (Erden et al., 2008). It features a knowledge base that provides information about modeling tools and facilitates data transfer and consistency across domain-specific models. VRS can be seen as an example of the integration of tools in the European shipbuilding industry, such as the physical testing platform; however, few details are provided about its implementation.

The Comprehensive Object-Oriented Requirements Engineering (CORE) software suite is a model-based systems engineering method. It enables designers to develop models in the form of requirements, functional decomposition, and system component interfaces. It does, however, have no direct relationships to detailed design and verification tools and needs to be integrated with other models, typically at the component level, using manual intervention. One of the proposed component-oriented methodologies is presented by Peak et al. (2007), which relies on the composable object (COB) paradigm (Paredis et al., 2001) and the systems modeling language (SysML) (Object Management Group, 2001), to integrate information from tools such as CATIA, Ansys, and MATLAB/Simulink. COBs consist of structural and behavioral

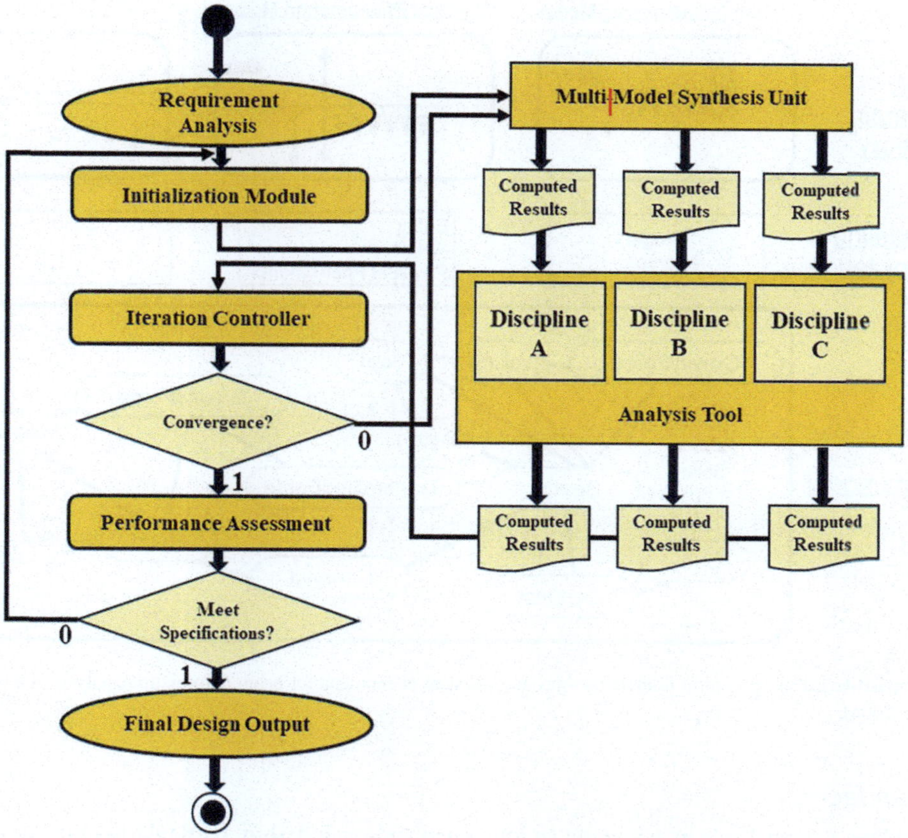

FIGURE 34.2 Overview of the design and engineering engine (DEE) illustrating multimodel synthesis and multidisciplinary analysis integration.

descriptions of systems, which are both human- and computer-readable and which provide a basis for synthesizing different views of a system (Peak et al., 2007). Fig. 34.2 illustrates the design and engineering engine (DEE) proposed by La Rocca and van Tooren (2007), which represents a knowledge-based platform designed to support multidisciplinary design and optimization while remaining independent of specific engineering domains. The main component in DEE is the multimodel generator (MMG), which serves as a product representation of parametric and object-oriented primitives. These primitives are designed to produce input models for different domain-specific analysis tools, which enable the assessment of structural and aerodynamic, as well as other disciplinary behaviors. Information sharing among these tools is managed by an agent-based communication network (Berends et al., 2008), providing loosely coupled interaction. Both CORE and DEE utilize component-based product models to integrate multiple views of the system. Nevertheless, external design tools typically require manual mapping at the component level when linked with CORE. The goal of object-oriented modeling is to overcome this limitation by allowing components with standardized parameters to be assembled into composite entities, which leads to the formation of more sophisticated systems. Experience in the industry also reveals that the mere use of domain-specific tools in isolation may obstruct communication. Still, integration enabled by these models may improve collaboration and design productivity (Jackson, 2006b).

The basic principle of the metamodeling concept is illustrated in Fig. 34.3, which shows the interrelations between the various design models used in the deformation analysis of a beam in the head design of a magnetic hard disk. At the highest level, known as the domain theory level, various design modeling systems incorporate background theoretical knowledge specific to the respective discipline. The bottom of Fig. 34.3 shows the metamodel of the design object, which has been defined using a vocabulary based on a fundamental ontology. This metamodel serves as an abstraction layer, encompassing key concepts about the system without reference to a specific modeling tool. The modeling of knowledge is then employed to maintain formal relationships between the concepts declared in the metamodel and those found in the domain-specific design environments. A case in point is the KIEF, in which the metamodel is associated with system concepts, physical phenomena, and governing laws, thereby enhancing the reuse and applicability of modeling objects across multiple domains.

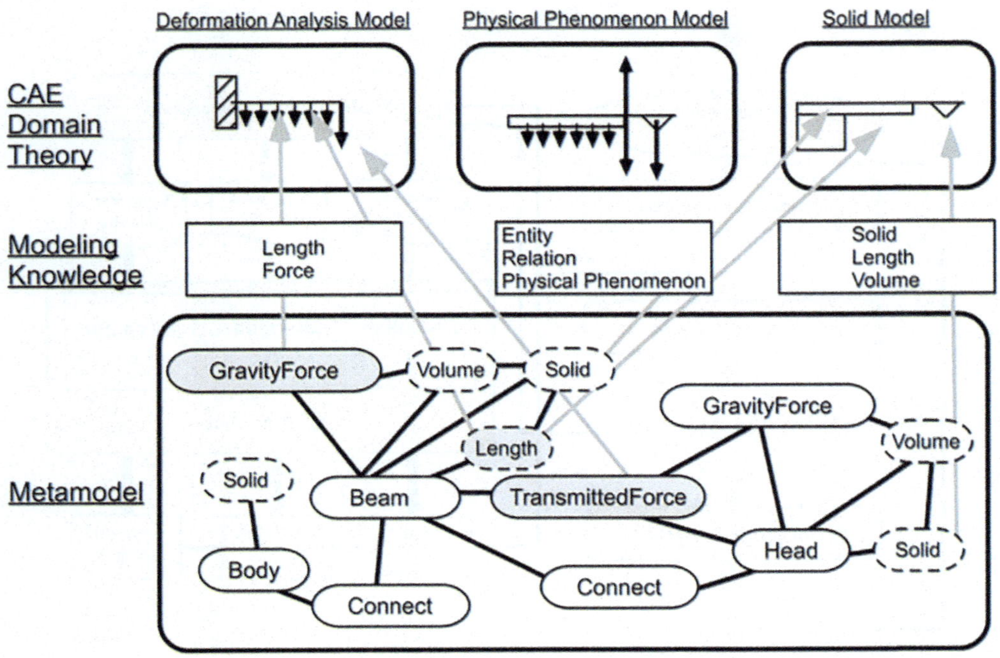

FIGURE 34.3 Metamodel-based integration of domain theories and physical phenomena for beam deformation analysis in a magnetic hard disk head design (Yoshioka et al., 2004).

34.3.1.3 *Human factors*

The human factor plays a critical role in achieving an integrated approach design, particularly in the design of mechatronic systems. Practical cooperation requires effective communication among all stakeholders, including engineers, designers, and other relevant parties. The research states that efficient communication is essential between engineers and the rest of the world and that this is becoming a crucial factor, especially with multidisciplinary teams undertaking contemporary design procedures. The importance of communication and cooperation between engineers is also evident in industrial studies, as it enables the efficient workflow of the design (Perrin, 2009). According to Pahl et al. (Wikander et al., 2001), information exchange and communication among designers are the core elements of systematic design methodology that have a direct impact on the division of labor and collaborative efficiency. The brainstorming and group evaluation techniques are recommended to promote the exchange of information at the initial stage. These methods are instrumental at the conceptual stage, when exploring solutions and developing ideas are the key goals. They are, however, less applicable at the later stages of design, where detailed and domain-specific information needs to be shared. Although the critical role of human interaction and information flow is widely accepted, the environments of modern design are not very supportive of incorporating these factors into the workflow. It is not well-equipped with tools that can combine the personal efforts of the designers operating in their domain-specific tools without losing a system-level perspective of the design goals. The evolution of design frameworks should focus on both mechanisms to ensure that individual contributions are captured and offer integrated visibility of the system, allowing not only communication but also coordination toward the achievement of shared aims.

34.3.2 Lack of interdisciplinary verification

The use of DSM tools only limits verification and design to specific perspectives. For example, finite element method (FEM) models are often used to determine mechanical strength and stiffness, whereas CACSD tools are used to develop and test controllers. There is extensive manual insight between tools, which results in a fragmented verification procedure. Domain-wide verification is typically performed at a later stage in the design process, when physical models can be created. Methods such as hardware-in-the-loop (HIL) and rapid control prototyping rely on such prototypes, so they are not as applicable to a concurrent design environment. In response to this, simulation models can be used as an alternative to physical prototypes. Multidomain models can be represented as transfer functions, which are modeled using block diagrams (as in MATLAB/Simulink) in control design. Other methods include physical modeling, where system behavior is modeled using differential equations and energy flows, and geometric modeling, which utilizes a finite-element mesh in combination with a multibody dynamics solver (Controllab Products B.V., n.d.; Ferretti et al., 2004). Physical modeling languages, including Modelica, bond graphs, and Simscape, enable users to create graphical representations of components in various fields, such as mechanical gears, electrical motors, and resistors. During the design evolution, simplified design models may be replaced by more detailed ones using these tools. Several modeling and simulation packages have been scaled to mechatronic systems, among them 20-sim, Dymola, gPROMS (Process Systems Enterprise Limited, n.d.), SABER (Synopsys, n.d.), HyBrSim (Mosterman, 2002), and Smile (Technical University of Berlin, n.d.). These facilitate modular, object-oriented modeling and component reuse, which is already available in libraries, making cross-domain interactivity easier (Ferretti et al., 2004). The weakness is that these models can be based on assumptions about expected behavior, and they require a significant amount of expertise from designers. It may be advisable to minimize reliance on assumptions by using first-principles simulations, such as FEM-based analysis.

34.3.3 Lack of automation in control design

Within the current work, automation of control software encompasses the process of generating code based on specific software models, as well as generating these models. Environments that utilize several commercial code generators include MATLAB/Simulink and UML-based modeling tools. For example, the Gene-Auto project has developed techniques for automatic model transformations based on a philosophy of correctness by construction (Toom et al., 2008). Integrated design notations are applied to both continuous and discrete-event control formalisms in the projects of PiCSi (Jackson, 2006a) and Flexicon. In this case, UML will serve as a common language, where both sequential flowcharts and Simulink models will be translated. Based on this single model, platform-independent Java code can be developed. This style tends to utilize existing domain-specific tools, such as model translation, rather than creating a new, general-purpose language. Another option is DSM languages, which increase the level of abstraction of control software design. DSM will not need to map modeling entities to UML before generating code, thus saving on development time (Kelly & Tolvanen, 2008). Nonetheless, the effort required to implement DSM prompts domain specialists to create the language itself and code generators. There is also considerable potential for automation in early design phases, where system requirements are converted into control structures and logic. The message sequencing charts and UML sequence diagrams are more limited in their expressiveness and offer specifications (Harel, 2001). More expressive live sequence charts can be automatically derived to control logic and software structure, which can be converted into UML models. Again, high-level functional requirements are translated into software using further techniques, such as requirements-based programming (RBP) (Rash et al., 2006) and the Functional Block CAD environment (Sakao et al., 1997), which enable designers to produce reusable, high-level control software components and executable code that is runnable on distributed systems.

34.4 An integrated approach for control software development

The application of domain-specific design methodologies and tools to realize an integrated and multidisciplinary system implies a set of constraints, especially in terms of multidomain modeling and interdesigner and intertool communication. The hardware and software components must be adequately modeled and are subject to multidisciplinary design, optimization, and verification. The framework discussed in this work is intended to support an integrated design process. It would not be prudent to design a single tool that can handle all the challenges of mechatronic system design. To begin with the design content of a total system is very cumbersome and voluminous. Building one model with all the required information would make it significantly larger, which can be viewed as a potential vulnerability (Cutkosky et al., 1993). Second the tools available today are specific to a domain and tailored to particular uses, so designers already have a good understanding of how to use them. Practically, a designer has the task of creating and maintaining models in their field (Browning, 2009).

Fig. 34.4 provides a diagram of the framework, where dashed-line boxes represent the existing domain-specific tools. Iterative design loops are not depicted to simplify the diagram. This can be categorized as an architecture framework, in the view of Browning (Browning, 2009), and is intended to facilitate interdeveloper communication and tool-to-tool model transformations. Fig. 34.4 will feature a high-level functional information model at the top level, allowing users to navigate the data, classify it, and obtain a summary of the detailed design data. Functions are used as elements of integration, allowing for the description of a system with varying levels of detail depending on the users' needs, while maintaining model coherence. The first design is defined in terms of the FBS modeling scheme (Umeda & Tomiyama, 1995), with system-level decisions regarding control design, such as strategies and goals. Formal specification languages can also be used to represent quantitative requirements and constraints. The first form of modeling the system architecture involves representing it as objects of software, hardware, or mechanical components. As this information is not well represented by standard

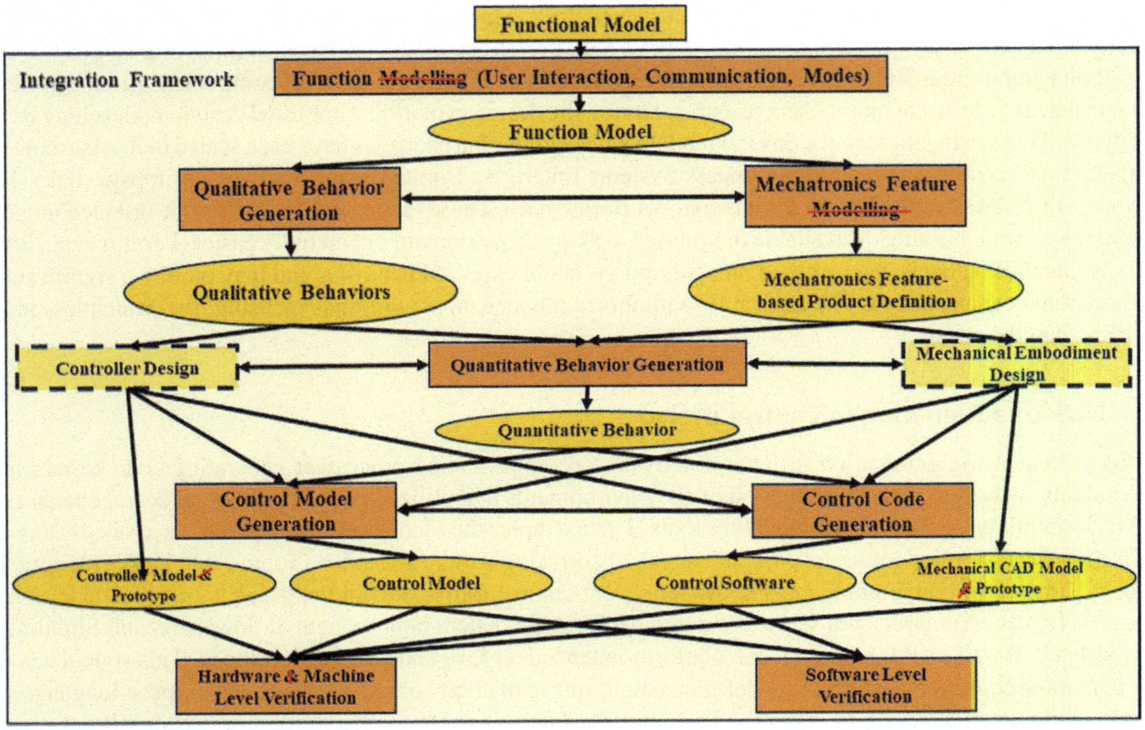

FIGURE 34.4 Conceptual layout of the integrated framework, highlighting *orange-colored* blocks as tool modules under development, whereas *dashed-outline* blocks denote preexisting commercial software integrated into the workflow.

FBS, a longer version of the scheme is used. In addition to an abstract functional description, detailed analysis output, and an implementation-oriented description, the model system should also include these elements to aid in the design being actualized. An example of such information includes control-oriented simulation output, verification data, and descriptive artifacts, such as manufacturing drawings or layout specifications. In the suggested system, these factors are clustered in a block, which is referred to as the mechatronic feature product definition in Fig. 34.4. The major difference between the current strategy and traditional component-based tools is the organization of the knowledge base. Instead of using only component descriptions, the framework takes a hybrid form that integrates process-oriented and component-oriented modeling paradigms. Such hybridization enables the capture of structural composition and behavioral evolution, thereby enhancing the expressiveness and reuse of design knowledge. To achieve this kind of integration in practice, domain-specific environments such as mechanical CAD systems and CACSD tools must be coupled together. This integration enables parallel design operations to be coordinated and part of synthesis, analysis, and model transformation to be partially automated. At the heart of this is an information manager that will assist in navigation, visualization, and checking of consistency throughout the system model and its related datasets. Quantitative behavior is used to display the outputs of the integrated domain-specific tools within the framework, as shown in Fig. 34.4. On the architectural level the system can be studied from several perspectives that address the particular concerns and needs of stakeholders, in accordance with existing practices in architectural description (Institute of Electrical and Electronics Engineers Standards Association, 2000). System-level requirements are systematically broken down, assigned, and traced to subsystems and the respective domain-specific design processes, ensuring transparency and traceability throughout the development process. To achieve continuous feedback control, parameters are defined in terms of performance indicators such as allowable overshoot, rise time, steady-state error, or power consumption. According to such specifications, a cost function may be developed. Then, controllers Proportional–Integral–Derivative (PID) or model predictive control (MPC) schemes can be synthesized and tuned semiautomatically using existing tools, e.g., the MATLAB Control System Toolbox. In contrast to the control design method (CDM), which assumes a strong connection between the functional design and the implementation process, the framework discussed in this chapter facilitates the simultaneous creation of the control system and the plant model. Control model generation and control code generation occur concurrently, as shown in Fig. 34.4. These opinions are in direct interaction with control analysis and design scenarios. Linearization and model-order reduction methods are employed to make the models amenable to control synthesis algorithms. The framework utilizes a knowledge-based engineering platform to reason about the interaction of components across physical domains. To sustain and extend the automation process of such activities, it is necessary to have a common conceptual foundation across design domains. This necessitates a formal statement of the kinds of models the framework can support, the modeling primitives to be used in constructing models, and the domain-specific vocabularies that give the abstractions meaning.

34.5 Conclusion

Mechatronic systems necessitate a paradigm shift in design, which entails collaboration and coordination, particularly during the initial phases of development. It is possible through a holistic approach where the interactions and interdependencies of various design domains are clearly considered. Currently, there are few tools to assist with combined architecture, and this is a clear sign that new constructions and solutions should be developed. System design issues have been identified in the chapter, and several critical issues in the design of mechatronic systems have been found, including the use of model tools and human interfaces, the lack of multidisciplinary system validation, and inadequate automation in the design of control software. According to the review the existing solutions to address the challenges are fragmented and, in most cases, target specific aspects. In the meantime a consistent design style also demands substantial exchange of models and information. In addition, formalizing architectural designs, characterizing functions, and specifying system requirements are essential for facilitating integration. The industry work is directed toward tool-level integration, whereas the research work performed in academic circles is aimed at developing integration techniques at the level of the underlying. Although these scholastic approaches provide a solid conceptual foundation, they are not typically well implemented in the discipline because of the abstract nature of the system. To seal this gap a conceptual integration framework for mechatronic system design is proposed. It is a combination of available tools and specifically constructed tools, with the help of which multidisciplinary teams can develop control software to generate software and ensure its correctness in a highly efficient manner. Through the use of the FBS modeling scheme, designers can create a high-level requirements specification model, providing a detailed picture of the system. Through the use of a domain-specific design and analysis tool, it is possible to design simulation and verification models that offer consistency and traceability across all design stages. The interaction between the tools and models occurs through the high-level functional models, which are the key integration component in this case. The suggested framework, compared to the CORE tool, is centered on incorporating available software tools for designers during the detailed design phase.

References

Barr, A., & Cohen, P. R. (1989). *The handbook of artificial intelligence*: 4. William Kaufmann, Inc. [chapter 21].

Berends, J., van Tooren, M. JL., & Schut, E. J. (2008). Design and implementation of a new generation multi-agent task environment framework. *49th AIAA/ASME/ASCE/AHS/ASC structures, structural dynamics, and materials conference, 4th AIAA multidisciplinary design optimization specialist conference*. American Institute of Aeronautics and Astronautics.

Boehm, B. (1986). A spiral model of software development and enhancement. *ACM SIGSOFT Software Engineering Notes, 11*(4), 14–24. doi:10.1145/12944.12948.

Bonnema, G. M. (2008). *FunKey architecting– an integrated approach to system architecting using functions, key drivers and system budgets* PhD thesis. University of Twente.

Browning, T. R. (2009). The many views of a process: Toward a process architecture framework for product development processes. *Systems Engineering, 12*(1), 69–90. http://www3.interscience.wiley.com/cgi-bin/fulltext/121395857/PDFSTART. doi:10.1002/sys.20109.

Citherlet, S., Clarke, J. A., & Hand, J. (2001). Integration in building physics simulation. *Energy and Buildings, 33*(5), 451–461. doi:10.1016/s0378-7788(00)00108-0.

Controllab Products B.V. (n.d.). 20-sim: Modeling and simulation software for mechatronic systems. (Accessed: July 13, 2026).

Cutkosky, M. R., Engelmore, R. S., Fikes, R. E., Genesereth, M. R., Gruber, T. R., Mark, W. S., Tenenbaum, J. M., & Weber, J. C. (1993). PACT: An experiment in integrating concurrent engineering systems. *Computer, 26*(1), 28–37. doi:10.1109/2.179153.

Dolk, D. R., & Kotterman, J. E. (1993). Model integration and theory of models. *Decision Support Systems, 9*, 51–63.

Erden, M. S., Komoto, H., Van Beek, T. J., D'amelio, V., Echavarria, E., & Tomiyama, T. (2008). A review of function modeling: Approaches and applications. *Artificial Intelligence for Engineering Design Analysis and Manufacturing, 22*, 147–169.

European Cooperation for Space Standardization. (1999). *Space engineering—Functional analysis (E-10-05A)* http://esapub.esrin.esa.it/pss/ecss-ct05.htm.

Ferretti, G., Magnani, G. A., & Rocco, P. (2004). Virtual prototyping of mechatronic systems. *Annual Reviews in Control, 28*(2), 193–206. doi:10.1016/j.arcontrol.2004.02.002.

Harel, D. (2001). From play-in scenarios to code: An achievable dream. *Computer, 34*(5), 53–60. doi:10.1109/2.895118.

Heck, H., Wills, W., & Vachtsevanos, V. (2003). Software technology for implementing reusable, distributed control systems. *IEEE Control Systems, 23*(1), 21–35. doi:10.1109/MCS.2003.1172827.

Institute of Electrical and Electronics Engineers Standards Association. (2000). *IEEE Std 1471–2000: Recommended practice for architectural description of software-intensive systems*. Institute of Electrical and Electronics Engineers Standards Association.

Jackson, C. K. (2006a). *Simulation driven design benchmark report*. Aberdeen Group.

Jackson, C. K. (2006b). *The mechatronic system design benchmark report*. Aberdeen Group.

Kelly, S., & Tolvanen, J.-P. (2008). *Domain-specific modeling: Enabling full code generation*. Wiley-IEEE Computer Society Press.

Knowledge Based Systems Inc. (n.d.). IDEF family of methods, IDEF. (Jul). 13, 2026. http://www.idef.com/.

La Rocca, G., & van Tooren, M. J. L. (2007). Enabling distributed multi-disciplinary design of complex products: A knowledge based engineering approach. *Journal of Design Research, 5*(3), 333–352. doi:10.1504/jdr.2007.014880.

Martin, J. N. (1997). *Systems engineering guidebook – A process for developing systems and products*. CRC Press.

McMahon, C. A., Caldwell, N. H. M., Darlington, M. J., Culley, S. J., Giess, M. D., & Clarkson, P. J. (2009). *The development of a set of principles for the through-life management of engineering information*. University of Bath http://www.bath.ac.uk/idmrc/themes/projects/kim/kim40rep007mjd10.doc.

Mosterman, P. J. (2002). HYBRSIM—A modelling and simulation environment for hybrid bond graphs. Proceedings of the Institution of Mechanical Engineers, Part I. *Journal of Systems and Control Engineering, 216*(1), 35–46. https://doi.org/10.1243/0959651021541417.

Muller, G. J. (2004). *CAFCR: A multi-view method for embedded systems architecting*. Delft University of Technology PhD thesis.

National Institute of Standards and Technology. (1993). *Integration definition for function modeling (IDEF0)* http://www.idef.com/pdf/idef0.pdf.

Object Management Group. (2001). *OMG systems modeling language, V1.0* http://www.omg.org/cgi-bin/apps/doc?formal/07-09-01.pdf.

Object Management Group. (2009). *Unified modeling language, V2, 2* http://www.omg.org/spec/UML/2.2/.

Pahl, G., Beitz, W., Feldhusen, J., & Grote, K. H. (2007). *Engineering design: A systematic approach* (pp. 1–617). Springer London. http://www.springerlink.com/openurl.asp?genre=book&isbn=978-1-84628-318-5. doi:10.1007/978-1-84628-319-2.

Paredis, C. J. J., Diaz-Calderon, A., Sinha, R., & Khosla, P. K. (2001). Composable models for simulation-based design. *Engineering with Computers, 17*(2), 112–128. doi:10.1007/pl00007197.

Peak, R. S., Burkhart, R. M., Friedenthal, S. A., Wilson, M. W., Bajaj, M., & Kim, I. (2007). Simulation based design using SysML part 1: A parametrics primer. In *Proceedings of INCOSE international symposium*. International Council on Systems Engineering.

Perrin, K. (2009). Digital prototyping in mechatronic design. *Project Mechatronics*. http://www.projectmechatronics.com/2009/07/13/digital-prototyping-in-mechatronic-design/.

Process Systems Enterprise Limited. (n.d.). *gPROMS advanced process modeling and process simulation*. http://www.psenterprise.com/gproms/index.html.

QFD, Institute. The official source for Quality Function Deployment (QFD). QFD Institute. https://www.qfdi.org/. (Accessed 13 July 2026).

Ramos-Hernandez, D. N., Fleming, P. J., & Bass, J. M. (2005). A novel object-oriented environment for distributed process control systems. *Control Engineering Practice, 13*(2), 213–230. doi:10.1016/j.conengprac.2004.03.007.

Rash, J. L., Hinchey, M. G., Rouff, C. A., Gračanin, D., & Erickson, J. (2006). A requirements-based programming approach to developing a NASA autonomous ground control system. *Artificial Intelligence Review, 25*(4), 285–297. doi:10.1007/s10462-007-9029-2.

Rzevski, G. (2003). On conceptual design of intelligent mechatronic systems. *Mechatronics, 13*(10), 1029–1044. doi:10.1016/s0957-4158(03)00041-2.

Sakao, T., Umeda, Y., Tomiyama, T., & Shimomura, Y. (1997). Model-based automatic generation of sequence-control programs from design information. *IEEE Expert, 12*, 54–61.

Schöner, H. P. (2004). Automotive mechatronics. *Control Engineering Practice, 12*(11), 1343–1351. doi:10.1016/j.conengprac.2003.10.004.

Stevens, R., Brook, P., Jackson, K., & Arnold, S. (1998). *System engineering: Coping with complexity*. Prentice Hall.

Suh, N. P. (1990). *The principles of design*. Oxford University Press.

Synopsys. (n.d). *Saber mixed-signal, mixed-technology simulation*. http://www.synopsys.com/Tools/SLD/MECHATRONICS/Saber/Pages/default.aspx.

Technical University of Berlin. (n.d.). *SMILE – The simulation environment for scientific computing*. http://www.smilenet.de.

Toom, A., Naks, T., Pantel, M., Gandriau, M., & Indrawati (2008). Gene-auto: An automatic code generator for a safe subset of Simulink/Stateflow and Scicos. *4th European congress on embedded real time software*. Toulouse, France: Association Aéronautique et Astronautique de France (3AF).

Umeda, Y., & Tomiyama, T. (1995). FBS modeling: Modeling scheme of function for conceptual design. In *Workshop on qualitative reasoning about physical systems* (pp. 271–278).

VRS ROPAX. (n.d.). *Virtual reality ship systems project webpage*. http://www.vrsproject.com/index.phtml.

Wikander, J., Törngren, M., & Hanson, M. (2001). The science and education of mechatronics engineering. *IEEE Robotics and Automation Magazine, 8*(2), 20–26. doi:10.1109/100.932753.

Yoshioka, M., Umeda, Y., Takeda, H., Shimomura, Y., Nomaguchi, Y., & Tomiyama, T. (2004). Physical concept ontology for the knowledge intensive engineering framework. *Advanced Engineering Informatics, 18*(2), 95–113. doi:10.1016/j.aei.2004.09.004.

Ziegler, J. G., & Nichols, N. B. (1943). Process lags in automatic control circuits. *Transactions of ASME, 65*, 433–444.

Chapter 35

Machine learning and artificial intelligence on HPC platforms for mechanical design—A review

Rishi J.P.[a] **and Sandeep B**[b]

[a] *Department of Mechanical Engineering, Vidyavardhaka College of Engineering, Mysuru, Karnataka, India,* [b] *Department of Mechanical Engineering ,Vidya Vikas Institute of Engineering & Technology, Mysuru, Karnataka, India*

35.1 Introduction

The modern mechanical systems are more and more performance-based, and the designs become more complicated, the connections between systems become more intertwined, and the constructions are more elaborate. Whether for lightweight composite design, advanced propulsion systems, precision manufacturing, predictive maintenance, or cyber-physical production, the new engineering applications demand design methods that are quick and efficient, yet highly intelligent and fundamentally reliable. Although well-proven analysis and modeling techniques are known and reliable, they do not always succeed in projects like the one of mechanical engineering because of its large scale, fast time scale, and complex uncertainties. Computational models have become increasingly useful in bridging the gap between raw engineering data and useful design intelligence, as complexity has grown.

Industry is being revolutionized by the convergence of machine learning (ML), artificial intelligence (AI), and high-performance computing (HPC). When it comes to pattern recognition, understanding system behavior, adapting to changing conditions, and predicting failures, mechanical workflows can harness the power of ML and AI and embed them within their processes. Through AI-enabled predictive maintenance, industrial operations are being revolutionized by reducing downtime and enhancing reliability (Abdelillah et al., 2023; Ates et al., 2023; Odili et al., 2024). Meanwhile, the AI-based design frameworks are allowing the creation of sophisticated materials (Basaran, 2020; Döllner, 2020) and scalable engineering components (Ibrahim et al., 2025; Kumbhar, 2025).

Apart from high-fidelity simulations, large data processing, training of deep learning algorithms, and design optimization, HPC also delivers the computation power for intelligent algorithms. With the latest HPC infrastructures, it is possible to run large-scale scientific workflows, advanced numerical simulations, and modeling of systems in real-time (Chetsa et al., 2013; Lingerfelt et al., 2016; Lusk & Mattsson, 2011; Park et al., 2012). Convergence of AI and HPC has led to new, energy-efficient architectures for computing; containerized computational pipelines; and models for execution of the workflow and associated computational processes that are designed to achieve sustainable and scalable scientific discovery (Álvarez Cid-Fuentes et al., 2019;Chillapalli, 2024; Merzky et al., 2025; Sukel, 2021). Model-driven engineering, design automation, and intelligent optimization are facilitated by integrating ML and AI with HPC. Several works have shown that deep reinforcement learning (RL), hybrid neural architectures, and physics-based AI models can facilitate a quicker mechanical design exploration process without losing fidelity and robustness (Chen et al., 2025; Sun et al., 2025; Zhang et al., 2025a). Last, but not least, in the field of aerospace, energy, transportation, and manufacturing (Chang et al., 2018; Pasupuleti, 2025; Pimenidis & Jayne, 2020), AI-powered digital twins provide real-time fault detection, system-level forecasting, and automated decision-making.

The combination of mechanical design methodology and ML, AI, and HPC (Fig. 35.1) is changing the game. These technologies will help engineering decision-making to become more intelligent, faster, and more reliable, with the help of predictive modeling, large-scale computational experimentation, and automation. This chapter brings together the latest

Computational Intelligence in Mechatronics: Solving Real-World Problems in Electronic Systems Design. DOI: https://doi.org/10.1016/B978-0-443-27580-7.00026-3

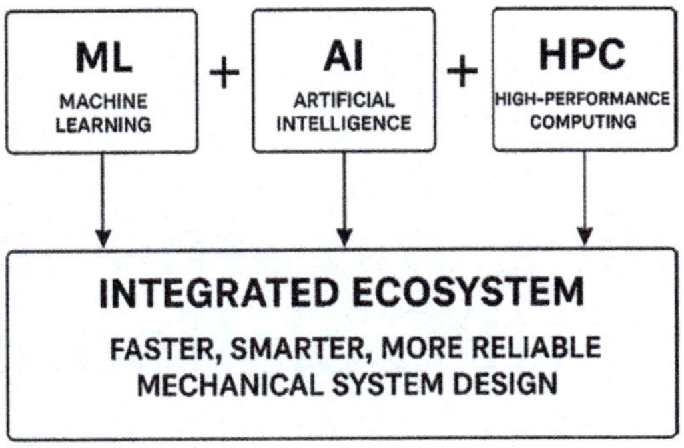

FIGURE 35.1 ML–AI–HPC integration framework for mechanical design. *AI*, Artificial intelligence; *ML*, machine learning; *HPC*, high-performance computing.

research, explores existing computational tools, and provides an overview of the state-of-the-art research on ML–AI–HPC convergence for mechanical system design.

35.2 Role of HPC in mechanical design

Today, HPC is a critical part of mechanical engineering and enables engineers to explore complex geometries, multiphysics phenomena, and large optimization challenges that are impractical to solve on traditional hardware. Simulations become more complex and accurate as mechanical systems are striving for higher degrees of efficiency, lower weight, tighter tolerances, and sensing capability. To meet this requirement, HPC platforms offer high levels of parallelism, increased data throughput, and increased computation speed as well as scalable frameworks, specifically engineered to satisfy engineering demands.

The initial developments in HPC helped speed up the design and simulation of materials. Lusk and Mattsson (2011) showed that energy-scale simulations using HPC were able to enable the understanding of the material behavior at the energy scale. Modern HPC infrastructures with special-purpose interconnects, distributed memory, and optimized compilers allow engineers to perform intensive workloads such as finite element analysis (FEA), computational fluid dynamics (CFD), and multibody dynamics (Park et al., 2012).

The complexity of mechanical design challenges has increased over the course of the evolution of HPC systems. HPC machines can perform large scientific simulations more efficiently with performance-counter-driven energy modeling compared to models with millions of degrees of freedom, as shown by Chetsa et al. (2013). Thanks to the automation of complex pipelines, e.g., through Blast Effects and Materials (BEAM) (Lingerfelt et al., 2016), engineers can automate whole meshes, solvers, data extraction, and postprocessing. HPC container frameworks (Zhou et al., 2022) allow for reproducible and portable simulation workflows on different systems such as clusters, cloud-HPC systems, and specialized accelerators. Key concepts in contemporary HPC environments are containerization and modular execution. Furthermore, these developed innovations directly impact high-fidelity optimizations such as structural-thermal codesign and digital prototyping.

HPC has now evolved into using ML/AI for applications as diverse as training surrogate models, deep learning architectures, and hybrid physics-ML systems, in addition to Partial Differential Equation(PDE)-based models. AI/ML-enabled HPC-driven workflows have significantly accelerated engineering simulations, particularly in turbulence modeling, uncertainty quantification, and real-time decision support (Fox & Jha, 2019; Jiang et al., 2019). A further study by Brewer et al. (2024) demonstrated that coupling AI with HPC increases predictive accuracy while reducing simulation costs.

Mechanical design is significantly enhanced by AI–HPC convergence, which enables high-resolution, time-dependent data to be processed quickly. For defect detection, structural assessment, and microstructural characterization of large-scale images, HPC systems are now equipped with parallel pipelines that enable large-scale image processing. The use of HPC-accelerated digital twins allows for real-time system monitoring, fault detection, and predictive control in mechanical environments such as turbomachinery, composite structures, and manufacturing systems (Ares de Parga et al., 2025; Pasupuleti, 2025). A hybrid AI–HPC architecture enhances scientific discovery further. Using AI-coupled runtimes in HPC-enabled systems (Merzky et al., 2025), mechanical designers can run optimization loops, RL, and multiscale simulations simultaneously across diverse processors, transforming the way they explore, assess, and analyze designs.

Last, mechanical design is becoming increasingly complex and needs computational tools that can deal with large design spaces, uncertainty, and multiple conflicting goals on realistic constraints. The reliability-driven design methods described

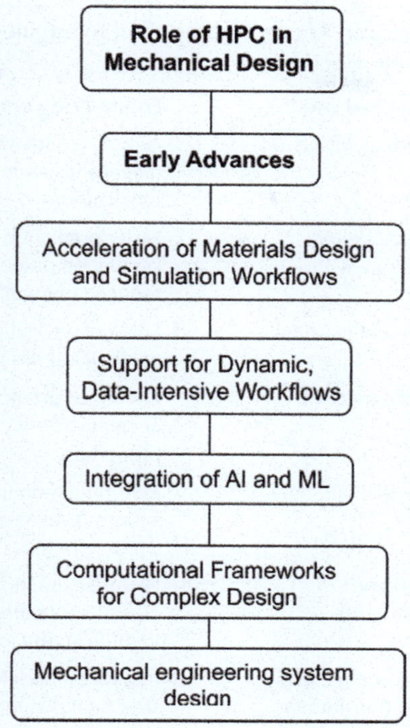

FIGURE 35.2 The evolution of HPC contributions in the field of mechanical engineering system design, from first principles to integrated computational frameworks (ICF). *AI*, Artificial intelligence; *ML*, machine learning; *HPC*, high-performance computing.

by Goh et al. (2009) and Ejakov et al. (2003) show that parametric sweeps, Monte Carlo analyses, and probabilistic design studies are possible with HPC platforms and are possible at scales that were not previously achievable.

In mechanical design, HPC has evolved from a numerical solver to a fully integrated computational ecosystem supporting data-driven modeling, AI-enhanced simulation, automated workflows, advanced optimization, and real-time engineering intelligence. It will become increasingly important to integrate HPC with ML and AI for next-generation engineering design methodologies as mechanical systems become more complex and demanding, as shown in Fig. 35.2.

35.3 Introduction to the foundations of machine learning and AI applications in engineering

ML and AI are transforming the formulation, modeling, optimization, and validation of engineering problems. Data-driven modeling improves physics-based understanding by recognizing patterns, minimizing uncertainty, and making predictions. Engineering systems can evolve beyond static architectures with AI as it expands these capabilities with autonomous reasoning, intelligent decision-making, and adaptive control.

By increasing the volume and complexity of sensor data, simulation data, imaging data, and digital manufacturing data, AI is gaining traction in engineering design. As demonstrated in engineering applications (Fox & Jha, 2019; Jiang et al., 2019), ML algorithms, especially supervised and unsupervised learning, are capable of extracting operational insights and system states from complex datasets. In addition to its pattern recognition capabilities, ML can also be used to detect damage, predict failures, classify materials, and monitor processes.

AI frameworks, especially reinforcement and metalearning, enhance engineering optimization by autonomously navigating complex design spaces. As a result, topology optimization, laminate sequencing, fault-tolerant control, and shape optimization for aerodynamics or thermal performance are improved (Ibrahim et al., 2025). In addition, AI models reduce computational costs through surrogate modeling, allowing engineers to use reduced-order approximations with maintained predictive accuracy (Brewer et al., 2024).

With the integration of physics constraints and principles into ML architectures, next-generation engineering design increasingly combines AI with domain knowledge. A robust and generalizable prediction is achieved through these physics-informed models, which bridge data-driven learning with classical engineering theories. In materials engineering, ML has accelerated microstructure discovery, performance prediction, and automated design (Basaran, 2020; Döllner, 2020).

TABLE 35.1 Summary of representative ML and AI methods applied to engineering domains, mapped to their functional contributions and supporting references.

Application area	ML/AI method used	Engineering contribution	Reference(s)
Defect detection and structural imaging	Deep learning, CNNs	Automated image-based diagnostics and defect localization	Huo et al. (2018)
Engineering simulation acceleration	Surrogate modeling, hybrid physics–ML	Faster prediction of simulation outputs with reduced computational cost	Brewer et al. (2024)
Autonomous design optimization	Reinforcement learning	Efficient navigation of large mechanical design spaces	Ibrahim et al. (2025)
Multiscale material behavior modeling	Physics-informed ML	Enhanced predictive accuracy for microstructural properties	Döllner (2020), Basaran (2020)
Digital twin for mechanical systems	Predictive ML, anomaly detection	Real-time monitoring and system-level forecasting	Pasupuleti (2025), Pimenidis and Jayne (2020)
Large-scale pattern recognition in engineering data	Supervised and unsupervised ML	Feature extraction, clustering, failure mode identification	Fox and Jha (2019), Jiang et al., 2019
Intelligent structural design	Hybrid AI architectures (GNN and transformer models)	Improved modeling of structural dependencies and performance	Chen et al. (2025), Sun et al. (2025), Zhang et al. (2025a)

AI, artificial intelligence; *ML*, machine learning; CNN; Convolutional Neural Network; GNN; Graph Neural Network

Through digital twins and cyber-physical systems, AI enhances modern engineering ecosystems beyond modeling and optimization. In complex environments, such as turbomachinery and automated manufacturing, it enables real-time monitoring, anomaly detection, and system forecasting (Pasupuleti, 2025; Pimenidis & Jayne, 2020). Through the integration of sensing data, ML prediction, and simulation, these systems enable self-correcting processes that adapt to changing environments. By enabling faster reasoning, richer pattern understanding, more efficient optimization, and deeper integration across various systems, AI/ML redefines engineering, as summarized in Table 35.1. Design workflows will depend on the effective integration of AI, ML, and HPC in the future.

35.4 Predictive analytics enabled by ML + HPC

ML and HPC enable engineers to run large predictive models at unprecedented speeds and resolutions, providing engineering with a solution that accurately forecasts system behavior under conditions that are traditionally costly or difficult to predict.

Through the use of computational power, ML and HPC can be combined to handle large datasets and complex simulations, enabling data-driven modeling, pattern recognition, anomaly detection, and surrogate modeling. High-fidelity simulations can be simulated for much lower computational cost using surrogate models and hybrid ML–physics approaches. These models can be used to simulate structural performance, thermal response, or dynamic behavior under a range of conditions rapidly, based on simulation data generated using HPCs. With HPC-enabled predictive analytics, mechanical systems that are safety-critical can be detected early. Based on deep learning models that recognize images, waveform signals, or sensor logs, cracking, delamination, or fatigue precursors can be accurately identified (Huo et al., 2018). When applied to HPC pipelines, these techniques allow the monitoring of large infrastructures, composite structures, or rotating machinery in close-to-real-time.

It is also possible to increase lifecycle estimation using ML–HPC coupling. The component lifespan and failure mode estimates are greatly enhanced using AI-driven predictive maintenance (Pasupuleti et al., 2025). Using ML models, degradation patterns are identified, temporal evolution is learned, and long-term reliability predictions are improved. With HPC simulations and lifecycle models, uncertainty, environment, and complex load conditions are taken into account.

Furthermore, ML–HPC frameworks facilitate the prediction of system behavior in real time, especially in digital twins. Areas such as AI-assisted system response prediction, ML-based prediction, data ingestion, and HPC-based simulation are utilized for forecasting system response with AI-enabled digital twins (Ares de Parga et al., 2025;

TABLE 35.2 Key engineering domains where ML combined with HPC enhances predictive accuracy and system forecasting.

Application area	ML/AI technique	Predictive capability	Reference(s)
Fatigue prediction	Surrogate models, deep learning	Crack growth forecasting, remaining life estimation	Huo et al. (2018), Brewer et al. (2024)
Thermal response modeling	Physics-informed ML, hybrid ML–CFD surrogates	Rapid thermal field prediction under variable boundary conditions	Jiang et al., 2019, Fox and Jha (2019)
Structural health monitoring	CNNs, image-based AI pipelines	Defect detection, anomaly classification, real-time damage assessment	Huo et al. (2018), Zhang et al. (2025a)
Digital twin forecasting	Predictive ML, anomaly detection	Real-time system behavior prediction and operational decision support	Pasupuleti (2025), Pimenidis and Jayne (2020)
Lifecycle estimation	Temporal ML models, pattern recognition	Failure mode prediction and lifespan estimation	Pasupuleti (2025), Ares de Parga et al. (2025)
Performance prediction for design	ML surrogates trained on HPC simulations	Fast structural, thermal, or flow performance evaluation	Brewer et al. (2024), Chen et al. (2025), Sun et al. (2025)

AI, artificial intelligence; *CFD*, computational fluid dynamics; *ML*, machine learning; *HPC*, high-performance computing.

Pimenidis & Jayne, 2020). Practical engineering examples illustrate how the solution ML–HPC predictive analytics is applied to turbomachinery, composites, and cyber-physical manufacturing.

Predictive analytics is undergoing a profound transformation with the advent of ML and HPC. Fatigue prediction models use HPC-generated stress–strain datasets to determine microcrack evolution and to estimate remaining useful life using pattern recognition. In thermal response modeling, ML-based surrogate models are trained on large simulations of CFD or finite element thermals. Using image-based and sensor-driven AI pipelines, structural health monitoring systems detect damage and forecast performance quickly (Chen et al., 2025; Sun et al., 2025; Zhang et al., 2025a). With this ML/HPC integration, mechanical systems can be designed, monitored, maintained, and optimized throughout their lifecycle with fast, scalable, and accurate forecasting models.

35.5 AI-driven design automation

Automating repetitive and computationally intensive tasks in engineering is transforming engineering design. As a result of this design automation, mechanical, electrical, and multidisciplinary design cycles are accelerated, efficiency is improved, and errors are reduced, and errors are reduced (Ibrahim et al., 2025; Kumbhar, 2025). The integration of AI optimizes system performance and enables predictive capabilities (Abdelillah et al., 2023; Adeleke et al., 2024).

35.5.1 Simulation automation

A workflow generation, parameter selection, and result interpretation process can be automated by AI-driven automation, simplifying engineering simulations such as FEA and CFD. ML models can be used to predict simulation results, which can help to lower the computational cost of simulations while maintaining accuracy (Chillapalli, 2024; Zhang et al., 2025a). AI surrogate models, for instance, enable developers to quickly prototype and make initial design choices without the need for repeated and time-consuming simulations (Abdelillah et al., 2023; Pasupuleti, 2025).

35.5.2 Parameter tuning and multiobjective optimization

An AI-driven design optimization tool automatically explores a large design space to satisfy design, safety, and cost goals (Dong et al., 2019). Techniques, such as RL and gradient-based optimization, can be used to efficiently identify optimized parameter sets. AI also supports engineers in making trade-offs between various design criteria, such as structural performance and weight, without resorting to manual iteration, which can be time-consuming and repetitive (Akçay Perdahcıoğlu et al., 2011; Efroni et al., 2025).

35.5.3 Geometry generation

By automating the generation of complex geometries based on functional and performance requirements, AI-assisted design tools enable designers to generate nonintuitive designs that are innovative and nonintuitive. A combination of AI and Computer-Aided Design (CAD) systems enables parametric modeling of adaptive geometries, shortened design cycles, and faster prototyping of mechanical parts (Ibrahim et al., 2025).

35.5.4 Automated meshing and preprocessing

To optimize element quality and distribution in accordance with geometry and analysis needs, AI algorithms can automate the time-consuming and expertise-dependent meshing process (Ates et al., 2023; Chillapalli, 2024). By automating the process, engineers can focus on higher level design requirements while minimizing manual intervention and errors.

35.5.5 Benefits to engineering workflows

With AI-driven design automation, engineering workflows are streamlined and design cycles are accelerated through

- The efficiency of repetitive tasks can be enhanced by automating repetitive tasks and simulations (Abdelillah et al., 2023; Zhang et al., 2025a).
- Accurate preprocessing, parameter tuning, and geometry generation: minimizing human error (Ibrahim et al., 2025; Kumbhar, 2025).
- Using generative and predictive AI to facilitate the exploration of new designs and materials (Bappy, 2024; Slimani et al., 2023).
- To provide predictive capability by forecasting system behavior under uncertainty (Magdum, 2023; Pasupuleti, 2025).

35.6 Intelligent optimization for mechanical systems

Intelligent optimization for mechanical systems is essential to maximize performance, reliability, and efficiency, and to reduce cost and resource usage. With the advent of AI and ML, optimization has been revolutionized, with the capacity to tackle complex high-dimensional problems with smarter, more adaptive methods. Simulations that are determined or gradient-based fail to provide a solution for complex and high-dimensional problems.

35.6.1 Surrogate-based optimization

Using surrogate-based optimization, costly simulations are replaced by an AI/ML model to reduce costs. These surrogate models help to predict system behavior over design space and provide rapid exploration of design options based on limited high-fidelity data (Gopakumar et al., 2018; Rahmani et al., 2021). Thus, in real time, it is possible to measure the structural, dynamic, or thermal properties of mechanical systems (Hashemi et al., 2025). It combines probabilistic modeling and the use of AI-based surrogates to estimate optimal configurations even in the scenario of high-dimensional spaces (Zhang et al., 2025a).

35.6.2 The evolutionary and gradient-free methods

Evolutionary algorithms are suitable for optimization problems in nondifferentiable or discontinuous design space domains, as are gradient-free methods. The use of ML combined with other optimization methods, such as genetic algorithms, particle swarm optimization, and differential evolution, helps in achieving convergence faster and improving the quality of solutions (Akçay Perdahçıoğlu et al., 2011; Xie et al., 2023). These AI-assisted strategies learn interesting search space dynamically, thus reducing the need for function evaluations to design optimal or near-optimal systems. This approach is specifically beneficial for mechanical engineering applications, where systems consist of multiple components, such as turbine blades, automotive drivetrains, and composite structures (Rajput et al., 2024).

35.6.3 Multiobjective optimization

AI/ML-driven Multi-Objective Optimization (MOO) frameworks can help optimize mechanical systems by simultaneously addressing multiple objectives, such as reducing weight and improving strength. These frameworks offer trade-off

TABLE 35.3 Integration framework for ML, AI, and HPC in mechanical system design: Workflow stages, roles, and computational support.

Stage	Description	ML/AI role	HPC role	References
1. Problem definition and data acquisition	Define objectives, collect data	Identify relevant features for modeling	Store/Handle large datasets	Rahmani et al. (2021), Gopakumar et al. (2018)
2. Data preprocessing and feature engineering	Clean, normalize, extract features	Dimensionality reduction, feature selection	Accelerate large-scale preprocessing	Hashemi et al. (2025)
3. Surrogate modeling and prediction	Approximate high-fidelity simulations	Train predictive models for system response	Parallel model training	Zhang et al. (2025a), Dong et al. (2019)
4. AI-driven optimization	Explore design space	Reinforcement learning, evolutionary algorithms, multiobjective optimization	Execute large candidate evaluations in parallel	Xie et al. (2023), Rajput et al. (2024)
5. HPC integration	Run simulations and ML training	Enable rapid model feedback	High-speed computation and resource scaling	Merzky et al. (2025), Brewer et al. (2024)
6. Validation and deployment	Verify optimal designs	Update AI/ML models based on feedback	Execute validation simulations quickly	Hashemi et al. (2025)

AI, artificial intelligence; *ML*, machine learning; *HPC*, high-performance computing.

information to designers by delivering Pareto-optimal solutions (Dong et al., 2019; Efroni et al., 2025). Evolutionary MOO, in conjunction with surrogate models, can be used to rapidly evaluate a large number of designs. To optimize the search process, RL and probabilistic modeling are applied to high-dimensional spaces as their degree of complexity increases (Hashemi et al., 2025; Xie et al., 2023).

35.6.4 When design exploration is performed using reinforcement learning

RL is a sound method for sequential design exploration. RL agents are used in mechanical system optimization (Zhang et al., 2025a) to propose modifications to the system in an iterative manner, with the performance of the modified system given as feedback. It is extremely effective in complex, adaptive systems with nonlinear and unpredictable interactions. The exploration approach by using surrogate modeling focuses on promising design spaces, thereby converging RL faster.

35.6.5 HPC for intelligent optimization

Its combination with AI-based optimization (Merzky et al., 2025; Wang et al., 2025) has revolutionized the way complex design spaces can be explored. The simultaneous evaluation of multiple designs and training of AI models is essential to speeding up optimization cycles, which HPC does greatly. AI–HPC convergence allows engineers to solve previously unsolvable problems and optimize mechanical systems in real-time with data-driven design (Brewer et al., 2024; Fox & Jha, 2019).

35.7 Integration framework: ML + AI + HPC in a unified workflow

In modern mechanical engineering design, ML, AI, and HPC can be combined to speed up design, optimize performance, and manage complex design spaces. Traditional engineering turns into an adaptive, data-driven, and intelligent system by offering the fusion of these technologies (approximating system behavior using ML, decision-making using AI, and parallel simulations using HPC) (Fox & Jha, 2019; Wang et al., 2025).

35.8 Workflow overview

A typical ML + AI + HPC workflow for mechanical system design is shown in Table 35.3, which consists of the following stages:

TABLE 35.4 Case studies of AI/ML/HPC applications in mechanical and structural.

Domain	Application area	AI/ML technique used	HPC role	Practical outcomes	References
Structural design	Composite laminate optimization	Surrogate modeling, multiobjective optimization	Parallel evaluation of multiple design variants	Reduced design iterations, improved material performance	
Manufacturing	Machining and process control	Multiobjective optimization, predictive modeling	Real-time HPC-enabled simulation	Reduced tool wear, improved surface quality, cost savings	Brown et al. (2021)
Aerospace	Wing and turbine optimization	Surrogate-based optimization, reinforcement learning	Large-scale parametric studies	Weight reduction, fuel efficiency, faster development	
Automotive	EV battery and crash analysis	Reinforcement learning, predictive maintenance	High-dimensional simulation analysis	Improved safety, predictive maintenance, reduced development time	
Industrial IoT	Predictive maintenance for machinery	Hybrid AI/ML frameworks	Real-time data processing and decision support	Minimized downtime, extended component lifetime	Brown et al. (2021)

AI, artificial intelligence; *IoT*, Internet of Things; *ML*, machine learning; *HPC*, high-performance computing; EV, Electric Vehicle.

35.9 Case study example

Hybrid composite beam structures for aerospace applications are optimized using an integrated AI/ML approach. The material composition and the orientation of the layers determine the stiffness and strength, which is predicted by a neural network based on experiments and simulations of mechanical and thermomechanical properties. The evolutionary algorithms and HPC, to achieve a balance between weight reduction and load capacity, lead to a number of designs. The designs are perfected using finite element simulations, and the models are continually updated, gaining in performance. In contrast to traditional iterative approaches, this approach drastically cuts down design cycle time and provides better results (Fox & Jha, 2019; Wang et al., 2025).

35.10 Case studies and real-world applications

Mechanical design and engineering is transformed thanks to the combination of ML, AI, and HPC. Case studies (Table 35.4) demonstrate that AI, ML, and HPC enhance design speed, optimize materials, increase prediction accuracy, and lower costs. The technologies make it possible to explore the complex design space efficiently, transforming mechanical, aerospace, and automotive design practices (Ibrahim et al., 2025; Kumbhar, 2025; Merzky et al., 2025). The application of these technologies in structural engineering, composites, manufacturing, aerospace, and automotive engineering accelerates design cycles, reduces costs, and improves accuracy.

35.10.1 Structural design and composites

In structural engineering, AI optimization and ML-based modeling increase material efficiency and structural performance. In composite material design, ML algorithms can predict mechanical properties and failure modes, allowing rapid exploration of material combinations, reducing the need for physical testing (Ibrahim et al., 2025; Kumbhar, 2025). By simultaneously evaluating multiple designs at the same time, HPC platforms also reduce computational time (Wang et al., 2025).

35.10.2 Manufacturing

In advanced manufacturing, AI and ML enable real-time process optimization, predictive maintenance, and adaptive control. To improve surface quality and reduce tool wear, ML-based surrogate modeling and multiobjective optimization can be used to adjust machining parameters in real-time (Chen et al., 2025; Rahmani et al., 2021). HPC frameworks that are AI-coupled allow high-fidelity process models to be implemented with minimal delay (Merzky et al., 2025).

35.10.3 Aerospace

Under multiple constraints, AI-driven aerospace design automation optimizes complex structures. Efroni et al. (2025) and Xie et al. (2023) applied multiobjective optimization and surrogate-based models to wing design and turbine components resulting in weight reduction, improved fuel efficiency, and shorter development cycles. By utilizing HPC workflows, engineers can explore high-dimensional design spaces and conduct parametric tests previously unimaginable.

35.10.4 Automotive

With crash simulations, fatigue life predictions, and EV battery management, AI and ML are transforming the automotive industry. By minimizing costly finite element simulations, surrogate models and RL improve predictive accuracy and speed up design (Akbar et al., 2022; Ibrahim et al., 2025). As a result of AI-driven predictive maintenance, component lifespans are further optimized and operational costs are reduced (Abdelillah et al., 2023).

Integration of AI, ML, and HPC accelerates design, optimizes material usage, improves prediction, and reduces costs, according to case studies (Table 35.4). The technologies enable efficient exploration of complex design spaces, revolutionizing mechanical, aerospace, and automotive applications (Ibrahim et al., 2025; Kumbhar, 2025; Merzky et al., 2025).

35.11 Challenges and future directions

Design automation, optimization, and predictive modeling have been made possible thanks to the use of ML, AI, and HPC. However, there are still hurdles that will keep the technology from being widely adopted.

35.11.1 Data quality and availability

To build strong ML/AI models, it is crucial to have representative, high-quality datasets. In engineering applications such as composites and aerospace, however, experimental data is often sparse, noisy, and inconsistent, and synthetic or simulated data must be used for these cases. This can lead to bias in the model, or it can reduce the accuracy of the model, particularly for predictive maintenance and failure modeling, which rely heavily on labeled datasets.

35.11.2 Model interpretability

Some of the more complicated models, including deep neural networks, are black-box models that present another difficulty in the use of AI/ML. Although these models are very accurate, the challenge arises in that they are not easy to explain and understand, meaning that they are not transparent enough for safety-critical engineering applications, where this is key for trust and compliance with regulatory requirements.

35.11.3 HPC access and cost

Despite its promise to enable large-scale design exploration, surrogate modeling, and multiobjective optimization, small and medium enterprises do not have access to HPC easily because of high operational costs and infrastructure requirements.

35.11.4 Integration complexity

To incorporate AI/ML models into traditional simulation-based engineering workflows, there are significant software and hardware challenges. Leveraging a multidisciplinary perspective is essential for coordinating use of predictive models, optimization algorithms, and HPC simulations (Brown et al., 2021).

35.11.5 Future directions

It is crucial to note that while these challenges exist, there are also indications of promising advancements in the integration of AI and ML into mechanical and structural engineering.

Physics-informed ML (PIML) combines data-driven models with physical laws, improving prediction accuracy and generalizability, especially with limited data. Besides its ability to reduce reliance on empirical data, PIML enhances interpretability.

In recent years, there has been a great deal of interest in the use of autonomous design systems that explore design options, optimize solutions, and generate valid design outcomes using AI. To shorten the design cycle, RL and surrogate-based optimization are used.

Adaptive control and predictive maintenance can be optimized and simulated in real time with HPC and real-time data processing in low latency. This coupling allows high-fidelity models to be continually updated with live data. Hybrid digital twins are AI, ML and HPC-enabled dynamic systems monitoring, virtual prototyping, and predictive maintenance applications as part of the aerospace, automotive, and industrial Internet of Things (IoT) (Brown et al., 2021).

35.12 Conclusion

Mechanical design is becoming a predictive, automated, and optimization-driven discipline as ML, AI, and HPC converge. Data-driven modeling is enabled by ML, autonomous decisions are enabled by AI, and simulation and optimization can be performed by HPC. In a collaborative effort, they enhance the speed and intelligence of engineering workflows through an integrated ecosystem.

- With ML–AI–HPC integrated into the design process, mechanical design becomes predictive, automated, and optimized (Fox & Jha, 2019; Wang et al., 2025).
- By using ML and AI in predictive design, it is possible to accurately predict system behavior, lifecycle performance, and failure modes in mechanical engineering (Brewer et al., 2024; Pasupuleti, 2025).
- AI-driven tools automate repetitive engineering tasks such as simulation setup, geometry generation, parameter tuning, and meshing, reducing manual effort and accelerating design cycles (Chillapalli, 2024; Ibrahim et al., 2025) while maintaining fidelity.
- With intelligent optimization, high-dimensional design spaces can be explored using surrogate modeling, RL, evolutionary algorithms, and physics-informed ML. Multiobjective optimization and data-driven decision-making can then be achieved (Xie et al., 2023; Zhang et al., 2025a).
- The computing power of HPC enables complex predictive and optimization tasks, such as large-scale simulations, parallel design evaluation, and training AI/ML models in real-time (Brewer et al., 2024; Merzky et al., 2025).
- In structural design, composite materials, manufacturing, aerospace, and automotive industries, it produces benefits such as reduced design time, enhanced performance, and cost savings (Chen et al., 2025; Kumbhar, 2025).
- The development of physics-based ML, autonomous design pipelines, HPC–AI integration in real-time, and hybrid digital twins is expected to advance mechanical system design by enhancing interpretability, adaptive learning, and operational intelligence.
- With improved interpretability, adaptive learning, and operational intelligence, physics-informed ML, autonomous design pipelines, HPC–AI integration in real time, and hybrid digital twins are poised to enhance mechanical system design.
- By combining the power of ML, AI, and HPC, mechanical design can become faster, smarter, more reliable, more innovative, and more resilient.

This integration enables prediction of lifecycle performance, evaluation of reliability, and system behavior. With AI-powered design automation, tasks are streamlined, designs are optimized, and geometry is generated more easily, and HPC guarantees efficient execution at scale. Engineers can quickly and accurately investigate complex design spaces using the convergence of ML, AI, and HPC, besides surrogate modeling and RL, as well as physics-based ML. Predictive modeling and automatic optimization in structural and composite design optimize materials and performance. Real-time process control, adaptive optimization, and predictive maintenance help enhance manufacturing. The aerospace and automotive industries are characterized by faster design cycles, lower computational costs, and more reliable predictions because of this integration. The challenges of data availability, model interpretability, accessibility of HPC, and integration complexity, however, are offset by current trends including physics-informed ML, autonomous design systems, HPC-AI coupling in real time, and hybrid digital twins. When combined with ML, AI, and HPC, the resulting new engineering methodologies will be faster, more reliable, and more innovative, enabling intelligent automation, predictive foresight, and strong optimization.

Acknowledgments

All authors contributed greatly to the development and writing of this chapter. The conceptualization of the chapter is done by Dr. Rishi J.P., Dr. Sandeep B. and Rishi J.P., who integrated the concepts of machine learning (ML), artificial intelligence (AI), and high-performance computing (HPC) for mechanical design. All the authors conceptualized the combination of machine learning (ML), artificial intelligence (AI), and high-performance computing (HPC) in mechanical design. Sandeep B. and Rishi J.P. conducted literature review and research synthesis for predictive analytics, AI-based design automation, and smart optimization. Rishi J.P. contributed the sections on HPC infrastructure, computational workflows, and case studies. Data curation, figure preparation, and workflow schematics were the main tasks for Dr. Sandeep B. and Rishi J..P. All the authors wrote, reviewed, and edited the manuscript.

Conflict of interest

The authors have no conflicts of interest to declare.

Funding received

No.

References

Abdelillah, F. M., Nora, H., Ouchani, S., & Benslimane, S. (2023). Hybrid data-driven and knowledge-based predictive maintenance framework in the context of industry 4. 0. In M. Mosbah, T. Kechadi, L. Bellatreche, & F. Gargouri (Eds.), *Model and data engineering. MEDI 2023. Lecture notes in computer science, vol 14396* (pp. 319–337). Springer. doi:10.1007/978-3-031-49333-1_23.

Adeleke, A., Ani, E., Olu-lawal, K., Olajiga, O., & Montero, D. (2024). Future of precision manufacturing: Integrating advanced metrology and intelligent monitoring for process optimization. *International Journal of Science and Research Archive, 11*(1), 2346–2355. doi:10.30574/ijsra.2024.11.1.0335.

Akbar, K., Zou, Y., Awais, Q., Baig, M. J. A., & Jamil, M. (2022). A machine learning-based robust state of health (SOH) prediction model for electric vehicle batteries. *Electronics, 11*(8), 1216. doi:10.3390/electronics11081216.

Akçay Perdahcıoğlu, D., Geijselaers, H. J. M., Ellenbroek, M., & de Boer, A. (2011). Dynamic substructuring and reanalysis methods in a surrogate-based design optimization environment. *Structural and Multidisciplinary Optimization, 45*(1), 129–138. doi:10.1007/s00158-011-0681-4.

Álvarez Cid-Fuentes, J., Álvarez, P., Amela, R., Ishii, K., Morizawa, R. K., & Badía, R. M. (2019). Efficient development of high performance data analytics in Python. *Future Generation Computer Systems, 111*, 570–581. doi:10.1016/j.future.2019.09.051.

Ares de Parga, S., Bravo, J. R., Sibuet, N., Hernandez, J. A., Rossi, R., Boschert, S., Quintana-Ortí, E. S., Tomás, A. E., Tatu, S. S., Vázquez-Novoa, F., Ejarque, J., & Badia, R. M. (2025). Parallel reduced order modeling for digital twins using high-performance computing workflows. *Computers & Structures, 316*, 107867.

Ates, C., Höfchen, T., Witt, M., Koch, R., & Bauer, H. (2023). Vibration-based wear condition estimation of journal bearings using convolutional autoencoders. *Sensors, 23*(22), 9212. doi:10.3390/s23229212.

Bappy, M. (2024). Exploring the integration of informed machine learning in engineering applications: A comprehensive review. *AJSLD, 3*(2), 11–21. doi:10.51699/ajsld.v3i2.3459.

Basaran, C. (2020). Entropy based fatigue, fracture, failure prediction and structural health monitoring. *Entropy, 22*(10), 1178 –1178. doi:10.3390/e22101178.

Brewer, W., Gainaru, A., Suter, F., Wang, F., Emani, M., & Jha, S. (2024). *AI-coupled HPC workflow applications*. Middleware and Performance.

Brown, O., Curtis, A., & Goodwin, J. (2021). Principles for evaluation of AI/ML model performance and robustness. *arXiv*. doi:10.48550/arXiv.2107.02868.

Chang, C. W., Lee, H.-W., & Liu, C.-H. (2018). A review of artificial intelligence algorithms used for smart machine tools. *Inventions, 3*(3), 41 –41. doi:10.3390/inventions3030041.

Chen, J., Wang, M., & Hsia, C. (2025). Artificial intelligence and machine learning in sensing and image processing. *Sensors, 25*(6), 1870. doi:10.3390/s25061870.

Chillapalli, N. (2024). AI and machine learning solutions for real-time fault-tolerant system design in critical applications. *Advances in Nonlinear Variational Inequalities, 28*(1s), 199–207. doi:10.52783/anvi.v28.2239.

Döllner, J. (2020). Geospatial artificial intelligence: Potentials of machine learning for 3D point clouds and geospatial digital twins. *PFG – Journal of Photogrammetry Remote Sensing and Geoinformation Science, 88*(1), 15–24. doi:10.1007/s41064-020-00102-3.

Dong, J., Qin, W., & Mo, J. (2019). Low-cost multi-objective optimization of multiparameter antenna structures based on the l1 optimization BPNN surrogate model. *Electronics, 8*(8), 839 –839. doi:10.3390/electronics8080839.

Efroni, Y., Kretzu, B., Jiang, D., Bhandari, J., Zheqing, B., Zhu, Z., & Ullrich, K. (2025). Aligned multi objective optimization. *arXiv*. doi:10.48550/arXiv.2502.14096.

Ejakov, M., Yeager, D., & Kayupov, M. (2003). Reliability and robustness analysis of engine ring-pack performance. In *SAE 2003 World Congress & Exhibition, Detroit, Michigan, United States, March 3, 2003*. doi:10.4271/2003-01-1221.

Fox, G., & Jha, S. (2019). *Understanding ML driven HPC: Applications and infrastructure*. Cornell University. doi:10.48550/arxiv.1909.02363.

Chetsa, G. L. T., Lefèvre, L., Pierson, J. M., Stolf, P., & Da Costa, G. (2013). Exploiting performance counters to predict and improve energy performance of HPC systems. *Future Generation Computer Systems, 36*, 287–298. doi:10.1016/j.future.2013.07.010.

Goh, Y., McMahon, C., & Booker, J. (2009). Improved utility and application of probabilistic methods for reliable mechanical design. *Proceedings of the Institution of Mechanical Engineers O: Journal of Risk and Reliability, 223*(3), 199–214. doi:10.1243/1748006xjrr244.

Gopakumar, A., Balachandran, P. V., Xue, D., Gubernatis, J. E., & Lookman, T. (2018). Multi-objective optimization for materials discovery via adaptive design. *Scientific Reports, 8*(1). doi:10.1038/s41598-018-21936-3.

Hashemi, A., Beheshti, J., & Mohammadi, M. (2025). Physics-based, AI-driven surrogate modeling for structural displacement prediction in mechanical systems with limited sensor data. *IEEE Access*. doi:10.1109/access.2025.3590664.

Huo, Y., Blaber, J., Damon, S. M., Boyd, B. D., Bao, S., Parvathaneni, P., Noguera, C. B., Chaganti, S., Nath, V., Greer, J. M., Lyu, I., French, W. R., Newton, A. T., Rogers, B. P., & Landman, B. A. (2018). Towards portable large-scale image processing with high-performance computing. *Journal of Digital Imaging, 31*(3), 304–314. doi:10.1007/s10278-018-0080-0.

Ibrahim, I., Thomas, P., & Adhikari, P. (2025). The integration of artificial intelligence in engineering design, manufacturing, and cad: A triangular revolution of innovation, efficiency, and automation. *Global Journal of Engineering and Technology Advances, 24*(2), 269–278. doi:10.30574/gjeta.2025.24.2.0256.

Jiang, Z., Gao, W., Wang, L., Xiong, X., Zhang, Y., Wen, X., & Zhan, J. (2019). HPC AI500: A benchmark suite for HPC AI SYstems. *Lecture notes in computer science* (pp. 10–22). doi:10.1007/978-3-030-32813-9_2.

Kumbhar, G. (2025). Application of artificial intelligence in predictive mechanical design and optimization. *International Journal for Multidisciplinary Research, 7*(5). doi:10.36948/ijfmr.2025.v07i05.55499.

Lingerfelt, E. J., Belianinov, A., Endeve, E., Ovchinnikov, O., Somnath, S., Borreguero, J. M., Grodowitz, N., Park, B., Archibald, R. K., Symons, C. T., Kalinin, S. V., Messer, O. E. B., Shankar, M., & Jesse, S. (2016). BEAM: A computational workflow system for managing and modeling material characterization data in HPC environments. *Procedia Computer Science, 80*, 2276–2280. doi:10.1016/j.procs.2016.05.410.

Lusk, Mark T., & Mattsson, Ann E. (2011). High-performance computing for materials design to advance energy science. *MRS Bulletin, 36*(3), 169–174. doi:10.1557/mrs.2011.30.

Magdum, S. (2023). Predictive maintenance and preventive measures for calibration devices: A mobile application approach. *International Journal of Engineering Technology and Management Sciences, 7*(4), 252–256. doi:10.46647/ijetms.2023.v07i04.034.

Merzky, A., Titov, M., Turilli, M., Kilic, O., Wang, T., & Jha, S. (2025). Scalable runtime architecture for data-driven, hybrid HPC and ML workflow applications. *arXiv*. doi:10.48550/arXiv.2503.13343.

Odili, P., Daudu, C., Adefemi, A., Ekemezie, I., & Usiagu, G. (2024). Operational readiness and assurance (OR&A) in asset integrity: Strategies for excellence in facility management. *Engineering Science & Technology Journal, 5*(2), 581–596. doi:10.51594/estj.v5i2.834.

Park, B. H., Kim, Y., Kim, B.-D., Hong, T., Kim, S.-J., & Lee, J. (2012). High performance computing: Infrastructure, application, and operation. *Journal of Computing Science and Engineering, 6*(4), 280–286. doi:10.5626/jcse.2012.6.4.280.

Pasupuleti, S. (2025). Using digital twins for fault detection and root cause analysis in mechanical systems. *International Scientific Journal of Engineering and Management, 04*(01), 1–6. doi:10.55041/isjem01640.

Pimenidis, E., & Jayne, C. (2020). Special issue on engineering applications of neural networks. *Neural Computing and Applications, 32*(16), 12241–12242. doi:10.1007/s00521-020-05028-w.

Rahmani, A. M., Yousefpoor, E., Yousefpoor, M. S., Mehmood, Z., Haider, A., Hosseinzadeh, M., & Naqvi, R. A. (2021). Machine learning (ML) in medicine: Review, applications, and challenges. *Mathematics, 9*(22), 2970 –2970. doi:10.3390/math9222970.

Rajput, K., Schram, M., Edelen, A., Colen, J., Kasparian, A., Roussel, R., Carpenter, A., Zhang, H., & Benesch, J. (2024). Harnessing the power of gradient-based simulations for multi-objective optimization in particle accelerators. *arXiv*. doi:10.48550/arXiv.2411.04817.

Slimani, K., Khoulji, S., & Kerkeb, M. (2023). Advancements and challenges in energy-efficient 6G mobile communication network. *E3S Web of Conferences, 412*, 01036. doi:10.1051/e3sconf/202341201036.

Sukel, K. (2021). Partners in design. *Mechanical Engineering, 143*(2), 34–39. doi:10.1115/1.2021-mar2.

Sun, P., Seyedi, A., & Gantz, L. (2025). Software- and hardware-accelerated inverse design of photonic devices. In *Proceeding SPIE 13601, Optical Design Automation, 136010J* (p. 28). doi:10.1117/12.3065466.

Wang, X., Zhang, G., Yin, J., Liu, S., Reshniak, V., Morgan, Z., Pawledzio, S., Proffen, T., Cao, H., Chakoumakos, B., Hoffmann, C., & Liu, Y. (2025). Integrated AI/ML and HPC framework for multidimensional neutron crystallography. *Structural Dynamics, 12*. doi:10.1063/4.0000339.

Xie, Z., Yu, C., & Qiao, W. (2023). Dropout strategy in reinforcement learning: Limiting the surrogate objective variance in policy optimization methods. *arXiv*. doi:10.48550/arXiv.2310.20380.

Zhang, X., Zhang, X., Qiu, L., Ruan, J., & Li, W. (2025a). A novel deep reinforcement learning framework fusing knowledge and probabilistic modeling for prognostic and maintenance of aircraft engines. *Engineering Research Express, 7*(2), 025570. doi:10.1088/2631-8695/ade04f.

Zhou, N., Zhou, H., & Hoppe, D. (2022). Containerization for high performance computing systems: Survey and prospects. *IEEE Transactions on Software Engineering, 49*(4), 2722–2740. doi:10.1109/tse.2022.3229221.

Index

Page numbers followed by "*f*" and "*t*" indicate, figures and tables respectively.

V

W

Printed and bound by CPI Group (UK) Ltd, Croydon, CR0 4YY
14/09/2026
02216388-0014